NETWORKS

Networks

An Introduction

M. E. J. Newman

University of Michigan
and
Santa Fe Institute

OXFORD

UNIVERSITY PRESS

OXFORD
UNIVERSITY PRESS

Great Clarendon Street, Oxford OX2 6DP

Oxford University Press is a department of the University of Oxford.
It furthers the University's objective of excellence in research, scholarship,
and education by publishing worldwide in

Oxford New York

Auckland Cape Town Dar es Salaam Hong Kong Karachi
Kuala Lumpur Madrid Melbourne Mexico City Nairobi
New Delhi Shanghai Taipei Toronto

With offices in

Argentina Austria Brazil Chile Czech Republic France Greece
Guatemala Hungary Italy Japan Poland Portugal Singapore
South Korea Switzerland Thailand Turkey Ukraine Vietnam

Oxford is a registered trade mark of Oxford University Press
in the UK and in certain other countries

Published in the United States
by Oxford University Press Inc., New York

British Library Cataloguing in Publication Data

Data available

Library of Congress Cataloging in Publication Data

Data available

Typeset by SPI Publisher Services, Pondicherry, India
Printed in Great Britain
on acid-free paper by
CPI Antony Rowe, Chippenham, Wiltshire

ISBN 978–0–19–920665–0 (Hbk.)

1 3 5 7 9 10 8 6 4 2

CONTENTS

PREFACE

The scientific study of networks, such as computer networks, biological networks, and social networks, is an interdisciplinary field that combines ideas from mathematics, physics, biology, computer science, the social sciences, and many other areas. The field has benefited enormously from the wide range of viewpoints brought to it by practitioners from so many different disciplines, but it has also suffered because human knowledge about networks is dispersed across the scientific community and researchers in one area often do not have ready access to discoveries made in another. The goal of this book is to bring our knowledge of networks together and present it in consistent language and notation, so that it becomes a coherent whole whose elements complement one another and in combination teach us more than any single element can alone.

The book is divided into five parts. Following a short introductory chapter, Part I describes the basic types of networks studied by present-day science and the empirical techniques used to determine their structure. Part II introduces the fundamental mathematical tools used in the study of networks as well as measures and statistics for quantifying network structure. Part III describes computer algorithms for the efficient analysis of network data, while Part IV describes mathematical models of network structure that can help us predict the behavior of networked systems and understand their formation and growth. Finally, Part V describes theories of processes taking place on networks, such as epidemics on social networks or search processes on computer networks.

The technical level of the presentation varies among the parts, Part I requiring virtually no mathematical knowledge for its comprehension, while Parts II and III require a grasp of linear algebra and calculus at the undergraduate level. Parts IV and V are mathematically more advanced and suitable for advanced undergraduates, postgraduates, and researchers working in the field. The book could thus be used as the basis of a taught course at more than one level. A less technical course suitable for those with moderate mathematical knowledge might cover the material of Chapters 1 to 8, while a more technical course for advanced students might cover the material of Chapters 6 to 14 and selected material thereafter. Each chapter from Part II onward is accompanied by a selection of exercises that can be used to test the reader's understanding of the material.

This book has been some years in the making and many people have helped me with it during that time. I must thank my ever-patient editor Sonke Adlung, with whom I have worked on various book projects for more than 15 years now, and whose constant encouragement and kind words have made working with him and Oxford University Press a real pleasure. Thanks are also due to Melanie Johnstone, Alison Lees, Emma Lonie, and April Warman for their help with the final stages of bringing the book to print.

I have benefited greatly during the writing of this book from the conversation, comments, suggestions, and encouragement of many colleagues and friends. They are, sadly, too numerous to mention exhaustively, but special thanks must go to Steve Borgatti, Duncan Callaway, Aaron Clauset, Betsy Foxman, Linton Freeman, Michelle Girvan, Martin Gould, Mark Handcock, Petter Holme, Jon Kleinberg, Alden Klovdahl, Liza Levina, Lauren Meyers, Cris Moore, Lou Pecora, Mason Porter, Sidney Redner, Puck Rombach, Cosma Shalizi, Steve Strogatz, Duncan Watts, Doug White, Lenka Zdeborova, and Bob Ziff, as well as to the many students, particularly Michelle Adan, Alejandro Balbin, Chris Fink, Ruthi Hortsch, and Jane Wang, whose feedback helped iron out a lot of rough spots. I would also especially like to thank Brian Karrer, who read the entire book in draft form and gave me many pages of thoughtful and thought-provoking comments, as well as spotting a number of mistakes and typos. Responsibility for any remaining mistakes in the book of course rests entirely with myself, and I welcome corrections from readers.

Finally, my profound thanks go to my wife Carrie for her continual encouragement and support during the writing of this book. Without her the book would still have been written but I would have smiled a lot less.

Mark Newman
Ann Arbor, Michigan
February 24, 2010

CHAPTER 1

INTRODUCTION

*A short introduction to networks
and why we study them*

A NETWORK is, in its simplest form, a collection of points joined together in pairs by lines. In the jargon of the field the points are referred to as *vertices*[1] or *nodes* and the lines are referred to as *edges*. Many objects of interest in the physical, biological, and social sciences can be thought of as networks and, as this book aims to show, thinking of them in this way can often lead to new and useful insights.

We begin, in this introductory chapter, with a discussion of why we are interested in networks and a brief description of some specific networks of note. All the topics in this chapter are covered in greater depth elsewhere in the book.

A small network composed of eight vertices and ten edges.

WHY ARE WE INTERESTED IN NETWORKS?

There are many systems of interest to scientists that are composed of individual parts or components linked together in some way. Examples include the Internet, a collection of computers linked by data connections, and human societies, which are collections of people linked by acquaintance or social interaction.

Many aspects of these systems are worthy of study. Some people study the nature of the individual components—how a computer works, for instance, or how a human being feels or acts—while others study the nature of the connections or interactions—the communication protocols used on the Internet or the dynamics of human friendships. But there is a third aspect to these interacting

---

[1]Singular: vertex.

1

systems, sometimes neglected but almost always crucial to the behavior of the system, which is the *pattern* of connections between components.

The pattern of connections in a given system can be represented as a network, the components of the system being the network vertices and the connections the edges. Upon reflection it should come as no surprise (although in some fields it is a relatively recent realization) that the structure of such networks, the particular pattern of interactions, can have a big effect on the behavior of the system. The pattern of connections between computers on the Internet, for instance, affects the routes that data take over the network and the efficiency with which the network transports those data. The connections in a social network affect how people learn, form opinions, and gather news, as well as affecting other less obvious phenomena, such as the spread of disease. Unless we know something about the structure of these networks, we cannot hope to understand fully how the corresponding systems work.

A network is a simplified representation that reduces a system to an abstract structure capturing only the basics of connection patterns and little else. Vertices and edges in a network can be labeled with additional information, such as names or strengths, to capture more details of the system, but even so a lot of information is usually lost in the process of reducing a full system to a network representation. This certainly has its disadvantages but it has advantages as well.

The most common network variants are discussed in detail in Chapter 6.

Scientists in a wide variety of fields have, over the years, developed an extensive set of tools—mathematical, computational, and statistical—for analyzing, modeling, and understanding networks. Many of these tools start from a simple network representation, a set of vertices and edges, and after suitable calculations tell you something about the network that might well be useful to you: which is the best connected vertex, say, or the length of a path from one vertex to another. Other tools take the form of network models that can make mathematical predictions about processes taking place on networks, such as the way traffic will flow over the Internet or the way a disease will spread through a community. Because they work with networks in their abstract form, these tools can in theory be applied to almost any system represented as a network. Thus if there is a system you are interested in, and it can usefully be represented as a network, then there are hundreds of different tools out there, already developed and well understood, that you can immediately apply to the analysis of your system. Certainly not all of them will give useful results—which measurements or calculations are useful for a particular system depends on what the system is and does and on what specific questions you are trying to answer about it. Still, if you have a well-posed question about a networked system there will, in many cases, already be a tool available that will help you address it.

Networks are thus a general yet powerful means of representing patterns of connections or interactions between the parts of a system. In this book, we discuss many examples of specific networks in different fields, along with techniques for their analysis drawn from mathematics, physics, the computer and information sciences, the social sciences, biology, and elsewhere. In doing so, we bring together a wide range of ideas and expertise from many disciplines to give a comprehensive introduction to the science of networks.

SOME EXAMPLES OF NETWORKS

One of the best known and most widely studied examples of a network is the Internet, the computer data network in which the vertices are computers and the edges are physical data connections between them, such as optical fiber cables or telephone lines. Figure 1.1 shows a picture of the structure of the Internet, a snapshot of the network as it was in 2003, reconstructed by observing the paths taken across the network by a large number of Internet data packets traveling between different sources and destinations. It is a curious fact that although the Internet is a man-made and carefully engineered network we don't know exactly what its structure is, since it was built by many different groups of people with only limited knowledge of each other's actions and little centralized control. Our best current data on its structure are derived from experimental studies, such as the one that produced this figure, rather than from any central repository of knowledge or coordinating authority.

We look at the Internet in more detail in Section 2.1.

There are a number of excellent practical reasons why we might want to study the network structure of the Internet. The function of the Internet is to transport data between computers (and other devices) in different parts of the world, which it does by dividing the data into pieces or *packets* and shipping them from vertex to vertex across the network until they reach their intended destination. Certainly the structure of the network will affect how efficiently it accomplishes this function and if we know the network structure we can address many questions of practical relevance. How should we choose the route by which data are transported? Is the shortest route always necessarily the fastest? If not, then what is, and how can we find it? How can we avoid bottlenecks in the traffic flow that might slow things down? What happens when a vertex or an edge fails (which they do with some regularity)? How can we devise schemes to route around such failures? If we have the opportunity to add new capacity to the network, where should it be added?

Knowledge of Internet structure also plays a central role in the development of new communications standards. New standards and protocols are

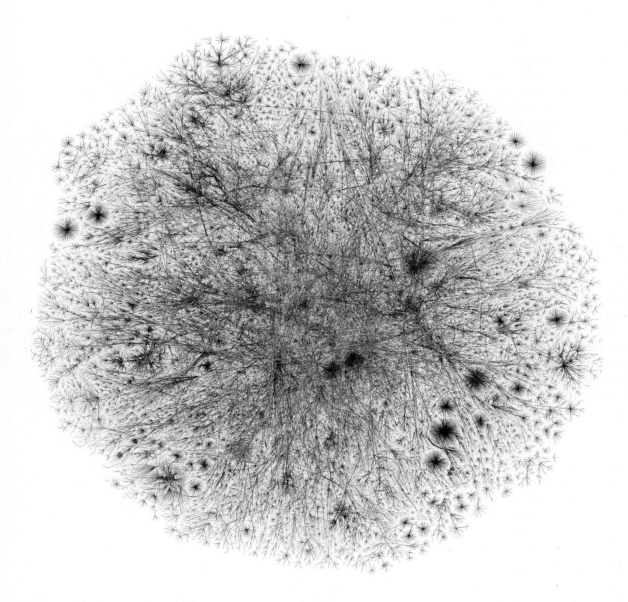

Figure 1.1: The network structure of the Internet. (See Plate I for color version.) The vertices in this representation of the Internet are "class C subnets"—groups of computers with similar Internet addresses that are usually under the management of a single organization—and the connections between them represent the routes taken by Internet data packets as they hop between subnets. The geometric positions of the vertices in the picture have no special meaning; they are chosen simply to give a pleasing layout and are not related, for instance, to geographic position of the vertices. The structure of the Internet is discussed in detail in Section 2.1. Figure created by the Opte Project (www.opte.org). Reproduced with permission.

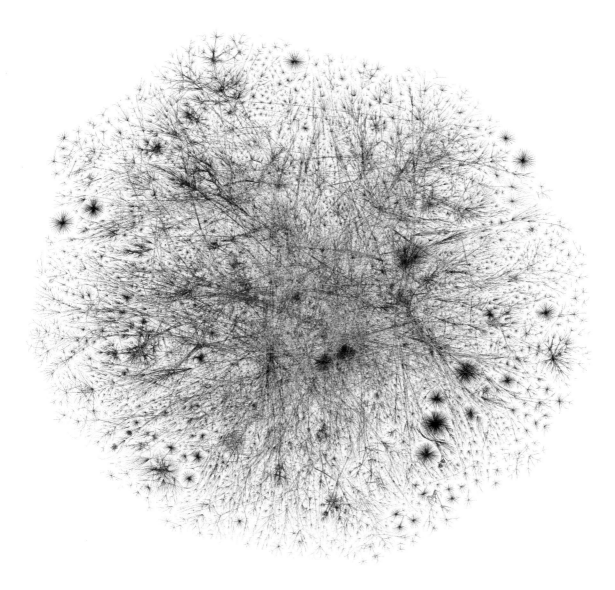

Plate I: The network structure of the Internet. The vertices in this representation of the Internet are "class C subnets"— groups of computers with similar Internet addresses that are usually under the management of a single organization— and the connections between them represent the routes taken by Internet data packets as they hop between subnets. The geometric positions of the vertices in the picture have no special meaning; they are chosen simply to give a pleasing layout and are not related, for instance, to geographic position of the vertices. The structure of the Internet is discussed in detail in Section 2.1. Figure created by the Opte Project (www.opte.org). Reproduced with permission.

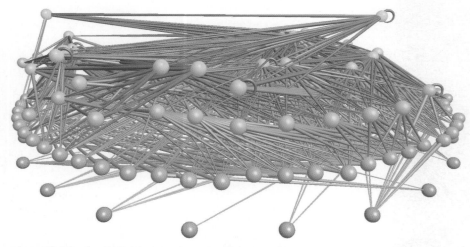

Plate II: The food web of Little Rock Lake, Wisconsin. This elegant picture summarizes the known predatory interactions between species in a freshwater lake in the northern United States. The vertices represent the species and the edges run between predator–prey species pairs. The vertical position of the vertices represents, roughly speaking, the trophic level of the corresponding species. The figure was created by Richard Williams and Neo Martinez [210].

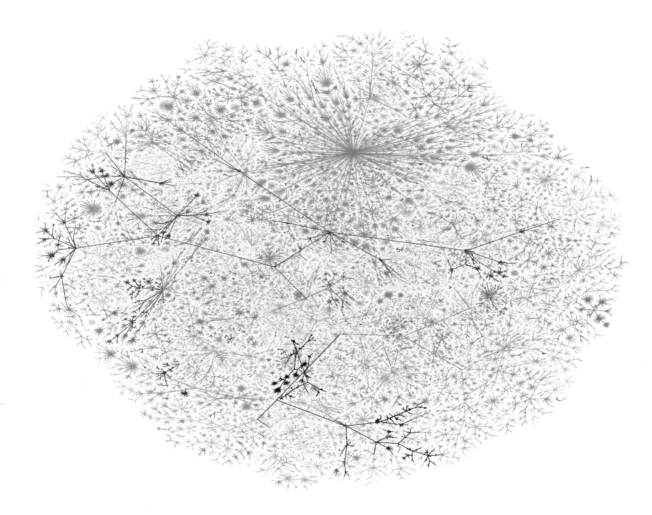

Plate III: The structure of the Internet at the level of autonomous systems. The vertices in this network representation of the Internet are autonomous systems and the edges show the routes taken by data traveling between them. This figure is different from Plate I, which shows the network at the level of class C subnets. The picture was created by Hal Burch and Bill Cheswick. Patent(s) pending and Copyright Lumeta Corporation 2009. Reproduced with permission.

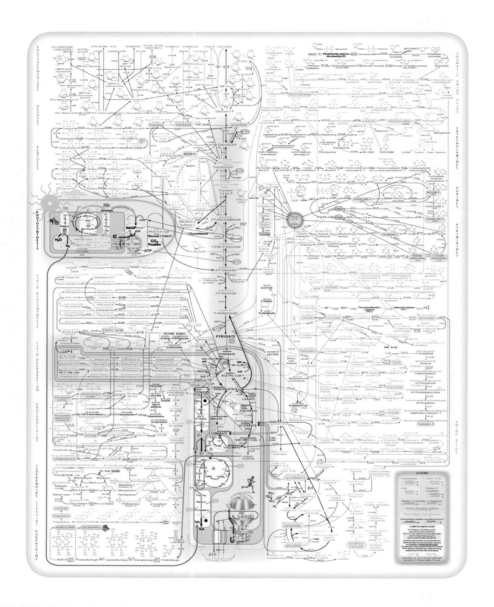

Plate IV: A metabolic network. A wallchart showing the network formed by the major metabolic pathways. Created by Donald Nicholson. Copyright of the International Union of Biochemistry and Molecular Biology. Reproduced with permission.

continually being devised for communication over the Internet, and old ones are revised. The parameters of these protocols are tuned for optimal performance with the structure of the Internet in mind. In the early days of the network, rather primitive models of network structure were employed in the tuning process, but as better structural data become available it becomes possible to better understand and improve performance.

A more abstract example of a network is the World Wide Web. In common parlance the words "Web" and "Internet" are often used interchangeably, but technically the two are quite distinct. The Internet is a physical network of computers linked by actual cables (or sometimes radio links) running between them. The Web, on the other hand, is a network of information stored on web pages. The vertices of the World Wide Web are web pages and the edges are "hyperlinks," the highlighted snippets of text or push-buttons on web pages that we click on to navigate from one page to another. A hyperlink is purely a software construct; you can link from your web page to a page that lives on a computer on the other side of the world just as easily as you can link to a friend down the hall. There is no physical structure, like an optical fiber, that needs to be built when you make a new link. The link is merely an address that tells the computer where to look next when you click on it.

The World Wide Web is discussed in more detail in Section 4.1.

Abstract though it may be, the World Wide Web, with its billions of pages and links, has proved enormously useful, not to mention profitable, to many people, and the structure of the network of links is of substantial interest. Since people tend to add hyperlinks between pages with related content, the link structure of the Web reveals something about the content structure. What's more, people tend to link more often to pages that they find useful than to those they do not, so that the number of links pointing to a page can be used as a measure of its usefulness. A more sophisticated version of this idea lies behind the operation of the popular Web search engine *Google*, as well as some others.

The Web also illustrates another concept of network theory, the *directed network*. Hyperlinks on the Web run in one specific direction, from one web page to another. Given an appropriate link on page A, you can click and arrive at page B. But there is no requirement that B contains a link back to A again. (It may contain such a link, but there is no law that says that it must and much of the time it will not.) One says that the edges in the World Wide Web are *directed*, running from the linking page to the linked.

The mechanics of Web search are discussed in Section 19.1.

Moving away from the technological realm, another type of network of scientific interest is the social network. A social network is, usually, a network of people, although it may sometimes be a network of groups of people, such as companies. The people or groups form the vertices of the network and the

Social networks are discussed in more depth in Chapter 3.

edges represent connections of some kind between them, such as friendship between individuals or business relationships between companies. The field of sociology has perhaps the longest and best developed tradition of the empirical study of networks as they occur in the real world, and many of the mathematical and statistical tools that are used in the study of networks are borrowed, directly or indirectly, from sociologists.

Figure 1.2 shows a famous example of a social network from the sociology literature, Wayne Zachary's "karate club" network. This network represents the pattern of friendships among members of a karate club at a north American university. The network was constructed by direct observation of interactions between the club's members. As is typical of such studies the network is small, having, in this case, only 34 vertices. Network representations of the Internet or the World Wide Web, by contrast, can have thousands or millions of vertices. In principle there is no reason why social networks cannot be similarly large. The entire population of the world, for example, can be regarded as a very large social network. But in practice social network data are limited to relatively small groups because of the effort involved in compiling them. The network of Fig. 1.2, for instance, was the product of two years of observations by one experimenter. In recent years a few larger social networks have been constructed by dint of enormous effort on the part of large groups of researchers. And online social networking services, such as Facebook or instant message "buddy lists," can provide network data on a previously unreachable scale. Studies are just beginning to emerge of the structure and properties of these larger networks.

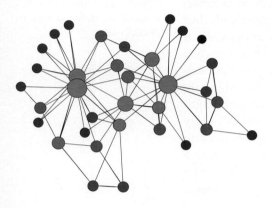

Figure 1.2: Friendship network between members of a club. This social network from a study conducted in the 1970s shows the pattern of friendships between the members of a karate club at an American university. The data were collected and published by Zachary [334].

Neural networks are discussed in Section 5.2 and food webs in Section 5.3.

A third realm in which networks have become important in recent years is biology. Networks occur in a number of situations in biology. Some are concrete physical networks like neural networks—the networks of connections between neurons in the brain—while others are more abstract. In Fig. 1.3 we show a picture of a "food web," an ecological network in which the vertices are species in an ecosystem and the edges represent predator–prey relationships between them. That is, pairs of species are connected by edges in this network if one species eats the other. The study of food webs forms a substantial branch of ecology and helps us to understand and quantify many ecological phenomena, particularly concerning energy and carbon flows in ecosystems. Food webs also provide us with another example of a directed network, like the World Wide Web discussed previously. The edges in a food web are

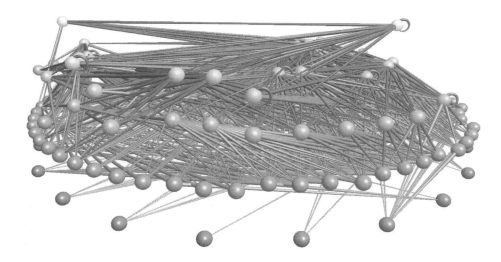

Figure 1.3: The food web of Little Rock Lake, Wisconsin. (See Plate II for color version.) This elegant picture summarizes the known predatory interactions between species in a freshwater lake in the northern United States. The vertices represent the species and the edges run between predator–prey species pairs. The vertical position of the vertices represents, roughly speaking, the trophic level of the corresponding species. The figure was created by Richard Williams and Neo Martinez [209].

asymmetric and are conventionally thought of as pointing from the prey to the predator, indicating the direction of the flow of energy when the prey is eaten. (This choice of direction is only a convention and one could certainly make the reverse choice. The important point is the asymmetry of the predator–prey interaction.)

Another class of biological networks is that of biochemical networks, such as metabolic networks, protein–protein interaction networks, and genetic regulatory networks. A metabolic network, for instance, is a representation of the chemical reactions that fuel cells and organisms. The reader may have seen the wallcharts of metabolic reactions that adorn the offices of some biochemists, incredibly detailed maps with hundreds of tiny inscriptions linked by a maze of arrows.[2] The inscriptions—the vertices in this network—are metabolites, the substrates and products of metabolism, and the arrows—directed edges—are reactions that turn one metabolite into another. The depiction of reactions as a

Biochemical networks are discussed in detail in Section 5.1.

---

[2] An example appears as Fig. 5.2 on page 83.

network is one of the first steps towards making sense of the bewildering array of biochemical data generated by recent and ongoing experiments in molecular genetics.

These are just a few examples of the types of network whose study is the focus of this book. There are many others that we will come across in later pages. Among them some of the best known are telephone networks, road, rail, and air networks, the power grid, citation networks, recommender networks, peer-to-peer networks, email networks, collaboration networks, disease transmission networks, river networks, and word networks.

PROPERTIES OF NETWORKS

We have seen that a variety of systems can be represented as networks. If we can gather data on the structure of one of these networks, what then can we do with those data? What can they tell us about the form and function of the system the network represents? What properties of networked systems can we measure or model and how are those properties related to the practical issues we care about? This, essentially, is the topic of this entire book, and we are not going to answer it in this chapter alone. Let us, however, look briefly here at a few representative concepts, to get a feel for the kinds of ideas we will be dealing with.

A first step in analyzing the structure of a network is often to make a picture of it. Figures 1.1, 1.2, and 1.3 are typical examples. Each of these was generated by a specialized computer program designed for network visualization and there are many such programs available, both commercially and for free, if you want to produce pictures like these for yourself. Visualization can be an extraordinarily useful tool in the analysis of network data, allowing one to see instantly important structural features of a network that would otherwise be difficult to pick out of the raw data. The human eye is enormously gifted at picking out patterns, and visualizations allow us to put this gift to work on our network problems. On the other hand, direct visualization of networks is only really useful for networks up to a few hundreds or thousands of vertices, and for networks that are relatively sparse, meaning that the number of edges is quite small. If there are too many vertices or edges in a network then pictures of the network will be too complicated for the eye to comprehend and their usefulness becomes limited. Many of the networks that scientists are interested in today have hundreds of thousands or even millions of vertices, which means that visualization is not of much help in their analysis and we need to employ other techniques to determine their structural features. In response to this need, network theory has developed a large toolchest of measures and metrics that can help us understand what our network data are telling us, even in cases where useful visualization is impossible.

An example of an important and useful class of network measures is that of measures of *centrality*. Centrality quantifies how important vertices (or edges) are in a networked system, and social network analysts in particular have expended considerable effort studying it. There are a wide variety of mathematical measures of vertex centrality that focus on different concepts and definitions of what it means to be central in a network. A simple but very useful example is the measure called *degree*. The degree of a vertex in a network is the number of edges attached to it. In a social network of friendships between individuals, for instance, such as the network of Fig. 1.2, the degree of an individual is the number of friends he or she has within the network. In the Internet degree would be the number of data connections a computer, router, or other device has. In many cases the vertices with the highest degrees in a network, those with the most connections, also play important roles in the functioning of the system, and hence degree can be a useful guide for focusing our attention on the system's most crucial elements.

In undirected networks degree is just a single number, but in directed networks vertices have two different degrees, *in-degree* and *out-degree*, corresponding to the number of edges pointing inward to and outward from those vertices. For example, the in-degree of a web page is the number of other pages that link to it and the out-degree is the number of pages to which it links. We have already mentioned one example of how centrality can be put to use on the Web to answer an important practical question: by counting the number of links a web page gets—the in-degree of the page—we (or a search engine operating on our behalf) can make a guess about which pages are most likely to contain information that might be of use to us.

It is an interesting observation that many networks are found to contain a small but significant number of "hubs"—vertices with unusually high degree. Social networks often contain a few central individuals with very many acquaintances; there are a few websites with an extraordinarily large number of links; there are a few metabolites that take part in almost all metabolic processes. A major topic of research in recent years has been the investigation of the effects of hubs on the performance and behavior of networked systems. Both empirical and theoretical results indicate that hubs can have a quite disproportionate effect, playing a central role particularly in network transport phenomena and resilience, despite being few in number.

Another example of a network concept that arises repeatedly and has real practical implications is the so-called *small-world effect*. One can define a distance, called the *geodesic distance*, between two vertices in a network to be the

See Chapter 7 for further discussion of centrality measures.

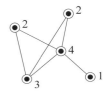

The number beside each vertex in this small network indicates the vertex's degree.

Hubs are discussed further in Section 8.3.

minimum number of edges one would have to traverse in order to get from one vertex to the other. For instance, two friends would have geodesic distance 1 in a friendship network because there is a single edge connecting them directly, while the friend of your friend would have distance 2 from you. As discussed in Sections 3.6 and 8.2, it is found empirically (and can be proven mathematically in some cases) that the mean geodesic distance, appropriately defined,[3] between vertex pairs is very short, typically increasing only as the logarithm of the number of vertices in the network. Although first studied in the context of friendship networks, this small-world effect appears to be very widespread, occurring in essentially all types of networks. In popular culture it is referred to as the "six degrees of separation," after a successful stage play and film of the same name. The semi-mythological claim is that you can get from anyone in the world to anyone else via a sequence of no more than five intermediate acquaintances—six steps in all.

The small-world effect can have interesting repercussions. For example, news and gossip spread over social networks. If you hear an interesting rumor from a friend, you may pass it on to your other friends, and they in turn pass it on to theirs, and so forth. Clearly the rumor will spread further and faster if it only takes six steps to reach anyone in the world than if it takes a hundred, or a million. It is a matter of common experience that indeed a suitably scandalous rumor can reach the ears of an entire community in what seems like the blink of an eye, and the structure of social networks has a lot to do with it.

And consider the Internet. One of the reasons the Internet functions at all is because any computer on the network is only a few "hops" over optical and other data lines from any other. In practice the paths taken by packets over the Internet are typically in the range of about ten to twenty hops long. Certainly the performance of the network would be much worse if packets had to make a thousand hops instead.

A third example of a network concept of practical importance is provided by clusters or communities in networks. We are most of us familiar with the idea that social networks break up into subcommunities—tightly knit groups of friends or acquaintances within the larger, looser network. Friendship networks, for instance, tend to contain cliques, circles, and gangs of friends within which connections are strong and frequent but between which they are weaker or rarer. The same is true of other kinds of social network also. For instance, in a network of business relationships between companies one often finds clusters formed of sets of companies that operate in particular sections of the econ-

---

[3]One must be careful when there are vertex pairs in the network that are connected by no path at all. Such issues are dealt with in Section 8.2.

omy. Connections might be stronger, for instance, between a pair of computer companies or a pair of biotech companies than between a computer company and a biotech company. And if it is the case that communities correspond to genuine divisions of interest or purpose in this way, then we may well learn something by taking a network and examining it to determine what communities it contains. The way a network breaks down into communities can reveal levels and concepts of organization that are not easy to see without network data, and can help us to understand how a system is structured. There is a substantial research literature in social network analysis as well as in other fields concerned with precisely these kinds of questions, and a large number of techniques have been developed to help us extract and analyze subcommunities within larger networks. These are highly active topics of research at present, and hold promise for exciting applications in the future.

OUTLINE OF THIS BOOK

This book is divided into five parts. In the first part, consisting of Chapters 2 to 5, we introduce the various types of network encountered in the real world, including technological, social, and biological networks, and the empirical techniques used to discover their structure. Although it is not the purpose of this book to describe any one particular network in great detail, the study of networks is nonetheless firmly founded on empirical observations and a good understanding of what data are available and how they are obtained is immensely helpful in understanding the science of networks as it is practiced today.

The second part of the book, Chapters 6 to 8, introduces the fundamental theoretical ideas on which our current understanding of networks is based. Chapter 6 describes the basic mathematics used to capture network ideas, Chapter 7 describes the measures and metrics we use to quantify network structure, and Chapter 8 describes some of the intriguing patterns and principles that emerge when we apply our mathematics and our metrics to real-world network data.

In the third part of the book, Chapters 9 to 11, we discuss computer algorithms for analyzing and understanding network data. Measurements of network properties, such as those described in Chapter 7, are typically only possible with the help of fast computers and much effort has been devoted over the years to the development of efficient algorithms for analyzing network data. This part of the book describes in detail some of the most important of these algorithms. A knowledge of this material will be of use to anyone who wants to work with network data.

In the fourth part of the book, Chapters 12 to 15, we look at mathematical models of networks. The material in these chapters forms a central part of the canon of the field and has been the subject of a vast amount of published scientific research. We study both traditional models, such as random graphs and their extensions, and newer models, such as models of growing networks and the "small-world model."

Finally, in the fifth and last part of the book, Chapters 16 to 19, we look at processes taking place on networks, including failure processes and resilience, network epidemiology, dynamical systems, and network search processes. The theory of these processes is less well developed than other aspects of the theory of networks and there is much work still to be done. The last chapters of the book probably raise at least as many questions as they answer, but this, surely, is a good thing. With luck readers will feel inspired to answer some of those questions themselves and the author looks forward to the new and exciting results they generate when they do.

Part I

The empirical study of networks

Chapter 2

Technological networks

A discussion of engineered networks like the Internet and
the power grid and how we determine their structure

In the next four chapters we define and describe some of the most commonly studied networks, dividing them into four general classes—technological networks, social networks, information networks, and biological networks. We will list the most important examples in each class and then describe the techniques used to measure their structure. (The classes are not rigorously defined and there is, as we will see, some overlap between them, with some networks belonging to more than one class. Nonetheless, the division into classes is a useful one, since networks in the same class are often treated using similar techniques or ideas.)

It is not our intention in this book to study any one network in great detail. Plenty of other books exist that do that. Nonetheless, network science is concerned with understanding and modeling the behavior of real-world networked systems and observational data are the starting point for essentially all the developments of the field, so the reader will find it useful to have a grasp of the types of data that are available, their strengths and limitations, and the means used to acquire them. In this chapter we look at technological networks, the physical infrastructure networks that have grown up over the last century or so and form the backbone of modern technological societies. Perhaps the most celebrated such network—and a relatively recent entry in the field—is the Internet, the global network of data connections, electrical, optical, and wireless, that links computers and other information systems together. Section 2.1 is devoted to a discussion of the Internet. A number of other important examples of technological networks, including power grids, transportation networks, delivery and distribution networks, and telephone networks, are discussed in subsequent sections.

2.1 THE INTERNET

The Internet is the worldwide network of physical data connections between computers and related devices. The Internet is a *packet switched* data network, meaning that messages sent over it are broken up into *packets*, small chunks of data, that are sent separately over the network and reassembled into a complete message again at the other end. The format of the packets follows a standard known as the *Internet Protocol* (IP) and includes an *IP address* in each packet that specifies the packet's destination, so that it can be routed correctly across the network.

The alternative to a packet switched network is a *circuit switched* network, the classic example of which is the telephone system. In a circuit switched network, vertices request connections when needed, such as when a telephone call is placed, and the network allocates a separate circuit for each connection, reserved for the sole use of that connection until the connection is ended. This works well for voice traffic, which consists of discrete phone calls each with a definite beginning and end, but it would be a poor model for a data network, in which data transmission typically occurs in brief, intermittent bursts. Using a packet switched model for the Internet allows computers to transmit and receive data intermittently or at varying rates without tying up capacity on the network. By making packets reasonably small, we also allow for a certain amount of unreliability in the network. It is not uncommon for packets to disappear on the Internet and never reach their destination, sometimes because of hardware or software failure, but more often because packets are deliberately deleted to reduce congestion in the busiest parts of the network. If a message is divided into several packets before transmission and a few packets are lost, then only those that are lost need be resent to complete the message. A software protocol called *Transport Control Protocol* or TCP, which runs on top of IP, performs the necessary error checking and retransmission automatically, without the need for intervention from computer users or other software.[1]

> The Internet should not be confused with the World Wide Web, a virtual network of web pages and hyperlinks, which we discuss in Section 4.1.

> The telephone network is discussed in Section 2.2.

---

[1]Most of the well-known communications protocols of the Internet are themselves built on top of TCP, including HTTP (the World Wide Web), SMTP (email), and FTP (file transfer). Thus communication is a three-layer process with a user-level protocol running on top of TCP, which in turn runs on top of IP, and the user protocols automatically benefit from the error-checking features and guaranteed transmission offered by TCP. (There are lower-level transport protocols as well, such as Ethernet, PPP, and ATM, but these will not concern us.) There are however also some applications of Internet technology that do not require guaranteed transmission. Most of the common examples are streaming media, such as audio and video transmissions, voice and

The simplest network representation of the Internet (there are others, as we will shortly see) is one in which the vertices of the network represent computers and other devices, and the edges represent physical connections between them, such as optical fiber lines. In fact, ordinary computers mostly occupy only the vertices on the "outside" of the network, those that data flows to and from, but they do not act as intermediate points for the flow of data between others. (Indeed, most computers only have a single connection to the net, so it would not be possible for them to lie on the path between any others.) The "interior" nodes of the Internet are primarily *routers*, powerful special-purpose computers at the junctions between data lines that receive data packets and forward them in one direction or another towards their intended destination.

The general overall shape of the Internet is shown, in schematic form, in Fig. 2.1. The network is composed of three levels or circles of vertices. The innermost circle, the core of the network, is the *backbone* of the network, the trunk lines that provide long-distance high-bandwidth data transport across the globe, along with the high-performance routers and switching centers that link them together. These trunk lines are the highways of the Internet, built with the fastest fiber optic connections available (and improving all the time). The backbone is operated by *network backbone providers* (NBPs), who are primarily national governments and communications companies such as AT&T, Global Crossing, British Telecom, and others.

The second circle of the Internet is composed of *Internet service providers* or ISPs—commercial companies, governments, universities, and others who contract with NBPs for connection to the backbone and then resell or otherwise provide that connection to end users, the ultimate consumers of Internet bandwidth, who form the third circle—businesses, government offices, academics, people in their homes, and so forth. In fact, as Fig. 2.1 shows, the ISPs are further subdivided into *regional ISPs* and *local* or *consumer ISPs*, the former being larger organizations whose primary customers are the local ISPs, who in turn sell network connections to the end users. This distinction is somewhat blurred however, because large consumer ISPs, such as America Online or British Telecom, often act as their own regional ISPs (and some may be backbone providers as well).

The network structure of the Internet is not dictated by any central authority. Protocols and guidelines are developed by an informal volunteer organization called the Internet Engineering Task Force, but one does not have to apply to any central Internet authority for permission to build a new spur on

---

teleconferencing, and online games. An alternative protocol to TCP called *User Datagram Protocol* (UDP), which provides no transmission guarantees, is used in such cases.

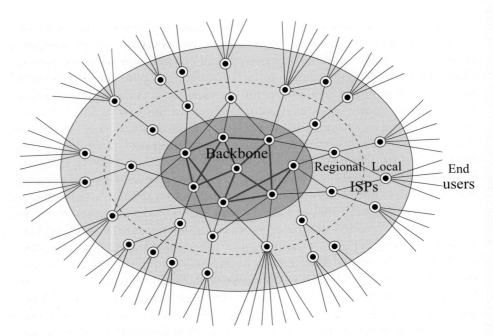

Figure 2.1: A schematic depiction of the structure of the Internet. The vertices and edges of the Internet fall into a number of different classes: the "backbone" of high-bandwidth long-distance connections; the ISPs, who connect to the backbone and who are divided roughly into regional (larger) and local (smaller) ISPs; and the end users—home users, companies, and so forth—who connect to the ISPs.

the Internet, or to take one out of service.

One of the remarkable features of the Internet is that the scheme used for the routing of packets from one destination to another is arrived at by automated negotiation among Internet routers using a system called the *Border Gateway Protocol* (BGP). BGP is designed in such a way that if new vertices or edges are added to the network, old ones disappear, or existing ones fail either permanently or temporarily, routers will take note and adjust their routing policy appropriately. Some human oversight is required to keep the system running smoothly, but no "Internet government" is needed to steer things from on high; the system organizes itself by the combined actions of many local and essentially autonomous computer systems.

While this is an excellent feature of the system from the point of view of robustness and flexibility, it is a problem for those who want to study the structure of the Internet, because there is no central registry from which one can determine that structure. There is no one whose job it is to maintain an official map of the network. Instead the network's structure must be determined by experimental measurements. There are two primary methods for doing this. The first uses something called "traceroute"; the second uses BGP.

2.1.1 MEASURING INTERNET STRUCTURE USING TRACEROUTE

It is not, at least for most of us, possible to probe the network structure of the Internet directly. We can, however, quite easily discover the particular path taken by data packets traveling between our own computer (or any computer to which we have access) and most others on the Internet. The standard tool for doing this is called *traceroute*.

In addition to a destination address, which says where it is going, each Internet packet also contains a source address, which says where it started from, and a *time-to-live* (TTL). The TTL is a number that specifies the maximum number of "hops" that the packet can make to get to its destination, a hop being the traversal of one edge in the network. At every hop, the TTL is decreased by one, and if ever it reaches zero the packet is discarded, meaning it is deleted and not forwarded any further over the network. If we are using TCP, a message is also then sent back to the sender informing them that the packet was discarded and where it got to. (This is a part of TCP's mechanism for guaranteeing the reliable transmission of data—see above.) The TTL exists mainly as a safeguard to prevent packets from getting lost on the Internet and wandering around forever, but we can make use of it to track packet progress as well. The idea is as follows.

First, we send out a TCP packet with the destination address of the network vertex we are interested in and a TTL of 1. The packet makes a single hop to the first router along the way, its TTL is decreased to zero, the packet is discarded by the router and a message is returned to us telling us, among other things, the IP address of the router. We record this address and then repeat the process with a TTL of 2. This time the packet makes two hops before dying and the returned message tells us the IP address of the second router. The process is repeated with larger and larger TTL until the destination is reached, and the set of IP addresses received as a result specifies the entire route taken to get there.[2] There are standard software tools that will perform the entire procedure

---

[2]We are assuming that each packet takes the same route to the destination. It is possible, but relatively rare, for different packets to take different routes, in which case the set of IP addresses returned by the traceroute procedure will not give a correct path through the network. This can happen, for instance, if congestion patterns along the route vary significantly while the procedure

automatically and print out the list of IP addresses for us. On most computers the tool that does this is called "traceroute."

We can use traceroute (or a similar tool) to probe the network structure of the Internet. The idea is to assemble a large data set of traceroute paths between many different pairs of points on the Internet. With luck, most of the edges in the network (though usually not all of them) will appear at least once in this set, and the union of all of them should give a reasonably complete picture of the network. Early studies, for the sake of expediency, limited themselves to just a few source computers, but more recent ones, such as the DIMES Project,[3] make use of distributed collections of thousands of sources to develop a very complete picture of the network.

See Section 6.7 for a discussion of tree networks.

The paths from any single source to a set of destinations form a tree-like structure as shown schematically in Fig. 2.2a, b, and c.[4] The source computers should, ideally, be well distributed over the network. If they are close together, then there may be a substantial overlap between the traceroute paths to distant vertices, which means that they will duplicate needlessly each other's efforts, rather than returning independent measurements.

Once one has a suitable set of traceroute data, a simple union of all the paths appearing in the data set gives us our snapshot of the network structure—see Fig. 2.2d. That is, we go through each path and record a vertex for every IP address that appears in the path and an edge between every pair of addresses that appear in adjacent positions. As hinted above, it is unlikely that such a procedure will find all the edges in the network (see Fig. 2.2d again), and for studies based on small numbers of sources there can be quite severe biases in the sampling of edges [3,192]. However, better and better data sets are becoming available as time passes, and it is believed that we now have a reasonably complete picture of the shape of the Internet.

In fact, it is rarely, if ever, done to record *every* IP address on the Internet as a separate vertex. There are believed to be about 2 billion unique IP addresses in use on the Internet at any one time, with many of those corresponding to end-user computers that appear and disappear as the computers are turned

---

is being performed, causing the network to reroute packets along less congested connections. Serious Internet mapping experiments perform repeated traceroute measurements to minimize the errors introduced by effects such as these.

[3]See www.netdimes.org.

[4]If there were a unique best path to every vertex, then the set of paths would be precisely a tree, i.e., it would contain no loops. Because of the way routing algorithms work, however, this is not in practice always the case—two routes that originate at the same point and pass through the same vertex on the way to their final destination can still take different routes to get to that vertex, so that the set of paths can contain loops.

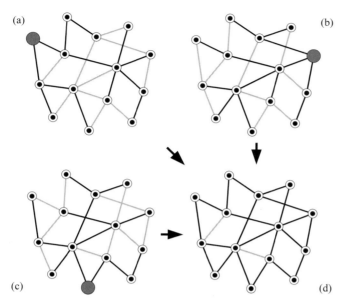

(a) (b) (c) (d)

Figure 2.2: Reconstruction of the topology of the Internet from traceroute data. In panels (a), (b), and (c) we show in bold the edges in three sets of traceroute paths starting from each of the three highlighted source vertices. In panel (d) we form the union of these edges to make a picture of the overall network topology. Note that a few edges are missing from this picture (the remaining gray edges in panel (d)) because, by chance, they happen not to appear in any of the three individual traceroute data sets.

on or off or connections to the Internet are made or broken. Most studies of the Internet ignore end-user computers and restrict themselves to just the routers, in effect concentrating on the inner zones in Fig. 2.1 and ignoring the outermost one. We will refer to such maps of the Internet as representations at the *router level*. The vertices in the network are routers, and the edges between them are network connections.

It may appear strange to ignore end-user computers, since the end users are, after all, the entire reason for the Internet's existence in the first place. However, it is the structure of the network at the router level that is responsible for most aspects of the performance, robustness, and efficiency of the network, that dictates the patterns of traffic flow on the network, and that forms the focus of most work on Internet structure and design. To the extent that these are the issues of scientific interest, therefore, it makes sense to concentrate our efforts on the router-level structure.

An example of a study of the topology of the Internet at the router level is that of Faloutsos *et al.* [111], who looked at the "degree distribution" of the network and discovered it to follow, approximately, a power law. We discuss degree distributions and power laws in networks in more detail in Section 8.4.

Even after removing all or most end-user computers from the network, the network structure at the router level may still be too detailed for our purposes. Often we would like a more coarse-grained representation of the network that gives us a broader overall picture of network structure. Such representations are created by grouping sets of IP addresses together into single vertices. Three different ways of grouping addresses are in common use giving rise to three different coarse-grained representations, at the level of subnets, domains, and autonomous systems.

A *subnet* is a group of IP addresses defined as follows. IP addresses consist of four numbers, each one in the range from 0 to 255 (eight bits in binary) and typically written in a string separated by periods or dots. For example, the IP address of the main web server at the author's home institution, the University of Michigan, is 141.211.144.190. IP addresses are allocated to organizations in blocks. The University of Michigan, for instance, owns (among others) all the addresses of the form 141.211.144.xxx, where "xxx" can be any number between 0 and 255. Such a block, where the first three numbers in the address are fixed and the last can be anything, is called a *class C subnet*. There are also class B subnets, which have the form 141.211.xxx.yyy, and class A subnets, which have the form 141.xxx.yyy.zzz.

Since all the addresses in a class C subnet are usually allocated to the same organization, a reasonable way of coarse-graining Internet network data is to group vertices into class C subnets. In most cases this will group together vertices in the same organization, although larger organizations, like the University of Michigan, own more than one class C subnet, so there will still be more than one vertex in the coarse-grained network corresponding to such organizations. Given the topology of the network at the router level, the level of individual IP addresses, it is easy to lump together into a single vertex all addresses in each class C subnet and place an edge between any two subnets if any router in one has a network connection to any router in the other. Figure 1.1 on page 4 shows an example of the network structure of the Internet represented at the level of class C subnets.

The second common type of coarse-graining is coarse-graining at the domain level. A *domain* is a group of computers and routers under, usually, the control of a single organization and identified by a single *domain name*, normally the last two or three parts of a computer's address when the address is written in human-readable text form (as opposed to the raw IP addresses considered above). For example, "umich.edu" is the domain name for the

University of Michigan and "oup.co.uk" is the domain name for Oxford University Press. The name of the domain to which a computer belongs can be determined in a straightforward manner from the computer's IP address by a "reverse DNS lookup," a network service set up to provide precisely this type of information. Thus, given the router-level network topology, it is a simple task to determine the domain to which each router belongs and group vertices in the network according to their domain. An edge is then placed between two vertices if any router in one has a direct network connection to any router in the other. The study by Faloutsos *et al.* [111] mentioned earlier looked at the domain-level structure of the Internet as well as the router-level structure.

The third common coarse-graining of the network is coarse-graining at the level of autonomous systems. An autonomous system is similar to a domain: it is a group of computers, usually under single administrative control, and it often (though not always) coincides with a domain. Coarse-graining at the autonomous system level is not usually used with data derived from trace-route sampling but with data derived using an alternative method based on BGP routing tables, for which it forms the most natural unit of representation. The BGP method and autonomous systems are discussed in detail in the next section.

2.1.2 MEASURING INTERNET STRUCTURE USING ROUTING TABLES

Internet routers maintain *routing tables* that allow them to decide in which direction incoming packets should be sent to best reach their destination. Routing tables are constructed from information shared between routers using the Border Gateway Protocol (BGP). They consist of lists of complete paths from the router in question to destinations on the Internet. When a packet arrives at a router, the router examines it to determine its destination and looks up that destination in the routing table. The first step of the path in the appropriate table entry tells the router how the packet should be sent on its way. Indeed, in theory routers need store only the first step on each path in order to route packets correctly. However, for efficient calculation of routes using BGP (the techniques of which we will not go into here) it is highly desirable that routers be aware of the entire path to each destination, and since the earliest days of the Internet all routers have operated in this way. We can make use of this fact to measure the structure of the Internet.

Routing tables in routers are represented at the level of *autonomous systems* (ASes). An autonomous system is a collection of computers and routers, usually under single administrative control, within which data routing is handled independently of the wider Internet, hence the name "autonomous system."

That is, when a data packet arrives at a router within an autonomous system, destined for a specific computer within that same autonomous system, it is the responsibility of the autonomous system to get the packet the last few steps to its final destination. Data passing between autonomous systems, however, is handled by the Internet-wide mechanisms of BGP. Thus it's necessary for BGP to know about routing only down to the level of autonomous systems and hence BGP tables are most conveniently represented in autonomous system terms. In practice, autonomous systems, of which there are (at the time of writing) about twenty thousand on the Internet, often coincide with domains, or nearly so.

Autonomous systems are assigned unique identification numbers. A routing path consists of a sequence of these AS numbers and since router tables consist of paths to a large number of destinations, we can construct a picture of the Internet at the autonomous system level by examining them. The process is very similar to that used for the traceroute method described in the previous section and depicted in Fig. 2.2. We first obtain a number of router tables. This is normally done simply by the gracious cooperation of router operators at a variety of organizations. Each router table contains a large number of paths starting from a single source (the router), and the union of these paths gives a good but not complete network snapshot in which the vertices are autonomous systems and the edges are the connections between autonomous systems. As with traceroute, it is important that the routers used be well scattered over the network to avoid too much duplication of results, and the number of routers used should be as large as possible to make the sampling of network edges as complete as possible. For example, the Routeviews Project,[5] a large BGP-based Internet mapping effort based at the University of Oregon, uses (again at the time of writing) a total of 223 source computers around the world to measure the structure of the entire network every two hours.

Figure 2.3 shows a picture of the Internet at the AS level derived from routing tables. Qualitatively, the picture is similar to Fig. 1.1 for the class C subnet structure, but there are differences arising because class C subnets are smaller units than many autonomous systems and so Fig. 1.1 is effectively a finer-grained representation than Fig. 2.3.

Using router-, subnet-, domain-, or AS-level structural data for the Internet, many intriguing features of the net's topology have been discovered in recent years [57, 66, 111, 211, 262, 265], many of which are discussed in later chapters of this book.

---

[5]See www.routeviews.org.

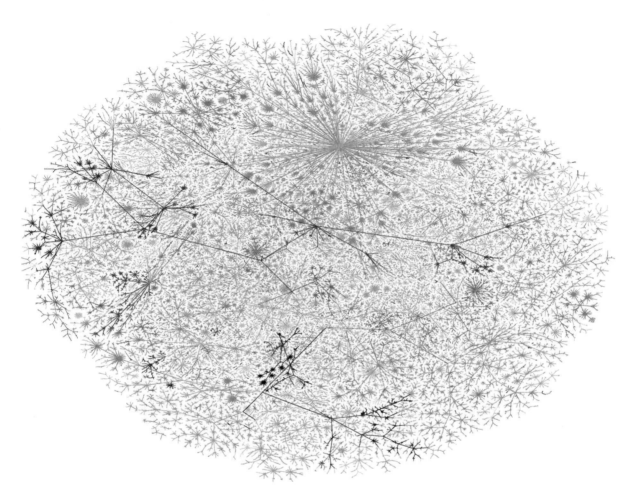

Figure 2.3: The structure of the Internet at the level of autonomous systems. (See Plate III for color version.) The vertices in this network representation of the Internet are autonomous systems and the edges show the routes taken by data traveling between them. This figure is different from Fig. 1.1, which shows the network at the level of class C subnets. The picture was created by Hal Burch and Bill Cheswick. Patent(s) pending and Copyright Lumeta Corporation 2009. Reproduced with permission.

One further aspect of the Internet worth mentioning here is the geographic location of its vertices on the surface of the Earth. In many of the networks that we will study in this book, vertices do not exist at any particular position in real space—the vertices of a citation network for instance are not located on any particular continent or in any particular town. Not so the Internet; its vertices, by and large, are quite well localized in space. Your computer sits on your desk, a router sits in the basement of an office building, and so forth. Things become more blurry once the network is coarse-grained. The domain `umich.edu` covers large parts of the state of Michigan. The domain `aol.com` covers most of North America. These are somewhat special cases, however, being unusually large domains. The majority of domains have a well-defined location at least to within a few miles. Furthermore, tools now exist for determining, at least approximately, the geographic location of a given IP address, domain, or autonomous system. Examples include *NetGeo*, *NetAcuity*, *GeoNetMap*, and many others. Geographic locations are determined primarily by looking them up in one of several registries that record the official addresses of the registered owners of domains or autonomous systems. These addresses need not in all cases correspond to the actual location of the corresponding computer hardware. For instance, the domain `ibm.com` is registered in New York City, but IBM's principal operations are in California. Nonetheless, an approximate picture of the geographic distribution of the Internet can be derived by these methods, and there has been some interest in the results [332].

Geographic localization is a feature the Internet shares with several other technological networks, as we will see in the following sections, but rarely with networks of other kinds.[6]

2.2 THE TELEPHONE NETWORK

The Internet is the best studied example of a technological network, at least as measured by volume of recent academic work. This is partly because data on Internet structure are relatively easy to come by and partly because of intense interest among engineers and computer scientists and among the public at large. Several other technological networks however are worthy of mention here. In this and the following sections of the chapter we look briefly at the telephone network and various distribution and transportation networks. A few other networks, such as software call graphs and electronic circuits, could also be considered technological networks and have been studied occasionally, but are beyond the scope of this book.

---

[6]Social networks are perhaps the main exception—in many cases people or groups of people can be considered to have reasonably well-defined geographic locations. Relatively little work has been done however on the effects of geographic distribution, perhaps because most social network studies have concentrated on populations in local neighborhoods, rather than ones spread out over significant geographic areas.

The telephone network—meaning the network of landlines and wireless links[7] that transmits telephone calls—is one of the oldest communication networks still in use (although the postal network is certainly older), but it has been little studied by network theorists, primarily because of a lack of good data about its structure. Of course, the structure of the phone network is known, but the data are largely proprietary to the telephone companies that own the network and, while not precisely secret, they are not openly shared with the research community in the same way that Internet data are. We hope that this situation will change, although the issue may become moot in the not too distant future, as telephone companies are sending an increasing amount of voice traffic over the Internet rather than over dedicated telephone lines, and it may not be long before the two networks merge into one.

Some general principles of operation of the telephone network are clear however. By contrast with the Internet, the traditional telephone network is, as mentioned in Section 2.1, not packet switched. Signals sent over the phone network are not disassembled and sent as sets of discrete packets. Instead the telephone network is *circuit switched*, which means that the telephone company has a number of lines or circuits available to carry telephone calls between different points and it assigns them to individual callers when those callers place phone calls. In the earliest days of the telephone systems in the United States and Europe the "lines" actually were individual wires, one each for each call the company could carry. Increasing the capacity of the network to carry more calls meant putting in more wires. Since the early part of the twentieth century, however, phone companies have employed techniques for *multiplexing* phone signals, i.e., sending many calls down the same wire simultaneously. The exception is the "last mile" of connection to the individual subscriber. The phone cable entering a house usually only carries one phone call at a time, although even that has changed in recent years as new technology has made it possible for households to have more than one telephone number and place more than one call at a time.

The basic form of the telephone network is relatively simple. Most countries with a mature landline (as opposed to wireless) telephone network use

---

[7]For most of its existence, the telephone network has connected together stationary telephones in fixed locations such as houses and offices using landlines. In the last twenty years or so fixed telephones have started to be replaced by wireless phones ("mobile phones" or "cell phones"), but it is important to realize that even calls made on wireless phones are still primarily carried over the traditional landline telephone network. The signal from a wireless phone makes the first step of its journey wirelessly to a nearby transmission tower, but from there it travels over ordinary phone lines. Thus, while the advent of wireless phones has had an extraordinary impact on society, it has had rather less impact on the nature of the telephone network.

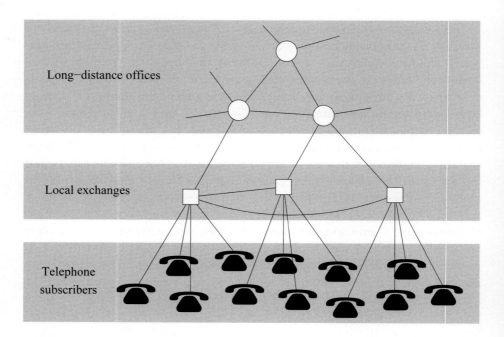

Long–distance offices

Local exchanges

Telephone
subscribers

Figure 2.4: A sketch of the three-tiered structure of a traditional telephone network.
In a telephone network individual subscriber telephones are connected to local exchanges, which are connected in turn to long-distance offices. The long-distance offices are connected amongst themselves by further lines, and there may be some connections between local exchanges as well.

a three-tiered design. Individual telephone subscribers are connected over local lines to local telephone exchanges, which are then connected over shared "trunk" lines to long-distance offices, sometimes also called toll-switching offices. The long-distance offices are then connected among themselves by further trunk lines. See Fig. 2.4 for a sketch of the network structure. The structure is, in many ways, rather similar to that of the Internet (Fig. 2.1), even though the underlying principles on which the two networks operate are quite different.

The three-level topology of the phone network is designed to exploit the fact that most telephone calls in most countries are local, meaning they connect subscribers in the same town or region. Phone calls between subscribers connected to the same local exchange can be handled by that exchange alone and do not need to make use of any trunk lines at all. Such calls are usually referred to as local calls, while calls that pass over trunk lines are referred to as trunk or long-distance calls. In many cases there may also be direct connections between nearby local exchanges that allow calls to be handled locally even when two subscribers are not technically attached to the same exchange.

The telephone network has had roughly this same topology for most of the last hundred years and still has it today, but many of the details about how the network works have changed. In particular, at the trunk level some telephone networks are no longer circuit switched. Instead they are now digital packet switched networks that work in a manner not dissimilar from the Internet, with voice calls digitized, broken into packets, and transmitted over optical fiber links. Only the "last mile" to the subscriber's telephone is still carried on an old-fashioned dedicated circuit, and even that is changing with the advent of digital and Internet telephone services. Nonetheless, in terms of geometry and topology the structure of the phone network is much the same as it has always been, being dictated in large part by the constraints of geography and the propensity for people to talk more often to others in their geographic vicinity than to those further away.

2.3 POWER GRIDS

The topology of power grids has received occasional study in the networks literature [16, 323]. A power grid, in this context, is the network of high-voltage transmission lines that provide long-distance transport of electric power within and between countries. Low-voltage local power delivery lines are normally excluded. The vertices in a power grid correspond to generating stations and switching substations, and the edges correspond to the high-voltage lines. The topology of power grids is not difficult to determine. The networks are usually overseen by a single authority and complete maps of grids are readily available. Indeed, very comprehensive data on power grids (as well as other energy-related networks such as oil and gas pipelines) are available from specialist publishers, either on paper or in electronic form, if one is willing to pay for them.

There is much of interest to be learned by looking at the structure of power grids. Like the Internet, power grids have a spatial aspect; the individual vertices each have a location somewhere on the globe, and their distribution in space is interesting from geographic, social, and economic points of view. Network statistics, both geographic and topological, may provide insight into the global constraints governing the shape and growth of grids. Power grids also display some unusual behaviors, such as cascading failures, which can give rise to surprising results such as the observed power-law distribution in the sizes of power outages [92].

However, while there is a temptation to apply simple models of the kind described in this book to try to explain these and other results, it is wise to be cautious. Power grids are very complicated systems. The flow of power is governed not only by simple physical laws, but also by precise and detailed control of the phases and voltages across transmission lines, monitored and adjusted on rapid timescales by sophisticated computer systems and on slower timescales by human operators. It turns out that power failures and other power-grid phenomena are influenced relatively little by the raw topology of the network and much more by operator actions and software design, and as a result network theory has not, so far, been very successful at shedding light on the behavior of power grids.

2.4 TRANSPORTATION NETWORKS

A moderate amount of work has been done on the structure and function of transportation networks such as airline routes and road and rail networks. The structure of these networks is not usually hard to determine, although compiling the data may be laborious. Airline networks can be reconstructed from published airline timetables, road and rail networks from maps. Geographic information systems (GIS) software can be useful for speeding the compilation of transportation data, and there are also a variety of online resources providing useful information such as latitude and longitude of airports.

One of the earliest examples of a study of a transportation network is the study by Pitts [268] of waterborne transport on Russian rivers in the Middle Ages. There was also a movement among geographers in the 1960s and 70s to study road and rail networks, particularly focusing on the interplay between their economics and their physical structure. The most prominent name in the movement was that of Karel Kansky, and his book on transportation networks is a good point of entry into that body of literature [168].

More recently a number of authors have produced studies applying new network analysis ideas to road, rail, and air networks [16, 136, 294]. In most of the networks studied the vertices represent geographic locations and the edges routes between them. For instance, in studies of road networks the vertices usually represent road intersections and the edges roads. The study by Sen *et al.* [294] of the rail network of India provides an interesting counterexample. Sen *et al.* argue, plausibly, that in the context of rail travel what matters to most people is whether there is a direct train to their destination or, if there is not, how many trains they will have to take to get there. People do not care so much about how many stops there are along the way, so long as they don't have to change trains. Thus, Sen *et al.* argue, a useful network representation in the case of rail travel is one in which the vertices represent locations and two vertices are connected by an edge if a single train runs between them. Then the distance between two vertices in the network—the number of edges you need to traverse to get from A to B—is equal to the number of trains you would have to take. A better representation still (although Sen *et al.* did not consider it) would be a "bipartite network," a network containing two types of vertex, one representing the locations and the other representing train routes. Edges in the network would then join locations to the routes that run through them. The first, simpler representation of Sen *et al.* can be derived from the bipartite one by making a "one-mode projection" onto the locations only. Bipartite networks and their projections are discussed in greater detail in Section 6.6.

2.5 DELIVERY AND DISTRIBUTION NETWORKS

Falling somewhere between transportation networks and power grids are the distribution networks, about which relatively little has been written within the field of networks research. Distribution networks include things like oil and gas pipelines, water and sewerage lines, and the routes used by the post office and package delivery and cargo companies. Figure 2.5 shows one example, the European gas distribution network, taken from a study by Carvalho *et al.* [64], who constructed the figure from data purchased from industry sources. In this network the edges are gas pipelines and the vertices are their intersections, including pumping, switching, and storage facilities and refineries.

If one is willing to interpret "distribution" in a loose sense, then one class of distribution networks that has been relatively well studied is river networks, though if one wants to be precise river networks are really collection networks, rather than distribution networks. In a river network the edges are rivers or streams and the vertices are their intersections. Like road networks no special techniques are necessary to gather data on the structure of river networks—the hard work of surveying the land has already been done for us by surveyors and cartographers, and all we need do is copy the results off their maps. See Fig. 2.6 for an example of a river network.

The topological and geographic properties of river networks have been studied in some detail [94, 208, 284]. Of particular note is the fact that river networks, to an excellent approximation, take the form of trees. That is, they contain no loops (if one disregards the occasional island midstream), a point that we discuss in more detail in Section 6.7.

Similar in some respects to river networks are networks of blood vessels in animals, and their equivalents in plants, such as root networks. These too

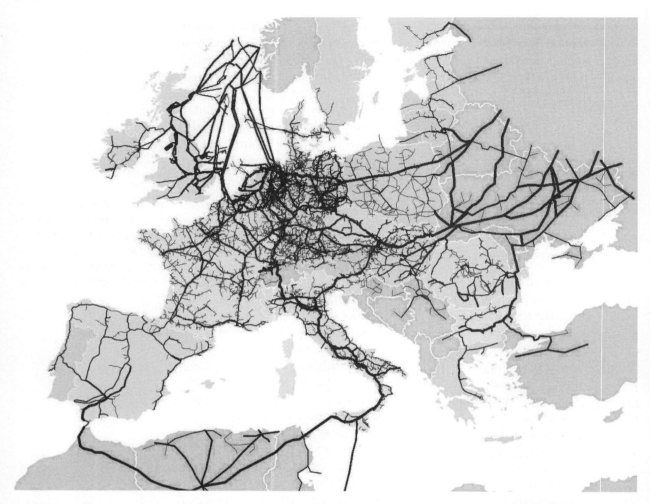

Figure 2.5: The network of natural gas pipelines in Europe. Thickness of lines indicates the sizes of the pipes. Figure created by R. Carvalho *et al.* [64]. Copyright 2009 American Physical Society. Reproduced with permission.

have been studied at some length. An early example of a mathematical result in this area is the formula for estimating the total geometric length of all edges in such a network by observing the number of times they intersect a regular array of straight lines [231]. This formula, whose derivation is related to the well-known "Buffon's needle" experiment for determining the value of π, is most often applied to root systems, but there is no reason it could not also be useful in the study of river networks or, with suitable modification, any other

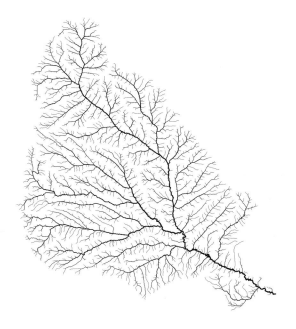

Figure 2.6: Drainage basin of the Loess Plateau. The network of rivers and streams on the Loess Plateau in the Shanxi province of China. The tree-like structure of the network is clearly visible—there are no loops in the network, so water at any point in the network drains off the plateau via a single path. Reproduced from Pelletier [266] by permission of the American Geophysical Union.

type of geographic network.

Also of note in this area is work on the scaling relationships between the structure of branching vascular networks in organisms and metabolic processes [26, 325, 326], an impressive example of the way in which an understanding of network structure can be parlayed into an understanding of the functioning of the systems the networks represent. We will see many more examples during the course of this book.

CHAPTER 3

SOCIAL NETWORKS

A discussion of social networks and the empirical
techniques used to probe their structure

SOCIAL networks are networks in which the vertices are people, or sometimes groups of people, and the edges represent some form of social interaction between them, such as friendship. Sociologists have developed their
own language for discussing networks: they refer to the vertices, the people,
as *actors* and the edges as *ties*. We will sometimes use these words when discussing social networks.

   We begin this chapter with a short summary of the origins and research focus of the field of social networks, before describing in detail some of the techniques used to discover social network structure. The material in this chapter
forms the basis for understanding many of the social network examples that
appear in the rest of the book.

3.1   THE EMPIRICAL STUDY OF SOCIAL NETWORKS

To most people the words "social network," if they mean anything, refer to
online social networking services such as *Facebook* and *MySpace*. The study of
social networks, however, goes back far farther than the networks' modern-
day computer incarnations. Indeed, among researchers who study networks,
sociologists have perhaps the longest and best established tradition of quantitative, empirical work. There are clear antecedents of social network analysis
to be found in the literature as far back as the end of the nineteenth century.
The true foundation of the field, however, is usually attributed to psychiatrist
Jacob Moreno, a Romanian immigrant to America who in the 1930s became
interested in the dynamics of social interactions within groups of people. At a

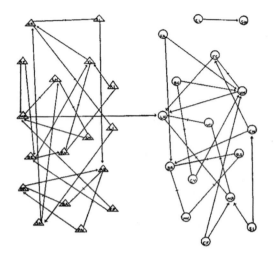

Figure 3.1: Friendships between schoolchildren. This early hand-drawn image of a social network, taken from the work of psychiatrist Jacob Moreno, depicts friendship patterns between the boys (triangles) and girls (circles) in a class of schoolchildren in the 1930s. Reproduced from [228] by kind permission of the American Society of Group Psychotherapy and Psychodrama.

medical conference in New York City in March 1933 he presented the results of a set of studies he had performed that may have been the first true social network studies, and the work attracted enough attention to merit a column in the *New York Times* a few days later. A year after that Moreno published a book entitled *Who Shall Survive?* [228] which, though not a rigorous work by modern standards, contained the seeds of the field of *sociometry*, which later became social network analysis.

Moreno called his diagrams of human interaction *sociograms*, rather than social networks (a term not coined until about twenty years later), but in everything but name they are clearly what we now know as networks. Figure 3.1, for instance, shows a hand-drawn figure from Moreno's book, depicting friendships within a group of schoolchildren. The triangles and circles represent boys and girls respectively and the figure reveals, among other things, that there are many friendships between two boys or two girls, but few between a boy and a girl. It is simple conclusions like this, that are both sociologically interesting and easy to see once one draws a picture of the network, that rapidly persuaded social scientists that there was merit in Moreno's methods.

One of the most important things to appreciate about social networks is that there are many different possible definitions of an edge in such a network and the particular definition one uses will depend on what questions one is interested in answering. Edges might represent friendship between individuals, but they could also represent professional relationships, exchange of goods or money, communication patterns, romantic or sexual relationships, or many other types of connection. If one is interested, say, in professional interactions between the boards of directors of Fortune 500 companies, then a network of who is dating whom or who looks at who else's Facebook page is probably not of much use. Moreover, the techniques one uses to probe different types of social interaction can also be quite different, so that different kinds of social network studies are typically needed to address different kinds of questions.

Direct questioning of experimental subjects is probably the most common method of determining the structure of social networks. We discuss it in detail in Section 3.2. Another important technique, the use of archival records (Sections 3.4 and 3.5), is illustrated by a different early example of a social network study. It was, apparently, a common practice in the US in the 1930s for newspapers to report on the public appearances of society women, and Davis, Gardner, and Gardner made use of this in a study of a social network of 18 women in a city in the American south. This study, often referred to in the literature as the "Southern Women Study," was described in a book by the researchers published in 1941 [86], although it was based on data from 1939. They took a sample of 14 social events attended by the women in question and recorded which women attended which events. Women in this network may be considered connected if they attended a common event. An alternative and more complete representation of the data is as an "affiliation network" or "bipartite graph," a network with two types of vertex, representing the women and the events, with edges connecting each woman to the events she attended. A visualization of the affiliation network for the Southern Women Study is shown in Fig. 3.2. One reason why this study has become so well known, in addition to its antiquity, is that the women were found by the researchers to split into two subgroups, tightly knit clusters of acquaintances with only rather loose between-cluster interaction. A classic problem in social network analysis is to devise a method or algorithm that can discover and extract such clustering from raw network data, and quite a number of researchers have made use of the Southern Women data as a test case for the development of such methods. Affiliation networks receive further attention in Section 3.5.

Such is the power of social network analysis that its techniques have, since Moreno and Davis *et al.*, been applied to an extraordinary variety of different communities, issues, and problems, including friendship and acquaintance patterns in local communities and in the population at large [36, 37, 175, 219, 311], and among students [334] and schoolchildren [112, 225, 277], contacts between business people and other professionals [78, 134], boards of directors of companies [87, 88, 207], collaborations of scientists [145, 146, 236], movie actors [16, 323], and musicians [139], sexual contact networks [183, 198, 272, 285] and dating patterns [34], covert and criminal networks such as networks of drug users [289] or terrorists [191], historical networks [259], online commu-

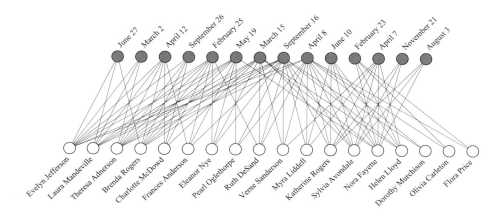

Figure 3.2: The affiliation network of the "Southern Women Study." This network (like all affiliation networks) has two types of vertex, the open circles at the bottom representing the 18 women who were the subjects of the study and the shaded circles at the top representing the social events they attended. The edges connect each woman to the events she attended, as deduced from newspaper reports. Data courtesy of L. Freeman and originally from Davis *et al.* [86].

nities such as Usenet [204, 300, 312] or Facebook [196], and social networks of animals [205, 286, 287].

We will see some examples of these and other networks throughout this book and we will give details as needed as we go along. The rest of the present chapter is devoted to a discussion of the different empirical methods used to measure social networks. The two techniques described above, namely direct questioning of subjects and the use of archival records, are two of the most important, but there are several others that find regular use. This chapter does not give a complete review of the subject—for that we refer the reader to specialized texts such as those of Wasserman and Faust [320] and Scott [293]—but we introduce as much material as will be needed for the later chapters of the book, while at the same time, we hope, giving some flavor for the challenges of empirical study in the field of social networks.

3.2 INTERVIEWS AND QUESTIONNAIRES

The most common general method for accumulating data on social networks is simply to ask people questions. If you are interested in friendship networks, then you ask people who their friends are. If you are interested in business relationships you ask people who they do business with, and so forth. The asking may take the form of direct interviews with participants or the completion by participants of questionnaires, either on paper or electronically. Indeed many modern studies, particularly surveys conducted by telephone, employ a combination of both interviews and questionnaires, wherein a professional interviewer reads questions from a questionnaire to a participant. By using a questionnaire, the designers of the study can guarantee that questions are asked, to a good approximation, in a consistent order and with consistent wording. By employing an interviewer to do the asking the study gains flexibility and reliability: interviewees often take studies more seriously when answering questions put to them by a human being, and interviewers may be given some latitude to probe interviewees when they are unclear, unresponsive, or confused. These are important considerations, since misunderstanding and inconsistent interpretation of survey questions are substantial sources of error. By making questions as uniform as possible and giving respondents personal help in understanding them, these errors can be reduced. A good introduction to social survey design and implementation has been given by Rea and Parker [279].

To find out about social networks, surveys typically employ a *name generator*, an item or series of items that invite respondents to name others with whom they have contact of a specified kind. For example, in their classic study of friendship networks among schoolchildren, Rapoport and Horvath [277] asked children to complete a questionnaire that included items worded as follows:[1]

My best friend at ＿＿ Junior High School is:
My second-best friend at ＿＿ Junior High School is:
My third-best friend at ＿＿ Junior High School is:

．
．
．

My eighth-best friend at ＿＿ Junior High School is:

The blanks "＿＿" in the questionnaire were filled in with the appropriate school name. The list stopped at the eighth-best friend and many children did not complete all eight.

Ideally all students within the school would be surveyed, though Rapoport and Horvath reported that in their case a few were absent on the day the survey was conducted. Note that the survey specifically asks children to name

---

[1] A junior high school in the United States is a school for children aged approximately 12 to 14 years.

only friends within the school. The resulting network will therefore record friendship ties within the school but none to individuals outside. Since all social network studies are limited to some community or portion of the population, and since it is highly unlikely that such a community will have ties solely within the community and none outside, all surveys must make some decision about how to deal with ties to outside individuals. Sometimes they are recorded. Sometimes, as here, they are not. Such details can be important since statistics derived from the survey results will often depend on the decisions made.

There are some points to notice about the data produced by name generators. First, the network ties, friendships in the case above, are determined by one respondent nominating another by name. This is a fundamentally asymmetric process. Individual A identifies individual B as their friend. In many cases B will also identify A as *their* friend, but there is no guarantee that this will happen and it is not uncommon for nomination to go only one way. We normally think of friendship as a two-way type of relationship, but surveys suggest that this not always the case. As a result, data derived from name generators are often best represented as directed networks, networks in which edges run in a particular direction from one vertex to another. If two individuals nominate each other then we have two directed edges, one pointing in either direction. Each vertex in the network also has two degrees, an out-degree—the number of friends identified by the corresponding individual—and an in-degree—the number of others who identified the individual as a friend.

We encountered directed networks previously in Chapter 1, in our discussion of the World Wide Web, and they are discussed in more detail in Section 6.4.

This brings us to a second point about name generators. It is common, as in the example above, for the experimenter to place a limit on the number of names a respondent can give. In the study of Rapoport and Horvath, this limit was eight. Studies that impose such a limit are called *fixed choice* studies. The alternative is to impose no limit. Studies that do this are called *free choice* studies.

Limits are often imposed purely for practical purposes, to reduce the work the experimenter must do. However, they may also help respondents understand what is required of them. In surveys of schoolchildren, for instance, there are some children who, when asked to name all their friends, will patiently name all the other children in the entire school, even if there are hundreds of them. Such responses are not particularly helpful in surveys—almost certainly the children in question are employing a definition of friendship different from that employed by most of their peers and by the investigators.

However, limiting the number of responses is for most purposes undesirable. In particular, it clearly limits the out-degree of the vertices in the network, imposing an artificial and possibly unrealistic cut-off. As discussed in Chapter 1, an interesting property of many networks is the existence of a small number of vertices with unusually high degree, and it is known that in some cases these vertices, though few in number, can have a dominant effect on the behavior of the network as a whole. By employing a name generator that artificially cuts off the degree, any information about the existence of such vertices is lost.

It is worth noticing, however, that even in a fixed-choice study there is normally no limit on the *in*-degree of vertices in the network; there is no limit to the number of times an individual can be nominated by others. And indeed in many networks it is found that a small number of individuals are nominated an unusually large number of times. Rapoport and Horvath [277] observed this in their friendship networks: while most children in a school are nominated as a friend of only a few others, a small number of popular children are nominated very many times. Rapoport and Horvath were some of the first scientists in any field to study quantitatively the degree distribution of a network, reporting and commenting extensively on the in-degrees in their friendship networks.

Not all surveys employing name generators produce directed networks. Sometimes we are interested in ties that are intrinsically symmetric between the two parties involved, in which case the edges in the network are properly represented as undirected. An example is networks of sexual contact, which are widely studied to help us understand the spread of sexually transmitted diseases [183,198,272,285]. In such networks a tie between individuals A and B means that A and B had sex. While participants in studies sometimes do not remember who they had sex with or may be unwilling to talk about it, it is at least in principal a straightforward yes-or-no question whether two people had sex, and the answer should not depend on which of the two you ask.[2] In such networks therefore, ties are normally represented as undirected.

Surveys can and often do ask respondents not just to name those with whom they have ties but to describe the nature of those ties as well. For instance, questions may ask respondents to name people they both like and dislike, or to name those with whom they have certain types of contact, such as socializing together, working together, or asking for advice. For example, in a study of the social network of a group of medical doctors, Coleman *et al.* [78] asked respondents the following questions:

---

[2]One can, by asking both, make some estimate of the accuracy of the survey. If individuals' responses disagree too often, it is a clear sign that the reliability of the responses is poor.

Who among your colleagues do you turn to most often for advice?

With whom do you most often discuss your cases in the course of an ordinary week?

Who are the friends among your colleagues who you see most often socially?

The names of a maximum of three doctors could be given in response to each question. A survey such as this, which asks about several types of interactions, effectively generates data on several different networks at once—the network of advice, the discussion network, and so forth.

Surveys may also pose questions aimed at measuring the strength of ties, asking for instance how often people interact or for how long, and they may ask individuals to give a basic description of themselves: their age, income, education, and so forth. Some of the most interesting results of social network studies concern the extent to which people's choice of whom they associate with reflects their own background and that of their associates. For instance, you might choose to socialize primarily with others of a similar age to yourself, but turn for advice to those who are older than you.

The main disadvantages of network studies based on direct questioning of participants are that they are first laborious and second inaccurate. The administering of interviews or questionnaires and the collation of responses is a demanding job that has been only somewhat helped in recent years by the increasing availability of computers and the use of online survey tools. Most studies have been limited to a few tens or at most hundreds of respondents— the 34-vertex social network of Fig. 1.2 is a typical example. It is a rare study that contains more than a thousand actors, and studies such as the National Longitudinal Study of Adolescent Health,[3] which compiled responses from over 90 000 participants, are very unusual and extraordinarily costly. Only a substantial public interest such as, in that case, the control of disease, can justify their funding.

Data based on direct questioning are also plagued by uncontrolled biases. Answers given by respondents are always, to some extent, subjective. If you ask people who their friends are, different people will interpret "friend" in different ways and thus give different kinds of answers. Investigators do their best to pose questions and record answers in a uniform fashion, but it is inevitable that inconsistencies will be present in the final data and anyone who has ever conducted a survey knows this well. This problem is not unique to social network studies. Virtually all social surveys suffer from such problems

---

[3]See www.cpc.unc.edu/projects/addhealth.

and a large body of expertise has been developed concerning techniques for dealing with them. Nonetheless, one should bear in mind when dealing with any social network data derived from interviews or questionnaires the possibility of uncontrolled experimental bias in the results.

3.2.1   EGO-CENTERED NETWORKS

Studies of the type described in the previous section, in which all or nearly all of the individuals in a community are surveyed, are called *sociometric* studies, a term coined by Jacob Moreno himself (see the discussion at the beginning of this chapter). For the purposes of determining network structure, sociometric studies are desirable; unless we survey all or nearly all of the population of interest, there is no way we can reconstruct the complete network of ties within that population. However, as discussed at the end of the preceding section, sociometric studies also require a lot of work and for large populations may simply be infeasible.

At the other end of the spectrum lie studies of *personal networks* or *ego-centered networks*.[4] An ego-centered network is the network surrounding one particular individual, meaning, usually, the individual surveyed and his or her immediate contacts. The individual surveyed is referred to as the *ego* and the contacts as *alters*.

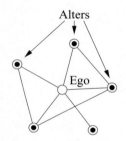

An ego-centered network consisting of an ego and five alters.

The typical survey of this kind is conducted using direct questioning techniques similar to those discussed in Section 3.2, with interviews, questionnaires, or a combination of both being the instruments of choice. One might, for instance, select a sample of the target population at random,[5] and ask them to identify all those with whom they have a certain type of contact. Participants might also be asked to describe some characteristics both of themselves and of their alters, and perhaps to answer some other simple questions, such as which alters also have contact with one another.

Obviously surveys of this type, and studies of ego-centered networks in general, cannot reveal the structure of an entire network. One receives snapshots of small local regions of the network, but in general those regions will not join together to form a complete social network. There are cases, however, where we are primarily interested in local network properties, and ego-

---

[4]Such networks are also called *egocentric* networks, although this term, which has its origins in social science and psychology, has taken on a different lay meaning which prompts us to avoid its use here.

[5]This can be done, for example, by *random-digit dialing*, the practice of calling random telephone numbers in the target area and surveying those who answer.

centered network studies can give us good data about these. For example, if we wish to know about the degrees of vertices in a network then a study in which a random sample of people are each asked to list their contacts can give us reasonable degree statistics. (Studies probing vertex degrees are discussed more below.) If we also gather data on the contacts between alters, we can estimate clustering coefficients (see Section 7.9). If we have data on characteristics of egos and alters we can measure assortative mixing (Sections 7.13 and 8.7).

An example of a study gathering ego-centered network data is the General Social Survey (GSS) [59], a large-scale survey conducted every year in the United States since 1972 (every two years since 1994). The GSS is not primarily a social network study. The purpose of the study is to gather data about life in the United States, how it is changing, and how it differs from or relates to life in other societies. The study contains a large number of items ranging from general questions probing the demographics and attitudes of the participants, to specific questions about recent events, political topics, or quality of life. However, among these many items there are in each iteration of the survey a few questions about social networks. The precise number and wording of these questions changes from one year to another, but here some examples from the survey of 1998, which was fairly typical:

> From time to time, most people discuss important matters with other people. Looking back over the last six months, who are the people with whom you discussed matters important to you? Do you feel equally close to all these people?

> Thinking now of close friends—not your husband or wife or partner or family members, but people you feel fairly close to—how many close friends would you say you have? How many of these close friends are people you work with now? How many of these close friends are your neighbors now?

And so on. By their nature these questions are of a "free choice" type, the number of friends or acquaintances the respondent can name being unlimited, although (and this is a criticism that has been leveled at the survey) they are also quite vague in their definitions of friends and acquaintances, so people may give answers of widely varying kinds.

Another example of an ego-centered network study is the study by Bernard *et al.* [36, 37, 175, 213] of the degree of individuals in acquaintance networks (i.e., the number of people that people know). It is quite difficult to estimate how many people a person knows because most people cannot recall at will all those with whom they are acquainted and there is besides a large amount of variation in people's subjective definition of "knowing." Bernard *et al.* came up with an elegant experimental technique to circumvent these difficulties. They

45

asked people to read through a list containing a sample of several hundred family names drawn from a telephone directory.[6] Participants counted up how many people they knew with names appearing on the list. Each person with a listed name was counted separately, so that two acquaintances called "Smith" would count as two people. They were instructed to use the following precise definition of acquaintance:

> You know the person and they know you by sight or by name; you can contact them in person by telephone or by mail; and you have had contact with the person in the past two years.

(Of course, many other definitions are possible. By varying the definition, one could probe different social networks.) Bernard *et al.* then fed the counts reported by participants into a statistical formula to estimate the total number of acquaintances of each participant.

Bernard *et al.* repeated their study with populations drawn from several different cities and the results varied somewhat from city to city, but overall they found that the typical number of acquaintances, in the sense defined above, of the average person in the United States is on the order of about 2000. In the city of Jacksonville, Florida, for instance, they found a figure of 1700, while in Orange County, California they found a figure of 2025. Many people find these numbers surprisingly high upon first encountering them, perhaps precisely because we are poor at recalling all of the many people we know. But repeated studies have confirmed figures of the same order of magnitude, at least in the United States. In some other countries the figures are lower. In Mexico City, for instance, Bernard *et al.* estimated that the average person knows about 570 others.

3.3 DIRECT OBSERVATION

An obvious method for constructing social networks is direct observation. Simply by watching interactions between individuals one can, over a period of time, form a picture of the networks of unseen ties that exist between those individuals. Most of us, for instance, will be at least somewhat aware of friendships or enmities that exist between our friends or coworkers. In direct observation studies, researchers attempt to develop similar insights about the members of the population of interest.

---

[6]Some care must be taken in the selection of the names, since the frequency of occurrence of names varies considerably, both from name to name, and geographically and culturally.

Direct observation tends to be a rather labor-intensive method of study, so its use is usually restricted to rather small groups, primarily ones with extensive face-to-face interactions in public settings. In Chapter 1 we saw one such example, the "karate club" network of Zachary [334]. Another example is the study by Freeman *et al.* [131, 132] of the social interactions of windsurfers on a beach. The experimenters simply watched the individuals in question and recorded the length in minutes of every pairwise interaction among them. A large number of direct-observation network data sets were compiled by Bernard and co-workers during the 1970s and 80s as part of a lengthy study of the accuracy of individuals' perception of their own social situation [38, 40, 41, 173]. These include data sets on interactions between students, faculty, and staff in a university department, on members of a university fraternity,[7] on users of a teletype service for the deaf, and several other examples.

One arena in which direct observation is essentially the only viable experimental technique is in studies of the social networks of animals—clearly animals cannot be surveyed using interviews or questionnaires. One method is to record instances of animal pairs engaging in recognizable social behaviors such as mutual grooming, courting, or close association and then to declare ties to exist between the pairs that engage in these behaviors most often [205]. Not all animal species form interesting or useful social networks, but informative studies have been performed of, amongst others, monkeys [121, 286, 287], kangaroos [143], and dolphins [80, 205]. Networks in which the ties represent aggressive behaviors have also been reported, such as networks of baboons [214], wolves [163, 316], and ants [77]. In cases where aggressive behaviors normally result in one animal's establishing dominance over another the resulting networks can be regarded as directed and are sometimes called *dominance hierarchies* [90, 91, 101].

3.4 DATA FROM ARCHIVAL OR THIRD-PARTY RECORDS

An increasingly important, voluminous, and often highly reliable source of social network data is archival records. Such records are, sometimes at least, relatively free from the vagaries of human memory and are often impressive in their scale, allowing us to construct networks of a size that would require far more effort were other techniques used.

---

[7]In American universities a "fraternity" is a semi-independent boarding house for male students.

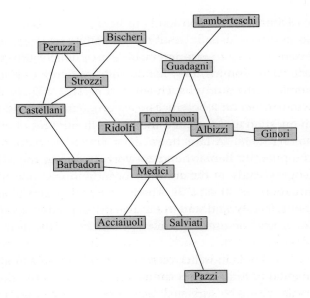

Figure 3.3: Intermarriage network of the ruling families of Florence. In this network the vertices represent fifteenth century Florentine families and the edges represent ties of marriage between them. After Padgett and Ansell [259].

A well-known small example of a study based on archival records is the study by Padgett and Ansell of the ruling families of Florence in the fifteenth century [259]. In this work, the investigators looked at contemporaneous historical records to determine which among the families had trade relations, marriage ties, or other forms of social contact with one another. Figure 3.3 shows one of the resulting networks, a network of intermarriages between 15 of the families. It is notable that the Medici family occupies a central position in this network, having marriage ties with members of no fewer than six other families. Padgett and Ansell conjectured that it was by shrewd manipulation of social ties such as these that the Medici rose to a position of dominance in Florentine society.

In recent years, with the widespread availability of computers and online databases, many more networks have been constructed from records of various types. A number of authors, for example, have looked at email networks [103, 313]. Drawing on email logs—automatic records kept by email servers of messages sent—it is possible to construct networks in which the vertices are people (or more correctly email addresses) and the directed edges between them are email messages. Exchange of email in such a network can be taken as a proxy for acquaintance between individuals, or we may be interested in the patterns of email exchange for some other reason. For instance, email messages can carry computer viruses and a knowledge of the structure of the network of messages may help us to predict and control the spread of those viruses.

Another form of email network is the network formed by email address books. An email address book is a computer file in which a computer user stores, for convenience, the email addresses of his or her regular correspondents. The set of all such address books can be regarded as defining a network in which the vertices represent the owners of the address books, and there is a directed edge from vertex A to vertex B if person B's address appears in person A's address book. This network is again of interest in the study of computer viruses, since some viruses search address books for the addresses of new victims to infect and hence spread over the address book network. Similar networks can also be constructed for other forms of electronic communication that use address books, such as instant messaging [301].

A form of network similar to but older than the email network is the *telephone call graph*. In such a network the vertices represent telephone numbers and directed edges between them represent telephone calls from one number to another. Call graphs can be reconstructed from call logs kept by telephone companies, although such logs are generally proprietary and not easily available outside of those companies, and call graphs have as a result only occasionally been examined in the scientific literature [1,9,258].

Telephone call graphs are quite distinct from the physical network of telephone cables discussed in Section 2.2. Indeed, a call graph is to the physical telephone network roughly as an email network is to the Internet.

Recent years have seen the rapid emergence of online social networking services, such as *Facebook* and *LinkedIn*, which exist primarily to promote, document, and exploit the networks of contacts between individuals. As a natural part of their operation, these services build records of connections between their participants and hence provide, at least in principle, a rich source of archival network data. These data, however, like those for telephone calls, are largely proprietary to the companies operating the services and hence quite difficult to get hold of. So far only a few studies have been published of online social networks [53], but internal studies have no doubt been performed by the companies themselves and it is only a matter of time before more data become publicly available.

A few other online communities, not explicitly oriented towards networks, have been studied using network techniques. For instance, Holme *et al.* [158] took records of interactions between members of a Swedish dating website and reconstructed from them the network of interactions between the site's members. This study was unusual in that the network was time-resolved— the date and time of each interaction were recorded, allowing investigators to reconstruct after the fact the timing and duration of contacts between individuals. Most of the sources of network data considered in this book are not time-resolved, but many of the networks they correspond to do nonetheless change over time. Time-resolved network studies, or *longitudinal* studies, as they are called in sociology, are certainly a growth area to watch for in the future.

Another source of network data representing online communities is the Internet newsgroup system *Usenet*, a worldwide online message-board system that allows users to post messages on a large variety of topics. Messages are date and time stamped and identified with the name or email address of the poster along with a unique reference number that allows a poster to indicate when a posting is a reply or follow-on to a previous posting. Thus one can reconstruct the thread of the conversation taking place in a newsgroup, and in particular assemble a network in which the vertices are posters and the edges represent a response by one poster to a posting by another. Studies of newsgroup networks of this kind have been performed by a number of authors [204, 300, 312].

Weblogs and online journals are another source of online social network data. Online journals of various kinds have become popular on the World Wide Web since around the turn of the century. On these websites the proprietor posts whatever thoughts he or she cares to make public, along with links to sites maintained by others. These links form a directed network that lies, in terms of semantic content, somewhere between a social network and the World Wide Web; the links are often informational—the linker wishes to bring to his or her readers' attention the contents of the linked site—but there is a strong social element as well, since people often link to sites operated by their friends or acquaintances. This trend is particularly noticeable within journal communities such as *LiveJournal* and among weblogs devoted to specific topics, such as science or politics. The structure of the networks of links can be extracted using "crawlers" similar to those used to search the Web—see Section 4.1. Studies of journals and weblogs have been performed for example by Adamic and Glance [4] and MacKinnon and Warren [206].

An interesting network that has features of both a social and a technological network is the network of trust formed by a set of cryptographic keys. Cryptosystems or cyphers (i.e., secret codes), long associated in the public mind with spies and skulduggery, have become a crucial part of the twenty-first-century economy, used to protect important data, particularly financial data such as credit card numbers, from theft and misuse. An important advance, central to the widespread and convenient use of cryptography, was the development in the 1970s of *public-key cryptography*. In traditional cryptosystems, two parties wishing to exchange messages must share a *key* that they use to encode and decode the messages. The key is typically a large number, which is used in combination with the chosen cryptosystem to dictate exactly how messages are converted from their original "plain text" form into code and back again. This key, which allows anyone possessing it to decode the messages, must be kept secret from any malicious snoopers, and this raises the difficult problem of how the parties in question agree on the key in the first place. Usually the key is generated by a computer program run by one party, but then it must be transmitted securely to the other party without anyone else seeing it. Sending the key over the Internet unencrypted would pose a significant risk of detection. Physical transmission, for example by conventional mail, would be reasonably secure, but would take a long time. Most customers buying goods over the Internet would not want to wait a week for a secret key to arrive by mail from their vendor.

These problems were solved with the invention of public-key cryptography in the 1970s. Public-key cryptosystems make use of any of several different *asymmetric cyphers* in which two different keys are used. One key, called the *public key*, is employed in the computer algorithm that converts the message from plain text into its encrypted form, but a different key, the *private key*, is needed to decrypt the message. The public key cannot be used for decryption.[8] The two keys are generated as a pair by one of the two parties wishing to exchange information and the public key is sent openly over the Internet or other channel to the second party. The private key remains the secret property of the first party and is not shared. The second party can then send messages to the first by encoding them with the public key and only the first party can decode them.[9] Although the public key can be intercepted by a third party in transmission, it will do the third party no good, since the public key cannot be used to decode messages, only to encode them. Indeed, in many cases, users of public-key systems deliberately broadcast their public keys to anyone who might want them, inviting the world to send them encoded messages, messages which only they can decode. It is from such practices that the name "public-key cryptography" arises.

Some asymmetric cyphers can also be used in the reverse direction. That is,

---

[8]Technically, the public key can be used to decrypt the message, but the calculation involved is extraordinarily complex and would take years or even centuries of effort on the fastest modern computers. For practical purposes, therefore, one can only decrypt the message if one has the private key.

[9]In practice it is a little more complicated than this. Asymmetric cyphers are computationally demanding to implement, far more so than the traditional (but less secure) symmetric cyphers in which the same key is used by both parties. To reduce demands on computer time, therefore, one usually uses the asymmetric cypher only to transmit from one party to the other a key for use in a symmetric cypher, and then the symmetric cypher, with that key, is used for all subsequent communications. In this way one benefits from the security of public-key cryptography without the computational overhead. For our purposes in this section, however, this is just a technical detail.

one can encode a message with the private key and it can only be decoded with the public key. Why would one want to do this, when everyone has the public key? The answer is that you can use it to prove your identity. Someone talking to you over the Internet, say, may want to be certain that you are who you claim to be (rather than some nefarious interloper) before they trust you with, for instance, their credit card number. So they send you a specific message that they choose, usually just a random string of digits, and ask you to encrypt it using your private key. Having done so, you send the encrypted message back to them and they decode it with the public key. If the decoded message matches the original one then they know that you are who you say you are, since no one else has your private key and hence no one else could have encrypted a message that decodes correctly with the public key.[10] This "digital signature" process is a crucial part of electronic commerce, allowing buyers and sellers to confirm each other's identities before doing business, and is used millions of times every day in transactions of every sort.

But there is still a fundamental problem with public-key encryption, namely the problem of knowing that the public key you are given really was created by the person you think it was created by. Some malicious person could create a public/private key pair and broadcast the public key to the world, labeled with, say, the name of a bank or retail trader, then use that key in a digital signature scheme to persuade some unsuspecting victim that they are the trader and that the victim should send them a credit card number.

One way around this problem is to have people sign each other's public keys [267]. That is, party A takes a public key that claims to belong to party B, and that A knows in fact to be genuine, and encrypts it with their own private key. Now if you have A's public key and you believe it to be genuine, then you can take the encrypted key and decode it with A's public key, thereby recovering B's public key, which A says is genuine. If you trust A to make this statement, then you can now also trust that the key you have is B's true public key.

But now one can repeat the process. Now that you have a genuine public key for party B, and if you trust B, then B can now sign the keys that *they* know to be genuine and you will be able to verify that they are genuine also. In this way, parties who trust each other can securely represent to one another that keys are genuine.

---

[10] Again, this is not completely true. One can encode a message using the public key that will decode with the same key, but again the calculations necessary to do this are extraordinarily lengthy, much lengthier than those using the private key, and hence for practical purposes only the person with the private key could have created the encrypted message.

The act of digitally signing someone else's public key is equivalent to saying that you know, or at least believe, the public key to be genuine, belonging to the person it claims to belong to. That act can be represented by a directed edge in a network. The vertices in the network represent the parties involved and a directed edge from party A to party B indicates that A has signed B's public key. The resulting directed network certainly has technological aspects but is in many ways more of a social network than anything else. People tend to vouch for the keys of other people they know, people they have communicated with or worked with frequently, so that they have both a good idea that the key in question is indeed genuine and a personal reason for making the effort to sign it.

Since public keys and the digital signatures of the people who sign them are, by necessity, public, it is relatively easy to construct a key-signing network from widely available data. There are a number of widely used key-signing networks associated, usually, with particular commercial cryptography products. One of the largest, for instance, is the network associated with the cryptography program PGP [267]. There have been only a small number of studies so far of the properties of key signing networks [47, 148] but there are certainly interesting questions awaiting answers in this area.

3.5 AFFILIATION NETWORKS

An important special case of the reconstruction of networks from archival records is the *affiliation network*. An affiliation network is a network in which actors are connected via comembership of groups of some kind. We saw one example in the introduction to this chapter, the Southern Women Study of Davis *et al.* [86], in which the authors drew their data from newspaper reports of social events and the "groups" were the sets of individuals who attended particular events. As we saw, the most complete representation of an affiliation network is as a network with two types of vertex representing the actors and the groups, with edges connecting actors to the groups to which they belong— see Fig. 3.2 on page 39. In such a representation, called a "bipartite network" or "two-mode network," there are no edges connecting actors directly to other actors (or groups to other groups), only actors to groups.

We study bipartite networks in more detail in Section 6.6.

Many examples of affiliation networks can be found in the literature. Another famous case is the study by Galaskiewicz [134] of the CEOs of companies in Chicago in the 1970s and their social interaction via clubs that they attended. In this network the CEOs are the actors and the clubs are the groups. Also in the business domain, quite a number of studies have been conducted of the boards of directors of companies [87, 88, 207]. In these networks the actors are company directors and the groups are the boards on which they sit. In addition to looking at the connections between directors in such networks, which arise as a result of their sitting on boards together, a considerable amount of attention has also been focused on the connections between boards (and hence between companies) that arise as a result of their sharing a common director, a so-called board "interlock."

More recently, some extremely large affiliation networks have been studied in the mathematics and physics literature. Perhaps the best known example is the network of collaborations of film actors, in which the "actors" in the network sense are actors in the dramatic sense also, and the groups to which they belong are the casts of films. This network is the basis, among other things, for a well-known parlor game, sometimes called the "Six Degrees of Kevin Bacon," in which one attempts to connect pairs of actors via chains of intermediate costars in a manner reminiscent of the small-world experiments of Stanley Milgram, which we discuss in Section 3.6. The film actor network has, with the advent of the Internet, become very thoroughly documented and has attracted the attention of many network analysts in recent years [16, 27, 323], although it is not clear whether there are any conclusions of real scientific interest to be drawn from its study.

Another example of a large affiliation network, one that holds more promise of providing useful results, is the coauthorship network of academics. In this network an actor is an academic author and a group is the set of authors of a learned paper. Like the film actor network, this network has become well documented in the last few years with the appearance of extensive online bibliographic resources covering many areas of human endeavor. Whether one is interested in papers published in journals or in more informal forums such as online preprint servers, excellent records now exist in most academic fields of authors and the papers they write, and a number of studies of the corresponding affiliation networks have been published [29, 89, 145, 146, 234–236].

3.6   THE SMALL-WORLD EXPERIMENT

An unusual contribution to the social networks literature was made by the experimental psychologist Stanley Milgram in the 1960s with his now-famous "small-world" experiments [219, 311]. Milgram was interested in quantifying the typical distance between actors in social networks. As discussed in Chapter 1, the "geodesic distance" between two vertices in a network is the minimum number of edges that must be traversed to travel from one vertex to the other through the network. Mathematical arguments suggest (as we will see later in this book) that this distance should be quite small for most pairs of vertices in most networks, a fact that was already well known in Milgram's time.[11] Milgram wanted to test this conjecture in real networks and to do this he concocted the following experiment.[12]

Milgram sent a set of packages, 96 in all, to recipients randomly chosen from the telephone directory in the US town of Omaha, Nebraska. The packages contained an official-looking booklet, or "passport," emblazoned with the crest of Milgram's home institution, Harvard University. Written instructions were included asking the recipients to attempt to get the passport to a specified target individual, a friend of Milgram's who lived in Boston, Massachusetts, over a thousand miles away. The only information supplied about the target was his name (and hence indirectly the fact that he was male), his address, and his occupation as a stockbroker. But the passport holders were not allowed simply to send their passport to the given address. Instead they were asked to pass it to someone they knew on a first-name basis and more specifically the person in this category who they felt would stand the best chance of getting the passport to the intended target. Thus they might decide to send it to someone they knew who lived in Massachusetts, or maybe someone who worked in the financial industry. The choice was up to them. Whoever they did send the passport to was then asked to repeat the process, sending it on to one of *their* acquaintances, so that after a succession of such steps the passport would, with luck, find its way into the hands of its intended recipient. Since every step of the process corresponded to the passport's changing hands between a pair of first-name acquaintances, the entire path taken corresponded to a path along the edges of the social network formed by the set of all such acquaintanceships. Thus the length of the path taken provided an upper bound on the geodesic distance in this network between the starting and ending individuals in the chain.

Of the 96 passports sent out, 18 found their way to the stockbroker target in Boston. While this may at first sound like a low figure, it is actually remarkably high—recent attempts to repeat Milgram's work have resulted in response rates orders of magnitude lower [93]. Milgram asked participants to record in the passport each step of the path taken, so he knew, among other things, how long each path was, and he found that the mean length of completed paths

---

[11]Milgram was particularly influenced in his work by a mathematical paper by Pool and Kochen [270] that dealt with the small-world phenomenon and had circulated in preprint form in the social science community for some years when Milgram started thinking about the problem, although the paper was not officially published until many years later.

[12]In fact Milgram conducted several sets of small-world experiments. The one described here is the first and most famous, but there were others [186, 311].

from Omaha to the target was just 5.9 steps. This result is the origin of the idea of the "six degrees of separation," the popular belief that there are only about six steps between any two people in the world.[13]

There are of course many reasons why this result is only approximate. Milgram used only a single target in Boston, and there is no guarantee the target was in any way typical of the population as a whole. And all the initial recipients in the study were in a single town in the same country.[14] (None of the completed chains that reached the target went outside the country.) Also there is no guarantee that chains took the shortest possible route to the target. Probably they did not, at least in some cases, so that the lengths of the paths found provide, as we have said, only an upper bound on the actual geodesic distance between vertices. And most of the chains of course were never completed. The passports were discarded or lost and never made their way to the target. It is reasonable to suppose that the chances of getting lost were greater for passports that took longer paths, and hence that the paths that were completed were a biased sample, having typical lengths shorter than the average.

For all these reasons and several others, Milgram's experiments should be taken with a large pinch of salt. Even so, the fundamental result that vertex pairs in social networks tend on average to be connected by short paths is now widely accepted, and has moreover been shown to extend to many other kinds of networks as well. Enough experiments have confirmed the effect in enough networks that, whatever misgivings we may have about Milgram's particular technique, the general result is not seriously called into question.

Milgram's experiments also, as a bonus, revealed some other interesting features of acquaintance networks. For instance, Milgram found that most of the passports that did find their way to the stockbroker target did so via just three of the target's friends. That is, a large fraction of the target's connections to the outside world seemed to be through only a few of his acquaintances, a phenomenon sometimes referred to as the "funneling" effect. Milgram called such well-connected acquaintances "sociometric superstars," and their existence has occasionally been noted in other networks also, such as collaboration networks [234], although not in some others [93].

Funneling is discussed further in Section 8.2.

A further interesting corollary of Milgram's experiment has been high-

---

[13]The phrase "six degrees of separation" did not appear in Milgram's writing. It is more recent and comes from the title of a popular Broadway play by John Guare [149], later made into a film, in which the lead character discusses Milgram's work.

[14]Furthermore, it appears that some of the initial recipients may have been selected not at random but by advertising for volunteers in the local newspaper [181], a procedure unlikely to produce a truly random sample of the population.

lighted by Kleinberg [177, 178]. (Milgram himself seems not to have appreciated the point.) The fact that a moderate number of the passports did find their way to the intended target person shows not only that short paths exist in the acquaintance network, but also that people are good at finding those paths. Upon reflection this is quite a surprising result. As Kleinberg has shown, it is possible and indeed common for a network to possess short paths between vertices but for them to be hard to find unless one has complete information about the structure of the entire network, which the participants in Milgram's studies did not. Kleinberg has suggested a possible explanation for how participants found the paths they did, based on conjectures about the structure of the network. We discuss his ideas in detail in Section 19.3.

Recently the small-world experiment has been repeated by Dodds *et al.* [93] using the modern medium of email. In this version of the experiment participants forwarded email messages to acquaintances in an effort to get them ultimately to a specified target person about whom they were told a few basic facts. The experiment improved on that of Milgram in terms of sheer volume, and also by having much more numerous and diverse target individuals and starting points for messages: 24 000 chains were started, most (though not all) with unique starting individuals, and with 18 different participating targets in 13 different countries. On the other hand, the experiment experienced enormously lower rates of participation than Milgram's, perhaps because the public is by now quite jaded in its attitude towards unsolicited mail. Of the 24 000 chains, only 384, or 1.5%, reached their intended targets, compared with 19% in Milgram's case. Still, the basic results were similar to those of Milgram. Completed chains had an average length of just over four steps. Because of their better data and considerably more careful statistical analysis, Dodds *et al.* were also able to compensate for biases due to unfinished chains and estimated that the true average path length for the experiment was somewhere between five and seven steps—very similar to Milgram's result. However, Dodds *et al.* observed no equivalent of the "sociometric superstars" of Milgram's experiment, raising the question of whether their appearance in Milgram's case was merely a fluke of the particular target individual he chose rather than a generic property of social networks.

An interesting variant on the small-world experiment has been proposed by Killworth and Bernard [39, 174], who were interested in how people "navigate" through social networks, and specifically how participants in the small-world experiments decide whom to forward messages to in the effort to reach a specified target. They conducted what they called "reverse small-world" ex-

The mechanisms of network search and message passing are discussed in greater detail in Section 19.3.

periments[15] in which they asked participants to *imagine* that they were taking part in a small-world experiment. A (fictitious) message was to be communicated to a target individual and participants were asked what they wanted to know about the target in order to make a decision about whom to forward the message to. The actual passing of the message never took place; the experimenters merely recorded what questions participants asked about the target. They found that three characteristics were sought overwhelmingly more often than any others, namely the name of the target, their geographic location, and their occupation—the same three pieces of information that Milgram provided in his original experiment. Some other characteristics came up with moderate frequency, particularly when the experiment was conducted in non-Western cultures or among minorities: in some cultures, for instance, parentage or religion were considered important identifying characteristics of the target.

While the reverse small-world experiments do not directly tell us about the structure of social networks, they do give us information about how people perceive and deal with social networks.

3.7   SNOWBALL SAMPLING, CONTACT TRACING, AND RANDOM WALKS

Finally in this chapter on social networks we take a look at a class of network-based techniques for sampling hidden populations.

Studies of some populations, such as drug users or illegal immigrants, present special problems to the investigator because the members of these populations do not usually want to be found and are often wary of giving interviews. Techniques have been developed, however, to sample these populations by making use of the social network that connects their members together. The most widely used such technique is *snowball sampling* [108, 127, 310].

Note that, unlike the other experimental techniques discussed in this chapter, snowball sampling is not intended as a technique for probing the structure of social networks. Rather, it is a technique for studying hidden populations that relies on social networks for its operation. It is important to keep this distinction clear. To judge by the literature, some professional social network analysts do not, and the results are often erroneous conclusions and bad science.

---

[15]Also sometimes called "INDEX" experiments, which is an abbreviation for "informant-defined experiment."

Standard techniques such as telephone surveys often do not work well when sampling hidden populations. An investigator calling a random telephone number and asking if anyone on the other end of the line uses drugs is unlikely to receive a useful answer. The target population in such cases is small, so the chances of finding one of its members by random search are also small, and when you do find one they will very likely be unwilling to discuss the highly personal and possibly illicit topic of the survey with an investigator they have never met before and have no reason to trust.

So investigators probe the population instead by getting some of its members to provide contact details for others. The typical survey starts off rather like a standard ego-centered network study (Section 3.2.1). You find one initial member of the population of interest and interview them about themselves. Then, upon gaining their confidence, you invite them also to name other members of the target population with whom they are acquainted. Then you go and find those acquaintances and interview *them* asking them also to name further contacts, and so forth through a succession of "waves" of sampling. Pretty soon the process "snowballs" and you have a large sample of your target population to work with.

Clearly this is a better way of finding a hidden population than random surveys, since each named individual is likely to be a member of the population, and you also have the advantage of an introduction to them from one of their acquaintances, which may make it more likely that they will talk to you. However, there are some serious problems with the method as well. In particular, snowball sampling gives highly biased samples. In the limit of a large number of waves, snowball sampling samples actors with probability proportional to their "eigenvector centrality" (see Section 7.2). Unfortunately, this limit is rarely reached in practice, and in any case the eigenvector centrality cannot be calculated without knowledge of the complete contact network, which by definition we don't have, making correction for the sampling bias difficult. In short, snowball sampling gives biased samples of populations and there is little we can do about it. Nonetheless, the technique is sufficiently useful for finding populations that are otherwise hard to pin down that it has been widely used, biases and all, in studies over the last few decades.

Sometimes, in the case of small target populations, a few waves of snowball sampling may find essentially all members of a local population, in which case the method can be regarded as returning data about the structure of the social network. If the contacts of each interviewed participant are recorded in the study, it should be possible to reconstruct the contact network when the study is complete. This has occasionally been done in such studies, although as noted above the object is more often to exploit the social network to find the population than to study the network itself.

A technique closely related to snowball sampling is *contact tracing*, which is essentially a form of snowball sampling applied to disease incidence. Some diseases, such as tuberculosis and HIV, are considered sufficiently serious that, when someone is discovered to be carrying them, an effort must be made to track down all those who might also have been infected. Thus, in most Western countries, when a patient tests positive for HIV, for instance, he or she will be questioned about recent sexual contacts, and possibly about other types of potentially disease-carrying contacts, such as needle sharing if the patient is an injection drug user. Then health authorities will make an effort to track down those contacts and test them also for HIV. The process is repeated with any who test positive, tracing their contacts as well, and so forth, until all leads have been exhausted. While the primary purpose of contract tracing is to curtail disease outbreaks and safeguard the health of the population, the process also produces data about the community through which a disease is spreading and such data have sometimes been used in scientific studies, particularly of sexually transmitted diseases, for which data may otherwise be hard to come by. Population samples derived from contact tracing studies display biases similar in type and magnitude to those seen in snowball sampling and should be treated with the same caution. Indeed, they contain extra biases as well, since contacts are rarely pursued when an individual tests negative for the disease in question, so the sample is necessarily dominated by carriers of the disease, who are themselves usually a biased sample of the population at large. Also, as with snowball sampling, contact tracing data can provide us with an experimental window on the structure of the contact network itself, but again we expect the data to be strongly biased, except in cases of small target populations for which the sampling process saturates.

There is another variant of snowball sampling that deals to some extent with the problems of bias in the sample. This is *random-walk sampling* [182,310]. In this method one again starts with a single member of the target community and interviews them and determines their contacts. Then, however, instead of interviewing all of those contacts, one chooses one of them at random and interviews only that one at the next step. If the person in question cannot be found or declines to be interviewed, one simply chooses another contact, and the process is repeated. Initially it appears that this will be a more laborious process than standard snowball sampling, since one spends a lot of time determining the names of individuals one never interviews, but this is not the case. In either method one has to determine the contacts of each person interviewed, so the total amount of work for a sample of a given size is the same. It is however very important that one really does determine all the contacts of each individual, even though most of the time only one of them is pursued. This is because for the method to work correctly one must make a random choice among those contacts, for example by rolling a die (or some modern electronic version thereof). To do this one must know the full set of contacts one is choosing between.

The advantage of the random-walk sampling method is that, as shown in Section 6.14, the asymptotic sampling probability of vertices in a random walk is simply proportional to vertex degree (see Eq. (6.60)). What's more, the asymptotic regime in such studies is, unlike snowball sampling, reached quite quickly for relatively small sample sizes.[16]

Knowing this, and given that we determine degree (i.e., the number of contacts an individual has) as a part of the interview process, we can easily compensate for sampling bias and make population estimates of quantities in a way that is, in theory at least, unbiased. In practice, many sources of bias remain, particularly those associated with participant subjectivity, inability to recall contacts, and non-participation of named contacts. Still, random-walk sampling is a great improvement on standard snowball sampling, and should be used more than it is. Its principal disadvantage is that it is relatively slow. Since the participants are interviewed serially, in a chain, rather than in parallel waves, a strict implementation of the method can take a long time to develop a large sample. One can get around this obstacle to some extent by running several short random walks in parallel instead of one long one, but the walks cannot be too short or they will not reach the asymptotic regime in which sampling is proportional to degree.

Another variant of the random-walk sampling idea is used to deal with a different problem, that of enrolling study participants. In some cases it is considered unethical to get participants to name their contacts, particularly when the topic of the study is one of dubious legality, and permission to perform such studies may be withheld by the authorities. To circumvent this problem one can make use of *respondent-driven sampling* [289]. In this technique, participants are usually paid to take part, and enrollment is achieved by handing out tickets to interviewees. Rather than asking people to name their contacts, the interviewees are simply told that they should give the tickets to their friends, and that both they and the friends will receive payment if the friend brings the ticket to the investigator and agrees to participate in the survey. In this

---

[16]In snowball sampling the sample size grows exponentially with the number of sampling waves and hence one typically only performs a logarithmic number of waves, which is not enough for the sampling process to reach equilibrium. In random walk sampling the sample size grows only linearly.

way, no one is ever asked to name names and all participants have actively volunteered their participation. In the case where a single ticket is given to each participant, the method is roughly equivalent to random-walk sampling and should in theory give a less biased sample than snowball sampling for the same reasons. In practice, a new bias is introduced because the recipient of the ticket is not necessarily chosen at random from an individual's acquaintances. Also, tickets frequently get lost or their recipients decline to participate, remuneration notwithstanding, so one would normally give out more than one ticket to each participant, which complicates the sampling process. Even so, it is believed that respondent-driven sampling provides superior population samples to snowball sampling, and it is the method of choice for studies in which one cannot ask people to name their contacts.

CHAPTER 4

NETWORKS OF INFORMATION

A description of networks of information or data, with a particular focus on the World Wide Web and citation networks

THIS CHAPTER focuses on networks of information, networks consisting of items of data linked together in some way. Information networks are all, so far as we know, man-made, with perhaps the best known example being the World Wide Web, though many others exist and are worthy of study, particularly citation networks of various kinds. These and several other types of information networks are discussed in this chapter.

In addition, there are some networks which could be considered information networks but which also have social aspects to them. Examples include networks of email communications, networks on social-networking websites such as *Facebook* or *LinkedIn*, and networks of weblogs and online journals. These and similar examples were discussed in the previous chapter on social networks, in Section 3.4, but they would have fitted perfectly well in the present chapter also. The classification of networks as social networks, information networks, and so forth is a fuzzy one, and there are plenty of examples that, like these, straddle the boundaries.

4.1 THE WORLD WIDE WEB

Although by no means the first information network created, the World Wide Web is probably the example best known to most people and a good place to start our discussion in this chapter.

As described in Chapter 1, the Web is a network in which the vertices are web pages consisting of text, pictures, or other information and the edges are the hyperlinks that allow us to navigate from page to page. Since hyperlinks

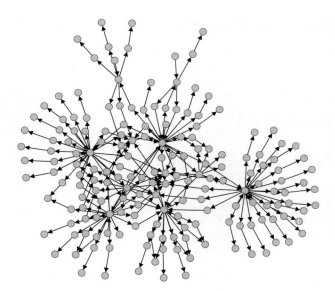

Figure 4.1: A network of pages on a corporate website. The vertices in this network represent pages on a website and the directed edges between them represent hyperlinks.

run in one direction only, the Web is a directed network. We can picture the network with an arrow on each edge indicating which way it runs. Some pairs of web pages are connected by hyperlinks running in both directions, which can be represented by two directed edges, one in each direction between the corresponding vertices. Figure 4.1 shows a picture of a small portion of the Web network, representing the connections between a set of web pages on a single website.

The World Wide Web was invented in the 1980s by scientists at the CERN high-energy physics laboratory in Geneva as a means of exchanging information among themselves and their coworkers, but it rapidly became clear that its potential was much greater [159]. At that time there were several similar ideas competing for dominance of the rapidly growing Internet, but the Web won the battle, largely because its inventors decided to give away for free the software technologies on which it was based—the Hypertext Markup Language (HTML) used to specify the appearance of pages and the Hypertext Transport Protocol (HTTP) used to transmit pages over the Internet. The Web's extraordinary rise is now a familiar story and most of us use its facilities at least occasionally, and in some cases daily. A crude estimate of the number of pages on the Web puts that number at over 25 billion at the time of the writing of this book.[1] The network structure of the Web has only been studied in detail

---

[1]This is only the number of reachable static pages. The number of unreachable pages is dif-

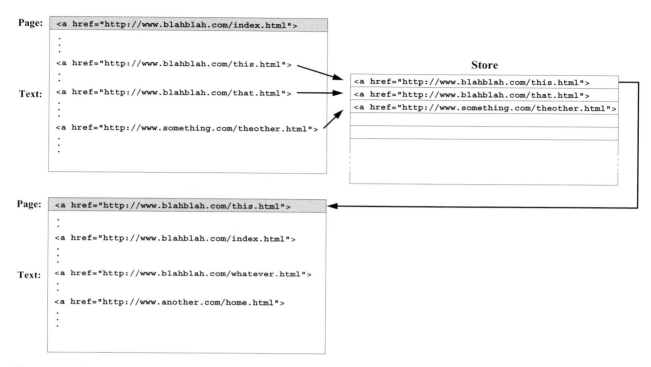

Figure 4.2: The operation of a web crawler. A web crawler iteratively downloads pages from the Web, starting from a given initial page. URLs are copied from the link tags in that initial page into a store. Once all links have been copied from the initial page, the crawler takes a URL from the store and downloads the corresponding page, then copies links from that, and so on.

relatively recently however.

The structure of the Web can be measured using a *crawler*, a computer program that automatically surfs the Web looking for pages. In its simplest form, the crawler performs a so-called breadth-first search on the Web network, as shown schematically in Fig. 4.2. One starts from any initial web page, downloads the text of that page over the Internet, and finds all the links in the text. Functionally, a link consists of an identifying "tag"—a short piece of text marking the link as a link—and a *Uniform Resource Locator* or URL, a standardized computer address that says how and where the linked web page can be found. By scanning for the tags and then copying the adjacent URLs a web crawler can

Breadth-first search is discussed at length in Section 10.3.

___

ficult to estimate, and dynamic pages (see later) are essentially infinite in number, although this may not be a very meaningful statement since these pages don't exist until someone asks for them.

rapidly extract URLs for all the links on a web page, storing them in memory or on a disk drive. When it is done with the current page, it removes one of the URLs from its store, uses it to locate a new page on the Web, and downloads the text of that page, and so the process repeats. If at any point the crawler encounters a URL that is the same as one already in its store, then that URL is ignored and not added to the store again, to avoid duplicate entries. Only URLs that are different from all those seen before are added to the store.

By repeating the process of downloading and URL extraction for a suitably long period of time one can find a significant portion of the pages on the entire Web. In practice, however, no web crawler actually finds all the pages on the Web. There are a number of reasons for this. First, some websites forbid crawlers to crawl their pages. Websites can place a file called `robots.txt` in their root directory that specifies which files, if any, crawlers can look at and may optionally specify that some crawlers are allowed to look at files while others are not. Compliance with the restrictions specified in a `robots.txt` file is voluntary, but in practice many crawlers do comply.

Second, many pages on the Web are dynamically generated: they are created on the fly by special software using, for instance, data from a database. Many corporate websites, as well as the web pages generated by search engines or directory services, fall into this category. The number of possible web pages that can be displayed as a result of a search using the *Google* search engine, for example, is so large as to be effectively infinite; it would not be possible (or sensible) for a crawler to crawl all of these pages. The crawler therefore has to make some choice about what counts as a web page and what does not. One choice would be to restrict oneself to static web pages—ones that are not generated on the fly. But it's not always simple to tell which pages are static, and besides, much useful information resides in the dynamic pages. In practice, the decisions made by crawlers about which pages to include tend to be fairly arbitrary, and it is not easy to guess which pages will be included in a crawl and which will not. But one can say with certainty that many will not and in this sense the crawl is always incomplete.

However, perhaps the most important reason why web crawls do not reach all the pages on the Web is that the network structure of the Web does not allow it. Since the Web is a directed network, not all pages are reachable from a given starting point. In particular, it is clear that pages that have no incoming hyperlinks—pages that no one links to—can never be found by a crawler that blindly follows links. Taking that idea one step further, it is also the case that a page will never be found if it is only linked to by pages that themselves have no incoming links. And so forth. In fact, the Web, and directed networks in general, have a special "component" structure, which we will examine in detail in Section 6.11.1, and most crawlers only find one part of that structure, the "giant out-component." In the case of the World Wide Web the giant out-component constitutes only about a half of all web pages and the other half of the Web is unreachable.[2]

Although we are interested in web crawlers as a tool for probing the structure of the Web so that we can study its network properties, this is not their main purpose. The primary use of web crawlers is to construct directories of web pages for search purposes. Web search engines such as *Google* indulge in web crawling on a massive scale to find web pages, parse their content, and construct indexes of the words and pictures they contain that can later be searched offline by fast database engines to find pages of interest to searchers. Because their primary interest is in indexing, rather than in reconstructing the network structure of the Web, search engine companies don't have any particular reason to take a good statistical sample of the Web and in network terms their crawls are probably quite biased. Still, many of them have graciously made their data available to academic researchers interested in web structure, and the data are good enough to give us a rough picture of what is going on. We will study a variety of features of the Web network in subsequent chapters.

Web searching, which itself raises some interesting network questions, is discussed in Section 19.1.

It isn't entirely necessary that we rely on search engine companies or other web enterprises for data on the structure of the Web. One can also perform one's own web crawls. There are a number of excellent web crawlers available for free, including *wget*, *Nutch*, *GRUB*, *Larbin*, *WebSPHINX*, and *ht://Dig*. While most of us don't have the time and network bandwidth to crawl billions of web pages, these programs can be useful for crawling single websites, and much useful insight and information can be acquired by doing so.

4.2 CITATION NETWORKS

A less well-known but much older information network is the network of citations between academic papers. Most papers refer to one or more other previous papers, usually in a bibliography at the end of the paper, and one can construct a network in which the vertices are papers and there is a directed edge from paper A to paper B if A cites B in its bibliography. There are many reasons why one paper might cite another—to point out information that may

---

[2]Which web pages a crawler finds does depend on where the crawl starts. A crawler *can* find a web page with no incoming links, for instance, if (and only if) it starts at that page. In practice, however, the starting point has remarkably little effect on what a crawler finds, since most of what is found consists of the giant out-component mentioned above, whose content does not depend on the starting point.

be useful to the reader, to give credit for prior work, to indicate influences on current work, or to disagree with the content of a paper. In general, however, if one paper cites another it is usually an indication that the contents of the earlier paper are relevant in some way to those of the later one, and hence citation networks are networks of relatedness of subject matter.

Quantitative studies of citation networks go back to the 1960s. The earliest seems to be the 1965 study by Price [274] (which is also the earliest study we know of to find a "power-law degree distribution," of which we talk in detail in Section 8.4). Studies such as this usually fall within the field formerly known as "library science" but now more often called "information science." The branch of information science dealing specifically with the statistical study of publications and citations is called *bibliometrics*.

The most common way to assemble citation data is to do it by hand, simply typing in all the entries in the bibliographies of papers to create a database that can then be used to assemble the network. In the 1960s when Price carried out his study, such databases were just starting to be created and he made use of an early version of what would later become the Science Citation Index. The Science Citation Index (along with its sister publications, the Social Science Citation Index and the Arts and Humanities Citation Index) is now one of the primary and most widely used sources of citation data. Another database, Scopus, provides a competing but largely similar service. Both are hand-maintained by professional staff and their coverage of the literature is reasonably complete and accurate, although the data are also quite expensive to purchase. Still, if one has the money, creating a citation network is only a matter of deciding which papers one wishes to include, using one of the databases to find the citations between those papers, and adding the appropriate directed edges to the network until it is complete.

See Section 4.1 for a discussion of web crawlers.

More recently, automated citation indexing by computer has started to become more common. For instance, the website *Citeseer*, maintained by Pennsylvania State University, performs citation indexing of papers in computer science and information science by crawling the Web to find freely available manuscripts of papers in electronic form, and then searching through those manuscripts to identify citations to other papers. This is a somewhat hit-or-miss operation because many papers are not on the Web or are not freely available, citations in papers have a wide variety of different formats and may include errors, and the same paper may exist in more than one place on the Web as well as in journals or books, and possibly in more than one different version. Nonetheless, enough progress has been made for Citeseer to become a useful tool in the computer science community. Other automatic citation indexing projects include *Citebase*, which indexes physics papers, and *Google Scholar*.

As with web crawls, the primary purpose of citation indexes is not to allow us to study the network structure of citation. Citation indexes are primarily research tools that allow researchers to discover by whom a paper has been cited, and hence to find research related to a topic of interest. Nonetheless, data from citation indices have been widely used to reconstruct the underlying networks and study their properties.

Citation networks are in many ways similar to the World Wide Web. The vertices of the network hold information in the form of text and pictures, just as web pages do, and the links from one paper to another play a role similar to hyperlinks on web pages, alerting the reader when information relevant to the topic of one paper can be found in another.[3] Papers with many citations are often more influential and widely read than those with few, just as is the case with web pages, and one can "surf" the citation network by following a succession of citations from paper to paper just as computer users surf the Web.

There is, however, at least one important difference between a citation network and the Web: a citation network is *acyclic*, while the Web is not. An acyclic network is one in which there are no closed loops of directed edges. On the World Wide Web, it is entirely possible to follow a succession of hyperlinks and end up back at the page you started at. Indeed this happens often. On a citation network, by contrast, it is essentially impossible. The reason is that in order to cite a paper, that paper must already have been written. One cannot cite a paper that doesn't exist yet. Thus all the directed edges in a citation network point backward in time, from newer papers to older ones. If we follow a path of such edges from paper to paper, we will therefore find ourselves going backward in time, but there is no way to go forward again, so we cannot close the loop and return to where we started.[4]

Acyclic networks are discussed further in Section 6.4.2.

See Fig. 6.3 for an illustration of a small acyclic network.

Citation networks have some surprising statistics. About 47% of all papers in the Science Citation Index have never been cited at all. Of the remainder, 9% have one citation, 6% have two, and it goes down quickly after that. Only 21% of all papers have 10 or more citations, and just 1% have 100 or more. These figures are a consequence of the power-law degree distribution of the network mentioned above and discussed more in Section 8.4.

---

[3]Indeed, academic studies of the Web within the information sciences sometimes refer to hyperlinks as "citations," a nomenclature that emphasizes the close similarities.

[4]On rare occasions it occurs that an author or authors will publish two papers simultaneously in the same volume of a journal and, with the help of the printers, arrange for each paper to cite the other, creating a cycle of length two in the network. Thus, the citation network is not strictly acyclic, having a small number of short cycles scattered about it.

The most highly cited paper in the Science Citation Index is a paper by Lowry *et al.* [202], which has been cited more than a quarter of a million times.[5] Like most very highly cited papers, it is a methodological paper in molecular biology.

Citation networks of the type described so far are the simplest but not the only possible network representation of citation patterns. An alternative and widely studied representation is the *cocitation network*. Two papers are said to be cocited if they are both cited by the same third paper. Cocitation is often taken as an indicator that papers deal with related topics and there is good evidence that this is a reasonable assumption in many cases.

A cocitation network is a network in which the vertices represent papers and the edges represent cocitation of pairs of papers. By contrast with ordinary citation networks, the edges in a cocitation network are normally considered undirected, since cocitation is a symmetric relationship. One can also define a strength for the cocitation between two papers as the number of other papers that cite both and one can create weighted cocitation networks in which the strengths of the edges are equal to this cocitation strength.

Another related concept, although one that is less often used, is *bibliographic coupling*. Two papers are said to be bibliographically coupled if they *cite* the same other papers (rather than being cited by the same papers). Bibliographic coupling, like cocitation, can be taken as an indicator that papers deal with related material and one can define a strength or weight of coupling by the number of common citations between two papers. From the bibliographic coupling figures one can then assemble a *bibliographic coupling network*, either weighted or not, in which the vertices are papers and the undirected edges indicate bibliographic coupling.

Cocitation and bibliographic coupling are discussed in more detail in Section 6.4.1.

4.2.1 PATENT AND LEGAL CITATIONS

Our discussions of citation networks have so far focused on citations between academic papers, but there are other types of citation also. Two of particular importance are citations between patents and between legal opinions.

Patents are temporary grants of ownership for inventions, which give their holders the right to take legal action against others who attempt to profit without permission from the protected inventions. They are typically issued to inventors—either individuals or corporations—by national governments after

---

[5] And it's been cited one more time now.

a review process to determine whether the invention in question is original and has not previously been invented by someone else. In applying for a patent, an inventor must describe his or her invention in sufficient detail to make adequate review possible and present the case that the invention is worthy of patent protection. A part of this case typically involves detailing the relationship between the invention and previously patented inventions, and in doing so the inventor will usually cite one or more previous patents. Citations may highlight dependencies between technologies, such as one invention depending for its operation on another, but more often patent citations are "defensive," meaning that the inventor cites the patent for a previous technology and then presents an argument for why the new technology is sufficiently different from the old one to merit its own patent. Governments, in the process of examining patent applications, will routinely consider their similarity to previous inventions, and defensive citations are one way in which an inventor can fend off in advance possible objections that might be raised. Typically there are a number of rounds of communication, back and forth between the government patent examiner and the inventor, before a patent application is finally accepted or rejected. During this process extra citations are often added to the application, either by the inventor or by the examiner, to document the further points discussed in their communications.

If and when a patent is finally granted, it is published, citations and all, so that the public may know which technologies have patent protection. These published patents provide a source of citation data that we can use to construct networks similar to the networks for citations between papers. In these networks the vertices are patents, each identified by a unique patent number, and the directed edges between them are citations of one patent by another. Like academic citation networks, patent networks are mostly acyclic, with edges running from more recent patents to older ones, although short loops can arise in the network in the not uncommon case that an inventor simultaneously patents a number of mutually dependent technologies. The structure of patent networks reflects the organization of human technology in much the same way that academic citations reflect the structure of research knowledge. Patent networks have been studied far less than academic citation networks, but studies have been growing in the last few years with the appearance of high-quality data sets, particularly for US patents [161], and there are a number of important technological and legal questions, for instance concerning antitrust policy, that can be addressed by examining their structure [69].

Another class of citation network that has begun to attract attention in recent years is that of legal citation networks. In countries where law cases can be heard by judges rather than juries, such as civil cases or appeals in Europe or the US, a judge will frequently issue an "opinion" after deciding a case, a narrative essay explaining his or her reasoning and conclusions. It is common practice in writing such an opinion to cite previous opinions issued in other cases in order to establish precedent, or occasionally to argue against it. Thus, like academic papers and patents, legal opinions form a citation network, with opinions being the vertices and citations being the directed edges. Again the network is approximately acyclic, as with the other networks in this section. The legal profession has long maintained indexes of citations between opinions for use by lawyers, judges, scholars, and others, and in recent years those indexes have made the jump to electronic form and are now available online. In the United States, for instance, two commercial services, LexisNexis and Westlaw,[6] provide thorough and detailed data on legal opinions and their citations via online services. In the last few years a number of studies have been published of the structure of legal citation networks using data derived from these services [125, 126, 194].

In principle it would be possible also to construct networks of cocitation or bibliographic coupling between either patents or legal opinions, but the author is not aware of any studies yet published of such networks.

4.3 OTHER INFORMATION NETWORKS

There are many other networks of information, although none have received the same level of study as the Web and citation networks. In the remainder of this chapter we briefly discuss a few examples of other networks.

4.3.1 PEER-TO-PEER NETWORKS

Peer-to-peer (P2P) file-sharing networks have become popular and widespread in the last decade or so. A *peer-to-peer network* is a network in which the nodes are computers containing information in the form, usually, of discrete files, and the edges between them are virtual links established for the purpose of sharing the contents of those files. The links exist only in software—they indicate only the intention of one computer to communicate with another should the need arise.

Peer-to-peer networks are typically used as a vehicle for distributed databases, particularly for the storage and distribution, often illegally, of music and movies, although there are substantial legal uses as well, such as local sharing

---

[6]Westlaw is owned and operated by Thomson Reuters, the same company that owns the Science Citation Index.

of files on corporate networks or the distribution of open-source software. (The network of router-to-router communications using the Border Gateway Protocol described in Section 2.1 is another less obvious example of a legitimate and useful peer-to-peer network.)

The point of a peer-to-peer network is that data is transferred directly between computers belonging to two end users of the network, two "peers." This contrasts with the more common server–client model, such as that used by the World Wide Web, in which central server computers supply requested data to a large number of client machines. The peer-to-peer model is favored particularly for illicit sharing of copyrighted material because the owners of a centralized server can easily be obliged to turn off the server by legal or law-enforcement action, but such actions are much more difficult when no central server exists.

On most peer-to-peer networks each computer is home to some information, but no computer has all the information in the network. If the user of a computer requires information stored on another computer, that information can be transmitted simply and directly over the Internet or over a local area network. This is a peer-to-peer transfer, but no special infrastructure is necessary to accomplish it—standard Internet protocols are perfectly adequate to the task. Things get interesting, however, when one wants to *find* which other computer has the desired information. One way to do that is to have a central server containing none of the information but just an index of which information is on which computers. Such a system was employed by the early file-sharing network *Napster*, but the central index server is, once again, susceptible to legal and other challenges, and such challenges were in the end responsible for shutting Napster down.[7]

To avoid this problem, developers have turned to distributed schemes for searching and this is where network concepts come into play. An illustrative example of a peer-to-peer system with distributed search is the *Gnutella* network, which underlies a number of popular file-sharing programs including *LimeWire* and the now-defunct *Morpheus*. In the simplest incarnation of this system (more sophisticated ones are in use now) computers form links to some number of their peers in such a way that all the computers form a connected network. Again, a link here is purely a software construct—a computer's network neighbors in the peer-to-peer sense are merely those others with which it intends to communicate when the need arises.

When a user instructs his or her computer to search the network for a spe-

---

[7]The Napster name was later bought up by the recording industry and is now the name of a legitimate online music service, although one that does not make use of peer-to-peer technology.

cific file the computer sends out a message to its network neighbors asking whether they have that file. If they do, they arrange to transmit it back to the first computer. If they do not, they pass the message on to *their* neighbors, and so forth until the file is found. As pointed out in Section 19.2, where we discuss search strategies on peer-to-peer networks at some length, this algorithm works, but only on relatively small networks. Since it requires messages to be passed between many computers for each individual search, the algorithm does not scale well as the network becomes large, the volume of network traffic eventually swamping the available data bandwidth. To get around this problem, modern peer-to-peer networks, including recent versions of Gnutella, employ a two-tiered network topology of nodes and "supernodes," in which searches are performed only among the supernodes and ordinary nodes contact them directly to request searches be performed. More details are given in Section 19.2.

So what is the structure of a peer-to-peer network like? In many cases, unfortunately, not a lot is known since the software is proprietary and its owners are reluctant to share operational details. The Gnutella system is more promising, being so-called open-source software, meaning that the original computer code for the software and the specification of the protocols it uses are freely available. By exploiting certain details of these protocols, particularly the ability for computers in the Gnutella network to "ping" one another (i.e., ask each other to identify themselves), a number of authors have been able to discover structures for Gnutella networks [282, 308]. The networks appear to have approximately power-law degree distributions and it has been suggested that this property could be exploited to improve search performance [6].

4.3.2 RECOMMENDER NETWORKS

A type of information network important for technology and commerce is the *recommender network*. Recommender networks represent people's preferences for things, such as for certain products sold by a retailer. Online merchants, for instance, usually keep records of which customers bought which products and sometimes ask them whether they liked the products or not. Many large supermarket chains record the purchases made by each of their regular customers (usually identified by a small card with a barcode on it that is scanned when purchases are made) and so can work out which products each customer buys frequently.

The fundamental representation of a recommender network is as a "bipartite network," a network with two types of vertex, one representing the products or other items and the other representing the people, with edges connecting people to the items they buy or like. One can also add strengths or weights to the edges to indicate, for instance, how often a person has bought an item or how much he or she likes it, or the strengths could be made negative to indicate dislikes.

Recommender networks have been studied for many types of goods and products, including books, music, films, and others. The primary commercial interest in recommender networks arises from their use in *collaborative filtering systems*, also sometimes called *recommender systems*, which are computer algorithms that attempt to guess items that people will like by comparing a person's known preferences with those of other people. If person A likes many of the same things as persons B, C, and D, and if persons B, C, and D all like some further item that A has never expressed an opinion about, then maybe (the theory goes) A would like that item too. A wide variety of computer algorithms have been developed for extracting conclusions of this type from recommender networks and are used extensively by retailers to suggest possible purchases to their customers, in the hope of drumming up business. The website of the online bookseller *Amazon.com*, for instance, has a feature that lists recommended book titles to customers based on their previously expressed preferences and purchases. And many supermarkets now print out discount coupons after customers have completed their purchases, coupons for products that the customer has not bought in the past but might be interested to try.

Research on recommender networks has in the past focused mainly on the development of new collaborative filtering algorithms, but it is reasonable to suppose that the success of these algorithms should depend to some extent on the structure of the recommender network itself, and there is therefore good reason to also study that structure. A few such studies have been published in the scientific literature [63, 147], but there is clearly room for further work.

4.3.3 KEYWORD INDEXES

Another type of information network, also bipartite in form, is the *keyword index*. Consider, for instance, a set of documents containing information on various topics. One can construct an index to that set so that one can look up words in that index and the index will list important occurrences of those words in the documents. Such indexes have historically appeared, of course, in books, as guides to their content, but more recently indexes have regularly been constructed as guides to other information collections, including sets of academic papers and the World Wide Web. The index constructed by a web search engine, as discussed in Section 4.1, is a good example; it consists, at a

We encountered bipartite networks previously in Section 3.5 and will study them further in Section 6.6.

minimum, of a list of words or phrases, with each word or phrase accompanied by a list of the web pages on which it occurs.

Such indexes can be represented as a bipartite network in which one of the two types of vertex represents words in the index and the other represents documents or pages. Then one places an edge between each word and the documents in which it occurs. Although such networks can be constructed for, amongst other things, the Web or collections of academic papers, they should not be confused with the networks of web links or citations discussed earlier in this chapter. Those are also networks of web pages and documents, but they are different from a keyword index. Those networks were networks of direct links between documents. An index is a network of links between index entries and the documents they point to.

Indexes are of practical importance as a method for searching large bodies of information. Web search engines, for example, rely heavily on them to quickly find web pages that correspond to a particular query. However, indexes also have other, more sophisticated applications. They are used, for example, as a basis for techniques that attempt to find documents or pages that are similar to one another. If one has a keyword index to a set of documents and finds that two documents share a lot of the same keywords, it may be an indication that the two cover similar topics. A variety of computer algorithms for spotting such connections have been developed, typically making use of ideas very similar to those used in the recommender systems discussed above—the problem of finding documents with similar keywords is in many ways similar to the problem of finding buyers who like similar products.

The identification of similar documents can be useful, for example, when searching through a body of knowledge. In a standard index search, one typically types in a set of keywords and gets back a list of documents containing those words. Search engines that can tell when documents are similar to each other may be able to respond more usefully to such queries because they can return documents that do not in fact contain the keywords entered, but which are similar to documents that do. In cases where a single concept is called by more than one name, this may be a very effective strategy for finding all of the relevant documents.

In the context of document retrieval, the classic method for determining document similarity and performing generalized searches of this type is *latent semantic indexing*, which is based on the application of the matrix technique known as singular value decomposition to the bipartite network of keywords and documents. The interested reader can find a discussion of latent semantic indexing in Ref. [193].

As with recommender systems, it is reasonable to suppose that the success of methods for finding similar documents or improving searches using similarity information depends on the structure of the bipartite keyword/document network, and hence that studies of that structure could generate useful insights. There has been relatively little interest in the problem within the network community so far and again there is plenty of room for future work.

CHAPTER 5

BIOLOGICAL NETWORKS

A discussion of various networks of interest in biology, including biochemical networks, neural networks, and ecological networks

NETWORKS are widely used in many branches of biology as a convenient representation of patterns of interaction between appropriate biological elements. Molecular biologists, for example, use networks to represent the patterns of chemical reactions among chemicals in the cell, while neuroscientists use them to represent patterns of connections between brain cells, and ecologists study the networks of interactions between species in ecosystems, such as predation or cooperation. In this chapter we describe the commonest kinds of biological networks and discuss methods for determining their structure.

5.1 BIOCHEMICAL NETWORKS

Among the biological networks those attracting the most attention in recent years have been biochemical networks, networks that represent the molecular-level patterns of interaction and mechanisms of control in the biological cell. The principal types of networks studied in this area are metabolic networks, protein–protein interaction networks, and genetic regulatory networks.

5.1.1 METABOLIC NETWORKS

Metabolism is the chemical process by which cells break down food or nutrients into usable building blocks (so-called *catabolic metabolism*) and then reassemble those building blocks to form the biological molecules the cell needs to complete its other tasks (*anabolic metabolism*). Typically this breakdown and reassembly involves chains or *pathways*, sets of successive chemical reactions that convert initial inputs into useful end products by a series of steps. The complete set of all reactions in all pathways forms a *metabolic network*.

The vertices in a metabolic network are the chemicals produced and consumed by the reactions. These chemicals are known generically as *metabolites*. By convention the definition of a metabolite is limited to small molecules, meaning things like carbohydrates (such as sugars) and lipids (such as fats), as well as amino acids and nucleotides. Amino acids and nucleotides are themselves the building blocks for larger polymerized macromolecules such as DNA, RNA, and proteins, but the macromolecules are not themselves considered metabolites—they are not produced by simple chemical reactions but by more complex molecular machinery within the cell, and hence are treated separately. (We discuss some of the mechanisms by which macromolecules are produced in Section 5.1.3.)

Although the fundamental purpose of metabolism is to turn food into useful biomolecules, one should be wary of thinking of it simply as an assembly line, even a very complicated one. Metabolism is not just a network of conveyor belts in which one reaction feeds another until the final products fall out the end; it is a dynamic process in which the concentrations of metabolites can change widely and rapidly, and the cell has mechanisms for turning on and off the production of particular metabolites or even entire portions of the network. Metabolism is a complex machine that reacts to conditions both within and outside the cell and generates a broad variety of chemical responses. A primary reason for the high level of scientific interest in metabolic networks is their importance as a stepping stone on the path towards an understanding of the chemical dynamics of the cell.

Generically, an individual chemical reaction in the cell involves the consumption of one or more metabolites that are broken down or combined to produce one or more others. The metabolites consumed are called the *substrates* of the reaction, while those produced are called the *products*.

The situation is complicated by the fact that most metabolic reactions do not occur spontaneously, or do so only at a very low rate. To make reactions occur at a usable rate, the cell employs an array of chemical catalysts, referred to as *enzymes*. Unlike metabolites, enzymes are mostly macromolecules, usually proteins but occasionally RNAs. Like all catalysts, enzymes are not consumed in the reactions they catalyze but they play an important role in metabolism nonetheless. Not only do they enable reactions that would otherwise be thermodynamically disfavored or too slow to be useful, but they also provide one of the mechanisms by which the cell controls its metabolism. By increasing or decreasing the concentration of the enzyme that catalyzes a particular reaction, the cell can turn that reaction on or off, or moderate its speed. Enzymes tend to be highly specific to the reactions they catalyze, each one enabling only one or a small number of reactions. Thousands of enzymes are known and many more are no doubt waiting to be discovered, and this large array of highly specific catalysts allows for a fine degree of control over the processes of the cell.

The details of metabolic networks vary between different species of organisms but, amongst animals at least, large parts are common to all or most species. Many important pathways, cycles, or other subportions of metabolic networks are essentially unchanged across the entire animal kingdom. For this reason one often refers simply to "metabolism" without specifying a particular species of interest; with minor variations, observations made in one species often apply to others.

The most correct representation of a metabolic network is as a bipartite network. We encountered bipartite networks previously in Section 3.5 on social affiliation networks and in Section 4.3.2 on recommender networks. A bipartite network has two distinct types of vertex, with edges running only between vertices of unlike kinds. In the case of affiliation networks, for example, the two types of vertex represented people and the groups they belonged to. In the case of a metabolic network they represent metabolites and metabolic reactions, with edges joining each metabolite to the reactions in which it participates. In fact, a metabolic network is really a *directed* bipartite network, since some metabolites go into the reaction (the substrates) and some come out of it (the products). By placing arrows on the edges we can distinguish between the ingoing and outgoing metabolites. An example is sketched in Fig. 5.1a.[1]

This bipartite representation of a metabolic network does not include any way of representing enzymes, which, though not metabolites themselves, are still an important part of the metabolism. Although it's not often done, one can in principle incorporate the enzymes by introducing a third class of vertex to represent them, with edges connecting them to the reactions they catalyze. Since enzymes are not consumed in reactions, these edges are undirected—running neither into nor out of the reactions they participate in. An example of such a network is sketched in Fig. 5.1b. Technically this is now a *tripartite network*, partly directed and partly undirected.[2]

Correct and potentially useful though they may be, however, neither of these representations is very often used for metabolic networks. The most

---

[1] The metabolic network is the only example of a directed bipartite network appearing in this book, and indeed the only naturally occurring example of such a network the author has come across, although no doubt there are others to be discovered if one looks hard enough.

[2] Also the only such network in the book.

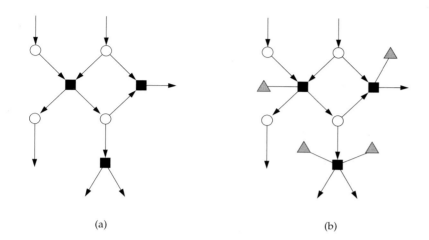

(a)   (b)

Figure 5.1: Bipartite and tripartite representations of a portion of a metabolic network. (a) A metabolic network can be represented as a directed bipartite network with vertices for the metabolites (circles) and reactions (squares) and directed edges indicating which metabolites are substrates (inputs) and products (outputs) of which reactions. (b) A third type of vertex (triangles) can be introduced to represent enzymes, with undirected edges linking them to the reactions they catalyze. The resulting network is a mixed directed/undirected tripartite network.

common representations of metabolic networks project the network onto just one set of vertices, either the metabolites or the reactions, with the former being the more popular choice. In one approach the vertices in the network represent metabolites and there is an undirected edge between any two metabolites that participate in the same reaction, either as substrates or as products. Clearly this projection loses much of the information contained in the full bipartite network, but, as we have said, it is nonetheless widely used. Another approach, probably the most common, is to represent the network as a directed network with a single type of vertex representing metabolites and a directed edge from one metabolite to another if there is a reaction in which the first metabolite appears as a substrate and the second as a product. This representation contains more of the information from the full network, but is still somewhat unsatisfactory since a reaction with many substrates or many products appears as many edges, with no easy way to tell that these edges represent aspects of the same reaction. The popularity of this representation arises from the fact that for many metabolic reactions only one product and one substrate are known

Projections of bipartite networks and the associated loss of information are discussed further in Section 6.6.

81

or are considered important, and therefore the reaction can be represented by only a single directed edge with no confusion arising. A number of companies produce large charts showing the most important parts of the metabolic network in this representation. An example is shown in Fig. 5.2. Such charts have become quite popular as wall decorations in the offices of molecular biologists and biochemists, although whether they are actually useful in practice is unclear.

The experimental measurement of metabolic networks is a complex and laborious process, although it has been made somewhat easier in recent years with the introduction of new techniques from molecular genetics. Experiments tend to focus neither on whole networks nor on individual reactions but on metabolic pathways. A number of tools are available to probe the details of individual pathways. Perhaps the most common is the use of radioactive isotopes to trace the intermediate products along a pathway. In this technique, the organism or cell under study is injected with a substrate for the pathway of interest in which one or more of the atoms has been replaced by a radioisotope. Typically this has little or no effect on the metabolic chemistry, but as the reactions of the pathway proceed, the radioactive atoms move from metabolite to metabolite. Metabolites can then be refined, for example by mass spectroscopy or chromatography, and tested for radioactivity. Those that show it can be assumed to be "downstream" products in the pathway fed by the initial radioactive substrate.

This method tells us the products along a metabolic pathway, but of itself does not tell us the order of the reactions making up the pathway. Knowledge of the relevant biochemistry—which metabolites can be transformed into which others by some chemical reaction—can often identify the ordering or at least narrow down the possibilities. Careful measurement of the strength of radioactivity of different metabolites, coupled with a knowledge of the half-life of the isotope used, can also give some information about pathway structure as well as rates of reactions.

Notice, however, that there is no way to tell if any of the reactions discovered have substrates other than those tagged with the radioisotope. If new substrates enter the pathway at intermediate steps (that is, they are not produced by earlier reactions in the pathway) they will not be radioactive and so will not be measured. Similarly, if there are reaction products that by chance do not contain the radioactive marker they too will not be measured.

An alternative approach to probing metabolic pathways is simply to increase the level of a substrate or enzyme for a particular reaction in the cell, thereby increasing the levels of the products of that reaction and those downstream of it in the relevant pathway or pathways, increases that can be mea-

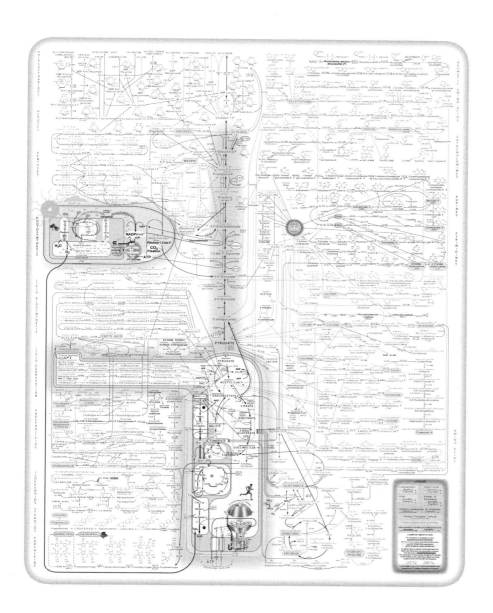

Figure 5.2: A metabolic network. (See Plate IV for color version.) A wallchart showing the network formed by the major metabolic pathways. Created by Donald Nicholson. Copyright of the International Union of Biochemistry and Molecular Biology. Reproduced with permission.

sured to determine the constituents of the pathway. This technique has the advantage of being able to detect products other than those that carry a particular radioactive marker inherited from a substrate, but it is still incapable of identifying substrates other than those produced as products along the pathway.

A complementary experimental technique that can probe the substrates of reactions is reaction inhibition, in which a reaction in a pathway is prevented from taking place or its rate is reduced. Over time, this results in a build-up in the cell of the substrates for that reaction, since they are no longer being used up. By watching for this build-up one can determine the reaction substrates. In principle the same method could also be used to determine the products of the reaction, since their concentration would decrease because they are not being produced any longer, but in practice this turns out to be a difficult measurement and is rarely done.

The inhibition of a reaction is usually achieved by disabling or removing an enzyme necessary for the reaction. This can be done in a couple of different ways. One can use *enzyme inhibitors*, which are chemicals that bind to an enzyme and prevent it from performing its normal function as a catalyst, or one can genetically alter the organism under study to remove or impair its ability to produce the enzyme (a so-called *knockout* experiment). The same techniques can also be used to determine which reactions are catalyzed by which enzymes in the first place, and hence to discover the structure of the third, enzymatic part of the tripartite metabolic network pictured in Fig. 5.1b.

The construction of a complete or partial picture of a metabolic network involves the combination of data from many different pathways, almost certainly derived from experiments performed by many different experimenters using many different techniques. There are now a number of public databases of metabolic pathway data from which one can draw to assemble networks, the best known being KEGG and MetaCyc. Assembling the network itself is a non-trivial task. Because the data are drawn from many sources, careful checking against the experimental literature (or "curation," as the lingo goes) is necessary to insure consistent and reliable inputs to the process, and missing steps in metabolic pathways must often be filled in by guesswork based on biochemistry and a knowledge of the genetics. A number of computer software packages have been developed that can reconstruct networks from raw metabolic data in an automated fashion, but the quality of the networks they create is generally thought to be poorer than that of networks created by knowledgeable human scientists (although the computers are much faster).

5.1.2 PROTEIN–PROTEIN INTERACTION NETWORKS

The metabolic networks of the previous section describe the patterns of chemical reactions that turn one chemical into another in the cell. As we have noted, the traditional definition of metabolism is restricted to small molecules and does not include proteins or other large molecules, except in the role of enzymes, in which they catalyze metabolic reactions but do not take part as reactants themselves.

Proteins do however interact with one another and with other biomolecules, both large and small, but the interactions are not purely chemical. Proteins sometimes interact chemically with other molecules—exchanging small subgroups, for example, such as the exchange of a phosphate group in the process known as phosphorylation. But the primary mode of protein–protein interaction—interactions of proteins with other proteins—is physical, their complicated folded shapes interlocking to create so-called *protein complexes* (see Fig. 5.3) but without the exchange of particles or subunits that defines chemical reactions.

The set of all protein–protein interactions forms a *protein–protein interaction network*, in which the vertices are proteins and two vertices are connected by an undirected edge if the corresponding proteins interact. Although this representation of the network is the one commonly used, it omits much useful information about the interactions. Interactions that involve three or more proteins, for instance, are represented by multiple edges, and there is no way to tell from the network itself that such edges represent aspects of the same interaction. This problem could be addressed by adopting a bipartite representation of the network similar to the one we sketched for metabolic networks in Fig. 5.1, with two kinds of vertex representing proteins and interactions, and undirected edges connecting proteins to the interactions in which they participate. Such representations, however, are rarely used.

There are a number of experimental techniques available to probe for interactions between proteins. One of the most reliable and trusted is *co-immunoprecipitation*. Immunoprecipitation (without the "co-") is a technique for extracting a single protein species from a sample containing more than one. The technique borrows from the immune system, which produces *antibodies*, specialized proteins that attach or *bind* to a specific other target protein when the two encounter each other. The immune system uses antibodies to neutralize proteins, complexes, or larger structures that are harmful to the body, but experimentalists have appropriated them for use in the laboratory. Immunopre-

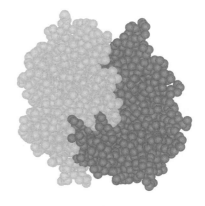

Figure 5.3: Two proteins joined to form a protein complex. Protein molecules can have complicated shapes that interlock with one another to form protein complexes.

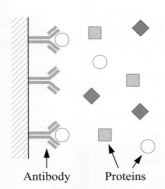

Antibody   Proteins

In immunoprecipitation, antibodies attached to a solid surface bind to a specific protein, represented here by the circles, pulling it out of the solution.

Transcription factors are discussed in more detail in Section 5.1.3.

cipitation involves attaching an antibody to a solid surface, such as the surface of a glass bead, then passing a solution containing the target protein (as well as others, in most cases) over the surface. The antibody and the target protein bind together, effectively attaching the protein to the surface via the antibody. The rest of the solution is then washed away, leaving the target protein to be recovered from the surface.

There are known naturally occurring antibodies for many proteins of scientific interest, but researchers also routinely create antibodies for specific proteins by injecting those proteins (or more often a portion of a protein) into an animal to provoke its immune system to generate the appropriate antibody.

Co-immunoprecipitation is an extension of the same method to the identification of protein interactions. An antibody is again attached to a suitable solid surface and binds to a known protein in a sample. If that protein is attached to others, forming a protein complex, then the entire complex will end up attached to the surface and will remain after the solution is washed away. Then the complex can be recovered from the surface and the different proteins that make it up individually identified, typically by testing to see if they bind to other known antibodies (a technique known as a *Western blot*).

Although well-established and reliable, co-immunoprecipitation is an impractical approach for reconstructing entire interaction networks, since individual experiments, each taking days, have to be performed for every interaction identified. If appropriate antibodies also have to be created the process would take even longer; the creation of a single antibody involves weeks or months of work, and costs a considerable amount of money too. As a result, the large-scale study of protein–protein interaction networks did not really take off until the adoption in the 1990s and early 2000s of so-called *high-throughput* methods for discovering interactions, methods that can identify interactions quickly and in a semi-automated fashion.

The oldest and best established of the high-throughput methods for protein interactions is the *two-hybrid screen*, invented by Fields and Song in 1989 [119].[3] This method relies on the actions of a specialized protein known as a *transcription factor*, which, if present in a cell, turns on the production of another protein, referred to as a *reporter*. The presence of the reporter can be detected by the experimenter by any of a number of relatively simple means. The idea of the two-hybrid screen is to arrange things so that the transcription factor is created when two proteins of interest interact, thereby turning on the reporter, which tells us that the interaction has taken place.

---

[3] Also called a *yeast two-hybrid screen* or Y2HS for short, in recognition of the fact that the technique is usually implemented inside yeast cells, as discussed later.

The two-hybrid screen relies on the fact that transcription factors are typically composed of two distinct parts, a so-called *binding domain* and an *activation domain*. It turns out that most transcription factors do not require the binding and activation domains to be actually attached to one another for the transcription factor to work. If they are merely in close enough proximity production of the reporter will be activated.

In a two-hybrid screen, a cell, usually a yeast cell, is persuaded to produce two proteins of interest, each with one of the domains of the transcription factor attached to it. This is done by introducing *plasmids* into the cell, fragments of DNA that code for the proteins and domains. Then, if the two proteins in question interact and form a complex, the two domains of the transcription factor will be brought together and, with luck, will activate production of the reporter.

See Section 5.1.3 for a discussion of DNA coding of proteins.

In a typical two-hybrid experiment, the protein attached to the binding domain of the transcription factor is a known protein (called the *bait* protein) whose interactions the experimenter wants to probe. Plasmids coding for a large number of other proteins (called *prey*) attached to copies of the activation domain are created, resulting in a so-called *library* of possible interaction targets for the bait. The plasmids for the bait and the library of prey are then introduced into a culture of yeast cells, with the concentration of prey carefully calibrated so that at most one prey plasmid enters each cell in most cases. Cells observed to produce the reporter are then assumed to contain plasmids coding for prey proteins that interact with the bait and the plasmids are recovered from those cells and analyzed to determine the proteins they correspond to.

The two-hybrid screen has two important advantages over older methods like co-immunoprecipitation. First, one can employ a large library of prey and hence test for interactions with many proteins in a single experiment, and second, the method is substantially cheaper and faster than co-immunoprecipitation per interaction detected. Where co-immunoprecipitation requires one to obtain or create antibodies for every protein tested, the two-hybrid screen requires only the creation of DNA plasmids and their later sequence analysis, both relatively simple operations for an experimenter armed with the machinery of modern genetic engineering.

One disadvantage of the two-hybrid screen is that the presence of the two domains of the transcription factor attached to the bait and prey proteins can get in the way of their interacting with one another and prevent the formation of a protein complex, meaning that some legitimate protein–protein interactions will not take place under the conditions of the experiment.

The principal disadvantage of the method, however, is that it is simply unreliable. It produces high rates of both false positive results—apparent interactions between proteins that in fact do not interact—and false negative results—failure to detect true interactions. By some estimates the rate of false positives may be as high as 50%, meaning that fully half of all interactions detected by the method are not real. This has not stopped a number of researchers from performing analyses on the interaction networks reconstructed from two-hybrid screen data, but the results should be viewed with caution. It is certainly possible that many or even most of the conclusions of such studies are substantially inaccurate.

An alternative and more accurate class of methods for high-throughput detection of protein interactions are the *affinity purification* methods (also sometimes called *affinity precipitation* methods). These methods are in some ways similar to the co-immunoprecipition method described previously, but avoid the need to develop antibodies for each protein probed. In an affinity purification method, a protein of interest is "tagged" by adding a portion of another protein to it, typically by introducing a plasmid that codes for the protein plus tag, in a manner similar to the introduction of plasmids in the two-hybrid screen. Then the protein is given the opportunity to interact with a suitable library of other proteins and a solution containing the resulting protein complexes (if any) passed over a surface to which are attached antibodies that bind to the tag. As a result, the tag, the attached protein, and its interaction partners are bound to the surface while the rest of the solution is washed away. Then, as in co-immunoprecipitation, the resulting complex or complexes can be analyzed to determine the identities of the interaction partners.

The advantage of this method is that it requires only a single antibody that binds to a known tag, and the same tag–antibody pair can be used in different experiments to bind different proteins. Thus, as with the two-hybrid screen, one need only generate new plasmids for each experiment, which is relatively easy, as opposed to generating new antibodies, which is slow and difficult. Some implementations of the method have a reliability comparable to that of co-immunoprecipitation. Of particular note is the method known as *tandem affinity purification*, which combines two separate purification stages and generates correspondingly higher-quality results. Tandem affinity purification is the source for some of the most reliable current data for protein–protein interaction networks.

As with metabolic reactions, there are now substantial databases of protein interactions available online, of which the most extensive are IntAct, MINT, and DIP, and from these databases interaction networks can be constructed for analysis. An example is shown in Fig. 5.4.

Figure 5.4: A protein–protein interaction network for yeast. A network of interactions between proteins in the single-celled organism *Saccharomyces cerevisiae* (baker's yeast), as determined using, primarily, two-hybrid screen experiments. From Jeong *et al.* [164]. Copyright Macmillan Publishers Ltd. Reproduced by permission.

5.1.3 GENETIC REGULATORY NETWORKS

As discussed in Section 5.1.1, the small molecules needed by biological organisms, such as sugars and fats, are manufactured in the cell by the chemical reactions of metabolism. Proteins, however, which are much larger molecules, are manufactured in a different manner, following recipes recorded in the cell's genetic material, DNA.

Proteins are biological polymers, long-chain molecules formed by the concatenation of a series of basic units called *amino acids*. The individual amino acids themselves are manufactured by metabolic processes, but their assembly into complete proteins is accomplished by the machinery of genetics. There are

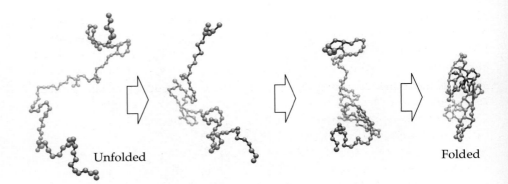

Unfolded

Folded

Figure 5.5: Protein folding. Proteins, which are long-chain polymers of amino acids, do not naturally remain in an open state (left), but collapse upon themselves to form a more compact folded state (right).

20 distinct amino acids that are used by all living organisms to build proteins, and different species of proteins are distinguished from one another by the particular sequence of amino acids that make them up. Once created, a protein does not stay in a loose chain-like form, but folds up on itself under the influence of thermodynamic forces and mechanical constraints, reliably producing a specific folded form or *conformation* whose detailed shape depends on the amino acid sequence—see Fig. 5.5. A protein's conformation dictates the physical interactions it can have with other molecules and can expose particular chemical groups or active sites on the surface of the protein that contribute to its biological function within the organism.

A protein's amino acid sequence is determined by a corresponding sequence stored in the DNA of the cell in which the protein is synthesized. This is the primary function of DNA in living matter, to act as an information storage medium containing the sequences of proteins needed by the cell. DNA is itself a long-chain polymer made up of units called *nucleotides*, of which there are four distinct species, adenine, cytosine, guanine, and thymine, commonly denoted A, C, G, and T, respectively.[4] The amino acids in proteins are encoded in DNA as trios of consecutive nucleotides called *codons*, such as ACG

---

[4]Technically, DNA is a double-stranded polymer, having two parallel chains of nucleotides forming the famous double helix shape. However, the two strands contain essentially the same sequence of nucleotides and so for our purposes the fact that there are two is not important (although it is very important in other circumstances, such as in the reproduction of a cell by cellular division and in the repair of damaged DNA).

or TTT, and a succession of such codons spells out the complete sequence of amino acids in a protein. A single strand of DNA can code for many proteins—hundreds or thousands of them—and two special codons, called the start and stop codons, are used to signal the beginning and end within the larger DNA strand of the sequence coding for a protein. The DNA code for a single protein, from start codon to stop codon, is called a *gene*.

Proteins are created in the cell by a mechanism that operates in two stages. In the first stage, known as *transcription*, an enzyme called *RNA polymerase* makes a copy of the coding sequence of a single gene. The copy is made of RNA, another information-bearing biopolymer, chemically similar but not identical to DNA. RNA copies of this type are called *messenger RNAs*. In the second stage, called *translation*, the protein is assembled, step by step, from the RNA sequence by an ingenious piece of molecular machinery known as a *ribosome*, a complex of interacting proteins and RNA. The translation process involves the use of *transfer RNAs*, short molecules of RNA that have a region at one end that recognizes and binds to a codon in the messenger RNA and a region at the other end that pulls the required amino acid into the correct place in the growing protein. The end result is a protein, assembled following the exact prescription spelled out in the corresponding gene. In the jargon of molecular biology, one says that the gene has been *expressed*.

The cell does not, in general, need to produce at all times every possible protein for which it contains a gene. Individual proteins serve specific purposes, such as catalyzing metabolic reactions, and it is important for the cell to be able to respond to its environment and circumstances by turning on or off the production of individual proteins as required. It does this by the use of *transcription factors*, which are themselves proteins and whose job is to control the transcription process by which DNA sequences are copied to RNA.

Transcription is performed by the enzyme RNA polymerase, which works by attaching to a DNA strand and moving along it, copying nucleotides one by one. The RNA polymerase doesn't just attach spontaneously, however, but is aided by a transcription factor. Transcription factors are specific to particular genes or sets of genes and regulate transcription in a variety of ways, but most commonly by binding to a recognized sub-sequence in the DNA, called a *promoter region*, which is adjacent to the beginning of the gene. The binding of the transcription factor to the promoter region makes it thermodynamically favorable for the RNA polymerase to attach to the DNA at that point and start transcribing the gene. (The end of the gene is marked by a stop codon and upon encountering this codon the RNA polymerase automatically detaches from the DNA strand and transcription ends.) Thus the presence in the cell of the transcription factor for the gene turns on or enhances the expression of that gene. We encountered an example of a transcription factor previously in our discussion of the two-hybrid screen in Section 5.1.2.

There are also transcription factors that inhibit expression by binding to a DNA strand in such a way as to prevent RNA polymerase from attaching to the strand and hence prevent transcription and the production of the corresponding protein.

But now here is the interesting point: being proteins, transcription factors are themselves produced by transcription from genes. Thus the protein encoded in a given gene can act as a transcription factor promoting or inhibiting production of one or more other proteins, which themselves can act as transcription factors for further proteins and so forth. The complete set of such interactions forms a *genetic regulatory network*. The vertices in this network are proteins or equivalently the genes that code for them and a directed edge from gene A to gene B indicates that A regulates the expression of B. A slightly more sophisticated representation of the network distinguishes between promoting and inhibiting transcription factors, giving the network two distinct types of edge.

The experimental determination of the structure of genetic regulatory networks involves identifying transcription factors and the genes that they regulate. The process has several steps. To begin with, one first confirms that a given candidate protein does bind to DNA roughly in the region of a gene of interest. The commonest technique for establishing the occurrence of such a binding is the *electrophoretic mobility shift assay*.[5] In this technique one creates strands of DNA containing the sequence to be tested and mixes them in solution with the candidate protein. If the two indeed bind, then the combined DNA/protein complex can be detected by *gel electrophoresis*, a technique in which one measures the speed of migration of electrically charged molecules or complexes through an agarose or polyacrylamide gel in an imposed electric field. In the present case the binding of the DNA and protein hinders the motion of the resulting complex through the gel, measurably reducing its speed when compared with unbound DNA strands. Typically one runs two experiments side by side, one with protein and one without, and compares the rate of migration to determine whether the protein binds to the DNA. One can also run parallel experiments using many different DNA sequences to test which (if any) bind to the protein.

An alternative though less sensitive technique for detecting binding is the *deoxyribonuclease footprinting assay*. Deoxyribonucleases (also called DNases

---

[5] "Assay" is biological jargon for an experimental test.

for short) are enzymes that, upon encountering DNA strands, cut them into shorter strands. There are many different DNases, some of which cut DNA only in particular places according to the sequence of nucleotides, but the footprinting technique uses a relatively indiscriminate DNase that will cut DNA at any point. If, however, a protein binds to a DNA strand at a particular location it will often (though not always) prevent the DNase from cutting the DNA at or close to that location. Footprinting makes use of this by mixing strands of DNA containing the sequence to be tested with the DNase and observing the resulting mix of strand lengths after the DNase has cut the DNA samples into pieces in a variety of different ways. Repeating the experiment with the protein present will result in a different mix of strand length if the protein binds to the DNA and prevents it from being cut in certain places. The mix is usually determined again by gel electrophoresis (strands of different lengths move at different speeds under the influence of the electric field) and one again runs side-by-side gel experiments with and without the protein to look for the effects of binding.

Both the mobility shift and footprinting assays can tell us if a protein binds somewhere on a given DNA sequence. To pin down exactly where it binds one typically must do some further work. For instance, one can create short strands of DNA, called *oligonucleotides*, containing possible sequences that the protein might bind to, and add them to the mix. If they bind to the protein then this will reduce the extent to which the longer DNAs bind and visibly affect the outcome of the experiment. By a combination of such experiments, along with computer-aided guesswork about which oligonucleotides are likely to work best, one can determine the precise sub-sequence to which a particular protein binds.

While these techniques can tell us the DNA sequence to which a protein binds, they cannot tell us which gene's promoter region that sequence belongs to (if any), whether the protein actually affects transcription of that gene, or, if it does, whether the transcription is promoted or inhibited. Further investigations are needed to address these issues.

Identification of the gene is typically done not by experiment but by computational means and requires a knowledge of the sequence of the DNA in the region where the protein binds. If we know the DNA sequence then we can search it for occurrences of the sub-sequence to which our protein binds, and then examine the vicinity to determine what gene or genes are there, looking for example for start and stop codons in the region and then recording the sequence of other codons that falls between them. Complete DNA sequences are now known for a number of organisms as a result of sequencing experiments starting in the late 1990s, and the identification of genes is as a result a relatively straightforward task.

Finally, we need to establish whether or not our protein actually acts as a transcription factor, which can be done either computationally or experimentally. The computational approach involves determining whether the subsequence to which the protein binds is indeed a promoter region for the identified gene. (It is possible for a protein to bind near a gene but not act as a transcription factor because the point at which it binds has no effect on transcription.) This is a substantially harder task than simply identifying nearby genes. The structure of promoter regions is, unfortunately, quite complex and varies widely, but computer algorithms have been developed that can identify them with some reliability.

Alternatively, one can perform an experiment to measure directly the concentration of the messenger RNA produced when the gene is transcribed. This can be achieved for example by using a *microarray* (colloquially known as a "DNA chip"), tiny dots of DNA strands attached in a grid-like array to a solid surface. RNA will bind to a dot if a part of its sequence matches the sequence of the dot's DNA and this binding can be measured using a fluorescence technique. By observing the simultaneous changes in binding on all the dots of the microarray, one can determine with some accuracy the change in concentration of any specific RNA and hence quantify the effect of the transcription factor. This technique can also be used to determine whether a transcription factor is a promoter or an inhibitor, something that is currently not easy using computational methods.

As with metabolic pathways and protein–protein interactions, there now exist electronic databases of genes and transcription factors, such as EcoCyc, from which it is possible to assemble snapshots of genetic regulatory networks. Current data on gene regulation are substantially incomplete and hence so are our networks, but more data are being added to the databases all the time.

5.2 NEURAL NETWORKS

A completely different use of networks in biology arises in the study of the brain and central nervous system in animals. One of the main functions of the brain is to process information and the primary information processing element is the *neuron*, a specialized brain cell that combines (usually) several inputs to generate a single output. Depending on the animal, an entire brain can contain anywhere from a handful of neurons to more than a hundred billion, wired together, the output of one cell feeding the input of another, to create a *neural network* capable of remarkable feats of calculation and decision making.

Figure 5.6 shows a sketch of a typical neuron, which consists of a cell body

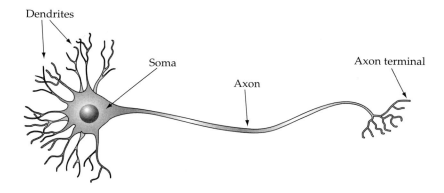

Dendrites

Soma

Axon

Axon terminal

Figure 5.6: The structure of a neuron. A typical neuron is composed of a cell body or soma with many dendrites that act as inputs and a single axon that acts as an output. Towards its tip, the axon branches to allow it to connect to the inputs of several other neurons.

or *soma*, along with a number of protruding tentacles, which are essentially wires for carrying signals in and out of the cell. Most of the wires are inputs, called *dendrites*, of which a neuron may have just one or two, or as many as a thousand or more. Most neurons have only one main output, called the *axon*, which is typically longer than the dendrites and may in some cases extend over large distances to connect the cell to others some way away. Although there is just one axon, it usually branches near its end to allow the output of the cell to feed the inputs of several others. The tip of each branch ends at an *axon terminal* that abuts the tip of the input dendrite of another neuron. There is a small gap, called a *synapse*, between terminal and dendrite across which the output signal of the first (presynaptic) neuron must be conveyed in order to reach the second (postsynaptic) neuron. The synapse plays an important role in the function of the brain, allowing transmission from cell to cell to be regulated by chemically modifying the properties of the gap.[6]

The actual signals that travel within neurons are electrochemical in nature. They consist of traveling waves of electrical voltage created by the motion of positively charged sodium and potassium ions in and out of the cell. These

---

[6]Neurons do sometimes have direct connections between them without synapses. These direct connections are called *gap junctions*, a confusing name, since it sounds like it might be a description of a synapse but is in reality quite different. In our brief treatment of neural networks, however, we will ignore gap junctions.

waves are called *action potentials* and typically consist of voltages on the order of tens of millivolts traveling at tens of meters per second. When an action potential reaches a synapse, it cannot cross the gap between the axon terminal and the opposing dendrite and the signal is instead transmitted chemically; the arrival of the action potential stimulates the production of a chemical neurotransmitter by the terminal, and the neurotransmitter diffuses across the gap and is detected by receptor molecules on the dendrite at the other side. This in turn causes ions to move in and out of the dendrite, changing its voltage.

These voltage changes, however, do not yet give rise to another traveling wave. The soma of the postsynaptic neuron sums the inputs from its dendrites and as a result may (or may not) send an output signal down its own axon. The neuron is stable against perturbations caused by voltages at a small number of its inputs, but if enough inputs are excited they can collectively drive the neuron into an unstable runaway state in which it "fires," generating a large electrochemical pulse that starts a new action potential traveling down the cell's axon and so a signal is passed on to the next neuron or neurons in the network. Thus the neuron acts as a switch or gate that aggregates the signals at its inputs and only fires when enough inputs are excited.

As described, inputs to neurons are excitatory, increasing the chance of firing of the neuron, but inputs can also be inhibiting—signals received at inhibiting inputs make the receiving neuron less likely to fire. Excitatory and inhibiting inputs can be combined in a single neuron and the combination allows neurons to perform quite complex information processing tasks all on their own, while an entire brain or brain region consisting of many neurons can perform tasks of extraordinary complexity. Current science cannot yet tell us exactly how the brain performs the more sophisticated cognitive tasks that allow animals to survive and thrive, but it is known that the brain constantly changes the pattern of wiring between neurons in response to inputs and experiences, and it is presumed that this pattern—the neural network—holds much of the secret. An understanding of the structure of neural networks is thus crucial if we are ever to explain the higher-level functions of the brain.

At the simplest level, a neuron can be thought of as a unit that accepts a number of inputs, either excitatory or inhibiting, combines them, and generates an output result that is sent to one or more further neurons. In network terms, a neural network can thus be represented as a set of vertices—the neurons—connected by two types of directed edges, one for excitatory inputs and one for inhibiting inputs. By convention, excitatory connections are denoted by an edge ending with an arrow "——▶", while inhibiting connections are denoted by an edge ending with a bar "———⊣".

In practice, neurons are not all the same. They come in a variety of differ-

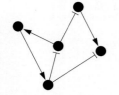

A wiring diagram for a small neural network.

ent types and even relatively small regions or circuits in the brain may contain many types. This variation can be encoded in our network representation by different types of vertex. Visually the types are often denoted by using different shapes for the vertices or by labeling. In functional terms, neurons can differ in a variety of ways, including the number and type of their inputs and outputs, the nature and speed of their response to their inputs, whether and to what extent they can fire spontaneously without receiving inputs, and many other things besides.

Experimental determination of the structure of neural networks is difficult and the lack of straightforward experimental techniques for probing network structure is a major impediment to current progress in neuroscience. Some useful techniques do exist, however, although their application can be extremely laborious.

The basic tool for structure determination is microscopy, either optical or electronic. One relatively simple approach works with cultured neurons on flat dishes. Neurons taken from animal brains at an early stage of embryonic development can be successfully cultured in a suitable nutrient medium and will, without prompting, grow synaptic connections to form a network. If cultured on a flat surface, the network is then roughly two-dimensional and its structure can be determined with reasonable reliability by simple optical microscopy. The advantage of this approach is that it is quick and inexpensive, but it has the substantial disadvantage that the networks studied are not the networks of real living animals and their structure is probably not very similar to that of a functional brain circuit.

In this respect, studies of real brains are much more satisfactory and likely to lead to greater insight, but they are also far harder, because real brains are three-dimensional and we do not currently have any form of microscopy suitable for probing such three-dimensional structures. Instead, therefore, researchers have resorted to cutting suitably preserved brains or brain regions into thin slices, whose structure is then determined by electron microscopy. Given the structure of an entire set of consecutive slices, one can, at least in principle, reconstruct the three-dimensional structure, identifying different types of neurons by their appearance, where possible. In the early days of such studies, most reconstruction was done by hand but more recently researchers have developed computer programs that can significantly speed the reconstruction process. Nonetheless, studies of this kind are very laborious and can take months or years to complete, depending on the size and complexity of the network studied.

Figure 5.7 shows an example of a "wiring diagram" of a neural network, reconstructed by hand from electron microscope studies of this type. The net-

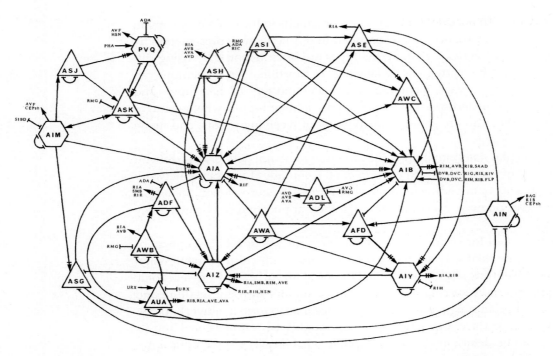

Figure 5.7: A diagram of a part of the brain circuitry of a worm. A portion of the neural circuitry of the nematode *Caenorhabditis elegans*, reconstructed by hand from electron micrographs of slices through the worm's brain. Reproduced from White *et al.* [328]. Copyright of the Royal Society. Reproduced by permission.

work in question is the neural network of the worm *Caenorhabditis elegans*, one of the best studied organisms in biology. The brain of *C. elegans* is simple—it has less than 300 neurons and essentially every specimen of the worm has the same wiring pattern. Several types of neuron, denoted by shapes and labels, are shown in the figure, along with a number of different types of connection, both excitatory and inhibiting. Some of the connections run out of the figure or enter from somewhere off the page. These are connections that run to or from other parts of the network not shown. The original experimenters determined the structure of the entire network and presented it as set of interconnected wiring diagrams like this one [328].

5.3 ECOLOGICAL NETWORKS

The final class of biological network that we consider in this chapter is networks of ecological interactions between species. Species in an ecosystem can interact in a number of different ways. They can eat one another, they can parasitize one another, they can compete for resources, or they can have any of a variety of mutually advantageous interactions, such as pollination or seed dispersal. Although in principle the patterns of interactions of all of these types could be represented in a combined "interaction network" with several different edge types, ecologists have traditionally separated interaction types into different networks. Food webs, for example—networks of predator–prey interactions (i.e., who eats whom)—have a long history of study. Networks of hosts and parasites or of mutualistic interactions are less well studied, but have nonetheless received significant attention in recent years.

5.3.1 FOOD WEBS

The biological organisms on our planet can be divided into *ecosystems*, groups of organisms that interact with one another and with elements of their environment such as sources of material, nutrients, and energy. Mountains, valleys, lakes, islands, and larger regions of land or water can all be home to ecosystems composed of many organisms each. Within ecological theory, ecosystems are usually treated as self-contained units with no outside interactions, although in reality perfect isolation is rare and many ecosystems are only approximately self-contained. Nonetheless, the ecosystem concept is one of significant practical utility for understanding ecological dynamics.

A *food web* is a directed network that represents which species prey on which others in a given ecosystem.[7] The vertices in the network correspond to species and the directed edges to predator–prey interactions. Figure 5.8 shows a small example, representing predation among species living in Antarctica. There are several points worth noticing about this figure. First, notice that not all of the vertices actually represent single species in this case. Some of them do—the vertices for sperm whales and humans, for instance. But some of them represent collections of species, such as birds or fish. This is common practice

---

[7]In common parlance, one refers to a *food chain*, meaning a chain of predator–prey relations between organisms starting with some lowly organism at the bottom of the chain, such as a microbe of some kind, and working all the way up to some ultimate predator at the top, such as a lion or a human being. Only a moment's reflection, however, is enough to convince us that real ecosystems cannot be represented by single chains, and a complete network of interactions is needed in most cases.

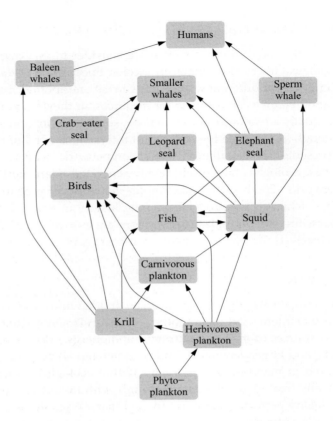

Figure 5.8: A food web of species in Antarctica. Vertices in a food web represent species or sometimes, as with some of the vertices in this diagram, groups of related species, such as fish or birds. Directed edges represent predator–prey interactions and run in the direction of energy flow, i.e., from prey to predator.

in the network representation of food webs. If a set of species such as birds all prey upon and are preyed on by the same other species, then the network can be simplified by representing them as a single vertex, without losing any information about who preys on whom. Indeed, even in cases where a set of species only have mostly, but not exactly, the same predators and prey we still sometimes group them, if we feel the benefits of the resulting simplification are worth a small loss of information. A set of species with the same or similar predators and prey is sometimes referred to as a *trophic species*.

Second, notice the direction of the edges in the network. One might imagine that the edges would point from predators to prey, but ecologists conventionally draw them in the opposite direction, from prey to predator. Thus the edge representing the eating of fish by birds runs *from* the fish vertex *to* the bird vertex. The reason for this apparently odd choice is that ecologists view food webs as representations of the flow of energy (or sometimes carbon) within ecosystems. The arrow from fish to birds indicates that the population of birds gains energy from the population of fish when the birds eat the fish.

Third, notice that almost all the arrows in the figure run up the page. Directed networks with this property—that they can be drawn so that the edges all run in one direction—are called *acyclic networks*. We encountered acyclic networks previously in our discussion of citation networks in Section 4.2. Food webs are usually only approximately acyclic. There are usually a few edges that do not run in the right direction,[8] but it is often a useful approximation to assume that the network is acyclic.

Acyclic networks are discussed in more detail in Section 6.4.2.

The acyclic nature of food webs indicates that there is an intrinsic pecking order among the species in ecosystems. Those higher up the order (which means higher up the page in Fig. 5.8) prey on those lower down, but not vice versa. A species' position in this pecking order is called by ecologists its *trophic level*. Species at the very bottom of the food web, of which there is just one in our example—the phytoplankton—have trophic level 1. Those that prey on them—krill, herbivorous plankton—have trophic level 2, and so forth all the way up to the species at the top of the web, which have no predators at all. In our antarctic example there are two species that have no predators, humans and small whales. (Note however that although such species are all, in a sense, at "the top of the food chain" they need not have the same trophic level.)

Trophic level is a useful general guide to the roles that species play in ecosystems, those in lower trophic levels tending to be smaller, more abundant species that are prey to other species higher up the food web, while those in higher trophic levels are predators, usually larger-bodied and less numerous. Calculating a species' trophic level, however, is not always easy. In principle, the rule is simple: a species' trophic level is 1 greater than the trophic level of its prey. Thus the herbivorous plankton and krill in our example have trophic level 2, because their prey has trophic level 1, and the carnivorous plankton have trophic level 3. On the other hand, the squid in our example prey on species at two different levels, levels 2 and 3, so it is unclear what level they belong to. A variety of mathematical definitions have been proposed to resolve this issue. One strategy is to define trophic level to be 1 greater than the mean of the trophic levels of the prey. There is, however, no accepted standard

---

[8]In Fig. 5.8, for example, there are edges in both directions between the fish and squid vertices, which makes it impossible to draw the network with all edges running in the same direction.

definition, and the only indisputable statement one can make is that in most food webs some species have ill-defined or mixed trophic level.

The food webs appearing in the ecological literature come in two basic types. *Community food webs* are complete webs for an entire ecosystem, as in Fig. 5.8—they represent, at least in theory, every predator–prey interaction in the system. *Source food webs* and *sink food webs* are subsets of complete webs that focus on species connected, directly or indirectly, to a specific prey or predator. In a source food web, for instance, one records all species that derive energy from a particular source species, such as grass. Our food web of antarctic species is, in fact, both a community food web and a source food web, since all of the species in the network derive their energy ultimately from phytoplankton. Phytoplankton is the source in this example, and the species above it (all of the species in this case) form the corresponding source web. A sink food web is the equivalent construct for a particular top predator in the network. In the antarctic example, for instance, humans consume the sperm and baleen whales and elephant seals, which in turn derive their energy from fish, squid, plankton, krill, and ultimately phytoplankton. This subset of species, therefore, constitutes the sink food web for humans—the web that specifies through which species or species groups the energy consumed by humans passes.

The experimental determination of the structure of food webs is typically done in one of two different ways, or sometimes a mixture of both. The first and most straightforward method is direct measurement. Having settled on the ecosystem to be studied, one first assembles a list of the species in that ecosystem and then determines their predator–prey interactions. For large-bodied animals such as mammals, birds, or larger fish, some predation can be established simply by observation in the field—we see a bird eating a fish and the presence of the corresponding edge is thereby established. More often, however, and particularly with smaller-bodied animals, interactions are established by catching and dissecting the animals in question and examining the contents of their stomachs to determine what they have been eating.

The second primary method of constructing food webs is by compilation from existing literature. Many predator–prey interactions are already known and have been recorded in the scientific literature, but not in the context of the larger food web, and one can often reconstruct a complete or partial picture of a food web by searching the literature for such records. Many of the currently available food web data sets were assembled in this way from pre-existing data, and some others were assembled by a combination of experimental measurement and literature searches.

In some cases attempts have also been made to measure not merely the presence (or absence) of interactions between species but also the strength of those interactions. One can quantify interaction strength by the fraction of its energy a species derives from each of its predators, or by the total rate of energy flow between a prey species and a predator. The result is a weighted directed network that sheds considerably more light on the flow of energy through an ecosystem than the more conventional unweighted food web. Measurements of interaction strength are, however, time-consuming, difficult, and yield uncertain results, so the current data on weighted food webs should be treated with caution.

Food web data from a variety of sources have been assembled into publicly available databases, starting in the late 1980s. Examples include the Ecoweb database [73] and the web-based collection at www.foodwebs.org.

5.3.2 OTHER ECOLOGICAL NETWORKS

Two other types of ecological network have received significant attention in the scientific literature (although less than has been paid to food webs). *Host–parasite networks* are networks of parasitic relationships between organisms, such as the relationship between a large-bodied animal and the insects and microorganisms that live on and inside it. In a sense parasitic relations are a form of predation—one species eating another—but in practical terms they are quite distinct from traditional predator–prey interactions. Parasites, for example, tend to be smaller-bodied than their hosts where predators tend to be larger, and parasites can live off their hosts for long, sometimes indefinite, periods of time without killing them, where predation usually results in the death of the prey.

Parasitic interactions, however, do form networks that are somewhat similar to traditional food webs. Parasites themselves frequently play host to still smaller parasites (called "hyperparasites"), which may have their own still smaller ones, and so forth through several levels.[9] There is a modest but growing literature on host–parasite networks, much of it based on research within the agriculture community, a primary reason for interest in parasites being their prevalence in and effects on livestock and crop species.

The other main class of ecological networks is that of *mutualistic networks*, meaning networks of mutually beneficial interactions between species. Three

---

[9]One is reminded of the schoolhouse rhyme by Augustus de Morgan:

> Great fleas have little fleas upon their backs to bite 'em,
> And little fleas have lesser fleas, and so ad infinitum.

specific types of mutualistic network that have received attention in the ecological literature are networks of plants and the animals (primarily insects) that pollinate them, networks of plants and the animals (such as birds) that disperse their seeds, and networks of ant species and the plants that they protect and eat. Since the benefit of a mutualistic interaction runs, by definition, in both directions between a pair of species, mutualistic networks are undirected networks (or bidirectional, if you prefer), in contrast with the directed interactions of food webs and host–parasite networks. Most mutualistic networks studied are also bipartite, consisting of two distinct, non-overlapping sets of species (such as plants and ants), with interactions only between members of different sets. In principle, however, non-bipartite mutualistic networks are also possible.

See Section 6.6 for a discussion of bipartite networks.

PART II

FUNDAMENTALS OF NETWORK THEORY

CHAPTER 6

MATHEMATICS OF NETWORKS

An introduction to the mathematical tools used in the
study of networks, tools that will be important to many
subsequent developments

IN THE next three chapters we introduce the fundamental quantitative foundations of the study of networks, concepts that are crucial for essentially all later developments in this book. In this chapter we introduce the basic theoretical tools used to describe and analyze networks, most of which come from graph theory, the branch of mathematics that deals with networks. Graph theory is a large field containing many results and we describe only a small fraction of those results here, focusing on the ones most relevant to the study of real-world networks. Readers interested in pursuing the study of graph theory further might like to look at the books by Harary [155] or West [324].

In the two chapters after this one we look first at measures and metrics for quantifying network structure (Chapter 7) and then at some of the remarkable patterns revealed in real-world networks when we apply the mathematics and metrics we have developed to their analysis (Chapter 8).

6.1 NETWORKS AND THEIR REPRESENTATION

To begin at the beginning, a *network*—also called a *graph* in the mathematical literature—is, as we have said, a collection of vertices joined by edges. Vertices and edges are also called *nodes* and *links* in computer science, *sites* and *bonds* in physics, and *actors*[1] and *ties* in sociology. Table 6.1 gives some examples of vertices and edges in particular networks.

---

[1]This use of the word "actor" sometimes leads to confusion: an actor need not be a person who actually acts, and need not even be a person. In a social network of business relationships

Network	Vertex	Edge
Internet	Computer or router	Cable or wireless data connection
World Wide Web	Web page	Hyperlink
Citation network	Article, patent, or legal case	Citation
Power grid	Generating station or substation	Transmission line
Friendship network	Person	Friendship
Metabolic network	Metabolite	Metabolic reaction
Neural network	Neuron	Synapse
Food web	Species	Predation

Table 6.1: Vertices and edges in networks. Some examples of vertices and edges in particular networks.

Throughout this book we will normally denote the number of vertices in a network by n and the number of edges by m, which is a common notation in the mathematical literature.

Most of the networks we will study in this book have at most a single edge between any pair of vertices. In the rare cases where there can be more than one edge between the same pair of vertices we refer to those edges collectively as a *multiedge*. In most of the networks we will study there are also no edges that connect vertices to themselves, although such edges will occur in a few instances. Such edges are called *self-edges* or *self-loops*.

A network that has neither self-edges nor multiedges is called a *simple network* or *simple graph*. A network with multiedges is called a *multigraph*.[2] Figure 6.1 shows examples of (a) a simple graph and (b) a non-simple graph having both multiedges and self-edges.

6.2 THE ADJACENCY MATRIX

There are a number of different ways to represent a network mathematically. Consider an undirected network with n vertices and let us label the vertices with integer labels $1 \ldots n$, as we have, for instance, for the network in Fig. 6.1a. It does not matter which vertex gets which label, only that each label is unique, so that we can use the labels to refer to any vertex unambiguously.

If we denote an edge between vertices i and j by (i, j) then the complete

---

between companies, for instance, the actors are the companies (and the ties are the business relationships).

[2]There does not seem to be a special name given to networks with self-edges. They are just called "networks with self-edges."

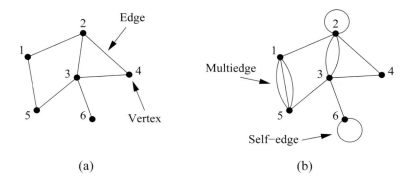

$$(a) \qquad\qquad (b)$$

Figure 6.1: Two small networks. (a) A simple graph, i.e., one having no multiedges or self-edges. (b) A network with both multiedges and self-edges.

network can be specified by giving the value of n and a list of all the edges. For example, the network in Fig. 6.1a has $n = 6$ vertices and edges $(1,2)$, $(1,5)$, $(2,3)$, $(2,4)$, $(3,4)$, $(3,5)$, and $(3,6)$. Such a specification is called an *edge list*. Edge lists are sometimes used to store the structure of networks on computers, but for mathematical developments like those in this chapter they are rather cumbersome.

A better representation of a network for present purposes is the *adjacency matrix*. The adjacency matrix $\mathbf{A}$ of a simple graph is the matrix with elements A_{ij} such that

$$A_{ij} = \begin{cases} 1 & \text{if there is an edge between vertices } i \text{ and } j, \\ 0 & \text{otherwise.} \end{cases} \qquad (6.1)$$

For example, the adjacency matrix of the network in Fig. 6.1a is

$$\mathbf{A} = \begin{pmatrix} 0 & 1 & 0 & 0 & 1 & 0 \\ 1 & 0 & 1 & 1 & 0 & 0 \\ 0 & 1 & 0 & 1 & 1 & 1 \\ 0 & 1 & 1 & 0 & 0 & 0 \\ 1 & 0 & 1 & 0 & 0 & 0 \\ 0 & 0 & 1 & 0 & 0 & 0 \end{pmatrix}. \qquad (6.2)$$

Two points to notice about the adjacency matrix are that, first, for a network with no self-edges such as this one the diagonal matrix elements are all zero, and second that it is symmetric, since if there is an edge between i and j then there is an edge between j and i.

It is also possible to represent multiedges and self-edges using an adjacency matrix. A multiedge is represented by setting the corresponding matrix element A_{ij} equal to the multiplicity of the edge. For example, a double edge between vertices i and j is represented by $A_{ij} = A_{ji} = 2$.

Self-edges are a little more complicated. A single self-edge from vertex i to itself is represented by setting the corresponding diagonal element A_{ii} of the matrix equal to 2. Why 2 and not 1? Essentially it is because every self-edge from i to i has two ends, both of which are connected to vertex i. We will find that many of our mathematical results concerning the adjacency matrix work equally well for networks with and without self-edges, but only if we are careful to count both ends of every edge, including the self-edges, by making the diagonal matrix elements equal to 2 rather than 1.[3]

Another way to look at this is that non-self-edges appear twice in the adjacency matrix—an edge from i to j means that both A_{ij} and A_{ji} are 1. To count edges equally, self-edges should also appear twice, and since there is only one diagonal matrix element A_{ii}, we need to record both appearances there.

To give an example, the adjacency matrix for the multigraph in Fig. 6.1b is

$$
\mathbf{A} = \begin{pmatrix}
0 & 1 & 0 & 0 & 3 & 0 \\
1 & 2 & 2 & 1 & 0 & 0 \\
0 & 2 & 0 & 1 & 1 & 1 \\
0 & 1 & 1 & 0 & 0 & 0 \\
3 & 0 & 1 & 0 & 0 & 0 \\
0 & 0 & 1 & 0 & 0 & 2
\end{pmatrix}. \tag{6.3}
$$

One can also have multiple self-edges (or "multi-self-edges" perhaps). Such edges are represented by setting the corresponding diagonal element of the adjacency matrix equal to twice the multiplicity of the edge.

6.3   WEIGHTED NETWORKS

Many of the networks we will study have edges that form simple on/off connections between vertices. Either they are there or they are not. In some situations, however, it is useful to represent edges as having a strength, weight, or value to them, usually a real number. Thus in the Internet edges might have weights representing the amount of data flowing along them or their bandwidth. In a food web predator–prey interactions might have weights measur-

---

[3]As discussed in the next section, this is not the case for directed networks. In directed networks, self-edges are represented by a 1 in the corresponding diagonal element of the adjacency matrix.

ing total energy flow between prey and predator. In a social network connections might have weights representing frequency of contact between actors. Such *weighted* or *valued networks* can be represented by giving the elements of the adjacency matrix values equal to the weights of the corresponding connections. Thus the adjacency matrix

$$\mathbf{A} = \begin{pmatrix} 0 & 2 & 1 \\ 2 & 0 & 0.5 \\ 1 & 0.5 & 0 \end{pmatrix} \tag{6.4}$$

represents a weighted network in which the connection between vertices 1 and 2 is twice as strong as that between 1 and 3, which in turn is twice as strong as that between 2 and 3.[4]

We have now seen two different types of network where the adjacency matrix can have off-diagonal elements with values other than 0 and 1, networks with weighted edges and networks with multiedges.[5] Indeed, if the weights in a weighted network are all integers it is possible to create a network with multiedges that has the exact same adjacency matrix, by simply choosing the multiplicities of the multiedges equal to the corresponding weights. This connection comes in handy sometimes. In some circumstances it is easier to reason about the behavior of a multigraph than a weighted network, or vice versa, and switching between the two can be a useful aid to analysis [242].

The weights in a weighted network are usually positive numbers, but there is no reason in theory why they should not be negative. For example, it is common in social network theory to construct networks of social relations between people in which positive edge weights denote friendship or other cordial relationships and negative ones represent animosity. We discuss such networks further in Section 7.11 when we consider the concept of structural balance.

Given that edges can have weights on them, it is not a huge leap to consider weights on vertices too, or to consider more exotic variables on either edges or

---

[4]The values on edges also sometimes represent lengths of some kind. On a road or airline network, for instance, edge values could represent the number of kilometers or miles the edges cover, or they could represent travel time along the edges, which can be regarded as a kind of length—one denominated in units of time rather than distance. Edge lengths are, in a sense, the inverse of edge weights, since two vertices that are strongly connected can be regarded as "close" to one another and two that are weakly connected can be regarded as far apart. Thus one could convert between weights and lengths by taking reciprocals, although this should be regarded as only an approximate procedure; in most cases there is no formal sense in which edge weights and lengths are equivalent.

[5]The diagonal elements are a special case, since they are equal to 0 or 2 in an undirected network even when there are no multiedges or weighted edges.

vertices, such as vectors or discrete enumerative variables like colors. Many such variations have been considered in the networks literature and we will discuss some of them later in the book. There is one case of variables on edges, however, that is so central to the study of networks that we discuss it straight away.

6.4 DIRECTED NETWORKS

A *directed network* or *directed graph*, also called a *digraph* for short, is a network in which each edge has a direction, pointing *from* one vertex *to* another. Such edges are themselves called *directed edges*, and can be represented by lines with arrows on them—see Fig. 6.2.

We encountered a number of examples of directed networks in previous chapters, including the World Wide Web, in which hyperlinks run in one direction from one web page to another, food webs, in which energy flows from prey to predators, and citation networks, in which citations point from one paper to another.

The adjacency matrix of a directed network has matrix elements

$$A_{ij} = \begin{cases} 1 & \text{if there is an edge } \textit{from } j \textit{ to } i, \\ 0 & \text{otherwise.} \end{cases} \tag{6.5}$$

Notice the direction of the edge here—it runs *from* the second index *to* the first. This is slightly counter-intuitive, but it turns out to be convenient mathematically and it is the convention we adopt in this book.

As an example, the adjacency matrix of the small network in Fig. 6.2 is

$$\mathbf{A} = \begin{pmatrix} 0 & 0 & 0 & 1 & 0 & 0 \\ 0 & 0 & 1 & 0 & 0 & 0 \\ 1 & 0 & 0 & 0 & 1 & 0 \\ 0 & 0 & 0 & 0 & 0 & 1 \\ 0 & 0 & 0 & 1 & 0 & 1 \\ 0 & 1 & 0 & 0 & 0 & 0 \end{pmatrix}. \tag{6.6}$$

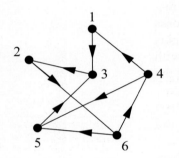

Figure 6.2: A directed network. A small directed network with arrows indicating the directions of the edges.

Note that this matrix is not symmetric. In general the adjacency matrix of a directed network is asymmetric.

We can, if we wish, think of undirected networks as directed networks in which each undirected edge has been replaced with two directed ones running in opposite directions between the same pair of vertices. The adjacency matrix for such a network is then symmetric and exactly the same as for the original undirected network.

Like their undirected counterparts, directed networks can have multiedges and self-edges, which are represented in the adjacency matrix by elements with values greater than 1 and by non-zero diagonal elements, respectively. An important point however is that self-edges in a directed network are represented by setting the corresponding diagonal element of the adjacency matrix to 1, not 2 as in the undirected case.[6] With this choice the same formulas and results, in terms of the adjacency matrix, apply for networks with and without self-edges.

6.4.1 COCITATION AND BIBLIOGRAPHIC COUPLING

It is sometimes convenient to turn a directed network into an undirected one for the purposes of analysis—there are many useful analytic techniques for undirected networks that do not have directed counterparts (or at least not yet).

One simple way to make a directed network undirected is just to ignore the edge directions entirely, an approach that can work in some cases, but inevitably throws out a lot of potentially useful information about the network's structure. A more sophisticated approach is to use "cocitation" or "bibliographic coupling," two different but related ideas that derive their names from their widespread use in the analysis of citation networks.

We briefly discussed cocitation in the context of citation networks in Section 4.2.

The *cocitation* of two vertices i and j in a directed network is the number of vertices that have outgoing edges pointing to both i and j. In the language of citation networks, for instance, the cocitation of two papers is the number of other papers that cite both. Given the definition above of the adjacency matrix of a directed network (Eq. (6.5)), we can see that $A_{ik}A_{jk} = 1$ if i and j are both cited by k and zero otherwise. Summing over all k, the cocitation C_{ij} of i and j is

$$C_{ij} = \sum_{k=1}^{n} A_{ik}A_{jk} = \sum_{k=1}^{n} A_{ik}A_{kj}^{T}, \qquad (6.7)$$

where A_{kj}^{T} is an element of the transpose of $\mathbf{A}$. We can define the *cocitation matrix* $\mathbf{C}$ to be the $n \times n$ matrix with elements C_{ij}, which is thus given by

$$\mathbf{C} = \mathbf{A}\mathbf{A}^{T}. \qquad (6.8)$$

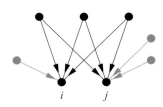

Vertices i and j are cited by three common papers, so their cocitation is 3.

---

[6]Indeed, one can understand the appearance of the 2 in the undirected case as a consequence of the equivalence between undirected and directed networks mentioned above: an undirected self-edge can be thought of as two directed self-edges at the same vertex, each of which contributes 1 to the corresponding element of the adjacency matrix.

Note that $\mathbf{C}$ is a symmetric matrix, since $\mathbf{C}^T = (\mathbf{A}\mathbf{A}^T)^T = \mathbf{A}\mathbf{A}^T = \mathbf{C}$.

Now we can define a *cocitation network* in which there is an edge between i and j if $C_{ij} > 0$, for $i \neq j$, i.e., an edge between any two vertices that are cocited in the original directed network. (We enforce the constraint that $i \neq j$ because the cocitation network is conventionally defined to have no self-edges, even though the diagonal elements of the cocitation matrix are in general non-zero—see below.) Better still, we can make the cocitation network a weighted network with positive integer weights on the edges equal to the corresponding elements C_{ij}. Then vertex pairs cited by more common neighbors have a stronger connection than those cited by fewer. Since the cocitation matrix is symmetric, the cocitation network is undirected, making it easier to deal with in many respects than the original directed network from which it was constructed.

The cocitation network turns out to make a lot of sense in many cases. In citation networks of academic papers, for instance, strong cocitation between papers is often a good indicator of papers that deal with related topics—if two papers are often cited together in the same bibliography they probably have something in common. And the more often they are cited together, the more likely it is that they are related.

The cocitation matrix thus plays a role similar to an adjacency matrix for the cocitation network. There is however one aspect in which the cocitation matrix differs from an adjacency matrix: its diagonal elements. The diagonal elements of the cocitation matrix are given by

$$C_{ii} = \sum_{k=1}^{n} A_{ik}^2 = \sum_{k=1}^{n} A_{ik}, \tag{6.9}$$

where we have assumed that the directed network is a simple graph, with no multiedges, so that all elements A_{ik} of the adjacency matrix are zero or one. Thus C_{ii} is equal to the total number of edges pointing to i—the total number of papers citing i in the citation network language. In constructing the cocitation network we ignore these diagonal elements, meaning that the network's adjacency matrix is equal to the cocitation matrix but with all the diagonal elements set to zero.

Bibliographic coupling is similar to cocitation. The *bibliographic coupling* of two vertices in a directed network is the number of other vertices to which both point. In a citation network, for instance, the bibliographic coupling of two papers i and j is the number of other papers that are cited by both i and j. Noting that $A_{ki}A_{kj} = 1$ if i and j both cite k and zero otherwise, the biblio-

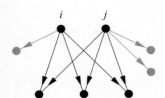

Vertices i and j cite three of the same papers and so have a bibliographic coupling of 3.

graphic coupling of i and j is

$$B_{ij} = \sum_{k=1}^{n} A_{ki} A_{kj} = \sum_{k=1}^{n} A_{ik}^{T} A_{kj},$$ (6.10)

and we define the *bibliographic coupling matrix* $\mathbf{B}$ to be the $n \times n$ matrix with elements B_{ij} so that

$$\mathbf{B} = \mathbf{A}^{T}\mathbf{A}.$$ (6.11)

The bibliographic coupling matrix is again a symmetric matrix and the off-diagonal elements can be used to define a weighted undirected network, the *bibliographic coupling network*, in which there is an edge with weight B_{ij} between any vertex pair i, j for which $B_{ij} > 0$. The diagonal elements of $\mathbf{B}$ are

$$B_{ii} = \sum_{k=1}^{n} A_{ki}^2 = \sum_{k=1}^{n} A_{ki}.$$ (6.12)

Thus B_{ii} is equal to the number of other vertices that vertex i points to—the number of papers i cites, in the citation language.

Bibliographic coupling, like cocitation, can be a useful measure of connection between vertices. In a citation network, for example, if two papers cite many of the same other papers it is often a good indication that they deal with similar subject matter, and the number of common papers cited can be an indicator of how strongly they overlap.

Although cocitation and bibliographic coupling are mathematically similar measures they can in practice give noticeably different results. In particular, they are affected strongly by the number of ingoing and outgoing edges that vertices have. For two vertices to have strong cocitation—to be pointed to by many of the same other vertices—they must both have a lot of incoming edges in the first place. In a citation network, for instance, two papers can only have strong cocitation if they are both well cited and hence strong cocitation is limited to influential papers, review articles, books, and similar highly cited items. Conversely, two papers can only have strong bibliographic coupling if they both cite many others, i.e., if they have large bibliographies. In practice, the sizes of bibliographies vary less than the number of citations papers receive, and hence bibliographic coupling is a more uniform indicator of similarity between papers than cocitation. The Science Citation Index, for example, makes use of bibliographic coupling in its "Related Records" feature, which allows users to find papers similar to a given paper. Cocitation would be less appropriate in this situation, since it tends not to work well for papers with few citations.

Bibliographic coupling also has the advantage that it can be computed as soon as a paper is published and the contents of the paper's bibliography are known. Cocitation, on the other hand, cannot be computed until a paper has been cited by other papers, which usually doesn't happen until at least a few months after publication, and sometimes years. Furthermore, the cocitation of two papers can change over time as the papers receive new citations, whereas bibliographic coupling is fixed from the moment the papers are published. (This could be an advantage or a disadvantage—there are situations in which changes in cocitation could reveal interesting information about the papers that cannot be gleaned from an unchanging measure like bibliographic coupling.)

In addition to their use as measures of vertex similarity, the cocitation and bibliographic coupling matrices are also used in search algorithms for directed networks, and in particular in the so-called HITS algorithm, which we describe in Section 7.5.

6.4.2 ACYCLIC DIRECTED NETWORKS

A *cycle* in a directed network is a closed loop of edges with the arrows on each of the edges pointing the same way around the loop. Networks like the World Wide Web have many such cycles in them. Some directed networks however have no cycles and these are called *acyclic* networks.[7] Ones with cycles are called *cyclic*. A self-edge—an edge connecting a vertex to itself—counts as a cycle, and so an acyclic network also has no self-edges.

A cycle in a directed network.

The classic example of an acyclic directed network is a citation network of papers, as discussed in Section 4.2. When writing a paper you can only cite another paper if it has already been written, which means that all the directed edges in a citation network point backward in time. Graphically we can depict such a network as in Fig. 6.3, with the vertices time-ordered—running from bottom to top of the picture in this case—so that all the edges representing the citations point downward in the picture.[8] There can be no closed cycles in such a network because any cycle would have to go down the picture and then come back up again to get back to where it started and there are no upward

---

[7] In the mathematical literature one often sees the abbreviation DAG, which is short for *directed acyclic graph*.

[8] As discussed in Section 4.2, there are in real citation networks rare instances in which two papers both cite each other, forming a cycle of length two in the citation network, for instance if an author publishes two related papers in the same issue of a journal. Citation networks are, nonetheless, acyclic to a good approximation.

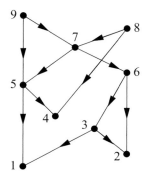

Figure 6.3: An acyclic directed network. In this network the vertices are laid out in such a way that all edges point downward. Networks that can be laid out in this way are called acyclic, since they possess no closed cycles of edges. An example of an acyclic network is a citation network of citations between papers, in which the vertical axis would represent date of publication, running up the figure, and all citations would necessarily point from later papers to earlier ones.

edges with which to achieve this.

It is less obvious but still true that if a network is acyclic it can be drawn in the manner of Fig. 6.3 with all edges pointing downward. The proof that this can be done turns out to be useful, because it also provides us with a method for determining whether a given network is acyclic.

Suppose we have an acyclic directed network of n vertices. There must be at least one vertex somewhere on the network that has ingoing edges only and no outgoing ones. To see this consider starting from any vertex in the network and making a path across the network by following edges, each in the correct direction denoted by its arrow. Either such a path will eventually encounter a vertex with no outgoing edges, in which case we are done, or each vertex it encounters has one or more outgoing edges, in which case we choose one such edge and continue our path. If the path never reaches a vertex with no outgoing edges, then it must eventually arrive back at a vertex that has been visited previously—at most we can visit all n vertices in the network once before the path either terminates or we are forced to revisit a vertex. However if we revisit a vertex then we have gone around a cycle in the network, which cannot be since the network is acyclic. Thus we must always in the end find a vertex with no outgoing edges and hence at least one such vertex always exists.

In practice, it is not necessary to actually construct the paths through the network to find a vertex with no outgoing edges—since we know that such a vertex exists, we can simply look through each vertex in turn until we find one.

We now take this vertex with no outgoing edges and draw it at the bottom of our picture. We remove this vertex from the network, along with any edges attached to it, and repeat the process, finding another vertex with no outgoing edges in the remaining network. We draw this second vertex above the first one in the figure, remove it from the network and repeat again. And so forth.

When we have drawn all vertices, we then add the directed edges between them to the picture. Since each edge, by definition, has incoming edges only from vertices drawn after it—and therefore above it—all edges in the final picture must be pointing downward. Note that the particular order in which we draw the vertices, and hence the picture we produce, is not necessarily unique. If at any stage in the process of drawing the vertices there is more than one vertex with no outgoing edges then we have a choice about which one we pick and hence a choice between overall vertex orders.

This process is a useful one for visualizing acyclic networks. Most computer algorithms for drawing such networks work by arranging the vertices in order along one axis in just this way, and then moving them around along the other axis to make the network structure as clear and visually pleasing as possible (which usually means minimizing the number of times that edges cross).

The process is useful for another reason too: it will break down if the network is cyclic, and therefore it gives us a way to test whether a given network is acyclic. If a network contains a cycle, then none of the vertices in that cycle will ever be removed during our process: none of them will be without outgoing edges until one of the others in the cycle is removed, and hence none of them can ever be removed. Thus, if the network contains a cycle there must come a point in our process where there are still vertices left in the network but all of them have outgoing edges. So a simple algorithm for determining whether a network is acyclic is:

1. Find a vertex with no outgoing edges.
2. If no such vertex exists, the network is *cyclic*. Otherwise, if such a vertex does exist, remove it and all its ingoing edges from the network.
3. If all vertices have been removed, the network is *acyclic*. Otherwise go back to step 1.

The adjacency matrix of an acyclic directed network has interesting properties. Suppose we construct an ordering of the vertices of an acyclic network as described above, so that all edges point in one direction, and suppose we then label the vertices in that order. Then there can be an edge from vertex j to vertex i only if $j > i$. Put another way, the adjacency matrix $\mathbf{A}$ (whose element A_{ij} records the presence of an edge *from j to i*) has all its non-zero elements above the diagonal—it is upper triangular. For instance, the adjacency matrix of the network shown in Fig. 6.3 is

$$
\mathbf{A} = \begin{pmatrix}
0 & 0 & 1 & 0 & 1 & 0 & 0 & 0 & 0 \\
0 & 0 & 1 & 0 & 0 & 1 & 0 & 0 & 0 \\
0 & 0 & 0 & 0 & 0 & 1 & 0 & 0 & 0 \\
0 & 0 & 0 & 0 & 1 & 0 & 0 & 1 & 0 \\
0 & 0 & 0 & 0 & 0 & 0 & 1 & 0 & 1 \\
0 & 0 & 0 & 0 & 0 & 0 & 1 & 0 & 0 \\
0 & 0 & 0 & 0 & 0 & 0 & 0 & 1 & 1 \\
0 & 0 & 0 & 0 & 0 & 0 & 0 & 0 & 0 \\
0 & 0 & 0 & 0 & 0 & 0 & 0 & 0 & 0
\end{pmatrix}.
\tag{6.13}
$$

Note also that the diagonal elements of the adjacency matrix are necessarily zero, since an acyclic network has no self-edges. Triangular matrices with zeros on the diagonal are called *strictly triangular*.

If the vertices of an acyclic network are not numbered in order as described above, then the adjacency matrix will not be triangular. (Imagine swapping rows and columns of the matrix above, for instance.) However, we can say that for every acyclic directed network there exists at least one labeling of the vertices such that the adjacency matrix will be strictly upper triangular.

The adjacency matrix also has the property that all of its eigenvalues are zero if and only if the network is acyclic. To demonstrate this, we must demonstrate the correspondence in both directions, i.e., that the adjacency matrix of an acyclic network has all eigenvalues zero and also that a network is acyclic if its adjacency matrix has all eigenvalues zero.

The former is the easier to prove. If a network is acyclic then we can order and label the vertices as described above and hence write the adjacency matrix in strictly upper triangular form. The diagonal elements of a triangular matrix, however, are its eigenvalues, and since these are all zero it follows immediately that all eigenvalues are zero for an acyclic network.

To show the converse, that the network is acyclic if the eigenvalues are all zero, it suffices to demonstrate that any cyclic network must have at least one non-zero eigenvalue. To demonstrate this we make use of a result derived in Section 6.10. There we show that the total number L_r of cycles of length r in a network is

$$
L_r = \sum_{i=1}^{n} \kappa_i^r,
\tag{6.14}
$$

where κ_i is the ith eigenvalue of the adjacency matrix. Suppose a network is cyclic. Let r be the length of one of the cycles it contains. Then by definition $L_r > 0$ for this network. However, this can only be the case if at least one of the terms in the sum on the right-hand side of Eq. (6.14) is greater than zero,

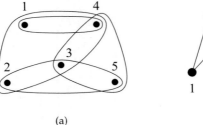 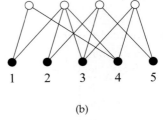

(a)   (b)

Figure 6.4: A hypergraph and corresponding bipartite graph. These two networks show the same information—the membership of five vertices in four different groups. (a) The hypergraph representation in which the groups are represented as hyperedges, denoted by the loops circling sets of vertices. (b) The bipartite representation in which we introduce four new vertices (open circles) representing the four groups, with edges connecting each vertex to the groups to which it belongs.

and hence the adjacency matrix has at least one non-zero eigenvalue. If all eigenvalues are zero, therefore, the network cannot be cyclic.

Matrices with all eigenvalues zero are called *nilpotent matrices*. Thus one could also say that a network is acyclic if and only if it has a nilpotent adjacency matrix.

6.5 HYPERGRAPHS

In some kinds of network the links join more than two vertices at a time. For example, we might want to create a social network representing families in a larger community of people. Families can have more than two people in them and the best way to represent family ties in such families is to use a generalized kind of edge that joins more than two vertices.[9] Such an edge is called a *hyperedge* and a network with hyperedges is called a *hypergraph*. Figure 6.4a shows a small example of a hypergraph in which the hyperedges are denoted by loops.

---

[9]We could just use ordinary edges joining vertex pairs to represent our family ties, placing an edge between any two vertices that correspond to individuals in the same family. This, however, doesn't explicitly tell us when two edges correspond to ties within the same family, and there is no single object in the network that corresponds to a family the way a hyperedge does in the hypergraph. In a number of ways, therefore, the hypergraph is more convenient.

Network	Vertex	Group	Section
Film actors	Actor	Cast of a film	3.5
Coauthorship	Author	Authors of an article	3.5
Boards of directors	Director	Board of a company	3.5
Social events	People	Participants at social event	3.1
Recommender system	People	Those who like a book, film, etc.	4.3.2
Keyword index	Keywords	Pages where words appear	4.3.3
Rail connections	Stations	Train routes	2.4
Metabolic reactions	Metabolites	Participants in a reaction	5.1.1

Table 6.2: Hypergraphs and bipartite graphs. Examples of networks that can be represented as hypergraphs or equivalently as bipartite graphs. The last column gives the section of this book in which each network is discussed.

Many of the networks that we will encounter in this book can be presented as hypergraphs. In particular, any network in which the vertices are connected together by common membership of groups of some kind can be represented in this way. In sociology such networks are called "affiliation networks" and we saw several examples of them in Section 3.5. Directors sitting on the boards of companies, scientists coauthoring papers, and film actors appearing together in films are all examples of such networks (see Table 6.2).

We will however talk very little about hypergraphs in this book, because there is another way of representing the same information that is more convenient for our purposes—the bipartite network.

6.6 BIPARTITE NETWORKS

The membership of vertices in groups represented by hyperedges in a hypergraph can equally and often more conveniently be represented as a *bipartite network*, also called a *two-mode network* in the sociology literature. In such a network there are two kinds of vertices, one representing the original vertices and the other representing the groups to which they belong. We discussed bipartite networks previously in the context of affiliation networks in Section 3.5 and of recommender networks in Section 4.3.2. For example, we can represent the network of film actors discussed in Section 3.5 as a bipartite network in which the two types of vertex are the actors themselves and the films in which they appear. The edges in a bipartite network run only between vertices of unlike types: in the film network they would run only between actors and films, and each actor would be connected by an edge to each film in which he or she appeared. A small example of a bipartite network is shown in Fig. 6.4b. This example network in fact portrays exactly the same set of group memberships as the hypergraph of Fig. 6.4a; the two are entirely equivalent.

Bipartite networks occur occasionally in contexts other than membership of groups. For example, if we were to construct a network of who is or has been married to whom within a population, that network would be bipartite, the two kinds of vertex corresponding to men and women and the edges between them marriages.[10]

The equivalent of an adjacency matrix for a bipartite network is a rectangular matrix called an *incidence matrix*. If n is the number of people or other participants in the network and g is the number of groups, then the incidence matrix $\mathbf{B}$ is a $g \times n$ matrix having elements B_{ij} such that

$$B_{ij} = \begin{cases} 1 & \text{if vertex } j \text{ belongs to group } i, \\ 0 & \text{otherwise.} \end{cases} \tag{6.15}$$

For instance, the 4×5 incidence matrix of the network shown in Fig. 6.4b is

$$\mathbf{B} = \begin{pmatrix} 1 & 0 & 0 & 1 & 0 \\ 1 & 1 & 1 & 1 & 0 \\ 0 & 1 & 1 & 0 & 1 \\ 0 & 0 & 1 & 1 & 1 \end{pmatrix}. \tag{6.16}$$

Although a bipartite network may give the most complete representation of a particular network it is often convenient to work with direct connections between vertices of just one type. We can use the bipartite network to infer such connections, creating a *one-mode projection* from the two-mode bipartite form. As an example, consider again the case of the films and actors. We can perform a projection onto the actors alone by constructing the n-vertex network in which the vertices represent actors and two actors are connected by an edge if they have appeared together in a film. The corresponding one-mode projection onto the films would be the g-vertex network where the vertices represent films and two films are connected if they share a common actor. Figure 6.5 shows the two one-mode projections of a small bipartite network.

When we form a one-mode projection each group in the bipartite network results in a cluster of vertices in the one-mode projection that are all connected to each other—a "clique" in network jargon (see Section 7.8.1). For instance, if a group contains four members in the bipartite network, then each of those four is connected to each of the others in the one-mode projection by virtue of

---

[10]In countries such as Spain or Canada, where same-sex marriages are permitted, the network would not be truly bipartite because there would be some edges between like kinds of vertex.

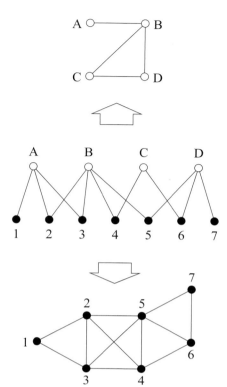

Figure 6.5: The two one-mode projections of a bipartite network. The central portion of this figure shows a bipartite network with four vertices of one type (open circles labeled A to D) and seven of another (filled circles, 1 to 7). At the top and bottom we show the one-mode projections of the network onto the two sets of vertices.

common membership in that group. (Such a clique of four vertices is visible in the center of the lower projection in Fig. 6.5.) Thus the projection is, generically, the union of a number of cliques, one for each group in the original bipartite network. The same goes for the other projection onto the groups.

The one-mode projection, as we have described it, is often useful and is widely employed, but its construction discards a lot of the information present in the structure of the original bipartite network and hence it is, in a sense, a less powerful representation of our data. For example, the projection loses any information about how many groups two vertices share in common. In the case of the actors and films, for instance, there are some pairs of actors who have appeared in many films together—Fred Astaire and Ginger Rogers, say, or William Shatner and Leonard Nimoy—and it's reasonable to suppose this indicates a stronger connection than between actors who appeared together only once.

We can capture information of this kind in our projection by making the projection weighted, giving each edge between two vertices in the projected network a weight equal to the number of common groups the vertices share. This weighted network still does not capture all the information in the bipartite original—it doesn't record the number of groups or the exact membership of each group for instance—but it is an improvement on the unweighted version and is quite widely used.

Mathematically the projection can be written in terms of the incidence matrix $\mathbf{B}$ as follows. The product $B_{ki}B_{kj}$ will be 1 if and only if i and j both belong to the same group k in the bipartite network. Thus, the total number P_{ij} of groups to which both i and j belong is

$$P_{ij} = \sum_{k=1}^{g} B_{ki}B_{kj} = \sum_{k=1}^{g} B_{ik}^{T}B_{kj}, \tag{6.17}$$

where B_{ik}^{T} is an element of the transpose $\mathbf{B}^{T}$ of $\mathbf{B}$. The $n \times n$ matrix $\mathbf{P} = \mathbf{B}^{T}\mathbf{B}$ is similar to an adjacency matrix for the weighted one-mode projection onto the n vertices. Its off-diagonal elements are equal to the weights in that network, the number of common groups shared by each vertex pair. $\mathbf{P}$ is not quite an adjacency matrix, however, since its diagonal elements are non-zero, even though the network itself, by definition, has no self-edges. (In this respect $\mathbf{P}$ is somewhat similar to the cocitation matrix of Section 6.4.1.) The diagonal elements have values

$$P_{ii} = \sum_{k=1}^{g} B_{ki}^{2} = \sum_{k=1}^{g} B_{ki}, \tag{6.18}$$

where we have made use of the fact that B_{ki} only takes the values 0 or 1. Thus P_{ii} is equal to the number of groups to which vertex i belongs.

Thus to derive the adjacency matrix of the weighted one-mode projection, we would calculate the matrix $\mathbf{P} = \mathbf{B}^{T}\mathbf{B}$ and set the diagonal elements equal to zero. And to derive the adjacency matrix of the unweighted projection, we would take the adjacency matrix of the weighted version and replace every non-zero matrix element with a 1.

The other one-mode projection, onto the groups, can be represented by a $g \times g$ matrix $\mathbf{P}' = \mathbf{B}\mathbf{B}^{T}$, whose off-diagonal element P'_{ij} gives the number of common members of groups i and j, and whose diagonal element P'_{ii} gives the number of members of group i.

One occasionally also comes across bipartite networks that are directed. For example, the metabolic networks discussed in Section 5.1.1 can be represented as directed bipartite networks—see Fig. 5.1a. A variety of more complex types of projection are possible in this case, although their use is rare and we won't spend time on them here. Weighted bipartite networks are also possible in principle, although no examples will come up in this book.

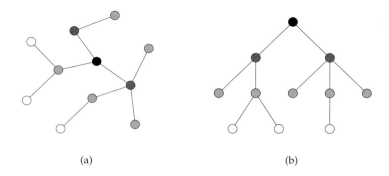

(a)   (b)

Figure 6.6: Two sketches of the same tree. The two panels here show two different depictions of a tree, a network with no closed loops. In (a) the vertices are positioned on the page in any convenient position. In (b) the tree is a laid out in a "rooted" fashion, with a root node at the top and branches leading down to "leaves" at the bottom.

6.7 TREES

A *tree* is a connected, undirected network that contains no closed loops—see Fig. 6.6a.[11] By "connected" we mean that every vertex in the network is reachable from every other via some path through the network. A network can also consist of two or more parts, disconnected from one another,[12] and if an individual part has no loops it is also called a tree. If all the parts of the network are trees, the complete network is called a *forest*.

Trees are often drawn in a *rooted* manner, as shown in Fig. 6.6b, with a *root node* at the top and a branching structure going down. The vertices at the bottom that are connected to only one other vertex are called *leaves*.[13] Topologically, a tree has no particular root—the same tree can be drawn with any node,

---

[11]In principle, one could put directions on the edges of a tree and make it a directed network, but the definition of a tree as a loopless network ignores directions if there are any. This means that a tree is not the same thing as a directed acyclic graph (Section 6.4.2). A directed tree is always a directed acyclic graph, but the reverse is not also true, since the definition of a loop in a directed acyclic graph takes the directions of the edges into account. A directed acyclic graph may well have loops if we ignore directions (see for example Fig. 6.3).

[12]Such parts are called "components"—see Section 6.11.

[13]This is a slightly odd way of drawing trees, with the root at the top and the leaves at the bottom. The more familiar trees of the wooden kind are, of course, the other way up. The upside-down orientation has, however, become conventional in mathematics and computer science, and we here bow to this convention.

including a leaf, as the root node, but in some applications there are other reasons for designating a root. A dendrogram is one example (see below).

Not very many of the real-world networks that we will encounter in this book are trees, although a few are. A river network is an example of a naturally occurring tree (see Fig. 2.6, for instance). Trees do nonetheless play several important roles in the study of networks. In Chapter 12 for instance we will study the network model known as the "random graph." In this model local groups of vertices—the so-called small components in the network—form trees, and we can exploit this property to derive a variety of mathematical results about random graphs. In Section 11.11.1 we introduce the "dendrogram," a useful tool that portrays a hierarchical decomposition of a network as a tree. Trees also occur commonly in computer science, where they are used as a basic building block for data structures such as AVL trees and heaps (see Sections 9.5 and 9.7 and Refs. [8, 81]) and in other theoretical contexts like minimum spanning trees [81], Cayley trees or Bethe lattices [269], and hierarchical models of networks (see Section 19.3.2 and Refs. [70, 179, 322]).

Perhaps the most important property of trees for our purposes is that, since they have no closed loops, there is exactly one path between any pair of vertices. (In a forest there is at most one path, but there may be none.) This is clear since if there were two paths between a pair of vertices A and B then we could go from A to B along one path and back along the other, making a loop, which is forbidden.

This property of trees makes certain kinds of calculation particularly simple, and trees are sometimes used as a basic model of a network for this reason. For instance, the calculation of a network's diameter (Section 6.10.1), the betweenness centrality of a vertex (Section 7.7), and certain other properties based on shortest paths are all relatively easy with a tree.

Another useful property of trees is that a tree of n vertices always has exactly $n - 1$ edges. To see this, consider building up a tree by adding vertices one by one. Starting with a single vertex and no edges, we add a second vertex and one edge to connect it to the first. Similarly when we add a third vertex we need at least one edge to connect it one of the others, and so forth. For every vertex we must add at least one edge to keep the network connected. This means that the number of edges must always be at least one less than the number of vertices. In mathematical terms, $n - 1$ is a lower bound on the number of edges.

But it is also an upper bound, because if we add more than one edge when we add a new vertex then we create a loop: the first edge connects the added vertex to the rest of the network and the second then connects together two vertices that are already part of the network. But adding an edge between two vertices that are already connected via the network necessarily creates a loop. Hence we are not allowed to add more than one edge per vertex if the network is to remain free of loops.

Thus the number of edges in a tree cannot be either more or less than $n - 1$, and hence is exactly $n - 1$.

The reverse is also true, that any connected network with n vertices and $n - 1$ edges is a tree. If such a network were not a tree then there must be a loop in the network somewhere, implying that we could remove an edge without disconnecting any part of the network. Doing this repeatedly until no loops are left, we would end up with a tree, but one with less than $n - 1$ edges, which cannot be. Hence we must have had a tree to begin with. As a corollary, this implies that the connected network on n vertices with the minimum number of edges is always a tree, since no connected network has less than $n - 1$ edges and all networks with $n - 1$ edges are trees.

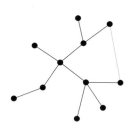

Adding an extra edge (gray) between any two vertices of a tree creates a loop.

6.8   PLANAR NETWORKS

A *planar network* is a network that can be drawn on a plane without having any edges cross.[14] Figure 6.7a shows a small planar network. Note that it is in most cases possible to find a way to draw a planar network so that some edges do cross (Fig. 6.7b). The definition of planarity only specifies that at least one arrangement of the vertices exists that results in no crossing.

Most of the networks we will encounter in this book are not planar, either because there is no relevant two-dimensional geometry to which the network is confined (e.g., citation networks, metabolic networks, collaboration networks), or else there is but there is nothing to stop edges from crossing on it (e.g., the Internet, airline route maps, email networks). However, there are a few important examples of networks that are planar. First of all, all trees are planar. For some trees, such as river networks, this is obvious. Rivers never cross one another; they only flow together. In other cases, such as the trees used in computer data structures, there is no obvious two-dimensional surface onto which the network falls but it is planar nonetheless.

Among non-tree-like networks, some are planar for physical reasons. A good example is a road network. Because roads are confined to the Earth's surface they form a roughly planar network. It does happen sometimes that roads meet without intersecting, one passing over the other on a bridge, so that

---

[14]A plane is a flat surface with open boundaries. One can define a generalization of a planar network for other types of two-dimensional surface, such as the torus, which wraps around on itself. A standard planar network, however, does not wrap around.

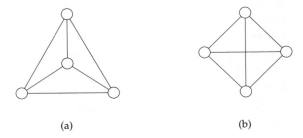

(a)   (b)

Figure 6.7: Two drawings of a planar graph. (a) A small planar graph with four vertices and six edges. It is self-evident that the graph is planar, since in this depiction it has no edges that cross. (b) The same graph redrawn with two of its edges crossing. Even though the edges cross, the graph is still planar—a graph is planar if it *can* be drawn without crossing edges.

in fact, if one wishes to be precise, the road network is not planar. However, such instances are rare (in the sense that there are far more places where roads intersect than there are bridges where they don't) and the network is planar to a good approximation.

Another example is the network of shared borders between countries, states, or provinces—see Fig. 6.8. We can take a map depicting any set of contiguous regions, represent each by a vertex, and draw an edge between any two that share a border. It is easy to see that the resulting network can always be drawn without crossing edges provided the regions in question are formed of contiguous landmasses.

Networks of this type, representing regions on a map, play an important role in the *four-color theorem*, a theorem that states that it is possible to color any set of regions on a two-dimensional map, real or imagined, with at most four colors such that no two adjacent regions have the same color, no matter how many regions there are or of what size or shape.[15] By constructing the network corresponding to the map in question, this problem can be converted into a problem of coloring the vertices of a planar graph in such a way that no two vertices connected by an edge have the same color. The number of colors required to color a graph in this way is called the *chromatic number* of the graph and many mathematical results are known about chromatic numbers.

---

[15]The theorem only applies for a map on a surface with topological genus zero, such as a flat plane or a sphere. A map on a torus (which has genus 1) can require as many as seven colors.

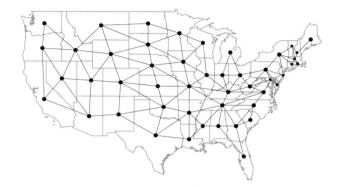

Figure 6.8: Graph of the adjacencies of the lower 48 United States. In this network each of the lower 48 states in the US is represented as a vertex and there is an edge between any two vertices if the corresponding states share a border. The resulting graph is planar, and indeed any set of states, countries, or other regions on a two-dimensional map can be turned into a planar graph in this way.

The proof of the four-color theorem—the proof that the chromatic number of a planar graph is always four or less—is one of the triumphs of traditional graph theory and was first given by Appel and Haken [20–22] in 1976 after more than a hundred years of valiant effort within the mathematics community.[16]

An important question that arises in graph theory is how to determine, given a particular network, whether that network is planar or not. For a small network it is a straightforward matter to draw a picture and play around with the positions of the vertices to see if one can find an arrangement in which no edges cross, but for a large network this is impractical and a more general method of determining planarity is needed. Luckily a straightforward one exists. We will only describe the method here, not prove why it works, since the proof is long and technical and not particularly relevant to the study of real-world networks. For those interested in seeing a proof, one is given by West [324].

Figure 6.9 shows two small networks, conventionally denoted K_5 and UG,

---

[16] Appel and Haken's proof is an interesting one and was controversial at the time of its publication because it made extensive use of a computer to check large numbers of special cases. On the one hand, the proof was revolutionary for being the first proof of a major mathematical result generated in this fashion. On the other hand a number of people questioned whether it could really be considered a proof at all, given that it was far too large for a human being to check its correctness by hand.

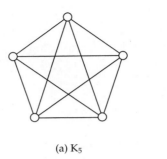
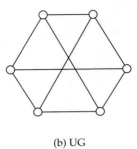

(a) K₅   (b) UG

Figure 6.9: The fundamental non-planar graphs K₅ and UG employed in Kuratow-ski's theorem. These two small graphs are non-planar and Kuratowski's theorem states that any non-planar graph contains at least one subgraph that is an expansion of K₅ or UG.

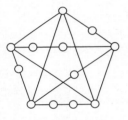

An expansion of K₅.

that are definitely not planar.[17] Neither of these networks can be drawn without edges crossing. It immediately follows that any network that contains a subset of vertices, or subgraph, in the form of K₅ or UG, is also not planar.

An *expansion* is a network derived by adding extra vertices in the middle of edges in another network. No such added vertices, however numerous, will ever make a non-planar network planar, so it is also the case that any expansion of K₅ or UG is non-planar, and hence that any network containing an expansion of K₅ or UG, is also non-planar.

Kuratowski's theorem (sometimes also called the *Kuratowski reduction theorem*) states that the converse is also true:

> Every non-planar network contains at least one subgraph that is an expansion of K₅ or UG.

"Expansion" should be taken here to include the null expansions, i.e., the graphs K₅ and UG themselves.

This theorem, first proved by Pontryagin in 1927 but named after Kuratow-ski who gave an independent proof a few years later,[18] provides us with a way of determining whether a graph is planar. If it contains a subgraph that is an

---

[17]In graph theory K$_n$ denotes the complete graph with n vertices, i.e., the graph of n vertices with all $\binom{n}{2}$ possible single edges present. UG stands for "utility graph." UG is the complete bipartite graph on two groups of three vertices.

[18]See Kennedy *et al.* [170] for an interesting history of the theorem and references to the original papers.

expansion of K$_5$ or UG it is not, otherwise it is.

Kuratowski's theorem is not, however, particularly useful for the analysis of real-world networks, because such networks are rarely precisely planar. (And if they are, then, as in the case of the shared border network of countries or states, it is usually clear for other reasons that they are planar and hence Kuratowski's theorem is unnecessary.) More often, like the road network, they are very nearly planar, but have a few edge crossings somewhere in the network. For such a network, Kuratowski's theorem would tell us, correctly, that the network was not planar, but we would be missing the point.

What we would really like is some measure of the degree of planarity of a network, a measure that could tell us, for example, that the road network of a country is 99% planar, even though there are a few bridges or tunnels here and there. One possible such measure is the minimum number of edge crossings with which the network can be drawn. This however would be a difficult measure to determine since, at least in the simplest approach, its evaluation would require us to try every possible way of drawing the network. Perhaps another approach would be to look at the number of subgraphs in a network that are expansions of K$_5$ or UG. So far, however, no widely accepted metric for degree of planarity has emerged. If such a measure were to gain currency it might well find occasional use in the study of real-world networks.

6.9 DEGREE

The *degree* of a vertex in a graph is the number of edges connected to it. We will denote the degree of vertex i by k_i. For an undirected graph of n vertices the degree can be written in terms of the adjacency matrix as[19]

$$k_i = \sum_{j=1}^{n} A_{ij}. \tag{6.19}$$

Every edge in an undirected graph has two ends and if there are m edges in total then there are $2m$ ends of edges. But the number of ends of edges is also equal to the sum of the degrees of all the vertices, so

$$2m = \sum_{i=1}^{n} k_i, \tag{6.20}$$

---

[19]Notice that this expression still gives the correct result if there are self-edges in the graph, provided each such edge is represented by a diagonal element $A_{ii} = 2$ as discussed earlier, and not 1.

or

$$m = \tfrac{1}{2} \sum_{i=1}^{n} k_i = \tfrac{1}{2} \sum_{ij} A_{ij}, \tag{6.21}$$

a result that we will use many times throughout this book.

The mean degree c of a vertex in an undirected graph is

$$c = \frac{1}{n} \sum_{i=1}^{n} k_i, \tag{6.22}$$

and combining this with Eq. (6.20) we get

$$c = \frac{2m}{n}. \tag{6.23}$$

This relation too will come up repeatedly throughout the book.

The maximum possible number of edges in a simple graph (i.e., one with no multiedges or self-edges) is $\binom{n}{2} = \tfrac{1}{2}n(n-1)$. The *connectance* or *density* ρ of a graph is the fraction of these edges that are actually present:

$$\rho = \frac{m}{\binom{n}{2}} = \frac{2m}{n(n-1)} = \frac{c}{n-1}, \tag{6.24}$$

where we have made use of Eq. (6.23).[20] The density lies strictly in the range $0 \leq \rho \leq 1$. Most of the networks we are interested in are sufficiently large that Eq. (6.24) can be safely approximated as $\rho = c/n$.

A network for which the density ρ tends to a constant as $n \to \infty$ is said to be *dense*. In such a network the fraction of non-zero elements in the adjacency matrix remains constant as the network becomes large. A network in which $\rho \to 0$ as $n \to \infty$ is said to be *sparse*, and the fraction of non-zero elements in the adjacency matrix also tends to zero. In particular, a network is sparse if c tends to a constant as n becomes large. These definitions of dense and sparse networks can, however, be applied only if one can actually take the limit $n \to \infty$, which is fine for theoretical model networks but doesn't work in most practical situations. You cannot for example take the limit as an empirical metabolic network or food web becomes large—you are stuck with the network nature gives you and it can't easily be changed.

In some cases real-world networks do change their sizes and by making measurements for different sizes we can make a guess as to whether they are best regarded as sparse or dense. The Internet and the World Wide Web are two

---

[20]Occasionally connectance is defined as $\rho = m/n^2$, which for large networks differs from Eq. (6.24) by about a factor of 2. With that definition $0 \leq \rho < \tfrac{1}{2}$.

examples of networks whose growth over time allows us to say with some conviction that they are best regarded as sparse. In other cases there may be independent reasons for regarding a network to be sparse or dense. In a friendship network, for instance, it seems unlikely that the number of a person's friends will double solely because the population of the world doubles. How many friends a person has is more a function of how much time they have to devote to the maintenance of friendships than it is a function of how many people are being born. Friendship networks therefore are usually regarded as sparse.

In fact, almost of all of the networks we consider in this book are considered to be sparse networks. This will be important when we look at the expected running time of network algorithms in Chapters 9 to 11 and when we construct mathematical models of networks in Chapters 12 to 15. One possible exception to the pattern is food webs. Studies comparing ecosystems of different sizes seem to show that the density of food webs is roughly constant, regardless of their size, indicating that food webs may be dense networks [102, 210].

Occasionally we will come across networks in which all vertices have the same degree. In graph theory, such networks are called *regular graphs*. A regular graph in which all vertices have degree k is sometimes called *k-regular*. An example of a regular graph is a periodic lattice such as a square or triangular lattice. On the square lattice, for instance, every vertex has degree four.

Vertex degrees are more complicated in directed networks. In a directed network each vertex has two degrees. The *in-degree* is the number of ingoing edges connected to a vertex and the *out-degree* is the number of outgoing edges. Bearing in mind that the adjacency matrix of a directed network has element $A_{ij} = 1$ if there is an edge from j to i, in- and out-degrees can be written

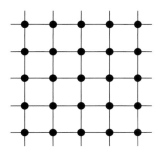

An infinite square lattice is an example of a 4-regular graph.

$$k_i^{\text{in}} = \sum_{j=1}^{n} A_{ij}, \qquad k_j^{\text{out}} = \sum_{i=1}^{n} A_{ij}. \tag{6.25}$$

The number of edges m in a directed network is equal to the total number of ingoing ends of edges at all vertices, or equivalently to the total number of outgoing ends of edges, so

$$m = \sum_{i=1}^{n} k_i^{\text{in}} = \sum_{j=1}^{n} k_j^{\text{out}} = \sum_{ij} A_{ij}. \tag{6.26}$$

Thus the mean in-degree c_{in} and the mean out-degree c_{out} of every directed network are equal:

$$c_{\text{in}} = \frac{1}{n} \sum_{i=1}^{n} k_i^{\text{in}} = \frac{1}{n} \sum_{j=1}^{n} k_j^{\text{out}} = c_{\text{out}}. \tag{6.27}$$

135

For simplicity we will just denote both by c, and combining Eqs. (6.26) and (6.27) we get

$$c = \frac{m}{n}. \tag{6.28}$$

Note that this differs by a factor of two from the equivalent result for an undirected network, Eq. (6.23).

6.10 PATHS

A *path* in a network is any sequence of vertices such that every consecutive pair of vertices in the sequence is connected by an edge in the network. In layman's terms a path is a route across the network that runs from vertex to vertex along the edges of the network. Paths can be defined for both directed and undirected networks. In a directed network, each edge traversed by a path must be traversed in the correct direction for that edge. In an undirected network edges can be traversed in either direction.

In general a path can intersect itself, visiting again a vertex it has visited before, or even running along an edge or set of edges more than once. Paths that do not intersect themselves are called *self-avoiding paths* and are important in some areas of network theory. Geodesic paths and Hamiltonian paths are two special cases of self-avoiding paths that we will study in this book.

The *length* of a path in a network is the number of edges traversed along the path (not the number of vertices). Edges can be traversed more than once, and if they are they are counted separately each time they are traversed. Again in layman's terms, the length of a path is the number of "hops" the path makes from vertex to adjacent vertex.

It is straightforward to calculate the number of paths of a given length r on a network. For either a directed or an undirected simple graph the element A_{ij} is 1 if there is an edge from vertex j to vertex i, and 0 otherwise. (We will consider only simple graphs for now, although the developments generalize easily to non-simple graphs.) Then the product $A_{ik}A_{kj}$ is 1 if there is a path of length 2 from j to i via k, and 0 otherwise. And the total number $N_{ij}^{(2)}$ of paths of length two from j to i, via any other vertex, is

$$N_{ij}^{(2)} = \sum_{k=1}^{n} A_{ik}A_{kj} = \left[\mathbf{A}^2\right]_{ij}, \tag{6.29}$$

where $[\ldots]_{ij}$ denotes the ijth element of a matrix.

Similarly the product $A_{ik}A_{kl}A_{lj}$ is 1 if there is a path of length three from j to i via l and k, in that order, and 0 otherwise, and hence the total number of

A path of length three in a network.

paths of length three is

$$N_{ij}^{(3)} = \sum_{k,l=1}^{n} A_{ik} A_{kl} A_{lj} = \left[\mathbf{A}^3\right]_{ij}. \tag{6.30}$$

Generalizing to paths of arbitrary length r, we see that[21]

$$N_{ij}^{(r)} = \left[\mathbf{A}^r\right]_{ij}. \tag{6.31}$$

A special case of this result is that the number of paths of length r that start and end at the same vertex i is $\left[\mathbf{A}^r\right]_{ii}$. These paths are just loops in the network, what we called "cycles" in our discussion of acyclic graphs in Section 6.1. The total number L_r of loops of length r anywhere in a network is the sum of this quantity over all possible starting points i:

$$L_r = \sum_{i=1}^{n} \left[\mathbf{A}^r\right]_{ii} = \text{Tr } \mathbf{A}^r. \tag{6.32}$$

Note that this expression counts separately loops consisting of the same vertices in the same order but with different starting points.[22] Thus the loop $1 \rightarrow 2 \rightarrow 3 \rightarrow 1$ is considered different from the loop $2 \rightarrow 3 \rightarrow 1 \rightarrow 2$. The expression also counts separately loops that consist of the same vertices but traversed in opposite directions, so that $1 \rightarrow 2 \rightarrow 3 \rightarrow 1$ and $1 \rightarrow 3 \rightarrow 2 \rightarrow 1$ are distinct.

Equation (6.32) can also be expressed in terms of the eigenvalues of the adjacency matrix. Let us consider the case of an undirected graph first. In this case, the adjacency matrix is symmetric, which means that it has n real non-negative eigenvalues, the eigenvectors have real elements, and the matrix can always be written in the form $\mathbf{A} = \mathbf{U}\mathbf{K}\mathbf{U}^T$, where $\mathbf{U}$ is the orthogonal matrix of eigenvectors and $\mathbf{K}$ is the diagonal matrix of eigenvalues. Then $\mathbf{A}^r = (\mathbf{U}\mathbf{K}\mathbf{U}^T)^r = \mathbf{U}\mathbf{K}^r\mathbf{U}^T$ and the number of loops is

$$L_r = \text{Tr}(\mathbf{U}\mathbf{K}^r\mathbf{U}^T) = \text{Tr}(\mathbf{U}^T\mathbf{U}\mathbf{K}^r) = \text{Tr } K^r$$
$$= \sum_i \kappa_i^r, \tag{6.33}$$

---

[21] For a rigorous proof we can use induction. If there are $N_{ik}^{(r-1)}$ paths of length $r-1$ from i to k, then by arguments similar to those above there are $N_{ij}^{(r)} = \sum_k N_{ik}^{(r-1)} A_{kj}$ paths of length r from i to j, or in matrix notation $\mathbf{N}^{(r)} = \mathbf{N}^{(r-1)}\mathbf{A}$, where $\mathbf{N}^{(r)}$ is the matrix with elements $N_{ij}^{(r)}$. This implies that if $\mathbf{N}^{(r-1)} = \mathbf{A}^{r-1}$ then $\mathbf{N}^{(r)} = \mathbf{A}^r$ and with the initial case $\mathbf{N}^{(1)} = \mathbf{A}$ we have $\mathbf{N}^{(r)} = \mathbf{A}^r$ for all r. Taking the ijth element of both sides then gives Eq. (6.31).

[22] If we wish to count each loop only once, we should roughly speaking divide by r, but this does not allow for paths that have symmetries under a change of starting points, such as paths that consist of the same subloop traversed repeatedly. Counting such symmetric paths properly is a complex problem that can be solved exactly in only a few cases.

where κ_i is the ith eigenvalue of the adjacency matrix and we have made use of the fact that the trace of a matrix product is invariant under cyclic permutations of the product.

For directed networks the situation is more complicated. In some cases the same line of proof works and we can again demonstrate that Eq. (6.33) is true, but in other cases the proof breaks down. Recall that directed graphs have, in general, asymmetric adjacency matrices, and some asymmetric matrices cannot be diagonalized.[23] An example is the matrix

$$\begin{pmatrix} 1 & 1 \\ 0 & 1 \end{pmatrix}, \quad \text{which describes the graph} \quad \bigcirc\!\!\!\longrightarrow\!\!\!\bigcirc .$$

This matrix has only a single (right) eigenvector $(1,0)$, and thus one cannot form an orthogonal matrix of eigenvectors with which to diagonalize it. Nonetheless Eq. (6.33) is still true even in such cases, but a different method of proof is needed, as follows.

Every real matrix, whether diagonalizable or not, can be written in the form $\mathbf{A} = \mathbf{QTQ}^T$, where $\mathbf{Q}$ is an orthogonal matrix and $\mathbf{T}$ is an upper triangular matrix. This form is called the *Schur decomposition* of $\mathbf{A}$ [217].

Since $\mathbf{T}$ is triangular, its diagonal elements are its eigenvalues. Furthermore those eigenvalues are the same as the eigenvalues of $\mathbf{A}$. To see this, let $\mathbf{x}$ be a right eigenvector of $\mathbf{A}$ with eigenvalue κ. Then $\mathbf{QTQ}^T\mathbf{x} = \mathbf{Ax} = \kappa\mathbf{x}$, and multiplying throughout by $\mathbf{Q}^T$, bearing in mind that $\mathbf{Q}$ is orthogonal, gives

$$\mathbf{TQ}^T\mathbf{x} = \kappa\mathbf{Q}^T\mathbf{x}, \tag{6.34}$$

and hence $\mathbf{Q}^T\mathbf{x}$ is an eigenvector of $\mathbf{T}$ with the same eigenvalue κ as the adjacency matrix.[24] Then

$$L_r = \text{Tr}\,\mathbf{A}^r = \text{Tr}(\mathbf{QT}^r\mathbf{Q}^T) = \text{Tr}(\mathbf{Q}^T\mathbf{QT}^r) = \text{Tr}\,\mathbf{T}^r$$
$$= \sum_i \kappa_i^r, \tag{6.35}$$

the final equality following because the diagonal elements of any power of a triangular matrix $\mathbf{T}$ are $\mathbf{T}$'s diagonal elements raised to the same power.

This demonstration works for any graph, whatever the properties of its adjacency matrix, and hence Eq. (6.35) is always true. We used this result in

---

[23]Such matrices have multiple or "degenerate" eigenvalues and technically have a non-zero nilpotent part in their Jordan decomposition.

[24]Indeed, any mapping $\mathbf{A} \to \mathbf{Q}^{-1}\mathbf{AQ}$ of a matrix preserves its eigenvalues. Such mappings are called *similarity transformations*.

Eq. (6.14) to show that the graph described by a nilpotent adjacency matrix (i.e., a matrix whose eigenvalues are all zero) must be acyclic. (All such matrices are non-diagonalizable, so one must use Eq. (6.35) in that case.)

Since the adjacency matrix of a directed graph is, in general, asymmetric it may have complex eigenvalues. But the number of loops L_r above is nonetheless always real, as it must be. The eigenvalues of the adjacency matrix are the roots of the characteristic polynomial $\det(\kappa\mathbf{I} - \mathbf{A})$, which has real coefficients, and all roots of such a polynomial are either themselves real or come in complex-conjugate pairs. Thus, while there may be complex terms in the sum in Eq. (6.33), each such term is complemented by another that is its complex conjugate and the sum itself is always real.

6.10.1 Geodesic paths

A *geodesic path*, also called simply a *shortest path*, is a path between two vertices such that no shorter path exists:

A geodesic path of length two between two vertices.

The length of a geodesic path, often called the *geodesic distance* or *shortest distance*, is thus the shortest network distance between the vertices in question. In mathematical terms, the geodesic distance between vertices i and j is the smallest value of r such that $[\mathbf{A}^r]_{ij} > 0$. In practice however there are much better ways of calculating geodesic distances than by employing this formula. We will study some of them in Section 10.3.

It is possible for there to be no geodesic path between two vertices if the vertices are not connected together by any route though the network (i.e., if they are in different "components"—see Section 6.11). In this case one sometimes says that the geodesic distance between the vertices is infinite, although this is mostly just convention—it doesn't really mean very much beyond the fact that the vertices are not connected.

Geodesic paths are necessarily self-avoiding. If a path intersects itself then it contains a loop and can be shortened by removing that loop while still connecting the same start and end points, and hence self-intersecting paths are never geodesic paths.

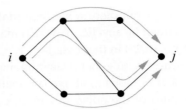

Figure 6.10: Vertices i and j have three geodesic paths between them of length three.

Geodesic paths are not necessarily unique, however. It is perfectly possible to have two or more paths of equal length between a given pair of vertices. The paths may even overlap along some portion of their length—see Fig. 6.10.

The *diameter* of a graph is the length of the longest geodesic path between any pair of vertices in the network for which a path actually exists. (If the diameter were merely the length of the longest geodesic path then it would be formally infinite in a network with more than one component if we adopted the convention above that vertices connected by no path have infinite geodesic distance. One can also talk about the diameters of the individual components separately, this being a perfectly well-defined concept whatever convention we adopt for unconnected vertices.)

6.10.2 EULERIAN AND HAMILTONIAN PATHS

An *Eulerian path* is a path that traverses each edge in a network exactly once. A *Hamiltonian path* is a path that visits each vertex exactly once. A network can have one or many Eulerian or Hamiltonian paths, or none. A Hamiltonian path is by definition self-avoiding, but an Eulerian path need not be. Indeed if there are any vertices of degree greater than two in a network an Eulerian path will have to visit those vertices more than once in order to traverse all their edges.

Eulerian paths form the basis of one of the oldest proofs in graph theory, which dates from 1736. Around that time the great mathematician Leonard Euler became interested the mathematical riddle now known as the *Königsberg Bridge Problem*. The city of Königsberg (now Kaliningrad) was built on the banks of the river Pregel, and on two islands that lie midstream. Seven bridges connected the land masses, as shown in Fig. 6.11a. The riddle asked, "Does there exist any walking route that crosses all seven bridges exactly once each?" Legend has it that the people of Königsberg spent many fruitless hours trying to find such a route, before Euler proved the impossibility of its exis-

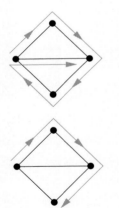

Examples of Eulerian and Hamiltonian paths in a small network.

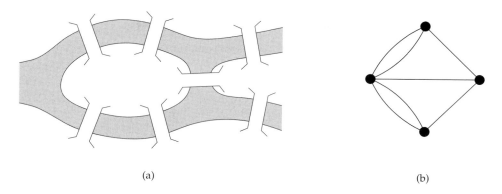

Figure 6.11: The Königsberg bridges. (a) In the eighteenth century the Prussian city of Königsberg, built on four landmasses around the river Pregel, was connected by seven bridges as shown. (b) The topology of the landmasses and bridges can be represented as a multigraph with four vertices and seven edges.

tence.[25] The proof, which perhaps seems rather trivial now, but which apparently wasn't obvious in 1736, involved constructing a network (technically a multigraph) with four vertices representing the four land masses and seven edges joining them in the pattern of the Königsberg bridges (Fig. 6.11b). Then the bridge problem becomes a problem of finding an Eulerian path on this network (and indeed the Eulerian path is named in honor of Euler for his work on this problem). Euler observed that, since any Eulerian path must both enter and leave every vertex it passes through except the first and last, there can at most be two vertices in the network with odd degree if such a path is to exist. Since all four vertices in the Königsberg network have odd degree, the bridge problem necessarily has no solution.

More precisely a network can have an Eulerian path only if there are exactly two or zero vertices of odd degree—zero in the case where the path starts and ends at the same vertex. This is not a sufficient condition for an Eulerian path, however. One can easily find networks that satisfy it and yet have no Eulerian path. The general problem of finding either an Eulerian or Hamiltonian path on a network, or proving that none exists, is a hard one and significant work is still being done on particular cases.

Eulerian and Hamiltonian paths have a number of practical applications in computer science, in job sequencing, "garbage collection," and parallel pro-

---

[25]No cheating: you're not allowed to swim or use a boat.

gramming [81]. A Hamiltonian path problem was also, famously, the first problem solved using a DNA-based computer [7].

6.11 COMPONENTS

It is possible for there to be no path at all between a given pair of vertices in a network. The network shown in Fig. 6.12, for example, is divided into two subgroups of vertices, with no connections between the two, so that there is no path from any vertex in the left subgroup to any vertex in the right. For instance, there is no path from the vertex labeled A to the vertex labeled B. A network of this kind is said to be *disconnected*. Conversely, if there is a path from every vertex in a network to every other the network is *connected*.

The subgroups in a network like that of Fig. 6.12 are called *components*. Technically a component is a subset of the vertices of a network such that there exists at least one path from each member of that subset to each other member, and such that no other vertex in the network can be added to the subset while preserving this property. (Subsets like this, to which no other vertex can be added while preserving a given property, are called *maximal subsets*.) The network in Fig. 6.12 has two components of three and four vertices respectively. A connected network necessarily has only one component. A singleton vertex that is connected to no others is considered to be a component of size one, and every vertex belongs to exactly one component.

The adjacency matrix of a network with more than one component can be written in block diagonal form, meaning that the non-zero elements of the matrix are confined to square blocks along the diagonal of the matrix, with all other elements being zero:

$$\mathbf{A} = \begin{pmatrix} \boxed{} & 0 & \cdots \\ 0 & \boxed{} & \cdots \\ \vdots & \vdots & \ddots \end{pmatrix}. \tag{6.36}$$

Note, however, that the vertex labels must be chosen correctly to produce this form. The visual appearance of blocks in the adjacency matrix depends on the vertices of each component being represented by adjacent rows and columns and choices of labels that don't achieve this will produce non-block-diagonal matrices, even though the choice of labels has no effect on the structure of the network itself. Thus, depending on the labeling, it may not always be imme-

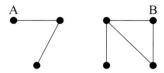

Figure 6.12: A network with two components. This undirected network contains two components of three and four vertices respectively. There is no path between pairs of vertices like A and B that lie in different components.

diately obvious from the adjacency matrix that a network has separate components. There do, however, exist computer algorithms, such as the "breadth-first search" algorithm described in Section 10.3, that can take a network with arbitrary vertex labels and quickly determine its components.

6.11.1 Components in directed networks

When we look at directed networks the definition of components becomes more complicated. The situation is worth looking at in some detail, because it assumes some practical importance in networks like the World Wide Web. Consider the directed network shown in Fig. 6.13. If we ignore the directed nature of the edges, considering them instead to be undirected, then the network has two components of four vertices each. In the jargon of graph theory these are called *weakly connected components*. Two vertices are in the same weakly connected component if they are connected by one or more paths through the network, where paths are allowed to go either way along any edge.

In many practical situations, however, this is not what we care about. For example, the edges in the World Wide Web are directed hyperlinks that allow Web users to surf from one page to another, but only in one direction. This means it is possible to reach one web page from another by surfing only if there is a directed path between them, i.e., a path in which we follow edges only in the forward direction. It would be useful to define components for directed networks based on such directed paths, but this raises some problems. It is certainly possible for there to be a directed path from vertex A to vertex B but no path back from B to A. Should we then consider A and B to be connected?

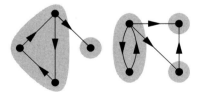

Figure 6.13: Components in a directed network. This network has two weakly connected components of four vertices each, and five strongly connected components (shaded).

Are they in the same component or not?

Clearly there are various answers one could give to these questions. One possibility is that we define A and B to be connected if and only if there exists both a directed path from A to B and a directed path from B to A. A and B are then said to be *strongly connected*. We can define components for a directed network using this definition of connection and these are called *strongly connected components*. Technically, a strongly connected component is a maximal subset of vertices such that there is a directed path in both directions between every pair in the subset. The strongly connected components of the network in Fig. 6.13 are highlighted by the shaded regions. Note that there can be strongly connected components consisting of just a single vertex and, as with the undirected case, each vertex belongs to exactly one strongly connected component. Note also that every strongly connected component with more than one vertex must contain at least one cycle. Indeed every *vertex* in such a component must belong to at least one cycle, since there is by definition a directed path from that vertex to every other in the component and a directed path back again, and the two paths together constitute a cycle. (A corollary of this observation is that acyclic directed graphs have no strongly connected components with more than one vertex, since if they did they wouldn't be acyclic.)

Strongly and weakly connected components are not the only useful definitions of components in a directed network. On the Web it could be useful to know what pages you can reach by surfing from a given starting point, but you might not care so much whether it's possible to surf back the other way. Considerations of this kind lead us to define the *out-component*, which is the set of vertices that are reachable via directed paths starting at a specified vertex A, and including A itself.

An out-component has the property that edges connecting it to other vertices (ones not in the out-component) only point inward towards the members of component, and never outward (since if they pointed outward then the vertices they connected to would by definition be members of the out-component).

Note that the members of an out-component depend on the choice of the starting vertex. Choose a different starting vertex and the set of reachable vertices may change. Thus an out-component is a property of the network structure and the starting vertex, and not (as with strongly and weakly connected components) of the network structure alone. This means, among other things, that a vertex can belong to more than one different out-component. In Fig. 6.14, for instance, we show the out-components of two different starting vertices, A and B. Vertices X and Y belong to both.

A few other points are worth noticing. First, it is self-evident that all the

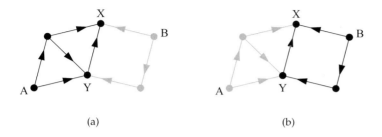

Figure 6.14: Out-components in a directed network. (a) The out-component of vertex A, which is the subset of vertices reachable by directed paths from A. (b) The out-component of vertex B. Vertices X and Y belong to both out-components.

members of the strongly connected component to which a vertex A belongs are also members of A's out-component. Furthermore, all vertices that are reachable from A are necessarily also reachable from all the other vertices in the strongly connected component. Thus it follows that the out-components of all members of a strongly connected component are identical. It would be reasonable to say that out-components really "belong" not to individual vertices, but to strongly connected components.

Very similar arguments apply to vertices *from which* a particular vertex can be reached. The *in-component* of a specified vertex A is the set of all vertices from which there is a directed path to A, including A itself. In-components depend on the choice of the specified vertex, and a vertex can belong to more than one in-component, but all vertices in the same strongly connected component have the same in-component. Furthermore, the strongly connected component to which a vertex belongs is a subset of its in-component, and indeed a vertex that is in both the in- and out-components of A is necessarily in the same strongly connected component as A (since paths exist in both directions) and hence A's strongly connected component is equal to the intersection of its in- and out-components.

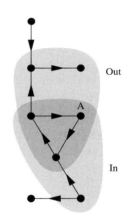

The in- and out-components of a vertex A in a small directed network.

6.12 INDEPENDENT PATHS, CONNECTIVITY, AND CUT SETS

A pair of vertices in a network will typically be connected by many paths of many different lengths. These paths will usually not be independent however. That is, they will share some vertices or edges, as in Fig. 6.10 for instance (page 140). If we restrict ourselves to independent paths, then the number of

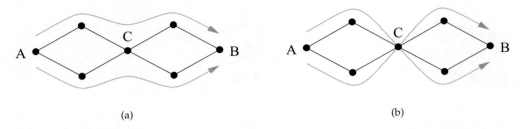

(a)   (b)

Figure 6.15: Edge independent paths. (a) There are two edge-independent paths from A to B in this figure, as denoted by the arrows, but there is only one vertex-independent path, because all paths must pass through the center vertex C. (b) The edge-independent paths are not unique; there are two different ways of choosing two independent paths from A to B in this case.

paths between a given pair of vertices is much smaller. The number of independent paths between vertices gives a simple measure of how strongly the vertices are connected to one another, and has been the topic of much study in the graph theory literature.

There are two species of independent path: edge-independent and vertex-independent. Two paths connecting a given pair of vertices are *edge-independent* if they share no edges. Two paths are *vertex-independent* (or *node-independent*) if they share no vertices other than the starting and ending vertices. If two paths are vertex-independent then they are also edge-independent, but the reverse is not true: it is possible to be edge-independent but not vertex-independent. For instance, the network shown in Fig. 6.15a has two edge-independent paths from A to B, as denoted by the arrows, but only one vertex-independent path— the two edge-independent paths are not vertex-independent because they share the intermediate vertex C.

Independent paths are also sometimes called *disjoint paths*, primarily in the mathematical literature. One also sees the terms *edge-disjoint* and *vertex-disjoint*, describing edge and vertex independence.

The edge- or vertex-independent paths between two vertices are not necessarily unique. There may be more than one way of choosing a set of independent paths. For instance Fig. 6.15b shows the same network as Fig. 6.15a, but with the two paths chosen a different way, so that they cross over as they pass through the central vertex C.

It takes only a moment's reflection to convince oneself that there can be only a finite number of independent paths between any two vertices in a finite network. Each path must contain at least one edge and no two paths can share an edge, so the number of independent paths cannot exceed the number of edges in the network.

The number of independent paths between a pair of vertices is called the *connectivity* of the vertices.[26] If we wish to be explicit about whether we are considering edge- or vertex-independence, we refer to *edge* or *vertex connectivity*. The vertices A and B in Fig. 6.15 have edge connectivity 2 but vertex connectivity 1 (since there is only one vertex-independent path between them).

The connectivity of a pair of vertices can be thought of as a measure of how strongly connected those vertices are. A pair that have only a single independent path between them are perhaps more tenuously connected than a pair that have many paths. This idea is sometimes exploited in the analysis of networks, for instance in algorithmic methods for discovering clusters or communities of strongly linked vertices within networks [122].

Connectivity can also be visualized in terms of "bottlenecks" between vertices. Vertices A and B in Fig. 6.15, for instance, are connected by only one vertex-independent path because vertex C forms a bottleneck through which only one path can go. This idea of bottlenecks is formalized by the notion of cut sets as follows.

Consider an undirected network. (In fact the developments here apply equally to directed ones, but for simplicity let us stick with the undirected case for now.) A *cut set*, or more properly a *vertex cut set*, is a set of vertices whose removal will disconnect a specified pair of vertices. For example, the central vertex C in Fig. 6.15 forms a cut set of size 1 for the vertices A and B. If it is removed, there will be no path from A to B. There are also other cut sets for A and B in this network, although all the others are larger than size 1.

An *edge cut set* is the equivalent construct for edges—it is a set of edges whose removal will disconnect a specified pair of vertices.

A *minimum cut set* is the smallest cut set that will disconnect a specified pair of vertices. In Fig. 6.15 the single vertex C is a minimum vertex cut set for vertices A and B. A minimum cut set need not be unique. For instance, there is a variety of minimum vertex cut sets of size two between the vertices A and B in this network:

---

[26]The word "connectivity" is occasionally also used in the networks literature as a synonym for "degree." Given that the word also has the older meaning discussed here, however, this seems an imprudent thing to do, and we avoid it in this book.

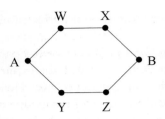

{W,Y}, {W,Z}, {X,Y}, and {X,Z} are all minimum cut sets for this network. (There are also many different minimum edge cut sets.) Of course all the minimum cut sets must have the same size.

An important early theorem in graph theory addresses the size of cut sets. *Menger's theorem* states:

> If there is no cut set of size less than n between a given pair of vertices, then there are at least n independent paths between the same vertices.

The theorem applies both to edges and to vertices and was first proved by Karl Menger [216] for the vertex case, although many other proofs have been given since. A simple one can be found in Ref. [324].

To understand why Menger's theorem is important, consider the following argument. If the minimum vertex cut set between two vertices has size n, Menger's theorem tells us that there must be at least n vertex-independent paths between those vertices. That is, the number of vertex-independent paths is greater than or equal to the size of the minimum cut set. Conversely, if we know there to be exactly n vertex-independent paths between two vertices, then, at the very least, we have to remove one vertex from each path in order to disconnect the two vertices, so the size of the minimum cut set must be at least n. We thus conclude that the number of vertex-independent paths must be both greater than or equal to *and* less than or equal to the size of the minimum cut set, which can only be true if the two are in fact equal. Thus Menger's theorem implies that:

> The size of the minimum vertex cut set that disconnects a given pair of vertices in a network is equal to the vertex connectivity of the same vertices.

Given that Menger's theorem also applies for edges, a similar argument can be used to show that the same result also applies for edge cut sets and edge connectivity.

The edge version of Menger's theorem has a further corollary that will be of some importance to us when we come to study computer algorithms for analyzing networks. It concerns the idea of *maximum flow*. Imagine a network of water pipes in the shape of some network of interest. The edges of the network correspond to the pipes and the vertices to junctions between pipes. Suppose that there is a maximum rate r, in terms of volume per unit time, at which water can flow through any pipe. What then is the maximum rate at which water than can flow through the network from one vertex, A, to another, B? The answer is that this maximum flow is equal to the number of edge-independent paths times the pipe capacity r.

We can construct a proof of this result starting from Menger's theorem. First, we observe that if there are n independent paths between A and B, each of which can carry water at rate r, then the network as a whole can carry a flow of at least nr between A and B, i.e., nr is a lower bound on the maximum flow.

At the same time, by Menger's theorem, we know that there exists a cut set of n edges between A and B. If we push the maximum flow (whatever it is) through the network from A to B and then remove one of the edges in this cut set, the maximum flow will be reduced by at most r, since that is the maximum flow an edge can carry. Thus if we remove all n edges in the cut set one by one, we remove at most nr of flow. But, since the cut set disconnects the vertices A and B, this removal must stop all of the flow. Hence the total capacity is at most nr, i.e., nr is an upper bound on the maximum flow.

Thus nr is both an upper and a lower bound on the maximum flow, and hence the maximum flow must in fact be exactly equal to nr.

This in outline is a proof of the *max-flow/min-cut theorem*, in the special case in which each pipe can carry the same fixed flow. The theorem says that the maximum flow between two vertices is always equal to the size of the minimum cut set times the capacity of a single pipe. The full max-flow/min-cut theorem applies also to weighted networks in which individual pipes can have different capacities. We look at this more general case in the following section.

In combination, Menger's theorem for edges and the max-flow/min-cut theorem show that for a pair of vertices in an undirected network three quantities are all numerically equal to each other: the edge connectivity of the pair (i.e., the number of edge-independent paths connecting them), the size of the minimum edge cut set (i.e., the number of edges that must be removed to disconnect them), and the maximum flow between the vertices if each edge in the network can carry at most one unit of flow. Although we have stated these results for the undirected case, nothing in any of the proofs demands an undirected network, and these three quantities are equal for directed networks as well.

The equality of the maximum flow, the connectivity, and the cut set size has an important practical consequence. There are simple computer algorithms, such as the augmenting path algorithm of Section 10.5.1, that can calculate maximum flows quite quickly (in polynomial time) for any given network, and the equality means that we can use these same algorithms to quickly calculate a connectivity or the size of a cut set as well. Maximum flow algorithms are now the standard numerical method for connectivities and cut sets.

6.12.1 MAXIMUM FLOWS AND CUT SETS ON WEIGHTED NETWORKS

As discussed in Section 6.3, networks can have weights on their edges that indicate that some edges are stronger or more prominent than others. In some cases these weights can represent capacities of the edges to conduct a flow of some kind. For example, they might represent maximum traffic throughput on the roads of a road network or maximum data capacity of Internet lines. We can ask questions about network flows on such networks similar to those we asked in the last section, but with the added twist that different edges can now have different capacities. For example, we can ask what the maximum possible flow is between a specified pair of vertices. We can also ask about cut sets. An edge cut set is defined as before to be a set of edges whose removal from the network would disconnect the specified pair of vertices. A *minimum edge cut set* is defined as being a cut set such that the sum of the weights on the edges of the set has the minimum possible value. Note that it is not now the number of edges that is minimized, but their weight. Nonetheless, this definition is a proper generalization of the one we had before—we can think of the unweighted case as being a special case of the weighted one in which the weights on all edges are equal, and the sum of the weights in the cut set is then simply proportional to the number of edges in the set.

Maximum flows and minimum cut sets on weighted networks are related by the max-flow/min-cut theorem in its most general form:

> The maximum flow between a given pair of vertices in a network is equal to the sum of the weights on the edges of the minimum edge cut set that separates the same two vertices.

We can prove this theorem using the results of the previous section.[27]

Consider first the special case in which the capacities of all the edges in our network are integer multiples of some fixed capacity r. We then transform our network by replacing each edge of capacity kr (with k integer) by k parallel edges of capacity r each. For instance, if $r = 1$ we would have something like this:

---

[27]For a first principles proof that is not based on Menger's theorem see, for instance, Ahuja *et al.* [8].

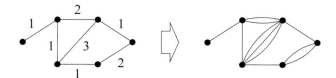

It is clear that the maximum flow between any two vertices in the transformed network is the same as that between the corresponding vertices in the original. At the same time the transformed network now has the form of a simple unweighted network of the type considered in Section 6.12, and hence, from the results of that section, we can immediately say that the maximum flow in the network is equal to the size *in unit edges* of the minimum edge cut set.

We note also that the minimum cut set on the transformed network must include either all or none of the parallel edges between any adjacent pair of vertices; there is no point cutting one such edge unless you cut all of the others as well—you have to cut all of them to disconnect the vertices. Thus the minimum cut set on the transformed network is also a cut set on the original network. And it is a minimum cut set on the original network, because every cut set on the original network is also a cut set with the same weight on the transformed network, and if there were any smaller cut set on the original network then there would be a corresponding one on the transformed network, which, by hypothesis, there is not.

Thus the maximum flows on the two networks are the same, the minimum cuts are also the same, and the maximum flow and minimum cut are equal on the transformed network. It therefore follows that the maximum flow and minimum cut are equal on the original network.

This demonstrates the theorem for the case where all edges are constrained to have weights that are integer multiples of r. This constraint can now be removed, however, by simply allowing r to tend to zero. This makes the units in which we measure edge weights smaller and smaller, and in the limit $r \rightarrow 0$ the edges can have any weight—any weight can be represented as a (very large) integer multiple of r—and hence the max-flow/min-cut theorem in the form presented above must be generally true.

Again there exist efficient computer algorithms for calculating maximum flows on weighted networks, so the max-flow/min-cut theorem allows us to calculate minimum cut weights efficiently also, and this is now the standard way of performing such calculations.[28]

---

[28] Another interesting and slightly surprising computational use of the max-flow/min-cut the-

6.13　THE GRAPH LAPLACIAN

Section 6.2 introduced an important quantity, the adjacency matrix, which captures the entire structure of a network and whose matrix properties can tell us a variety of useful things about networks. There is another matrix, closely related to the adjacency matrix but differing in some important respects, that can also tell us much about network structure. This is the graph Laplacian.

6.13.1　DIFFUSION

Diffusion is, among other things, the process by which gas moves from regions of high density to regions of low, driven by the relative pressure (or partial pressure) of the different regions. One can also consider diffusion processes on networks, and such processes are sometimes used as a simple model of spread across a network, such as the spread of an idea or the spread of a disease. Suppose we have some commodity or substance of some kind on the vertices of a network and there is an amount ψ_i of it at vertex i. And suppose that the commodity moves along the edges, flowing from one vertex j to an adjacent one i at a rate $C(\psi_j - \psi_i)$ where C is a constant called the *diffusion constant*. That is, in a small interval of time the amount of fluid flowing from j to i is $C(\psi_j - \psi_i)\,\mathrm{d}t$. Then the rate at which ψ_i is changing is given by

$$\frac{\mathrm{d}\psi_i}{\mathrm{d}t} = C\sum_j A_{ij}(\psi_j - \psi_i). \tag{6.37}$$

The adjacency matrix in this expression insures that the only terms appearing in the sum are those that correspond to vertex pairs that are actually connected by an edge. Equation (6.37) works equally well for both undirected and directed networks, but let us focus here on undirected ones.[29] We will also consider our networks to be simple (i.e., to have at most a single edge between any pair of vertices and no self-edges).

---

orem is in the polynomial-time algorithm for finding ground states of the thermal random-field Ising model [257], an interesting cross-fertilization between network theory and physics: it is relatively common for physics ideas to find application in network theory, but the reverse has been considerably rarer.

[29]In fact, the graph Laplacian matrix for undirected networks defined in this section does not have a clean generalization for directed networks, although several possible candidates have been suggested. Most of the results in the remaining sections of this chapter do not generalize easily to the directed case.

Splitting the two terms in Eq. (6.37), we can write

$$
\frac{d\psi_i}{dt} = C\sum_j A_{ij}\psi_j - C\psi_i\sum_j A_{ij} = C\sum_j A_{ij}\psi_j - C\psi_i k_i
$$

$$
= C\sum_j (A_{ij} - \delta_{ij}k_i)\psi_j, \tag{6.38}
$$

where k_i is the degree of vertex i as usual and we have made use of the result $k_i = \sum_j A_{ij}$—see Eq. (6.19). (And δ_{ij} is the Kronecker delta, which is 1 if $i = j$ and 0 otherwise.)

Equation (6.38) can be written in matrix form as

$$
\frac{d\psi}{dt} = C(\mathbf{A} - \mathbf{D})\psi, \tag{6.39}
$$

where ψ is the vector whose components are numbers ψ_i, $\mathbf{A}$ is the adjacency matrix, and $\mathbf{D}$ is the diagonal matrix with the vertex degrees along its diagonal:

$$
\mathbf{D} = \begin{pmatrix} k_1 & 0 & 0 & \cdots \\ 0 & k_2 & 0 & \cdots \\ 0 & 0 & k_3 & \cdots \\ \vdots & \vdots & \vdots & \ddots \end{pmatrix}. \tag{6.40}
$$

It is common to define the new matrix

$$
\mathbf{L} = \mathbf{D} - \mathbf{A}, \tag{6.41}
$$

so that Eq. (6.38) takes the form

$$
\frac{d\psi}{dt} + C\mathbf{L}\psi = 0, \tag{6.42}
$$

which has the same form as the ordinary diffusion equation for a gas, except that the Laplacian operator ∇^2 that appears in that equation has been replaced by the matrix $\mathbf{L}$. The matrix $\mathbf{L}$ is for this reason called the *graph Laplacian*, although its importance stretches much further than just diffusion processes. The graph Laplacian, as we will see, turns up in a variety of different places, including random walks on networks, resistor networks, graph partitioning, and network connectivity.[30]

---

[30]In fact the graph Laplacian doesn't occupy *quite* the same position as ∇^2 does in the normal diffusion equation—there is a plus sign in Eq. (6.42) where a minus sign appears in the normal equation. We could easily get rid of this discrepancy by reversing the sign of the definition in Eq. (6.41), but the definition as given has become the standard one and so for consistency we will stick with it.

Written out in full, the elements of the Laplacian matrix are

$$L_{ij} = \begin{cases} k_i & \text{if } i = j, \\ -1 & \text{if } i \neq j \text{ and there is an edge } (i,j), \\ 0 & \text{otherwise,} \end{cases} \quad (6.43)$$

so it has the degrees of the vertices down its diagonal and a -1 element for every edge. Alternatively we can write

$$L_{ij} = \delta_{ij} k_i - A_{ij}. \quad (6.44)$$

We can solve the diffusion equation (6.42) by writing the vector ψ as a linear combination of the eigenvectors $\mathbf{v}_i$ of the Laplacian thus:

$$\psi(t) = \sum_i a_i(t)\, \mathbf{v}_i, \quad (6.45)$$

with the coefficients $a_i(t)$ varying over time. Substituting this form into (6.42) and making use of $\mathbf{L}\mathbf{v}_i = \lambda_i \mathbf{v}_i$, where λ_i is the eigenvalue corresponding to the eigenvector $\mathbf{v}_i$, we get

$$\sum_i \left(\frac{da_i}{dt} + C\lambda_i a_i \right) \mathbf{v}_i = 0. \quad (6.46)$$

But the eigenvectors of a symmetric matrix such as the Laplacian are orthogonal, and so, taking the dot product of this equation with any eigenvector $\mathbf{v}_j$, we get

$$\frac{da_i}{dt} + C\lambda_i a_i = 0, \quad (6.47)$$

for all i, which has the solution

$$a_i(t) = a_i(0)\, e^{-C\lambda_i t}. \quad (6.48)$$

Given an initial condition for the system, as specified by the quantities $a_i(0)$, therefore, we can solve for the state at any later time, provided we know the eigenvalues and eigenvectors of the graph Laplacian.

6.13.2 EIGENVALUES OF THE GRAPH LAPLACIAN

This is the first of many instances in which the eigenvalues of the Laplacian will arise, so it is worth spending a little time understanding their properties. The Laplacian is a symmetric matrix, and so has real eigenvalues. However, we can say more than this about them. In fact, as we now show, all the eigenvalues of the Laplacian are also non-negative.

Consider an undirected network with n vertices and m edges and let us arbitrarily designate one end of each edge to be end 1 and the other to be end 2. It doesn't matter which end is which, only that they have different labels.

Now let us define an $m \times n$ matrix $\mathbf{B}$ with elements as follows:

$$B_{ij} = \begin{cases} +1 & \text{if end 1 of edge } i \text{ is attached to vertex } j, \\ -1 & \text{if end 2 of edge } i \text{ is attached to vertex } j, \\ 0 & \text{otherwise.} \end{cases} \qquad (6.49)$$

Thus each row of the matrix has exactly one $+1$ and one -1 element.

The matrix $\mathbf{B}$ is called the *edge incidence matrix*. It bears some relation to, but is distinct from, the incidence matrix for a bipartite graph defined in Section 6.6.

Now consider the sum $\sum_k B_{ki} B_{kj}$. If $i \neq j$, then the only non-zero terms in the sum will occur if both B_{ik} and B_{jk} are non-zero, i.e., if edge k connects vertices i and j, in which case the product will have value -1. For a simple network, there is at most one edge between any pair of vertices and hence at most one such non-zero term, so the value of the entire sum will be -1 if there is an edge between i and j and zero otherwise.

If $i = j$ then the sum is $\sum_k B_{ki}^2$, which has a term $+1$ for every edge connected to vertex i, so the whole sum is just equal to the degree k_i of vertex i.

Thus the sum $\sum_k B_{ki} B_{kj}$ is precisely equal to an element of the Laplacian $\sum_k B_{ki} B_{kj} = L_{ij}$—the diagonal terms L_{ii} are equal to the degrees k_i and the off-diagonal terms L_{ij} are -1 if there is an edge (i, j) and zero otherwise. (See Eq. (6.43).) In matrix form we can write

$$\mathbf{L} = \mathbf{B}^T \mathbf{B}, \qquad (6.50)$$

where $\mathbf{B}^T$ is the transpose of $\mathbf{B}$.

Now let $\mathbf{v}_i$ be an eigenvector of $\mathbf{L}$ with eigenvalue λ_i. Then

$$\mathbf{v}_i^T \mathbf{B}^T \mathbf{B} \mathbf{v}_i = \mathbf{v}_i^T \mathbf{L} \mathbf{v}_i = \lambda_i \mathbf{v}_i^T \mathbf{v}_i = \lambda_i, \qquad (6.51)$$

where we assume that the eigenvector $\mathbf{v}_i$ is normalized so that its inner product with itself is 1.

Thus any eigenvalue λ_i of the Laplacian is equal to $(\mathbf{v}_i^T \mathbf{B}^T)(\mathbf{B} \mathbf{v}_i)$. But this quantity is itself just the inner product of a real vector $(\mathbf{B} \mathbf{v}_i)$ with itself. In other words, it is the sum of the squares of the (real) elements of that vector and hence it cannot be negative. The smallest value it can have is zero:

$$\lambda_i \geq 0 \qquad (6.52)$$

for all i.

This is an important physical property of the Laplacian. It means, for instance, that the solution, Eq. (6.48), of the diffusion equation on any network contains only decaying exponentials or constants and not growing exponentials, so that the solution tends to an equilibrium value as $t \to \infty$, rather than diverging.[31]

While the eigenvalues of the Laplacian cannot be negative, they can be zero, and in fact the Laplacian always has at least one zero eigenvalue. Consider the vector $\mathbf{1} = (1, 1, 1, \ldots)$. If we multiply this vector by the Laplacian, the ith element of the result is given by

$$\sum_j L_{ij} \times 1 = \sum_j (\delta_{ij} k_i - A_{ij}) = k_i - \sum_j A_{ij} = k_i - k_i$$
$$= 0, \tag{6.53}$$

where we have made use of Eqs. (6.19) and (6.44). In vector notation, $\mathbf{L} \cdot \mathbf{1} = 0$. Thus the vector $\mathbf{1}$ is always an eigenvector of the graph Laplacian with eigenvalue zero.[32] Since there are no negative eigenvalues, this is the lowest of the eigenvalues of the Laplacian. Following convention, we number the n eigenvalues of the Laplacian in ascending order: $\lambda_1 \leq \lambda_2 \leq \ldots \leq \lambda_n$. So we always have $\lambda_1 = 0$.

Note that the presence of a zero eigenvalue implies that the Laplacian has no inverse: the determinant of the matrix is the product of its eigenvalues, and hence is always zero for the Laplacian, so that the matrix is singular.

6.13.3 COMPONENTS AND THE ALGEBRAIC CONNECTIVITY

Suppose we have a network that is divided up into c different components of sizes $n_1, n_2, \ldots, n_c$. To make the notation simple let us number the vertices of the network so that the first n_1 vertices are those of the first component, the next n_2 are those of the second component, and so forth. With this choice the Laplacian of the network will be block diagonal, looking something like this:

See the discussion of block diagonal matrices in Section 6.11.

---

[31] This is clearly the right answer from a physical point of view, since the fluid in our diffusion process is conserved—there is a fixed, finite amount of it—so it is impossible for the amount on any vertex to become infinite.

[32] It is not a properly normalized eigenvector. The properly normalized vector would be $(1, 1, 1, \ldots)/\sqrt{n}$.

$$\mathbf{L} = \begin{pmatrix} \boxed{} & 0 & \cdots \\ 0 & \boxed{} & \cdots \\ \vdots & \vdots & \ddots \end{pmatrix}.$$ (6.54)

What is more, each block in the Laplacian is, by definition, the Laplacian of the corresponding component: it has the degrees of the vertices in that component along its diagonal and -1 in each position corresponding to an edge within that component. Thus we can immediately write down c different vectors that are eigenvectors of $\mathbf{L}$ with eigenvalue zero: the vectors that have ones in all positions corresponding to vertices in a single component and zero elsewhere. For instance, the vector

$$\mathbf{v} = (\underbrace{1, 1, 1, \ldots,}_{n_1 \text{ ones}} \underbrace{0, 0, 0, \ldots}_{\text{zeros}}),$$ (6.55)

is an eigenvector with eigenvalue zero.

Thus in a network with c components there are always at least c eigenvectors with eigenvalue zero. In fact, it can be shown that the number of zero eigenvalues is always exactly equal to the number of components [324]. (Note that the vector $\mathbf{1}$ of all ones is just equal to the sum of the c other eigenvectors, so it is not an independent eigenvector.) An important corollary of this result is that the second eigenvalue of the graph Laplacian λ_2 is non-zero if and only if the network is connected, i.e., consists of a single component. The second eigenvalue of the Laplacian is called the *algebraic connectivity* of the network.[33] It will come up again in Section 11.5 when we look at the technique known as spectral partitioning.

6.14   RANDOM WALKS

Another context in which the graph Laplacian arises is in the study of random walks on networks. A *random walk* is a path across a network created by taking repeated random steps. Starting at some specified initial vertex, at each step of the walk we choose uniformly at random between the edges attached to the current vertex, move along the chosen edge to the vertex at its other end, and repeat. Random walks are normally allowed to go along edges more than

---

[33]Occasionally λ_2 is also called the *spectral gap*.

once, visit vertices more than once, or retrace their steps along an edge just traversed. *Self-avoiding walks*, which do none of these things, are also studied sometimes, but we will not discuss them here.

Random walks arise, for instance, in the random walk sampling method for social networks discussed in Section 3.7 and in the random walk betweenness measure of Section 7.7.

Consider a random walk that starts at a specified vertex and takes t random steps. Let $p_i(t)$ be the probability that the walk is at vertex i at time t. If the walk is at vertex j at time $t - 1$, the probability of taking a step along any particular one of the k_j edges attached to j is $1/k_j$, so on an undirected network $p_i(t)$ is given by

$$p_i(t) = \sum_j \frac{A_{ij}}{k_j} p_j(t - 1), \tag{6.56}$$

or $\mathbf{p}(t) = \mathbf{A}\mathbf{D}^{-1}\mathbf{p}(t - 1)$ in matrix form where $\mathbf{p}$ is the vector with elements p_i and, as before, $\mathbf{D}$ is the diagonal matrix with the degrees of the vertices down its diagonal.

There are a couple of other useful ways to write this relation. One is to define $\mathbf{D}^{1/2}$ to be the matrix with the square roots $\sqrt{k_i}$ of the degrees down the diagonal, so that

$$\mathbf{D}^{-1/2}\mathbf{p}(t) = \left[\mathbf{D}^{-1/2}\mathbf{A}\mathbf{D}^{-1/2}\right]\left[\mathbf{D}^{-1/2}\mathbf{p}(t - 1)\right]. \tag{6.57}$$

This form is convenient in some situations because the matrix $\mathbf{D}^{-1/2}\mathbf{A}\mathbf{D}^{-1/2}$ is a symmetric one. This matrix is called the *reduced adjacency matrix* and has elements equal to $1/\sqrt{k_i k_j}$ if there is an edge between i and j and zero otherwise. Equation (6.57) tells us that the vector $\mathbf{D}^{-1/2}\mathbf{p}$ gets multiplied by one factor of the reduced adjacency matrix at each step of the random walk, and so the problem of understanding the random walk can be reduced to one of understanding the effects of repeated multiplication by a simple symmetric matrix.

For our purposes, however, we take a different approach. In the limit as $t \to \infty$ the probability distribution over vertices is given by setting $t = \infty$: $p_i(\infty) = \sum_j A_{ij}p_j(\infty)/k_j$, or in matrix form:

$$\mathbf{p} = \mathbf{A}\mathbf{D}^{-1}\mathbf{p}. \tag{6.58}$$

Rearranging, this can also be written as

$$(\mathbf{I} - \mathbf{A}\mathbf{D}^{-1})\mathbf{p} = (\mathbf{D} - \mathbf{A})\mathbf{D}^{-1}\mathbf{p} = \mathbf{L}\mathbf{D}^{-1}\mathbf{p} = 0. \tag{6.59}$$

Thus $\mathbf{D}^{-1}\mathbf{p}$ is an eigenvector of the Laplacian with eigenvalue 0.

On a connected network, for instance—one with only a single component—we know (Section 6.13.3) that there is only a single eigenvector with eigenvalue zero, the vector whose components are all equal. Thus, $\mathbf{D}^{-1}\mathbf{p} = a\mathbf{1}$, where a is a constant and $\mathbf{1}$ is the vector whose components are all ones. Equivalently $\mathbf{p} = a\mathbf{D}\mathbf{1}$, so that $p_i = ak_i$. Then on a connected network the probability that a random walk will be found at vertex i in the limit of long time is simply proportional to the degree of that vertex. If we choose the value of a to normalize p_i properly, this gives

$$p_i = \frac{k_i}{\sum_j k_j} = \frac{k_i}{2m},$$ (6.60)

where we have used Eq. (6.20).

The simple way to understand this result is that vertices with high degree are more likely to be visited by the random walk because there are more ways of reaching them. We used Eq. (6.60) in Section 3.7 in our analysis of the random-walk sampling method for social networks.

An important question about random walks concerns the *first passage time*. The first passage time for a random walk from a vertex u to another vertex v is the number of steps before a walk starting at u first reaches v. Since the walk is random, the first passage time between two vertices is not fixed; if we repeat the random walk process more than once it can take different values on different occasions. But we can ask for example what the mean first passage time is.

To answer this question, we modify our random walk slightly to make it into an *absorbing random walk*. An absorbing walk is one that has one or more absorbing states, meaning vertices that the walk can move to, but not leave again. We will consider just the simplest case of a single absorbing vertex v. Any walk that arrives at vertex v must stay there ever afterwards, but on the rest of the network the walk is just a normal random walk. We can answer questions about the first passage time by considering the probability $p_v(t)$ that a walk is at vertex v after a given amount of time, since this is also the probability that the walk has a first passage time to v that is less than or equal to t. And the probability that a walk has first passage time exactly t is $p_v(t) - p_v(t-1)$, which means that the mean first passage time τ is[34]

$$\tau = \sum_{t=0}^{\infty} t\big[p_v(t) - p_v(t-1)\big].$$ (6.61)

---

[34]One might think that this equation could be simplified by reordering the terms so that most of them cancel out, but this is not allowed. The sum viewed as individual terms is not absolutely convergent and hence does not have a unique limit. Only the complete sum over t as written is meaningful and a reordering of the terms will give the wrong answer.

To calculate the probability $p_v(t)$ we could apply Eq. (6.56) (or (6.58)) repeatedly to find $\mathbf{p}(t)$ and substitute the result into Eq. (6.61). Note, however, that since the random walk can move *to* vertex v but not away from it, the adjacency matrix $\mathbf{A}$ has elements $A_{iv} = 0$ for all i but A_{vi} can still be non-zero. Thus in general $\mathbf{A}$ is asymmetric. Although we can work with such an asymmetric matrix, the computations are harder than for symmetric matrices and in this case there is no need. Instead we can use the following trick.

Consider Eq. (6.56) for any $i \neq v$:

$$p_i(t) = \sum_j \frac{A_{ij}}{k_j} p_j(t-1) = \sum_{j \neq v} \frac{A_{ij}}{k_j} p_j(t-1), \tag{6.62}$$

where the second equality applies since $A_{iv} = 0$ and hence the terms with $j = v$ don't contribute to the sum. But if $i \neq v$ then there are no terms in A_{vj} in the sum either. This allows us to write the equation in the matrix form

$$\mathbf{p}'(t) = \mathbf{A}'\mathbf{D}'^{-1}\mathbf{p}'(t-1), \tag{6.63}$$

where $\mathbf{p}'$ is $\mathbf{p}$ with the vth element removed and $\mathbf{A}'$ and $\mathbf{D}'$ are $\mathbf{A}$ and $\mathbf{D}$ with their vth row and column removed. Note that $\mathbf{A}'$ and $\mathbf{D}'$ are symmetric matrices, since the rows and columns containing the asymmetric elements have been removed. Iterating Eq. (6.63), we now get

$$\mathbf{p}'(t) = \left[\mathbf{A}'\mathbf{D}'^{-1}\right]^t \mathbf{p}'(0). \tag{6.64}$$

Since we have removed the element p_v from the vector $\mathbf{p}$, we cannot calculate its value directly using this equation, but we can calculate it indirectly by noting that $\sum_i p_i(t) = 1$ at all times. Thus

$$p_v(t) = 1 - \sum_{i \neq v} p_i(t) = 1 - \mathbf{1} \cdot \mathbf{p}'(t), \tag{6.65}$$

where again $\mathbf{1} = (1, 1, 1, \ldots)$. Using Eqs. (6.61), (6.64), and (6.65) we then have a mean first passage time of

$$\tau = \sum_{t=0}^{\infty} t\, \mathbf{1} \cdot \left[\mathbf{p}'(t-1) - \mathbf{p}'(t)\right] = \mathbf{1} \cdot \left[\mathbf{I} - \mathbf{A}'\mathbf{D}'^{-1}\right]^{-1} \cdot \mathbf{p}'(0), \tag{6.66}$$

where $\mathbf{I}$ is the identity matrix and we have made use of the result that

$$\sum_{t=0}^{\infty} t\left(\mathbf{M}^{t-1} - \mathbf{M}^t\right) = \left[\mathbf{I} - \mathbf{M}\right]^{-1}, \tag{6.67}$$

for any matrix $\mathbf{M}$ (assuming the sum actually converges).

We can simplify Eq. (6.66) by writing

$$[\mathbf{I} - \mathbf{A}'\mathbf{D}'^{-1}]^{-1} = \mathbf{D}'[\mathbf{D}' - \mathbf{A}']^{-1} = \mathbf{D}'\mathbf{L}'^{-1}, \qquad (6.68)$$

so that

$$\tau = \mathbf{1} \cdot \mathbf{D}'\mathbf{L}'^{-1} \cdot \mathbf{p}'(0), \qquad (6.69)$$

where the symmetric matrix $\mathbf{L}'$ is the graph Laplacian with the vth row and column removed. $\mathbf{L}'$ is called the *vth reduced Laplacian*. Note that, even though, as we noted in Section 6.13.2, the Laplacian has no finite inverse, the reduced Laplacian can have an inverse. The eigenvector $(1, 1, 1, \dots)$ whose zero eigenvalue causes the determinant of the Laplacian to be zero is, in general, not an eigenvector of the reduced matrix.

For convenience, we now introduce the symmetric matrix $\mathbf{\Lambda}^{(v)}$, which is equal to $\mathbf{L}'^{-1}$ with a vth row and column reintroduced having elements all zero:

$$\Lambda_{ij}^{(v)} = \begin{cases} 0 & \text{if } i = v \text{ or } j = v, \\ \left[\mathbf{L}'^{-1}\right]_{ij} & \text{if } i < v \text{ and } j < v, \\ \left[\mathbf{L}'^{-1}\right]_{i-1,j} & \text{if } i > v \text{ and } j < v, \\ \left[\mathbf{L}'^{-1}\right]_{i,j-1} & \text{if } i < v \text{ and } j > v, \\ \left[\mathbf{L}'^{-1}\right]_{i-1,j-1} & \text{if } i > v \text{ and } j > v. \end{cases} \qquad (6.70)$$

Then we observe that for a walk starting at vertex u, the initial probability distribution $\mathbf{p}'(0)$ has all elements 0 except the one corresponding to vertex u, which is 1. Thus, combining Eqs. (6.69) and (6.70), the mean first passage time for a random walk from u to v is given by

$$\tau = \sum_i k_i \Lambda_{iu}^{(v)}, \qquad (6.71)$$

where we have made use of the fact that the non-zero elements of the diagonal matrix $\mathbf{D}'$ are the degrees k_i of the vertices. Thus if we can calculate the inverse of the vth reduced Laplacian then a sum over the elements in the uth column immediately gives us the mean first passage time for a random walk from u to v. And sums over the other columns give us the first passage times from other starting vertices to the same target vertex v—we get n first passage times from a single matrix inversion.

6.14.1 RESISTOR NETWORKS

There are interesting mathematical connections between random walks on networks and the calculation of current flows in networks of resistors. Suppose

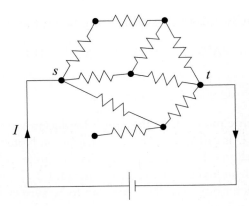

Figure 6.16: A resistor network with applied voltage. A network in which the edges are resistors and the vertices are electrical junctions between them, with a voltage applied between vertices s and t so as to generate a total current I.

we have a network in which the edges are identical resistors of resistance R and the vertices are junctions between resistors, as shown in Fig. 6.16, and suppose we apply a voltage between two vertices s and t such that a current I flows from s to t through the network. What then is the current flow through any given resistor in the network?

The currents in the network obey Kirchhoff's current law, which is essentially a statement that electricity is conserved, so that the net current flowing in or out of any vertex is zero. Let V_i be the voltage at vertex i, measured relative to any convenient reference potential. Then Kirchhoff's law says that

$$\sum_j A_{ij} \frac{V_j - V_i}{R} + I_i = 0, \tag{6.72}$$

where I_i represents any current injected into vertex i by an external current source. In our case this external current is non-zero only for the two vertices s and t connected to the external voltage:

$$I_i = \begin{cases} +I & \text{for } i = s, \\ -I & \text{for } i = t, \\ 0 & \text{otherwise.} \end{cases} \tag{6.73}$$

(In theory there's no reason why one could not impose more complex current source arrangements by applying additional voltages to the network and making more elements I_i non-zero, but let us stick to our simple case in this discussion.)

Noting that $\sum_j A_{ij} = k_i$, Eq. (6.72) can also be written as $k_i V_i - \sum_j A_{ij} V_j = RI_i$ or

$$\sum_j (\delta_{ij} k_i - A_{ij}) V_j = RI_i, \qquad (6.74)$$

which in matrix form is

$$\mathbf{LV} = R\mathbf{I}, \qquad (6.75)$$

where $\mathbf{L} = \mathbf{D} - \mathbf{A}$ is once again the graph Laplacian.

As discussed in Section 6.13.2, the Laplacian has no inverse because it always has at least one eigenvalue that is zero, so we cannot simply invert Eq. (6.75) to get the voltage vector $\mathbf{V}$. We can, however, solve for $\mathbf{V}$ by once again making use of the reduced Laplacian of Section 6.14.

The reason why we cannot invert Eq. (6.75) is that the equation does not in fact fix the absolute value of the voltages V_i. We can add any multiple of the vector $\mathbf{1} = (1, 1, 1, \ldots)$ to the solution of this equation and get another solution, since $\mathbf{1}$ is an eigenvector of $\mathbf{L}$ with eigenvalue zero:

$$\mathbf{L}(\mathbf{V} + c\mathbf{1}) = \mathbf{LV} + \mathbf{L1} = \mathbf{LV} = R\mathbf{I}. \qquad (6.76)$$

In physical terms these different solutions correspond to different choices of the reference potential against which we measure our voltages. The actual currents flowing around the system are identical no matter what reference potential we choose. If we fix our reference potential at a particular value, then we will fix the solution for the voltages as well, and our equation for $\mathbf{V}$ will become solvable.

Let us choose, arbitrarily, to set our reference potential equal to the potential at the target vertex t where the current exits the network. (We could choose any other vertex just as well, but this choice is the simplest.) That is, the voltage at this vertex is chosen to be zero and all others are measured in terms of their potential difference from vertex t. But now we can remove the element $V_t = 0$ from $\mathbf{V}$ in Eq. (6.75), along with the corresponding column t in the Laplacian, without affecting the result, since they contribute zero to the matrix multiplication anyway. And we can also remove row t from both sides of the equation, since we already know the value of V_t, so there's no need to calculate it. That leaves us with a modified equation $\mathbf{L'V'} = R\mathbf{I'}$, with $\mathbf{L'}$ being the tth reduced Laplacian, which in general has a well-defined inverse. Then

$$\mathbf{V'} = R\mathbf{L'}^{-1}\mathbf{I'}, \qquad (6.77)$$

and once we have the voltages we can calculate in a straightforward manner any other quantity of interest, such as the current along a given edge in the network.

Note that, for the simple case discussed here in which current is injected into the network at just one vertex and removed at another, $\mathbf{I}'$ has only one non-zero element. (The other one, I_t, has been removed.) Thus the vector $\mathbf{V}'$ on the left-hand side of Eq. (6.77) is simply proportional to the column of the inverse reduced Laplacian corresponding to vertex s. To use the notation of Section 6.14, if $\mathbf{\Lambda}^{(t)}$ is the inverse of the tth reduced Laplacian with the tth row and column reintroduced having elements all zero (see Eq. (6.70)), then

$$V_i = RI\Lambda_{is}^{(t)} . \tag{6.78}$$

PROBLEMS

6.1 Consider the following two networks:

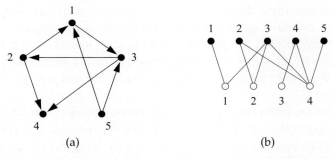

(a)   (b)

Network (a) is a directed network. Network (b) is undirected but bipartite. Write down:

  a) the adjacency matrix of network (a);

  b) the cocitation matrix of network (a);

  c) the incidence matrix of network (b);

  d) the projection matrix (Eq. (6.17)) for the projection of network (b) onto its black vertices.

6.2 Let $\mathbf{A}$ be the adjacency matrix of an undirected network and $\mathbf{1}$ be the column vector whose elements are all 1. In terms of these quantities write expressions for:

  a) the vector $\mathbf{k}$ whose elements are the degrees k_i of the vertices;

  b) the number m of edges in the network;

  c) the matrix $\mathbf{N}$ whose element N_{ij} is equal to the number of common neighbors of vertices i and j;

d) the total number of triangles in the network, where a triangle means three vertices, each connected by edges to both of the others.

6.3 Consider an acyclic directed network of n vertices, labeled $i = 1 \ldots n$, and suppose that the labels are assigned in the manner of Fig. 6.3 on page 119, such that all edges run from vertices with higher labels to vertices with lower.

a) Find an expression for the total number of edges ingoing to vertices $1 \ldots r$ and another for the total number of edges outgoing from vertices $1 \ldots r$, in terms of the in- and out-degrees k_i^{in} and k_i^{out} of the vertices.

b) Hence find an expression for the total number of edges running to vertices $1 \ldots r$ from vertices $r + 1 \ldots n$.

c) Show that in any acyclic network the in- and out-degrees must satisfy

$$k_r^{out} \leq \sum_{i=1}^{r-1} \left(k_i^{in} - k_i^{out} \right),$$

for all r.

6.4 Consider a bipartite network, with its two types of vertex, and suppose that there are n_1 vertices of type 1 and n_2 vertices of type 2. Show that the mean degrees c_1 and c_2 of the two types are related by

$$c_2 = \frac{n_1}{n_2} c_1.$$

6.5 Using Kuratowski's theorem, prove that this network is not planar:

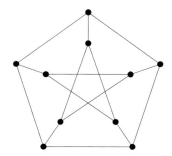

6.6 Consider a connected planar network with n vertices and m edges. Let f be the number of "faces" of the network, i.e., areas bounded by edges when the network is drawn in planar form. The "outside" of the network, the area extending to infinity on all sides, is also considered a face. The network can have multiedges and self-edges:

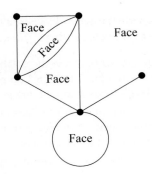

a) Write down the values of n, m, and f for a network with a single vertex and no edges.

b) How do n, m, and f change when we add a single vertex to the network along with a single edge attaching it to another vertex?

c) How do n, m, and f change when we add a single edge between two extant vertices (or a self-edge attached to just one vertex), in such a way as to maintain the planarity of the network?

d) Hence by induction prove a general relation between n, m, and f for all connected planar networks.

e) Now suppose that our network is simple (i.e., it contains no multiedges or self-edges). Show that the mean degree c of such a network is strictly less than six.

6.7 Consider the set of all paths from vertex s to vertex t on an undirected graph with adjacency matrix $\mathbf{A}$. Let us give each path a weight equal to α^r, where r is the length of the path.

a) Show that the sum of the weights of all the paths from s to t is given by Z_{st} which is the st element of the matrix $\mathbf{Z} = (\mathbf{I} - \alpha\mathbf{A})^{-1}$, where $\mathbf{I}$ is the identity matrix.

b) What condition must α satisfy for the sum to converge?

c) Hence, or otherwise, show that the length ℓ_{st} of a geodesic path from s to t, if there is one, is

$$\ell_{st} = \lim_{\alpha \to 0} \frac{\partial \log Z_{st}}{\partial \log \alpha}.$$

6.8 What is the difference between a 2-component and a 2-core? Draw a small network that has one 2-core but two 2-components.

6.9 In Section 5.3.1, we gave one possible definition of the trophic level x_i of a species in a (directed) food web as the mean of the trophic levels of the species' prey, plus one.

a) Show that x_i, when defined in this way, is the ith element of the vector

$$\mathbf{x} = (\mathbf{D} - \mathbf{A})^{-1}\mathbf{D} \cdot \mathbf{1},$$

where $\mathbf{D}$ is the diagonal matrix of in-degrees, $\mathbf{A}$ is the (asymmetric) adjacency matrix, and $\mathbf{1} = (1, 1, 1, \ldots)$.

b) This expression does not work for autotrophs—species with no prey—because the corresponding vector element diverges. Such species are usually given a trophic level of one. Suggest a modification of the calculation that will correctly assign trophic levels to these species, and hence to all species.

6.10 What is the size k of the minimum vertex cut set between s and t in this network?

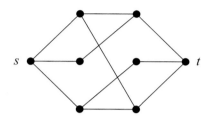

Prove your result by finding one possible cut set of size k and one possible set of k independent paths between s and t. Why do these two actions constitute a proof that the minimum cut set has size k?

CHAPTER 7

MEASURES AND METRICS

An introduction to some standard measures and metrics for quantifying network structure, many of which were introduced first in the study of social networks, although they are now in wide use in many other areas

I F WE KNOW the structure of a network we can calculate from it a variety of useful quantities or measures that capture particular features of the network topology. In this chapter we look at some of these measures. Many of the most important ideas in this area come from the social sciences, from the discipline of *social network analysis*, which was developed to aid our understanding of social network data such as those described in Chapter 3, and much of the language used to describe these ideas reflects their sociological origin. Nonetheless, the methods described are now widely used in areas outside the social sciences, including computer science, physics, and biology, and form an important part of the basic network toolbox.[1]

In the chapter following this one we will apply some of the measures developed here to the analysis of network data from a variety of fields and in the process reveal some intriguing features and patterns that will play an important role in later developments.

7.1 DEGREE CENTRALITY

A large volume of research on networks has been devoted to the concept of *centrality*. This research addresses the question, "Which are the most important or central vertices in a network?" There are of course many possible definitions

---

[1]For those interested in traditional social network analysis, introductions can be found in the books by Scott [293] and by Wasserman and Faust [320].

of importance, and correspondingly many centrality measures for networks. In this and the following several sections we describe some of the most widely used such measures.

Perhaps the simplest centrality measure in a network is just the degree of a vertex, the number of edges connected to it (see Section 6.9). Degree is sometimes called *degree centrality* in the social networks literature, to emphasize its use as a centrality measure. In directed networks, vertices have both an in-degree and an out-degree, and both may be useful as measures of centrality in the appropriate circumstances.

Although degree centrality is a simple centrality measure, it can be very illuminating. In a social network, for instance, it seems reasonable to suppose that individuals who have connections to many others might have more influence, more access to information, or more prestige than those who have fewer connections. A non-social network example is the use of citation counts in the evaluation of scientific papers. The number of citations a paper receives from other papers, which is simply its in-degree in the citation network, gives a crude measure of whether the paper has been influential or not and is widely used as a metric for judging the impact of scientific research.

7.2 EIGENVECTOR CENTRALITY

A natural extension of the simple degree centrality is *eigenvector centrality*. We can think of degree centrality as awarding one "centrality point" for every network neighbor a vertex has. But not all neighbors are equivalent. In many circumstances a vertex's importance in a network is increased by having connections to other vertices that are *themselves important*. This is the concept behind eigenvector centrality. Instead of awarding vertices just one point for each neighbor, eigenvector centrality gives each vertex a score proportional to the sum of the scores of its neighbors. Here's how it works.

Let us make some initial guess about the centrality x_i of each vertex i. For instance, we could start off by setting $x_i = 1$ for all i. Obviously this is not a useful measure of centrality, but we can use it to calculate a better one x_i', which we define to be the sum of the centralities of i's neighbors thus:

$$x_i' = \sum_j A_{ij} x_j, \tag{7.1}$$

where A_{ij} is an element of the adjacency matrix. We can also write this expression in matrix notation as $\mathbf{x}' = \mathbf{A}\mathbf{x}$, where $\mathbf{x}$ is the vector with elements x_i. Repeating this process to make better estimates, we have after t steps a vector of centralities $\mathbf{x}(t)$ given by

$$\mathbf{x}(t) = \mathbf{A}^t \mathbf{x}(0). \tag{7.2}$$

Now let us write $\mathbf{x}(0)$ as a linear combination of the eigenvectors $\mathbf{v}_i$ of the adjacency matrix thus:

$$\mathbf{x}(0) = \sum_i c_i \mathbf{v}_i , \tag{7.3}$$

for some appropriate choice of constants c_i. Then

$$\mathbf{x}(t) = \mathbf{A}^t \sum_i c_i \mathbf{v}_i = \sum_i c_i \kappa_i^t \mathbf{v}_i = \kappa_1^t \sum_i c_i \left[\frac{\kappa_i}{\kappa_1} \right]^t \mathbf{v}_i, \tag{7.4}$$

where the κ_i are the eigenvalues of $\mathbf{A}$, and κ_1 is the largest of them. Since $\kappa_i / \kappa_1 < 1$ for all $i \neq 1$, all terms in the sum other than the first decay exponentially as t becomes large, and hence in the limit $t \to \infty$ we get $\mathbf{x}(t) \to c_1 \kappa_1^t \mathbf{v}_1$. In other words, the limiting vector of centralities is simply proportional to the leading eigenvector of the adjacency matrix. Equivalently we could say that the centrality $\mathbf{x}$ satisfies

$$\mathbf{A}\mathbf{x} = \kappa_1 \mathbf{x}. \tag{7.5}$$

This then is the eigenvector centrality, first proposed by Bonacich [49] in 1987. As promised the centrality x_i of vertex i is proportional to the sum of the centralities of i's neighbors:

$$x_i = \kappa_1^{-1} \sum_j A_{ij} x_j, \tag{7.6}$$

which gives the eigenvector centrality the nice property that it can be large either because a vertex has many neighbors or because it has important neighbors (or both). An individual in a social network, for instance, can be important, by this measure, because he or she knows lots of people (even though those people may not be important themselves) or knows a few people in high places.

Note also that the eigenvector centralities of all vertices are non-negative. To see this, consider what happens if the initial vector $\mathbf{x}(0)$ happens to have only non-negative elements. Since all elements of the adjacency matrix are also non-negative, multiplication by $\mathbf{A}$ can never introduce any negative elements to the vector and $\mathbf{x}(t)$ in Eq. (7.2) must have all elements non-negative.[2]

---

[2]Technically, there could be more than one eigenvector with eigenvalue κ_1, only one of which need have all elements non-negative. It turns out, however, that this cannot happen: the adjacency matrix has only one eigenvector of eigenvalue κ_1. See footnote 2 on page 346 for a proof.

Equation (7.5) does not fix the normalization of the eigenvector centrality, although typically this doesn't matter because we care only about which vertices have high or low centrality and not about absolute values. If we wish, however, we can normalize the centralities by, for instance, requiring that they sum to n (which insures that average centrality stays constant as the network gets larger).

In theory eigenvector centrality can be calculated for either undirected or directed networks. It works best however for the undirected case. In the directed case other complications arise. First of all, a directed network has an adjacency matrix that is, in general, asymmetric (see Section 6.4). This means that it has two sets of eigenvectors, the left eigenvectors and the right eigenvectors, and hence two leading eigenvectors. So which of the two should we use to define the centrality? In most cases the correct answer is to use the right eigenvector. The reason is that centrality in directed networks is usually bestowed by other vertices pointing towards you, rather than by you pointing to others. On the World Wide Web, for instance, the number and stature of web pages that point to your page can give a reasonable indication of how important or useful your page is. On the other hand, the fact that your page might point to other important pages is neither here nor there. Anyone can set up a page that points to a thousand others, but that does not make the page important.[3] Similar considerations apply also to citation networks and other directed networks. Thus the correct definition of eigenvector centrality for a vertex i in a directed network makes it proportional to the centralities of the vertices that point to i thus:

$$x_i = \kappa_1^{-1} \sum_j A_{ij} x_j, \qquad (7.7)$$

which gives $\mathbf{Ax} = \kappa_1 \mathbf{x}$ in matrix notation, where $\mathbf{x}$ is the right leading eigenvector.

However, there are still problems with eigenvector centrality on directed networks. Consider Fig. 7.1. Vertex A in this figure is connected to the rest of the network, but has only outgoing edges and no incoming ones. Such a vertex will always have centrality zero because there are no terms in the sum

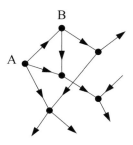

Figure 7.1: A portion of a directed network. Vertex A in this network has only outgoing edges and hence will have eigenvector centrality zero. Vertex B has outgoing edges and one ingoing edge, but the ingoing one originates at A, and hence vertex B will also have centrality zero.

---

[3]This is not entirely true, as we will see in Section 7.5. Web pages that point to many others are often directories of one sort or another and can be useful as starting points for web surfing. This is a different kind of importance, however, from that highlighted by the eigenvector centrality and a different, complementary centrality measure is needed to quantify it.

in Eq. (7.7). This might not seem to be a problem: perhaps a vertex that no one points to *should* have centrality zero. But then consider vertex B, which has one ingoing edge, but that edge originates at vertex A, and hence B also has centrality zero, because the one term in its sum in Eq. (7.7) is zero. Taking this argument further, we see that a vertex may be pointed to by others that themselves are pointed to by many more, and so on through many generations, but if the progression ends up at a vertex or vertices that have in-degree zero, it is all for nothing—the final value of the centrality will still be zero.

In mathematical terms, only vertices that are in a strongly connected component of two or more vertices, or the out-component of such a component, can have non-zero eigenvector centrality.[4] In many cases, however, it is appropriate for vertices with high in-degree to have high centrality even if they are not in a strongly-connected component or its out-component. Web pages with many links, for instance, can reasonably be considered important even if they are not in a strongly connected component. Recall also that acyclic networks, such as citation networks, have no strongly connected components of more than one vertex (see Section 6.11.1), so all vertices will have centrality zero. Clearly this make the standard eigenvector centrality completely useless for acyclic networks.

A variation on eigenvector centrality that addresses these problems is the Katz centrality, which is the subject of the next section.

7.3 KATZ CENTRALITY

One solution to the issues of the previous section is the following: we simply give each vertex a small amount of centrality "for free," regardless of its position in the network or the centrality of its neighbors. In other words, we define

$$x_i = \alpha \sum_j A_{ij} x_j + \beta, \tag{7.8}$$

where α and β are positive constants. The first term is the normal eigenvector centrality term in which the centralities of the vertices linking to i are summed, and the second term is the "free" part, the constant extra term that all vertices receive. By adding this second term, even vertices with zero in-degree still get centrality β, and once they have a non-zero centrality, then the vertices they point to derive some advantage from being pointed to. This means that any vertex that is pointed to by many others will have a high centrality, although

---

[4]For the left eigenvector it would be the in-component.

those that are pointed to by others with high centrality themselves will still do better.

In matrix terms, Eq. (7.8) can be written

$$\mathbf{x} = \alpha \mathbf{A} \mathbf{x} + \beta \mathbf{1}, \tag{7.9}$$

where $\mathbf{1}$ is the vector $(1, 1, 1 \ldots)$. Rearranging for $\mathbf{x}$, we find that $\mathbf{x} = \beta(\mathbf{I} - \alpha \mathbf{A})^{-1} \cdot \mathbf{1}$. As we have said, we normally don't care about the absolute magnitude of the centrality, only about which vertices have high or low centrality values, so the overall multiplier β is unimportant. For convenience we usually set $\beta = 1$, giving

$$\mathbf{x} = (\mathbf{I} - \alpha \mathbf{A})^{-1} \cdot \mathbf{1}. \tag{7.10}$$

This centrality measure was first proposed by Katz in 1953 [169] and we will refer to it as the *Katz centrality*.

The Katz centrality differs from ordinary eigenvector centrality in the important respect of having a free parameter α, which governs the balance between the eigenvector term and the constant term in Eq. (7.8). If we wish to make use of the Katz centrality we must first choose a value for this constant. In doing so it is important to understand that α cannot be arbitrarily large. If we let $\alpha \to 0$, then only the constant term survives in Eq. (7.8) and all vertices have the same centrality β (which we have set to 1). As we increase α from zero the centralities increase and eventually there comes a point at which they diverge. This happens at the point where $(\mathbf{I} - \alpha \mathbf{A})^{-1}$ diverges in Eq. (7.10), i.e., when $\det(\mathbf{I} - \alpha \mathbf{A})$ passes through zero. Rewriting this condition as

$$\det(\mathbf{A} - \alpha^{-1}\mathbf{I}) = 0, \tag{7.11}$$

we see that it is simply the characteristic equation whose roots α^{-1} are equal to the eigenvalues of the adjacency matrix.[5] As α increases, the determinant first crosses zero when $\alpha^{-1} = \kappa_1$, the largest eigenvalue of $\mathbf{A}$, or alternatively when $\alpha = 1/\kappa_1$. Thus, we should choose a value of α less than this if we wish the expression for the centrality to converge.[6]

Beyond this, however, there is little guidance to be had as to the value that α should take. Most researchers have employed values close to the maximum of $1/\kappa_1$, which places the maximum amount of weight on the eigenvector term

---

[5]The eigenvalues being defined by $\mathbf{A}\mathbf{v} = \kappa\mathbf{v}$, we see that $(\mathbf{A} - \kappa\mathbf{I})\mathbf{v} = 0$, which has non-zero solutions for $\mathbf{v}$ only if $(\mathbf{A} - \kappa\mathbf{I})$ cannot be inverted, i.e., if $\det(\mathbf{A} - \kappa\mathbf{I}) = 0$, and hence this equation gives the eigenvalues κ.

[6]Formally one recovers finite values again when one moves past $1/\kappa_1$ to higher α, but in practice these values are meaningless. The method returns good results only for $\alpha < 1/\kappa_1$.

and the smallest amount on the constant term. This returns a centrality that is numerically quite close to the ordinary eigenvector centrality, but gives small non-zero values to vertices that are not in the strongly connected components or their out-components.

The Katz centrality can be calculated directly from Eq. (7.10) by inverting the matrix on the right-hand side, but often this isn't the best way to do it. Inverting a matrix on a computer takes an amount of time proportional to n^3, where n is the number of vertices. This makes direct calculation of the Katz centrality prohibitively slow for large networks. Networks of more than a thousand vertices or so present serious problems.

A better approach in many cases is to evaluate the centrality directly from Eq. (7.8) (or equivalently, Eq. (7.9)). One makes an initial estimate of $\mathbf{x}$— probably a bad one, such as $\mathbf{x} = 0$—and uses that to calculate a better estimate

$$\mathbf{x}' = \alpha \mathbf{A}\mathbf{x} + \beta \mathbf{1}. \tag{7.12}$$

Repeating the process many times, $\mathbf{x}$ converges to a value close to the correct centrality. Since $\mathbf{A}$ has m non-zero elements, each iteration requires m multiplication operations and the total time for the calculation is proportional to rm, where r is the number of iterations necessary for the calculation to converge. Unfortunately, r depends on the details of the network and on the choice of α, so we cannot give a general guide to how many iterations will be necessary. Instead one must watch the values of x_i to observe when they converge to constant values. Nonetheless, for large networks it is almost always worthwhile to evaluate the centrality this way rather than by inverting the matrix.

We have presented the Katz centrality as a solution to the problems encountered with ordinary eigenvector centrality in directed networks. However, there is no reason in principle why one cannot use Katz centrality in undirected networks as well, and there are times when this might be useful. The idea of adding a constant term to the centrality so that each vertex gets some weight just by virtue of existing is a natural one. It allows a vertex that has many neighbors to have high centrality regardless of whether those neighbors themselves have high centrality, and this could be desirable in some applications.

A possible extension of the Katz centrality is to consider cases in which the additive constant term in Eq. (7.8) is not the same for all vertices. One could define a generalized centrality measure by

$$x_i = \alpha \sum_j A_{ij} x_j + \beta_i, \tag{7.13}$$

where β_i is some intrinsic, non-network contribution to the centrality for each vertex. For example, in a social network the importance of an individual might depend on non-network factors such as their age or income and if we had information about these factors we could incorporate it into the values of the β_i. Then the vector $\mathbf{x}$ of centralities is given by

$$\mathbf{x} = (\mathbf{I} - \alpha\mathbf{A})^{-1}\boldsymbol{\beta}, \qquad (7.14)$$

where $\boldsymbol{\beta}$ is the vector whose elements are the β_i. One nice feature of this approach is that the difficult part of the calculation—the inversion of the matrix—only has to be done once for a given network and choice of α. For difference choices of the β_i we need not recalculate the inverse, but simply multiply the inverse into different vectors $\boldsymbol{\beta}$.

7.4 PAGERANK

The Katz centrality of the previous section has one feature that can be undesirable. If a vertex with high Katz centrality points to many others then those others also get high centrality. A high-centrality vertex pointing to one million others gives all one million of them high centrality. One could argue—and many have—that this is not always appropriate. In many cases it means less if a vertex is only one among many that are pointed to. The centrality gained by virtue of receiving an edge from a prestigious vertex is diluted by being shared with so many others. For instance, the famous *Yahoo!* web directory might contain a link to my web page, but it also has links to millions of other pages. *Yahoo!* is an important website, and would have high centrality by any sensible measure, but should I therefore be considered very important by association? Most people would say not: the high centrality of *Yahoo!* will get diluted and its contribution to the centrality of my page should be small because my page is only one of millions.

We can allow for this by defining a variation on the Katz centrality in which the centrality I derive from my network neighbors is proportional to their centrality *divided by their out-degree*. Then vertices that point to many others pass only a small amount of centrality on to each of those others, even if their own centrality is high.

In mathematical terms this centrality is defined by

$$x_i = \alpha \sum_j A_{ij} \frac{x_j}{k_j^{\text{out}}} + \beta. \qquad (7.15)$$

This gives problems however if there are vertices in the network with out-degree $k_i^{\text{out}} = 0$. If there are any such vertices then the first term in Eq. (7.15) is indeterminate—it is equal to zero divided by zero (because $A_{ij} = 0$ for all i). This problem is easily fixed however. It is clear that vertices with no out-going edges should contribute zero to the centrality of any other vertex, which we can contrive by artificially setting $k_i^{out} = 1$ for all such vertices. (In fact, we could set k_i^{out} to any non-zero value and the calculation would give the same answer.)

In matrix terms, Eq. (7.15), is then

$$\mathbf{x} = \alpha \mathbf{A} \mathbf{D}^{-1} \mathbf{x} + \beta \mathbf{1}, \tag{7.16}$$

with $\mathbf{1}$ being again the vector $(1, 1, 1, \ldots)$ and $\mathbf{D}$ being the diagonal matrix with elements $D_{ii} = \max(k_i^{out}, 1)$. Rearranging, we find that $\mathbf{x} = \beta(\mathbf{I} - \alpha \mathbf{A} \mathbf{D}^{-1})^{-1} \cdot \mathbf{1}$, and thus, as before, β plays the role only of an unimportant overall multiplier for the centrality. Conventionally we set $\beta = 1$, giving

$$\mathbf{x} = (\mathbf{I} - \alpha \mathbf{A} \mathbf{D}^{-1})^{-1} \mathbf{1} = \mathbf{D}(\mathbf{D} - \alpha \mathbf{A})^{-1} \mathbf{1}. \tag{7.17}$$

This centrality measure is commonly known as *PageRank*, which is the trade name given it by the Google web search corporation, which uses it as a central part of their web ranking technology [55]. The aim of the Google web search engine is to generate lists of useful web pages from a preassembled index of pages in response to text queries. It does this by first searching the index for pages matching a given query using relatively simple criteria such as text matching, and then ranking the answers according to scores based on a combination of ingredients of which PageRank is one. Google returns useful answers to queries not because it is better at finding relevant pages, but because it is better at deciding what order to present its findings in: its perceived accuracy arises because the results at the top of the list of answers it returns are often highly relevant to the query, but it is possible and indeed likely that many irrelevant answers also appear on the list, lower down.

PageRank works on the Web precisely because having links to your page from important pages elsewhere is a good indication that your page may be important too. But the added ingredient of dividing by the out-degrees of pages insures that pages that simply point to an enormous number of others do not pass much centrality on to any of them, so that, for instance, network hubs like *Yahoo!* do not have a disproportionate influence on the rankings.

As with the Katz centrality, the formula for PageRank, Eq. (7.17), contains one free parameter α, whose value must be chosen somehow before the algorithm can be used. By analogy with Eq. (7.11) and the argument that follows it, we can see that the value of α should be less than the inverse of the largest eigenvalue of $\mathbf{A} \mathbf{D}^{-1}$. For an undirected network this largest eigenvalue turns

Web search is discussed in more detail in Section 19.1.

out to be 1 and the corresponding eigenvector is $(k_1, k_2, k_3, \ldots)$, where k_i is the degree of the ith vertex.[7] Thus α should be chosen less than 1. For a directed network, this result does not follow and in general the leading eigenvalue will be different from 1, although in practical cases it is usually still roughly of order 1.

The *Google* search engine uses a value of $\alpha = 0.85$ in its calculations, although it's not clear that there is any rigorous theory behind this choice. More likely it is just a shrewd guess based on experimentation to find out what works well.

As with the Katz centrality we can generalize PageRank to the case where the additive constant term in Eq. (7.15) is different for different vertices:

$$x_i = \alpha \sum_j A_{ij} \frac{x_j}{k_j^{\text{out}}} + \beta_i. \tag{7.18}$$

In matrix form this gives a solution for the centrality vector of

$$\mathbf{x} = \mathbf{D}(\mathbf{D} - \alpha \mathbf{A})^{-1} \boldsymbol{\beta}. \tag{7.19}$$

One could, for instance, use this for ranking web pages, giving β_i a value based perhaps on textual relevance to a search query. Pages that contained the word or words being searched for more often or in more prominent places could be given a higher intrinsic centrality than others, thereby pushing them up the rankings. The author is not aware, however, of any cases in which this technique has been implemented in practice.

Finally, one can also imagine a version of PageRank that did not have the additive constant term in it at all:

$$x_i = \alpha \sum_j A_{ij} \frac{x_j}{k_j}, \tag{7.20}$$

which is similar to the original eigenvector centrality introduced back in Section 7.2, but now with the extra division by k_j. For an undirected network, however, this measure is trivial: it is easy to see that it gives simply $x_i = k_i$

---

[7]It is easy to confirm that this vector is indeed an eigenvector with eigenvalue 1. That there is no eigenvalue larger than 1 is less obvious. It follows from a standard result in linear algebra, the Perron–Frobenius theorem, which states that the largest eigenvalue of a matrix such as $\mathbf{AD}^{-1}$ that has all elements non-negative is unique—there is only one eigenvector with this eigenvalue— that the eigenvector also has all elements non-negative, and that it is the only eigenvector with all elements non-negative. Combining these results, it is clear that the eigenvalue 1 above must be the largest eigenvalue of the matrix $\mathbf{AD}^{-1}$. For a discussion of the Perron–Frobenius theorem see Ref. [217] and the two footnotes on page 346 of this book.

	with constant term	without constant term
divide by out-degree	$\mathbf{x} = \mathbf{D}(\mathbf{D} - \alpha\mathbf{A})^{-1} \cdot \mathbf{1}$ PageRank	$\mathbf{x} = \mathbf{A}\mathbf{D}^{-1}\mathbf{x}$ degree centrality
no division	$\mathbf{x} = (\mathbf{I} - \alpha\mathbf{A})^{-1} \cdot \mathbf{1}$ Katz centrality	$\mathbf{x} = \kappa_1^{-1}\mathbf{A}\mathbf{x}$ eigenvector centrality

Table 7.1: **Four centrality measures.** The four matrix-based centrality measures discussed in the text are distinguished by whether or not they include an additive constant term in their definition and whether they are normalized by dividing by the degrees of neighboring vertices. Note that the diagonal matrix $\mathbf{D}$, which normally has elements $D_{ii} = k_i$, must be defined slightly differently for PageRank, as $D_{ii} = \max(1, k_i)$—see Eq. (7.15) and the following discussion. Each of the measures can be applied to directed networks as well as undirected ones, although only three of the four are commonly used in this way. (The measure that appears in the top right corner of the table is equivalent to degree centrality in the undirected case but takes more complicated values in the directed case and is not widely used.)

and therefore is just the same as ordinary degree centrality. For a directed network, on the other hand, it does not reduce to any equivalent simple value and it might potentially be of use, although it does not seem to have found use in any prominent application. (It does suffer from the same problem as the original eigenvector centrality, that it gives non-zero scores only to vertices that fall in a strongly connected component of two or more vertices or in the out-component of such a component. All other vertices get a zero score.)

In Table 7.1 we give a summary of the different matrix centrality measures we have discussed, organized according to their definitions and properties. If you want to use one of these measures in your own calculations and find the many alternatives bewildering, eigenvector centrality and PageRank are probably the two measures to focus on initially. They are the two most commonly used measures of this type. The Katz centrality has found widespread use in the past but has been favored less in recent work, while the PageRank measure without the constant term, Eq. (7.20), is the same as degree centrality for undirected networks and not in common use for directed ones.

7.5 HUBS AND AUTHORITIES

In the case of directed networks, there is another twist to the centrality measures introduced in this section. So far we have considered measures that accord a vertex high centrality if those that point to it have high centrality.

However, in some networks it is appropriate also to accord a vertex high centrality if it *points to* others with high centrality. For instance, in a citation network a paper such as a review article may cite other articles that are authoritative sources for information on a particular subject. The review itself may contain relatively little information on the subject, but it tells us where to find the information, and this on its own makes the review useful. Similarly, there are many examples of web pages that consist primarily of links to other pages on a given topic or topics and such a page of links could be very useful even if it does not itself contain explicit information on the topic in question.

Thus there are really two types of important node in these networks: *authorities* are nodes that contain useful information on a topic of interest; *hubs* are nodes that tell us where the best authorities are to be found. An authority may also be a hub, and vice versa: review articles often contain useful discussions of the topic at hand as well as citations to other discussions. Clearly hubs and authorities only exist in directed networks, since in the undirected case there is no distinction between pointing to a vertex and being pointed to.

One can imagine defining two different types of centrality for directed networks, the *authority centrality* and the *hub centrality*, which quantify vertices' prominence in the two roles. This idea was first put forward by Kleinberg [176] and developed by him into a centrality algorithm called *hyperlink-induced topic search* or *HITS*.

The HITS algorithm gives each vertex i in a network an authority centrality x_i and a hub centrality y_i. The defining characteristic of a vertex with high authority centrality is that it is pointed to by many hubs, i.e., by many other vertices with high hub centrality. And the defining characteristic of a vertex with high hub centrality is that it *points to* many vertices with high authority centrality.

Thus an important scientific paper (in the authority sense) would be one cited in many important reviews (in the hub sense). An important review is one that cites many important papers. Reviews, however, are not the only publications that can have high hub centrality. Ordinary papers can have high hub centrality too if they cite many other important papers, and papers can have both high authority and high hub centrality. Reviews too may be cited by other hubs and hence have high authority centrality as well as high hub centrality.

In Kleinberg's approach, the authority centrality of a vertex is defined to be proportional to the sum of the hub centralities of the vertices that point to it:

$$x_i = \alpha \sum_j A_{ij} y_j, \tag{7.21}$$

where α is a constant. Similarly the hub centrality of a vertex is proportional to the sum of the authority centralities of the vertices it points to:

$$y_i = \beta \sum_j A_{ji} x_j \, , \qquad (7.22)$$

with β another constant. Notice that the indices on the matrix element A_{ji} are swapped around in this second equation: it is the vertices that i points to that define its hub centrality.

In matrix terms these equations can be written as

$$\mathbf{x} = \alpha \mathbf{A} \mathbf{y}, \qquad \mathbf{y} = \beta \mathbf{A}^T \mathbf{x}, \qquad (7.23)$$

or, combining the two,

$$\mathbf{A}\mathbf{A}^T \mathbf{x} = \lambda \mathbf{x}, \qquad \mathbf{A}^T \mathbf{A} \mathbf{y} = \lambda \mathbf{y}, \qquad (7.24)$$

where $\lambda = (\alpha\beta)^{-1}$. Thus the authority and hub centralities are respectively given by eigenvectors of $\mathbf{A}\mathbf{A}^T$ and $\mathbf{A}^T\mathbf{A}$ with the same eigenvalue. By an argument similar to the one we used for the standard eigenvector centrality in Section 7.1 we can show that we should in each case take the eigenvector corresponding to the leading eigenvalue.

A crucial condition for this approach to work, is that $\mathbf{A}\mathbf{A}^T$ and $\mathbf{A}^T\mathbf{A}$ have the same leading eigenvalue λ, otherwise we cannot satisfy both conditions in Eq. (7.24). It is easily proved, however, that this is the case, and in fact that all eigenvalues are the same for the two matrices. If $\mathbf{A}\mathbf{A}^T\mathbf{x} = \lambda\mathbf{x}$ then multiplying both sides by $\mathbf{A}^T$ gives

$$\mathbf{A}^T\mathbf{A}(\mathbf{A}^T\mathbf{x}) = \lambda(\mathbf{A}^T\mathbf{x}), \qquad (7.25)$$

and hence $\mathbf{A}^T\mathbf{x}$ is an eigenvector of $\mathbf{A}^T\mathbf{A}$ with the same eigenvalue λ. Comparing with Eq. (7.24) this means that

$$\mathbf{y} = \mathbf{A}^T\mathbf{x}, \qquad (7.26)$$

which gives us a fast way of calculating the hub centralities once we have the authority ones—there is no need to solve both the eigenvalue equations in Eq. (7.24) separately.

Note that $\mathbf{A}\mathbf{A}^T$ is precisely the cocitation matrix defined in Section 6.4.1 (Eq. (6.8)) and the authority centrality is thus, roughly speaking, the eigenvector centrality for the cocitation network.[8] Similarly $\mathbf{A}^T\mathbf{A}$ is the bibliographic

---

[8]This statement is only approximately correct since, as discussed in Section 6.4.1, the cocitation matrix is not precisely equal to the adjacency matrix of the cocitation network, having non-zero elements along its diagonal where the adjacency matrix has none.

coupling matrix, Eq. (6.11), and hub centrality is the eigenvector centrality for the bibliographic coupling network.

A nice feature of the hub and authority centralities is that they circumvent the problems that ordinary eigenvector centrality has with directed networks, that vertices outside of strongly connected components or their out-components always have centrality zero. In the hubs and authorities approach vertices not cited by any others have authority centrality zero (which is reasonable), but they can still have non-zero hub centrality. And the vertices that *they* cite can then have non-zero authority centrality by virtue of being cited. This is perhaps a more elegant solution to the problems of eigenvector centrality in directed networks than the more ad hoc method of introducing an additive constant term as we did in Eq. (7.8). We can still introduce such a constant term into the HITS algorithm if we wish, or employ any of the other variations considered in previous sections, such as normalizing vertex centralities by the degrees of the vertices that point to them. Some variations along these lines are explored in Refs. [52, 256], but we leave the pursuit of such details to the enthusiastic reader.

The HITS algorithm is an elegant construction that should in theory provide more information about vertex centrality than the simpler measures of previous sections, but in practice it has not yet found much application. It is used as the basis for the web search engines *Teoma* and *Ask.com*, and will perhaps in future find further use, particularly in citation networks, where it holds clear advantages over other eigenvector measures.

7.6 CLOSENESS CENTRALITY

An entirely different measure of centrality is provided by the *closeness centrality*, which measures the mean distance from a vertex to other vertices. In Section 6.10.1 we encountered the concept of the geodesic path, the shortest path through a network between two vertices. Suppose d_{ij} is the length of a geodesic path from i to j, meaning the number of edges along the path.[9] Then the mean geodesic distance from i to j, averaged over all vertices j in the network, is

$$\ell_i = \frac{1}{n} \sum_j d_{ij}. \tag{7.27}$$

---

[9]Recall that geodesic paths need not be unique—vertices can be joined by several shortest paths of the same length. The length d_{ij} however is always well defined, being the length of any one of these paths.

This quantity takes low values for vertices that are separated from others by only a short geodesic distance on average. Such vertices might have better access to information at other vertices or more direct influence on other vertices. In a social network, for instance, a person with lower mean distance to others might find that their opinions reach others in the community more quickly than the opinions of someone with higher mean distance.

In calculating the average distance some authors exclude from the sum in (7.27) the term for $j = i$, so that

$$\ell_i = \frac{1}{n-1} \sum_{j(\neq i)} d_{ij}, \tag{7.28}$$

which is a reasonable strategy, since a vertex's influence on itself is usually not relevant to the working of the network. On the other hand, the distance d_{ii} from i to itself is zero by definition, so this term in fact contributes nothing to the sum. The only difference the change makes to ℓ_i is in the leading divisor, which becomes $1/(n-1)$ instead of $1/n$, meaning that ℓ_i changes by a factor of $n/(n-1)$. Since this factor is independent of i and since, as we have said, we usually care only about the relative centralities of different vertices and not about their absolute values, we can in most cases ignore the difference between Eqs. (7.27) and (7.28). In this book we use (7.27) because it tends to give slightly more elegant analytic results.

The mean distance ℓ_i is not a centrality measure in the same sense as the others in this chapter, since it gives *low* values for more central vertices and high values for less central ones, which is the opposite of our other measures. In the social networks literature, therefore, researchers commonly calculate the inverse of ℓ_i rather than ℓ_i itself. This inverse is called the *closeness centrality* C_i:

$$C_i = \frac{1}{\ell_i} = \frac{n}{\sum_j d_{ij}}. \tag{7.29}$$

Closeness centrality is a very natural measure of centrality and is often used in social and other network studies. But it has some problems. One issue is that its values tend to span a rather small dynamic range from largest to smallest. As discussed in Sections 3.6, 8.2, and 12.7, geodesic distances d_{ij} between vertices in most networks tend to be small, the typical distance increasing only logarithmically with the size of the entire network. This means that the ratio between the smallest distance, which is 1, and the largest, which is of order $\log n$, is itself only of order $\log n$, which is small. But the smallest and largest distances provide lower and upper bounds on the average distance ℓ_i, and hence the range of values of ℓ_i and similarly of C_i is also small. In a typical network the values of C_i might span a factor of five or less. What this means in practice is that it is difficult to distinguish between central and less central vertices using this measure: the values tend to be cramped together with the differences between adjacent values showing up only when you examine the trailing digits. This means that even small fluctuations in the structure of the network can change the order of the values substantially.

For example, it has become popular in recent years to rank film actors according to their closeness centrality in the network of who has appeared in films with who else [323]. Using data from the Internet Movie Database,[10] we find that in the largest component of the network, which includes more than 98% of all actors, the smallest closeness centrality of any actor is 2.4138 for the actor Christopher Lee,[11] while the largest is 8.6681 for an Iranian actress named Leia Zanganeh. The ratio of the two is just 3.6 and about half a million other actors lie in between. As we can immediately see, the values must be very closely spaced. The second best centrality score belongs to actor Donald Pleasence, who scores 2.4164, just a tenth of a percent less than winner Lee. Because of the close spacing of values, the leaders under this dubious measure of superiority change frequently as the small details of the film network shift when new films are made or old ones added to the database. In an analysis using an earlier version of the database, Watts and Strogatz [323] proclaimed Rod Steiger to be the actor with the lowest closeness centrality. Steiger falls in sixth place in our analysis and it is entirely possible that the rankings will have changed again by the time you read this. Other centrality measures, including degree centrality and eigenvector centrality, typically don't suffer from this problem because they have a wider dynamic range and the centrality values, particular those of the leaders, tend to be widely separated.

The closeness centrality has another problem too. If, as discussed in Section 6.10.1, we define the geodesic distance between two vertices to be infinite if the vertices fall in different components of the network, then ℓ_i is infinite for all i in any network with more than one component and C_i is zero. There are two strategies for getting around this. The most common one is simply to average over only those vertices in the same component as i. Then n in Eq. (7.29) becomes the number of vertices in the component and the sum is over only that component. This gives us a finite measure, but one that has its own problems. In particular, distances tend to be smaller between vertices in small components, so that vertices in such components get lower values of ℓ_i

---

[10]www.imdb.com

[11]Perhaps most famous for his role as the evil wizard Saruman in the film version of *The Lord of the Rings*.

and higher closeness centrality than their counterparts in larger components. This is usually undesirable: in most cases vertices in small components are considered *less* well connected than those in larger ones and should therefore be given lower centrality.

Perhaps a better solution, therefore, is to redefine closeness in terms of the harmonic mean distance between vertices, i.e., the average of the inverse distances:

$$C_i' = \frac{1}{n-1} \sum_{j(\neq i)} \frac{1}{d_{ij}}. \tag{7.30}$$

(Notice that we are obliged in this case to exclude from the sum the term for $j = i$, since $d_{ii} = 0$ which would make this term infinite. This means that the sum has only $n - 1$ terms in it, hence the leading factor of $1/(n-1)$.)

This definition has a couple of nice properties. First, if $d_{ij} = \infty$ because i and j are in different components, then the corresponding term in the sum is simply zero and drops out. Second, the measure naturally gives more weight to vertices that are close to i than to those far away. Intuitively we might imagine that the distance to close vertices is what matters in most practical situations— once a vertex is far away in a network it matters less exactly how far away it is, and Eq. (7.30) reflects this, having contributions close to zero from all such vertices.

Despite its desirable qualities, however, Eq. (7.30) is rarely used in practice. We have seen it employed only occasionally.

An interesting property of entire networks, which is related to the closeness centrality, is the mean geodesic distance between vertices. In Section 8.2 we will use measurements of mean distance in networks to study the so-called "small-world effect."

For a network with only one component, the mean distance between pairs of vertices, conventionally denoted just ℓ (now without the subscript), is

$$\ell = \frac{1}{n^2} \sum_{ij} d_{ij} = \frac{1}{n} \sum_i \ell_i. \tag{7.31}$$

In other words ℓ is just the mean of ℓ_i over all vertices.

For a network with more than one component we run into the same problems as before, that d_{ij} is infinite when i and j are in different components and hence ℓ is also infinite. The most common way around this problem is to average only over paths that run between vertices in the same component. Let $\{\mathscr{C}_m\}$ be the set of components of a network, with $m = 1, 2 \ldots$ Then we define

$$\ell = \frac{\sum_m \sum_{ij \in \mathscr{C}_m} d_{ij}}{\sum_m n_m^2}, \tag{7.32}$$

where n_m is the number of vertices in component $\mathscr{C}_m$. This measure is now finite for all networks, although it is not now equal to a simple average over the values of ℓ_i for each vertex.

An alternative and perhaps better approach would be to use the trick from Eq. (7.30) and define a harmonic mean distance ℓ' according to

$$\frac{1}{\ell'} = \frac{1}{n(n-1)} \sum_{i \neq j} \frac{1}{d_{ij}} = \frac{1}{n} \sum_i C_i', \tag{7.33}$$

or equivalently

$$\ell' = \frac{n}{\sum_i C_i'}, \tag{7.34}$$

where C_i' is the harmonic mean closeness of Eq. (7.30). (Note that, as in (7.30), we exclude from the first sum in (7.33) the terms for $i = j$, which would be infinite since $d_{ii} = 0$.)

Equation (7.34) automatically removes any contributions from vertex pairs for which $d_{ij} = \infty$. Despite its elegance, however, Eq. (7.34), like Eq. (7.30), is hardly ever used.

7.7 BETWEENNESS CENTRALITY

A very different concept of centrality is *betweenness centrality*, which measures the extent to which a vertex lies on paths between other vertices. The idea of betweenness is usually attributed to Freeman [128] in 1977, although as Freeman himself has pointed out [129], it was independently proposed some years earlier by Anthonisse [19] in an unpublished technical report.

Suppose we have a network with something flowing around it from vertex to vertex along the edges. For instance, in a social network we might have messages, news, information, or rumors being passed from one person to another. In the Internet we have data packets moving around. Let us initially make the simple assumption that every pair of vertices in the network exchanges a message with equal probability per unit time (more precisely every pair that is actually connected by a path) and that messages always take the shortest (geodesic) path though the network, or one such path, chosen at random, if there are several. Then let us ask the following question: if we wait a suitably long time until many messages have passed between each pair of vertices, how many messages, on average, will have passed through each vertex en route to their destination? The answer is that, since messages are passing down each geodesic path at the same rate, the number passing through each vertex is simply proportional to the number of geodesic paths the vertex lies on. This number of geodesic paths is what we call the betweenness centrality, or just betweenness for short.

Vertices with high betweenness centrality may have considerable influence within a network by virtue of their control over information passing between others. The vertices with highest betweenness in our message-passing scenario are the ones through which the largest number of messages pass, and if those vertices get to see the messages in question as they pass, or if they get paid for passing the messages along, they could derive a lot of power from their position within the network. The vertices with highest betweenness are also the ones whose removal from the network will most disrupt communications between other vertices because they lie on the largest number of paths taken by messages. In real-world situations, of course, not all vertices exchange communications with the same frequency, and in most cases communications do not always take the shortest path. Nonetheless, betweenness centrality may still be an approximate guide to the influence vertices have over the flow of information between others.

Having seen the basic idea of betweenness centrality, let us make things more precise. For the sake of simplicity, suppose for the moment that we have an undirected network in which there is at most one geodesic path between any pair of vertices. (There may be zero paths if the vertices in question are in different components.) Consider the set of all geodesic paths in such a network. Then the betweenness centrality of a vertex i is defined to be the number of those paths that pass through i.

Mathematically, let n_{st}^i be 1 if vertex i lies on the geodesic path from s to t and 0 if it does not or if there is no such path (because s and t lie in different components of the network). Then the betweenness centrality x_i is given by

$$x_i = \sum_{st} n_{st}^i. \tag{7.35}$$

Note that this definition counts separately the geodesic paths in either direction between each vertex pair. Since these paths are the same on an undirected network this effectively counts each path twice. One could compensate for this by dividing x_i by 2, and often this is done, but we prefer the definition given here for a couple of reasons. First, it makes little difference in practice whether one divides the centrality by 2, since one is usually concerned only with the relative magnitudes of the centralities and not with their absolute values. Second, as discussed below, Eq. (7.35) has the advantage that it can be applied unmodified to directed networks, in which the paths in either direction between a vertex pair can differ.

Note also that Eq. (7.35) includes paths from each vertex to itself. Some people prefer to exclude such paths from the definition, so that $x_i = \sum_{s \neq t} n_{st}^i$, but again the difference is typically not important. Each vertex lies on one path from itself to itself, so the inclusion of these terms simply increases the betweenness by 1, but does not change the rankings of the vertices—which ones have higher or lower betweenness—relative to one another.

There is also a choice to be made about whether the path from s to t should be considered to pass through the vertices s and t themselves. In the social networks literature it is usually assumed that it does not. We prefer the definition where it does: it seems reasonable to define a vertex to be on a path between itself and someone else, since normally a vertex has control over information flowing from itself to other vertices or vice versa. If, however, we exclude the endpoints of the path as sociologists commonly do, the only effect is to reduce the number of paths through each vertex by twice the size of the component to which the vertex belongs. Thus the betweennesses of all vertices within a single component are just reduced by an additive constant and the ranking of vertices within the component is again unchanged. (The rankings of vertices in different components can change relative to one another, but this is rarely an issue because betweenness centrality is not typically used to compare vertices in different components, since such vertices are not competing for influence in the same arena.)

These developments are all for the case in which there is at most one geodesic path between each vertex pair. More generally, however, there may be more than one. The standard extension of betweenness to this case gives each path a weight equal to the inverse of the number of paths. For instance, if there are two geodesic paths between a given pair of vertices, each of them gets weight $\frac{1}{2}$. Then the betweenness of a vertex is defined to be the sum of the weights of all geodesic paths passing through that vertex.

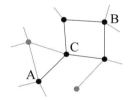

Vertices A and B are connected by two geodesic paths. Vertex C lies on both paths.

Note that the geodesic paths between a pair of vertices need not be vertex-independent, meaning they may pass through some of the same vertices (see figure). If two or more paths pass through the same vertex then the betweenness sum includes contributions from each of them. Thus if there are, say, three geodesic paths between a given pair of vertices and two of them pass through a particular vertex, then they contribute $\frac{2}{3}$ to that vertex's betweenness.

Formally, we can express the betweenness for a general network by redefining n_{st}^i to be the number of geodesic paths from s to t that pass through i. And we define g_{st} to be the total number of geodesic paths from s to t. Then the betweenness centrality of vertex i is

$$x_i = \sum_{st} \frac{n_{st}^i}{g_{st}}, \tag{7.36}$$

where we adopt the convention that $n_{st}^i/g_{st} = 0$ if both n_{st}^i and g_{st} are zero. This definition is equivalent to our message-passing thought experiment above, in which messages pass between all pairs of vertices in a network at the same average rate, traveling along shortest paths, and in the case of several shortest paths between a given pair of vertices they choose at random between those several paths. Then x_i is proportional to the average rate at which traffic passes though vertex i.

Betweenness centrality can be applied to directed networks as well. In a directed network the shortest path between two vertices depends, in general, on the direction you travel in. The shortest path from A to B is different from the shortest path from B to A. Indeed there may be a path in one direction and no path at all in the other. Thus it is important in a directed network explicitly to include the path counts in either direction between each vertex pair. The definition in Eq. (7.36) already does this and so, as mentioned above, we can use the same definition without modification for the directed case. This is one reason why we prefer this definition to other slight variants that are sometimes used.

Although the generalization of betweenness to directed networks is straightforward, however, it is rarely if ever used, so we won't discuss it further here, concentrating instead on the much more common undirected case.

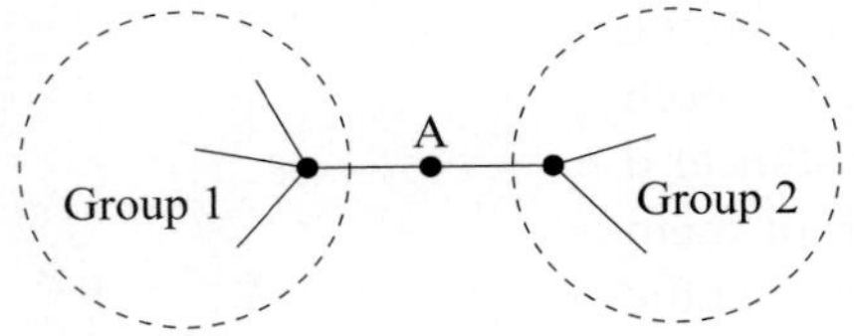

Betweenness centrality differs from the other centrality measures we have considered in being not principally a measure of how well-connected a vertex is. Instead it measures how much a vertex falls "between" others. Indeed a vertex can have quite low degree, be connected to others that have low degree, even be a long way from others on average, and still have high betweenness. Consider the situation depicted in Fig. 7.2. Vertex A lies on a bridge between two groups within a network. Since any shortest path (or indeed any path whatsoever) between a vertex in one group and a vertex in the other must pass along this bridge, A acquires very high betweenness, even though it is itself on the periphery of both groups and in other respects may be not well connected: probably A would not have particularly impressive values for eigenvector or closeness centrality, and its degree centrality is only 2, but nonetheless it might have a lot of influence in the network as a result of its control over the flow of information between others. Vertices in roles

Figure 7.2: A low-degree vertex with high betweenness. In this sketch of a network, vertex A lies on a bridge joining two groups of other vertices. All paths between the groups must pass through A, so it has a high betweenness even though its degree is low.

like this are sometimes referred to in the sociological literature as *brokers*.[12]

Betweenness centrality also has another interesting property: its values are typically distributed over a wide range. The maximum possible value for the betweenness of a vertex occurs when the vertex lies on the shortest path between every other pair of vertices. This occurs for the central vertex in a *star graph*, a network composed of a vertex attached to $n - 1$ others by single edges. In this situation the central vertex lies on all n^2 shortest paths between vertex pairs except for the $n - 1$ paths from the peripheral vertices to themselves. Thus the betweenness centrality of the central vertex is $n^2 - n + 1$. At the other end of the scale, the smallest possible value of betweenness in a network with a single component is $2n - 1$, since at a minimum each vertex lies on every path that starts or ends with itself. (There are $n - 1$ paths from a vertex to others, $n - 1$ paths from others to the vertex, and one path from the vertex to itself, for a total of $2(n - 1) + 1 = 2n - 1$.) This situation occurs, for instance, when a network has a "leaf" attached to it, a vertex connected to the rest of the network by just a single edge.

Thus the ratio of largest and smallest possible betweenness values is

$$\frac{n^2 - n + 1}{2n - 1} \simeq \tfrac{1}{2}n, \tag{7.37}$$

where the equality becomes exact in the limit of large n. Thus in theory there could be a factor of almost $\tfrac{1}{2}n$ between the largest and smallest betweenness centralities, which could become very large for large networks. In real networks the range is usually considerably smaller than this, but is nonetheless large and typically increasing with increasing n.

Taking again the example of the network of film actors from the previous section, the individual with the highest betweenness centrality in the largest component of the actor network is the great Spanish actor Fernando Rey, most famous in the English-speaking world for his 1971 starring role next to Gene Hackman in *The French Connection*.[13] Rey has a betweenness score of 7.47×10^8,

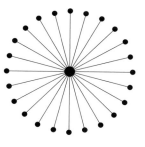

A star graph.

---

[12]Much of sociological literature concerns power or "social capital." It may seem ruthless to think of individuals exploiting their control over other people's information to gain the upper hand on them, but it may also be realistic. At least in situations where there is a significant pay-off to having such an upper hand (like business relationships, for example), it is reasonable to suppose that notions of power derived from network structure really do play into people's manipulations of the world around them.

[13]It is perhaps no coincidence that the highest betweenness belongs to an actor who appeared in both European and American films, played roles in several different languages, and worked extensively in both film and television, as well as on stage. Rey was the archetypal "broker," with a career that made him a central figure in several different arms of the entertainment business that otherwise overlap relatively little.

while the lowest score of any actor[14] in the large component is just 8.91×10^5. Thus there is a ratio of almost a thousand between the two limits—a much larger dynamic range than the ratio of 3.6 we saw in the case of closeness centrality. One consequence of this is that there are very clear winners and losers in the betweenness centrality competition. The second highest betweenness in the actor network is that of Christopher Lee (again), with 6.46×10^8, a 14% percent difference from winner Fernando Rey. Although betweenness values may shift a little as new movies are made and new actors added to the network, the changes are typically small compared with these large gaps between the leaders, so that the ordering at the top of the list changes relatively infrequently, giving betweenness centrality results a robustness not shared by those for closeness centrality.

The values of betweenness calculated here are raw path counts, but it is sometimes convenient to normalize betweenness in some way. Several of the standard computer programs for network analysis, such as Pajek and UCINET, perform such normalizations. One natural choice is to normalize the path count by dividing by the total number of (ordered) vertex pairs, which is n^2, so that betweenness becomes the fraction (rather than the number) of paths that run through a given vertex:[15]

$$x_i = \frac{1}{n^2} \sum_{st} \frac{n_{st}^i}{g_{st}}. \tag{7.38}$$

With this definition, the values of the betweenness lie strictly between zero and one.

Some other variations on the betweenness centrality idea are worth mentioning. Betweenness gets at an important idea in network analysis, that of the flow of information or other traffic and of the influence vertices might have over that flow. However, betweenness as defined by Freeman is based on counting only the shortest paths between vertex pairs, effectively assuming that all or at least most traffic passes along those shortest paths. In reality traf-

---

[14]This score is shared by many actors. It is the minimum possible score of $2n - 1$ as described above.

[15]Another possibility, proposed by Freeman [128] in his original paper on betweenness, is to divide by the maximum possible value that betweenness can take on any network of size n, which, as mentioned above, occurs for the central vertex in a star graph. The resulting expression for between is then

$$x_i = \frac{1}{n^2 - n + 1} \sum_{st} \frac{n_{st}^i}{g_{st}}.$$

We, however, prefer Eq. (7.38), which we find easier to interpret, although the difference between the two becomes small anyway in the limit of large n.

fic flows along paths other than the shortest in many networks. Most of us, for instance, will have had the experience of hearing news about one of our friends not from that friend directly but from another mutual acquaintance—the message has passed along a path of length two via the mutual acquaintance, rather than along the direct (geodesic) path of length one.

A version of betweenness centrality that makes some allowance for effects like this is the *flow betweenness*, which was proposed by Freeman *et al.* [130] and is based on the idea of maximum flow. Imagine each edge in a network as a pipe that can carry a unit flow of some fluid. We can ask what the maximum possible flow then is between a given source vertex s and target vertex t through these pipes. In general the answer is that more than a single unit of flow can be carried between source and target by making simultaneous use of several different paths through the network. The flow betweenness of a vertex i is defined according to Eq. (7.35), but with n_{st}^i being now the amount of flow through vertex i when the maximum flow is transmitted from s to t.

See Section 6.12 for a discussion of maximum flow in networks.

As we saw in Section 6.12, the maximum flow between vertices s and t is also equal to the number of edge-independent paths between them. Thus another way equivalent to look at the flow betweenness would be to consider n_{st}^i to be the number of independent paths between s and t that run through vertex i.

A slight problem arises because the independent paths between a given pair of vertices are not necessarily unique. For instance, the network shown in Fig. 7.3 has two edge-independent paths between s and t but we have two choices about what those paths are, either the paths denoted by the solid arrows, or those denoted by the dashed ones. Furthermore, our result for the flow betweenness will depend on which choice we make; the vertices labeled A and B fall on one set of paths but not the other. To get around this problem, Freeman *et al.* define the flow through a vertex for their purposes to be the *maximum* possible flow over all possible choices of paths, or equivalently the maximum number of independent paths. Thus in the network of Fig. 7.3, the contribution of the flow between s and t to the betweenness of vertex A would be 1, since this is the maximum value it takes over all possible choices of flow paths.

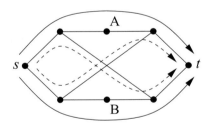

Figure 7.3: Edge-independent paths in a small network. The vertices s and t in this network have two independent paths between them, but there are two distinct ways of choosing those paths, represented by the solid and dashed curves.

In terms of our information analogy, one can think of flow betweenness as measuring the betweenness of vertices in a network in which a maximal amount of information is continuously pumped between all sources and targets. Flow betweenness takes account of more than just the geodesic paths between vertices, since flow can go along non-geodesic paths as well as geodesic ones. (For example, the paths through vertices A and B in the example above are not geodesic.) Indeed, in some cases *none* of the paths that appear in the solution of the maximum flow problem are geodesic paths, so geodesic paths may not be counted at all by this measure.

But this point highlights a problem with flow betweenness: although it typically counts more paths than the standard shortest-path betweenness, flow betweenness still only counts a subset of possible paths, and some important ones (such as geodesic paths) may be missed out altogether. One way to look at the issue is that both shortest-path betweenness and flow betweenness assume flows that are optimal in some sense—passing only along shortest paths in the first case and maximizing total flow in the second. Just as there is no reason to suppose that information or other traffic always takes the shortest path, there is no reason in general to suppose it should act to maximize flow (although of course there may be special cases in which it does).

A betweenness variant that does count all paths is the *random-walk betweenness* [243]. In this variant traffic between vertices s and t is thought of as performing an (absorbing) random walk that starts at vertex s and continues until it reaches vertex t. The betweenness is defined according to $x_i = \sum_{st} n^i_{st}$ but with n^i_{st} now being the number of times that the random walk from s to t passes through i on its journey, averaged over many repetitions of the walk.

See Section 6.14 for a discussion of random walks.

Note that in this case $n^i_{st} \neq n^i_{ts}$ in general, even on an undirected network. For instance, consider this portion of a network:

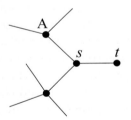

A random walk from s to t may pass through vertex A before returning to s and stepping thence to t, but a walk from t to s will never pass through A because its first step away from t will always take it to s and then the walk will finish.

Since every possible path from s to t occurs in a random walk with some probability (albeit a very small one) the random-walk betweenness includes contributions from all paths.[16] Note, however, that different paths appear in general with different probabilities, so paths do not contribute equally to the

---

[16] All paths, that is, that terminate at the target vertex t the first time they reach it. Since we use an absorbing random walk, paths that visit the target, move away again, and then return are not included in the random-walk betweenness.

betweenness scores, longer paths typically making smaller contributions than shorter ones, a bias that is plausible in some but by no means all cases.

Random walk betweenness would be an appropriate betweenness measure for traffic that traverses a network with no idea of where it is going—it simply wanders around at random until it reaches its destination. Shortest-path betweenness is the exact opposite. It is the appropriate measure for information that knows exactly where it is going and takes the most direct path to get there. It seems likely that most real-world situations fall somewhere in between these two extremes. However, it is found in practice [243] that the two measures often give quite similar results, in which case one can with reasonable justification assume that the "correct" answer, the one lying between the limits set by the shortest-path and random-walk measures, is similar to both. In cases where the two differ by a considerable margin, however, we should be wary of attributing too much authority to either measure—there is no guarantee that either is telling us a great deal about true information flow in the network.

Other generalizations of betweenness are also possible, based on other models of diffusion, transmission, or flow along network edges. We refer the interested reader to the article by Borgatti [51], which draws together many of the possibilities into a broad general framework for betweenness measures.

7.8 GROUPS OF VERTICES

Many networks, including social and other networks, divide naturally into groups or communities. Networks of people divide into groups of friends, coworkers, or business partners; the World Wide Web divides into groups of related web pages; biochemical networks divide into functional modules, and so forth. The definition and analysis of groups within networks is a large and fruitful area of network theory. In Chapter 11 we discuss some of the sophisticated computer methods that have been developed for detecting groups, such as hierarchical clustering and spectral partitioning. In this section we discuss some simpler concepts of network groups which can be useful for probing and describing the local structure of networks. The primary constructs we look at are cliques, plexes, cores, and components.

7.8.1 CLIQUES, PLEXES, AND CORES

A *clique* is a maximal subset of the vertices in an undirected network such that every member of the set is connected by an edge to every other. The word "maximal" here means that there is no other vertex in the network that can

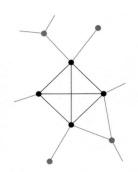

A clique of four vertices within a network.

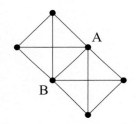

Two overlapping cliques. Vertices A and B in this network both belong to two cliques of four vertices.

be added to the subset while preserving the property that every vertex is connected to every other. Thus a set of four vertices in a network would be a clique if (and only if) each of the four is directly connected by edges to the other three *and* if there is no other vertex anywhere in the network that could be added to make a group of five vertices all connected to each other. Note that cliques can overlap, meaning that they can share one or more of the same vertices.

The occurrence of a clique in an otherwise sparse network is normally an indication of a highly cohesive subgroup. In a social network, for instance, one might encounter a set of individuals each of whom was acquainted with each of the others, and such a clique would probably indicate that the individuals in question are closely connected—a set of coworkers in an office for example or a group of classmates in a school.

However, it's also the case that many circles of friends form only near-cliques, rather than perfect cliques. There may be some members of the group who are unacquainted, even if most members know one another. The requirement that every possible edge be present within a clique is a very stringent one, and it seems natural to consider how we might relax this requirement. One construct that does this is the *k-plex*. A *k*-plex of size n is a maximal subset of n vertices within a network such that each vertex is connected to at least $n - k$ of the others. If $k = 1$, we recover the definition of an ordinary clique—a 1-plex is the same as a clique. If $k = 2$, then each vertex must be connected to all or all-but-one of the others. And so forth.[17] Like cliques, *k*-plexes can overlap one another; a single vertex can belong to more than one *k*-plex.

The *k*-plex is a useful concept for discovering groups within networks: in real life many groups in social and other networks form *k*-plexes. There is no solid rule about what value *k* should take. Experimentation starting from small values is the usual way to proceed. Smaller values of *k* tend to be meaningful for smaller groups, whereas in large groups the smaller values impose too stringent a constraint but larger values often give useful results. This suggests another possible generalization of the clique idea: one could specify that each member be connected to a certain *fraction* of the others, say 75% or 50%. (As far as we know, this variant doesn't have a name and it is not in wide use, but perhaps it should be.)

Many other variations on the clique idea have been proposed in the literature. For instance Flake *et al.* [122] proposed a definition of a group as a subset

---

[17]This definition is slightly awkward to remember, since the members of a *k*-plex are allowed to be unconnected to $k - 1$ other members and not *k*. It would perhaps have been more sensible to define *k* such that a 0-plex was equivalent to a normal clique, but for better or worse we are stuck with the definition we have.

of vertices such that each has at least as many connections to vertices inside the group as to vertices outside. Radicchi *et al.* [276] proposed a weaker definition of a group as a subset of vertices such that the total number of connections of all vertices in the group to others in the group is greater than the total number of connections to vertices outside.[18]

Another concept closely related to the k-plex is the k-core. A k-core is a maximal subset of vertices such that each is connected to at least k others in the subset.[19] It should be obvious (or you can easily prove it for yourself) that a k-core of n vertices is also an $(n - k)$-plex. However, the set of all k-cores for a given value of k is not the same as the set of all k-plexes for any value of k, since n, the size of the group, can vary from one k-core to another. Also, unlike k-plexes (and cliques), k-cores cannot overlap, since by their definition two k-cores that shared one or more vertices would just form a single larger k-core.

The k-core is of particular interest in network analysis for the practical reason that it is very easy to find the set of all k-cores in a network. A simple algorithm is to start with your whole network and remove from it any vertices that have degree less than k, since clearly such vertices cannot under any circumstances be members of a k-core. In so doing, one will normally also reduce the degrees of some other vertices in the network—those that were connected to the vertices just removed. So we then go through the network again to see if there are any more vertices that now have degree less than k and if there are we remove those too. And so we proceed, repeatedly pruning the network to remove vertices with degree less than k until no such vertices remain.[20] What is left over will, by definition, be a k-core or a set of k-cores, since each vertex is connected to at least k others. Note that we are not necessarily left with a *single* k-core—there's no guarantee that the network will be connected once we are done pruning it, even if it was connected to start with.

Two other generalizations of cliques merit a brief mention. A *k-clique* is a maximal subset of vertices such that each is no more than a distance k away from any of the others via the edges of the network. For $k = 1$ this just recovers

---

[18]Note that for the purposes of this latter definition, an edge between two vertices A and B within the group counts as *two* connections, one from A to B and one from B to A.

[19]We have to be careful about the meaning of the word "maximal" here. It is possible to have a group of vertices such that each is connected to at least k others and no *single* vertex can be added while retaining this property, but it may be possible to add more than one vertex. Such groups, however, are not considered to be k-cores. A group is only a k-core if it is not a subset of any larger group that is a k-core.

[20]A closely related process, *bootstrap percolation*, has also been studied in statistical physics, principally on regular lattices.

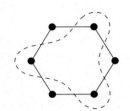

The outlined set of three vertices in this network constitute a 2-clique, but one that is not connected via paths within the 2-clique.

the definition of an ordinary clique. For larger k it constitutes a relaxation of the stringent requirements of the usual clique definition. Unfortunately it is not a very well-behaved one, since a k-clique by this definition need not be connected via paths that run within the subset (see figure). If we restrict ourselves to paths that run only within the subset then the resulting object is known as either a *k-clan* or a *k-club*. (The difference between the two lies in whether we impose the restriction that paths stay within the group from the outset, or whether we first find k-cliques and then discard those with outside paths. The end results can be different in the two cases. For more details see Wasserman and Faust [320].).

7.8.2 COMPONENTS AND k-COMPONENTS

In Section 6.11 we introduced the concept of a component. A component in an undirected network is a maximal subset of vertices such that each is reachable by some path from each of the others. A useful generalization of this concept is the k-component. A *k-component* (sometimes also called a *k-connected component*) is a maximal subset of vertices such that each is reachable from each of the others by at least k vertex-independent paths—see Fig. 7.4. (Recall that two paths are said to be vertex-independent if they share none of the same vertices, except the starting and ending vertices—see Section 6.12.) For the common special cases $k = 2$ and $k = 3$, k-components are also called *bicomponents* and *tricomponents* respectively.

A 1-component by this definition is just an ordinary component—there is at least one path between every pair of vertices—and k-components for $k \geq 2$ are nested within each other. A 2-component or bicomponent, for example, is necessarily a subset of a 1-component, since any pair of vertices that are connected by at least two paths are also connected by at least one path. Similarly a tricomponent is necessarily a subset of a bicomponent, and so forth. (See Fig. 7.4 again.)

As discussed in Section 6.12, the number of vertex-independent paths between two vertices is equal to the size of the vertex cut set between the same two vertices, i.e., the number of vertices that would have to be removed in order to disconnect the two. So another way of defining a k-component would be to say that it is a maximal subset of vertices such that no pair of vertices can be disconnected from each other by removing less than k vertices.

A variant of the k-component can also be defined using edge-independent paths, so that vertices are in the same k-component if they are connected by k or more edge-independent paths, or equivalently if they cannot be disconnected by the removal of less than k edges. In principal this variant could be useful in

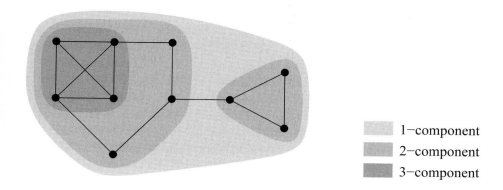

1–component
2–component
3–component

Figure 7.4: The k-components in a small network. The shaded regions denote the k-components in this small network, which has a single 1-component, two 2-components, one 3-component, and no k-components for any higher value of k. Note that the k-components are nested within one another, the 2-components falling inside the 1-component and the 3-component falling inside one of the 2-components.

certain circumstances but in practice it is rarely used.

The idea of a k-component is a natural one in network analysis, being connected with the idea of network robustness. For instance, in a data network such as the Internet, the number of vertex-independent paths between two vertices is also the number of independent routes that data might take between the same two vertices, and the size of the cut set between them is the number of vertices in the network—typically routers—that would have to fail or otherwise be knocked out to sever the data connection between the two endpoints. Thus a pair of vertices connected by two independent paths cannot be disconnected from one another by the failure of any single router. A pair of vertices connected by three paths cannot be disconnected by the failure of any two routers. And so forth. A k-component with $k \geq 2$ in a network like the Internet is a subset of the network that has robust connectivity in this sense. One would hope, for instance, that most of the network backbone—the system of high volume world-spanning links that carry long-distance data (see Section 2.1)—is a k-component with high k, so that it would be difficult for points on the backbone to lose connection with one another.

Note that for $k \geq 3$, the k-components in a network can be non-contiguous (see figure). Ordinary components (1-components) and bicomponents, by contrast, are always contiguous. Within the social networks literature, where non-contiguous components are often considered undesirable, k-components are

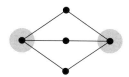

The two highlighted vertices in this network form a tricomponent, even though they are not directly connected to each other. The other three vertices are not in the tricomponent.

197

sometimes defined slightly differently: a k-component is defined to be a maximal subset of vertices such that every pair in the set is connected by at least k vertex-independent paths *that themselves are contained entirely within the subset*. This definition rules out non-contiguous k-components, but it is also mathematically and computationally more difficult to work with than the standard definition. For this reason, and because there are also plenty of cases in which it is appropriate to count non-contiguous k-components, the standard definition remains the most widely used one in fields other than sociology.

7.9 TRANSITIVITY

A property very important in social networks, and useful to a lesser degree in other networks too, is *transitivity*. In mathematics a relation "$\circ$" is said to be transitive if $a \circ b$ and $b \circ c$ together imply $a \circ c$. An example would be equality. If $a = b$ and $b = c$, then it follows that $a = c$ also, so "$=$" is a transitive relation. Other examples are "greater than," "less than," and "implies."

In a network there are various relations between pairs of vertices, the simplest of which is "connected by an edge." If the "connected by an edge" relation were transitive it would mean that if vertex u is connected to vertex v, and v is connected to w, then u is also connected to w. In common parlance, "the friend of my friend is also my friend." Although this is only one possible kind of network transitivity—other network relations could be transitive too—it is the only one that is commonly considered, and networks showing this property are themselves said to be transitive. This definition of network transitivity could apply to either directed or undirected networks, but let us take the undirected case first, since it's simpler.

Perfect transitivity only occurs in networks where each component is a fully connected subgraph or clique, i.e., a subgraph in which all vertices are connected to all others.[21] Perfect transitivity is therefore pretty much a useless concept in networks. However, *partial* transitivity can be very useful. In many networks, particularly social networks, the fact that u knows v and v knows w

---

[21]To see this suppose we have a component that is perfectly transitive but not a clique, i.e., there is at least one pair of vertices u, w in the component that are not directly connected by an edge. Since u and w are in the same component they must therefore be connected by some path of length greater than one, $u, v_1, v_2, v_3, \ldots, w$. Consider the first two links in this path. Since u is connected by an edge to v_1 and v_1 to v_2 it follows that u must be connected to v_2 if the network is perfectly transitive. Then consider the next two links. Since u is connected to v_2 and v_2 to v_3 it follows that u must be connected to v_3. Repeating the argument all the way along the path, we can then see that u must be connected by an edge to w. But this violates the hypothesis that u and w are not directly connected. Hence no perfectly transitive components exist that are not cliques.

doesn't *guarantee* that u knows w, but makes it much more likely. The friend of my friend is not necessarily my friend, but is far more likely to be my friend than some randomly chosen member of the population.

We can quantify the level of transitivity in a network as follows. If u knows v and v knows w, then we have a path uvw of two edges in the network. If u also knows w, we say that the path is *closed*—it forms a loop of length three, or a triangle, in the network. In the social network jargon, u, v, and w are said to form a *closed triad*. We define the *clustering coefficient*[22] to be the fraction of paths of length two in the network that are closed. That is, we count all paths of length two, and we count how many of them are closed, and we divide the second number by the first to get a clustering coefficient C that lies in the range from zero to one:

The path uvw (solid edges) is said to be closed if the third edge directly from u to w is present (dashed edge).

$$C = \frac{\text{(number of closed paths of length two)}}{\text{(number of paths of length two)}}. \tag{7.39}$$

$C = 1$ implies perfect transitivity, i.e., a network whose components are all cliques. $C = 0$ implies no closed triads, which happens for various topologies, such as a tree (which has no closed loops of any kind—see Section 6.7) or a square lattice (which has closed loops with even numbers of vertices only and no closed triads).

Note that paths in networks, as defined in Section 6.10 have a direction and two paths that traverse the same edges but in opposite directions are counted separately in Eq. (7.39). Thus uvw and wvu are distinct paths and are counted separately. Similarly, closed paths are counted separately in each direction.[23]

An alternative way to write the clustering coefficient is

$$C = \frac{\text{(number of triangles)} \times 6}{\text{(number of paths of length two)}}. \tag{7.40}$$

Why the factor of six? It arises because each triangle in the network gets counted six times over when we count up the number of closed paths of length two. Suppose we have a triangle uvw. Then there are six paths of length two

---

[22]It's not entirely clear why the clustering coefficient has the name it has. The name doesn't appear to be connected with the earlier use of the word clustering in social network analysis to describe groups or clusters of vertices (see Section 11.11.2). The reader should be careful to avoid confusing these two uses of the word.

[23]In fact, we could count each path just in one direction, provided we did it for both the numerator and denominator of Eq. (7.39). Doing so would decrease both counts by a factor of two, but the factors would cancel and the end result would be the same. In most cases, and particularly when writing computer programs, it is easier to count paths in both directions—it avoids having to remember which paths you have counted before.

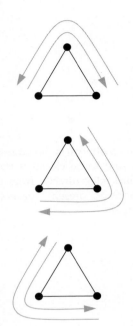

A triangle contains six distinct paths of length two, all of them closed.

in it: uvw, vwu, wuv, wvu, vuw, and uwv. Each of these six is closed, so the number of closed paths is six times the number of triangles.

Yet another way to write the clustering coefficient would be to note that if we have a path of length two, uvw, then it is also true to say that vertices u and w have a common neighbor in v—they share a mutual acquaintance in social network terms. If the triad uvw is closed then u and w are themselves acquainted, so the clustering coefficient can be thought of also as the fraction of pairs of people with a common friend who are themselves friends or equivalently as the mean probability that two people with a common friend are themselves friends. This is perhaps the most common way of defining the clustering coefficient. In mathematical notation:

$$C = \frac{(\text{number of triangles}) \times 3}{(\text{number of connected triples})}. \tag{7.41}$$

Here a "connected triple" means three vertices uvw with edges (u,v) and (v,w). (The edge (u,w) can be present or not.) The factor of three in the numerator arises because each triangle gets counted three times when we count the connected triples in the network. The triangle uvw for instance contains the triples uvw, vwu, and wuv. In the older social networks literature the clustering coefficient is sometimes referred to as the "fraction of transitive triples," which is a reference to this definition of the coefficient.

Social networks tend to have quite high values of the clustering coefficient. For example, the network of film actor collaborations discussed earlier has been found to have $C = 0.20$ [241]; a network of collaborations between biologists has been found to have $C = 0.09$ [236]; a network of who sends email to whom in a large university has $C = 0.16$ [103]. These are typical values for social networks. Some denser networks have even higher values, as high as 0.5 or 0.6. (Technological and biological networks by contrast tend to have somewhat lower values. The Internet at the autonomous system level, for instance, has a clustering coefficient of only about 0.01. This point is discussed in more detail in Section 8.6.)

In what sense are these clustering coefficients for social networks high? Well, let us assume, to make things simple, that everyone in a network has about the same number c of friends. Consider one of my friends in this network and suppose they pick *their* friends completely at random from the whole population. Then the chance that one of their c friends happens to be a particular one of my other friends would be c/n, where n is the size of the network. Thus in this network the probability of two of my friends being acquainted, which is by definition the clustering coefficient, would be just c/n. Of course it is not the case that everyone in a network has the same number of friends, and we will see how to perform better calculations of the clustering coefficient later (Section 13.4), but this crude calculation will serve our purposes for the moment.

For the networks cited above, the value of c/n is 0.0003 (film actors), 0.00001 (biology collaborations), and 0.00002 (email messages). Thus the measured clustering coefficients are *much* larger than this estimate based on the assumption of random network connections. Even though the estimate ignores, as we have said, any variation in the number of friends people have, the disparity between the calculated and observed values of the clustering coefficient is so large that it seems unlikely it could be eliminated just by allowing the number of friends to vary. A much more likely explanation is that our other assumption, that people pick their friends at random, is seriously flawed. The numbers suggest that there is a much greater chance that two people will be acquainted if they have another common acquaintance than if they don't.

Although this argument is admittedly crude, we will see in Section 8.6 how to make it more accurate and so show that our basic conclusion is indeed correct.

Some social networks, such as the email network above, are directed networks. In calculating clustering coefficients for direct networks, scientists have typically just ignored their directed nature and applied Eq. (7.41) as if the edges were undirected. It is however possible to generalize transitivity to take account of directed links. If we have a directed relation between vertices such as "u likes v" then we can say that a triple of vertices is closed or transitive if u likes v, v likes w, and also u likes w. (Note that there are many distinct ways for such a triple to be transitive, depending on the directions of the edges. The example given here is only one of six different possibilities.) One can calculate a clustering coefficient or fraction of transitive triples in the obvious fashion for the directed case, counting all directed paths of length two that are closed and dividing by the total number of directed paths of length two. For some reason, however, such measurements have not often appeared in the literature.

A transitive triple of vertices in a directed network.

7.9.1 LOCAL CLUSTERING AND REDUNDANCY

We can also define a clustering coefficient for a single vertex. For a vertex i, we define[24]

$$C_i = \frac{\text{(number of pairs of neighbors of } i \text{ that are connected)}}{\text{(number of pairs of neighbors of } i\text{)}}. \tag{7.42}$$

---

[24]The notation C_i is used for both the local clustering coefficient and the closeness centrality and we should be careful not to confuse the two.

That is, to calculate C_i we go through all distinct pairs of vertices that are neighbors of i in the network, count the number of such pairs that are connected to each other, and divide by the total number of pairs, which is $\frac{1}{2}k_i(k_i - 1)$ where k_i is the degree of i,. C_i is sometimes called the *local clustering coefficient* and it represents the average probability that a pair of i's friends are friends of one another.

Local clustering is interesting for several reasons. First, in many networks it is found empirically to have a rough dependence on degree, vertices with higher degree having a lower local clustering coefficient on average. This point is discussed in detail in Section 8.6.1.

Second, local clustering can be used as a probe for the existence of so-called "structural holes" in a network. While it is common in many networks, especially social networks, for the neighbors of a vertex to be connected among themselves, it happens sometimes that these expected connections between neighbors are missing. The missing links are called *structural holes* and were first studied in this context by Burt [60]. If we are interested in efficient spread of information or other traffic around a network, as we were in Section 7.7, then structural holes are a bad thing—they reduce the number of alternative routes information can take through the network. On the other hand structural holes can be a good thing for the central vertex i whose friends lack connections, because they give i power over information flow between those friends. If two friends of i are not connected directly and their information about one another comes instead via their mutual connection with i then i can control the flow of that information. The local clustering coefficient measures how influential i is in this sense, taking lower values the more structural holes there are in the network around i. Thus local clustering can be regarded as a type of centrality measure, albeit one that takes small values for powerful individuals rather than large ones.

In this sense, local clustering can also be thought of as akin to the betweenness centrality of Section 7.7. Where betweenness measures a vertex's control over information flowing between all pairs of vertices in its component, local clustering is like a local version of betweenness that measures control over flows between just the immediate neighbors of a vertex. One measure is not necessarily better than another. There may be cases in which we want to take all vertices into account and others where we want to consider only immediate neighbors—the choice will depend on the particular questions we want to answer. It is worth pointing out however that betweenness is much more computationally intensive to calculate than local clustering (see Section 10.3.6), and that in practice betweenness and local clustering are strongly correlated [60]. There may in many cases be little to be gained by performing the more costly

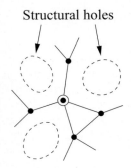

Structural holes

When the neighbors of a node are not connected to one another we say the network contains "structural holes."

full calculation of betweenness and much to be saved by sticking with clustering, given that the two contain much the same information.[25]

In his original studies of structural holes, Burt [60] did not in fact make use of the local clustering coefficient as a measure of the presence of holes.[26] Instead, he used another measure, which he called *redundancy*. The original definition of redundancy was rather complicated, but Borgatti [50] has shown that it can be simplified to the following: the redundancy R_i of a vertex i is the mean number of connections from a neighbor of i to other neighbors of i. Consider the example shown in Fig. 7.5 in which vertex i has four neighbors. Each of those four *could* be acquainted with any of the three others, but in this case none of them is connected to all three. One is connected to none of the others, two are connected to one other, and the last is connected to two others. The redundancy is the average of these numbers $R_i = \frac{1}{4}(0+1+1+2) = 1$. The minimum possible value of the redundancy of a vertex is zero and the maximum is $k_i - 1$, where k_i is the degree of vertex i.

Figure 7.5: Redundancy. The neighbors of the central vertex in this figure have 0, 1, 1, and 2 connections to other neighbors respectively. The redundancy is the mean of these values: $R_i = \frac{1}{4}(0+1+1+2) = 1$.

It's probably obvious that R_i is related to the local clustering C_i. To see precisely what the relation is, we note that if the average number of connections from a friend of i to other friends is R_i, then the total number of connections between friends is $\frac{1}{2}k_i R_i$. And the total number of pairs of friends of i is $\frac{1}{2}k_i(k_i - 1)$. The local clustering coefficient, Eq. (7.42), is the ratio of these two quantities:

$$C_i = \frac{\frac{1}{2}k_i R_i}{\frac{1}{2}k_i(k_i - 1)} = \frac{R_i}{k_i - 1}. \tag{7.43}$$

Given that $k_i - 1$ is the maximum value of R_i, the local clustering coefficient can be thought of as simply a version of the redundancy rescaled to have a maximum value of 1. Applying Eq. (7.43) to the example of Fig. 7.5 implies that the local clustering coefficient for the central vertex should be $C_i = \frac{1}{3}$, and the reader can easily verify that this is indeed the case.

A third context in which the local clustering coefficient arises is in the calculation of the global clustering coefficient itself. Watts and Strogatz [323] proposed calculating a clustering coefficient for an entire network as the mean of

---

[25] As an example, in Section 11.11.1 we study methods for partitioning networks into clusters or communities and we will see that effective computer algorithms for this task can be created based on betweenness measures, but that almost equally effective and much faster algorithms can be created based on local clustering.

[26] Actually, the local clustering coefficient hadn't yet been invented. It was first proposed to this author's knowledge by Watts [321] a few years later.

the local clustering coefficients for each vertex:

$$C_{WS} = \frac{1}{n} \sum_{i=1}^{n} C_i, \tag{7.44}$$

where n is the number of vertices in the network. This is a different definition for the clustering coefficient from the one given earlier, Eq. (7.41), and the two definitions are not equivalent. Furthermore, they can give substantially different numbers for a given network and because both definitions are in reasonably common use this can give rise to confusion. We favor our first definition for C, Eq. (7.41), because it has a simple interpretation and because it is normally easier to calculate. Also the second definition, Eq. (7.44), tends to be dominated by vertices with low degree, since they have small denominators in Eq. (7.42), and the measure thus gives a rather poor picture of the overall properties of any network with a significant number of such vertices.[27] It's worth noting, however, that the definition of Eq. (7.44) was actually proposed before Eq. (7.41) and, perhaps because of this, it finds moderately wide use in network studies. So you need at least to be aware of both definitions and clear which is being used in any particular situation.

7.10 RECIPROCITY

The clustering coefficient of Section 7.9 measures the frequency with which loops of length three—triangles—appear in a network. Of course, there is no reason why one should concentrate only on loops of length three, and people have occasionally looked at the frequency of loops of length four or more [44,61,133,140,238]. Triangles occupy a special place however because in an undirected simple graph the triangle is the shortest loop we can have (and usually the most commonly occurring). However, in a *directed* network this is not the case. In a directed network, we can have loops of length two—a pair of vertices between which there are directed edges running in both directions—and it is interesting to ask about the frequency of occurrence of these loops also.

A loop of length two in a directed network.

The frequency of loops of length two is measured by the *reciprocity*, and tells you how likely it is that a vertex that you point to also points back at you. For instance, on the World Wide Web if my web page links to your web page, how likely is it, on average, that yours link back again to mine? In general, it's found

---

[27] As discussed in Section 8.6.1, vertices with low degree tend to have high values of C_i in most networks and this means that C_{WS} is usually larger than the value given by Eq. (7.41), sometimes much larger.

that you are much more likely to link to me if I link to you than if I don't. (That probably isn't an Earth-shattering surprise, but it's good to know when the data bear out one's intuitions.) Similarly in friendship networks, such as the networks of schoolchildren described in Section 3.2 where respondents were asked to name their friends, it is much more likely that you will name me if I name you than if I do not.

If there is a directed edge from vertex i to vertex j in a directed network and there is also an edge from j to i then we say the edge from i to j is *reciprocated*. (Obviously the edge from j to i is also reciprocated.) Pairs of edges like this are also sometimes called *co-links*, particularly in the context of the World Wide Web [104].

The reciprocity r is defined as the fraction of edges that are reciprocated. Noting that the product of adjacency matrix elements $A_{ij}A_{ji}$ is 1 if and only if there is an edge from i to j and an edge from j to i and is zero otherwise, we can sum over all vertex pairs i, j to get an expression for the reciprocity:

$$r = \frac{1}{m} \sum_{ij} A_{ij}A_{ji} = \frac{1}{m} \operatorname{Tr} \mathbf{A}^2, \tag{7.45}$$

where m is, as usual, the total number of (directed) edges in the network.

Consider for example this small network of four vertices:

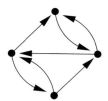

There are seven directed edges in this network and four of them are reciprocated, so the reciprocity is $r = \frac{4}{7} \simeq 0.57$. In fact, this is about the same value as seen on the World Wide Web. There is about a 57% percent chance that if web page A links to web page B then B also links back to A.[28] As another example, in a study of a network of who has whom in their email address book it was found that the reciprocity was about $r = 0.23$ [248].

---

[28]This figure is an unusually high one among directed networks, but there are reasons for it. One is that many of the links between web pages are between pages on the same website, and it is common for such pages to link to each other. If you exclude links between pages on the same site the value of the reciprocity is lower.

7.11 SIGNED EDGES AND STRUCTURAL BALANCE

In some social networks, and occasionally in other networks, edges are allowed to be either "positive" or "negative." For instance, in an acquaintance network we could denote friendship by a positive edge and animosity by a negative edge:

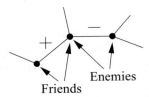

One could also consider varying degrees of friendship or animosity—networks with more strongly positive or negative edges in them—but for the moment let's stick to the simple case where each edge is in just one of two states, positive or negative, like or dislike. Such networks are called *signed networks* and their edges are called *signed edges*.

It is important to be clear here that a negative edge is not the same as the absence of an edge. A negative edge indicates, for example, two people who interact regularly but dislike each other. The absence of an edge represents two people who do not interact. Whether they would like one another if they did interact is not recorded.

Now consider the possible configurations of three edges in a triangle in a signed network, as depicted in Fig. 7.6. If "+" and "−" represent like and dislike, then we can imagine some of these configurations creating social problems if they were to arise between three people in the real world. Configuration (a) is fine: everyone likes everyone else. Configuration (b) is probably also fine, although the situation is more subtle than (a). Individuals u and v like one another and both dislike w, but the configuration can still be regarded as stable in the sense that u and v can agree over their dislike of w and get along just fine, while w hates both of them. No one is conflicted about their allegiances.

Put another way, w is u's enemy and v is w's enemy, but there is no problem with u and v being friends if one considers that the "enemy of my enemy is my friend."

Configuration (c) however could be problematic. Individual u likes individual v and v likes w, but u thinks w is an idiot. This is going to place a strain on the friendship between u and v because u thinks v's friend is an idiot. Alternatively, from the point of view of v, v has two friends, u and w and they don't get along, which puts v in an awkward position. In many real-life situations of this kind the tension would be resolved by one of the acquaintances being

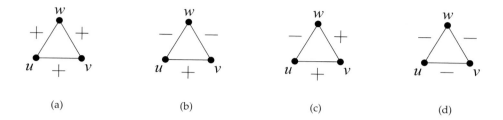

(a)   (b)   (c)   (d)

Figure 7.6: Possible triad configurations in a signed network. Configurations (a) and (b) are balanced and hence relatively stable, but configurations (c) and (d) are unbalanced and liable to break apart.

broken, i.e., the edge would be removed altogether. Perhaps v would simply stop talking to one of his friends, for instance.

Configuration (d) is somewhat ambiguous. On the one hand, it consists of three people who all dislike each other, so no one is in doubt about where things stand: everyone just hates everyone else. On the other hand, the "enemy of my enemy" rule does not apply here. Individuals u and v might like to form an alliance in recognition of their joint dislike of w, but find it difficult to do so because they also dislike each other. In some circumstances this might cause tension. (Think of the uneasy alliance of the US and Russia against Germany during World War II, for instance.) But what one can say definitely is that configuration (d) is often unstable. There may be little reason for the three to stay together when none of them likes the others. Quite probably three enemies such as these would simply sever their connections and go their separate ways.

The feature that distinguishes the two stable configurations in Fig. 7.6 from the unstable ones is that they have an even number of minus signs around the loop.[29] One can enumerate similar configurations for longer loops, of length four or greater, and again find that loops with even numbers of minus signs appear stable and those with odd numbers unstable.

This alone would be an observation of only slight interest, where it not for the intriguing fact that this type of stability really does appear have an effect on the structure of networks. In surveys it is found that the unstable configurations in Fig. 7.6, the ones with odd numbers of minus signs, occur

Two stable configurations in loops of length four.

---

[29]This is similar in spirit to the concept of "frustration" that arises in the physics of magnetic spin systems.

far less often in real social networks than the stable configurations with even numbers of minus signs.

Networks containing only loops with even numbers of minus signs are said to show *structural balance*, or sometimes just *balance*. An important consequence of balance in networks was proved by Harary [154]:

> A balanced network can be divided into connected groups of vertices such that all connections between members of the same group are positive and all connections between members of different groups are negative.

Note that the groups in question can consist of a single vertex or many vertices, and there may be only one group or there may be very many. Figure 7.7 shows a balanced network and its division into groups. Networks that can be divided into groups like this are said to be *clusterable*. Harary's theorem tells us that all balanced networks are clusterable.

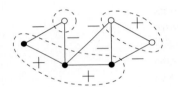

Figure 7.7: A balanced, clusterable network. Every loop in this network contains an even number of minus signs. The dotted lines indicate the division of the network into clusters such that all acquaintances within clusters have positive connections and all acquaintances in different clusters have negative connections.

Harary's theorem is straightforward to prove, and the proof is "constructive," meaning that it shows not only when a network is clusterable but also tells us what the groups are.[30] We consider initially only networks that are connected—they have just one component. In a moment we will relax this condition. We will color in the vertices of the network each in one of two colors, denoted by the open and filled circles in Fig. 7.7, for instance. We start with any vertex we please and color it with whichever color we please. Then we color in the others according to the following algorithm:

1. A vertex v connected by a positive edge to another u that has already been colored gets colored the same as u.
2. A vertex v connected by a negative edge to another u that has already been colored gets colored the opposite color from u.

For most networks it will happen in the course of this coloring process that we sometimes come upon a vertex whose color has already been assigned. When this happens there is the possibility of a conflict arising between the previously assigned color and the one that we would like to assign to it now according to the rules above. However, as we now show, this conflict only arises if the network as a whole is unbalanced.

If in coloring in a network we come upon a vertex that has already been colored in, it immediately implies that there must be another path by which that vertex can be reached from our starting point and hence that there is at least one, and possibly more than one, loop in the network to which this ver-

---

[30]The proof we give is not Harary's proof, which was quite different and not constructive.

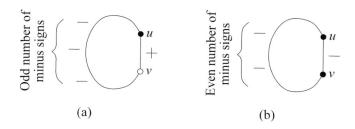

Figure 7.8: Proof that a balanced network is clusterable. If we fail to color a network in two colors as described in the text, then there must exist a loop in the network that has one or other of the two configurations shown here, both of which have an odd number of minus signs around them (counting the one between the vertices u and v), and hence the network is not balanced.

tex belongs—the loop consisting of the two paths between the starting point and the vertex. Since the network is balanced, every loop to which our vertex belongs must have an even number of negative edges around it. Now let us suppose that the color already assigned to the vertex is in conflict with the one we would like to assign it now. There are two ways in which this could happen, as illustrated in Fig. 7.8. In case (a), we color in a vertex u and then move onto its neighbor v, only to find that v has already been colored the opposite color to u, even though the edge between them is positive. This presents a problem. But if u and v are opposite colors, then around any loop containing them both there must be an *odd* number of minus signs, so that the color changes an odd number of times and ends up the opposite of what it started out as. And if there is an odd number of minus signs around the loop, then the network is not balanced.

In case (b) vertices u and v have the same color but the edge between them is negative. Again we have a problem. But if u and v are the same color then there must be an *even* number of negative edges around the rest of the loop connecting them which, along with the negative edge between u and v, gives us again an odd total number of negative edges around the entire loop, and hence the network is again not balanced.

Either way, if we ever encounter a conflict about what color a vertex should have then the network must be unbalanced. If the network is balanced, therefore, we will never encounter such a conflict and we will be able to color the entire network with just two colors while obeying the rules.

Once we have colored the network in this way, we can immediately deduce the identity of the groups that satisfy Harary's theorem: we simply divide the network into contiguous clusters of vertices that have the same color—see Fig. 7.7 again. In every such cluster, since all vertices have the same color, they must be joined by positive edges. Conversely, all edges that connected different clusters must be negative, since the clusters have different colors. (If they did not have different colors they would be considered the same cluster.)

Thus Harary's theorem is proved and at the same time we have deduced a method for constructing the clusters.[31] It only remains to extend the proof to networks that have more than one component, but this is trivial, since we can simply repeat the proof above for each component separately.

The practical importance of Harary's result rests on the fact that, as mentioned earlier, many real social networks are found naturally to be in a balanced or mostly balanced state. In such cases it would be possible, therefore, for the network to form into groups such that everyone likes others within their group with whom they have contact and dislikes those in other groups. It is widely assumed in social network theory that this does indeed often happen. Structural balance and clusterability in networks are thus a model for cliquishness or insularity, with people tending to stick together in like-minded groups and disdaining everyone outside their immediate community.

It is worth asking whether the inverse of Harary's clusterability theorem is also true. Is it also the case that a network that is clusterable is necessarily balanced? The answer is no, as this simple counter-example shows:

---

[31] As an interesting historical note, we observe that while Harary's proof of his theorem is perfectly correct, his interpretation of it was, in this author's opinion, erroneous. In his 1953 paper [154], he describes the meaning of the theorem in the following words: "A psychological interpretation of Theorem 1 is that a 'balanced group' consists of two highly cohesive cliques which dislike each other." (Harary is using the word "clique" in a non-technical sense here to mean a closed group of people, rather than in the graph theoretical sense of Section 7.8.1.) However, just because it is possible to color the network in two colors as described above does not mean the network forms two groups. Since the vertices of a single color are not necessarily contiguous, there are in general many groups of each color, and it seems unreasonable to describe these groups as forming a single "highly cohesive clique" when in fact they have no contact at all. Moreover, it is neither possible nor correct to conclude that the members of two groups of opposite colors dislike each other unless there is at least one edge connecting the two. If two groups of opposite colors never actually have any contact then it might be that they would get along just fine if they met. It's straightforward to prove that such an occurrence would lead to an unbalanced network, but Harary's statement says that the *present* balanced network implies dislike, and this is untrue. Only if the network were to remain balanced upon addition of one or more edges between groups of unlike colors would his conclusion be accurate.

In this network all three vertices dislike each other, so there is an odd number of minus signs around the loop, but there is no problem dividing the network into three clusters of one vertex each such that everyone dislikes the members of the other clusters. This network is clusterable but not balanced.

7.12 SIMILARITY

Another central concept in social network analysis is that of similarity between vertices. In what ways can vertices in a network be similar, and how can we quantify that similarity? Which vertices in a given network are most similar to one another? Which vertex v is most similar to a given vertex u? Answers to questions like these can help us tease apart the types and relationships of vertices in social networks, information networks, and others. For instance, one could imagine that it might be useful to have a list of web pages that are similar—in some appropriate sense—to another page that we specify. In fact, several web search engines already provide a feature like this: "Click here for pages similar to this one."

Similarity can be determined in many different ways and most of them have nothing to do with networks. For example, commercial dating and matchmaking services try to match people with others to whom they are similar by using descriptions of people's interests, background, likes, and dislikes. In effect, these services are computing similarity measures between people based on personal characteristics. Our focus in this book, however, is on networks, so we will concentrate on the more limited problem of determining similarity between the vertices of a network using the information contained in the network structure.

There are two fundamental approaches to constructing measures of network similarity, called *structural equivalence* and *regular equivalence*. The names are rather opaque, but the ideas they represent are simple enough. Two vertices in a network are structurally equivalent if they share many of the same network neighbors. In Fig. 7.9a we show a sketch depicting structural equivalence between two vertices i and j—the two share, in this case, three of the same neighbors, although both also have other neighbors that are not shared.

Regular equivalence is more subtle. Two regularly equivalent vertices do not necessarily share the same neighbors, but they have neighbors *who are*

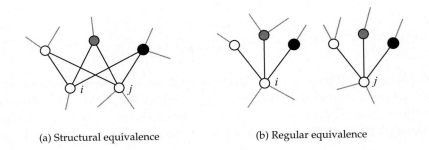

(a) Structural equivalence   (b) Regular equivalence

Figure 7.9: Structural equivalence and regular equivalence. (a) Vertices i and j are structurally equivalent if they share many of the same neighbors. (b) Vertices i and j are regularly equivalent if their neighbors are themselves equivalent (indicated here by the different shades of vertices).

themselves similar. Two history students at different universities, for example, may not have any friends in common, but they can still be similar in the sense that they both know a lot of other history students, history instructors, and so forth. Similarly, two CEOs at two different companies may have no colleagues in common, but they are similar in the sense that they have professional ties to their respective CFO, CIO, members of the board, company president, and so forth. Regular equivalence is illustrated in Fig. 7.9b.

In the next few sections we describe some mathematical measures that quantify these ideas of similarity. As we will see, measures for structural equivalence are considerably better developed than those for regular equivalence.

7.12.1 COSINE SIMILARITY

We start by looking at measures of structural equivalence and we will concentrate on undirected networks. Perhaps the simplest and most obvious measure of structural equivalence would be just a count of the number of common neighbors two vertices have. In an undirected network the number n_{ij} of common neighbors of vertices i and j is given by

$$n_{ij} = \sum_k A_{ik}A_{kj}, \tag{7.46}$$

which is the ijth element of $\mathbf{A}^2$. This quantity is closely related to the "cocitation" measure introduced in Section 6.4.1. Cocitation is defined for directed networks whereas we are here considering undirected ones, but otherwise it is essentially the same thing.

However, a simple count of common neighbors for two vertices is not on its own a very good measure of similarity. If two vertices have three common neighbors is that a lot or a little? It's hard to tell unless we know, for instance, what the degrees of the vertices are, or how many common neighbors other pairs of vertices share. What we need is some sort of normalization that places the similarity value on some easily understood scale. One strategy might be simply to divide by the total number of vertices in the network n, since this is the maximum number of common neighbors two vertices can have in a simple graph. (Technically the maximum is actually $n - 2$, but the difference is small when n is large.) However, this unduly penalizes vertices with low degree: if a vertex has degree three, then it can have at most three neighbors in common with another vertex, but the two vertices would still receive a small similarity value if the divisor n were very large. A better measure would allow for the varying degrees of vertices. Such a measure is the *cosine similarity*, sometimes also called *Salton's cosine*.

In geometry, the inner or dot product of two vectors $\mathbf{x}$ and $\mathbf{y}$ is given by $\mathbf{x} \cdot \mathbf{y} = |\mathbf{x}| \, |\mathbf{y}| \cos \theta$, where $|\mathbf{x}|$ is the magnitude of $\mathbf{x}$ and θ is the angle between the two vectors. Rearranging, we can write the cosine of the angle as

$$\cos \theta = \frac{\mathbf{x} \cdot \mathbf{y}}{|\mathbf{x}| \, |\mathbf{y}|}. \tag{7.47}$$

Salton [290] proposed that we regard the ith and jth rows (or columns) of the adjacency matrix as two vectors and use the cosine of the angle between them as our similarity measure. Noting that the dot product of two rows is simply $\sum_k A_{ik} A_{kj}$ for an undirected network, this gives us a similarity

$$\sigma_{ij} = \cos \theta = \frac{\sum_k A_{ik} A_{kj}}{\sqrt{\sum_k A_{ik}^2} \sqrt{\sum_k A_{jk}^2}}. \tag{7.48}$$

Assuming our network is an unweighted simple graph, the elements of the adjacency matrix take only the values 0 and 1, so that $A_{ij}^2 = A_{ij}$ for all i, j. Then $\sum_k A_{ik}^2 = \sum_k A_{ik} = k_i$, where k_i is the degree of vertex i (see Eq. (6.19)). Thus

$$\sigma_{ij} = \frac{\sum_k A_{ik} A_{kj}}{\sqrt{k_i k_j}} = \frac{n_{ij}}{\sqrt{k_i k_j}}. \tag{7.49}$$

The cosine similarity of i and j is therefore the number of common neighbors of the two vertices divided by the geometric mean of their degrees. For the vertices i and j depicted in Fig. 7.9a, for instance, the cosine similarity would be

$$\sigma_{ij} = \frac{3}{\sqrt{4 \times 5}} = 0.671\ldots \tag{7.50}$$

Notice that the cosine similarity is technically undefined if one or both of the vertices has degree zero, but by convention we normally say in that case that $\sigma_{ij} = 0$.

The cosine similarity provides a natural scale for our similarity measure. Its value always lies in the range from 0 to 1. A cosine similarity of 1 indicates that two vertices have exactly the same neighbors. A cosine similarity of zero indicates that they have none of the same neighbors. Notice that the cosine similarity can never be negative, being a sum of positive terms, even though cosines in general can of course be negative.

7.12.2 PEARSON COEFFICIENTS

An alternative way to normalize the count of common neighbors is to compare it with the expected value that count would take on a network in which vertices choose their neighbors at random. This line of argument leads us to the *Pearson correlation coefficient*.

Suppose vertices i and j have degrees k_i and k_j respectively. How many common neighbors should we expect them to have? This is straightforward to calculate if they choose their neighbors purely at random. Imagine that vertex i chooses k_i neighbors uniformly at random from the n possibilities open to it (or $n - 1$ on a network without self-loops, but the distinction is slight for a large network), and vertex j similarly chooses k_j neighbors at random. For the first neighbor that j chooses there is a probability of k_i/n that it will choose one of the ones k_i chose, and similarly for each succeeding choice. (We neglect the possibility of choosing the same neighbor twice, since it is small for a large network.) Then in total the expected number of common neighbors between the two vertices will be k_j times this, or $k_i k_j/n$.

A reasonable measure of similarity between two vertices is the actual number of common neighbors they have minus the expected number that they *would* have if they chose their neighbors at random:

$$\sum_k A_{ik}A_{jk} - \frac{k_ik_j}{n} = \sum_k A_{ik}A_{jk} - \frac{1}{n}\sum_k A_{ik}\sum_l A_{jl}$$
$$= \sum_k A_{ik}A_{jk} - n\langle A_i\rangle\langle A_j\rangle$$
$$= \sum_k [A_{ik}A_{jk} - \langle A_i\rangle\langle A_j\rangle]$$
$$= \sum_k (A_{ik} - \langle A_i\rangle)(A_{jk} - \langle A_j\rangle), \qquad (7.51)$$

where $\langle A_i\rangle$ denotes the mean $n^{-1}\sum_k A_{ik}$ of the elements of the ith row of the adjacency matrix. Equation (7.51) will be zero if the number of common neighbors of i and j is exactly what we would expect on the basis of random chance. If it is positive, then i and j have more neighbors than we would expect by chance, which we take as an indication of similarity between the two. Equation (7.51) can also be negative, indicating that i and j have fewer neighbors than we would expect, a possible sign of dissimilarity.

Equation (7.51) is simply n times the covariance $\text{cov}(A_i, A_j)$ of the two rows of the adjacency matrix. It is common to normalize the covariance, as we did with the cosine similarity, so that its maximum value is 1. The maximum value of the covariance of any two sets of quantities occurs when the sets are exactly the same, in which case their covariance is just equal to the variance of either set, which we could write as σ_i^2 or σ_j^2, or in symmetric form as $\sigma_i\sigma_j$. Normalizing by this quantity then gives us the standard Pearson correlation coefficient:

$$r_{ij} = \frac{\text{cov}(A_i, A_j)}{\sigma_i\sigma_j} = \frac{\sum_k(A_{ik} - \langle A_i\rangle)(A_{jk} - \langle A_j\rangle)}{\sqrt{\sum_k(A_{ik} - \langle A_i\rangle)^2}\sqrt{\sum_k(A_{jk} - \langle A_j\rangle)^2}}. \qquad (7.52)$$

This quantity lies strictly in the range $-1 \leq r_{ij} \leq 1$.

The Pearson coefficient is a widely used measure of similarity. It allows us to say when vertices are both similar or dissimilar compared with what we would expect if connections in the network were formed at random.

7.12.3 OTHER MEASURES OF STRUCTURAL EQUIVALENCE

There are many other possible measures of structural equivalence. For instance, one could also normalize the number n_{ij} of common neighbors by dividing by (rather than subtracting) the expected value of k_ik_j/n. That would give us a similarity of

$$\frac{n_{ij}}{k_ik_j/n} = n\frac{\sum_k A_{ik}A_{jk}}{\sum_k A_{ik}\sum_k A_{jk}}. \qquad (7.53)$$

This quantity will be 1 if the number of common neighbors is exactly as expected on the basis of chance, greater than one if there are more common neighbors than that, and less than one for dissimilar vertices with fewer common neighbors than we would expect by chance. It is never negative and has the nice property that it is zero when the vertices in question have no common neighbors. This measure could be looked upon as an alternative to the cosine similarity: the two differ in that one has the product of the degrees $k_i k_j$ in the denominator while the other has the square root of the product $\sqrt{k_i k_j}$. It has been suggested that Eq. (7.53) may in some cases be a superior measure to the cosine similarity because, by normalizing with respect to the expected number of common neighbors rather than the maximum number, it allows us to easily identify statistically surprising coincidences between the neighborhoods of vertices, which cosine similarity does not [195].

Another measure of structural equivalence is the so-called *Euclidean distance*,[32] which is equal to the number of neighbors that differ between two vertices. That is, it is the number of vertices that are neighbors of i but not of j, or vice versa. Euclidean distance is really a dissimilarity measure, since it is larger for vertices that differ more.

In terms of the adjacency matrix the Euclidean distance d_{ij} between two vertices can be written

$$d_{ij} = \sum_k (A_{ik} - A_{jk})^2. \tag{7.54}$$

As with our other measures it is sometimes convenient to normalize the Euclidean distance by dividing by its possible maximum value. The maximum value of d_{ij} occurs when two vertices have no neighbors in common, in which case the distance is equal to the sum of the degrees of the vertices: $d_{ij} = k_i + k_j$. Dividing by this maximum value the normalized distance is

$$\frac{\sum_k (A_{ik} - A_{jk})^2}{k_i + k_j} = \frac{\sum_k (A_{ik} + A_{jk} - 2A_{ik}A_{jk})}{k_i + k_j} = 1 - 2\frac{n_{ij}}{k_i + k_j}, \tag{7.55}$$

where we have made use of the fact that $A_{ij}^2 = A_{ij}$ because A_{ij} is always zero or one, and n_{ij} is again the number of neighbors that i and j have in common. To within additive and multiplicative constants, this normalized Euclidean distance can thus be regarded as just another alternative normalization of the number of common neighbors.

---

[32]This is actually a bad name for it—it should be called *Hamming distance*, since it is essentially the same as the Hamming distance of computer science and has nothing to do with Euclid.

7.12.4 REGULAR EQUIVALENCE

The similarity measures discussed in the preceding sections are all measures of structural equivalence, i.e., they are measures of the extent to which two vertices share the same neighbors. The other main type of similarity considered in social network analysis is regular equivalence. As described above, regularly equivalent vertices are vertices that, while they do not necessarily share neighbors, have neighbors who are themselves similar—see Fig. 7.9b again.

Quantitative measures of regular equivalence are less well developed than measures of structural equivalence. In the 1970s social network analysts came up with some rather complicated computer algorithms, such as the "REGE" algorithm of White and Reitz [320,327], that were intended to discover regular equivalence in networks, but the operation of these algorithms is involved and not easy to interpret. More recently, however, some simpler algebraic measures have been developed that appear to work reasonably well. The basic idea [45, 162,195] is to define a similarity score σ_{ij} such that i and j have high similarity if they have neighbors k and l that themselves have high similarity. For an undirected network we can write this as

$$\sigma_{ij} = \alpha \sum_{kl} A_{ik} A_{jl} \sigma_{kl}, \tag{7.56}$$

or in matrix terms $\boldsymbol{\sigma} = \alpha \mathbf{A} \boldsymbol{\sigma} \mathbf{A}$. Although it may not be immediately clear, this expression is a type of eigenvector equation, where the entire matrix $\boldsymbol{\sigma}$ of similarities is the eigenvector. The parameter α is the eigenvalue (or more correctly, its inverse) and, as with the eigenvector centrality of Section 7.2, we are normally interested in the leading eigenvalue, which can be found by standard methods.

This formula however has some problems. First, it doesn't necessarily give a high value for the "self-similarity" σ_{ii} of a vertex to itself, which is counterintuitive. Presumably, all vertices are highly similar to themselves! As a consequence of this, Eq. (7.56) also doesn't necessarily give a high similarity score to vertex pairs that have a lot of common neighbors, which in the light of our examination of structural equivalence in the preceding few sections we perhaps feel it should. If we had high self-similarity scores for all vertices, on the other hand, then Eq. (7.56) would automatically give high similarity also to vertices with many common neighbors.

We can fix these problems by introducing an extra diagonal term in the similarity thus:

$$\sigma_{ij} = \alpha \sum_{kl} A_{ik} A_{jl} \sigma_{kl} + \delta_{ij}, \tag{7.57}$$

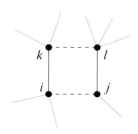

Vertices i and j are considered similar (dashed line) if they have respective neighbors k and l that are themselves similar.

See Section 11.1 for a discussion of computer algorithms for finding eigenvectors.

or in matrix notation

$$\sigma = \alpha \mathbf{A} \sigma \mathbf{A} + \mathbf{I}. \tag{7.58}$$

However, while expressions like this have been proposed as similarity measures, they still suffer from some problems. Suppose we evaluate Eq. (7.58) by repeated iteration, taking a starting value, for example, of $\sigma^{(0)} = 0$ and using it to compute $\sigma^{(1)} = \alpha \mathbf{A} \sigma \mathbf{A} + \mathbf{I}$, and then repeating the process many times until σ converges. On the first few iterations we will get the following results:

$$\sigma^{(1)} = \mathbf{I}, \tag{7.59a}$$

$$\sigma^{(2)} = \alpha \mathbf{A}^2 + \mathbf{I}, \tag{7.59b}$$

$$\sigma^{(3)} = \alpha^2 \mathbf{A}^4 + \alpha \mathbf{A}^2 + \mathbf{I}. \tag{7.59c}$$

The pattern is clear: in the limit of many iterations, we will get a sum over even powers of the adjacency matrix. However, as discussed in Section 6.10, the elements of the rth power of the adjacency matrix count paths of length r between vertices, and hence this measure of similarity is a weighted sum over the numbers of paths of even length between pairs of vertices.

But why should we consider only paths of even length? Why not consider paths of all lengths? These questions lead us to a better definition of regular equivalence as follows: vertices i and j are similar if i has a neighbor k that is itself similar to j.[33] Again we assume that vertices are similar to themselves, which we can represent with a diagonal δ_{ij} term in the similarity, and our similarity measure then looks like

$$\sigma_{ij} = \alpha \sum_k A_{ik} \sigma_{kj} + \delta_{ij}, \tag{7.60}$$

or

$$\sigma = \alpha \mathbf{A} \sigma + \mathbf{I}, \tag{7.61}$$

in matrix notation. Evaluating this expression by iterating again starting from $\sigma^{(0)} = 0$, we get

$$\sigma^{(1)} = \mathbf{I}, \tag{7.62a}$$

$$\sigma^{(2)} = \alpha \mathbf{A} + \mathbf{I}, \tag{7.62b}$$

$$\sigma^{(3)} = \alpha^2 \mathbf{A}^2 + \alpha \mathbf{A} + \mathbf{I}. \tag{7.62c}$$

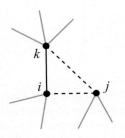

In the modified definition of regular equivalence vertex i is considered similar to vertex j (dashed line) if it has a neighbor k that is itself similar to j.

---

[33]This definition is not obviously symmetric with respect to i and j but, as we see, does in fact give rise to an expression for the similarity that is symmetric.

In the limit of a large number of iterations this gives

$$\sigma = \sum_{m=0}^{\infty} (\alpha \mathbf{A})^m = (\mathbf{I} - \alpha \mathbf{A})^{-1}, \tag{7.63}$$

which we could also have deduced directly by rearranging Eq. (7.61). Now our similarity measure includes counts of paths at all lengths, not just even paths. In fact, we can see now that this similarity measure could be defined a completely different way, as a weighted count of all the paths between the vertices i and j with paths of length r getting weight α^r. So long as $\alpha < 1$, longer paths will get less weight than shorter ones, which seems sensible: in effect we are saying that vertices are similar if they are connected either by a few short paths or by very many long ones.

Equation (7.63) is reminiscent of the formula for the Katz centrality, Eq. (7.10). We could call Eq. (7.63) the "Katz similarity" perhaps, although Katz himself never discussed it. The Katz centrality of a vertex would then be simply the sum of the Katz similarities of that vertex to all others. Vertices that are similar to many others would get high centrality, a concept that certainly makes intuitive sense. As with the Katz centrality, the value of the parameter α is undetermined—we are free to choose it as we see fit—but it must satisfy $\alpha < 1/\kappa_1$ if the sum in Eq. (7.63) is to converge, where κ_1 is the largest eigenvalue of the adjacency matrix.

In a sense, this regular equivalence measure can be seen as a generalization of our structural equivalence measures in earlier sections. With those measures we were counting the common neighbors of a pair of vertices, but the number of common neighbors is also of course the number of paths of length two between the vertices. Our "Katz similarity" measure merely extends this concept to counting paths of all lengths.

Some variations of this similarity measure are possible. As defined it tends to give high similarity to vertices that have high degree, because if a vertex has many neighbors it tends to increase the number of those neighbors that are similar to any other given vertex and hence increases the total similarity to that vertex. In some cases this might be desirable: maybe the person with many friends *should* be considered more similar to others than the person with few. However, in other cases it gives an unwanted bias in favor of high-degree nodes. Who is to say that two hermits are not "similar" in an interesting sense? If we wish, we can remove the bias in favor of high degree by dividing by vertex degree thus:

$$\sigma_{ij} = \frac{\alpha}{k_i} \sum_k A_{ik} \sigma_{kj} + \delta_{ij}, \tag{7.64}$$

or in matrix notation $\sigma = \alpha \mathbf{D}^{-1} \mathbf{A} \sigma + \mathbf{I}$, where, as previously, $\mathbf{D}$ is the diagonal matrix with elements $D_{ii} = k_i$. This expression can be rearranged to read:[34]

$$\sigma = (\mathbf{I} - \alpha \mathbf{D}^{-1} \mathbf{A})^{-1} = (\mathbf{D} - \alpha \mathbf{A})^{-1} \mathbf{D}. \qquad (7.65)$$

Another useful variant is to consider cases where the last term in Eqs. (7.60) or (7.64) is not simply diagonal, but includes off-diagonal terms too. Such a generalization would allow us to specify explicitly that particular pairs of vertices are similar, based on some other (probably non-network) information that we have at our disposal. Going back to the example of CEOs at companies that we gave at the beginning of Section 7.12, we might, for example, want to state explicitly that the CFOs and CIOs and so forth at different companies are similar, and then our similarity measure would, we hope, correctly deduce from the network structure that the CEOs are similar also. This kind of approach is particularly useful in the case of networks that consist of more than one component, so that some pairs of vertices are not connected at all. If, for instance, we have two separate components representing people in two different companies, then there will be no paths of any length between individuals in different companies, and hence a measure like (7.60) or (7.64) will never assign a non-zero similarity to such individuals. If however, we explicitly insert some similarities between members of the different companies, our measure will then be able to generalize and extend those inputs to deduce similarities between other members.

This idea of generalizing from a few given similarities arises in other contexts too. For example, in the fields of machine learning and information retrieval there is a considerable literature on how to generalize known similarities between a subset of the objects in a collection of, say, text documents to the rest of the collection, based on network data or other information.

7.13 HOMOPHILY AND ASSORTATIVE MIXING

Consider Fig. 7.10, which shows a friendship network of children at an American school, determined from a questionnaire of the type discussed in Section 3.2.[35] One very clear feature that emerges from the figure is the division of

---

[34]It is interesting to note that when we expand this measure in powers of the adjacency matrix, as we did in Eq. (7.63), the second-order (i.e., path-length two) term is the same as the structural equivalence measure of Eq. (7.53), which perhaps lends further credence to both expressions as natural measures of similarity.

[35]The study used a "name generator"—students were asked to list the names of others they considered to be their friends. This results in a directed network, but we have neglected the edge

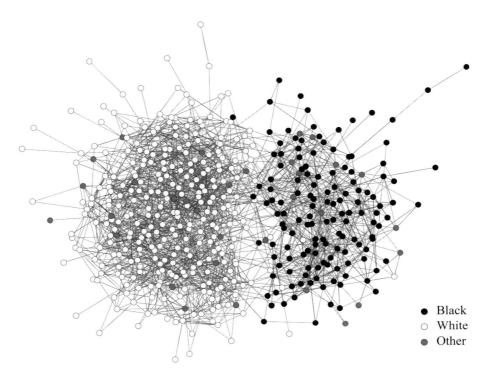

○ Black
○ White
● Other

Figure 7.10: Friendship network at a US high school. The vertices in this network represent 470 students at a US high school (ages 14 to 18 years). The vertices are color coded by race as indicated in the key. Data from the National Longitudinal Study of Adolescent Health [34, 314].

the network into two groups. It turns out that this division is principally along lines of race. The different shades of the vertices in the picture correspond to students of different race as denoted in the legend, and reveal that the school is sharply divided between a group composed principally of black children and a group composed principally of white.

This is not news to sociologists, who have long observed and discussed such divisions [225]. Nor is the effect specific to race. People are found to form friendships, acquaintances, business relations, and many other types of tie based on all sorts of characteristics, including age, nationality, language, income, educational level, and many others. Almost any social parameter you

directions in the figure. In our representation there is an undirected edge between vertices *i* and *j* if either of the pair considers the other to be their friend (or both).

can imagine plays into people's selection of their friends. People have, it appears, a strong tendency to associate with others whom they perceive as being similar to themselves in some way. This tendency is called *homophily* or *assortative mixing*.

More rarely, one also encounters *disassortative mixing*, the tendency for people to associate with others who are *unlike* them. Probably the most widespread and familiar example of disassortative mixing is mixing by gender in sexual contact networks. The majority of sexual partnerships are between individuals of opposite sex, so they represent connections between people who differ in their gender. Of course, same-sex partnerships do also occur, but they are a much smaller fraction of the ties in the network.

Assortative (or disassortative) mixing is also seen in some nonsocial networks. Papers in a citation network, for instance, tend to cite other papers in the same field more than they do papers in different fields. Web pages written in a particular language tend to link to others in the same language.

In this section we look at how assortative mixing can be quantified. Assortative mixing by discrete characteristics such as race, gender, or nationality is fundamentally different from mixing by a scalar characteristic like age or income, so we treat the two cases separately.

7.13.1 ASSORTATIVE MIXING BY ENUMERATIVE CHARACTERISTICS

Suppose we have a network in which the vertices are classified according to some characteristic that has a finite set of possible values. The values are merely enumerative—they don't fall in any particular order. For instance, the vertices could represent people and be classified according to nationality, race, or gender. Or they could be web pages classified by what language they are written in, or biological species classified by habitat, or any of many other possibilities.

The network is assortative if a significant fraction of the edges in the network run between vertices of the same type, and a simple way to quantify assortativity would be to measure that fraction. However, this is not a very good measure because, for instance, it is 1 if all vertices belong to the same single type. This is a trivial sort of assortativity: all friends of a human being, for example, are also human beings,[36] but this is not really an interesting statement. What we would like instead is a measure that is large in non-trivial cases but small in trivial ones.

A good measure turns out to be the following. We find the fraction of edges

---

[36]Ignoring, for the purposes of argument, dogs, cats, imaginary friends, and so forth.

that run between vertices of the same type, and then we subtract from that figure the fraction of such edges we would *expect* to find if edges were positioned at random without regard for vertex type. For the trivial case in which all vertices are of a single type, for instance, 100% of edges run between vertices of the same type, but this is also the expected figure, since there is nowhere else for the edges to fall. The difference of the two numbers is then zero, telling us that there is no non-trivial assortativity in this case. Only when the fraction of edges between vertices of the same type is significantly greater than we would expect on the basis of chance will our measure give a positive score.

In mathematical terms, let us denote by c_i the class or type of vertex i, which is an integer $1 \ldots n_c$, with n_c being the total number of classes. Then the total number of edges that run between vertices of the same type is

$$\sum_{\text{edges }(i,j)} \delta(c_i, c_j) = \tfrac{1}{2} \sum_{ij} A_{ij} \delta(c_i, c_j), \tag{7.66}$$

where $\delta(m, n)$ is the Kronecker delta and the factor of $\tfrac{1}{2}$ accounts for the fact that every vertex pair i, j is counted twice in the second sum.

Calculating the expected number of edges between vertices if edges are placed at random takes a little more work. Consider a particular edge attached to vertex i, which has degree k_i. There are by definition $2m$ ends of edges in the entire network, where m is as usual the total number of edges, and the chances that the other end of our particular edge is one of the k_j ends attached to vertex j is thus $k_j/2m$ if connections are made purely at random.[37] Counting all k_i edges attached to i, the total expected number of edges between vertices i and j is then $k_i k_j/2m$, and the expected number of edges between all pairs of vertices of the same type is

$$\tfrac{1}{2} \sum_{ij} \frac{k_i k_j}{2m} \delta(c_i, c_j), \tag{7.67}$$

where the factor of $\tfrac{1}{2}$, as before, prevents us from double-counting vertex pairs. Taking the difference of (7.66) and (7.67) then gives us an expression for the difference between the actual and expected number of edges in the network

---

[37] Technically, we are making connections at random while preserving the vertex degrees. We could in principle ignore vertex degrees and make connections truly at random, but in practice this is found to give much poorer results.

that join vertices of like types:

$$\tfrac{1}{2} \sum_{ij} A_{ij}\, \delta(c_i, c_j) - \tfrac{1}{2} \sum_{ij} \frac{k_i k_j}{2m}\, \delta(c_i, c_j) = \tfrac{1}{2} \sum_{ij} \left(A_{ij} - \frac{k_i k_j}{2m} \right) \delta(c_i, c_j).$$

(7.68)

Conventionally, one calculates not the number of such edges but the fraction, which is given by this same expression divided by the number m of edges:

$$Q = \frac{1}{2m} \sum_{ij} \left(A_{ij} - \frac{k_i k_j}{2m} \right) \delta(c_i, c_j).$$

(7.69)

This quantity Q is called the *modularity* [239,250] and is a measure of the extent to which like is connected to like in a network. It is strictly less than 1, takes positive values if there are more edges between vertices of the same type than we would expect by chance, and negative ones if there are less.

For Fig. 7.10, for instance, where the types are the three ethnic classifications "black," "white," and "other," we find a modularity value of $Q = 0.305$, indicating (positive) assortative mixing by race in this particular network.[38] Negative values of the modularity indicate disassortative mixing. We might see a negative modularity, for example, in a network of sexual partnerships where most partnerships were between individuals of opposite sex.

The quantity

$$B_{ij} = A_{ij} - \frac{k_i k_j}{2m}$$

(7.70)

in Eq. (7.69) appears in a number of situations in the study of networks. We will encounter it, for instance, in Section 11.8 when we study community detection in networks. In some contexts it is useful to consider B_{ij} to be an element of a matrix **B**, which itself is called the *modularity matrix*.

The modularity, Eq. (7.69), is always less than 1 but in general it does not achieve the value $Q = 1$ even for a perfectly mixed network, one in which every vertex is connected only to others of the same type. Depending on the sizes of the groups and the degrees of vertices, the maximum value of Q can be considerably less than 1. This is in some ways unsatisfactory: how is one to

---

[38] An alternative measure of assortativity has been proposed by Gupta *et al.* [152]. That measure however gives equal weight to each group of vertices, rather than to each edge as the modularity does. With this measure if one had a million vertices of each of two types, which mixed with one another entirely randomly, and ten more vertices of a third type that connected only among themselves, one would end up with a score of about 0.5 [239], which appears to imply strong assortativity when in fact almost all of the network mixes randomly. For most purposes therefore, the measure of Eq. (7.69) gives results more in line with our intuitions.

know when one has strong assortative mixing and when one doesn't? To rectify the problem, we can normalize Q by dividing by its value for the perfectly mixed network. With perfect mixing all edges fall between vertices of the same type and hence $\delta(c_i, c_j) = 1$ whenever $A_{ij} = 1$. This means that the first term in the sum in Eq. (7.69) sums to $2m$ and the modularity for the perfectly mixed network is

$$Q_{\max} = \frac{1}{2m}\left(2m - \sum_{ij}\frac{k_i k_j}{2m}\delta(c_i, c_j)\right). \tag{7.71}$$

Then the normalized value of the modularity is given by

$$\frac{Q}{Q_{\max}} = \frac{\sum_{ij}(A_{ij} - k_i k_j/2m)\delta(c_i, c_j)}{2m - \sum_{ij}(k_i k_j/2m)\delta(c_i, c_j)}. \tag{7.72}$$

This quantity, sometimes called an *assortativity coefficient*, now takes a maximum value of 1 on a perfectly mixed network.

Although it can be a useful measure in some circumstances, however, Eq. (7.72) is only rarely used. Most often, the modularity is used in its unnormalized form, Eq. (7.69).

An alternative form for the modularity, which is sometimes useful in practical situations, can be derived in terms of the quantities

$$e_{rs} = \frac{1}{2m}\sum_{ij} A_{ij}\,\delta(c_i, r)\,\delta(c_j, s), \tag{7.73}$$

which is the fraction of edges that join vertices of type r to vertices of type s, and

$$a_r = \frac{1}{2m}\sum_{i} k_i\,\delta(c_i, r), \tag{7.74}$$

which is the fraction of ends of edges attached to vertices of type r. Then, noting that

$$\delta(c_i, c_j) = \sum_r \delta(c_i, r)\delta(c_j, r), \tag{7.75}$$

we have, from Eq. (7.69)

$$
\begin{aligned}
Q &= \frac{1}{2m}\sum_{ij}\left(A_{ij} - \frac{k_i k_j}{2m}\right)\sum_r \delta(c_i, r)\delta(c_j, r)\\
&= \sum_r\left[\frac{1}{2m}\sum_{ij}A_{ij}\,\delta(c_i, r)\delta(c_j, r) - \frac{1}{2m}\sum_i k_i\,\delta(c_i, r)\frac{1}{2m}\sum_j k_j\delta(c_j, r)\right]\\
&= \sum_r\left(e_{rr} - a_r^2\right).
\end{aligned}
\tag{7.76}
$$

This form can be useful, for instance, when we have network data in the form of a list of edges and the types of the vertices at their ends, but no explicit data on vertex degrees. In such a case e_{rs} and a_r are relatively easy to calculate, while Eq. (7.69) is quite awkward.

7.13.2 ASSORTATIVE MIXING BY SCALAR CHARACTERISTICS

We can also have homophily in a network according to scalar characteristics like age or income. These are characteristics whose values come in a particular order, so that it is possible say not only when two vertices are exactly the same according to the characteristic but also when they are approximately the same. For instance, while two people can certainly be of exactly the same age—born on the same day even—they can also be approximately the same age—born within a couple of years of one another, say—and people could (and in fact often do) choose who they associate with on the basis of such approximate ages. There is no equivalent approximate similarity for the enumerative characteristics of the previous section: there is no sense in which people from France and Germany, say, are more nearly of the same nationality than people from France and Spain.[39]

If network vertices with similar values of a scalar characteristic tend to be connected together more often that those with different values then the network is considered assortatively mixed according to that characteristic. If, for example, people are friends with others around the same age as them, then the network is assortatively mixed by age. Sometimes you may also hear it said that the network is *stratified* by age, which means the same thing—one can think of age as a one-dimensional scale or axis, with individuals of different ages forming connected "strata" within the network.

Consider Fig. 7.11, which shows friendship data for the same set of US schoolchildren as Fig. 7.10 but now as a function of age. Each dot in the figure corresponds to one pair of friends and the position of the dot along the two axes gives the ages of the friends, with ages measured by school grades.[40] As the figure shows, there is substantial assortative mixing by age among the students: many dots lie within the boxes close to the diagonal line that represent

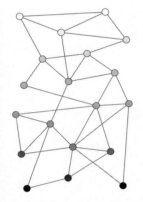

A sketch of stratified network in which most connections run between vertices at or near the same "level" in the network, with level along the vertical axis in this case and also denoted by the shades of the vertices.

---

[39]Of course, one could make up some measure of national differences, based say on geographic distance, but if the question we are asked is, "Are these two people of the same nationality?" then under normal circumstances the only answers are "yes" and "no." There is nothing in between.

[40]In the US school system there are 12 grades of one year each and to begin grade g students normally must be at least of age $g + 5$. Thus the 9th grade corresponds to children of age 14 and 15.

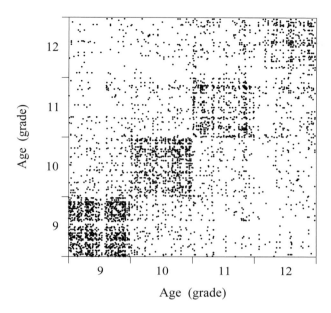

Figure 7.11: Ages of pairs of friends in high school. In this scatter plot each dot corresponds to one of the edges in Fig. 7.10, and its position along the horizontal and vertical axes gives the ages of the two individuals at either end of that edge. The ages are measured in terms of the grades of the students, which run from 9 to 12. In fact, grades in the US school system don't correspond precisely to age since students can start or end their high-school careers early or late, and can repeat grades. (Each student is positioned at random within the interval representing their grade, so as to spread the points out on the plot. Note also that each friendship appears twice, above and below the diagonal.)

friendships between students in the same grade. There is also, in this case, a notable tendency for students to have more friends of a wider range of ages as their age increases so there is a lower density of points in the top right box than in the lower left one.

One could make a crude measure of assortative mixing by scalar characteristics by adapting the ideas of the previous section. One could group the vertices into bins according to the characteristic of interest (say age) and then treat the bins as separate "types" of vertex in the sense of Section 7.13.1. For instance, we might group people by age in ranges of one year or ten years. This however misses much of the point about scalar characteristics, since it considers vertices falling in the same bin to be of identical types when they may

227

be only approximately so, and vertices falling in different bins to be entirely different when in fact they may be quite similar.

A better approach is to use a covariance measure as follows. Let x_i be the value for vertex i of the scalar quantity (age, income, etc.) that we are interested in. Consider the pairs of values (x_i, x_j) for the vertices at the ends of each edge (i, j) in the network and let us calculate their covariance over all edges as follows. We define the mean μ of the value of x_i at the end of an edge thus:

$$\mu = \frac{\sum_{ij} A_{ij} x_i}{\sum_{ij} A_{ij}} = \frac{\sum_i k_i x_i}{\sum_i k_i} = \frac{1}{2m} \sum_i k_i x_i. \tag{7.77}$$

Note that this is not simply the mean value of x_i averaged over all vertices. It is an average over edges, and since a vertex with degree k_i lies at the ends of k_i edges it appears k_i times in the average (hence the factor of k_i in the sum).

Then the covariance of x_i and x_j over edges is

$$
\begin{aligned}
\mathrm{cov}(x_i, x_j) &= \frac{\sum_{ij} A_{ij}(x_i - \mu)(x_j - \mu)}{\sum_{ij} A_{ij}} \\
&= \frac{1}{2m} \sum_{ij} A_{ij}(x_i x_j - \mu x_i - \mu x_j + \mu^2) \\
&= \frac{1}{2m} \sum_{ij} A_{ij} x_i x_j - \mu^2 \\
&= \frac{1}{2m} \sum_{ij} A_{ij} x_i x_j - \frac{1}{(2m)^2} \sum_{ij} k_i k_j x_i x_j \\
&= \frac{1}{2m} \sum_{ij} \left(A_{ij} - \frac{k_i k_j}{2m} \right) x_i x_j, \tag{7.78}
\end{aligned}
$$

where we have made use of Eqs. (6.21) and (7.77). Note the strong similarity between this expression and Eq. (7.69) for the modularity—only the delta function $\delta(c_i, c_j)$ in (7.69) has changed, being replaced by $x_i x_j$.

The covariance will be positive if, on balance, values x_i, x_j at either end of an edge tend to be both large or both small and negative if they tend to vary in opposite directions. In other words, the covariance will be positive when we have assortative mixing and negative for disassortative mixing.

Just as with the modularity measure of Section 7.13.1, it is sometimes convenient to normalize the covariance so that it takes the value 1 in a perfectly mixed network—one in which all edges fall between vertices with precisely equal values of x_i (although in most cases such an occurrence would be extremely unlikely in practice). Putting $x_j = x_i$ in Eq. (7.78) gives a perfect mixing value of

$$\frac{1}{2m}\sum_{ij}\left(A_{ij} - \frac{k_i k_j}{2m}\right)x_i^2 = \frac{1}{2m}\sum_{ij}\left(k_i\delta_{ij} - \frac{k_i k_j}{2m}\right)x_i x_j, \qquad (7.79)$$

and the normalized measure, sometimes called an *assortativity coefficient*, is the ratio of the two:

$$r = \frac{\sum_{ij}(A_{ij} - k_i k_j/2m)x_i x_j}{\sum_{ij}(k_i\delta_{ij} - k_i k_j/2m)x_i x_j}. \qquad (7.80)$$

Although it may not be immediately obvious, this is in fact an example of a (Pearson) correlation coefficient, having a covariance in its numerator and a variance in the denominator. We encountered another example in a different context in Section 7.12.2. The correlation coefficient varies in value between a maximum of 1 for a perfectly assortative network and a minimum of -1 for a perfectly disassortative one. A value of zero implies that the values of x_i at the ends of edges are uncorrelated.[41]

For the data of Fig. 7.11 the correlation coefficient is found to take a value of $r = 0.616$, indicating that the student friendship network has significant assortative mixing by age—students tend to be friends with others who have ages close to theirs.

It would be possible in principle also to have assortative (or disassortative) mixing according to vector characteristics, with vertices whose vectors have similar values, as measured by some appropriate metric, being more (or less) likely to be connected by an edge. One example of such mixing is the formation of friendships between individuals according to their geographic locations, location being specified by a two-dimensional vector of, for example, latitude/longitude coordinates. It is certainly the case that in general people tend to be friends with others who live geographically close to them, so one would expect mixing of this type to be assortative. Formal treatments of vector assortative mixing, however, have not been much pursued in the network literature so far.

---

[41] There could be non-linear correlations in such a network and we could still have $r = 0$; the correlation coefficient detects only linear correlations. For instance, we could have vertices with high and low values of x_i connected predominantly to vertices with intermediate values. This is neither assortative nor disassortative by the conventional definition and would give a small value of r, but might nonetheless be of interest. Such non-linear correlations could be discovered by examining a plot such as Fig. 7.11 or by using alternative measures of correlation such as information theoretic measures. Thus it is perhaps wise not to rely solely on the value of r in investigating assortative mixing.

7.13.3 ASSORTATIVE MIXING BY DEGREE

A special case of assortative mixing according to a scalar quantity, and one of particular interest, is that of mixing by degree. In a network that shows assortative mixing by degree the high-degree vertices will be preferentially connected to other high-degree vertices, and the low to low. In a social network, for example, we have assortative mixing by degree if the gregarious people are friends with other gregarious people and the hermits with other hermits. Conversely, we could have disassortative mixing by degree, which would mean that the gregarious people were hanging out with hermits and vice versa.

The reason this particular case is interesting is because, unlike age or income, degree is itself a property of the network structure. Having one structural property (the degrees) dictate another (the positions of the edges) gives rise to some interesting features in networks. In particular, in an assortative network, where the high-degree nodes tend to stick together, one expects to get a clump or *core* of such high-degree nodes in the network surrounded by a less dense *periphery* of nodes with lower-degree. This *core/periphery structure* is a common feature of social networks, many of which are found to be assortatively mixed by degree. Figure 7.12a shows a small assortatively mixed network in which the core/periphery structure is clearly visible.

On the other hand, if a network is disassortatively mixed by degree then high-degree vertices tend to connected to low-degree ones, creating star-like features in the network that are often readily visible. Figure 7.12b shows an example of a small disassortative network. Disassortatively networks do not usually have a core/periphery split but are instead more uniform.

Assortative mixing by degree can be measured in the same way as mixing according to any other scalar quantity. We define a covariance of the type described by Eq. (7.78), but with x_i now equal to the degree k_i:

$$\text{cov}(k_i, k_j) = \frac{1}{2m} \sum_{ij} \left(A_{ij} - \frac{k_i k_j}{2m} \right) k_i k_j, \tag{7.81}$$

or if we wish we can normalize by the maximum value of the covariance to get a correlation coefficient or assortativity coefficient:

$$r = \frac{\sum_{ij}(A_{ij} - k_i k_j / 2m) k_i k_j}{\sum_{ij}(k_i \delta_{ij} - k_i k_j / 2m) k_i k_j}. \tag{7.82}$$

We give examples of the application of this formula to a number of networks in Section 8.7.

One point to notice is that the evaluation of Eq. (7.81) or Eq. (7.82) requires only the structure of the network and no other information (unlike the calcu-

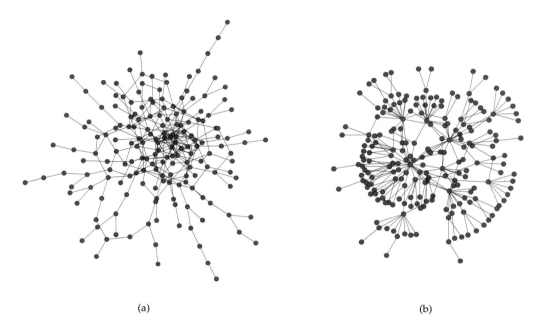

Figure 7.12: Assortative and disassortative networks. These two small networks are not real networks—they were computer generated to display the phenomenon of assortativity by degree. (a) A network that is assortative by degree, displaying the characteristic dense core of high-degree vertices surrounded by a periphery of lower-degree ones. (b) A disassortative network, displaying the star-like structures characteristic of this case. Figure from Newman and Girvan [249]. Copyright 2003 Springer-Verlag Berlin Heidelberg. Reproduced with kind permission of Springer Science and Business Media.

lations for other forms of assortative mixing). Once we know the adjacency matrix (and hence the degrees) of all vertices we can calculate r. Perhaps for this reason mixing by degree is one of the most frequently studied types of assortative mixing.

PROBLEMS

7.1 Consider a k-regular undirected network (i.e., a network in which every vertex has degree k).

a) Show that the vector $\mathbf{1} = (1,1,1,\ldots)$ is an eigenvector of the adjacency matrix with eigenvalue k.

b) By making use of the fact that eigenvectors are orthogonal (or otherwise), show that there is no other eigenvector that has all elements positive. The Perron–Frobenius theorem says that the eigenvector with the largest eigenvalue always has all elements non-negative (see footnote 2 on page 346), and hence the eigenvector $\mathbf{1}$ gives, by definition, the eigenvector centrality of our k-regular network and the centralities are the same for every vertex.

c) Find the Katz centralities of all vertices in a k-regular network.

d) You should have found that, as with the eigenvector centrality, the Katz centralities of all vertices in the network are the same. Name a centrality measure that could give different centrality values for different vertices in a regular network.

7.2 Suppose a directed network takes the form of a tree with all edges pointing inward towards a central vertex:

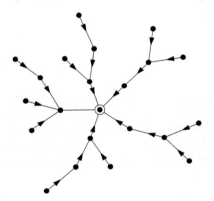

What is the PageRank centrality of the central vertex in terms of the single parameter α appearing in the definition of PageRank and the geodesic distances d_i from each vertex i to the central vertex?

7.3 Consider an undirected tree of n vertices. A particular edge in the tree joins vertices 1 and 2 and divides the tree into two disjoint regions of n_1 and n_2 vertices as sketched here:

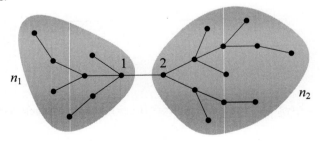

Show that the closeness centralities C_1 and C_2 of the two vertices, defined according to Eq. (7.29), are related by

$$\frac{1}{C_1} + \frac{n_1}{n} = \frac{1}{C_2} + \frac{n_2}{n}.$$

7.4 Consider an undirected (connected) tree of n vertices. Suppose that a particular vertex in the tree has degree k, so that its removal would divide the tree into k disjoint regions, and suppose that the sizes of those regions are $n_1 \ldots n_k$.

    a) Show that the unnormalized betweenness centrality x of the vertex, as defined in Eq. (7.36), is

$$x = n^2 - \sum_{m=1}^{k} n_m^2.$$

    b) Hence, or otherwise, calculate the betweenness of the ith vertex from the end of a "line graph" of n vertices, i.e., n vertices in a row like this:

7.5 Consider these three networks:

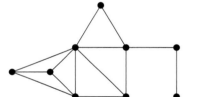 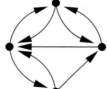 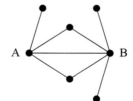

    a) Find a 3-core in the first network.

    b) What is the reciprocity of the second network?

    c) What is the cosine similarity of vertices A and B in the third network?

7.6 Among all pairs of vertices in a directed network that are connected by an edge or edges, suppose that half are connected in only one direction and the rest are connected in both directions. What is the reciprocity of the network?

7.7 In this network + and − indicate pairs of people who like each other or don't, respectively:

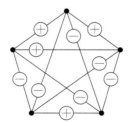

a) Is the network structurally balanced and why?

b) Is it clusterable and, if so, what are the clusters?

7.8 In a survey of couples in the US city of San Francisco, Catania *et al.* [65] recorded, among other things, the ethnicity of their interviewees and calculated the fraction of couples whose members were from each possible pairing of ethnic groups. The fractions were as follows:

		Women				
		Black	Hispanic	White	Other	Total
Men	Black	0.258	0.016	0.035	0.013	0.323
	Hispanic	0.012	0.157	0.058	0.019	0.247
	White	0.013	0.023	0.306	0.035	0.377
	Other	0.005	0.007	0.024	0.016	0.053
	Total	0.289	0.204	0.423	0.084	

Assuming the couples interviewed to be a representative sample of the edges in the undirected network of relationships for the community studied, and treating the vertices as being of four types—black, Hispanic, white, and other—calculate the numbers e_{rr} and a_r that appear in Eq. (7.76) for each type. Hence calculate the modularity of the network with respect to ethnicity.

CHAPTER 8

THE LARGE-SCALE STRUCTURE OF NETWORKS

*A discussion of some of the recurring patterns and
structures revealed when we apply the concepts
developed in previous chapters to the study of real-world
networks*

IN PREVIOUS chapters of this book we have looked at different types of natural and man-made networks and techniques for determining their structure (Chapters 2 to 5), the mathematics used to represent networks formally (Chapter 6), and the measures and metrics used to quantify network structure (Chapter 7). In this chapter we combine what we have learned so far, applying our theoretical ideas and measures to empirical network data to get a picture of what networks look like in the real world.

As we will see, there are a number of common recurring patterns seen in network structures, patterns that can have a profound effect on the way networked systems work. Among other things, we discuss in this chapter component sizes, path lengths and the small-world effect, degree distributions and power laws, and clustering coefficients.

8.1   COMPONENTS

We begin our discussion of the structure of real-world networks with a look at component sizes. In an undirected network, we typically find that there is a large component that fills most of the network—usually more than half and not infrequently over 90%—while the rest of the network is divided into a large number of small components disconnected from the rest. This situation is sketched in Fig. 8.1. (The large component is often referred to as the "giant

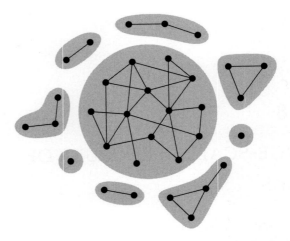

Figure 8.1: Components in an undirected network. In most undirected networks there is a single large component occupying a majority, or at least a significant fraction, of the network, along with a number of small components, typically consisting of only a handful of vertices each.

component," although this is a slightly sloppy usage. As discussed in Section 12.5, the words "giant component" have a specific meaning in network theory and are not precisely synonymous with "largest component." In this book we will be careful to distinguish between "largest" and "giant.")

A typical example of this kind of behavior is the network of film actors discussed in Section 3.5. In this network the vertices represent actors in movies and there is an edge between two actors if they have ever appeared in the same movie. In a version of the network from May 2000 [253], it was found that 440 971 out of 449 913 actors were connected together in the largest component, or about 98%. Thus just 2% of actors were not part of the largest component.

Table 8.1 summarizes the properties of many of the networks discussed in this chapter, and gives, among other things, the size S of the largest component in each case as a fraction of total network size. (For the directed networks in the table it is the size of the largest weakly connected component that is quoted. Component sizes in directed networks are discussed further in the following section.) As we can see from the table our figure for the actor network is quite typical for the networks listed and not unusually large.

See Section 6.11.1 for the definition of a weakly connected component.

As the table also shows, there are quite a few networks for which the largest component fills the entire network so that $S = 1$, i.e., the network has only a single component and no smaller components. In the cases where this happens

	Network	Type	n	m	c	S	ℓ	α	C	C_{ws}	r	Ref(s).
Social	Film actors	Undirected	449 913	25 516 482	113.43	0.980	3.48	2.3	0.20	0.78	0.208	16,323
	Company directors	Undirected	7 673	55 392	14.44	0.876	4.60	–	0.59	0.88	0.276	88,253
	Math coauthorship	Undirected	253 339	496 489	3.92	0.822	7.57	–	0.15	0.34	0.120	89,146
	Physics coauthorship	Undirected	52 909	245 300	9.27	0.838	6.19	–	0.45	0.56	0.363	234,236
	Biology coauthorship	Undirected	1 520 251	11 803 064	15.53	0.918	4.92	–	0.088	0.60	0.127	234,236
	Telephone call graph	Undirected	47 000 000	80 000 000	3.16			2.1				9,10
	Email messages	Directed	59 812	86 300	1.44	0.952	4.95	1.5/2.0		0.16		103
	Email address books	Directed	16 881	57 029	3.38	0.590	5.22	–	0.17	0.13	0.092	248
	Student dating	Undirected	573	477	1.66	0.503	16.01	–	0.005	0.001	−0.029	34
	Sexual contacts	Undirected	2 810					3.2				197,198
Information	WWW nd.edu	Directed	269 504	1 497 135	5.55	1.000	11.27	2.1/2.4	0.11	0.29	−0.067	13,28
	WWW AltaVista	Directed	203 549 046	1 466 000 000	7.20	0.914	16.18	2.1/2.7				56
	Citation network	Directed	783 339	6 716 198	8.57			3.0/–				280
	Roget's Thesaurus	Directed	1 022	5 103	4.99	0.977	4.87	–	0.13	0.15	0.157	184
	Word co-occurrence	Undirected	460 902	16 100 000	66.96	1.000		2.7		0.44		97,116
Technological	Internet	Undirected	10 697	31 992	5.98	1.000	3.31	2.5	0.035	0.39	−0.189	66,111
	Power grid	Undirected	4 941	6 594	2.67	1.000	18.99	–	0.10	0.080	−0.003	323
	Train routes	Undirected	587	19 603	66.79	1.000	2.16	–		0.69	−0.033	294
	Software packages	Directed	1 439	1 723	1.20	0.998	2.42	1.6/1.4	0.070	0.082	−0.016	239
	Software classes	Directed	1 376	2 213	1.61	1.000	5.40	–	0.033	0.012	−0.119	315
	Electronic circuits	Undirected	24 097	53 248	4.34	1.000	11.05	3.0	0.010	0.030	−0.154	115
	Peer-to-peer network	Undirected	880	1 296	1.47	0.805	4.28	2.1	0.012	0.011	−0.366	6,282
Biological	Metabolic network	Undirected	765	3 686	9.64	0.996	2.56	2.2	0.090	0.67	−0.240	166
	Protein interactions	Undirected	2 115	2 240	2.12	0.689	6.80	2.4	0.072	0.071	−0.156	164
	Marine food web	Directed	134	598	4.46	1.000	2.05	–	0.16	0.23	−0.263	160
	Freshwater food web	Directed	92	997	10.84	1.000	1.90	–	0.20	0.087	−0.326	209
	Neural network	Directed	307	2 359	7.68	0.967	3.97	–	0.18	0.28	−0.226	323,328

Table 8.1: Basic statistics for a number of networks. The properties measured are: type of network, directed or undirected; total number of vertices n; total number of edges m; mean degree c; fraction of vertices in the largest component S (or the largest weakly connected component in the case of a directed network); mean geodesic distance between connected vertex pairs ℓ; exponent α of the degree distribution if the distribution follows a power law (or "−" if not; in/out-degree exponents are given for directed graphs); clustering coefficient C from Eq. (7.41); clustering coefficient C_{ws} from the alternative definition of Eq. (7.44); and the degree correlation coefficient r from Eq. (7.82). The last column gives the citation(s) for each network in the bibliography. Blank entries indicate unavailable data.

there is usually a good reason. For instance, the Internet is a communication network—its reason for existence is to provide connections between its nodes. There must be at least one path from your vertex to your friend's vertex if the network is to serve its purpose of allowing your and your friend to communicate. To put it another way, there would be no point in being a part of the Internet if you are not part of its largest component, since that would mean that you are disconnected from and unable to communicate with almost everyone else. Thus there is a strong pressure on every vertex of the Internet to be part of the largest component and thus for the largest component to fill the entire network. In other cases the largest component fills the network because of the way the network is measured. The first Web network listed in the table, for instance, is derived from a single web crawl, as described in Section 4.1. Since a crawler can only find a web page if that page is linked to by another page, it follows automatically that all pages found by a single crawl will be connected into a single component. A Web network may, however, have more than one component if, like the "AltaVista" network in the table, it is assembled using several web crawls starting from different locations.

Can a network have two or more large components that fill a sizable fraction of the entire graph? Usually the answer to this question is no. We will study this point in more detail in Section 12.6, but the basic argument is this. If we had a network of n vertices that was divided into two large components of about $\frac{1}{2}n$ vertices each, then there would be $\frac{1}{4}n^2$ possible pairs of vertices such that one vertex was in one large component and the other vertex in the other large component. If there is an edge between *any* of these pairs of vertices, then the two components are joined together and are in fact just one component. For example, in our network of movie actors, with half a million vertices, there would about 50 billion pairs, only one of which would have to be joined by an edge to join the two large components into one. Except in very special cases, it is highly unlikely that not one such pair would be connected, and hence also highly unlikely that we will have two large components.

And what about networks with no large component? It is certainly possible for networks to consist only of small components, small groups of vertices connected among themselves but not connected to the rest of the world. An example would be the network of immediate family ties, in which two people are considered connected if they are family members living under the same roof. Such a network is clearly broken into many small components consisting of individual families, with no large component at all. In practice, however, situations like this arise rather infrequently in the study of networks for the anthropocentric reason that people don't usually bother to represent such situations by networks at all. Network representations of systems are normally only useful if most of the network is connected together. If a network is so sparse as to be made only of small components, then there is normally little to be gained by applying techniques like those described in this book. Thus, essentially all of the networks we will be looking at do contain a large component (and certainly all those in Table 8.1, although for some of them the size of that component has not been measured and the relevant entry in the table is blank).

So the basic picture we have of the structure of most networks is that of Fig. 8.1, of a large component filling most of the network, sometimes all of it, and perhaps some other small components that are not connected to the bulk of the network.

8.1.1 COMPONENTS IN DIRECTED NETWORKS

As discussed in Section 6.11, the component structure of directed networks is more complicated than for undirected ones. Directed graphs have weakly and strongly connected components. The weakly connected components correspond closely to the concept of a component in an undirected graph, and the typical situation for weakly connected components is similar to that for undirected graphs: there is usually one large weakly connected component plus, optionally, other small ones. Figures for the sizes of the largest weakly connected components in several directed network are given in Table 8.1.

A strongly connected component, as described in Section 6.11, is a maximal subset of vertices in a network such that each can reach and is reachable from all of the others along a directed path. As with weakly connected components, there is typically one large strongly connected component in a directed network and a selection of small ones. The largest strongly connected component of the World Wide Web, for instance, fills about a quarter of network [56].

Associated with each strongly connected component is an out-component (the set of all vertices that can be reached from any starting point in the strongly connected component along a directed path) and an in-component (the set of vertices from which the strongly connected component can be reached). By their definition, in- and out-components are supersets of the strongly connected component to which they belong and if there is a large strongly connected component then the corresponding in- and out-components will often contain many vertices that lie outside the strongly connected component. In the Web, for example, the portion of the in- and out-components that lie outside the largest strongly connected component each also occupy about a quarter of the network [56].

Each of the small strongly connected components will have its own in- and

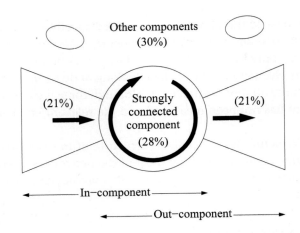

Figure 8.2: The "bow tie" diagram of components in a directed network. The typical directed network consists of one large strongly connected component and many small ones, each with an in-component and an out-component. Note that by definition each in-component includes the corresponding strongly connected component as a subset, as does each out-component. The largest strongly connected component and its in- and out-components typically occupy a significant fraction of the whole network. The percentages shown here indicate how much of the network is taken up by each part of the bow tie in the case of the World Wide Web. After Broder *et al.* [56].

out-components also. Often these will themselves be small, but they need not be. It can happen that a small strongly connected component $\mathscr{C}$ is connected by a directed path to the large strongly connected component, in which case the out-component of the large strongly connected component belongs to (and probably forms the bulk of) $\mathscr{C}$'s out-component. Notice that the large out-component can be reachable from many small components in this way—the out-components of different strongly connected components can overlap in directed networks and any vertex can and usually does belong to many out-components. Similar arguments apply, of course, for in-components as well.

The overall picture for a directed network can be represented using the "bow tie" diagram introduced by Broder and co-workers [56]. In Fig. 8.2 we show the bow tie for the case of the World Wide Web, including percentages (from Ref. [56]) for the fraction of the network occupied by its different parts.

Not all directed networks have a large strongly connected component. In particular, any acyclic directed network has no strongly connected components of size greater than one since if two vertices belong to the same strongly connected component then by definition there exists a directed path through the network in both directions between them, and hence there is a cycle from one vertex to the other and back. Thus if there are no cycles in a network there can be no strongly connected components with two or more vertices. Real-life networks are not usually perfectly acyclic, but some, such as citation networks (Section 4.2) are approximately so. Such networks typically have a few small strongly connected components of two or perhaps three vertices each, but no large ones.

8.2 SHORTEST PATHS AND THE SMALL-WORLD EFFECT

One of the most remarkable and widely discussed of network phenomena is the *small-world effect*, the finding that in many—perhaps most—networks the typical network distances between vertices are surprisingly small. In Section 3.6 we discussed Stanley Milgram's letter-passing experiment in the 1960s, in which people were asked to get a letter from an initial holder to a distant target person by passing it from acquaintance to acquaintance through the social network. The letters that made it to the target did so in a remarkably small number of steps, around six on average. Milgram's experiment is a beautiful and powerful demonstration of the small-world effect, although also a rather poorly controlled one. But with the very complete network data we have for many networks these days it is now possible to measure directly the path lengths between vertices and verify the small-world effect explicitly.

In Section 7.6 we defined the mean distance ℓ between vertices in a network (see Eqs. (7.31) and (7.32)). In mathematical terms, the small-world effect is the hypothesis that this mean distance is small, in a sense that will be defined shortly. In Table 8.1 we list the value of ℓ for each of the networks in the table, and we see that indeed it takes quite small values, always less than 20 and usually less than 10, even though some of the networks have millions of vertices.

One can well imagine that the small-world effect could have substantial implications for networked systems. Suppose a rumor is spread over a social network for instance (or a disease for that matter). Clearly it will reach people much faster if it is only about six steps from any person to any other than if it is a hundred, or a million. Similarly, the speed with which one can get a response from another computer on the Internet depends on how many steps or "hops" data packets have to make as they traverse the network. Clearly a network in which the typical number of hops is only ten or twenty will perform much better than one in which it is ten times as much. (While this point was not articulated by the original designers of the Internet in the 1960s, they must have had some idea of its truth, even if only vaguely, to believe that a network like the Internet could be built and made to work.)

In fact, once one looks more deeply into the mathematics of networks, which we will do in later chapters, one discovers that the small-world effect is not so surprising after all. As we will see in Section 12.7, mathematical models of networks suggest that path lengths in networks should typically scale as $\log n$ with the number n of network vertices, and should therefore tend to remain small even for large networks because the logarithm is a slowly growing function of its argument.

The shortest path from i to j in this network has length 1, but the shortest path from j to i has length 2.

One can ask about path lengths on directed networks as well, although the situation is more complicated there. Since in general the path from vertex i to vertex j is different in a directed network from the path from j to i, the two paths can have different lengths. Our average distance ℓ should therefore include terms for both distances separately. It's also possible for there to be no path in one direction between two vertices, which we would conventionally denote by setting $d_{ij} = \infty$. As before we could get around the problems caused by the infinite values by defining ℓ as an average over only the finite ones, as in Eq. (7.32). Values calculated in this way are given for the directed networks in Table 8.1. One could also (and perhaps more elegantly) use a harmonic mean as in Eq. (7.34), although this is rarely done.

One can also examine the diameter of a network, which, as described in Section 6.10.1, is the length of the longest finite geodesic path anywhere in the network. The diameter is usually found to be relatively small as well and calculations using network models suggest that it should scale logarithmically with n just as the average distance does. The diameter is in general a less useful measure of real-world network behavior than mean distance, since it really only measures the distance between one specific pair of vertices at the extreme end of the distribution of distances. Moreover, the diameter of a network could be affected substantially by a small change to only a single vertex or a few vertices, which makes it a poor indicator of the behavior of the network as a whole. Nonetheless, there are cases where it is of interest. In Section 8.4 we discuss so-called "scale-free" networks, i.e., networks with power-law degree distributions. Such networks are believed to have an unusual structure consisting of a central "core" to the network that contains most of the vertices and has a mean geodesic distance between vertex pairs that scales only as $\log \log n$ with network size, and not as $\log n$, making the mean distance for the whole network scale as $\log \log n$ also. Outside of this core there are longer "streamers" or "tendrils" of vertices attached to the core like hair, which have length typically of order $\log n$, making the *diameter* of the network of order $\log n$ [67,75]. This sort of behavior could be detected by measuring separately the mean geodesic distance and diameter of networks of various sizes to confirm that they vary differently with n. (It's worth noting, however, that behavior of the form $\log \log n$ is very difficult to confirm in real-world data because $\log \log n$ is a *very* slowly varying function of n.)

Another interesting twist on the small-world effect was discussed by Milgram in his original paper on the problem. He noticed, in the course of his letter-passing experiments, that most of the letters destined for a given target person passed through just one or two acquaintances of the target. Thus, it appeared, most people who knew the target person knew him through these one or two people. This idea, that one or two of your acquaintances are especially well connected and responsible for most of the connection between you and the rest of the world has been dubbed *funneling*, and it too is something we can test against complete networks with the copious data available to us today. If, for instance, we focus on geodesic paths between vertices, as we have been doing in this section, then we could measure what fraction of the shortest paths between a vertex i and every other reachable vertex go through each of i's neighbors in the network. For many networks, this measurement does reveal a funneling effect. For instance, in the coauthorship network of physicists from Table 8.1 it is found that, for physicists having five or more collaborators, 48% of geodesic paths go through one neighbor of the average vertex, the remaining 52% being distributed over the other four or more neighbors. A similar result is seen in the Internet. Among nodes having degree five or greater in a May 2005 snapshot of Internet structure at the autonomous system level, an average of 49% of geodesic paths go through one neighbor of the average vertex. It is tempting to draw conclusions about the routing of Internet packets from this latter result—perhaps that the network will tend to overload a small number of well-connected nodes rather than distributing load more evenly—but it is worth noticing that, although Internet packets tended to be routed along shortest paths during the early days of the Internet, much more sophisticated routing strategies are in place today, so statistics for shortest paths may not reflect actual packet flows very closely.

Milgram referred to these people as "sociometric superstars." We discussed them previously in Section 3.6.

8.3 DEGREE DISTRIBUTIONS

In this section, we look at one of the most fundamental of network properties, the frequency distribution of vertex degrees. This distribution will come up time and again throughout this book as a defining characteristic of network structure.

As described in Section 6.9, the degree of a vertex is the number of edges attached to it. Let us first consider undirected networks. We define p_k to be the fraction of vertices in such a network that have degree k. For example, consider this network:

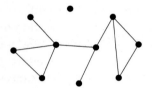

It has $n = 10$ vertices, of which 1 has degree 0, 2 have degree 1, 4 have degree 2, 2 have degree 3, and 1 has degree 4. Thus the values of p_k for $k = 0, \ldots, 4$ are

$$p_0 = \frac{1}{10}, \quad p_1 = \frac{2}{10}, \quad p_2 = \frac{4}{10}, \quad p_3 = \frac{2}{10}, \quad p_4 = \frac{1}{10}, \tag{8.1}$$

and $p_k = 0$ for all $k > 4$. The quantities p_k represent the *degree distribution* of the network.

The value p_k can also be thought of as a probability: it is the probability that a randomly chosen vertex in the network has degree k. This will be a useful viewpoint when we study theoretical models of networks in Chapters 12 to 15.

Sometimes, rather than the fraction of vertices with a given degree, we will want the total number of such vertices. This is easily calculated from the degree distribution, being given simply by np_k, where n is as usual the total number of vertices.

Another construct containing essentially the same information as the degree distribution is the *degree sequence*, which is the set $\{k_1, k_2, k_3, \ldots\}$ of degrees for all the vertices. For instance, the degree sequence of the small graph above is $\{0, 1, 1, 2, 2, 2, 2, 3, 3, 4\}$. (The degree sequence need not necessarily be given in ascending order of degrees as here. For instance, in many cases the vertices are given numeric labels and their degrees are then listed in the order of the labels.)

It is probably obvious, but bears saying anyway, that a knowledge of the degree distribution (or degree sequence) does not, in most cases, tell us the complete structure of a network. For most choices of vertex degrees there is more than one network with those degrees. These two networks, for instance, are different but have the same degrees:

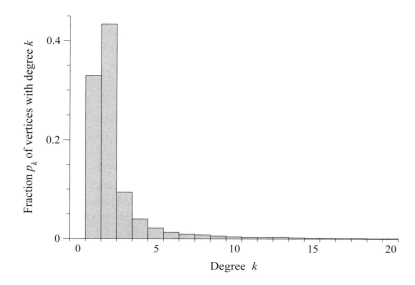

Figure 8.3: The degree distribution of the Internet. A histogram of the degree distribution of the vertices of the Internet graph at the level of autonomous systems.

Thus we cannot tell the complete structure of a network from its degrees alone. The degree sequence certainly gives us very important information about a network, but it doesn't give us complete information.

It is often illuminating to make a plot of the degree distribution of a large network as a function of k. Figure 8.3 shows an example of such a plot for the Internet at the level of autonomous systems. The figure reveals something interesting: most of the vertices in the network have low degree—one or two or three—but there is a significant "tail" to the distribution, corresponding to vertices with substantially higher degree.[1] The plot cuts off at degree 20, but in fact the tail goes much further than this. The highest degree vertex in the network has degree 2407. Since there are, for this particular data set, a total of 19 956 vertices in the network, that means that the most highly connected vertex is connected to about 12% of all other vertices in the network. We call such a well-connected vertex a *hub*.[2] Hubs will play an important role in the

---

[1] For the Internet there are no vertices of degree zero, since a vertex is not considered part of the Internet unless it is connected to at least one other.

[2] We used the word hub in a different and more technical sense in Section 7.5 to describe vertices in directed networks that point to many "authorities." Both senses are common in the net-

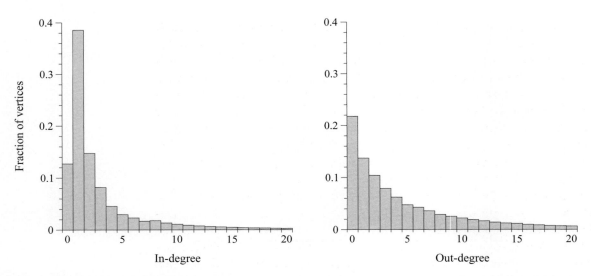

Figure 8.4: The degree distributions of the World Wide Web. Histograms of the distributions of in- and out-degrees of pages on the World Wide Web. Data are from the study by Broder *et al.* [56].

developments of the following chapters.

In fact, it turns out that almost all real-world networks have degree distributions with a tail of high-degree hubs like this. In the language of statistics we say that the degree distribution is *right-skewed*. Right-skewed degree distributions are discussed further in Section 8.4, and will reappear repeatedly throughout this book.

One can also calculate degree distributions for directed networks. As discussed in Section 6.9, directed networks have two different degrees for each vertex, the in-degree and the out-degree, which are, respectively, the number of edges ingoing and outgoing at the vertex of interest. There are, correspondingly, two different degree distributions in a directed network, the in-degree and out-degree distributions, and one can make a plot of either, or both. Figure 8.4, for example, shows the degree distributions for the World Wide Web.

If we wish to be more sophisticated, we might observe that the true degree distribution of a directed network is really a joint distribution of in- and out-

works literature, and in many cases the reader must deduce from the context which is being used. In this book we will mostly use the word in the less technical sense introduced here, of a vertex with unusually high degree. When we use it in the other sense of Section 7.5 we will say so explicitly.

degrees. We can define p_{jk} to be the fraction of vertices having simultaneously an in-degree j and an out-degree k. This is a two-dimensional distribution that cannot be plotted as a simple histogram, although it could be plotted as a two-dimensional density plot or as a surface plot. By using a joint distribution in this way we can allow for the possibility that the in- and out-degrees of vertices might be correlated. For instance, if vertices with high in-degree also tended to have high out-degree, then we would see this reflected in large values of p_{jk} when both j and k were large. If we only have the separate distributions of in- and out-degree individually, but not the joint distribution, then there is no way of telling whether the network contains such correlations.

In practice, the joint in/out degree distribution of directed networks has rarely been measured or studied, so there is relatively little data on it. This is, in some ways, a pity, since many of our theories of directed networks depend on a knowledge of the joint distribution to give accurate answers (see Section 13.11), while others make predictions about the joint distribution that we would like to test against empirical data. For the moment, however, this is an area awaiting more thorough exploration.

8.4 POWER LAWS AND SCALE-FREE NETWORKS

Returning to the Internet, another interesting feature of its degree distribution is shown in Fig. 8.5, where we have replotted the histogram of Fig. 8.3 using logarithmic scales. (That is, both axes are logarithmic. We have also made the range of the bins bigger in the histogram to make the effect clearer—they are of width five in Fig. 8.5 where they were only of width one before.) As the figure shows, when viewed in this way, the degree distribution follows, roughly speaking, a straight line. In mathematical terms, the logarithm of the degree distribution p_k is a linear function of degree k thus:

$$\ln p_k = -\alpha \ln k + c, \tag{8.2}$$

where α and c are constants. The minus sign here is optional—we could have omitted it—but it is convenient, since the slope of the line in Fig. 8.5 is clearly negative, making α a positive constant equal to minus the slope in the figure. In this case, the slope gives us a value for α of about 2.1.

Taking the exponential of both sizes of Eq. (8.2), we can also write this logarithmic relation as

$$p_k = Ck^{-\alpha}, \tag{8.3}$$

where $C = e^c$ is another constant. Distributions of this form, varying as a power of k, are called *power laws*. Based on the evidence of Fig. 8.5 we can say

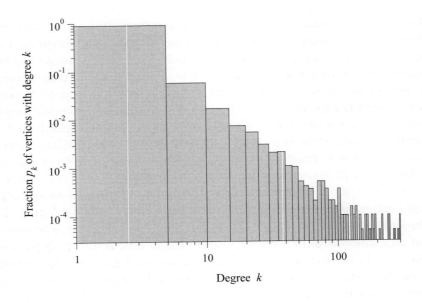

Figure 8.5: The power-law degree distribution of the Internet. Another histogram of the degree distribution of the Internet graph, plotted this time on logarithmic scales. The approximate straight-line form of the histogram indicates that the degree distribution roughly follows a power law of the form (8.3).

that, roughly speaking, the degree distribution of the Internet follows a power law.

This is, in fact, a common pattern seen in quite a few different networks. For instance, as shown in Fig. 8.8 on page 253, both the in- and out-degrees of the World Wide Web roughly follow power-law distributions, as do the in-degrees in many citation networks (but not the out-degrees).

The constant α is known as the *exponent* of the power law. Values in the range $2 \leq \alpha \leq 3$ are typical, although values slightly outside this range are possible and are observed occasionally. Table 8.1 gives the measured values of the exponents for a number of networks that have power-law or approximately power-law degree distributions, and we see that most of them fall in this range. The constant C in Eq. (8.3) is mostly uninteresting, being fixed by the requirement of normalization, as described in Section 8.4.2.

Degree distributions do not usually follow Eq. (8.3) over their entire range. Looking at Fig. 8.3, for example, we can see that the degree distribution is not monotonic for small k, even allowing for statistical fluctuations in the histogram. A true power-law distribution is monotonically decreasing over its entire range and hence the degree distribution must in this case deviate from the true power law in the small-k regime. This is typical. A common situation is that the power law is obeyed in the tail of the distribution, for large values of k, but not in the small-k regime. When one says that a particular network has a power-law degree distribution one normally means only that the tail of the distribution has this form. In some cases, the distribution may also deviate from the power-law form for high k as well. For instance, there is often a cut-off of some type that limits the maximum degree of vertices in the tail.

Networks with power-law degree distributions are sometimes called *scale-free networks*, and we will use this terminology occasionally. Of course, there are also many networks that are not scale-free, that have degree distributions with non-power-law forms, but the scale-free ones will be of particular interest to us because they have a number of intriguing properties. Telling the scale-free ones from the non-scale-free is not always easy however. The simplest strategy is to look at a histogram of the degree distribution on a log–log plot, as we did in Fig. 8.5, to see if we have a straight line. There are, however, a number of problems with this approach and where possible we recommend you use other methods, as we now explain.

8.4.1 DETECTING AND VISUALIZING POWER LAWS

As a tool for visualizing or detecting power-law behavior, a simple histogram like Fig. 8.5 presents some problems. One problem obvious from the figure is that the statistics of the histogram are poor in the tail of the distribution, the large-k region, which is precisely the region in which the power law is normally followed most closely. Each bin of the histogram in this region contains only a few samples, which means that statistical fluctuations in the number of samples from bin to bin are large. This is visible as a "noisy signal" at the right-hand end of Fig. 8.5 that makes it difficult to determine whether the histogram really follows a straight line or not, and what the slope of that line is.

There are a number of solutions to this problem. The simplest is to use a histogram with larger bins, so that more samples fall into each bin. In fact, we already did this in going from Fig. 8.3 to Fig. 8.5—we increased the bin width from one to five between the two figures. Larger bins contain more samples and hence give less noise in the tail of the histogram, but at the expense of less detail overall, since the number of bins is correspondingly reduced. Bin width in this situation is always something of a compromise: we would like to use very wide bins in the tail of the distribution where noise is a problem, but narrower ones at the left-hand end of the histogram where there are many samples and we would prefer to have more bins if possible.

Alternatively, we could try to get the best of both worlds by using bins of different sizes in different parts of the histogram. For example, we could use bins of width one for low degrees and switch to width five for higher degrees. In doing this we must be careful to normalize the bins correctly: a bin of width five will on average accrue five times as many samples as a similarly placed bin of width one, so if we wish to compare counts in the two we should divide the number of samples in the larger bin by five. More generally, we should divide sample counts by the width of their bins to make counts in bins of different widths comparable.

We need not restrict ourselves to only two different sizes of bin. We could use larger and larger bins as we go further out in the tail. We can even make every bin a different size, each one a little larger than the one before it. One commonly used version of this idea is called *logarithmic binning*. In this scheme, each bin is made wider than its predecessor by a constant factor a. For instance, if the first bin in a histogram covers the interval $1 \leq k < 2$ (meaning that all vertices of degree 1 fall in this bin) and $a = 2$, then the second would cover the interval $2 \leq k < 4$ (vertices of degrees 2 and 3), the third the interval $4 \leq k < 8$, and so forth. In general the nth bin would cover the interval $a^{n-1} \leq k < a^n$ and have width $a^n - a^{n-1} = (a - 1) a^{n-1}$. The most common choice for a is $a = 2$, since larger values tend to give bins that are too coarse while smaller ones give bins with non-integer limits.

Figure 8.6 shows the degree distribution of the Internet binned logarithmically in this way. We have been careful to normalize each bin by dividing by its width, as described above. As we can see, the histogram is now much less noisy in the tail and it is considerably easier to see the straight-line behavior of the degree distribution. The figure also reveals a nice property of logarithmically binned histograms, namely that when plotted on logarithmic scales as here, the bins in such a histogram appear to have equal width. This is, in fact, the principal reason for this particular choice of bins and also the origin of the name "logarithmic binning."

Note that on a logarithmically binned histogram there is never any bin that contains vertices of degree zero. Since there is no zero on logarithmic scales like those of Fig. 8.6, this doesn't usually make much difference, but if we do want to know how many vertices there are of degree zero we will have to measure this number separately.

A different solution to the problem of visualizing a power-law distribution

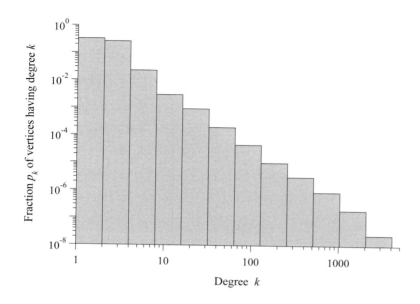

Figure 8.6: Histogram of the degree distribution if the Internet, created using logarithmic binning. In this histogram the widths of the bins are constant on a logarithmic scale, meaning that on a linear scale each bin is wider by a constant factor than the one to its left. The counts in the bins are normalized by dividing by bin width to make counts in different bins comparable.

is to construct the *cumulative distribution function*, which is defined by

$$P_k = \sum_{k'=k}^{\infty} p_{k'}. \tag{8.4}$$

In other words, P_k is the fraction of vertices that have degree k or greater. (Alternatively, it is the probability at a randomly chosen vertex has degree k or greater.)

Suppose the degree distribution p_k follows a power law in its tail. To be precise, let us say that $p_k = Ck^{-\alpha}$ for $k \geq k_{min}$ for some k_{min}. Then for $k \geq k_{min}$ we have

$$P_k = C \sum_{k'=k}^{\infty} k'^{-\alpha} \simeq C \int_k^{\infty} k'^{-\alpha} \, dk'$$
$$= \frac{C}{\alpha - 1} k^{-(\alpha-1)}, \tag{8.5}$$

where we have approximated the sum by an integral, which is reasonable since

251

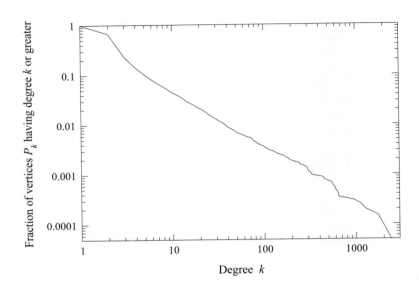

Figure 8.7: Cumulative distribution function for the degrees of vertices on the Internet. For a distribution with a power-law tail, as is approximately the case for the degree distribution of the Internet, the cumulative distribution function, Eq. (8.4), also follows a power law, but with a slope 1 less than that of the original distribution.

the power law is a slowly varying function for large k. (We are also assuming that $\alpha > 1$ so that the integral converges.) Thus we see that if the distribution p_k follows a power law, then so does the cumulative distribution function P_k, but with an exponent $\alpha - 1$ that is 1 less than the original exponent.

This gives us another way of visualizing a power-law distribution: we plot the cumulative distribution function on log–log scales, as we did for the original histogram, and again look for straight-line behavior. We have done this in Fig. 8.7 for the case of the Internet, and the (approximate) straight-line form is clearly visible. Three more examples are shown in Fig. 8.8, for the in- and out-degree distributions of the World Wide Web and for the in-degree distribution of a citation network.

This approach has some advantages. In particular, the calculation of P_k does not require us to bin the values of k as we do with a normal histogram. P_k is perfectly well defined for any value of k and can be plotted just as a normal function. When bins in a histogram contain more than one value of k—i.e., when their width is greater than 1—the binning of data necessarily throws away quite a lot of the information contained in the data, eliminating, as it

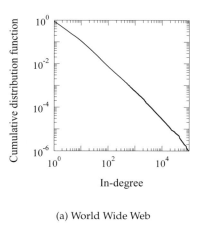
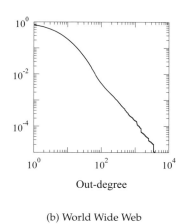
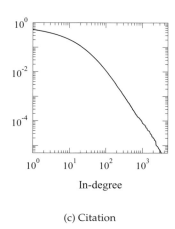

(a) World Wide Web   (b) World Wide Web   (c) Citation

Figure 8.8: Cumulative distribution functions for in- and out-degrees in three directed networks. (a) The in-degree distribution of the World Wide Web, from the data of Broder *et al.* [56]. (b) The out-degree distribution for the same Web data set. (c) The in-degree distribution of a citation network, from the data of Redner [280]. The distributions follow approximate power-law forms in each case.

does, the distinction between any two values that fall into the same bin. The cumulative distribution function on the other hand preserves all of the information contained in the data, because no bins are involved. The most obvious manifestation of this difference is that the number of points in a plot like Fig. 8.5 or Fig. 8.6 is relatively small, whereas in a cumulative distribution plot like Fig. 8.7 there are as many points along the k (horizontal) axis as there are distinct values of k.

The cumulative distribution function is also easy to calculate. The number of vertices with degree greater than or equal to that of the rth-highest-degree vertex in a network is, by definition, r. Thus the *fraction* with degree greater than or equal to that of the rth-highest-degree vertex in a network is $P_k = r/n$. So a simple way of finding P_k is to sort the degrees of the vertices in descending order and then number them from 1 to n in that order. These numbers are the so-called *ranks* r_i of the vertices. A plot of r_i/n as a function of degree k_i, with the vertices in rank order, then gives us our cumulative distribution plot.[3]

---

[3]Such plots are also sometimes called *rank/frequency* plots because one of their earliest uses was to detect power-law behavior in the frequency of occurrence of words in natural languages. If the data you are measuring are frequencies, then the cumulative distribution graph is a plot of rank against frequency. Since then such plots have been used to detect power-law behavior in

For instance, consider again the small example network we looked at at the beginning of Section 8.3, on page 244. The degrees of the vertices in that case were $\{0, 1, 1, 2, 2, 2, 2, 3, 3, 4\}$. Listing these in decreasing order and numbering them, we can easily calculate P_k as follows:

Degree k	Rank r	$P_k = r/n$
4	1	0.1
3	2	0.2
3	3	0.3
2	4	0.4
2	5	0.5
2	6	0.6
2	7	0.7
1	8	0.8
1	9	0.9
0	10	1.0

Then a plot of the last column as a function of the first gives us our cumulative distribution function.

Cumulative distributions do have some disadvantages. One is that they are less easy to interpret than ordinary histograms, since they are only indirectly related to the actual distribution of vertex degrees. A more serious disadvantage is that the successive points on a cumulative plot are correlated—the cumulative distribution function in general only changes a little from one point to the next, so adjacent values are not at all independent. This means that it is not valid for instance to extract the exponent of a power-law distribution by fitting the slope of the straight-line portion of a plot like Fig. 8.7 and equating the result with $\alpha - 1$, at least if the fitting is done using standard methods such as least squares that assume independence between the data points.

In fact, it is in general not good practice to evaluate exponents by performing straight-line fits to either cumulative distribution functions or ordinary histograms. Both are known to give biased answers, although for different reasons [72, 141]. Instead, it is usually better to calculate α directly from the data,

---

many quantities other than frequencies, but the name "rank/frequency plot" is still often used.

using the formula[4]

$$\alpha = 1 + N \left[\sum_i \ln \frac{k_i}{k_{min} - \frac{1}{2}} \right]^{-1}.$$ (8.6)

Here, k_{min} is the minimum degree for which the power law holds, as before, and N is the number of vertices with degree greater than or equal to k_{min}. The sum is performed over only those vertices with $k \geq k_{min}$, and not over all vertices.

We can also calculate the statistical error on α from the formula:

$$\sigma = \sqrt{N} \left[\sum_i \ln \frac{k_i}{k_{min} - \frac{1}{2}} \right]^{-1} = \frac{\alpha - 1}{\sqrt{N}}.$$ (8.7)

For example, applying Eqs. (8.6) and (8.7) to the degree sequence of the Internet from Fig. 8.3 gives an exponent value of $\alpha = 2.11 \pm 0.01$.

The derivation of these formulas, which makes use of maximum likelihood techniques, would take us some way from our primary topic of networks, so we will not go into it here. The interested reader can find a discussion in Ref. [72], along with many other details such as methods for determining the value of k_{min} and methods for telling whether a particular distribution follows a power law at all.

8.4.2 PROPERTIES OF POWER-LAW DISTRIBUTIONS

Quantities with power-law distributions behave in some surprising ways. We take a few pages here to look at some of the properties of power-law distributions, since the results will be of use to us later on.

Power laws turn up in a wide variety of places, not just in networks. They are found in the sizes of city populations [24, 336], earthquakes [153], moon craters [230], solar flares [203], computer files [84], and wars [283]; in the frequency of use of words in human languages [109, 336], the frequency of occurrence of personal names in most cultures [335], the numbers of papers scientists write [201], and the number of hits on web pages [5]; in the sales of books, music recordings, and almost every other branded commodity [83, 185]; and in the numbers of species in biological taxa [58, 330]. A review of the data and some mathematical properties of power laws can be found in Ref. [244]. Here we highlight just a few issues that will be relevant for our study of networks.

---

[4]In fact, this formula is only an approximation to the full formula for the exponent. The full formula, unfortunately, does not give a closed-form solution for α and is therefore hard to use. Equation (8.6) works well provided k_{min} is greater than about 6, which is true for many networks. In cases where it is not, however, the full formula must be used—see Ref. [72].

Normalization: The constant C appearing in Eq. (8.3) is fixed by the requirement that the degree distribution be normalized. That is, when we add up the total fraction of vertices having all possible degrees $k = 0 \ldots \infty$, we must get 1:

$$\sum_{k=0}^{\infty} p_k = 1. \tag{8.8}$$

If our degree distribution truly follows a pure power law, obeying Eq. (8.3) for all k, then no vertices of degree zero are allowed, because p_0 would then be infinite, which is impossible since it is a probability and must lie between 0 and 1. Let us suppose therefore that the distribution starts at $k = 1$. Substituting from Eq. (8.3) we then find that $C \sum_k k^{-\alpha} = 1$, or

$$C = \frac{1}{\sum_{k=1}^{\infty} k^{-\alpha}} = \frac{1}{\zeta(\alpha)}, \tag{8.9}$$

where $\zeta(\alpha)$ is the Riemann zeta function. Thus the correctly normalized power-law distribution is

$$p_k = \frac{k^{-\alpha}}{\zeta(\alpha)}, \tag{8.10}$$

for $k > 0$ with $p_0 = 0$.

This is a reasonable starting point for mathematical models of scale-free networks—we will use it in Chapter 13—but it's not a very good representation of most real-world networks, which deviate from pure power-law behavior for small k as described above and seen in Fig. 8.3. In that case, the normalization constant will take some other value dependent on the particular shape of the distribution, but nonetheless it is still fixed by the requirement of normalization and we must make sure we get it right in our calculations.

For some of our calculations we will be interested only in the tail of the distribution where the power-law behavior holds and can discard the rest of the data. In such cases, we normalize over only the tail, starting from the minimum value $k_{\min}$ for which the power law holds, as above. This gives

$$p_k = \frac{k^{-\alpha}}{\sum_{k=k_{\min}}^{\infty} k^{-\alpha}} = \frac{k^{-\alpha}}{\zeta(\alpha, k_{\min})}, \tag{8.11}$$

where $\zeta(\alpha, k_{\min})$ is the so-called generalized or incomplete zeta function.

Alternatively, we could observe, as we did for Eq. (8.5), that in the tail of the distribution the sum over k is well approximated by an integral, so that the normalization constant can written

$$C \simeq \frac{1}{\int_{k_{\min}}^{\infty} k^{-\alpha} \, dk} = (\alpha - 1) k_{\min}^{\alpha-1}, \tag{8.12}$$

or

$$p_k \simeq \frac{\alpha - 1}{k_{min}} \left(\frac{k}{k_{min}} \right)^{-\alpha}. \tag{8.13}$$

In the same approximation the cumulative distribution function, Eq. (8.5), is given by

$$P_k = \left(\frac{k}{k_{min}} \right)^{-(\alpha-1)}. \tag{8.14}$$

Moments: Of great interest to us will be the moments of the degree distribution. The first moment of a distribution is its mean:

$$\langle k \rangle = \sum_{k=0}^{\infty} k p_k. \tag{8.15}$$

The second moment is the mean square:

$$\langle k^2 \rangle = \sum_{k=0}^{\infty} k^2 p_k. \tag{8.16}$$

And the mth moment is

$$\langle k^m \rangle = \sum_{k=0}^{\infty} k^m p_k. \tag{8.17}$$

Suppose we have a degree distribution p_k that has a power-law tail for $k \geq k_{min}$, in the manner of the Internet or the World Wide Web. Then

$$\langle k^m \rangle = \sum_{k=0}^{k_{min}-1} k^m p_k + C \sum_{k=k_{min}}^{\infty} k^{m-\alpha}. \tag{8.18}$$

Since the power law is a slowly varying function of k for large k, we can again approximate the second sum by an integral thus:

$$\langle k^m \rangle \simeq \sum_{k=0}^{k_{min}-1} k^m p_k + C \int_{k_{min}}^{\infty} k^{m-\alpha} \, dk$$
$$= \sum_{k=0}^{k_{min}-1} k^m p_k + \frac{C}{m-\alpha+1} \left[k^{m-\alpha+1} \right]_{k_{min}}^{\infty}. \tag{8.19}$$

The first term here is some finite number whose value depends on the particular (non-power-law) form of the degree distribution for small k. The second term however depends on the values of m and α. If $m - \alpha + 1 < 0$, then the bracket has a finite value, and $\langle k^m \rangle$ is well-defined. But if $m - \alpha + 1 \geq 0$ then the bracket diverges and with it the value of $\langle k^m \rangle$. Thus, the mth moment of the degree distribution is finite if and only if $\alpha > m + 1$. Put another way, for a given value of α all moments will diverge for which $m \geq \alpha - 1$.

Of particular interest to us will be the second moment $\langle k^2 \rangle$, which arises in many calculations to do with networks (such as mean degree of neighbors, Section 13.3, robustness calculations, Section 16.2.1, epidemiological processes, Section 17.8.1, and many others). The second moment is finite if and only if $\alpha > 3$. As discussed above, however, most real-world networks with power-law degree distributions have values of α in the range $2 \leq \alpha \leq 3$, which means that the second moment should diverge, an observation that has a number of remarkable implications for the properties of scale-free networks, some of which we will explore in coming chapters. Notice that this applies even for networks where the power law only holds in the tail of the distribution—the distribution does not have to follow a power law everywhere for the second moment to diverge.

These conclusions, however, are slightly misleading. In any real network all the moments of the degree distribution will actually be finite. We can always calculate the mth moment directly from the degree sequence thus:

$$\langle k^m \rangle = \frac{1}{n} \sum_{i=1}^{n} k_i^m, \qquad (8.20)$$

and since all the k_i are finite, so must the sum be. When we say that the mth moment is infinite, what we mean is that if we were to calculate it for an arbitrarily large network with the same power-law degree distribution the value would be infinite. But for any finite network Eq. (8.20) applies and all moments are finite.

There is however another factor that limits the values of the higher moments of the degree distribution, namely that most real-world networks are simple graphs. That is, they have no multiedges and no self-loops, which means that a vertex can have, at most, one edge to every other vertex in the network, giving it a maximum degree of $n - 1$, where n is the total number of vertices. In practice, the power-law behavior of the degree distribution may be cut off for other reasons before we reach this limit, but in the worst case, an integral such as that of Eq. (8.19) will be cut off in a simple graph at $k = n$ so that

$$\langle k^m \rangle \sim \left[k^{m-\alpha+1} \right]_{k_{\min}}^{n} \sim n^{m-\alpha+1}, \qquad (8.21)$$

as $n \to \infty$ for $m > \alpha - 1$. This again gives moments that are finite on finite networks but become infinite as the size of the network becomes infinite. For instance, the second moment goes as

$$\langle k^2 \rangle \sim n^{3-\alpha}. \qquad (8.22)$$

In a network with $\alpha = \frac{5}{2}$, this diverges as $n^{1/2}$ as the network becomes large.

We will throughout this book derive results that depend on moments of the degree distributions of networks. Some of those results will show unusual behavior in power-law networks because of the divergence of the moments. On practical, finite networks that divergence is replaced by large finite values of the moments. In many cases, however, this produces similar results to a true divergence. On the Internet, for instance, with its power-law degree distribution and a total of about $n \simeq 20\,000$ autonomous systems as vertices, we can expect the second (and all higher moments) to take not infinite but very large values. For the Internet data we used in Figs. 8.3 and 8.5 the second moment has the value $\langle k^2 \rangle = 1159$, which can in practice be treated as infinite for many purposes.

Top-heavy distributions: Another interesting quantity is the fraction of edges in a network that connect to the vertices with the highest degrees. For a pure power-law degree distribution, it can be shown [244] that a fraction W of ends of edges attach to a fraction P of the highest-degree vertices in the network, where

$$W = P^{(\alpha-2)/(\alpha-1)}, \tag{8.23}$$

A set of curves of W against P is shown in Fig. 8.9 for various values of α. Curves of this kind are called *Lorenz curves*, after Max Lorenz, who first studied them around the turn of the twentieth century [200]. As the figure shows, the curves are concave downward for all values of α, and for values only a little above 2 they have a very fast initial increase, meaning that a large fraction of the edges are connected to a small fraction of the highest degree nodes.

Thus, for example, the in-degree distribution of the World Wide Web follows a power law above about $k_{min} = 20$ with exponent around $\alpha = 2.2$. Equation (8.23) with $P = \frac{1}{2}$ then tells us that we would expect that about $W = 0.89$ or 89% of all hyperlinks link to pages in the top half of the degree distribution, while the bottom half gets a mere 11%. Conversely, if we set $W = \frac{1}{2}$ in Eq. (8.23) we get $P = 0.015$, implying that 50% of all the links go to less than 2% of the "richest" vertices. Thus the degree distribution is in a sense "top-heavy," a large fraction of the "wealth"—meaning incoming hyperlinks in this case—falling to a small fraction of the vertices.

This calculation assumes a degree distribution that follows a perfect power law, whereas in reality, as we have seen, degree distributions usually only follow a power law in their high-degree tail. The basic principle still holds, however, and even if we cannot write an exact formula like Eq. (8.23) for a particular network we can easily evaluate W as a function of P directly from degree

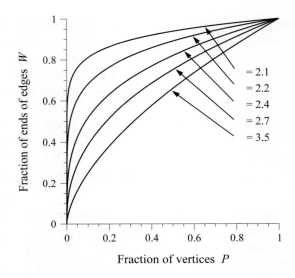

Figure 8.9: Lorenz curves for scale-free networks. The curves show the fraction W of the total number of ends of edges in a scale-free network that are attached to the fraction P of vertices with the highest degrees, for various values of the power-law exponent α.

data. For the real degree distribution of the Web[5] we find that 50% of the incoming hyperlinks point to just 1.1% of the richest vertices (so Eq. (8.23) was not too bad in this case).

Similarly, for paper citations 8.3% of the highest cited papers get 50% of all the citations[6] and on the Internet just 3.3% of the most highly connected nodes have 50% of the connections.[7]

In the remaining chapters of this book we will see many examples of networks with power-law degree distributions, and we will make use of the results of this section to develop an understanding of their behavior.

---

[5]Using the data of Broder *et al.* [56].

[6]Using the data of Redner [280].

[7]For the AS-level data of Fig. 8.3.

8.5 DISTRIBUTIONS OF OTHER CENTRALITY MEASURES

Vertex degree is just one of a variety of centrality measures for vertices in networks, as discussed in Chapter 7. Other centrality measures include eigenvector centrality and its variations (Sections 7.2 to 7.5), closeness centrality (Section 7.6), and betweenness centrality (Section 7.7). The distributions of these other measures, while of lesser importance in the study of networks than the degree distribution, are nonetheless of some interest.

Eigenvector centrality can be thought of as an extended form of degree centrality, in which we take into account not only how many neighbors a vertex has but also how central those neighbors themselves are (Section 7.2). Given its similarity to degree centrality, it is perhaps not surprising to learn that eigenvector centrality often has a highly right-skewed distribution. The left panel of Fig. 8.10 shows the cumulative distribution of eigenvector centralities for the vertices of the Internet, using again the autonomous-system-level data that we used in Section 8.3. As the figure shows, the tail of the distribution approximately follows a power law but the distribution rolls off for vertices with low centrality. Similar roughly power-law behavior is also seen in eigenvector centralities for other scale-free networks, such as the World Wide Web and citation networks, while other networks show right-skewed but non-power-law distributions.

Betweenness centrality (Section 7.7) also tends to have right-skewed distributions on most networks. The right panel of Fig. 8.10 shows the cumulative distribution of betweenness for the vertices of the Internet and, as we can see, this distribution is again roughly power-law in form. Again there are some other networks that also have power-law betweenness distributions and others still that have skewed but non-power-law distributions.

An exception to this pattern is the closeness centrality (Section 7.6), which is the mean geodesic distance from a vertex to all other reachable vertices. As discussed in Section 7.6 the values of the closeness centrality are typically limited to a rather small range from a lower bound of 1 to an upper bound of order $\log n$, and this means that their distribution cannot have a long tail. In Fig. 8.11, for instance, we show the distributions of closeness centralities for our snapshot of the Internet, and the distribution spans well under an order of magnitude from a minimum of 2.30 to a maximum of 7.32. There is no long tail to the distribution, and the distribution is not even roughly monotonically decreasing (as our others have been) but shows clear peaks and dips.

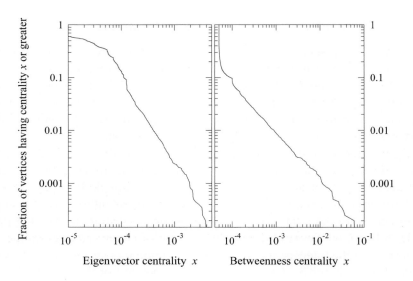

Figure 8.10: Cumulative distribution functions for centralities of vertices on the Internet. Left panel: eigenvector centrality. Right panel: betweenness centrality.

8.6 CLUSTERING COEFFICIENTS

See Section 7.9 for a discussion of clustering coefficients.

The clustering coefficient measures the average probability that two neighbors of a vertex are themselves neighbors. In effect it measures the density of triangles in the networks and it is of interest because in many cases it is found to have values sharply different from what one would expect on the basis of chance. To see what we mean by this, look again at Table 8.1 on page 237, which gives measured values of the clustering coefficient for a variety of networks. (Look at the column denoted C, which gives values for the coefficient defined by Eq. (7.41).) Most of the values are of the order of tens of percent—there is typically a probability between about 10% and maybe 60% that two neighbors of a vertex will be neighbors themselves. However, as we will see in Section 13.4, if we consider a network with a given degree distribution in which connections between vertices are made at random, the clustering coefficient takes the value

$$C = \frac{1}{n} \frac{\left[\langle k^2 \rangle - \langle k \rangle \right]^2}{\langle k \rangle^3}.$$

(8.24)

In networks where $\langle k^2 \rangle$ and $\langle k \rangle$ have fixed finite values, this quantity becomes small as $n \to \infty$ and hence we expect the clustering coefficient to be very small on large networks. This makes the values in Table 8.1, which are of order 1,

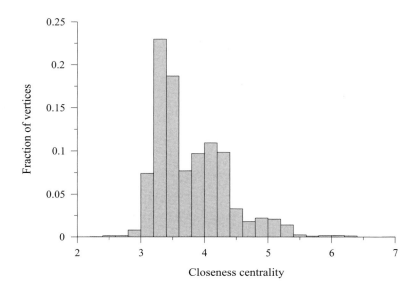

Figure 8.11: Histogram of closeness centralities of vertices on the Internet. Unlike Fig. 8.10 this is a normal non-cumulative histogram showing the actual distribution of closeness centralities. This distribution does not follow a power law.

quite surprising, and indeed many of them turn out to be much larger than the estimate given by Eq. (8.24). For instance, the collaboration network of physicists is measured to have a clustering coefficient of 0.45. Plugging the appropriate values for n, $\langle k \rangle$, and $\langle k^2 \rangle$ into Eq. (8.24) on the other hand gives $C = 0.0023$. Thus the measured value is more than a hundred times greater than the value we would expect if physicists chose their collaborators at random.

Presumably this large difference is indicative of real social effects at work. There are a number of reasons why a real collaboration network might contain more triangles than one would expect by chance, but for example it might be that people introduce pairs of their collaborators to one another and those pairs then go on to collaborate themselves. This is an example of the process that social network analysts call *triadic closure*: an "open" triad of vertices (i.e., a triad in which one vertex is linked to the other two, but the third possible edge is absent) is "closed" by the addition of the last edge, forming a triangle.

One can study triadic closure processes directly if one has time-resolved data on the formation of a network. The network of physics collaborators discussed here was studied in this way in Ref. [233], where it was shown that pairs of individuals who have not previously collaborated, but who have another mutual collaborator, are enormously more likely to collaborate in future than pairs who do not—a factor of 45 times as likely in that particular study. Furthermore, the probability of future collaboration also goes up sharply as the number of mutual collaborators increases, with pairs having two mutual collaborators being more than twice as likely to collaborate in future as those having just one.

However, it is not always the case that the measured clustering coefficient greatly exceeds the expected value given by Eq. (8.24). Take the example of the Internet again. For the data set we examined earlier the measured clustering coefficient is just 0.012. The expected value, if connections were made at random, is 0.84. (The large value arises because, as discussed in Section 8.4, the Internet has a highly right-skewed degree distribution, which makes $\langle k^2 \rangle$ large.) Clearly in this case the clustering is far *less* than one would expect on the basis of chance, suggesting that in the Internet there are forces at work that shy away from the creation of triangles.[8]

In some other networks, such as food webs or the World Wide Web, clustering is neither higher nor lower than expected, taking values roughly comparable with those given by Eq. (8.24). It is not yet well understood why clustering coefficients take such different values in different types of network, although one theory is that it may be connected with the formation of groups or communities in networks [252].

The clustering coefficient measures the density of triangles in a network. There is no reason, however, for us to limit ourselves to studying only triangles. We can also look at the densities of other small groups of vertices, or *motifs*, as they are often called. One can define coefficients similar to the clustering coefficient to measure the densities of different motifs, although more often one simply counts the numbers of the motifs of interest in a network. And, as with triangles, one can compare the results with the values one would expect to find if connections in the network are made at random. In general, one can

---

[8]It is sometimes claimed that essentially all networks show clustering higher than expected [12, 323], which is at odds with the results given here. There seem to be two reasons for the disagreement. First, the claims are based primarily on comparisons of measured clustering coefficients against values calculated on the Poisson random graph, a simple model network with a Poisson degree distribution, which we study in Chapter 12.1. Many networks, however, have right-skewed degree distributions which are very far from Poissonian, and hence the random graph is a poor model against which to compare measurements and probably gives misleading results. Second, the clustering coefficients in these comparisons are mostly calculated as an average of the local clustering, following Eq. (7.44). On networks with highly skewed degree distributions this definition can give very different results from the definition, Eq. (7.41), used in our calculations. Usually Eq. (7.44) gives much larger numbers than Eq. (7.41), which could explain the discrepancies in the findings.

find counts that are higher, lower, or about the same as the expected values, all of which can have implications for the understanding of the networks in question. For example, Milo *et al.* [221] looked at motif counts in genetic regulatory networks and neural networks and found certain small motifs that occurred far more often than was expected on the basis of chance. They conjectured that these motifs were playing the role of functional "circuit elements," such as filters or pulse generators, and that their frequent occurrence in these networks might be an evolutionary result of their usefulness to the organisms involved.

8.6.1 LOCAL CLUSTERING COEFFICIENT

In Section 7.9.1 we introduced the local clustering coefficient for a vertex:

$$C_i = \frac{\text{(number of pairs of neighbors of } i \text{ that are connected)}}{\text{(number of pairs of neighbors of } i\text{)}}, \tag{8.25}$$

which is the fraction of pairs of neighbors of vertex i that are themselves neighbors. If we calculate the local clustering coefficient for all vertices in a network, an interesting pattern emerges in many cases: we find that on average vertices of higher degree tend to have lower local clustering [278, 318]. Figure 8.12, for example, shows the average value of the local clustering coefficient for vertices of degree k on the Internet as a function of k. The decrease of the average C_i with k is clear. It has been conjectured that plots of this type take either the form $C_i \sim k^{-0.75}$ [318] or the form $C_i \sim k^{-1}$ [278]. In this particular case neither of these conjectures matches the data very well, but for some other networks they appear reasonable.

On possible explanation for the decrease in C_i with increasing degree is that vertices group together into tightly knit groups or communities, with vertices being connected mostly to others within their own group. In a network showing this kind of behavior vertices that belong to small groups are constrained to have low degree, because they have relatively few fellow group members to connect to, while those in larger groups can have higher degree. (They don't have to have higher degree, but they can.) At the same time, the local clustering coefficient of vertices in small groups will tend to be larger. This occurs because each group, being mostly detached from the rest of the network, functions roughly as its own small network and, as discussed in Section 8.6, smaller networks are expected to have higher clustering. When averaged over many groups of different sizes, therefore, we would expect vertices of lower degree to have higher clustering on average, as in Fig. 8.12.[9]

Community structure in networks is discussed at some length in Chapter 11.

---

[9]An alternative and more complex proposal is that the behavior of the local clustering co-

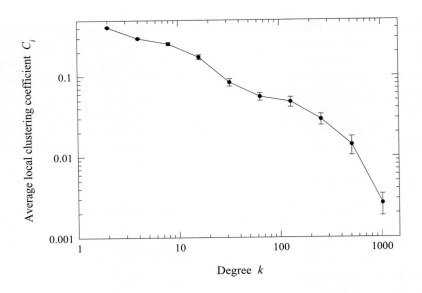

Figure 8.12: Local clustering as a function of degree on the Internet. A plot of the measured mean local clustering coefficient of vertices on the Internet (at the level of autonomous systems) averaged over all vertices with the given degree.

8.7 ASSORTATIVE MIXING

Assortative mixing or homophily is the tendency of vertices to connect to others that are like them in some way. We discussed assortative mixing in Section 7.13, where we gave some examples from social networks, such as the high school friendships depicted in Figs. 7.10 and 7.11 in which school students tend to associate more with others of the same ethnicity or age as themselves.

Of particular interest is assortative mixing by degree, the tendency of vertices to connect others with degrees that are similar to their own. We can also have disassortative mixing by degree, in which vertices connect to others with very different degrees. As we saw in Section 7.13.3, assortative mixing can have substantial effects on the structure of a network (see particularly Fig. 7.12).

Assortative mixing by degree can be quantified in a number of different

---

efficient arises through hierarchical structure in a network—that not only are there groups, but that the groups are divided into smaller groups, and those into still smaller ones, and so on. See Refs. [95, 278, 309].

ways. One of them is to use the correlation coefficient defined in Eq. (7.82):

$$r = \frac{\sum_{ij}(A_{ij} - k_i k_j/2m)k_i k_j}{\sum_{ij}(k_i \delta_{ij} - k_i k_j/2m)k_i k_j}. \tag{8.26}$$

If we were going to calculate the value of this coefficient, however, we should not do it directly from this equation, because the double sum over vertices i and j has a lot of terms (n^2 of them) and is slow to evaluate on a computer. Instead we write

$$r = \frac{S_1 S_e - S_2^2}{S_1 S_3 - S_2^2}, \tag{8.27}$$

with

$$S_e = \sum_{ij} A_{ij} k_i k_j = 2 \sum_{\text{edges }(i,j)} k_i k_j, \tag{8.28}$$

where the second sum is over all distinct (unordered) pairs of vertices (i, j) connected by an edge, and

$$S_1 = \sum_i k_i, \qquad S_2 = \sum_i k_i^2, \qquad S_3 = \sum_i k_i^3. \tag{8.29}$$

The sum in (8.28) has m terms, where m is the number of edges in the network and the sums in (8.29) have n terms each, so Eq. (8.27) is usually a lot faster to evaluate than Eq. (8.26).

In Table 8.1 we show the values of r for a range of networks and the results reveal an interesting pattern. While none of the values are of very large magnitude—the correlations between degrees are not especially strong—there is a clear tendency for the social networks to have positive r, indicating assortative mixing by degree, while the rest of the networks—technological, information, biological—have negative r, indicating disassortative mixing.

The reasons for this pattern are not known for certain, but it appears that many networks have a tendency to negative values of r because they are simple graphs. As shown by Maslov *et al.* [211], graphs that have only single edges between vertices tend in the absence of other biases to show disassortative mixing by degree because the number of edges that can fall between high-degree vertex pairs is limited. Since most networks are represented as simple graphs this implies that most should be disassortative, as indeed Table 8.1 indicates they are.

And what about the social networks? One suggestion is that social networks are assortatively mixed because they tend to be divided into groups, as discussed in Section 8.6.1. If a network is divided up into tightly knit groups of vertices that are mostly disconnected from the rest of the network, then, as

The computer time needed to calculate network quantities is an important topic in its own right. We discuss the main issues in Chapter 9.

we have said, vertices in small groups tend to have lower degree than vertices in larger groups. But since the members of small groups are in groups with other members of the same small groups, it follows that the low-degree vertices will tend to be connected to other low-degree vertices, and similarly for high-degree ones. This simple idea can be turned into a quantitative calculation [252] and indeed it appears that, at least under some circumstances, this mechanism does produce positive values of r.

Thus a possible explanation of the pattern of r-values seen in Table 8.1 is that most networks are naturally disassortative by degree because they are simple graphs while social networks (and perhaps a few others) override this natural bias and become assortative by virtue of their group structure.

PROBLEMS

8.1 One can calculate the diameter of certain types of network exactly.

a) What is the diameter of a clique?

b) What is the diameter of a square portion of square lattice, with L edges (or equivalently $L + 1$ vertices) along each side, like this:

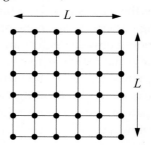

What is the diameter of the corresponding hypercubic lattice in d dimensions with L edges along each side? Hence what is the diameter of such a lattice as a function of the number n of vertices?

c) A Cayley tree is a symmetric regular tree in which each vertex is connected to the same number k of others, until we get out to the leaves, like this:

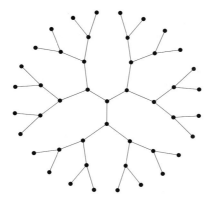

(We have $k = 3$ in this picture.)

Show that the number of vertices reachable in d steps from the central vertex is $k(k - 1)^{d-1}$ for $d \geq 1$. Hence find an expression for the diameter of the network in terms of k and the number of vertices n.

d) Which of the networks in parts (i), (ii), and (iii) displays the small-world effect, defined as having a diameter that increases as $\log n$ or slower?

8.2 Suppose that a network has a degree distribution that follows the exponential form $p_k = Ce^{-\lambda k}$, where C and λ are constants.

a) Find C as a function of λ.

b) Calculate the fraction P of vertices that have degree k or greater.

c) Calculate the fraction W of ends of edges that are attached to vertices of degree k or greater.

d) Hence show that for the exponential degree distribution with exponential parameter λ, the Lorenz curve—the equivalent of Eq. (8.23)—is given by

$$W = P - \frac{1 - e^\lambda}{\lambda} P \ln P.$$

e) Show that the value of W is greater than one for some values of P in the range $0 \leq P \leq 1$. What is the meaning of these "unphysical" values?

8.3 A particular network is believed to have a degree distribution that follows a power law. Among a random sample of vertices in the network, the degrees of the first 20 vertices with degree 10 or greater are:

16	17	10	26	13
14	28	45	10	12
12	10	136	16	25
36	12	14	22	10

Estimate the exponent α of the power law and the error on that estimate using Eqs. (8.6) and (8.7).

8.4 Consider the following simple and rather unrealistic mathematical model of a network. Each of n vertices belongs to one of several groups. The mth group has n_m vertices and each vertex in that group is connected to others in the group with independent probability $p_m = A(n_m - 1)^{-\beta}$, where A and β are constants, but not to any vertices in other groups. Thus this network takes the form of a set of disjoint clusters or communities.

a) Calculate the expected degree $\langle k \rangle$ of a vertex in group m.

b) Calculate the expected value $\overline{C}_m$ of the local clustering coefficient for vertices in group m.

c) Hence show that $\overline{C}_m \propto \langle k \rangle^{-\beta/(1-\beta)}$.

d) What value would β have to have for the expected value of the local clustering to fall off with increasing degree as $\langle k \rangle^{-3/4}$?

PART III

COMPUTER ALGORITHMS

CHAPTER 9

BASIC CONCEPTS OF ALGORITHMS

*An introduction to some of the basic concepts of
computer algorithms for network calculations,
particularly data structures for storing networks and
methods for estimating the time computations will take*

IN THE preceding chapters of this book we have introduced various types of
networks encountered in scientific study, methods for collecting data about
those networks, and some of the basic theoretical tools used to describe and
quantify networks. Then in the last chapter we combined these ideas in an
analysis of the structural features of a variety of real-world networks, revealing
in the process a number of interesting patterns that will be important to our
further studies in the remainder of the book.

Analysis of this kind, and most analysis involved in the contemporary
study of networks, is primarily performed using computers. In the early days
of network analysis in the first part of the twentieth century, calculations were
mostly performed by hand, partly out of necessity, since computers were slow,
expensive, and rare, but also because the networks studied were typically quite
small, consisting of perhaps just a few dozen vertices or even less. These days
we are concerned with networks that have thousands or even millions of vertices. Gathering and analyzing the data for networks like these is only possible
because of the advent of fast cheap computing.

Some networks calculations are simple enough that it is obvious how one
would get a computer to carry them out, but many are not and performing
them efficiently requires careful consideration and thoughtful programming.
Even merely storing a network in a computer requires some thought, since
there are many methods for doing it and the choice of method can make a
substantial difference to the performance of subsequent calculations.

In this chapter and the following two we discuss some of the techniques

275

and algorithms used for network calculations on computers. A good understanding of the material discussed here will form a solid foundation for writing software to perform a wide variety of calculations with network data.

In this chapter we describe some simple but important ideas about the running time of algorithms and data structures for the storage of networks. We will not describe any actual algorithms in this chapter, but the ideas introduced form a foundation for understanding the algorithms that appear in the following chapters.

In Chapter 10 we describe a selection of basic network algorithms, including many of the classics of the field, such as algorithms for calculating centrality indices, finding components, and calculating shortest paths and maximum flows. We continue our study of algorithms in Chapter 11, where we look at matrix-based algorithms and particularly at methods for network "partitioning."

Understanding the content of these chapters does not require that you know how to program a computer. We will not, for instance, discuss particular programming languages. However, some experience with programming will certainly help enormously in understanding the material, and the reader who has none will in any case probably not have very much use for the methods we describe.

Conversely, readers who already have a thorough knowledge of computer algorithms may well find some of the material here too basic for them, particularly the material on run times and data structures in the present chapter. Such readers should feel free to skip material as appropriate and move quickly on to the possibly less familiar subject matter of Chapters 10 and 11. For very advanced readers for whom all the material covered here is already familiar, or readers who simply wish to go into the subject in greater detail, we recommend the books by Cormen *et al.* [81], which is a general computer science text on algorithms, and by Ahuja *et al.* [8], which is specifically on network algorithms.

Before we leap into the study of algorithms, one further word of advice is worthwhile. Many of the standard algorithms for the study of networks are already available, ready-made, in the form of professional network analysis software packages. Many of these packages are of very high quality, produced by excellent and knowledgeable programmers, and if they are adequate for your needs then there is no reason not to use them. Writing and debugging your own software for the analysis of network data can take hours or days, and there is little reason to expend that time when someone else has already done it for you. Table 9.1 lists some of the most widely used current software packages for the analysis of network data along with a brief description of

Name	Availability	Platform	Description
Pajek	Free	W	Interactive social network analysis and visualization
Net Workbench	Free	WML	Interactive network analysis and visualization
Netminer	Commercial	W	Interactive social network analysis and visualization
InFlow	Commercial	W	Interactive social network analysis and visualization
UCINET	Commercial	W	Interactive social network analysis
yEd	Free	WML	Interactive visualization
Visone	Free	WL	Interactive visualization
Graphviz	Free	L	Visualization
NetworkX	Free	WML	Interactive network analysis and Python library
JUNG	Free	WML	JAVA library for network analysis and visualization
igraph	Free	WML	C/R/Python libraries for network analysis
GTL	Free	WML	C++ library for network analysis
LEDA/AGD	Commercial	WL	C++ library for network analysis

Table 9.1: A selection of software implementing common network algorithms. Platforms are Microsoft Windows (W), Apple Macintosh (M), and Linux (L). Most Linux programs also run under Unix and Unix-like systems such as BSD, and many Windows programs can run on Macs and Linux systems using emulation software.

what they do. The present author, for instance, has made considerable use of Graphviz, Pajek, and yEd, all of which provide useful features that could save you a lot of time in your work. Some other network calculations, especially the matrix-based methods of Chapter 11 and calculations using the models of Chapters 12 to 15, can be performed using standard mathematical software such as Matlab, Mathematica, or Maple, and again there is no reason not to make use of these resources if they are adequate for the particular task before you.

That said there are still some excellent reasons for studying network algorithms and computer methods. First of all, even when you are making use of pre-packaged software to do your calculations, it helps greatly if you understand how the algorithms work and what the software is doing. Much time can be wasted when people fail to understand how a program works or misunderstand the kinds of answers the program can give them. Furthermore, if you are going to undertake a substantial amount of work using network data, you will sooner or later find that you need to do something that cannot be done with standard software and you'll have to write some programs of your own.

Second, there is a marked tendency in the current networks literature for some researchers to restrict their calculations to those that can be carried out using the standard software. By relying on pre-packaged programs to do their calculations for them, researchers have become limited in what types of analysis they can perform. In this way, the standard packages have, in effect, shaped the research agenda of the empirical side of the field, which is completely the reverse of what it should be. Good research decides the interesting questions first and then goes in search of answers. Research that restricts itself only to the questions it already knows how to answer will be narrowly focused indeed. By following the developments in this and the following chapters, and, if you wish, reading further in the algorithms literature, you give yourself the opportunity to pursue whatever network questions are of interest to you, without having to rely on others to produce software to tackle those questions.

9.1 RUNNING TIME AND COMPUTATIONAL COMPLEXITY

Before we can look at exactly how network algorithms work, there is an important issue we need to tackle, that of *computational complexity*. If you have programmed computers before, you may well have had the experience of writing a program to perform a particular calculation and setting it running, only to find that it is still running an hour or even a day later. Performing a quick back-of-the-envelope calculation, you discover to your dismay that the calculation you have started will take a thousand years to finish, and hence that the program you wrote is basically useless.

The concept of computational complexity (or just "complexity" for short) is essentially a more formal version of back-of-the-envelope calculations like this one, and is useful precisely because it helps us to avoid wasting our energies on programs that will not finish running in any reasonable amount time. By considering the complexity of an algorithm before we even start to write a computer program, we can be sure we are writing one that will actually finish.

Computational complexity is a measure of the running time of a computer algorithm. Consider a simple example: how long does it take to find the largest number in a list of n numbers? Assuming the numbers are not given to us in some special order (such as largest first), then there is no quicker way to find the largest than simply to go through the whole list, item by item, keeping a running record of the largest number we have seen, until we get to the end.

This is a very simple example of a computer algorithm. We could use it, for instance, to find the vertex in a network that has the highest degree. The algorithm consists of a number of steps, one for each number in the list. On each step, we examine the next number in the list and ask whether it is larger than the largest we have seen so far. If it is, it becomes the new largest-number-seen-so-far, otherwise nothing happens and we move on to the next step.

Now here is the crucial point: in the *worst possible case* the most work we will have to do for this algorithm is on each step to (1) examine the next number, (2) compare it with our previous record holder, and (3) replace the previous record holder with the new number. That is, the largest amount of work we have to do happens when every number is bigger than all the ones before it.

In this case the amount of work we do is the same on every step and hence the total time taken to complete the algorithm, its running time, is just $n\tau$, where τ is the time taken on each individual step. If we are lucky, the actual time taken may be less than this, but it will never be more. Thus we say that the running time or *time complexity* of this algorithm is order n, or just $O(n)$ for short. Technically the notation $O(n)$ means that the running time varies as a constant times n *or less*, to leading order in n.[1] We say "to leading-order" because it is possible that there may be contributions to the running time that increase with system size more slowly than this leading-order term. For instance, there might be some initial start-up time for the algorithm, such as time taken initializing variables, that is a constant independent of n. We would denote this time as being $O(1)$, i.e., a constant times 1. By convention, however, one drops such sub-leading terms when citing the complexity of an algorithm, because if n is large enough that the running time of the program becomes a serious issue then the sub-leading terms will usually be small enough by comparison with the leading ones that they can be safely neglected.[2] Thus the time complexity of our simple largest-number algorithm is just $O(n)$.

Technically, the computational complexity of an algorithm is an indication of how the algorithm's running time scales *with the size of its input*. In our example, the input to the algorithm is the list of numbers and the size of that input is the length n of the list. If this algorithm were used to find the highest degree node in a network, then the size of the input would be the number of vertices in the network. In many of the network algorithms we will look at this will be the case—the number of vertices n will be the important parameter we consider. In other cases, the important parameter will be the number of edges m in the network, while in others still we will need both m and n to fully specify the size of the input—there could be different parts to an algorithm, for instance, that operate separately on the vertices and the edges, so that the total running time depends on both. Thus, for example, we will see in Section 10.3 that the algorithm known as "breadth-first search," which is used for finding geodesic paths in networks, has a computational complexity $O(m) + O(n)$ for

---

[1] If we wish to say that the running time is exactly proportional to n, we can use the notation $\Theta(n)$.

[2] There are occasional instances where this is not true, so it is worth just bearing in mind the possibility of sub-leading terms.

a network with m edges and n vertices, meaning that it runs in time $am + bn$ where a and b are constants, or quicker. Very often one writes this, in shorthand and slightly sloppy form, as $O(m + n)$. This latter notation is not meant to imply that the constants in front of m and n are the same.

In a lot of networks research we are concerned with sparse graphs (see Section 6.9) and particularly with graphs for which m increases in proportion to n as n becomes large. To put that another way, the mean degree of the network $c = 2m/n$ remains constant (see Eq. (6.23)). In such networks, $O(m + n) \equiv O(n)$ and we can drop the m from our notation.

The importance of the computational complexity lies in its use for estimating the actual running time of real algorithms. If a particular algorithm is going to take a month to solve a problem of interest, or a year or a century, we'd like to know that in advance. We want to estimate how long the calculation is going to take before we start it, so we can make a decision about whether the wait is justified. A knowledge of the computational complexity allows us to do that by measuring run-time on a small problem and then scaling up appropriately to the size of the real problem.

For example, suppose we wish to run the breadth-first search algorithm mentioned above on a network with a million vertices and ten million edges. Knowing that the algorithm has time complexity $O(m + n)$, we could start out with a small test-run of the program on a network with $n = 1000$ vertices, say, and $m = 10\,000$ edges. Often we artificially create small networks just for the purposes of such tests. Perhaps we find that the program finishes in a second on the test network. We then scale up this result knowing that the running time varies as $am + bn$. On the full network with $n = 1\,000\,000$ and $m = 10\,000\,000$ both n and m are a thousand times larger than on the test network, so the program should take about a thousand times longer to finish, i.e., a thousand seconds or about a quarter of an hour. Armed with this information we can safely start our program working on the larger problem and step out for a cup of tea or a phone call while we wait for it finish.

Conversely, suppose we had an algorithm with computational complexity $O(n^4)$. That means that if we increase the number of vertices n in our network by a factor of a thousand the running time will increase by a trillion. In such a case it almost does not matter what the run time of the algorithm is on our small test network; the run time on the full network is going to be prohibitively long. For instance, if the test network takes a second again, then the full network would take a trillion seconds, which is around $30\,000$ years. In this case, we would certainly abandon the calculation, or at least look for a faster algorithm that can complete it in reasonable time.

Finding the computational complexity of an algorithm, generating test networks, performing small runs, and doing scaling calculations of this type all require some work—additional work on top of the work of developing and programming the computer algorithm in the first place. Nonetheless, this extra work is well worth the effort involved and one should always perform this type of analysis, at least in some rough manner, before embarking on any major numerical calculations. Computational complexity will be one of our major concerns throughout of the discussions of algorithms in this chapter and the following two. An algorithm is next to useless if its running time scales poorly with the size of a network. In practice, any algorithm that scales with system size as $O(n^3)$ or greater is useless for large networks, although such algorithms still find some use for the smaller cases. In the world of computer science, where many researchers have devoted their entire careers to the invention of new algorithms for solving particular problems, the calculation of the computational complexity of an algorithm is a primary goal—often *the* primary goal—of research. Plenty of papers are published whose sole contribution is to provide a calculation of the complexity of some algorithm.

It is worth mentioning that calculations of the run time of algorithms based on their complexity, as above, do not always give very accurate answers. We have mentioned already that standard measures of time complexity neglect sub-leading contributions to the run time, which may introduce inaccuracies in practical situations. But in addition there are, for technical reasons, many cases where the behavior of the run time is considerably poorer than a simple scaling argument would suggest. For instance, in calculations on networks it is important that the entire network fit in the main memory (RAM) of our computer if the algorithm is to run quickly. If the network is so large that at least part of it must be stored on a disk or some other slow form of storage, then the performance of the algorithm may be substantially hindered.[3] Even if the entire network fits in the main memory, there may be additional space required for the operation of the algorithm, and that must fit in the memory too. Also, not all kinds of memory are equally fast. Modern computers have a small amount of extra-fast "cache" memory that the computer can use for storing small quantities of frequently used data. If all or most of the data for a calculation fit in the cache, then the program will run far faster than if it does not.

There are also cases in which a program will perform better than the es-

---

[3]There are whole subfields in computer science devoted to the development of algorithms that run quickly even when part of the data is stored on a slow disk. Usually such algorithms work by reordering operations so that many operations can be performed on the same data, stored in the main memory, before swapping those data for others on the disk.

timate based on its complexity would indicate. In particular, the complexity is usually calculated by considering the behavior of the program in the worst case. But for some programs the worst-case behavior is relatively rare, occurring only for certain special values of the program inputs or particularly unlucky parameter choices, and the typical behavior is significantly better than the worst case. For such programs the complexity can give an unreasonably pessimistic estimate of running time.

For all of these reasons, and some others as well, programs can show unexpected behaviors as the size of their input increases, sometimes slowing down substantially more than we would expect given their theoretical time complexity and sometimes running faster. Nonetheless, computational complexity is still a useful general guide to program performance and an indispensable tool in the computer analysis of large networks.

9.2   STORING NETWORK DATA

The first task of most programs that work with network data is to read the data, usually from a computer file, and store it in some form in the memory of the computer. Network data stored in files can be in any of a large number of different formats, some standard, some not, but typically the file contains an entry containing information for each vertex or for each edge, or sometimes both. The way the data are stored in the computer memory after they are read from the file can, as we will see, make a substantial difference to both the speed of a program and the amount of memory it uses. Here we discuss some of the commonest ways to store network data.

The first step in representing a network in a computer is to label the vertices so that each can be uniquely identified. The most common way of doing this is to give each a numeric label, usually an integer, just as we have been doing in our mathematical treatment of networks in Chapters 6 and 7. In the simplest case, we can number the n vertices of a network by the consecutive integers $i = 1 \ldots n$, although in some cases we might wish to use non-consecutive integers for some reason or to start the numbering from a different point. (For instance, in the C programming language it is conventional for numbering to start at zero and run through $i = 0 \ldots n - 1$.) Most, though not all, file formats for storing networks already specify integer labels for vertices, which may simplify things. For those that don't, one typically just labels vertices consecutively in the order they are read from the file. In what follows, we will assume that vertices are numbered $1 \ldots n$.

Often the vertices in a network have other notations or values attached to them in addition to their integer labels. The vertices in a social network, for instance, might have names; vertices in the World Wide Web might have URLs; vertices on the Internet might have IP addresses or AS numbers. Vertices could also have properties like age, capacity, or weight represented by other numbers, integer or not. All of these other notations and values can be stored straightforwardly in the memory of the computer by defining an array of a suitable type with n elements, one for each vertex, and filling it with the appropriate values in order. For example, we might have an array of n text strings to store the names of the individuals in a social network, and another array of integers to store their ages in years.

Having devised a suitable scheme for storing the properties of vertices, we then need a way to represent the edges in the network. This is where things get more complicated.

9.3 THE ADJACENCY MATRIX

In most of the mathematical developments of previous chapters we have represented networks by their adjacency matrix A_{ij}—see Section 6.2. The adjacency matrix also provides one of the simplest ways to represent a network on a computer. Most computer languages provide two-dimensional arrays that can be used to store an adjacency matrix directly in memory. An array of integers can be used if the adjacency matrix consists only of integers, as it does for unweighted simple graphs or multigraphs. An array of floating-point numbers is needed for an adjacency matrix that may have reals (non-integers) as its elements, as does the adjacency matrix of some weighted networks—see Section 6.3.

Storing a network in the form of an adjacency matrix is convenient in many ways. Most of the formulas and calculations described in this book are written out in terms of adjacency matrices. So if we have that matrix stored in our computer it is usually a trivial matter to turn those formulas into computer code to calculate the corresponding quantities.

The adjacency matrix can be highly advantageous for other reasons too. For instance, if one wishes to add or remove an edge between a given pair of vertices, this can be achieved very quickly with an adjacency matrix. To add an edge between vertices i and j one simply increases the ijth element of the adjacency matrix by one. To remove an edge between the same vertices one decreases the element by one. These operations take a constant amount of time regardless of the size of the network, so their computational complexity is $O(1)$. Similarly if we want to test whether there is an edge between a given pair of vertices i and j we need only inspect the value of the appropriate matrix element, which can also be done in $O(1)$ time.

Undirected networks give a slight twist to the issue since they are represented by symmetric matrices. If we want to add an undirected edge between vertices i and j, then in principle we should increase both the ijth and jith elements of the adjacency matrix by one, but in practice this is a waste of time. A better approach is to update only elements in the upper triangle of the matrix and leave the lower one empty, knowing that its correct value is just the mirror image of the upper triangle.[4] To put this another way, we only update elements (i, j) of the matrix for which $i < j$. (For networks in which self-edges are allowed, we would use the diagonal elements as well, so we would update elements with $i \leq j$—see Section 6.2.) For instance, if we wish to create an edge between vertex 2 and vertex 1, this means in principle that we want to increase both the $(2, 1)$ element and the $(1, 2)$ element of the adjacency matrix by one. But, since we are only updating elements with $i < j$, we would increase only the $(1, 2)$ element and leave the other alone.[5]

Taking this idea one step further, we could not bother to store the lower triangle of the adjacency matrix in memory at all. If we are not going to update it, why waste memory storing it? Unfortunately, dropping the lower triangle of the matrix makes our remaining matrix triangular itself, and most computer languages don't contain arrays designed to hold triangular sets of quantities. One can, by dint of a certain amount of work, arrange to store triangular matrices using, for example, the dynamic memory allocation facilities provided by languages like C and JAVA, but this is only worth the effort if memory space is the limiting factor in performing your calculation.

The adjacency matrix is not always a convenient representation, however. It is cumbersome if, for instance, we want to run quickly through the neighbors of a particular vertex, at least on a sparse graph. The neighbors of vertex i are denoted by non-zero elements in the ith row of the adjacency matrix and to find them all we would have to go through all the elements of the row one by one looking for those that are non-zero. This takes time $O(n)$ (since that is the length of the row), which could be a lot of time in a large network, and yet on a sparse graph most of that time is wasted because each vertex is connected to only a small fraction of the others and most of the elements in the adjacency matrix are zero. As we will see in this chapter, many network algorithms do indeed require us to find all neighbors of a vertex, often repeatedly, and for

---

[4]For directed networks, which are represented by asymmetric adjacency matrices, this issue does not arise—the full matrix, both the upper and lower triangles, is used to store the structure of the network.

[5]Of course we could equally well store the edges in the lower triangle of the matrix and neglect the upper triangle. Either choice works fine.

Operation	Adjacency matrix	Adjacency list	Adjacency tree
Insert	$O(1)$	$O(1)$	$O(\log(m/n))$
Delete	$O(1)$	$O(m/n)$	$O(\log(m/n))$
Find	$O(1)$	$O(m/n)$	$O(\log(m/n))$
Enumerate	$O(n)$	$O(m/n)$	$O(m/n)$

Table 9.2: The leading-order time complexity of four operations for various representations of a network of n vertices and m edges. The operations are adding an edge to the network (insert), removing an edge from the network (delete), testing whether a given pair of vertices are connected by an edge (find), and listing the neighbors of a given vertex (enumerate).

such algorithms the adjacency matrix is not an ideal tool.

The computational complexity of the network operations discussed here for an adjacency matrix is summarized in Table 9.2.

Another disadvantage of the adjacency matrix representation is that for sparse graphs it makes inefficient use of computer memory. In a network in which most elements of the adjacency matrix are zero, most of the memory occupied by the matrix is used for storing those zeros. As we will see, other representations exist, such as the "adjacency list," that avoid storing the zeros and thereby take up much less space.[6]

It is a simple matter to work out how much memory is consumed in storing the adjacency matrix of a network. The matrix has n^2 elements. If each of them is an integer (which requires 4 bytes for its storage on most modern computers) then the entire matrix will take $4n^2$ bytes. At the time of writing, a typical computer has about 10^{10} bytes of RAM (10 GB), and hence the largest network that can be stored in adjacency matrix format satisfies $4n^2 = 10^{10}$, or $n = 50\,000$. This is not nearly large enough to store the largest networks of interest today, such as large subsets of the Web graph or large social networks, and is not even big enough for some of the medium-sized ones.

The disadvantages of the adjacency matrix representation described here

---

[6]One advantage of the adjacency matrix is that the amount of space it consumes is independent of the number of edges in the network. (It still depends on the number of vertices, of course.) As we will see in Section 9.4, other data formats such as the adjacency list use varying amounts of memory, even for networks with the same number of vertices, depending on how many edges there are. In calculations where edges are frequently added or removed it may be convenient—and increase the speed of our algorithms—to have the size of our data structures remain constant, although this advantage must be weighed against the substantial space savings of using the adjacency list or other memory-efficient formats.

apply primarily to sparse networks. If one is interested in dense networks—those in which a significant fraction of all possible edges are present—then the adjacency matrix format may be appropriate. It will still use a lot of memory in such cases, but so will any data format, since there is simply a lot of information that needs to be stored, so the advantages of other formats are less significant. The adjacency matrix may also be a good choice if you are only interested in relatively small networks. For instance, the social network analysis package UCINET, which is targeted primarily at sociological practitioners working with smaller networks, uses the adjacency matrix format exclusively. Most current research on networks however is focused on larger data sets, and for these another representation is needed.

9.4 THE ADJACENCY LIST

The simplest alternative to storing the complete adjacency matrix of a network is to use an *adjacency list*. The adjacency list is, in fact, probably the most widely used network representation for computer algorithms.

An adjacency list is actually not just a single list but a set of lists, one for each vertex i. Each list contains the labels of the other vertices to which i is connected by an edge. Thus, for example, this small network:

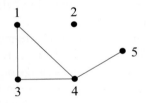

would be represented by this adjacency list:

Vertex	Neighbors
1	3, 4
2	
3	4, 1
4	5, 1, 3
5	4

An adjacency list can be stored in a series of integer arrays, one for each vertex, or as a two-dimensional array with one row for each vertex.[7] It is common to also store somewhere the degree of each vertex, so that we know how many

---

[7]Note that the number of entries in the list of neighbors for a vertex varies from one vertex

entries there are in the list of neighbors for each vertex; this can be done using a separate array of n integers. Note also that there is usually no requirement that the neighbors of a vertex in the adjacency list appear in numerical order. Normally they are allowed to appear in any order.

In the example adjacency list above, each edge appears twice. For instance, the existence of an edge between vertices 1 and 3 means that vertex 3 is listed as a neighbor of vertex 1 and vertex 1 is also listed as a neighbor of vertex 3. To represent m edges, therefore, we need to store $2m$ integers. This is much better than the n^2 integers used to store the full adjacency matrix.[8] For instance, on a computer where each integer occupies 4 bytes of memory, a network with $n = 10\,000$ vertices and $m = 100\,000$ edges would occupy 800 kB in adjacency list form, as opposed to 400 MB in matrix format. The double storage of the edges is slightly wasteful—we could save an additional factor of two if we only stored each edge once. However, the double storage turns out to have other advantages, making our algorithms substantially faster and easier to program in many cases, and these benefits are normally worth the extra cost in terms of space. In these days of cheap memory, not many networks are large enough that space to store an adjacency list is a serious consideration.

An adjacency list can store networks with multiedges or self-edges. A multiedge is represented by multiple identical entries in the list of neighbors of a vertex, all pointing to the same adjacent vertex. A self-edge is represented by an entry identifying a vertex as its own neighbor. In fact, a self-edge is most correctly represented by *two* such entries in the list, so that the total number of entries in the list is still equal to the degree of the vertex. (Recall that a self-edge adds 2 to the degree of the vertex it is connected to.)

The example adjacency list above is for an undirected network, but adjacency lists can be used with directed networks as well. For instance, this network:

---

to another, and may even be zero, being equal to the degree of the corresponding vertex. Most modern computer languages, including C and its derivatives and JAVA, allow the creation of two-dimensional matrices with rows having varying numbers of elements in this fashion. Some older languages, like FORTRAN 77, do not allow this, making it more difficult to store adjacency lists in a memory-efficient way.

[8]Note that the amount of memory used is now a function of m rather than than n. For algorithms in which edges are added or removed from a network during the course of a calculation this means that the size of the adjacency list can change, which can complicate the programming and potentially slow down the calculation. Normally, however, this added complication is not enough to outweigh the considerable benefits of the adjacency list format.

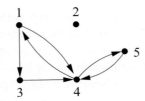

could be represented by this adjacency list:[9]

Vertex	Outgoing edges
1	3, 4
2	
3	4
4	5, 1
5	4

Here we have listed the outgoing edges for each vertex. Since each edge is outgoing from *some* vertex, this approach is guaranteed to capture every edge in the network, but each edge now appears only once in the adjacency list, not twice as in the undirected case.

Alternatively, we could represent the same network by listing the ingoing edges for each vertex thus:

Vertex	Incoming edges
1	4
2	
3	1
4	3, 1, 5
5	4

In principle these two representations are equivalent. Both include all the edges and either of them can be constructed from a knowledge of the other. When creating computer programs, however, the crucial point is to have the information you need for your calculations easily available, so that the program runs fast. Different calculations require different information and some might need ingoing edges while others need outgoing ones. The choice of which adjacency list to use thus depends on the particular calculations being

---

[9]Indeed, the adjacency list for an undirected network such as that given above could be viewed as a special case of the directed adjacency list for a network in which each undirected edge is replaced by two directed ones, one in each direction. It takes only a moment to convince oneself that this results precisely in the sort of double representation of each edge that we saw in the undirected case.

performed. Some calculations even require both ingoing and outgoing edges, in which case we could create a double adjacency list like this:

Vertex	Incoming edges	Outgoing edges
1	4	3, 4
2		
3	1	4
4	3, 1, 5	5, 1
5	4	4

Note that, as in the undirected case considered above, this double adjacency list stores each edge twice, once as an incoming edge and once as an outgoing one, and is thus in some respects wasteful of space, although not to an extent that is normally a problem.

As with the adjacency matrix it is important also to ask how fast our calculations will run if we use an adjacency list. Will they still run at a reasonable pace? If the answer is no, then the adjacency list is not a useful representation, no matter what its other advantages may be.

Consider the undirected case[10] and the four basic network operations that we considered previously for the adjacency matrix, addition and removal of edges, finding whether an edge is present, and enumeration of all edges connected to a vertex—see Table 9.2.

We can add an edge to our network very quickly: to add an edge (i, j) we need only add one new entry each to the ends of the neighbor lists for vertices i and j, which takes time $O(1)$.

Finding or removing an edge is a little harder. To find whether an edge exists between vertices i and j we need to go through the list of neighbors of i to see whether j appears in that list, or vice versa. Since the list of neighbors is in no particular order, there is no quicker way of doing this than simply going through the entire list step by step from the beginning. In the worst case, we will have check all elements to find our edge or to confirm that it does not exist, and on average[11] this will take time of order the mean number c of elements in the list, which is given by the mean degree $c = 2m/n$ (Eq. 6.23).

---

[10]The answers are essentially the same in the directed case. The demonstration is left as an exercise.

[11]We are thus calculating a sort of "average worst-case" behavior, allowing for the worst case in which we have to look through the entire list, but then averaging that worst case over many different lists. This is a reasonable (and standard) approach because almost all of the algorithms we will be considering do many successive "find" operations during a single run, but it does mean that we are technically not computing the complexity of the absolute worst case situation.

Thus the "find" operation takes time $O(m/n)$ for a network in adjacency list form. This is a bit slower than the same operation using an adjacency matrix, which takes time $O(1)$ (Section 9.3). On a sparse graph with constant mean degree, so that $m \propto n$ (see Sections 6.9 and 9.1), $O(m/n) \equiv O(1)$, so technically the complexity of the adjacency list is as good as that of the adjacency matrix, but in practice the former will be slower than the latter by a constant factor which could become large if the average degree is large.

Removing an edge involves first finding it, which takes time $O(m/n)$, and then deleting it. The deletion operation can be achieved in $O(1)$ time by simply moving the last entry in the list of neighbors to overwrite the entry for the deleted edge and decreasing the degree of the vertex by one (see figure). (If there is no last element, then we need do nothing other than decreasing the degree by one.) Thus the leading-order running time for the edge removal operation is $O(m/n)$.

However, the adjacency list really comes into its own when we need to run quickly through the neighbors of a vertex, a common operation in many network calculations, as discussed in Section 9.3. We can do this very easily by simply running through the stored list of neighbors for the vertex in question, which takes time proportional to the number of neighbors, which on average is $c = 2m/n$. The leading-order time complexity of the operation is thus $O(m/n)$, much better than the $O(n)$ of the adjacency matrix for the same operation.

The computational complexity of operations on the adjacency list is summarized in Table 9.2.

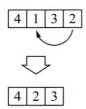

The element "1" is deleted from a list by moving the last element "2" to overwrite it.

9.5 TREES

The adjacency list is, as we have said, probably the most commonly used format for the storage of networks. Its main disadvantage is the comparatively long time it takes to find or remove edges—$O(m/n)$ time, compared with the $O(1)$ of the adjacency matrix. In many of the most common network algorithms, such as the breadth-first search of Section 10.3 or the augmenting path algorithm of Section 10.5, this is not a problem, since one never needs to perform these operations. Some algorithms, however, such as the algorithm for calculating the clustering coefficient given in Section 10.2, do require these operations and can be slowed by the use of an adjacency list. On a network with mean degree 100, for instance, the edge-finding operations in the clustering coefficient algorithm can be expected to slow the calculation down by about a factor of 100, which could make the difference between a calculation that takes an hour to finish and one that takes a week.

A data structure that has most of the advantages of the adjacency list, while being considerably faster in many situations, is the *adjacency tree*.[12] An adjacency tree is identical to an adjacency list except that each "row"—the set of neighbors of each vertex—is stored in a tree rather than a simple array. If you already know what a tree is then you probably don't need to read the rest of this section—you'll have got the idea already. If you don't know what a tree is, read on.[13]

A tree is a data structure for storing values of some kind, such as integers or real numbers, in a way that allows them to be added, removed, and located quickly. Trees come in a number of types. We consider the case where the values stored are all distinct, which is the relevant case for our purposes, although it is only slightly more complicated to allow for identical values.

Figure 9.1a shows an example of a tree that is being used to store the integers $\{1, 3, 4, 6, 7, 10, 12, 14\}$. The basic structure is one of a set of nodes (not to be confused with the nodes of the network that we are storing), which correspond to memory locations in the computer, each containing one of the integers. The tree nodes are arranged in a top-down fashion with a *root node* at the top.[14] Each node can have zero, one, or two *child nodes* that are drawn immediately below it and connected to it by lines to indicate the child relationship. One can create trees with nodes that have more than two children, but for our purposes it works better to limit the number to two. A tree with at most two children per node is called a *binary tree*.

Each node in the tree, except for the root, has exactly one *parent node*. The root has no parent. Nodes with no children are called *leaves*. On a computer a tree is typically implemented with dynamically allocated memory locations to store the numbers, and pointers from one node to another to indicate the child and parent relationships.[15]

---

[12] This is not a standard name. As far as the present author is aware, this data structure doesn't have a standard name, since it is not used very often. Moreover, the name isn't even entirely accurate. The structure is technically not a tree but a forest, i.e., a collection of many trees. Still, "adjacency tree" is simple and descriptive, and analogous to "adjacency list," which is also, technically, not a single list but a collection of lists.

[13] The word "tree" has a different meaning here from the one it had in Section 6.7, although the two are related. There a tree meant a network with no loops in it. Here it refers to a data structure, although as we will see, the tree data structure can also be regarded as a network with no loops. In effect we are using a network to store information about another network, which is a nice touch.

[14] As pointed out in a previous footnote on page 127, it is slightly odd to put the "root" the top of the tree. Most of us are more familiar with trees that have their roots at the bottom. We could of course draw the tree the other way up—it would have the same meaning—but it has become conventional to draw the root at the top of the picture, and we bow to that convention here.

[15] A "pointer" in computer programming is a special variable that holds the address in memory of another variable.

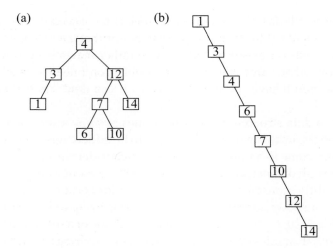

Figure 9.1: Two trees containing the same set of numbers. The two trees depicted here both store the numbers $\{1, 3, 4, 6, 7, 10, 12, 14\}$. (a) A balanced tree in which the depths of the subtrees below each node differ by no more than 1. (b) An unbalanced tree.

A defining property of our binary tree will be that the values stored in the left child of node i (if there is one) and in all other nodes descended from that child, are *less than* the value stored in i itself. Conversely, the values stored in the right child of i (if there is one) and all nodes descended from it are *greater than* the value stored in i. The reader might like to confirm that the values in Fig. 9.1a satisfy these requirements at every node in the tree.

Our goal is to use trees to store the lists of neighbors of each vertex in a network. We will show that if we do so we can perform our four basic network operations—addition, removal, and finding of edges, and enumeration of the complete set of a vertex's neighbors—very quickly on average. Below we first explain in the general language of the binary tree how these operations are achieved. At the end of the section we discuss how they are used in the specific context of the adjacency tree format for networks.

The find operation: The first tree operation we consider is the "find" operation, which is the operation of determining whether a particular value is present in our tree. We accomplish this operation as follows. Starting at the root node:

1. Examine the value x in the current node. If it is the value we are looking for, stop—our task is done.
2. If not and the value we are looking for is less than x, then by the prop-

erties described above, the value we are looking for must be in the left child of the current node or one of its descendants. So we now move to the left child, if there is one, which becomes our new current node. If there is no left child, then the value we are looking for does not exist in the tree and our task is done.

3. Conversely if the value we are looking for is greater than x, we move to the right child, if there is one, which becomes our new current node. If there is no right child, the value we are looking for does not exist in the tree and our task is done.

4. Repeat from step 1.

Taking the example of the tree in Fig. 9.1a, suppose that we are trying to determine whether the number 7 appears in the tree. Starting at the root note we find a 4, which is not the number 7 that we are looking for. Since 7 is greater than 4, we move to the right child of the root and there find a 12. Since 7 is less than 12 we move to the left child and there we find our number 7, and our search is over.

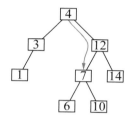

The path taken through our binary tree to locate the number 7.

On the other hand, suppose we wish to determine whether the number 9 appears in the tree. Again we would start at the root and move right, then left, then right as we went down the tree, arriving after three steps at the node containing the number 10. This number is not equal to the number 9 and since 9 is less than 10 we would normally now move to the left child down the tree. In this case, however, there is no left child, and hence we conclude that the number we are looking for does not exist in the tree.

How long does the "find" operation take? That depends on how many steps we have to take through the tree: we have to perform the same operations on each step, so the time taken is simply proportional to the number of steps. The maximum number of steps to reach any node from the root is called the *depth* of the tree. Unfortunately, the depth doesn't have a simple fixed value. There are many possible ways to store the same numbers in a tree while still obeying the conditions on child and parent nodes. For instance, both of the trees in Fig. 9.1 are valid ways to store the same set of numbers. The one in panel (a) has depth 4, while the one in panel (b) has depth 8. In general the maximum possible depth is given by an arrangement like (b) and is equal to the number k of values stored in the tree. For such a tree the find operation would, in the worst case, take $O(k)$ time, which is no better than what we get if we store the values in a simple array—in either case we just end up going through the values one by one until we find the one we want.

On the other hand, if we can make the tree reasonably *balanced*, like the one in Fig. 9.1a, then the depth can be a lot less than k. Let us calculate the minimum depth required to store k values. If we start at the top of the tree and fill in the levels one by one, we can put one value at the root, two values in the second level, four in the third, and so forth, each level holding twice as many values as the previous one. In general, level l can hold 2^{l-1} values and the total number of values in L levels is

$$\sum_{l=1}^{L} 2^{l-1} = 2^L - 1. \tag{9.1}$$

Setting this equal to k and rearranging, we find that $L = \log_2(k+1)$. However, L must be an integer, so, rounding up, we conclude that a minimum of

$$L = \lceil \log_2(k+1) \rceil \tag{9.2}$$

levels are needed to store k numbers, where $\lceil x \rceil$ denotes the smallest integer not less than x. If we can pack the values into the tree like this, filling each level completely, then our find operation will only take $O(\lceil \log_2(k+1) \rceil)$ to complete. In fact, since $\lceil \log_2(k+1) \rceil \leq \log_2(k+1) + 1$ we could also say that it will take $O(\log_2(k+1))$ time, where as usual we have kept only the leading-order term and dropped the sub-leading $+1$ term. We can also neglect the base of the logarithm, since all logs are proportional to one another, regardless of their base, and we can replace $\log(k+1)$ by $\log k$ since again we are only interested in the leading-order scaling.

Thus, we conventionally say that the find operation in a balanced tree containing k values can be completed in time $O(\log k)$. This is much better than the $O(k)$ of the simple list or the unbalanced tree. For a tree with $k = 100$ values stored in it we have $\log_2 k \simeq 7$, so the find operation should be about $100/7 \simeq 14$ times faster than for the simple list, and the speed advantage increases further the larger the value of k.

The addition operation: And how do we add a new value to a tree? The crucial point to notice is that in adding a new value we must preserve the relations between parent and child nodes, that lower values are stored in the left child of a node and its descendants and higher ones in the right child and descendants. These relations were crucial to the speedy performance of the find operation above, so we must make sure they are maintained when new items are added to the tree.

This, however, turns out not to be difficult. To add an item to the tree we first perform a "find" operation as above, to check if the value in question already exists in the tree. If the value already exists, then we don't need to add it. (Imagine for example that we are adding an edge to a network. If that edge already exists then we don't need to add it.)

On the other hand, if the value does not exist in the tree then the find operation will work its way down the tree until it gets to the leaf node that *would*

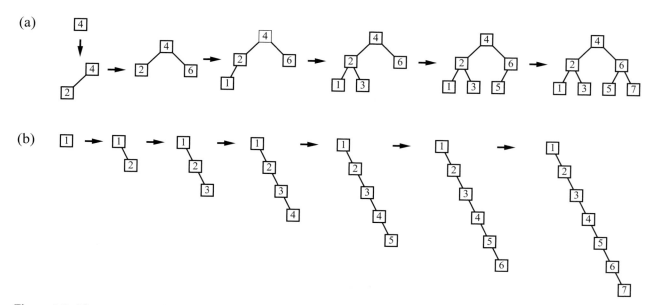

Figure 9.2: The structure of a tree depends on the order in which elements are added to it. Here the values 1 to 7 are added to a tree in two different orders. In (a) they are added in the order 4, 2, 6, 1, 3, 5, 7, resulting in a perfectly balanced tree with the minimum possible depth of three. In (b) they are added in the order 1, 2, 3, 4, 5, 6, 7, resulting in an unbalanced tree with the maximum possible depth of seven.

have been the parent of our value, but which does not have an appropriate child node. Then we simply add the appropriate child of that leaf node and store our value in it, thereby increasing the number of nodes in the tree by one. Since the find operation takes $O(\log k)$ time and the creation of the new node takes constant time, the leading-order complexity of the addition operation is $O(\log k)$.

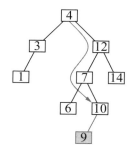

Addition of the number 9 to the tree.

Balancing the tree: This is satisfying and simple, but it immediately raises another problem. The position a newly added value occupies in the tree depends on the values that are already in the tree, and hence the shape of the tree depends, in general, on the values we add to it and the order in which we add them. Indeed, it turns out a tree can take quite different shapes even for the same values, just as a result of adding those values in different orders. In particular a tree can end up balanced or not as a result of different orders of addition—see Fig. 9.2—and if we are unlucky and get an unbalanced tree then the speed of our tree operations can be severely affected. Obviously we would like to avoid this if possible.

For algorithms in which elements are added to the tree only at the beginning of the program and no elements are added or removed thereafter (the clustering coefficient algorithm of Section 10.2 is an example), a simple solution to the problem is just to randomize the order in which the elements are added. Although the resulting tree is not completely packed full, as in the case considered above, it will still have depth $O(\log k)$ on average.[16]

If our algorithm requires us to add or remove values as we go along, then this approach will not work and we must explicitly balance the tree by performing *rebalancing* operations whenever the tree becomes unbalanced. We typically don't attempt to keep the elements packed as tightly in the tree as possible, but we can still achieve $O(\log k)$ run times by adopting a looser definition of balance. One practical and simple definition (though not the only one) is that a tree is balanced if the depth of the two subtrees below any node differ by no more than 1. The tree in Fig. 9.1a satisfies this criterion. For instance, the depths of the subtrees of the node containing the value 12 are 2 on the left and 1 on the right.

It is straightforward to prove that a tree satisfying this criterion, which is

---

[16]The proof is as follows.

Consider the set of "empty nodes," meaning the missing children immediately below current nodes in the tree (gray boxes in the figure on the right). Suppose that when there are k nodes in total in the tree there are c_k such empty nodes and that their average depth, measured from the root, is d_k. When we add one new value to the tree at random it will occupy one of these empty nodes thereby decreasing their number by one. At the same time two new empty nodes will appear, the children of the newly added node. Overall therefore $c_{k+1} = c_k + 1$. Noting that $c_1 = 2$, this immediately implies that $c_k = k + 1$.

At the same time the sum of the lengths of all paths from the root to an empty node, which by definition is equal to $c_k d_k$, decreases (on average) by d_k when a new random node is added and increases by $2(d_k + 1)$ for the two new ones, so that $c_{k+1} d_{k+1} = c_k d_k + d_k + 2$. Eliminating c_k, we then find that $d_{k+1} = d_k + 2/(k + 2)$. Noting that $d_1 = 2$, this implies

$$d_k = 1 + \sum_{m=0}^{k-1} \frac{2}{m+2} = -1 + 2 \sum_{m=1}^{k+1} \frac{1}{m} = 2\gamma - 1 + 2\ln k + O(1/k),$$

where $\gamma = 0.5772\ldots$ is Euler's constant.

Thus the average depth of the empty nodes is $O(\log k)$. Since the find and addition operations on the tree both involve searching the tree until an empty node is encountered, it immediately follows that both operations have average complexity $O(\log k)$ on our randomized tree, just as they do on the optimally packed tree.

Note that this is only a statement about the average behavior of the tree and not about the worst-case behavior. If we are unlucky the depth of the tree could be much larger than the average, up to the maximum depth k. The randomized tree only provides a guarantee of average performance, meaning performance averaged over many possible runs of an algorithm. Individual runs will vary around the mean.

called an *AVL tree*, has depth $O(\log k)$. What's more it is possible to maintain this level of balance with fairly simple rebalancing operations called "pivots" that themselves take only $O(\log k)$ time to perform. As a result, we can both find elements in and add elements to an AVL tree in time $O(\log k)$, even if we have to rebalance the tree after every single addition (although usually we will not have to do this).

Details of the workings of the AVL tree can be found in most books on computer algorithms, such as Cormen *et al.* [81]. However, if you need to use such a tree in your own work it's probably not worth programming it yourself. There are many readily available standard implementations of AVL trees or other balanced trees that will do the job just fine and save you a lot of effort. A suitable one appears, for instance, in the standard C++ library STL, which is available with every installation of C++.

The deletion operation: The process of deleting a value from a tree is rather involved, but none of the algorithms described in this book require it, so we will not go into it in detail here. The basic idea is that you first find the value you want to delete in the normal way, which takes time $O(\log k)$, then delete it and perform a "pivot" operation of the type mentioned above to remove the hole left by the deletion. Each of these operations takes at most time $O(\log k)$ and hence the entire deletion can be complete in $O(\log k)$ time. The interested reader can find the details in Ref. [81].

Enumerating the items in a tree: We can also quickly run through the items stored in a tree, an operation that takes $O(k)$ time, just as it does for a simple list stored in an array. To do this we use an *Euler tour*, which is a circuit of the tree that visits each node at most twice and each edge exactly twice. An Euler tour starts at the root of the tree and circumnavigates the tree by following its outside edge all the way round (see figure). More precisely, starting from the root we move to the left child of the current node if we haven't already visited it and failing that we move to the right child. If we have already visited both children we move upward to the parent. We repeat these steps until we reach the root again, at which point the tour is complete. Since each edge is traversed twice and there are $k - 1$ edges in a tree with k nodes (see Section 6.7), this immediately tells us we can run through all elements in the tree in $O(k)$ time.

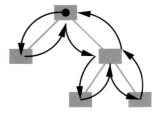

An Euler tour.

Given the above properties of trees, the adjacency tree for a network is now simple to define: we store the list of neighbors of each vertex as values in a binary tree. There is one tree for each vertex, or two for a directed network if we want to store both the incoming edges and the outgoing ones. Edges can be found, added, or deleted from the tree for a given vertex in time $O(\log k)$, where k is the degree of the vertex, or $O(\langle \log k \rangle)$ when averaged over all vertices. However, the average of the logs of a set of positive numbers is always less than the log of their average, so a running time of $O(\langle \log k \rangle)$ also implies a running time of $O(\log \langle k \rangle) \equiv O(\log(m/n))$.

The computational complexity of operations on the adjacency tree is summarized in Table 9.2. Comparing our three storage formats for networks, the adjacency matrix, list, and tree, we see that there is no overall winner; the choice of format is going to depend on what we want to do with it. For use in algorithms where we are only going to add and remove edges and check for their existence, the adjacency matrix is the fastest option. On the other hand, in algorithms where we are only adding edges or enumerating the neighbors of vertices, which includes most of the algorithms in this chapter, the adjacency list is the clear winner. The adjacency tree is the fastest format if we need to find or delete edges as well as create them. (Note that $\log(m/n)$ is usually much better than a mere m/n, particularly in the case where the mean degree $c \gg 1$.) In algorithms that only need to enumerate edges and not find, add, or delete them, the adjacency list and adjacency tree are equally good in principle. In practice, however, the additional complexity of the adjacency tree usually makes it somewhat slower and the adjacency list is the format of choice.

It's worth bearing in mind that speed isn't everything, and in particular that the adjacency matrix format uses a prohibitive amount of memory space for networks with large numbers of vertices (see Section 9.3). Overall, as we have said, the adjacency list is the format used most often, but the others certainly have their place and each will be useful to us more than once in the remainder of this book.

9.6 OTHER NETWORK REPRESENTATIONS

We have discussed three ways of representing network data in the memory of a computer. These are probably the most useful simple representations and the ones that you are most likely to need if you write your own programs to analyze networks, but there are a few other representations that it is worth knowing about.

Hybrid matrix/list representations: The representations of Table 9.2 all have their advantages and disadvantages, but none is optimal. In the best of all possible worlds, we would like a data structure that can insert, delete, and find edges in $O(1)$ time and enumerate the $O(m/n)$ neighbors (on average) of a given vertex in $O(m/n)$ time, but none of our representations can do this. It is possible to create a representation that *can* do this, however, if we are willing to sacrifice memory space: we can make a hybrid representation that consists of an adjacency matrix *and* an adjacency list. Non-zero elements in the adjacency matrix, those corresponding to edges, are accompanied by pointers that point to the corresponding elements in the adjacency list. Then we can find whether an edge exists between a specified pair of vertices in $O(1)$ time using the adjacency matrix as usual. And we can enumerate the neighbors of a vertex in $O(m/n)$ time using the adjacency list. We can add an edge in $O(1)$ time since both matrix and list allow this anyway (Table 9.2). And finally, we can delete an edge in $O(1)$ time by first locating it in the adjacency matrix and setting the corresponding element to zero, then following the pointers to the relevant elements of the adjacency list and deleting those too by moving the last element of the list to fill their place.

In terms of time complexity, i.e., scaling of run time with network size, this hybrid data structure is optimal.[17] Its main disadvantage is that it uses even more memory than the ordinary adjacency matrix, and hence is suitable only for relatively small networks, up to a few tens of thousands of vertices on a typical computer at the time of writing. If this is not an issue in your case, however, and speed is, then this hybrid representation may be worth considering.

Representations with variables on edges: In some networks the edges have values, weights, or labels on them. One can store additional properties like these using simple variants of the adjacency matrix or adjacency list representations. For instance, if edges come in several types we could define an additional $n \times n$ matrix to go with the adjacency matrix that has elements indicating the type of each extant edge. (For edges that do not exist the elements of such a matrix would have no meaning.) Or one could combine the two matrices into a single one that has a non-zero element for every extant edge whose value indicates the edge type. If there are many different variables associated with each edge, as there are for instance in some social network studies, then one could use many different matrices, one for each variable, or a matrix whose elements are themselves arrays of values or more complicated programming objects like structures. Similarly, with an adjacency list one could replace the elements of the list with arrays or structures that contain all the details of the edges they correspond to.

---

[17]It does place some overhead on the edge addition and deletion operations, meaning the complexity is still $O(1)$ but the operations take a constant factor longer to complete, since we have to update both adjacency matrix and list, where normally we would only have to update one or the other. Whether this makes an significant difference to the running of a program will depend on the particular algorithm under consideration.

However, these representations can be wasteful or clumsy. The matrix method can waste huge amounts of memory storing meaningless matrix elements in all the positions corresponding to edges that don't exist. The adjacency list (for an undirected network) contains two entries for each edge, both of which would have to be updated every time we modify the properties of that edge. If each edge has many properties this means a lot of extra work and wasted space.

In some cases, therefore, it is worthwhile to create an additional data structure that stores the properties of the edges separately. For instance, one might use a suitable array of m elements, one for each edge. This array can be linked to the main representation of the network structure: with an adjacency list we could store a pointer from each entry in the list to the corresponding element in the array of edge data. Then we can immediately find the properties of any edge we encounter in the main adjacency list. Similarly, each entry in the array of edge data could include pointers to the elements in the adjacency list that correspond to the edge in question. This would allow us to go through the array of edge data looking for edges with some particular property and, for example, delete them.

Edge lists: One very simple representation of a network that we have not yet mentioned is the *edge list*. This is simply a list of the labels of pairs of vertices that are connected by edges. Going back to this network, which we saw in Section 9.4:

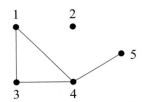

the edge list representation would be $(1,3), (4,1), (4,3), (4,5)$. The order of the edges is usually not important in an edge list, nor is the order of the vertices in the vertex pairs.

The edge list is a convenient and space-efficient way to store the structure of a network, and furthermore allows us easily to associate properties with the edges—we can simply store those properties along with the corresponding pairs of labels. It is not such a good representation if we wish to store properties of vertices. Indeed, the representation doesn't explicitly list vertices at all, so there is no way to tell that a vertex even exists if it is not connected to any edges. Vertex 2 in the network above is an example of this problem: it doesn't appear in the edge list because it has no edges. On the other hand, this problem and the problem of storing vertex properties can be remedied easily enough by creating a separate list of vertices and the data associated with them.

However, the edge list is a poor format for storing network data in computer memory for most of the algorithms we will be considering in this book. It does not, for instance, allow us to determine quickly whether a particular edge exists—we would have to go through the entire list to answer that question. And, crucially, it does not allow us easily to enumerate the neighbors of a given vertex, an operation that is central to many algorithms. For these reasons, the edge list is hardly ever used as a format for the representation of a network in memory.

Where it does find use is in file formats for networks. Being a fairly compact representation, edge lists are often used as a way to store network structure in computer files on a disk or other storage medium. When we wish to perform calculations on these networks we must read the file and convert its contents into a more suitable form for calculation, such as an adjacency matrix or adjacency list. This, however, is simple: we create an empty network in the memory of our computer, one with no edges initially, then run through the edges stored in the edge list and add them one by one to the network in the memory. Since the operation of adding an edge can be accomplished quickly in all of the formats we have considered (Table 9.2), this normally does not take a significant amount of time. When it is finished, we have a complete copy of the network in the format of our choice stored in the memory of the computer, and we are ready to continue with our computations.

9.7 HEAPS

The last data structure we describe in this chapter is a specialized structure called a *binary heap*. Unlike the structures introduced in the last few sections, heaps are not normally used for storing networks themselves, but are used for storing values *on* networks, usually values associated with a network's vertices. The definitive property of a heap is that it allows us to quickly find the entry in the heap with the minimum (or maximum) value.

We will make use of the binary heap in Section 10.4 when we study one of the most famous of network algorithms, Dijkstra's algorithm, which is an algorithm for finding shortest paths on weighted networks. This is the main place the heap will come up in this book, so if you are not interested in, or do not need to know, the detailed workings of Dijkstra's algorithm, you can safely skip the remainder of this chapter. Otherwise, read on.

Suppose, then, that we have some numerical value associated with every vertex in a network. That value might be, for instance, the distance from another vertex in the network, or a time until something happens. To give a concrete example, consider a disease spreading across a social network, as discussed in Chapters 1 and 3, and suppose we want to make a computer model of the spread. One simple and efficient way to do this is to associate with each vertex of the network a number representing our current estimate of the time at which that vertex will be infected by the disease (if it ever is). Initially each of these times is set to ∞, except for a single vertex representing the initial carrier of the disease, for which the time is set to zero. Then a simple algorithm for simulating the disease involves at each step finding the next vertex to be infected, i.e., the one with the earliest infection time, infecting it, and then calculating the time until it subsequently infects each of its neighbors. If any of those infection times is earlier than the current recorded time for the same neighbor, the new time supersedes the old one. Then we find the vertex in the network with the next earliest infection time and the process proceeds.

The crucial requirements for this algorithm to run efficiently are that we should be able to quickly find the smallest value of the infection time anywhere on the network and that we should be able to quickly decrease the value at any other given vertex. The binary heap is a data structure that allows us to do these things.

The binary heap is built upon a binary tree structure similar to the trees in Section 9.5, although the tree in a binary heap is arranged and used in a different fashion. Each of the items stored in a heap (items that will represent vertices in the network context) consists of two parts, an integer label that identifies the item and a numerical value. In the disease example above, for instance, the labels are the vertex indices $i = 1 \ldots n$ for vertices not yet infected and the values are the times at which the vertices are infected with the disease.

The items in the heap are stored at the nodes of a binary tree as depicted in Fig. 9.3. There are two important features to notice about this structure. First, the tree is always completely packed. We fill the tree row by row starting with the root node at the top and filling each row from left to right. Thus the tree is denser than the typical binary tree of Section 9.5 and always has a depth that is logarithmic in the number of items in the tree—see Eq. (9.2).

Second, the *values* associated with the items in the heap (the lower number at each node in Fig. 9.3) are *partially ordered*. This means that each value is greater than or equal to the one above it in the tree and less than or equal to both of the two below it. If we follow any branch down the tree—any path from top to bottom—the values grow larger along the branch or stay the same, but never decrease. The values are said to be "partially" ordered because the ordering only applies along branches and not between different branches. That is, there is no special relation between values on different branches of the tree;

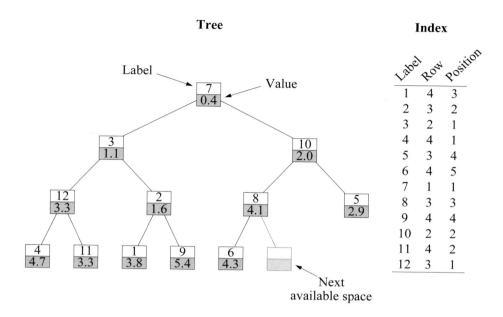

Figure 9.3: The structure of a binary heap. A binary heap consists of two parts, a tree and an index. The nodes of the tree each contain a label and a numerical value, and the tree is partially ordered so that the numerical value at each node is greater than or equal to the value above it in the tree and less than or equal to both of the values below it. The index is a separate array that lists by label the positions of each of the items in the tree, so that we can find a given item quickly. New items are added to the tree in the next available space at the bottom, starting a new row if necessary.

they may be larger or smaller than one another, whether they are on the same level in the tree or on different levels.

The property of partial ordering has the important result that the value stored at the root node of the tree is always the smallest value anywhere. Since values are non-decreasing along all branches of the tree starting from the root, it follows that no value can be smaller than the root value.

The binary heap also has another component to it, the *index*, which tells us the location of each item in the tree. The index is an array containing the coordinates in the tree of all the items, listed in order of their labels. It might, for instance, contain the row in the tree and position along that row of each item, starting with the item with label 1 and proceeding through each in turn—see Fig. 9.3 again.

A heap allows us to perform three operations on its contents: (1) adding

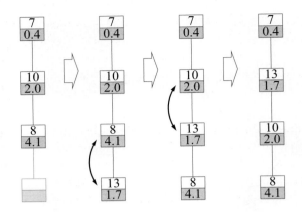

Figure 9.4: Sifting a value up the heap. A branch in the tree initially contains three items as shown. A new item with value 1.7 is added at the bottom. The upward sift repeatedly compares and swaps this value with the value above it until it reaches its correct place in the partial ordering. In this case the added value 1.7 gets swapped twice, ending up (correctly) between the values 0.4 and 2.0.

an item to the heap, (2) reducing the numerical value stored in the item with a specified label, and (3) finding and removing the item with the smallest numerical value from the heap. Let us look at how each of these operations is carried out.

Adding an item: To add an item to the heap we place it in the first available free space at the bottom of the tree as indicated in Fig. 9.3. If the bottom row of the tree is full, we start a new row. It is important in adding an item to the tree that we preserve the partially ordered structure, but a new item added at the bottom may well violate that structure by having a numerical value smaller than the item above it. We can fix this by *sifting* the tree as illustrated in Fig. 9.4. The newly added item is "sifted up" its branch of the tree by comparing its value with that of the item above it. If its value is smaller, the two are swapped, the new item moving up one row. We repeat this process until the new item either reaches a point where it is not smaller than the item above it, or it has risen to the root of the tree. If it rises to the root of the tree, then by definition it is has the smallest value in the tree because it is smaller than the value for the previous root item.

When we add a new item we also have to update the index of the heap. The coordinates of the new item are recorded at the position in the index corresponding to the item's label and then during the sifting operation, we simply

304

swap the index entries for every two items that are swapped in the tree.

Since the tree is completely packed it has depth given by Eq. (9.2), which is $O(\log k)$, where k is the number of items in the tree. Thus the maximum number of swaps we have to do during the sifting process is $O(\log k)$ and the total time to add an item to the heap, ignoring sub-leading terms, is also $O(\log k)$.

Reducing a value in the heap: Reducing the numerical value associated with a given labeled item is similar to the addition operation. We first use the index to locate the given item in the tree and we reduce its numerical value as desired. Since this may mean that the item is now smaller than one or more of those above it, violating the partial ordering, we sift up as before until the item reaches its correct place in the ordering, updating the index as we go. This operation, like the addition operation, takes time $O(\log k)$.

Finding and removing the smallest value: The item in the heap with the smallest numerical value is easily found since, as we have said, it is always located at the root of the tree. In Fig. 9.3, for example, the smallest value is 0.4 for the item with label 7.

We remove the smallest item by first deleting it from the tree and deleting its entry in the index. This leaves a hole in the tree which violates the condition that the tree be completely packed, but we can fix this problem by taking the last item from the bottom row of the tree and moving it up to the root, at the same time updating the relevant entry in the index. This, however, creates its own problem because in moving the item we will likely once again create a violation of the partial ordering of the tree. The item at the root of the tree is supposed to have the smallest numerical value and it's rather unlikely that the item we have moved satisfies this condition. This problem we can fix by "sifting down." Sifting down involves comparing the numerical value stored in the root item with *both* of those below it. If it is larger than either of them, we swap it with the smaller of the two and at the same time swap the corresponding entries in the index. We repeatedly perform such comparisons and swaps, moving our item down the tree until either it reaches a point at which it is smaller than both of the items below it, or it reaches the bottom of the tree.

Again the sifting process, and hence the entire process of removing the root item from the tree, takes time $O(\log k)$.

Thus the binary heap allows us to do all three of our operations—adding an item, reducing a value, or finding and removing the item with the smallest value—in time $O(\log k)$.

PROBLEMS

9.1 What (roughly) is the time complexity of:

a) Vacuuming a carpet if the size of the input to the operation is the number n of square feet of carpet?

b) Finding a word in a (paper) dictionary if the size of the input is the number n of words in the dictionary?

9.2 Suppose you have a sparse network with $m \propto n$. What is the time complexity of:

a) Multiplying an arbitrary n-element vector by the adjacency matrix, if the network is stored in the adjacency matrix format.

b) Performing the same multiplication if the network is in adjacency list format.

c) The "modularity matrix" **B** of a network is the matrix with elements

$$B_{ij} = A_{ij} - \frac{k_i k_j}{2m}.$$

(See Eq. (7.70) on page 224.) What is the time complexity of multiplying an arbitrary vector by the modularity matrix of our sparse network if the network is in adjacency list format?

d) In fact, if we are clever about it, this last operation can be performed in time $O(n)$ for the sparse network with $m \propto n$. Describe an algorithm that achieves this.

9.3 An interesting question, which is discussed in some detail in Chapter 16, concerns what happens to a network if you disable or remove its vertices one by one. The question is of relevance, for instance, to the vaccination of populations against the spread of disease. One typical approach is to remove vertices in order of their degrees, starting with the highest degrees first. Note that once you remove one vertex (along with its associated edges) the degrees of some of the other vertices may change.

In most cases it is not possible to do the experiment of removing vertices from a real network to see what effect it has, but we can simulate the process on a computer by taking a network stored in computer memory, removing its vertices, and then measuring various properties of the remaining network.

a) What is the time complexity of finding the highest-degree vertex in a network, assuming the vertices are given to you in no particular order?

b) If we perform the repeated vertex removal in a dumb way, searching exhaustively for the highest-degree vertex, removing it, then searching for the next highest, and so forth, what is the time complexity of the entire operation?

c) Describe how the same operation could be performed with the degrees of the vertices stored instead in a heap. You will need to modify the heap structure of Section 9.7 in a couple of ways to make the algorithm work. One modification is trivial: the heap needs to be sorted in the opposite order so that the largest element is at the root. What other modification is needed, and how would you do it? What now is the time complexity of the entire calculation?

d) Taking the same approach, describe in a sentence or two a method for taking n numbers in random order and sorting them into decreasing order using a heap. Show that the time complexity of this sorting algorithm is $O(n \log n)$.

e) The degrees of the vertices in a simple graph are integers between zero and n. It is possible to sort such a set of integers into numerical order, either increasing or decreasing, in time $O(n)$. Describe briefly an algorithm that achieves this feat.

CHAPTER 10

FUNDAMENTAL NETWORK ALGORITHMS

*A discussion of some of the most important and
fundamental algorithms for performing network
calculations on a computer*

A RMED WITH the tools and data structures of Chapter 9, we look in this
chapter at the algorithms that are used to perform network calculations.
We start with some simple algorithms for calculating quantities such as degrees, degree distributions, and clustering. In the later sections of the chapter we look at more sophisticated algorithms for shortest paths, betweenness,
maximum flows, and other non-local quantities.

In the following chapter we extend our examination of network algorithms
to algorithms based on matrix calculations and linear algebra, including algorithms for matrix-based centralities like eigenvector centrality and algorithms
for graph partitioning and community discovery in networks.

10.1 ALGORITHMS FOR DEGREES AND DEGREE DISTRIBUTIONS

Many network quantities are easy to calculate and require only the simplest
of algorithms, algorithms that are little more than translations into computer
code of the definitions of the quantities in question. Nonetheless, it is worth
looking at these algorithms at least briefly, for two reasons. First, there is in
some cases more than one simple algorithm for calculating a quantity, and one
algorithm may be much faster than another. It pays to evaluate one's algorithm at least momentarily before writing a computer program, to make sure
one is going about the calculation in the most sensible manner. Second, it is
worthwhile to calculate the computational complexity of even the simplest algorithm, so that one can make an estimate of how long a computation will take
to finish—see Section 9.1. Even simple algorithms can take a long time to run.

One of the most fundamental and important of network quantities is the degree of a vertex. Normally degrees are very simple to calculate. In fact, if a network is stored in the form of an adjacency list[1] then, as described in Section 9.4, we normally maintain an array containing the degree of each vertex so that we know how many entries there are in the list of neighbors for each vertex. That means that finding the degree of any particular vertex is a simple matter of looking it up in this array, which takes $O(1)$ time.

If the network is stored in an adjacency matrix, then the calculation takes longer. Calculating the degree of a vertex i in this case involves going through all elements of the ith row of the adjacency matrix and counting the number that are non-zero. Since there are n elements in each row of the matrix, where n is the number of vertices in the network, the calculation takes time $O(n)$, making the calculation far slower than for the adjacency list. If one needed to find the degrees of vertices frequently during the course of a larger calculation using an adjacency matrix, it might make good sense to calculate the degree of each vertex once and for all and store the results for later easy retrieval in a separate array.

In Section 8.3 we discussed degree distributions, which are of considerable interest in the study of networks for the effect they have on network structure and processes on networks (see Chapters 13 and 14). Calculating a degree distribution p_k is also very straightforward: once we have the degrees of all vertices, we make a histogram of them by creating an array to store the number of vertices of each degree up to the network maximum, setting all the array elements initially to zero, and then running through the vertices in turn, finding the degree k of each and incrementing by one the kth element of the array. This process trivially takes time $O(n)$ to complete. Once it is done the fraction p_k of vertices of degree k is given by the count in the kth array element divided by n.

The cumulative distribution function P_k of Section 8.4.1 requires a little more work. There are two common ways to calculate it. One is first to form a histogram of the degrees as described above and then to calculate the cumulative distribution directly from it using

$$P_k = \sum_{k'=k}^{\infty} p_{k'} = -p_{k-1} + \sum_{k'=k-1}^{\infty} p_{k'} = P_{k-1} - p_{k-1}. \tag{10.1}$$

Noting that $P_0 = \sum_{k'=0}^{\infty} p_{k'} = 1$, we can then start from P_0 and use Eq. (10.1) to calculate successive P_k up to any desired value of k. This process trivially

---

[1]Or an adjacency tree—see Section 9.5.

takes $O(n)$ time and, since the calculation of p_k also takes $O(n)$ time, the whole process is $O(n)$.

In fact, however, as described in Section 8.4.1, this is not usually how one calculates the cumulative distribution function. Although the method is fast, it's also moderately complicated and there is a simpler way of doing the calculation that involves taking the degrees of all the vertices, sorting them in descending order, and ranking them from 1 to n. A plot of the rank divided by n as a function of degree then gives the cumulative distribution. The most time-consuming part of this calculation is the sorting of the degrees. Sorting is a well-studied problem and the fastest general algorithms[2] run in time $O(n \log n)$. Thus the leading order scaling of this algorithm to calculate the cumulative distribution is $O(n \log n)$. This is slower than the first method described above, which was $O(n)$, but not much slower and the second method has the considerable advantage that almost all computers provide standard software for sorting numbers, which means that in most cases one doesn't have to write a program at all to calculate the cumulative distribution. All spreadsheet programs, for instance, include facilities for sorting numbers, so one can calculate cumulative distributions directly in a spreadsheet.

Another quantity of interest is the correlation coefficient r for vertex degrees, Eq. (8.26), which measures assortative mixing by degree. This too is straightforward to calculate—one uses Eq. (8.27) and the sums defined in Eqs. (8.28) and (8.29). Given the degrees of all vertices, the sum in Eq. (8.28) takes time $O(m)$ to evaluate, where m is the number of edges in the network, and the sums in Eq. (8.29) each take time $O(n)$, so the total time required to calculate r is $O(m + n)$. As mentioned in Section 9.1, we are often concerned with sparse networks in which the mean degree remains constant as the network gets larger, i.e., networks in which $m \propto n$. In such networks $O(m + n) \equiv O(n)$ and the time to calculate r taken just scales as the number of vertices. On the other hand, if the network is dense, meaning that $m \propto n^2$, then $O(m) \equiv O(n^2)$, which is considerably worse.

10.2 CLUSTERING COEFFICIENTS

The calculation of clustering coefficients is only slightly more complicated than the calculation of degrees. To see how it works, we start by calculating the local clustering coefficient C_i for a single vertex i on an undirected network,

---

[2]For integers, such as vertex degrees, it is under certain conditions possible to sort faster, in time $O(n)$, using the so-called *radix sort* algorithm. See, for example, Cormen *et al.* [81].

Eq. (7.42):[3]

$$C_i = \frac{(\text{number of pairs of neighbors of } i \text{ that are connected})}{(\text{number of pairs of neighbors of } i)}. \tag{10.2}$$

Calculating the numerator involves going through every pair of distinct neighbors of vertex i and counting how many are connected. We need only consider each pair once, which we can do conveniently by restricting ourselves to pairs (j, l) for which $j < l$. For each pair we determine whether an edge exists between them, which is done in various ways depending on the representation used for the network as described in Sections 9.3–9.5, and count up the number of such edges. Then we divide the result by the number of pairs, which is just $\frac{1}{2}k_i(k_i - 1)$, where k_i is the degree of the vertex.

To calculate the overall clustering coefficient for the entire network, which is given by

$$C = \frac{(\text{number of triangles}) \times 3}{(\text{number of connected triples})}. \tag{10.3}$$

(see Eq. (7.41)), we extend the same calculation to the whole network. That is we consider for every vertex each pair of neighbors (j, l) with $j < l$ and find whether they are connected by an edge.[4] We add up the total number of such edges over all vertices and then divide by the number of connected triples, which is $\sum_i \frac{1}{2}k_i(k_i - 1)$.

This last algorithm is simple and straightforward, a direct implementation of the formula (10.3) defining the clustering coefficient, but some interesting issues nonetheless come up when we consider its running time. Even without performing a full calculation of the complexity of the algorithm we can see that something unusual is going to happen because a vertex i with degree k_i has $\frac{1}{2}k_i(k_i - 1)$ pairs of neighbors. We have to check for the presence of an edge between each such pair on the entire network and hence the total number of checks we have to perform is

$$\sum_i \tfrac{1}{2}k_i(k_i - 1) = \tfrac{1}{2}n(\langle k^2 \rangle - \langle k \rangle), \tag{10.4}$$

where

$$\langle k \rangle = \frac{1}{n}\sum_i k_i, \qquad \langle k^2 \rangle = \frac{1}{n}\sum_i k_i^2, \tag{10.5}$$

---

[3]An equivalent to the clustering coefficient can be defined for a directed network (see Section 7.9) but we limit ourselves here to the much commoner undirected case.

[4]Note that this calculation automatically accounts for the factor of three appearing in the numerator of Eq. (10.3), since each triangle is counted three times, once each from the point of view of the three vertices it connects.

are the mean and mean square degree for the network. (We previously denoted the mean degree by c, but we use the alternate notation $\langle k \rangle$ here for clarity, and to highlight the distinction between the mean and the mean square.)

See Section 8.3 for a discussion of degree distributions.

The interesting point here is that Eq. (10.4) depends in a non-trivial way on the degree distribution of our network. The running times of other algorithms we have seen so far have depended on the number of vertices n and the number of edges m, and hence, indirectly, on the mean degree $\langle k \rangle = 2m/n$. For the clustering coefficient, however, we see that the amount of work we have to do, and hence also the running time, depends not only on n and $\langle k \rangle$, but on the second moment $\langle k^2 \rangle$, which is an additional independent parameter. Even if we suppose that the degree distribution remains the same with increasing n so that the quantities $\langle k \rangle$ and $\langle k^2 \rangle$ can be considered constant, strange things can happen. Consider the case of a network whose degree distribution follows a power law $p_k \sim k^{-\alpha}$, as described in Section 8.4. For such networks, the first moment is well behaved but the second moment $\langle k^2 \rangle$ formally diverges if $\alpha < 3$ (see Section 8.4.2) which implies that it will take an infinite amount of time to evaluate the clustering coefficient!

To understand better what is going on let us perform a more careful calculation of the time complexity of our clustering coefficient algorithm. We start by considering again a single vertex i. And let us assume that we have our network stored in adjacency list form. In that case, we can, as we have seen, easily enumerate all of the neighbors of our vertex in time that goes like k_i. For each neighbor j we run through each other neighbor $l > j$ that could be paired with it, for a total of $\frac{1}{2}k_i(k_i - 1)$ pairs and determine for each pair whether an edge exists between them. This latter operation takes a time proportional, to leading order, to either k_j or to k_l (see Table 9.2), depending on whether we find the edge by looking at the adjacency list for vertex j or for vertex l. Let us for the moment assume a simple algorithm that chooses at random between the two vertices, in which case the typical time taken will go as the average of the two degrees, i.e., it will be proportional to $k_j + k_l$.

Let Γ_i denote the set of neighbors of vertex i. Then the total time taken to check for edges between all pairs of neighboring vertices is proportional to

$$\sum_{j,l \in \Gamma_i : j < l} (k_j + k_l) = \frac{1}{2} \sum_{j,l \in \Gamma_i : j \neq l} (k_j + k_l) = \sum_{j,l \in \Gamma_i : j \neq l} k_j$$

$$= (k_i - 1) \sum_{j \in \Gamma_i} k_j. \tag{10.6}$$

The total time needed to calculate the numerator of Eq. (10.3) is then proportional to the sum of this quantity over all vertices i:

$$\sum_i (k_i - 1) \sum_{j \in \Gamma_i} k_j = \sum_{ij} A_{ij}(k_i - 1)k_j = \sum_{ij} A_{ij}k_i k_j - \sum_j k_j^2 \qquad (10.7)$$

where A_{ij} is an element of the adjacency matrix and we have made use of the result $\sum_i A_{ij} = k_j$ (Eq. (6.19)).

Compare this equation with our earlier expression for the correlation coefficient r between degrees in a network, Eq. (7.82), which quantifies assortativity by degree in networks:

$$r = \frac{\sum_{ij}(A_{ij} - k_i k_j / 2m)k_i k_j}{\sum_{ij}(k_i \delta_{ij} - k_i k_j / 2m)k_i k_j}. \qquad (10.8)$$

As we can see, the first term in Eq. (10.7) is the same as the first term in the numerator of the correlation coefficient. As a result, our estimate of the time to calculate the clustering coefficient depends on whether the degrees of vertices are correlated or not. This can lead to some interesting behaviors for specific networks, but for simplicity let us assume here that there is no correlation between degrees, that the network we are considering has no assortativity. In that case $r = 0$ in Eq. (10.8), which can only occur if the numerator is itself zero, or equivalently if

$$\sum_{ij} A_{ij}k_i k_j = \frac{1}{2m} \sum_{ij} k_i^2 k_j^2 = \frac{1}{2m}\left[\sum_i k_i^2\right]^2. \qquad (10.9)$$

Combining this result with Eq. (10.7), the running time for our calculation of the clustering coefficient on an uncorrelated network is proportional to

$$\frac{1}{2m}\left[\sum_i k_i^2\right]^2 - \sum_j k_j^2 = n\langle k^2 \rangle \left[\frac{\langle k^2 \rangle}{\langle k \rangle} - 1\right], \qquad (10.10)$$

where we have made use of the fact that $2m = \sum_i k_i = n\langle k \rangle$ (see Eq. (6.20)).

This is a measure of the time taken to evaluate the numerator of Eq. (10.3). The denominator is simply equal to $\sum_i k_i(k_i - 1)$ and so just takes $O(n)$ time to evaluate, given that, for a network stored in adjacency list format, we already have the degrees of all vertices available. This will never be longer than the time represented in Eq. (10.10), so Eq. (10.10) gives the leading-order time complexity of the calculation of the clustering coefficient.

So we see that the calculation of the clustering coefficient indeed takes a time that depends not only on n and m but also on the second moment $\langle k^2 \rangle$ of the degree distribution. In many cases this does not matter, since the second moment often tends to a modest constant value as the network becomes large. But for networks with highly skewed degree distributions $\langle k^2 \rangle$ can become very large and in the case of a power-law degree distribution with exponent $\alpha < 3$ it formally diverges (see Section 8.4.2) and with it so does the expected running time of our algorithm.

More realistically, if the network is a simple graph with no multiedges, then the maximum allowed degree is $k = n$ and the degree distribution is cut off, which means that the second moment scales at worst as $n^{3-\alpha}$ (Eq. (8.22)) while the first moment remains constant. This in turn implies that the running time of our clustering coefficient algorithm on a scale-free network would go as $n \times n^{3-\alpha} \times n^{3-\alpha} = n^{7-2\alpha}$. For values of α in the typical range of $2 \leq \alpha \leq 3$ (Table 8.1), this gives running times that vary from a minimum of $O(n)$ for $\alpha = 3$ to a maximum of $O(n^3)$ for $\alpha = 2$. For the lower values of α this makes the calculation of the clustering coefficient quite arduous, taking a time that increases sharply as the network gets larger.

So can we improve on this algorithm? There are various possibilities. Most of the work of the algorithm is in the "find" operation to determine whether there is an edge between a given pair of vertices, and the algorithm will be considerably faster if we can perform this operation more efficiently. One simple (though memory-inefficient) method is to make use of the hybrid matrix/list data structure of Section 9.6, which can perform the find operation in constant time.[5] Even in this case, however, the number of find operations that must be performed is still equal to the number of connected triples in the network, which means the running time is given by Eq. (10.4), and hence still formally diverges on a network with a power-law degree distribution. On a simple graph for which the power law is cut off at $k = n$, it will go as $n^{4-\alpha}$, which ranges from $O(n)$ to $O(n^2)$ for values of α in the interesting range $2 \leq \alpha \leq 3$. This is better than our earlier algorithm, but still relatively poor for the lower values of α.

These difficulties are specific to the case of scale-free networks. In other cases there is usually no problem calculating the clustering coefficient quickly. Some alternative algorithms have been proposed for calculating approximate values of the clustering coefficient rapidly, such as the algorithm of Schank and Wagner [292], and these may be worth considering if you need to perform calculations on very large networks.

---

[5]Other possible ways to improve the algorithm are to use the adjacency tree structure of Section 9.5, or to use the adjacency list but always test for the presence of an edge between two vertices by searching the neighbors of the lower-degree vertex to see if the higher is among them (rather than the algorithm described above, which chooses which one to search at random).

10.3 SHORTEST PATHS AND BREADTH-FIRST SEARCH

We now move on to some more complex algorithms, algorithms for calculating mostly non-local quantities on the networks, such as shortest paths between vertices. The study of each of these algorithms has three parts. Two are, as before, the description of the algorithm and the analysis of its running time. But now we also include a proof that the algorithm described performs the calculation it claims to. For the previous algorithms in this chapter such proofs were unnecessary; the algorithms were direct implementations of the equations defining the quantities calculated. As we move onto more complex algorithms, however, it will become much less obvious why those algorithms give the results they do, and to gain a full understanding we will need to examine their working in some detail.

The first algorithm we look at is the standard algorithm for finding shortest distances in a network, which is called *breadth-first search*.[6] A single run of the breadth-first search algorithm finds the shortest (geodesic) distance from a single source vertex s to *every other* vertex in the same component of the network as s. In some cases we want to know only the shortest distance between a single pair of vertices s, t, but there is no procedure known for calculating such a distance that is faster in the worst case than calculating the distances from s to every other vertex using breadth-first search and then throwing away all of the results except for the one we want.[7]

With only minor modifications, as we will describe, breadth-first search can also find the geodesic path one must take to realize each shortest distance and if there is more than one geodesic path, it can find all such paths. It works also on both directed and undirected networks, although our description will focus on the undirected case.

10.3.1 DESCRIPTION OF THE ALGORITHM

Breadth-first search finds the shortest distance from a given starting vertex s to every other vertex in the same component as s. The basic principle behind the algorithm is illustrated in Fig. 10.1. Initially we know only that s has distance 0 from itself and the distances to all other vertices are unknown. Now we find all the neighbors of s, which by definition have distance 1 from s. Then we find all the neighbors of *those* vertices. Excluding those we have already visited, these

---

[6]In physics, breadth-first search is sometimes called the "burning algorithm."

[7]In many cases we may find the result we want before calculating all distances, in which case we can save ourselves the effort of calculating the rest, but in the worst case we will have to calculate them all. See Section 10.3.4 for more discussion.

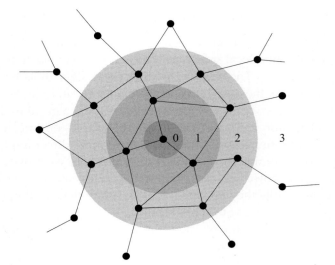

Figure 10.1: Breadth-first search. A breadth-first search starts at a given vertex, which by definition has distance 0 from itself, and grows outward in layers or waves. The vertices in the first wave, which are the immediate neighbors of the starting vertex, have distance 1. The neighbors of those neighbors have distance 2, and so forth.

A network path from s to t of length d (where $d = 3$ in this case) necessarily includes a path of length $d - 1$ (i.e., 2) from s to an immediate neighbor of t.

vertices must have distance 2. And *their* neighbors, excluding those we have already visited have distance 3, and so on. On every iteration, we grow the set of vertices visited by one step.

This is the basic idea of breadth-first search. Now let us go over it more carefully to see how it works in practice and show that it really does find correct geodesic distances. We begin by noting the following fact:

Every vertex whose shortest distance from s is d has a network neighbor whose shortest distance from s is $d - 1$.

This follows since if the shortest path from s to a vertex t is of length d then the penultimate vertex along that path, which is a neighbor of t, can, by definition, be reached in $d - 1$ steps and hence cannot have shortest distance greater than $d - 1$. It also cannot have shortest distance less than $d - 1$ because it if did there would be a path to t of length less than d.

Now suppose that we already know the distance to every vertex on the network that is d steps or less from our source vertex s. For example, we might know all the distances to vertices at distance 2 or less from the central vertex in Fig. 10.1. For every neighbor of one of the vertices at distance d there exists a path of length $d + 1$ to that neighbor: we can get to the vertex at distance d along a path of length d and then we take one more step to its neighbor. Thus every such neighbor is *at most* $d + 1$ steps from s, but it could be less than $d + 1$ from s if there is another shorter path through the network. However, we already know whether there is a shorter path to any particular vertex, since by hypothesis we know the distance to every vertex d steps or less from s.

Consider the set of all vertices that are neighbors of vertices at distance d but that are not already known to have distance d or less from s. We can say immediately that (1) all neighbors in this set have distance $d + 1$ from s, and (2) that there are no other vertices at distance $d + 1$. The latter follows from the property cited above: all vertices at distance $d + 1$ must be neighbors of vertices at distance d. Thus we have found the set of vertices at distance $d + 1$, and hence we now know the distances to all vertices that are $d + 1$ or less from s.

Now we just repeat the process. On each round of the algorithm we find all the vertices one step further out from s than on the last round. The algorithm continues until we reach a point at all the neighbors of vertices at distance d are found already to have known distances of d or less. This implies that there are no vertices of distance $d + 1$ and hence, by the property above, no vertices of any greater distance either, and so we must have found every vertex in the component containing s.

As a corollary of the process of finding distance, breadth-first search thus also finds the component to which vertex s belongs, and indeed breadth-first search is the algorithm of choice for finding components in networks.

10.3.2   A NAIVE IMPLEMENTATION

Let us now consider how we would implement breadth-first search on our computer. The simplest approach (but not, as we will see, the best) would go something like this. We create an array of n elements to store the distance of each vertex from the source vertex s, and initially set the distance of vertex s from itself to be zero while all other vertices have unknown distance from s. Unknown distances could be indicated, for instance, by setting the corresponding element of the array to -1, or some similar value that could never occur in reality.

We also create a distance variable d to keep track of where we are in the breadth-first search process and set its value initially to zero. Then we do the following:

1. Find all vertices that are distance d from s, by going through the distance array, element by element.
2. Find all the neighbors of those vertices and check each one to see if its distance from s is unknown (denoted, for example, by an entry -1 in the distance array).
3. If the number of neighbors with unknown distances is zero, the algorithm is over. Otherwise, if the number of neighbors with unknown dis-

tances is non-zero, set the distance of each of those neighbors to $d + 1$.

4. Increase the value of d by 1.
5. Repeat from step 1.

When the algorithm is finished we are left with an array that contains the distances to every vertex in the component of the network that contains s (and every vertex in every other component has unknown distance).

How long does this algorithm take? First of all we have to set up the distance array, which has one element for each vertex. We spend a constant amount of time setting up each element, so overall we spend $O(n)$ time setting up the distance array.

For the algorithm proper, on each iteration we go through all n vertices looking for those with distance d. Most will not have distance d in which case we pass over them, spending only $O(1)$ time on each. Thus there is a basic cost of $O(n)$ time for each iteration. The total number of iterations we will for the moment call r, and overall we thus spend $O(rn)$ time on this part of the algorithm, in the worst case.

However, when we *do* come across a vertex with distance d, we must pause at that vertex and spend an additional amount of time checking each of its neighbors to see if their distances are unknown and assigning them distance $d +$ 1 if they are. If we assume that the network is stored in adjacency list format (see Section 9.4) then we can go through the neighbors of a vertex in $O(m/n)$ on average, and during the whole course of the algorithm we pause like this at each vertex exactly once so that the total extra time we spend on checking neighbors of vertices is $n \times O(m/n) = O(m)$.

Thus the total running time of the algorithm, including set-up, is $O(n + rn + m)$.

And what is the value of the parameter r? The value of r is the maximum distance from our source vertex s to any other vertex. In the worst case, this distance is equal to the diameter of the network (Section 6.10.1) and the worst-case diameter is simply n, which is realized when the network is just a chain of n vertices strung one after another in a line. Thus in the worst case our algorithm will have running time $O(m + n^2)$ (where we have dropped the first n because we are keeping only the leading-order terms).

This is very pessimistic, however. As discussed in Sections 8.2 and 12.7 the diameter of most networks increases only as $\log n$, in which case our algorithm would run in time $O(m + n \log n)$ to leading order. This may be a moot point, however, since we can do significantly better than this if we use a little cunning in the implementation of our algorithm.

10.3.3 A BETTER IMPLEMENTATION

The time-consuming part of the implementation described in the previous section is step 1, in which we go through the list of distances to find vertices that are distance d from the starting vertex s. Since this operation involves checking the distances of all n vertices, only a small fraction of which will be at distance d, it wastes a lot of time. Observe, however, that in each wave of the breadth-first search process we find and label all vertices with a given distance $d + 1$. If we could store a list of these vertices, then on the next wave we wouldn't have to search through the whole network for vertices at distance $d + 1$; we could just use our list.

The most common implementation of this idea makes use of a *first-in/first-out buffer* or *queue*, which is nothing more than an array of (in this case) n elements that store a list of labels of vertices. On each sweep of the algorithm, we read the vertices with distance d from the list, we use these to find the vertices with distance $d + 1$, add those vertices with distance $d + 1$ to the list, and repeat.

To do this in practice, we fill up the queue array starting from the beginning. We keep a pointer, called the write pointer, which is a simple integer variable whose value indicates the next empty location at the end of the queue that has not been used yet. When we want to add an item to the queue, we store it in the element of the array pointed to by the write pointer and then increase the pointer by one to point to the next empty location.

At the same time we also keep another pointer, the read pointer, which points to the next item in the list that is to be read by our algorithm. Each item is read only once and once it is read the read pointer is increased by one to point to the next unread item.

Here is a sketch of the organization of the queue:

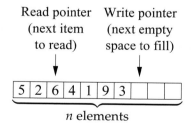

Our breadth-first search algorithm now uses two arrays of n elements, one for the queue and one for the distances from s to each other vertex. The algorithm is as follows.

1. Place the label of the source vertex s in the first element of the queue, set the read pointer to point to it, and set the write pointer to point to the second element, which is the first empty one. In the distance array, record the distance of vertex s from itself as being zero and the distances to all other vertices as "unknown" (for instance, by setting the corresponding elements of the distance array to -1, or some similar impossible value).

2. If the read and write pointers are pointing to the same element of the queue array then the algorithm is finished. Otherwise, read a vertex label from the element pointed to by the read pointer and increase that pointer by one.

3. Find the distance d for that vertex by looking in the distance array.

4. Go through each neighboring vertex in turn and look up its distance in the distance array as well. If it has a known distance, leave it alone. If it has an unknown distance, assign it distance $d + 1$, store its label in the queue array in the element pointed to by the write pointer, and increase the write pointer by one.

5. Repeat from step 2.

Note the test applied in step 2: if the read pointer points to the same element as the write pointer, then there is no vertex to be read from the queue (since the write pointer always points to an empty element). Thus this test tells us when there are no further vertices waiting to have their neighbors investigated.

Note also that this algorithm reads all the vertices with distance d from the queue array one after another and uses them to find all the vertices with distance $d + 1$. Thus all vertices with the same distance appear one after another in the queue array, with the vertices of distance $d + 1$ immediately after those of distance d. Furthermore, each vertex appears in the queue array at most once. A vertex may of course be a neighbor of more than one other, but a vertex is assigned a distance and put in the queue only on the first occasion on which it is encountered. If it is encountered again, its distance is known rather than unknown, and hence it is not again added to the queue. Of course, a vertex may not appear in the queue array at all if it is never reached by the breadth-first search process, i.e., if it belongs to a different component from s.

Thus the queue does exactly what we wanted it to: it stores all vertices with a specified distance for us so that we have the list handy on the next sweep of the algorithm. This spares us from having to search through the network for them and so saves us a lot of time. In all other respects the algorithm works exactly as in the simple implementation of Section 10.3.2 and gives the same answers.

How long does this implementation of the algorithm take to run? Again there is an initial time of $O(n)$ to set up the distance array (see Section 10.3.2).

Then, for each element in the queue, which means for each of the vertices in the same component as s, we do the following operations: we run through its neighbors, of which there are $O(m/n)$ on average, and either calculate their distance and add them to the queue, or do nothing if their distance is already known. Either way the operations take $O(1)$ time. Thus for each vertex in the component, of which there are in the worst case n, we spend time $O(m/n)$ and hence we require overall at most a time $n \times O(m/n) = O(m)$ to complete the algorithm for all n vertices.

Thus, including the time to set up the distance array, the whole algorithm takes time $O(m + n)$, which is better than the $O(m + n \log n)$ of the naive implementation (Section 10.3.2). For the common case of a sparse network with $m \propto n$, $O(m + n)$ is equivalent to $O(n)$ and our algorithm runs in time proportional to the number of vertices.[8] This seems just about optimal, since the algorithm is calculating the distance of all n vertices from the source vertex s. Thus it is assigning n numbers to the n elements of the distance array, which in the best possible case must take $O(n)$ time.

On a sparse network, therefore, the breadth-first search algorithm does as well as we can hope for in finding the distances from a single vertex to all others, and indeed it is the fastest known algorithm for performing this operation.

10.3.4 VARIANTS OF BREADTH-FIRST SEARCH

There are a number of minor variants of breadth-first search that merit a mention. First, one might wish to calculate the shortest distance between only a single pair of vertices s and t, rather than between s and all others. As mentioned in Section 10.3 there is no known way to do this faster than using breadth-first search. We can, however, improve the running time slightly by the obvious tactic of stopping the algorithm as soon as the distance to the target vertex t has been found. There is no point in continuing to calculate distances to the remaining vertices once we have the answer we want. In the worst case, the calculation still takes $O(m + n)$ time since, after all, our particular target vertex t might turn out to be the last one the algorithm finds. If we are lucky, however, and encounter the target early then the running time might be considerably shorter.

Conversely, we sometimes want to calculate the shortest distance between every pair of vertices in an entire network, which we can do by performing a breadth-first search starting at each vertex in the network in turn. The total

---

[8]On the other hand, for a dense network where $m \propto n^2$, we have a running time of $O(n^2)$.

running time for this "all-pairs shortest path" calculation is $n \times O(m + n) = O(n(m + n))$, or $O(n^2)$ on a sparse graph. As with the standard breadth-first search, this is optimal in the sense that we are calculating $O(n^2)$ quantities in $O(n^2)$ time, which is the best we can hope for.

As mentioned in the previous section, breadth-first search can also be used to identify the members of the component to which a vertex s belongs. At the end of the algorithm the distance array contains the distance from s to every vertex in its component, while distances to all other vertices are recorded as unknown. Thus we can find the size of the component just by counting the number of vertices with known distances. It takes time $O(n)$ to perform the count, so the operation of finding the component still takes $O(m + n)$ time in total.

The closeness centrality of Section 7.6 can also be calculated simply using breadth-first search. Recall that closeness is defined as the inverse of the mean distance from a vertex to all others in the same component. Since our breadth-first search calculates distances to all others in a component we need then only go through the distance array, calculate the sum of all known distances, divide by the size of the component, and take the inverse. Again the running time is $O(n + m)$. The variant closeness defined in terms of the harmonic mean in Eq. (7.30) can also be calculated, in the same running time, by a similar method.

10.3.5 FINDING SHORTEST PATHS

The breadth-first search algorithm as we have described it so far finds the shortest distance from a vertex s to all others in the same component of the network. It does not tell us the particular path or paths by which that shortest distance is achieved. With only a relatively small modification of the algorithm, however, we can calculate the paths as well. The trick is to construct another network on top of our original network, this one directed, that represents the shortest paths. This other network is often called the *shortest path tree*, although in the most general case it is a directed acyclic graph, not a tree.

The idea is as follows. At the start of our algorithm we create an extra network, which will become our shortest path tree, with the same number n of vertices as our original network and the same vertex labels, but with no edges at all. Then we start our breadth-first search algorithm from the specified source vertex s as before. The algorithm repeatedly pulls a vertex out of the queue and examines its neighbors, as described in Section 10.3.3, but now every time the neighbor j of some vertex i turns out to be a previously unseen vertex, one whose distance is recorded as "unknown," we not only assign j a distance and store it in the queue, we also add a directed edge to our shortest

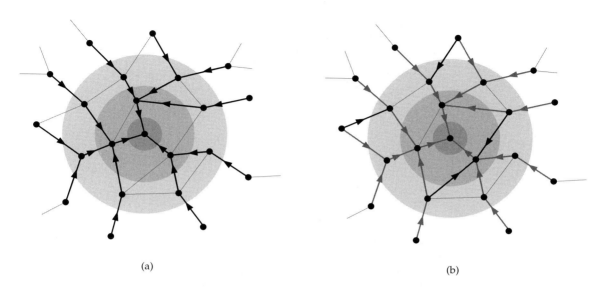

(a)   (b)

Figure 10.2: Shortest path trees. (a) A simple shortest path tree for the network of Fig. 10.1. Each vertex has a directed edge pointing to the vertex by which it was reached during the breadth-first search process. By following directed edges from any vertex we can find a shortest path to the starting vertex in the center. (b) The full shortest path tree (which is actually not a tree at all but a directed acyclic graph) contains extra directed edges that allow us to reconstruct all possible shortest paths.

path tree *from* vertex j *to* vertex i. This directed edge tells us that we found j by following a path from its neighbor i. However, vertex i will also have a directed edge leading from it to one of its neighbors, telling us that we found i by following *that* path, and so forth. Thus, by following a succession of these directed edges we eventually get all the way back to s, and so we can reconstruct the entire shortest path between j and s.

So, when our breadth-first search is finished, the shortest path tree contains the information we need to find the actual shortest path from every vertex in the component containing s to s itself. An example of a shortest path tree is shown in Fig. 10.2a for the same network as in Fig. 10.1.

This algorithm works well and the extra step of adding an edge to the shortest path tree can be accompanied quickly—in $O(1)$ time if we store the network in adjacency list format (see Table 9.2). Thus the overall running time of the algorithm is still $O(m + n)$ to find all distances from s and the corresponding shortest paths.

The algorithm does have one shortcoming, however, which is that it only finds *one* shortest path to each vertex. As pointed out in Section 6.10.1, a pair of vertices may have more than one shortest path between them (see Fig. 6.10). Another slight modification of the algorithm allows us to deal with this case.

Multiple shortest paths exist between any vertex and the source vertex s if the path to s splits in two or more directions at some point along its length. This occurs if there is a vertex j somewhere along that path, say at distance $d + 1$ from s, that has more than one neighbor at distance d—see Fig. 10.2b. We can record this circumstance in our shortest-path tree by adding more than one directed edge from j to each of the relevant neighbors. These directed edges tell us that we can find a shortest path to vertex s by taking a step to any of those neighboring vertices.

To do this we modify our algorithm as follows. We perform the breadth-first search starting from s as before, and add directed edges from newly found vertices to their neighbor as before. But we also add an extra step. If, in the process of examining the neighbors of a vertex i that has distance d from the source vertex, we discover a neighbor j that already has an assigned distance, and that distance is $d + 1$, then we know that a path of length $d + 1$ has already been found to j, but we also know that *another* path of length $d + 1$ must exist via the current vertex i. So we add an extra directed edge to the shortest path tree from j to i. This makes the shortest path tree no longer a tree but, as we have said, it's usually called a tree anyway. In any case, the algorithm gives exactly what we want. When it is finished running the shortest path "tree" allows us to reconstruct all shortest paths from every vertex in the component to the source vertex s. See Fig. 10.2b.

10.3.6 BETWEENNESS CENTRALITY

In Section 7.7 we described betweenness centrality, a widely used centrality index that measures the extent to which a vertex in a network lies on the paths between other vertices. The betweenness centrality of vertex v is the number of geodesic paths between pairs of vertices s, t that pass through v. (Sometimes it is normalized to be the fraction of such paths, rather than the total number. The difference is only a multiplicative constant—see Section 7.7.) Given that we have a method for finding the shortest path (or paths) between any two vertices (Section 10.3.5), we can with only a little more work now create an algorithm for calculating betweenness.

The simplest way to calculate betweenness would be to implement the definition of the measure directly: use breadth-first search to find the shortest path between s and t, as described in Section 10.3.5 (assuming such a path exists), and then work our way along that path checking the vertices it passes though to see if the vertex v we are interested in lies among them. Repeating this process for every distinct pair s, t, we can then count the total number of paths that pass through v. (Things are slightly more complicated for the case in which a pair of vertices are connected by more than one shortest path, but let us ignore this complication for the moment—we will come to it soon.)

This algorithm is certainly a correct algorithm and it would work, but it is also inefficient. As we have seen, breadth-first search takes time $O(m + n)$ to find a shortest path between two vertices, and there are $\frac{1}{2}n(n - 1)$ distinct pairs of vertices s, t. Thus the work of calculating betweenness for a single vertex would take $O(n^2(m + n))$ time, or $O(n^3)$ in the common case of a sparse graph for which $m \propto n$. (The operation of checking the vertices along each shortest path will take time of the order of the length of the path, which is typically $O(\log n)$ (Section 8.2), making it negligible compared with the time taken to find the path.) This is prohibitively slow: while one might be able to calculate the betweenness of a vertex on a given network in, say, an hour's work, the same calculation on a graph ten times larger would take $10^3 = 1000$ hours, or more than a month of computer time.

But we can do a lot better if we make use of some of our results about breadth-first search from previous sections. First, the standard breadth-first search can find paths between a source s and all other vertices (in the same component) in time $O(m + n)$, which means, as noted in Section 10.3.4, we can find paths between all pairs in the network in time $O(n(m + n))$, or $O(n^2)$ on a sparse network.

An improved algorithm for calculating the betweenness of a vertex v might work as follows. For each s we use breadth-first search to find shortest paths between s and all other vertices, constructing a shortest path tree as described in Section 10.3.5. Then we use that tree to trace the paths from each vertex back to s, counting in the process the number of paths that go through v. We repeat this calculation for all s and so end up with a count of the total number of shortest paths that pass through v.

Indeed, we can trivially extend this algorithm to calculate betweenness for *all* vertices at the same time—we simply maintain a count of the number of paths that go through every vertex, for example in an array.[9]

---

[9]Note that this actually counts each path twice (since the path between i and j is counted once when i is considered the source vertex and once when j is), except for the path from each vertex to itself, which is counted only once (when that vertex is the source). This, however, is correct: the betweenness centrality, as defined in Eq. (7.36), indeed counts each path twice, except for the path from a vertex to itself. As discussed in Section 7.7, some researchers define betweenness differently, counting paths only once, but that merely reduces all values by a factor of two.

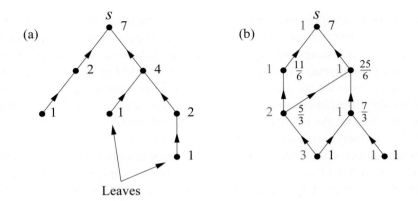

Figure 10.3: Calculation of betweenness centrality. (a) When there is only a single shortest path from a source vertex s (top) to all other reachable vertices, those paths necessarily form a tree, which makes the calculation of the contribution to betweenness from this set of paths particularly simple, as described in the text. (b) For cases in which there is more than one shortest path to some vertices, the calculation is more complex. First we must calculate the number of paths from the source s to each other vertex (numbers to the left of vertices), and then use these to weight the path counts appropriately and derive the betweenness scores (numbers to the right of vertices).

For any given s, this algorithm will take time $O(m + n)$ to find the shortest paths. Paths have length that by definition is less than or equal to the diameter of the network, which is typically of order $\log n$, and hence traversing the n paths from each vertex to s will take time $O(n \log n)$, for a running time of $O(m + n \log n)$ for each value of s. Repeating for all s, the whole algorithm will then take total time $O(n(m + n \log n))$ or $O(n^2 \log n)$ on a sparse network.

This is much better than our earlier $O(n^3)$ algorithm, but we can do better still. It is in fact possible to cut the running time down to just $O(n(m + n))$ by exploiting the fact that many of the shortest paths in the shortest path tree share many of the same edges. To understand this development, consider Fig. 10.3a, which shows a shortest path tree from a vertex s to all other vertices on a graph. In this case the shortest path tree really is a tree, meaning there is only one shortest path from s to any other vertex. This case is a good first example to study because of its simplicity, but we will consider the more general case in just a moment.

We use the tree to calculate a score for each vertex representing the number of shortest paths passing through that vertex. We find first the "leaves" of the tree, i.e., those vertices such that no shortest paths from other vertices to s pass through them. (In Fig. 10.3a the leaves are drawn at the bottom of the tree.) We assign a score of 1 to each of these leaves—the only path to s that passes through these vertices is the one that starts there.[10] Then, starting at the bottom of the tree we work upward, assigning to each vertex a score that is 1 plus the sum of the scores on the neighboring vertices immediately below it. That is, the number of paths through a vertex v is 1 for the path that starts at v plus the count of all paths that start below v in the tree and hence have to pass through it.

When we have worked all the way up the tree in this manner and reached vertex s, the scores at each vertex are equal to the betweenness counts for paths that end at vertex s. Repeating the process for all s and summing the scores, we arrive at the full betweenness scores for all paths.

In practice, the process of working up the tree can be accomplished by running through the vertices in order of decreasing distance from s. Conveniently, we already have a list of vertices in order of their distances, namely the entries in the queue array created by the breadth-first search process. Thus the betweenness algorithm in practice involves running backwards through the list of vertices in this array and calculating the number of paths through each vertex as above until the beginning of the array is reached.

In the worst case, this process involves going through all n vertices and checking every neighbor of every vertex, of which there are a total of $2m$, so that the overall running time is $O(m+n)$. The breadth-first search itself also takes time $O(m+n)$ (as usual) and hence the total time to count paths for each source vertex s is $O(m+n)$, which means the complete betweenness calculation takes time $O(n(m+n))$, as promised.

In general, however, we cannot assume that the shortest paths to a given vertex form a tree. As we saw in Section 10.3.5, often they do not. Consider, for instance, the "tree" shown in Fig. 10.3b. Following the definition of betweenness in Section 7.7, multiple shortest paths between the same pair of vertices are given equal weights summing to 1, so that for a vertex pair connected by three shortest paths, for example, we give each path weight $\frac{1}{3}$. Note that some of the paths may share vertices for part of their length, resulting in vertices with greater weight.

---

[10]In this case we are considering the first and last vertices on a path to be members of that path. As discussed in Section 7.7, the first and last vertices are sometimes excluded from the calculation, which means that the betweenness score of each vertex is smaller by an additive constant equal to twice the number of vertices in the component. If we wish to calculate betweenness according to this alternative definition, the simplest approach is to use the algorithm described here and then subtract the additive constant from each vertex's score at the end.

To calculate correctly the weights of the paths flowing through each vertex in a network, we need first to calculate the total number of shortest paths from each vertex to s. This is actually quite straightforward to do: the shortest paths from s to a vertex i must pass through one or more neighbors of i and the total number of shortest paths to i is simply the sum of the numbers of shortest paths to each of those neighbors. We can calculate these sums as part of a modified breadth-first search process as follows.

Consider Fig. 10.3b and suppose we are starting at vertex s. We carry out the following steps:

1. Assign vertex s distance zero, to indicate that it is zero steps from itself, and set $d = 0$. Also assign s a weight $w_s = 1$ (whose purpose will become clear shortly).

2. For each vertex i whose assigned distance is d, follow each attached edge to the vertex j at its other end and then do one of the following three things:

   a) If j has not yet been assigned a distance, assign it distance $d + 1$ and weight $w_j = w_i$.

   b) If j has already been assigned a distance and that distance is equal to $d + 1$, then the vertex's weight is increased by w_i, that is $w_j \leftarrow w_j + w_i$.

   c) If j has already been assigned a distance less than $d + 1$, do nothing.

3. Increase d by 1.

4. Repeat from step 2 until there are no vertices that have distance d.

The resulting weights for the example of Fig. 10.3b are shown to the left of each vertex in the figure. Each weight is the sum of the ones above it in the "tree." (It may be helpful to work through this example yourself by hand to see how the algorithm arrives at these values for the weights.) Physically, the weight on a vertex i represents the number of distinct geodesic paths between the source vertex s and i.

Now if two vertices i and j are connected by a directed edge in the shortest path "tree" pointing from j to i, then the fraction of the paths to s that pass through (or starting at) j and that also pass through i is given by w_i / w_j.

Thus, and finally, to calculate the contribution to the betweenness from shortest paths starting at all vertices and ending at s, we need only carry out the following steps:

1. Find every "leaf" vertex t, i.e., a vertex such that no paths from s to other vertices go though t, and assign it a score of $x_t = 1$.

2. Now, starting at the bottom of the tree, work up towards s and assign to each vertex i a score $x_i = 1 + \sum_j x_j w_i / w_j$, where the sum is over the neighbors j immediately below vertex i.

3. Repeat from step 2 until vertex s is reached.

The resulting scores are shown to the right of each vertex in Fig. 10.3b. Now repeating this process for all n source vertices s and summing the resulting scores on the vertices gives us the total betweenness scores for all vertices in time $O(n(m + n))$.[11]

This algorithm again takes time $O(n(m + n))$ in general or $O(n^2)$ on a sparse network, which is the best known running time for any betweenness algorithm at the time of writing, and moreover seems unlikely to be beaten by any future algorithm given that the calculation of the betweenness necessarily requires us to find shortest paths between all pairs of vertices, which operation also has time complexity $O(n(m + n))$. Indeed, even if we want to calculate the betweenness of only a single vertex it seems unlikely we can do better given that such a calculation still requires us to find all shortest paths.

10.4 SHORTEST PATHS IN NETWORKS WITH VARYING EDGE LENGTHS

In Section 6.3 we discussed weighted networks, networks in which the edges have values or strengths representing, for instance, the traffic capacities of connections on the Internet or the frequencies of contacts between acquaintances in a social network. In some cases the values on edges can be interpreted as lengths for the edges. The lengths could be real lengths, such as distances along roads in a road network, or they could represent quantities that act like lengths, such as transmission delays for packets traveling along Internet connections. In other cases they might just be approximately length-like measures: one might say, for instance, that a pair of acquaintances in a social network are twice as far apart as another pair if they see one another half as often.

Sometimes with networks such as these we would like to calculate the shortest path between two vertices taking the lengths of the edges into account. For instance, we might want to calculate the shortest driving route from A to B via a road network or we might want to calculate the route across the Internet that gets a data packet to its destination in the shortest time. (In fact, this is exactly what many Internet routers do when routing data packets.)

But now we notice a crucial—and annoying—fact. The shortest path across a network when we take edge lengths into account may not the be same as the shortest path in terms of number of edges. Consider Fig. 10.4. The shortest path between s and t in this small network traverses four edges, but is

---

[11] As discussed in footnote 9, these scores give the betweenness as defined in Eq. (7.36). To get true path counts one would have to divide by two and add a half (or equivalently add one then divide by two) to correct for the double counting of paths between distinct vertices.

still shorter, in terms of total edge length, than the competing path with just two edges. Thus we cannot find the shortest path in such a network using standard breadth-first search, which finds paths with the minimum number of edges. For problems like this we need a different algorithm. We need *Dijkstra's algorithm*.

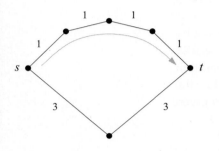

Dijkstra's algorithm, like breadth-first search, finds the shortest distance from a given source vertex s to every other vertex in the same component of a network, but does so taking the lengths of edges into account.[12] It works by keeping a record of the shortest distance it has found so far to each vertex and updating that record whenever a shorter one is found. It can be shown that, at the end of the algorithm, the shortest distance found to each vertex is in fact the shortest distance possible by any route. In detail the algorithm is as follows.

We start by creating an array of n elements to hold our current estimates of the distances from s to every vertex. At all times during the running of the algorithm these estimates are upper bounds on the true shortest distances. Initially we set our estimate of the distance from s to itself to be zero, which is trivially correct, and from s to every other vertex to be ∞, which is clearly a safe upper bound.

Figure 10.4: The shortest path in a network with varying edge lengths. The numbers on the edges in this network represent their lengths. The shortest path between s and t, taking the lengths into account, is the upper path marked with the arrow (which has total length 4), even though it traverses more edges than the alternative, lower path (which has length 6).

We also create another array of n elements in which we record whether we are certain that the distance we have to a given vertex is the smallest possible distance. For instance, we might use an integer array with 1s to indicate the distances we are sure about and 0s for the distances that are just our best current estimate. Initially, we put a 0 in every element of this array. (You might argue that we know for certain that the distance from s to itself is zero and hence that we should put a 1 in the element corresponding to vertex s. Let us, however, pretend that we don't know this to begin with, as it makes the algorithm work out more neatly.)

Now we do the following.

1. We find the vertex v in the network that has the smallest estimated distance from s, i.e., the smallest distance about which we are not yet certain.

---

[12] We assume that the lengths are all non-negative. If lengths can be negative, which happens in some cases, then the problem is much harder, falling in the class of "NP-complete" computational problems, for which even the best known algorithms take an amount of time exponential in n to finish, in the worst case [8]. Indeed, if edges are allowed to have negative lengths, there may not be any shortest path between a pair of vertices, since one can have a loop in the network that has negative length, so that one can reduce the length of a path arbitrarily by going around the loop repeatedly.

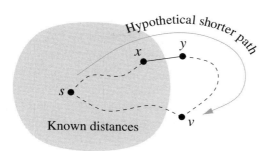

Figure 10.5: Paths in Dijkstra's algorithm. If v is the vertex with the smallest estimated (i.e., not certain) distance from s then that estimated distance must in fact be the true shortest distance to v. If it were not and there were a shorter path $s, \ldots, x, y, \ldots, v$ then all points along that path must have shorter distances from s than v's estimated distance, which means that y has a smaller estimated distance than v, which is impossible.

2. We mark this distance as being certain.
3. We calculate the distances from s via v to each of the neighbors of v by adding to v's distance the lengths of the edges connecting v to each neighbor. If any of the resulting distances is smaller than the current estimated distance to the same neighbor, the new distance replaces the older one.
4. We repeat from step 1 until the distances to all vertices are flagged as being certain.

Simple though it is to describe, it's not immediately obvious that this algorithm does what it is supposed to do and finds true shortest paths. The crucial step is step 2 where we declare the current smallest estimated distance in fact to be certain. That is, we claim that among vertices for which we don't yet definitely know the distance, the smallest distance recorded to any vertex is in fact the smallest possible distance to that vertex.

To see why this is true consider such a vertex, which we'll again call v, and consider a hypothetical path from s to v that has a shorter length than the current estimated distance recorded for v. The situation is illustrated in Fig. 10.5. Since this hypothetical path is shorter than the estimated distance to v, the distance along the path to each vertex in the path must also be less than that estimated distance.

Furthermore, there must exist somewhere along the path a pair of adjacent vertices x, y such that x's distance is known for certain and y's is not. Vertex x need not necessarily be distinct from vertex s (although we have drawn it that way in the figure), but vertex y must be distinct from v: if y and v were the same vertex, so that v was a neighbor of x, then we would already have found the shorter path to v when we explored the neighbors of x in step 3 above and we would accordingly have revised our estimate of v's distance downward. Since this hasn't happened, y and v must be distinct vertices.

But notice now that y's current estimated distance will be at most equal to its distance from s *along the path* because that distance is calculated in step 3 above when we explore x's neighbors. And since, as we have said, all distances along the path are necessarily less than the current estimated distance to v, it follows that y's estimated distance must be less than v's and we have a contradiction, because v is by hypothesis the vertex with the shortest estimated distance. Hence there is no path to vertex v with length less than v's current estimated distance, so we can safely mark that distance as being certain, as in step 2 above.

Thus on each step the algorithm correctly flags one additional distance as being known exactly and when all distances have been so flagged the algorithm has done its job.

As with breadth-first search, the running time of Dijkstra's algorithm depends on how it is implemented. The simplest implementation is one that searches through all vertices on each round of the algorithm to find the one that has the smallest estimated distance. This search takes time $O(n)$. Then we must calculate a new estimated distance to each of the neighbors of the vertex we find, of which there are $O(m/n)$ on average. To leading order, one round thus takes time $O(m/n + n)$ and the whole algorithm, which runs (in the worst case of a network with a single component) for n rounds, takes time $O(m + n^2)$ to find the distance from s to every other vertex.

But we can do better than this. If we store the estimated distances in a binary heap (see Section 9.7) then we can find the smallest one and remove it from the heap in time $O(\log n)$. The operation of replacing an estimated distance with a new and better estimate (which in the worst case we have to do an average of $O(m/n)$ times per round) also takes $O(\log n)$ time, and hence a complete round of the algorithm takes time $O((m/n)\log n + \log n)$ and all n rounds then take $O((m + n)\log n)$, or $O(n \log n)$ on a sparse network with $m \propto n$. This is very nearly the best running time known for this problem,[13] and close to, though not quite as good as, the $O(m + n)$ for the equivalent problem on an unweighted network (factors of $\log n$ being close to constant given that

---

[13] In theory one can achieve a slightly better running time of $O(m + n \log n)$ using a data structure known as a Fibonacci heap [81], but in practice the operation of the Fibonacci heap is so complicated that the calculation usually ends up running slower.

the logarithm is a very slowly growing function of its argument).

As we have described it, Dijkstra's algorithm finds the shortest distance from a vertex s to every other in the same component but, like breadth-first search, it can be modified also to find the actual paths that realize those distances. The modification is very similar to the one for breadth-first search. We maintain a shortest path tree, which is initially empty and to which we add directed edges pointing from the vertices along the first step of their shortest path to s. We create such a directed edge when we first assign a vertex an estimated distance less than ∞ and move the edge to point to a new vertex every time we find a new estimated distance that is less than the current one. The last position in which an edge comes to rest indicates the true first step in the shortest path. If a new estimate of the distance to a vertex is ever exactly the same as the current estimate then we put two directed edges in the shortest path tree indicating the two alternative paths that give the shortest distance. When the algorithm is finished the shortest path tree, like those in Fig. 10.2, can be used to reconstruct the shortest paths themselves, or to calculate other quantities such as a weighted version of betweenness centrality (which could be used for instance as a measure of traffic flow in a network where traffic always takes the shortest weighted path).

10.5 MAXIMUM FLOWS AND MINIMUM CUTS

In Section 6.12 we discussed the ideas of connectivity, independent paths, cut sets, and maximum flows in networks. In particular, we defined two paths that connect the same vertices s and t to be edge-independent if they share none of the same edges and vertex-independent if they share none of the same vertices except for s and t themselves. And the edge or vertex connectivity of the vertices is the number of edge- or vertex-independent paths between them. We also showed that the edge or vertex connectivity is equal to the size of the minimum edge or vertex cut set—the minimum number of edges or vertices that need to be removed from the network to disconnect s from t. Connectivity is thus a simple measure of the robustness of the connection between a pair of vertices. Finally, we showed that the edge-connectivity is also equal to the maximum flow that can pass from s to t if we think of the network as a network of pipes, each of which can carry one unit of flow.

In this section we look at algorithms for calculating maximum flows between vertices on networks. As we will see, there is a simple algorithm, the Ford–Fulkerson or augmenting path algorithm, that calculates the maximum flow between two vertices in average time $O((m + n)m/n)$. Once we have this maximum flow, then we also immediately know the number of edge-independent paths and the size of the minimum edge cut set between the same vertices. With a small extension, the algorithm can also find the particular edges that constitute the minimum edge cut set. A simple modification of the augmenting path algorithm allows us also to calculate vertex-independent paths and vertex cuts sets.

All the developments of this section are described for undirected networks, but in fact the algorithms work perfectly well, without modification, for directed networks as well. Readers who want to know more about maximum flow algorithms are recommended to look at the book by Ahuja *et al.* [8], which contains several hundred pages on the topic and covers almost every conceivable detail.

10.5.1 THE AUGMENTING PATH ALGORITHM

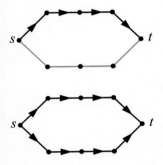

A simple breadth-first search finds a path from source s to target t (top) in this network. A second search using only the edges not used in the first finds a second path (bottom).

In this section we describe the augmenting path algorithm of Ford and Fulkerson for calculating maximum flows between vertices in a network.[14] The case of primary interest to us is the one where each edge in the network can carry the same single unit of flow. The algorithm can be used in the more general case where the edges have varying capacities, but we will not discuss that case here.[15]

The basic idea behind the augmenting path algorithm is a simple one. We first find a path from source s to target t using the breadth-first search algorithm of Section 10.3.[16] This "uses up" some of the edges in the network, filling them to capacity so that they can carry no more flow. Then we find another path from s to t among the remaining edges and we repeat this procedure until no more paths can be found.

Unfortunately, this does not yet give us a working algorithm, because as we have described it the procedure will not always find the maximum flow. Consider Fig. 10.6a. If we apply breadth-first search between s and t we find

---

[14]The augmenting path algorithm is not the only algorithm for calculating maximum flows. It is, however, the simplest and its average performance is about as good as any other, so it is a good choice for everyday calculations. It's worth noting, however, that the *worst-case* performance of the algorithm is quite poor: for pathological networks, the algorithm can take a very long time to run. Another algorithm, the *preflow-push algorithm* [8], has much better worst-case performance and comparable average-case performance, but is considerably more complicated to implement.

[15]See Ahuja *et al.* [8] or Cormen *et al.* [81] for details of the general case.

[16]Technically, the augmenting path algorithm doesn't specify how paths are to be found. Here we study the particular version in which paths are found using breadth-first search, which is known to be one of the better-performing variants. Sometimes this variant is called the *shortest augmenting path algorithm* or the *Edmonds–Karp algorithm*.

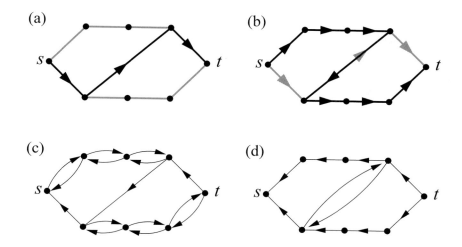

Figure 10.6: The augmenting path algorithm. (a) We find a first path from source s to target t using breadth-first search. This leaves no more independent paths from s to t among the remaining edges. (b) However, if we allow flows in both directions along an edge (such as the central edge in this network), then we can find another path. Panels (c) and (d) show the residual graphs corresponding to panels (a) and (b).

the path marked in bold. Unfortunately, once we have filled all the edges along this path to capacity there are no more paths from s to t that can be constructed with the remaining edges, so the algorithm stops after finding just one path. It is clear, however, that there are in fact two edge-independent paths from s to t—along the top and bottom of the network—and a maximum flow of two, so the algorithm has given the wrong answer.

There is however, a simple fix for this problem, which is to allow fluid to flow simultaneously *both ways* down an edge in our network. That is, we allow a state in which there is one unit of flow in each direction along any given edge. If the edges were real pipes, then this would not be possible: if a pipe is full of fluid flowing one way, then there is no room for fluid flowing the other way too. However, if fluid *were* flowing both ways down an edge, the net flow in and out of either end of that edge would be zero—the two flows would effectively cancel out giving zero net flow. And zero flow down an edge certainly is possible.

So we use a trick and allow our algorithm to place a unit of flow both ways down any edge, but declare this to mean in practice that there is no flow at all on that edge. This means that the paths we will find will no longer necessarily be independent paths, since two of them can share an edge so long as they pass along it in opposite directions. But this doesn't matter: the flows the paths represent are still allowed flows, since no pipe is ever required to carry more than one unit of flow, and we know that in the end our final, maximum flow will be numerically equal to the actual number of independent paths, even though those independent paths may be different from the paths picked out by the algorithm. Thus we create an algorithm that counts independent paths by counting a special class of *non*-independent paths: strange as this sounds, the max-flow/min-cut theorem tells us that it must work, and indeed it does.

More generally, since the maximum allowed flow down an edge is one unit in either direction, we can have any number of units flowing either way down an edge provided they cancel out so that net flow is no more than one unit. Thus, two units of flow in either direction would be allowed, or three units one way and four the other, and so forth. Three units one way and five the other would not be allowed, however.[17]

To see how this works in practice, consider Fig. 10.6 again. We begin by performing a breadth-first search that finds the path shown in panel (a). Now, however, there is a second path to be found, as shown in panel (b), making use of the fact that we are still allowed to send one unit of flow *backwards* along the edge in the center of the network. After this, however, there are no more paths left from s to t and so the algorithm stops and tells us that the maximum possible flow is two units, which is the correct answer.

This is merely one example of the algorithm: we still have to prove that it gives the correct answer in all cases, which we do in Section 10.5.3. To understand the proof, however, we first need to understand how the algorithm is implemented.

10.5.2 IMPLEMENTATION AND RUNNING TIME

Implementation of the augmenting path algorithm makes use of a *residual graph*, which is a directed network in which the edges connect the pairs of vertices on the original network between which we still have capacity available to carry one or more units of flow in the given direction. For instance, Figs. 10.6c and 10.6d show the residual graphs corresponding to the flow states in 10.6a and 10.6b.

The residual graph is constructed by first taking the initial network and

---

[17]On networks with directed edges, we allow either the same flow in both directions along an edge (i.e., zero net flow) or one more unit in the forward direction than in the backward direction, but not vice versa.

replacing each undirected edge with two directed ones, one in each direction. We now perform our breadth-first searches on this residual graph, rather than on the original network, respecting the directions of the edges. Every time our algorithm finds a new path through the network, we update the residual graph by adding a directed edge in the opposite direction to the path between every pair of vertices along the path, provided no such edge already exists. If a vertex pair already has such a backward-pointing edge, we instead take away a forward-pointing one. (There will always be such a forward-pointing edge, otherwise the path would not exist in the first place.) The largest number of edges we update during this process is m, the total number of edges in the original network, so the process takes time $O(m)$ and thus makes no difference to the $O(m + n)$ time complexity of the breadth-first search.

Now we find the next path by performing another breadth-first search on the updated residual graph. By always working on the residual graph in this way, we insure that we find only paths along edges that have not yet reached their maximum flow. Such paths are called *augmenting paths*. The process is repeated until our breadth-first search fails to find any augmenting path from s to t, at which point we have found all the paths there are and the number of paths found is equal to the number of units in the maximum flow from s to t.

Each breadth-first search, along with the corresponding updates to the residual graph, takes time $O(m + n)$ for a network stored in adjacency list format (see Sections 9.4 and 10.3). Moreover, the number of independent paths from s to t can be no greater than the smaller of the two degrees k_s and k_t of the source and target vertices (since each path must leave or enter one of those vertices along some edge, and that edge can carry at most one path). Thus the running time of the algorithm is $O\big(\min(k_s, k_t)(m + n)\big)$. If we are interested in the average running time over many pairs of vertices, then we can make use of the fact that $\langle \min(k_s, k_t) \rangle \leq \langle k \rangle$ (where the averages are over all vertices), and recalling that $\langle k \rangle = 2m/n$ (Eq. (6.23)), this implies that the average running time of the algorithm is $O\big((m + n)m/n\big)$, which is $O(n)$ on a sparse network with $m \propto n$. (On the other hand, on a dense graph where $m \propto n^2$, we would have $O(n^3)$, which is much worse.)

10.5.3 WHY THE ALGORITHM GIVES CORRECT ANSWERS

It is plausible but not immediately obvious that the augmenting path algorithm correctly finds maximum flows. We can prove that it does as follows.

Suppose at some point during the operation of the algorithm (including the very beginning) we have found some (or no) paths for flow from s to t, but any paths we have found do not yet constitute the maximum possible flow.

That is, there is still room in the network for more flow from s to t. If this is the case then, as we will now show, there must exist at least one augmenting path from s to t, which by definition carries one unit of flow. And if there exists an augmenting path, our breadth-first search will always find it, and so the algorithm will go on finding augmenting paths until there is no room in the network for more flow, i.e., we have reached the maximum flow, which is equal to the number of paths found.

Thus the proof that the algorithm is correct requires only that we prove the following theorem:

> If at some point in our algorithm the flow from s to t is less than the maximum possible flow, then there must exist at least one augmenting path on the current residual graph.

Consider such a point in the operation of the algorithm and consider the flows on the network as represented by f, the set of all individual net flows along the edges of the network. And consider also the maximum possible flow from s to t, represented by f_{max}, the corresponding set of individual net flows. By hypothesis, the total flow out of s and into t is greater in f_{max} than in f. Let us calculate the difference flow $\Delta f = f_{max} - f$, by which we mean we subtract the net flow along each edge in f from the net flow along the corresponding edge in f_{max}, respecting flow direction—see Fig. 10.7. (For instance, the difference of two unit flows in the same direction would be zero while the difference of two in opposite directions would be two in one direction or the other.)

Since the total flow is greater in f_{max} than in f, the difference flow Δf must have a net flow out of s and net flow into t. What's more, because the "fluid" composing the flow is conserved at vertices, every vertex except s and t must have zero net flow in or out of it in both f_{max} and f and hence also in Δf. But if each vertex other than s and t has zero net flow, then the flow from s to t must form at least one path across the network—it must leave every vertex it enters, except the last one, vertex t. Let us choose any one of these paths formed by the flow from s to t in Δf and let us call this path p.

Since there is a positive flow in Δf in the forward direction along each edge in p, there must have been no such flow in f along any of the same edges. If there were such a flow in f then when we performed the subtraction $\Delta f = f_{max} - f$ the flow in Δf would be either zero or negative on the edge in question (depending on the flow in f_{max}), but could not be positive. Thus we can always safely add to f a unit of flow forward along each edge in p without overloading any of the edges. But this immediately implies that p is an augmenting path for f.

Thus, for any flow that is not maximal, at least one augmenting path always

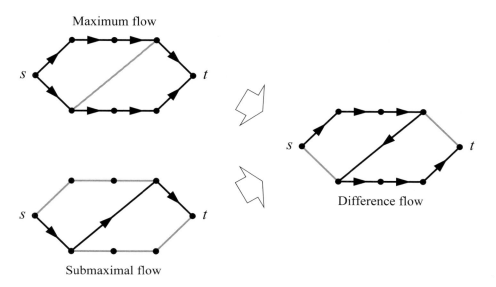

Figure 10.7: Correctness of the augmenting path algorithm. If we subtract from the maximum flow f_{max} (upper left) any submaximal flow f (lower left), the resulting difference flow (right) necessarily contains at least one path from s to t, and that path is necessarily an augmenting path for f.

exists, and hence it follows that the augmenting path algorithm as described above is correct and will always find the maximum flow.

10.5.4 FINDING INDEPENDENT PATHS AND MINIMUM CUT SETS

Once we have found the maximum possible flow between a given pair of vertices, we also automatically have the size of the minimum edge cut set and the number of edge-independent paths, which are both numerically equal to the number of units in the maximum flow (see Section 6.12).

We might also wish to know exactly where the independent paths run. The augmenting path algorithm does not give us this directly since, as we have seen, the augmenting paths it finds are not necessarily the same as the independent paths, but only a very small extension of the algorithm is necessary to find the independent paths: we take the final residual graph produced at the end of the algorithm and remove from it every pair of directed edges that joins the same two vertices in opposite directions—see Fig. 10.8. In other words we are removing all network edges that carry no net flow. The edges remaining

Figure 10.8: Reconstructing the independent paths from the residual graph. Deleting every pair of edges on the residual graph that join the same two vertices in opposite directions leaves a graph consisting of the independent paths only, spelled out in directed edges that point backwards along those paths from target to source.

after we have done this are necessarily those that actually carry the maximum flow and it is a straightforward matter to trace these edges from s to t to reconstruct the paths themselves.[18] (In fact, as Fig. 10.8 shows, the remaining directed edges in the residual graph point backwards from t to s, so it is often easier to reconstruct the paths backwards.)

Another thing we might want is the set of edges that constitutes the minimum cut set for the vertices s and t. In fact in most cases there is more than one cut set of the minimum size, so more generally we would like to find one of the minimum cut sets. Again we can do this by a small extension of the augmenting path algorithm. The procedure is illustrated in Fig. 10.9. We again consider the final residual graph generated at the end of the algorithm. By definition this graph has no directed path in it from s to t (since if it did the algorithm would not have stopped yet). Thus we can reach some subset of vertices by starting at vertex s, but we cannot reach all of them. (For example, we cannot reach t.) Let V_s be the subset of vertices reachable from s by some path on the residual graph and let V_t be the set of all the other vertices in the graph that are not in V_s. Then the set of edges on the original graph that connect vertices in V_s to vertices in V_t constitutes a minimum cut set for s and t.

Why does this work? Clearly if we removed all edges that connect vertices in V_s to those in V_t we disconnect s and t, since then there is no path at all between s and t. Thus the edges between V_s and V_t constitute a cut set. That it

---

[18]Note, however, that the independent paths are not necessarily unique: there can be more than one choice of paths and some of them may not be found by this algorithm. Furthermore, there can be points in the network where paths come together at a vertex and then part ways again. If such points exist, you will have to make a choice about which way to go at the parting point. It doesn't matter what choice you make in the sense that all choices lead to a correct set of paths, but different choices will give different sets of paths.

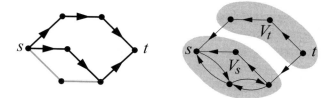

Figure 10.9: Finding a minimum cut set. Once we have found a set of maximum flows for a given s and t (left) we can find a corresponding minimum cut set by considering the residual graph (right). The set V_s is the set of vertices reachable from s by following directed edges on the residual graph and V_t is the rest of the vertices. The minimum cut set is the set of edges (two of them in this case) that connect V_s to V_t on the original network.

is a *minimum* cut set we can see by the following argument. Every edge from a vertex in V_s to a vertex in V_t must be carrying a unit of flow from V_s to V_t. If it were not, then it would have available capacity away from V_s, meaning that there would be a corresponding directed edge away from V_s in the residual graph. In that case, however, the vertex at the far end of that edge would be reachable from V_s on the residual graph and therefore would be a part of V_s. Since the vertex in question is, by hypothesis, in V_t and not in V_s, it follows that it must be carrying a unit of the maximum flow from s to t.

Now, since every edge in the cut set between V_s and V_t is carrying a unit of flow, the size of that cut set is numerically equal to the size of the flow from V_s to V_t, which is also the flow from s to t. And, by the max-flow/min-cut theorem, a cut set between s and t that is equal in size to the maximum flow between s and t is a minimum cut set, and hence our result is proved.

10.5.5 FINDING VERTEX-INDEPENDENT PATHS

Once we know how to find edge-independent paths it is straightforward to find vertex-independent paths as well. First, note that any set of vertex-independent paths between two vertices s and t is necessarily also a set of edge-independent paths: if two paths share none of the same vertices, then they also share none of the same edges. Thus, we can find vertex-independent paths using the same algorithm that we used to find edge-independent paths, but adding the restriction that no two paths may pass through the same vertex. One way to impose this restriction is the following. First, we replace our undirected network with a directed one, as shown in Fig. 10.10, with a directed

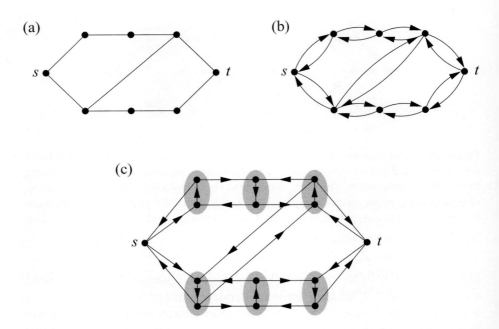

Figure 10.10: Mapping from the vertex-independent path problem to the edge-independent path problem. Starting with an undirected network (a), we (b) replace each edge by two directed edges, then (c) replace each vertex, except for s and t, with a pair of vertices with a directed edge between them (shaded) following the prescription in Fig. 10.11. Edge-independent paths on the final network then correspond to vertex-independent paths on the initial network.

edge in either direction between every connected pair of vertices. This does not change the maximum flow possible in the network and hence does not change the number of independent paths either.

Second, we replace each of the vertices in the network, except s and t, with a construct like that shown in Fig. 10.11. Each vertex is replaced with two vertices separated by a directed edge. All original incoming edges connect to the first of these two (on the left in Fig. 10.11) and all outgoing edges to the second. This new construct functions as the original vertex did, allowing flows to pass in along ingoing edges and out along outgoing ones, but with one important difference: assuming that the new edge joining the two vertices has unit capacity like all others, we are now limited to just one unit of flow through the entire construct, since every path through the construct must traverse this central edge. Thus every allowed flow on this network corresponds to a flow

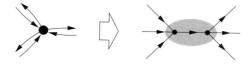

Figure 10.11: Vertex transformation for the vertex-independent path algorithm. Each vertex in the network is replaced by a pair of vertices joined by a single directed edge. All incoming edges are connected to one of the pair and all outgoing edges to the other as shown.

on the original network with at most a single unit passing though each vertex.

Transforming the entire network of Fig. 10.10a using this method gives us a network that looks like Fig. 10.10c. Now we simply apply the normal augmenting path algorithm to this directed network, and the number of *edge*-independent paths we find is equal to the number of *vertex*-independent paths on the original network.

PROBLEMS

10.1 What is the time complexity, as a function of the number n of vertices and m of edges, of the following network operations if the network in question is stored in adjacency list format?

a) Calculating the mean degree.

b) Calculating the median degree.

c) Calculating the air-travel route between two airports that has the shortest total flying time, assuming the flying time of each individual flight is known.

d) Calculating the minimum number of routers that would have to fail to disconnect two given routers on the Internet.

10.2 For an undirected network of n vertices stored in adjacency list format show that:

a) It takes time $O(n(n+m))$ to find the diameter of the network.

b) It takes time $O(\langle k \rangle)$ on average to list the neighbors of a vertex, where $\langle k \rangle$ is the average degree in the network, but time $O(\langle k^2 \rangle)$ to list the second neighbors.

10.3 For a directed network in which in- and out-degrees are uncorrelated, show that it takes time $O(m^2/n)$ to calculate the reciprocity of the network. Why is the restriction to uncorrelated degrees necessary? What could happen if they were correlated?

10.4 Suppose that we define a new centrality measure x_i for vertex i in a network to be a sum of contributions as follows: 1 for vertex i itself, α for each vertex at (geodesic) distance 1 from i, α^2 for each vertex at distance 2, and so forth, where $\alpha < 1$ is a given constant.

a) Write an expression for x_i in terms of α and the geodesic distances d_{ij} between vertex pairs.

b) Describe briefly an algorithm for calculating this centrality measure. What is the time complexity of calculating x_i for all i?

CHAPTER 11

MATRIX ALGORITHMS AND GRAPH PARTITIONING

A discussion of network algorithms that use matrix and linear algebra methods, including algorithms for partitioning network nodes into groups

IN THE preceding chapter we discussed a variety of computer algorithms for calculating quantities of interest on networks, including degrees, centralities, shortest paths, and connectivity. We continue our study of network algorithms in this chapter with algorithms based on matrix calculations and methods of linear algebra applied to the adjacency matrix or other network matrices such as the graph Laplacian. We begin with a simple example, the calculation of eigenvector centrality, which involves finding the leading eigenvector of the adjacency matrix, and then we move on to some more advanced examples, including Fiedler's spectral partitioning method and algorithms for network community detection.

11.1   LEADING EIGENVECTORS AND EIGENVECTOR CENTRALITY

As discussed in Section 7.2, the eigenvector centrality of a vertex i in a network is defined to be the ith element of the leading eigenvector of the adjacency matrix, meaning the eigenvector corresponding to the largest (most positive) eigenvalue. Eigenvector centrality is an example of a quantity that can be calculated by a computer in a number of different ways, but not all of them are equally efficient. One way to calculate it would be to use a standard linear algebra method to calculate the complete set of eigenvectors of the adjacency matrix, and then discard all of them except the one corresponding to the largest eigenvalue. This, however, would be a wasteful approach, since it involves cal-

345

culating a lot of things that we don't need. A simpler and faster method for calculating the eigenvector centrality is the *power method*.

If we start with essentially any initial vector $\mathbf{x}(0)$ and multiply it repeatedly by the adjacency matrix $\mathbf{A}$, we get

$$\mathbf{x}(t) = \mathbf{A}^t\mathbf{x}(0), \qquad (11.1)$$

and, as shown in Section 7.2, $\mathbf{x}(t)$ will converge[1] to the required leading eigenvector of $\mathbf{A}$ as $t \to \infty$. This is the power method, and, simple though it is, there is no faster method known for calculating the eigenvector centrality (or the leading eigenvector of any matrix). There are a few caveats, however:

1. The method will not work if the initial vector $\mathbf{x}(0)$ happens to be orthogonal to the leading eigenvector. One simple way to avoid this problem is to choose the initial vector to have all elements positive. This works because all elements of the leading eigenvector of a real matrix with non-negative elements have the same sign,[2] which means that any vector or-

---

[1]Technically the power method finds the eigenvector corresponding to the eigenvalue of largest absolute magnitude and hence the method would fail to find the eigenvector we want if the largest absolute magnitude belongs to a negative eigenvalue. For a matrix with all elements non-negative, however, such as the adjacency matrix, it turns out this can never happen. Here is a proof of this result for an undirected network where $\mathbf{A}$ is symmetric; the general case is covered, for example, in Ref. [217]. Let μ be the most negative eigenvalue of a real symmetric matrix $\mathbf{A}$ and let $\mathbf{w}$ be the corresponding eigenvector, with elements w_i. Then, given that $\mathbf{w}^T\mathbf{w} = \sum_i w_i^2 > 0$,

$$|\mu|\mathbf{w}^T\mathbf{w} = |\mu\mathbf{w}^T\mathbf{w}| = |\mathbf{w}^T\mathbf{A}\mathbf{w}| = \left|\sum_{ij} A_{ij}w_iw_j\right| \leq \sum_{ij}|A_{ij}w_iw_j| = \sum_{ij}A_{ij}|w_i||w_j| = \mathbf{x}^T\mathbf{A}\mathbf{x},$$

where $\mathbf{x}$ is the vector with components $|w_i|$. The inequality here follows from the so-called triangle inequality $|a + b| \leq |a| + |b|$, which is true for all real numbers a, b. Rearranging, we now find that

$$|\mu| \leq \frac{\mathbf{x}^T\mathbf{A}\mathbf{x}}{\mathbf{w}^T\mathbf{w}} = \frac{\mathbf{x}^T\mathbf{A}\mathbf{x}}{\mathbf{x}^T\mathbf{x}},$$

where we have made use of $\mathbf{x}^T\mathbf{x} = \sum_i |w_i|^2 = \mathbf{w}^T\mathbf{w}$. Now we write $\mathbf{x}$ as a linear combination of the normalized eigenvectors $\mathbf{v}_i$ of $\mathbf{A}$ thus: $\mathbf{x} = \sum_i c_i\mathbf{v}_i$, where the c_i are real coefficients whose exact values are not important for this proof. Then, if κ_i is the eigenvalue corresponding to $\mathbf{v}_i$ and κ_1 is the most positive eigenvalue, we have

$$\frac{\mathbf{x}^T\mathbf{A}\mathbf{x}}{\mathbf{x}^T\mathbf{x}} = \frac{\sum_j c_j\mathbf{v}_j^T\mathbf{A}\sum_i c_i\mathbf{v}_i}{\sum_j c_j\mathbf{v}_j^T\sum_i c_i\mathbf{v}_i} = \frac{\sum_j c_j\mathbf{v}_j^T\sum_i c_i\kappa_i\mathbf{v}_i}{\sum_j c_j\mathbf{v}_j^T\sum_i c_i\mathbf{v}_i} = \frac{\sum_i c_i^2\kappa_i}{\sum_i c_i^2} \leq \frac{\sum_i c_i^2\kappa_1}{\sum_i c_i^2} = \kappa_1,$$

where we have made use of the orthogonality property $\mathbf{v}_j^T\mathbf{v}_i = \delta_{ij}$. (The inequality is an exact equality if and only if $\mathbf{x}$ is an eigenvector with eigenvalue κ_1.) Putting these results together, we find that $|\mu| \leq \kappa_1$ and hence the most negative eigenvalue never has a magnitude greater than that of the most positive eigenvalue (although if we are unlucky the two magnitudes could be equal). The result proved here is one part of the *Perron–Frobenius theorem*. The other part, that the leading eigenvector has all elements non-negative, is proved in the following footnote.

[2]This result, like that in footnote 1, is a part of the Perron–Frobenius theorem. To prove it—

thogonal to the leading eigenvector must contain both positive and negative elements. Hence, if we choose all elements of our initial vector to be positive, we are guaranteed that the vector cannot be orthogonal to the leading eigenvector.

2. The elements of the vector have a tendency to grow on each iteration—they get multiplied by approximately a factor of the leading eigenvalue

---

at least for the case of symmetric $\mathbf{A}$—let κ_1 be the most positive eigenvalue of $\mathbf{A}$ and let $\mathbf{v}$ be a corresponding eigenvector. (We will allow, for the moment, the possibility that there is more than one eigenvector with eigenvalue κ_1, though we show below that in fact this cannot happen in a connected network.) Note that $\kappa_1 \geq 0$ since the sum of the eigenvalues of $\mathbf{A}$ is given by $\mathrm{Tr}\,\mathbf{A} \geq 0$, and hence at least one eigenvalue must be non-negative. Then, given that $\mathbf{v}^T\mathbf{v} = \sum_i v_i^2 > 0$ and all elements of $\mathbf{A}$ are non-negative, we have

$$\kappa_1\mathbf{v}^T\mathbf{v} = |\kappa_1\mathbf{v}^T\mathbf{v}| = |\mathbf{v}^T\mathbf{A}\mathbf{v}| = \left|\sum_{ij} A_{ij}v_iv_j\right| \leq \sum_{ij}|A_{ij}v_iv_j| = \sum_{ij}A_{ij}|v_i||v_j| = \mathbf{x}^T\mathbf{A}\mathbf{x},$$

where $\mathbf{x}$ is the vector with elements $|v_i|$. Rearranging this result, we find

$$\kappa_1 \leq \frac{\mathbf{x}^T\mathbf{A}\mathbf{x}}{\mathbf{v}^T\mathbf{v}} = \frac{\mathbf{x}^T\mathbf{A}\mathbf{x}}{\mathbf{x}^T\mathbf{x}},$$

where we have made use of $\mathbf{x}^T\mathbf{x} = \sum_i |v_i|^2 = \mathbf{v}^T\mathbf{v}$. As demonstrated in footnote 1 on page 346, for any vector $\mathbf{x}$ we have

$$\frac{\mathbf{x}^T\mathbf{A}\mathbf{x}}{\mathbf{x}^T\mathbf{x}} \leq \kappa_1,$$

with the equality being achieved only when $\mathbf{x}$ is an eigenvector corresponding to eigenvalue κ_1. The only way to reconcile the two inequalities above is if they are in fact equalities in this case, implying that $\mathbf{x}$ must indeed be an eigenvector with eigenvalue κ_1. But $\mathbf{x}$ has all elements non-negative, and hence there exists an eigenvector with eigenvalue κ_1 and all elements non-negative.

It is still possible that there might be more than one eigenvector with eigenvalue κ_1, and that one of the others might have negative elements. This, however, we can rule out as follows. Recall that eigenvectors with same eigenvalue can always be chosen orthogonal, and any eigenvector $\mathbf{v}$ that is orthogonal to the eigenvector with all elements non-negative would have to have both positive and negative elements in order that the product of the two vectors equal zero. Thus there is only one eigenvector with all elements non-negative.

Then, for eigenvector $\mathbf{v}$, by the results above, the vector $\mathbf{x}$ with elements $|v_i|$ is necessarily equal to the unique eigenvector with all elements non-negative. Thus if v_i is one of the positive elements of $\mathbf{v}$ then $v_i = x_i$ and

$$\sum_j A_{ij}|v_j| = \sum_j A_{ij}x_j = \kappa_1 x_i = \kappa_1 v_i = \sum_j A_{ij}v_j,$$

or, equivalently, $\sum_j A_{ij}(|v_j| - v_j) = 0$. But $|v_j| - v_j \geq 0$ so this last result can only be true if for all j we have either $A_{ij} = 0$ or $v_j - |v_j| = 0$, meaning that $v_j = |v_j| \geq 0$. Thus if $v_i > 0$ then $v_j > 0$ whenever $A_{ij} \neq 0$. In network terms, if $v_i > 0$ then $v_j > 0$ for every neighbor of i. But then we can start at i and work outwards, moving from neighbor to neighbor and so demonstrate that $v_j > 0$ for every vertex and hence $\mathbf{v} = \mathbf{x}$ and the leading eigenvector is unique.

The only exception to this last result is when the network has more than component, so that some vertices are not reachable from an initial vertex i. In that case, it is possible for the elements of the leading eigenvector corresponding to vertices in different components to have different signs. This, however, causes no problems for any of the results presented here.

each time, which is usually greater than 1. Computers however cannot handle arbitrarily large numbers. Eventually the variables storing the elements of the vector will overflow their allowed range. To obviate this problem, we must periodically renormalize the vector by dividing all the elements by the same value, which we are allowed to do since an eigenvector divided throughout by a constant is still an eigenvector. Any suitable divisor will do, but we might, for instance, divide by the magnitude of the vector, thereby normalizing it so that its new magnitude is 1.

3. How long do we need to go on multiplying by the adjacency matrix before the result converges to the leading eigenvalue? This will depend on how accurate an answer we require, but one simple way to gauge convergence is to perform the calculation in parallel for two different initial vectors and watch to see when they reach the same value, within some prescribed tolerance. This scheme works best if, for the particular initial vectors chosen, at least some elements of the vector converge to the final answer from opposite directions for the two vectors, one from above and one from below. (We must make the comparisons immediately after the renormalization of the vector described in (2) above—if we compare unnormalized vectors, then most likely all elements will increase on every iteration and no convergence will be visible.) If we can find some elements that do this (and we usually can), then it is a fairly safe bet that the difference between the two values for such an element is greater than the difference of either from the true value of the same element in the leading eigenvector.

The power method can also be used to calculate the leading eigen*value* κ_1 of the adjacency matrix. Once the algorithm has converged to the leading eigenvector, one more multiplication by the adjacency matrix will multiply that vector by exactly a factor of κ_1. Thus, we can take the ratio of the values of any element of the vector at two successive iterations of the algorithm after convergence and that ratio should equal κ_1. Or we could take the average of the ratios for several different elements to reduce numerical errors. (We should however avoid elements whose values are very small, since a small error in such an element could lead to a large fractional error in the ratio; our accuracy will be better if we take the average of some of the larger elements.)

11.1.1 COMPUTATIONAL COMPLEXITY

How long does the power method take to run? The answer comes in two parts. First, we need to know how long each multiplication by the adjacency matrix takes, and second we need to know how many multiplications are needed to get a required degree of accuracy in our answer.

If our network is stored in adjacency matrix form, then multiplying that matrix into a given vector is straightforward. Exactly n^2 multiplications are needed for one matrix multiplication—one for each element of the adjacency matrix. We can do better, however, if our network is in adjacency list form. Elements of the adjacency matrix that are zero contribute nothing to the matrix multiplication and so can be neglected. The adjacency list allows us to skip the zero terms automatically, since it stores only the non-zero ones anyway.

In an ordinary unweighted network each non-zero element of the adjacency matrix is equal to 1. Let $\{u_j\}$, $j = 1\ldots k_i$ be the set of neighbors of vertex i (where k_i is the degree of i). Then the ith element of $\mathbf{Ax}$, which we denote $[\mathbf{Ax}]_i$, is given by $[\mathbf{Ax}]_i = \sum_{j=1}^{k_i} x_{u_j}$. The evaluation of this sum involves only k_i operations, so one element of the matrix multiplication can be completed in time proportional to k_i and all elements can be completed in time proportional to $\sum_i k_i = 2m$, where m is the total number of edges in the network, or in other words in $O(m)$ time.

And how many such multiplications must we perform? Equation (7.4) tells us that after t iterations our vector is equal to

$$\mathbf{x}(t) = \kappa_1^t \sum_{i=1}^n c_i \left[\frac{\kappa_i}{\kappa_1} \right]^t \mathbf{v}_i, \tag{11.2}$$

where $\mathbf{v}_i$ is the normalized ith eigenvector, κ_i is the corresponding eigenvalue, and the c_i are constants whose values depend on the choice of initial vector. Rearranging slightly, we can write this as

$$\frac{\mathbf{x}(t)}{c_1 \kappa_1^t} = \mathbf{v}_1 + \frac{c_2}{c_1} \left(\frac{\kappa_2}{\kappa_1} \right)^t \mathbf{v}_2 + \ldots, \tag{11.3}$$

which gives us our estimate of the leading eigenvector $\mathbf{v}_1$ plus the dominant contribution to the error. Neglecting the smaller terms, the root-mean-square error on the eigenvector is then

$$\sqrt{\left| \frac{\mathbf{x}(t)}{c_1 \kappa_1^t} - \mathbf{v}_1 \right|^2} = \frac{c_2}{c_1} \left(\frac{\kappa_2}{\kappa_1} \right)^t, \tag{11.4}$$

and if we want this error to be at most ϵ then we require

$$t \geq \frac{\ln(1/\epsilon) + \ln(c_1/c_2)}{\ln(\kappa_1/\kappa_2)}. \tag{11.5}$$

Neither ϵ nor the constants c_1 and c_2 depend on the network size. All the variation in the run time comes from the eigenvalues κ_1 and κ_2. The eigenvalues range in value from a maximum of κ_1 to a minimum of $\kappa_n \geq -|\kappa_1|$ and hence have a mean spacing of at most $2\kappa_1/(n-1)$. Thus an order-of-magnitude estimate for the second eigenvalue is $\kappa_2 \simeq \kappa_1 - a\kappa_1/n$, where a is a constant of order unity, and hence

$$\ln \frac{\kappa_1}{\kappa_2} \simeq -\ln\left(1 - \frac{a}{n}\right) = \frac{a}{n} + O(n^{-2}). \tag{11.6}$$

Combining Eqs. (11.5) and (11.6), we find that the number of steps required for convergence of the power method is $t = O(n)$ to leading order.[3]

Overall therefore, the complete calculation of the eigenvector centralities of all n vertices of the network takes $O(n)$ multiplications which take $O(m)$ time each, or $O(mn)$ time overall, for a network stored in adjacency list format. If our network is sparse with $m \propto n$, a running time of $O(mn)$ is equivalent to $O(n^2)$. On the other hand, if the network is dense, with $m \propto n^2$, then $O(mn)$ is equivalent to $O(n^3)$.

Conversely, if our network is stored in adjacency matrix format the multiplications take $O(n^2)$ time, as noted above, so the complete calculation takes $O(n^3)$, regardless of whether the network is sparse or dense. Thus for the common case of a sparse matrix the adjacency list is the representation of choice for this calculation.

11.1.2 CALCULATING OTHER EIGENVALUES AND EIGENVECTORS

The power method of the previous section calculates the largest eigenvalue of a matrix and the corresponding eigenvector. This is probably the most common type of eigenvector calculation encountered in the study of networks, but there are cases where we wish to know other eigenvectors or eigenvalues as well. One example is the calculation of the so-called algebraic connectivity, which is the second smallest (or second most negative) eigenvalue of the graph Laplacian. As we saw in Section 6.13.3, the algebraic connectivity is non-zero if and only if a network is connected (i.e., has just a single component). The algebraic connectivity also appears in Section 11.5 as a measure of how easily a network can be bisected into two sets of vertices such that only a small number of edges run between the sets. Moreover, as we will see the elements of the corresponding eigenvector of the Laplacian tell us exactly how that bisection

---

[3]In fact, this estimate usually errs on the pessimistic side, since the spacing of the highest eigenvalues tends to be wider than the mean spacing, so that in practice the algorithm may be faster than the estimate would suggest.

should be performed. Thus it will be useful to us to have a method for calculating eigenvalues beyond the largest one and their accompanying eigenvectors.

There are a number of techniques that can be used to find non-leading eigenvalues and eigenvectors of matrices. For instance, we can calculate the eigenvector corresponding to the most negative eigenvalue by shifting all the eigenvalues by a constant amount so that the most negative one becomes the eigenvalue of largest magnitude. The eigenvalues of the graph Laplacian $\mathbf{L}$, for instance, are all non-negative. If we number them in ascending order as in Section 6.13.2, so that $\lambda_1 \leq \lambda_2 \ldots \leq \lambda_n$, with $\mathbf{v}_1, \mathbf{v}_2, \ldots, \mathbf{v}_n$ being the corresponding eigenvectors, then

$$(\lambda_n \mathbf{I} - \mathbf{L})\mathbf{v}_i = (\lambda_n - \lambda_i)\mathbf{v}_i, \tag{11.7}$$

and hence $\mathbf{v}_i$ is an eigenvector of $\lambda_n \mathbf{I} - \mathbf{L}$ with eigenvalue $\lambda_n - \lambda_i$. These eigenvalues are still all non-negative, but their order is reversed from those of the original Laplacian, so that the former smallest has become the new largest. Now we can calculate the eigenvector corresponding to the smallest eigenvalue of the Laplacian by finding the leading eigenvector of $\lambda_n \mathbf{I} - \mathbf{L}$ using the technique described in Section 11.1. We can also find the eigenvalue λ_1 by taking the measured value of $\lambda_n - \lambda_1$, subtracting λ_n, and reversing the sign. (Performing these calculations does require that we know the value of λ_n, so the complete calculation would be a two-stage process consisting of first finding the largest eigenvalue of $\mathbf{L}$, then using that to find the smallest.[4])

In this particular case, it would not in fact be very useful to calculate the smallest eigenvalue or its associated eigenvector since, as we saw in Section 6.13.2, the smallest eigenvalue of the Laplacian is always zero and the eigenvector is $(1, 1, 1, \ldots)$. However, if we can find the second-largest eigenvalue of a matrix we can use the same subtraction method also to find the second-smallest. And the second-smallest eigenvalue of the Laplacian is, as we have said, definitely of interest.

We can find the second-largest eigenvalue (and the corresponding eigenvector) using the following trick. Let $\mathbf{v}_1$ be the normalized eigenvector corresponding to the largest eigenvalue of a matrix $\mathbf{A}$, as found, for instance, by the power method of Section 11.1. Then we choose any starting vector $\mathbf{x}$ as before

---

[4]If we wish to be more sophisticated, we can note that it is sufficient to shift the eigenvalues by any amount greater than or equal to λ_n. Anderson and Morley [18] have shown that $\lambda_n \leq 2k_{max}$ where k_{max} is the largest degree in the network, which we can find in time $O(n)$, considerably faster than we can find λ_n itself. Thus a quicker way to find the smallest eigenvalue would be to find the largest eigenvalue of $2k_{max}\mathbf{I} - \mathbf{L}$.

and define

$$\mathbf{y} = \mathbf{x} - (\mathbf{v}_1^T \mathbf{x})\mathbf{v}_1. \tag{11.8}$$

This vector has the property that

$$\mathbf{v}_i^T \mathbf{y} = \mathbf{v}_i^T \mathbf{x} - (\mathbf{v}_1^T \mathbf{x})(\mathbf{v}_i^T \mathbf{v}_1) = \mathbf{v}_i^T \mathbf{x} - \mathbf{v}_1^T \mathbf{x} \, \delta_{i1}$$
$$= \begin{cases} 0 & \text{if } i = 1, \\ \mathbf{v}_i^T \mathbf{x} & \text{otherwise,} \end{cases} \tag{11.9}$$

where $\mathbf{v}_i$ is again the ith eigenvector of $\mathbf{A}$ and δ_{ij} is the Kronecker delta. In other words it is equal to $\mathbf{x}$ along the direction of every eigenvector of $\mathbf{A}$ except the leading eigenvector, in whose direction it has no component at all. This means that the expansion of $\mathbf{y}$ in terms of the eigenvectors of $\mathbf{A}$, which is given by $\mathbf{y} = \sum_{i=1}^{n} c_i \mathbf{v}_i$ with $c_i = \mathbf{v}_i^T \mathbf{y}$, has no term in $\mathbf{v}_1$, since $c_1 = \mathbf{v}_1^T \mathbf{y} = 0$. Thus

$$\mathbf{y} = \sum_{i=2}^{n} c_i \mathbf{v}_i, \tag{11.10}$$

with the sum starting at $i = 2$.

Now we use this vector $\mathbf{y}$ as the starting vector for repeated multiplication by $\mathbf{A}$, as before. After multiplying $\mathbf{y}$ by $\mathbf{A}$ a total of t times, we have

$$\mathbf{y}(t) = \mathbf{A}^t \mathbf{y}(0) = \kappa_2^t \sum_{i=2}^{n} c_i \left[\frac{\kappa_i}{\kappa_2} \right]^t \mathbf{v}_i. \tag{11.11}$$

The ratio κ_i / κ_2 is less than 1 for all $i > 2$ (assuming only a single eigenvalue of value κ_2) and hence in the limit of large t all terms in the sum disappear except the first so that $\mathbf{y}(t)$ tends to a multiple of $\mathbf{v}_2$ as $t \to \infty$. Normalizing this vector, we then have our result for $\mathbf{v}_2$.

This method has the same caveats as the original power method for the leading eigenvector, as well as one additional one: it is in practice possible for the vector $\mathbf{y}$, Eq. (11.8), to have a very small component in the direction of $\mathbf{v}_1$. This can happen as a result of numerical error in the subtraction, or because our value for $\mathbf{v}_1$ is not exactly correct. If $\mathbf{y}$ does have a component in the direction of $\mathbf{v}_1$, then although it may start out small it will get magnified relative to the others when we multiply repeatedly by $\mathbf{A}$ and eventually it may come to dominate $\mathbf{y}(t)$, Eq. (11.11), or at least to contribute a sufficiently large term as to make the calculation of $\mathbf{v}_2$ inaccurate. To prevent this happening, we periodically perform a subtraction similar to that of Eq. (11.8), removing any component in the direction of $\mathbf{v}_1$ from $\mathbf{y}(t)$, while leaving the components in all other directions untouched. (The subtraction process is sometimes referred to as *Gram–Schmidt orthogonalization*—a rather grand name for a simple procedure. The repeated application of the process to prevent the growth of unwanted terms is called *reorthogonalization*.)

We could in theory extend this method to find further eigenvectors and eigenvalues of our matrix, but in practice the approach does not work well beyond the first couple of eigenvectors because of cumulative numerical errors. Moreover it is also slow because for each additional eigenvector we calculate we must carry out the entire repeated multiplication process again. In practice, therefore, if we wish to calculate anything beyond the first eigenvector or two, other methods are used.

11.1.3 EFFICIENT ALGORITHMS FOR COMPUTING ALL EIGENVALUES AND EIGENVECTORS OF MATRICES

If we wish to calculate all or many of the eigenvalues or eigenvectors of a matrix $\mathbf{A}$ then specialized techniques are needed. The most widely used such techniques involve finding an orthogonal matrix $\mathbf{Q}$ such that the similarity transform $\mathbf{T} = \mathbf{Q}^T\mathbf{A}\mathbf{Q}$ gives either a tridiagonal matrix (if $\mathbf{A}$ is symmetric) or a Hessenberg matrix (if $\mathbf{A}$ is asymmetric). If we can find such a transformation and if $\mathbf{v}_i$ is an eigenvector of $\mathbf{A}$ with eigenvalue κ_i, then, bearing in mind that for an orthogonal matrix $\mathbf{Q}^{-1} = \mathbf{Q}^T$, we have

$$\kappa_i \mathbf{Q}^T \mathbf{v}_i = \mathbf{Q}^T \mathbf{A}\mathbf{v}_i = \mathbf{T}\mathbf{Q}^T\mathbf{v}_i. \tag{11.12}$$

In other words, the vector $\mathbf{w}_i = \mathbf{Q}^T\mathbf{v}_i$ is an eigenvector of $\mathbf{T}$ with eigenvalue κ_i. Thus if we can find the eigenvalues of $\mathbf{T}$ and the corresponding eigenvectors, we automatically have the eigenvalues of $\mathbf{A}$ as well, and the eigenvectors of $\mathbf{A}$ are simply $\mathbf{v}_i = \mathbf{Q}\mathbf{w}_i$. Luckily there exist efficient numerical methods for finding the eigenvalues and eigenvectors of tridiagonal and Hessenberg matrices, such as the *QL algorithm* [273]. The QL algorithm takes time O(n) to reach an answer for an $n \times n$ tridiagonal matrix and O(n^2) for a Hessenberg one.

The matrix $\mathbf{Q}$ can be found in various ways. For a general symmetric matrix the *Householder algorithm* [273] can find $\mathbf{Q}$ in time O(n^3). More often, however, we are concerned with sparse matrices, in which case there are faster methods. For a symmetric matrix, the *Lanczos algorithm* [217] can find $\mathbf{Q}$ in time O(mn), where m is the number of network edges in an adjacency matrix, or more generally the number of non-zero elements in the matrix. For sparse matrices with $m \propto n$ this gives a running time of O(n^2), considerably better than the Householder method. A similar method, the *Arnoldi algorithm* [217], can find $\mathbf{Q}$ for an asymmetric matrix.

Thus, combining the Lanczos and QL algorithms, we expect to be able to find all eigenvalues and eigenvectors of a sparse symmetric matrix in time

$O(mn)$, which is as good as the worst-case run time of our direct multiplication method for finding just the leading eigenvector. (To be fair, the direct multiplication is much simpler, so its overall run time will typically be better than that of the combined Lanczos/QL algorithm, although the scaling with system size is the same.)

While there is certainly much to be gained by learning about the details of these algorithms, one rarely implements them in practice. Their implementation is tricky (particularly in the asymmetric case), and has besides already been done in a careful and professional fashion by many software developers. In practice, therefore, if one wishes to solve eigensystem problems for large networks, one typically turns to commercial or freely available implementations in professionally written software packages. Examples of such packages include Matlab, LAPACK, and Mathematica. We will not go into more detail here about the operation of these algorithms.

11.2   DIVIDING NETWORKS INTO CLUSTERS

We now turn to the topics that will occupy us for much of the rest of the chapter, *graph partitioning* and *community detection*.[5] Both of these terms refer to the division of the vertices of a network into groups, clusters, or communities according to the pattern of edges in the network. Most commonly one divides the vertices so that the groups formed are tightly knit with many edges inside groups and only a few edges between groups.

Consider Fig. 11.1, for instance, which shows patterns of collaborations between scientists in a university department. Each vertex in this network represents a scientist and links between vertices indicate pairs of scientists who have coauthored one or more papers together. As we can see from the figure, this network contains a number of densely connected clusters of vertices, corresponding to groups of scientists who have worked closely together. Readers familiar with the organization of university departments will not be surprised to learn that in general these clusters correspond, at least approximately, to formal research groups within the department.

But suppose one did not know how university departments operate and wished to study them. By constructing a network like that in Fig. 11.1 and then observing its clustered structure, one would be able to deduce the existence of groups within the larger department and by further investigation could prob-

---

[5]Community detection is sometimes also called "clustering," although we largely avoid this term to prevent confusion with the other, and quite different, use of the word clustering introduced in Section 7.9.

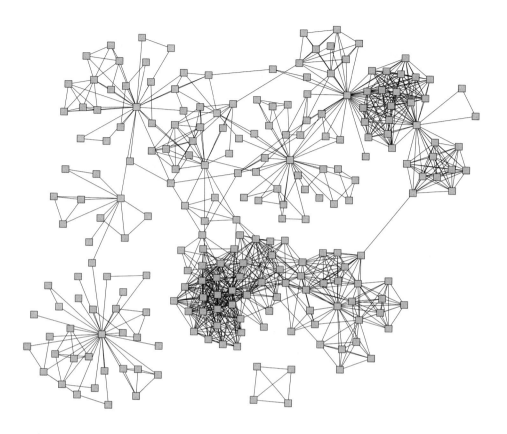

Figure 11.1: Network of coauthorships in a university department. The vertices in this network represent scientists in a university department, and edges links pairs of scientists who have coauthored scientific papers. The network has clear clusters or "community structure," presumably reflecting divisions of interests and research groups within the department.

ably quickly work out how the department was organized. Thus the ability to discover groups or clusters in a network can be a useful tool for revealing structure and organization within networks at a scale larger than that of a single vertex. In this particular case the network is small enough and sparse enough that the groups are easily visible by eye. Many of the networks that have engaged our interest in this book, however, are much larger or denser networks for which visual inspection is not a useful tool. Finding clusters in such networks is a task for computers and the algorithms that run on them.

11.2.1 PARTITIONING AND COMMUNITY DETECTION

There are a number of reasons why one might want to divide a network into groups or clusters, but they separate into two general classes that lead in turn to two corresponding types of computer algorithm. We will refer to these two types as *graph partitioning* and *community detection* algorithms. They are distinguished from one another by whether the number and size of the groups is fixed by the experimenter or whether it is unspecified.

Graph partitioning is a classic problem in computer science, studied since the 1960s. It is the problem of dividing the vertices of a network into a given number of non-overlapping groups of given sizes such that the number of edges between groups is minimized. The important point here is that the number and sizes of the groups are fixed. Sometimes the sizes are only fixed roughly—within a certain range, for instance—but they are fixed nonetheless. For instance, a simple and prototypical example of a graph partitioning problem is the problem of dividing a network into two groups of equal size, such that the number of edges between them is minimized.

Graph partitioning problems arise in a variety of circumstances, particularly in computer science, but also in pure and applied mathematics, physics, and of course in the study of networks themselves. A typical example is the numerical solution of network processes on a parallel computer.

In the last part of this book (Chapters 16 to 19) we will study processes that take place on networks, such as diffusion processes or the spread of diseases. These processes can be modeled mathematically by placing variables on the vertices of a network and evolving them according to equations that typically depend on the variables' current values and the values on neighboring vertices. The solution of such equations is often a laborious computational task, but it can be sped up by using a parallel computer, a computer with more than one processor or CPU. Many modern personal computers have two or more processors and large research organizations sometimes use parallel computers with very many processors. Solutions of network equations can be spread across several processors by assigning to each processor the task of solving the equations on a subset of the vertices. For instance, on a two-processor desktop computer we might give a half of the vertices to each processor.

The catch is that, unless the network consists of totally unconnected components, some vertices on one processor are always going to have neighbors that are on the other processor and hence the solution of their equations involves variables whose value is known only to the other processor. To complete the solution, therefore, those values have to be transmitted from the one processor to the other at regular intervals throughout the calculation and this is typically

Partition of a network into two groups of equal sizes.

a slow process (or at least it's slow compared to the dazzling speed of most other computer operations). The time spent sending messages between processors can, in fact, be the primary factor limiting the speed of calculations on parallel computers, so it is important to minimize interprocessor communication as much as possible. One way that we do this is by minimizing the number of pairs of neighboring vertices assigned to different processors.

Thus we want to divide up the vertices of the network into different groups, one for each processor, such that the number of edges between groups is minimized. Most often we want to assign an equal or roughly equal number of vertices to each processor so as to balance the workload among them. This is precisely a graph partitioning problem of the type described above.

The other type of cluster finding problem in networks is the problem we call community detection. Community detection problems differ from graph partitioning in that the number and size of the groups into which the network is divided are not specified by the experimenter. Instead they are determined by the network itself: the goal of community detection is to find the natural fault lines along which a network separates. The sizes of the groups are not merely unspecified but might in principle vary widely from one group to another. A given network might divide into a few large groups, many small ones, or a mixture of all different sizes.

The most common use for community detection is as a tool for the analysis and understanding of network data. We saw in Fig. 11.1 an example of a network for which a knowledge of the group structure might help us understand the organization of the underlying system. Figure 7.10 on page 221 shows another example of clusters of vertices, in a network of friendships between US high-school students. In this case the network splits into two clear groups, which, as described in Section 7.13, are primarily dictated by students' ethnicity, and this structure and others like it can give us clues about the nature of the social interactions within the community represented.

Community detection has uses in other types of networks as well. Clusters of nodes in a web graph for instance might indicate groups of related web pages. Clusters of nodes in a metabolic network might indicate functional units within the network.

Community detection is a less well-posed problem than graph partitioning. Loosely stated, it is the problem of finding the natural divisions of a network into groups of vertices such that there are many edges within groups and few edges between groups. What exactly we mean by "many" or "few," however, is debatable, and a wide variety of different definitions have been proposed, leading to a correspondingly wide variety of different algorithms for community detection. In this chapter we will focus mainly on the most widely used formulation of the problem, the formulation in terms of modularity optimization, but we will mention briefly a number of other approaches at the end of the chapter.

In summary, the fundamental difference between graph partitioning and community detection is that the number and size of the groups into which a network is divided is specified in graph partitioning but unspecified in community detection. However, there is also a difference between the goals of the two types of calculations. Graph partitioning is typically performed as a way of dividing up a network into smaller more manageable pieces, for example to perform numerical calculations. Community detection is more often used as a tool for understanding the structure of a network, for shedding light on large-scale patterns of connection that may not be easily visible in the raw network topology.

Notice also that in graph partitioning calculations the goal is usually to find the best division of a network, subject to certain conditions, regardless of whether any good division exists. If the performance of a calculation on a parallel computer, for example, requires us to divide a network into pieces, then we had better divide it up. If there are no good divisions, then we must make do with the least bad one. With community detection, on the other hand, where the goal is normally to understand the structure of the network, there is no need to divide the network if no good division exists. Indeed if a network has no good divisions then that in itself may be a useful piece of information, and it would be perfectly reasonable for a community detection algorithm only to divide up networks when good divisions exist and to leave them undivided the rest of the time.

11.3   GRAPH PARTITIONING

In the next few sections we consider the graph partitioning problem and look at two well-known methods for graph partitioning. The first, the Kernighan–Lin algorithm, is not based on matrix methods (and therefore doesn't strictly belong in this chapter) but it provides a simple introduction to the partitioning problem and is worth spending a little time on. In Section 11.5 we look at a more sophisticated partitioning method based on the spectral properties of the graph Laplacian. This spectral partitioning method both is important in its own right and will also provide a basis for our discussion of community detection later in the chapter.

First, however, we address an important preliminary question: why does one need fancy partitioning algorithms at all? Partitioning is an easy problem to state, so is it not just as easy to solve?

11.3.1 WHY PARTITIONING IS HARD

The simplest graph partitioning problem is the division of a network into just two parts. Division into two parts is sometimes called *graph bisection*. Most of the algorithms we consider in this chapter are in fact algorithms for bisecting networks rather than for dividing them into arbitrary numbers of parts. This may at first appear to be a drawback, but in practice it is not, since if we can divide a network into two parts, then we can divide it into more than two by further dividing one or both of those parts. This repeated bisection is the commonest approach to the partitioning of networks into arbitrary numbers of parts.

Formally the graph bisection problem is the problem of dividing the vertices of a network into two non-overlapping groups of given sizes such that the number of edges running between vertices in different groups is minimized. The number of edges between groups is called the *cut size*.[6]

Simple though it is to describe, this problem is not easy to solve. One might imagine that one could bisect a network simply by looking through all possible divisions of the network into two parts of the required sizes and choosing the one with the smallest cut size. For all but the smallest of networks, however, this so-called *exhaustive search* turns out to be prohibitively costly in terms of computer time.

The number of ways of dividing a network of n vertices into two groups of n_1 and n_2 vertices respectively is $n!/(n_1! n_2!)$. Approximating the factorials using Stirling's formula $n! \simeq \sqrt{2\pi n}(n/e)^n$ and making use of the fact that $n_1 + n_2 = n$, we get

$$\frac{n!}{n_1! n_2!} \simeq \frac{\sqrt{2\pi n}(n/e)^n}{\sqrt{2\pi n_1}(n_1/e)^{n_1}\sqrt{2\pi n_2}(n_2/e)^{n_2}} = \frac{n^{n+1/2}}{n_1^{n_1+1/2}n_2^{n_2+1/2}}. \qquad (11.13)$$

Thus, for instance, if we want to divide a network into two parts of equal size $\frac{1}{2}n$ the number of different ways to do it is roughly

$$\frac{n^{n+1/2}}{(n/2)^{n+1}} = \frac{2^{n+1}}{\sqrt{n}}. \qquad (11.14)$$

So the amount of time required to look through all of these divisions will go up roughly exponentially with the size of the network. Unfortunately, the exponential is a very rapidly growing function of its argument, which means the

---

[6]The problem is somewhat similar to the minimum cut problem of Section 6.12, but we are now searching for the minimum cut over all possible bisections of a network, rather than just between a given pair of vertices.

partitioning task quickly leaves the realm of the possible at quite moderate values of n. Values up to about $n = 30$ are feasible with current computers, but go much beyond that and the calculation becomes intractable.

One might wonder whether it is possible to find a way around this problem. After all, brute-force enumeration of all possible divisions of a network is not a very imaginative way to solve the partitioning problem. Perhaps one could find a way to limit one's search to only those divisions of the network that have a chance of being the best one? Unfortunately, there are some fundamental results in computer science that tell us that no such algorithm will ever be able to find the best division of the network in all cases. Either an algorithm can be clever and run quickly, but will fail to find the optimal answer in some (and perhaps most) cases, or it always finds the optimal answer but takes an impractical length of time to do it. These are the only options.[7]

This is not to say, however, that clever algorithms for partitioning networks do not exist or that they don't give useful answers. Even algorithms that fail to find the very best division of a network may still find a pretty good one, and for many practical purposes pretty good is good enough. The goal of essentially all practical partitioning algorithms is just to find a "pretty good" division in this sense. Algorithms that find approximate, but acceptable, solutions to problems in this way are called *heuristic algorithms* or just *heuristics*. All the algorithms for graph partitioning discussed in this chapter are heuristic algorithms.

11.4 THE KERNIGHAN–LIN ALGORITHM

The *Kernighan–Lin algorithm*, proposed by Brian Kernighan[8] and Shen Lin in 1970 [171], is one of the simplest and best known heuristic algorithms for the graph bisection problem. The algorithm is illustrated in Fig. 11.2.

We start by dividing the vertices of our network into two groups of the required sizes in any way we like. For instance, we could divide the vertices randomly. Then, for each pair (i, j) of vertices such that i lies in one of the groups

---

[7]Technically, this statement has not actually been proved. Its truth hinges on the assumption that two fundamental classes of computational problem, called P and NP, are not the same. Although this assumption is universally believed to be true—the world would pretty much fall apart if it weren't—no one has yet proved it, nor even has any idea about where to start. Readers interested in the fascinating branch of theoretical computer science that deals with problems of this kind are encouraged to look, for example, at the book by Moore and Mertens [227].

[8]Some readers may be familiar with Kernighan's name. He was one of the authors of the original book describing the C programming language [172]. "Kernighan" is pronounced "Kernihan"—the "g" is silent.

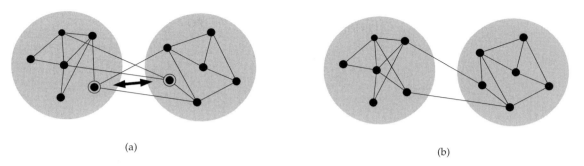

Figure 11.2: The Kernighan–Lin algorithm. (a) The Kernighan–Lin algorithm starts with any division of the vertices of a network into two groups (shaded) and then searches for pairs of vertices, such as the pair highlighted here, whose interchange would reduce the cut size between the groups. (b) The same network after interchange of the two vertices.

and j in the other, we calculate how much the cut size between the groups would change if we were to interchange i and j, so that each was placed in the other group. Among all pairs (i, j) we find the pair that reduces the cut size by the largest amount or, if no pair reduces it, we find the pair that increases it by the smallest amount. Then we swap that pair of vertices. Clearly this process preserves the sizes of the two groups of vertices, since one vertex leaves each group and another joins. Thus the algorithm respects the requirement that the groups take specified sizes.

The process is then repeated, but with the important restriction that each vertex in the network can only be moved once. Once a vertex has been swapped with another it is not swapped again (at least not in the current round of the algorithm—see below). Thus, on the second step of the algorithm we consider all pairs of vertices excluding the two vertices swapped on the first step.

And so the algorithm proceeds, swapping on each step that pair that most decreases, or least increases, the number of edges between our two groups, until eventually there are no pairs left to be swapped, at which point we stop. (If the sizes of the groups are unequal then there will be vertices in the larger group that never get swapped, equal in number to the difference between the sizes of the groups.)

When all swaps have been completed, we go back through every state that the network passed through during the swapping procedure and choose among them the state in which the cut size takes its smallest value.[9]

---

[9]One might imagine that an equivalent procedure would be to go on swapping vertex pairs

Finally, this entire process is performed repeatedly, starting each time with the best division of the network found on the last time around and continuing until no improvement in the cut size occurs. The division with the best cut size on the last round is the final division returned by the algorithm.

Once we can divide a network into two pieces of given size then, as we have said, we can divide into more than two simply by repeating the process. For instance, if we want to divide a network into three pieces of equal size, we would first divide into two pieces, one twice the size of the other, and then further divide the larger one into two equally sized halves. (Note, however, that even if the algorithm were able to find the optimal division of the network in each of these two steps, there would be no guarantee that we would end up with the optimal division of the network into three equal parts. Nonetheless, we do typically find a reasonably good division, which, as we have said, is often good enough. This point is discussed further in Section 11.9.)

Note that if we choose the initial assignment of vertices to groups randomly, then the Kernighan–Lin algorithm may not give the same answer if it is run twice on the same network. Two different random starting states could (though needn't necessarily) result in different divisions of the network. For this reason, people sometimes run the algorithm more than once to see if the results vary. If they do vary then among the divisions of the network returned on the different runs it makes sense to take the one with the smallest cut size.

As an example of the use of the Kernighan–Lin algorithm, consider Fig. 11.3, which shows an application of the algorithm to a mesh, a two-dimensional network of the type often used in parallel finite-element computations. Suppose we want to divide this network into two parts of equal size. Looking at the complete network in Fig. 11.3a there is no obvious division—there is no easy cut or bottleneck where the network separates naturally—but we must do the best we can. Figure 11.3b shows the best division found by the Kernighan–Lin algorithm, which involves cutting 40 edges in the network. Though it might not be the best possible division of the network, this is certainly good enough for many practical purposes.

The primary disadvantage of the Kernighan–Lin algorithm is that it is quite slow. The number of swaps performed during one round of the algorithm is

---

until no swap can be found that decreases the cut size. This, however, turns out to be wrong. It is perfectly possible for the cut size to decrease for a few steps of the algorithm, then increase, then decrease again. If we halt the algorithm the first time we see the cut size increasing, we run the risk of missing a later state with smaller cut size. Thus the correct algorithm is the one described here, with two separate processes, one of vertex swapping, and one of checking the states so generated to see which is optimal.

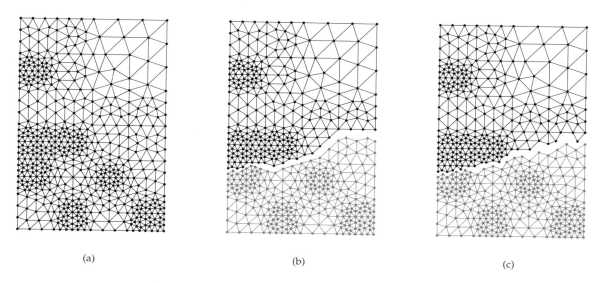

Figure 11.3: Graph partitioning applied to a small mesh network. (a) A mesh network of 547 vertices of the kind commonly used in finite element analysis. (b) The edges removed indicate the best division of the network into parts of 273 and 274 vertices found by the Kernighan–Lin algorithm. (c) The best division found by spectral partitioning. The network is from Bern *et al.* [35].

equal to the smaller of the sizes of the two groups, which lies between zero and $\frac{1}{2}n$ in a network of n vertices. Thus there are $O(n)$ swaps in the worst case. For each swap we have to examine all pairs of vertices in different groups, of which there are, in the worst case, $\frac{1}{2}n \times \frac{1}{2}n = \frac{1}{4}n^2 = O(n^2)$. And for each of these we need to determine the change in the cut size if the pair is swapped.

When a vertex i moves from one group to the other any edges connecting it to vertices in its current group become edges between groups after the swap. Let us suppose that there are k_i^{same} such edges. Similarly, any edges that i has to vertices in the other group, of which there are say k_i^{other}, become within-group edges after the swap, but with one exception. If i is being swapped with vertex j and there is an edge between i and j, then that edge lies between groups before the swap and still lies between groups after the swap. Thus the change in the cut size due to the movement of i is $k_i^{\text{other}} - k_i^{\text{same}} - A_{ij}$. A similar expression applies for vertex j also and the total change in cut size as a result of the swap is

$$\Delta = k_i^{\text{other}} - k_i^{\text{same}} + k_j^{\text{other}} - k_j^{\text{same}} - 2A_{ij}. \qquad (11.15)$$

For a network stored in adjacency list form, the evaluation of this expression

363

involves running through all the neighbors of i and j in turn, and hence takes time of order the average degree in the network, or $O(m/n)$, where m is, as usual, the total number of edges in the network.

Thus the total time for one round of the algorithm is $O(n \times n^2 \times m/n) = O(mn^2)$, which is $O(n^3)$ on a sparse network in which $m \propto n$ or $O(n^4)$ on a dense network. This in itself would already be quite bad, but we are not yet done. This time must be multiplied by the number of rounds the algorithm performs before the cut size stops decreasing. It is not well understood how the number of rounds required varies with network size. In typical applications the number is small, maybe five or ten for networks of up to a few thousand vertices, and larger networks are currently not possible because of the demands of the algorithm, so in practice the number of rounds is always small. Still, it seems quite unlikely that the number of rounds would actually increase as network size grows, and even if it remains constant the time complexity of the algorithm will still be $O(mn^2)$, which is relatively slow.

We can improve the running time of the algorithm a little by a couple of tricks. If we initially calculate and store the number of neighbors, k_i^{same} and k_i^{other}, that each vertex has within and between groups and update it every time a vertex is moved, then we save ourselves the time taken to recalculate these quantities on each step of the algorithm. And if we store our network in adjacency matrix form then we can tell whether two vertices are connected (and hence evaluate A_{ij}) in time $O(1)$. Together these two changes allow us to calculate Δ above in time $O(1)$ and improve the overall running time to $O(n^3)$. For a sparse graph this is the same as $O(mn^2)$, but for a dense one it gives us an extra factor of n.

Overall, however, the algorithm is quite slow. Even with $O(n^3)$ performance the algorithm is suitable only for networks up to a few hundreds or thousands of vertices, but not more.

11.5   SPECTRAL PARTITIONING

So are there faster methods for partitioning networks? There are indeed, although they are typically more complex than the simple Kernighan–Lin algorithm, and may be correspondingly more laborious to implement. In this section we discuss one of the most widely used methods, the *spectral partitioning* method of Fiedler [118, 271], which makes use of the matrix properties of the graph Laplacian. We describe the spectral partitioning method as applied to the graph bisection problem, the problem of dividing a graph into two parts of specified sizes. As discussed in the previous section, division into more than two groups is typically achieved by repeated bisection, dividing and subdividing the network to give groups of the desired number and size.

Consider a network of n vertices and m edges and a division of that network into two groups, which we will call group 1 and group 2. We can write the cut size for the division, i.e., the number of edges running between the two groups, as

$$R = \tfrac{1}{2} \sum_{\substack{i, j \text{ in} \\ \text{different} \\ \text{groups}}} A_{ij}, \tag{11.16}$$

where the factor of $\tfrac{1}{2}$ compensates for our counting each edge twice in the sum.

Let us define a set of quantities s_i, one for each vertex i, which represent the division of the network thus:

$$s_i = \begin{cases} +1 & \text{if vertex } i \text{ belongs to group 1,} \\ -1 & \text{if vertex } i \text{ belongs to group 2.} \end{cases} \tag{11.17}$$

Then

$$\tfrac{1}{2}(1 - s_i s_j) = \begin{cases} 1 & \text{if } i \text{ and } j \text{ are in different groups,} \\ 0 & \text{if } i \text{ and } j \text{ are in the same group,} \end{cases} \tag{11.18}$$

which allows us to rewrite Eq. (11.16) as

$$R = \tfrac{1}{4} \sum_{ij} A_{ij}(1 - s_i s_j), \tag{11.19}$$

with the sum now over all values of i and j. The first term in the sum is

$$\sum_{ij} A_{ij} = \sum_i k_i = \sum_i k_i s_i^2 = \sum_{ij} k_i \delta_{ij} s_i s_j, \tag{11.20}$$

where k_i is the degree of vertex i as usual, δ_{ij} is the Kronecker delta, and we have made use of the fact that $\sum_j A_{ij} = k_i$ (see Eq. (6.19)) and $s_i^2 = 1$ (since $s_i = \pm 1$). Substituting back into Eq. (11.19) we then find that

$$R = \tfrac{1}{4} \sum_{ij} (k_i \delta_{ij} - A_{ij}) s_i s_j = \tfrac{1}{4} \sum_{ij} L_{ij} s_i s_j, \tag{11.21}$$

where $L_{ij} = k_i \delta_{ij} - A_{ij}$ is the ijth element of the graph Laplacian matrix—see Eq. (6.44).

Equation (11.21) can be written in matrix form as

$$R = \tfrac{1}{4} \mathbf{s}^T \mathbf{L} \mathbf{s}, \tag{11.22}$$

where $\mathbf{s}$ is the vector with elements s_i. This expression gives us a matrix formulation of the graph partitioning problem. The matrix $\mathbf{L}$ specifies the structure of our network, the vector **s** defines a division of that network into groups, and our goal is to find the vector **s** that minimizes the cut size (11.22) for given **L**.

You will probably not be surprised to learn that, in general, this minimization problem is not an easy one. If it were easy then we would have a corresponding easy way to solve the partitioning problem and, as discussed in Section 11.3.1, there are good reasons to believe that partitioning has no easy solutions.

What makes our matrix version of the problem hard in practice is that the s_i cannot take just any values. They are restricted to the special values ± 1. If they were allowed to take any real values the problem would be much easier; we could just differentiate to the find the optimum.

This suggests a possible approximate approach to the minimization problem. Suppose we indeed allow the s_i to take any values (subject to a couple of basic constraints discussed below) and then find the values that minimize R. These values will only be approximately the correct ones, since they probably won't be ± 1, but they may nonetheless be good enough to give us a handle on the optimal partitioning. This idea leads us to the so-called *relaxation method*, which is one of the standard methods for the approximate solution of vector optimization problems such as this one. In the present context it works as follows.

The allowed values of the s_i are actually subject to two constraints. First, as we have said, each individual one is allowed to take only the values ± 1. If we regard **s** as a vector in a Euclidean space then this constraint means that the vector always points to one of the 2^n corners of an n-dimensional hypercube centered on the origin, and always has the same length, which is $\sqrt{n}$. Let us relax the constraint on the vector's direction, so that it can point in any direction in its n-dimensional space. We will however still keep its length the same. (It would not make sense to allow the length to vary. If we did that then the minimization of R would have the obvious trivial solution **s** = 0, which would tell us nothing.) So **s** will be allowed to take any value, but subject to the constraint that $|\mathbf{s}| = \sqrt{n}$, or equivalently

$$\sum_i s_i^2 = n. \tag{11.23}$$

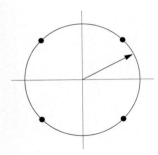

The relaxation of the constraint allows **s** to point to any position on a hypersphere circumscribing the original hypercube, rather than just the corners of the hypercube.

Another way of putting this is that **s** can now point to any location on the surface of a hypersphere of radius $\sqrt{n}$ in our n-dimensional Euclidean space. The hypersphere includes the original allowed values at the corners of the hypercube, but also includes other points in between.

The second constraint on the s_i is that the numbers of them that are equal to $+1$ and -1 respectively must equal the desired sizes of the two groups. If those two sizes are n_1 and n_2, this second constraint can be written as

$$\sum_i s_i = n_1 - n_2. \tag{11.24}$$

or in vector notation

$$\mathbf{1}^T\mathbf{s} = n_1 - n_2, \tag{11.25}$$

where $\mathbf{1}$ is the vector $(1, 1, 1, \ldots)$ whose elements are all 1. We keep this second constraint unchanged in our relaxed calculations, so that our partitioning problem, in its relaxed form, is a problem of minimizing the cut size, Eq. (11.22), subject to the two constraints (11.23) and (11.24).

This problem is now just a standard piece of algebra. We differentiate with respect to the elements s_i, enforcing the constraints using two Lagrange multipliers, which we denote λ and 2μ (the extra 2 being merely for notational convenience):

$$\frac{\partial}{\partial s_i}\left[\sum_{jk} L_{jk}s_j s_k + \lambda\left(n - \sum_j s_j^2\right) + 2\mu\left((n_1 - n_2) - \sum_j s_j\right)\right] = 0. \tag{11.26}$$

Performing the derivatives, we then find that

$$\sum_j L_{ij}s_j = \lambda s_i + \mu, \tag{11.27}$$

or, in matrix notation

$$\mathbf{Ls} = \lambda\mathbf{s} + \mu\mathbf{1}. \tag{11.28}$$

We can calculate the value of μ by recalling that $\mathbf{1}$ is an eigenvector of the Laplacian with eigenvalue zero so that $\mathbf{L} \cdot \mathbf{1} = 0$ (see Section 6.13.2). Multiplying (11.28) on the left by $\mathbf{1}^T$ and making use of Eq. (11.25), we then find that $\lambda(n_1 - n_2) + \mu n = 0$, or

$$\mu = -\frac{n_1 - n_2}{n}\lambda. \tag{11.29}$$

If we define the new vector

$$\mathbf{x} = \mathbf{s} + \frac{\mu}{\lambda}\mathbf{1} = \mathbf{s} - \frac{n_1 - n_2}{n}\mathbf{1}, \tag{11.30}$$

then Eq. (11.28) tells us that

$$\mathbf{Lx} = \mathbf{L}\left(\mathbf{s} + \frac{\mu}{\lambda}\mathbf{1}\right) = \mathbf{Ls} = \lambda\mathbf{s} + \mu\mathbf{1} = \lambda\mathbf{x}, \tag{11.31}$$

where we have used $\mathbf{L} \cdot \mathbf{1} = 0$ again.

In other words, $\mathbf{x}$ is an eigenvector of the Laplacian with eigenvalue λ. We are still free to choose which eigenvector it is—any eigenvector will satisfy

Eq. (11.31)—and clearly we should choose the one that gives the smallest value of the cut size R. Notice, however, that

$$\mathbf{1}^T \mathbf{x} = \mathbf{1}^T \mathbf{s} - \frac{\mu}{\lambda} \mathbf{1}^T \mathbf{1} = (n_1 - n_2) - \frac{n_1 - n_2}{n} n = 0, \qquad (11.32)$$

where we have used Eq. (11.25). Thus $\mathbf{x}$ is orthogonal to $\mathbf{1}$, which means that, while it should be an eigenvector of $\mathbf{L}$, it cannot be the eigenvector $(1, 1, 1, \ldots)$ that has eigenvalue zero.

So which eigenvector should we choose? To answer this question we note that

$$R = \tfrac{1}{4} \mathbf{s}^T \mathbf{L} \mathbf{s} = \tfrac{1}{4} \mathbf{x}^T \mathbf{L} \mathbf{x} = \tfrac{1}{4} \lambda \mathbf{x}^T \mathbf{x}. \qquad (11.33)$$

But from Eq. (11.30) we have

$$\begin{aligned}
\mathbf{x}^T \mathbf{x} &= \mathbf{s}^T \mathbf{s} + \frac{\mu}{\lambda} \left(\mathbf{s}^T \mathbf{1} + \mathbf{1}^T \mathbf{s} \right) + \frac{\mu^2}{\lambda^2} \mathbf{1}^T \mathbf{1} \\
&= n - 2 \frac{n_1 - n_2}{n} (n_1 - n_2) + \frac{(n_1 - n_2)^2}{n} n \\
&= 4 \frac{n_1 n_2}{n},
\end{aligned} \qquad (11.34)$$

and hence

$$R = \frac{n_1 n_2}{n} \lambda. \qquad (11.35)$$

Thus the cut size is proportional to the eigenvalue λ. Given that our goal is to minimize R, this means we should choose $\mathbf{x}$ to be the eigenvector corresponding to the smallest allowed eigenvalue of the Laplacian. All the eigenvalues of the Laplacian are non-negative (see Section 6.13.2). The smallest one is the zero eigenvalue that corresponds to the eigenvector $(1, 1, 1, \ldots)$ but we have already ruled this one out—$\mathbf{x}$ has to be orthogonal to this lowest eigenvector. Thus the best thing we can do is choose $\mathbf{x}$ proportional to the eigenvector $\mathbf{v}_2$ corresponding to the second lowest eigenvalue λ_2, with its normalization fixed by Eq. (11.34).

Finally, we recover the corresponding value of $\mathbf{s}$ from Eq. (11.30) thus:

$$\mathbf{s} = \mathbf{x} + \frac{n_1 - n_2}{n} \mathbf{1}, \qquad (11.36)$$

or equivalently

$$s_i = x_i + \frac{n_1 - n_2}{n}. \qquad (11.37)$$

This gives us the optimal relaxed value of $\mathbf{s}$.

As we have said, however, the real vector $\mathbf{s}$ is subject to the additional constraints that its elements take the values ± 1 and moreover that exactly n_1 of them are $+1$ and the other n_2 are -1. Typically these constraints will prevent **s** from taking exactly the value given by Eq. (11.37). Let us, however, do the best we can and choose **s** to be as close as possible to our ideal value subject to its constraints, which we do by making the product

$$\mathbf{s}^T\left(\mathbf{x} + \frac{n_1 - n_2}{n}\mathbf{1}\right) = \sum_i s_i\left(x_i + \frac{n_1 - n_2}{n}\right) \tag{11.38}$$

as large as possible. The maximum of this expression is achieved by assigning $s_i = +1$ for the vertices with the largest (i.e., most positive) values of $x_i + (n_1 - n_2)/n$ and $s_i = -1$ for the remainder.

Note however that the most positive values of $x_i + (n_1 - n_2)/n$ are also the most positive values of x_i, which are in turn also the most positive elements of the eigenvector $\mathbf{v}_2$ (to which, as we have said, **x** is proportional). So after this moderately lengthy derivation we actually arrive at a very simple final prescription for dividing our network. We calculate the eigenvector $\mathbf{v}_2$, which has n elements, one for each vertex in the network, and place the n_1 vertices with the most positive elements in group 1 and the rest in group 2.

There is one further small subtlety. It is arbitrary which group we call group 1 and which we call group 2, and hence which one we assign to the more positive elements of the eigenvector and which to the more negative. Thus, if the sizes of the two groups are different there are two different ways of making the split—either the larger or the smaller group could correspond to the more positive values. (In the geometrical language of our vectors, this is equivalent to saying our eigenvector calculation might find the vector **x** that we actually want, or minus that vector—both are good eigenvectors of the Laplacian.) To get around this problem, we simply compute the cut size for both splits of the network and choose the one with the smaller value.

Thus our final algorithm is as follows:

1. Calculate the eigenvector $\mathbf{v}_2$ corresponding to the second smallest eigenvalue λ_2 of the graph Laplacian.
2. Sort the elements of the eigenvector in order from largest to smallest.
3. Put the vertices corresponding to the n_1 largest elements in group 1, the rest in group 2, and calculate the cut size.
4. Then put the vertices corresponding to the n_1 smallest elements in group 1, the rest in group 2, and recalculate the cut size.
5. Between these two divisions of the network, choose the one that gives the smaller cut size.

In Fig. 11.3c we show the result of the application of this method to the same mesh network that we studied in conjunction with the Kernighan–Lin algorithm. In this case the spectral method finds a division of the network very similar to that given by the Kernighan–Lin algorithm, although the cut size is slightly worse—the spectral method cuts 46 edges in this case, where the Kernighan–Lin algorithm cut only 40. This is typical of the spectral method. It tends to find divisions of a network that have the right general shape, but are not perhaps quite as good as those returned by other methods.

An advantage of the spectral approach, however, is its speed. The time-consuming part of the algorithm is the calculation of the eigenvector $\mathbf{v}_2$, which takes time $O(mn)$ using either the orthogonalization method or the Lanczos method (see Section 11.1.2), or $O(n^2)$ on a sparse network having $m \propto n$. This is one factor of n better than the $O(n^3)$ of the Kernighan–Lin algorithm, which makes the algorithm feasible for much larger networks. Spectral partitioning can be extended to networks of hundreds of thousands of vertices, where the Kernighan–Lin algorithm is restricted to networks of a few thousand vertices at most.

The second eigenvalue of the Laplacian has come up previously in this book in Section 6.13.3, where we saw that it is non-zero if and only if a network is connected. The second eigenvalue is for this reason sometimes called the *algebraic connectivity* of a network. In this section we have seen it again in another context, that of partitioning. What happens if a network is not connected and the second eigenvalue is zero? In that case, the two lowest eigenvalues are the same, and the corresponding eigenvectors are indeterminate—any mixture of two eigenvectors with the same eigenvalue is also an eigenvector. This is not however a serious problem. If the network is not connected, having more than one component, then usually we are interested either in partitioning one particular component, such as the largest component, or in partitioning all components individually, and so we just treat the components separately as connected networks according to the algorithm above.

The algebraic connectivity itself appears in our expression for the cut size, Eq. (11.35), and indeed is a direct measure of the cut size, being directly proportional to it, at least within the "relaxed" approximation used to derive the equation. Thus the algebraic connectivity is a measure of *how easily* a network can be divided. It is small for networks that have good cuts and large for those that do not. This in a sense is a generalization of our earlier result that the algebraic connectivity is non-zero for connected networks and zero for unconnected ones—we now see that *how* non-zero it is is a measure of how connected the network is.

11.6 COMMUNITY DETECTION

In the last few sections we looked at the problem of graph partitioning, the division of network vertices into groups of given number and size, so as to minimize the number of edges running between groups. A complementary problem, introduced in Section 11.2.1, is that of community detection, the search for the naturally occurring groups in a network regardless of their number or size, which is used primarily as a tool for discovering and understanding the large-scale structure of networks.

The basic goal of community detection is similar to that of graph partitioning: we want to separate the network into groups of vertices that have few connections between them. The important difference is that the number or size of the groups is not fixed. Let us focus to begin with on a very simple example of a community detection problem, probably *the* simplest, which is analogous to the graph bisection problems we examined in previous sections. We will consider the problem of dividing a network into just two non-overlapping groups or communities, as previously, but now without any constraint on the sizes of the groups, other than that the sum of the sizes should equal the size n of the whole network. Thus, in this simple version of the problem, the number of groups is still specified but their sizes are not, and we wish to find the "natural" division of the network into two groups, the fault line (if any) along which the network inherently divides, although we haven't yet said precisely what we mean by that, so that the question we're asking is not yet well defined.

Our first guess at how to tackle this problem might be simply to find the division with minimum cut size, as in the corresponding graph partitioning problem, but without any constraint on the sizes of our groups. However, a moment's reflection reveals that this will not work. If we divide a network into two groups with any number of vertices allowed in the groups then the optimum division is simply to put all the vertices in one of the groups and none of them in the other. This trivial division insures that the cut size between the two groups will be zero—there will be no edges between groups because one of the groups contains no vertices! As an answer to our community detection problem, however, it is clearly not useful.

One way to do better would be to impose loose constraints of some kind on the sizes of the groups. That is, we could allow the sizes of the groups to vary, but not too much. An example of this type of approach is *ratio cut partitioning* in which, instead of minimizing the standard cut size R, we instead minimize the ratio $R/(n_1 n_2)$, where n_1 and n_2 are the sizes of the two groups. The denominator $n_1 n_2$ has its largest value, and hence reduces the ratio by the largest amount, when n_1 and n_2 are equal $n_1 = n_2 = \frac{1}{2}n$. For unequal group sizes the denominator becomes smaller the greater the inequality, and diverges when either group size becomes zero. This effectively eliminates solutions in which all vertices are placed in the same group, since such solutions never give the minimum value of the ratio, and biases the division towards those solutions in which the groups are of roughly equal size.

As a tool for discovering the natural divisions in a network, however, the ratio cut is not ideal. In particular, although it allows group sizes to vary it is still biased towards a particular choice, that of equally sized groups. More importantly, there is no principled rationale behind its definition. It works reasonably well in some circumstances, but there's no fundamental reason to believe it will give sensible answers or that some other approach will not give better ones.

An alternative strategy is to focus on a different measure of the quality of a division other than the simple cut size or its variants. It has been argued that the cut size is not itself a good measure because a good division of a network into communities is not merely one in which there are few edges between communities. On the contrary, the argument goes, a good division is one where there are *fewer than expected* such edges. If we find a division of a network that has few edges between its groups, but nonetheless the number of such edges is about what we would have expected were edges simply placed at random in the network, then most people would say we haven't found anything significant. It is not the total cut size that matters, but how that cut size compares with what we expect to see.

In fact, in the conventional development of this idea one considers not the number of edges between groups but the number within groups. The two approaches are equivalent, however, since every edge that lies within a group necessarily does not lie between groups, so one can calculate one number from the other given the total number of edges in the network as whole. We will follow convention here and base our calculations on the numbers of within-group edges.

Our goal therefore will be to find a measure that quantifies how many edges lie within groups in our network relative to the number of such edges expected on the basis of chance. This, however, is an idea we have encountered before. In Section 7.13.1 we considered the phenomenon of assortative mixing in networks, in which vertices with similar characteristics tend to be connected by edges. There we introduced the measure of assortative mixing known as modularity, which has a high value when many more edges in a network fall between vertices of the same type than one would expect by chance. This is precisely the type of measure we need to solve our current community detection problem. If we consider the vertices in our two groups to be vertices of two types then good divisions of the network into communities are precisely those that have high values of the corresponding modularity.

Thus one way to detect communities in networks is to look for the divisions that have the highest modularity scores and in fact this is the most commonly used method for community detection. Like graph partitioning, modularity maximization is a hard problem (see Section 11.3.1). It is believed that, as with partitioning, the only algorithms capable of always finding the division with maximum modularity take exponentially long to run and hence are useless for all but the smallest of networks [54]. Instead, therefore, we turn again to heuristic algorithms, algorithms that attempt to maximize the modularity in an intelligent way that gives reasonably good results most of the time.

11.7 SIMPLE MODULARITY MAXIMIZATION

One straightforward algorithm for maximizing modularity is the analog of the Kernighan–Lin algorithm [245]. This algorithm divides networks into two communities starting from some initial division, such as a random division into equally sized groups. The algorithm then considers each vertex in the network in turn and calculates how much the modularity would change if that vertex were moved to the other group. It then chooses among the vertices the one whose movement would most increase, or least decrease, the modularity and moves it. Then it repeats the process, but with the important constraint that a vertex once moved cannot be moved again, at least on this round of the algorithm.

And so the algorithm proceeds, repeatedly moving the vertices that most increase or least decrease the modularity. Notice that in this algorithm we are not swapping pairs as we did in the Kernighan–Lin algorithm. In that algorithm we were required to keep the sizes of the groups constant, so for every vertex removed from a group we also had to add one. Now we no longer have such a constraint and so we can move single vertices on each step.

When all vertices have been moved exactly once, we go back over the states through which the network has passed and select the one with the highest modularity. We then use that state as the starting condition for another round of the same algorithm, and we keep repeating the whole process until the modularity no longer improves.

Figure 11.4 shows an example application of this algorithm to the "karate club" network of Zachary, which we encountered previously in Chapter 1 (see Fig. 1.2 on page 6). This network represents the pattern of friendships between members of a karate club at a North American university, as determined by direct observation of the club's members by the experimenter over a period

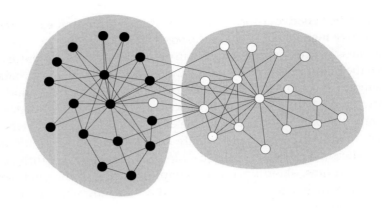

Figure 11.4: Modularity maximization applied to the karate club network. When we apply our vertex-moving modularity maximization algorithm to the karate club network, the best division found is the one indicated here by the two shaded regions, which split the network into two groups of 17 vertices each. This division is very nearly the same as the actual split of the network in real life (open and solid circles), following the dispute among the club's members. Just one vertex is classified incorrectly.

of about two years. The network is interesting because during the period of observation a dispute arose among the members of the club over whether to raise the club's fees and as a result the club eventually split into two parts, of 18 and 16 members respectively, the latter departing to form their own club. The colors of the vertices in Fig. 11.4 denote the members of the two factions, while the shaded regions show the communities identified in the network by our vertex-moving algorithm. As we can see from the figure, the communities identified correspond almost perfectly to the known groups in the network. Just one vertex on the border between the groups is incorrectly assigned. Thus in this case our algorithm appears to have picked out structure of genuine sociological interest from an analysis of network data alone. It is precisely for results of this kind, that shed light on potentially important structural features of networks, that community detection methods are of interest.

The vertex moving algorithm is also quite efficient. At each step of the algorithm we have to evaluate the modularity change due to the movement of each of $O(n)$ vertices, and each such evaluation, like the corresponding ones for the Kernighan–Lin algorithm, can be achieved in time $O(m/n)$ if the network is stored as an adjacency list. Thus each step takes time $O(m)$ and there are n steps in one complete round of the algorithm for a total time of $O(mn)$. This is considerably better than the $O(mn^2)$ of the Kernighan–Lin algorithm, and the algorithm is in fact one of the better of the many proposed algorithms for modularity maximization.[10] The fundamental reason for the algorithm's speed is that when moving single vertices we only have to consider $O(n)$ possible moves at each step, by contrast with the $O(n^2)$ possible swaps of vertex pairs that must be consider in a step of the Kernighan–Lin algorithm.

11.8 SPECTRAL MODULARITY MAXIMIZATION

Having seen in the previous section an algorithm for modularity maximization analogous to the Kernighan–Lin algorithm, it is natural to ask whether there also exists an analog for community detection of the spectral graph partitioning algorithm of Section 11.5. The answer is yes, there is indeed such an algorithm, as we now describe.

In Section 7.13.1 we wrote an expression for the modularity of a division of a network as follows (Eq. (7.69)):

$$Q = \frac{1}{2m} \sum_{ij} \left(A_{ij} - \frac{k_i k_j}{2m} \right) \delta(c_i, c_j) = \frac{1}{2m} \sum_{ij} B_{ij}\, \delta(c_i, c_j), \qquad (11.39)$$

where c_i is the group or community to which vertex i belongs, $\delta(m, n)$ is the Kronecker delta, and

$$B_{ij} = A_{ij} - \frac{k_i k_j}{2m}. \qquad (11.40)$$

Note that B_{ij} has the property

$$\sum_j B_{ij} = \sum_j A_{ij} - \frac{k_i}{2m} \sum_j k_j = k_i - \frac{k_i}{2m} 2m = 0, \qquad (11.41)$$

and similarly for sums over i. (We have made use of Eq. (6.20) in the second equality.) This property will be important shortly.

Let us again consider the division of a network into just two parts (we will consider the more general case later) and again represent such a division by the quantities

$$s_i = \begin{cases} +1 & \text{if vertex } i \text{ belongs to group 1,} \\ -1 & \text{if vertex } i \text{ belongs to group 2.} \end{cases} \qquad (11.42)$$

---

[10]If the network is stored in adjacency matrix form then the total run time can be improved further to $O(n^2)$, although for the common case of a sparse network this makes relatively little difference, and the adjacency matrix is costly in terms of memory space.

We note that the quantity $\frac{1}{2}(s_i s_j + 1)$ is 1 if i and j are in the same group and zero otherwise, so that

$$\delta(c_i, c_j) = \tfrac{1}{2}(s_i s_j + 1). \tag{11.43}$$

Substituting this expression into Eq. (11.39), we find

$$Q = \frac{1}{4m} \sum_{ij} B_{ij} \left(s_i s_j + 1 \right) = \frac{1}{4m} \sum_{ij} B_{ij}\, s_i s_j, \tag{11.44}$$

where we have used Eq. (11.41). In matrix terms we can write this as

$$Q = \frac{1}{4m}\, \mathbf{s}^T \mathbf{B} \mathbf{s}, \tag{11.45}$$

where $\mathbf{s}$ is, as before, the vector with elements s_i, and $\mathbf{B}$ is the $n \times n$ matrix with elements B_{ij}, also called the *modularity matrix*.

Equation (11.45) is similar in form to our expression, Eq. (11.22), for the cut size of a network in terms of the graph Laplacian. By exploiting this similarity we can derive a spectral algorithm for community detection that is closely analogous to the spectral partitioning method of Section 11.5.

We wish to find the division of a given network that maximizes the modularity Q. That is, we wish to find the value of $\mathbf{s}$ that maximizes Eq. (11.45) for a given modularity matrix $\mathbf{B}$. The elements of $\mathbf{s}$ are constrained to take values ± 1, so that the vector always points to one of the corners of an n-dimensional hypercube, but otherwise there are no constraints on the problem. In particular, the number of elements with value $+1$ or -1 is not fixed as it was in the corresponding graph partitioning problem—the sizes of our communities are unconstrained.

As before, this optimization problem is a hard one, but it can be tackled approximately—and effectively—by a relaxation method. We relax the constraint that $\mathbf{s}$ must point to a corner of the hypercube and allow it to point in any direction, though keeping its length the same, meaning that it can take any real value subject only to the constraint that

$$\mathbf{s}^T \mathbf{s} = \sum_i s_i^2 = n. \tag{11.46}$$

The maximization is now a straightforward problem. We maximize Eq. (11.44) by differentiating, imposing the constraint with a single Lagrange multiplier β:

$$\frac{\partial}{\partial s_i}\left[\sum_{jk} B_{jk} s_j s_k + \beta\left(n - \sum_j s_j^2 \right) \right] = 0. \tag{11.47}$$

When we perform the derivatives, this gives us

$$\sum_j B_{ij} s_j = \beta s_i, \tag{11.48}$$

or in matrix notation

$$\mathbf{B}\mathbf{s} = \beta\mathbf{s}. \tag{11.49}$$

In other words, $\mathbf{s}$ is one of the eigenvectors of the modularity matrix. Substituting (11.49) back into Eq. (11.45), we find that the modularity itself is given by

$$Q = \frac{1}{4m}\beta\mathbf{s}^T\mathbf{s} = \frac{n}{4m}\beta, \tag{11.50}$$

where we have used Eq. (11.46). For maximum modularity, therefore, we should choose $\mathbf{s}$ to be the eigenvector $\mathbf{u}_1$ corresponding to the largest eigenvalue of the modularity matrix.

As before, we typically cannot in fact choose $\mathbf{s} = \mathbf{u}_1$, since the elements of $\mathbf{s}$ are subject to the constraint $s_i = \pm 1$. But we do the best we can and choose it as close to $\mathbf{u}_1$ as possible, which means maximizing the product

$$\mathbf{s}^T\mathbf{u}_1 = \sum_i s_i [\mathbf{u}_1]_i, \tag{11.51}$$

where $[\mathbf{u}_1]_i$ is the ith element of $\mathbf{u}_1$. The maximum is achieved when each term in the sum is non-negative, i.e., when

$$s_i = \begin{cases} +1 & \text{if } [\mathbf{u}_1]_i > 0, \\ -1 & \text{if } [\mathbf{u}_1]_i < 0. \end{cases} \tag{11.52}$$

In the unlikely event that a vector element is exactly zero, either value of s_i is equally good and we can choose whichever we prefer.

And so we are led the following very simple algorithm. We calculate the eigenvector of the modularity matrix corresponding to the largest (most positive) eigenvalue and then assign vertices to communities according to the signs of the vector elements, positive signs in one group and negative signs in the other.

In practice this method works very well. For example, when applied to the karate club network of Fig. 11.4 it works perfectly, classifying every one of the 34 vertices into the correct group.

One potential problem with the algorithm is that the matrix $\mathbf{B}$ is, unlike the Laplacian, not sparse, and indeed usually has all elements non-zero. At first sight, this appears to make the algorithm's complexity significantly worse than that of the normal spectral bisection algorithm; as discussed in Section 11.1.1, finding the leading eigenvector of a matrix takes time $O(mn)$, which is equivalent to $O(n^3)$ in a dense matrix, as opposed to $O(n^2)$ in a sparse one. In fact, however, by exploiting special properties of the modularity matrix it is still possible to find the eigenvector in time $O(n^2)$ on a sparse network. The details can be found in [246].

Overall, this means that the spectral method is about as fast as, but not significantly faster than, the vertex-moving algorithm of Section 11.7. Both have time complexity $O(n^2)$ on sparse networks.[11] There is, however, merit to having both algorithms. Given that all practical modularity maximizing algorithms are merely heuristics—clever perhaps, but not by any means guaranteed to perform well in all cases—having more than one fast algorithm in our toolkit is always a good thing.

11.9   DIVISION INTO MORE THAN TWO GROUPS

The community detection algorithms of the previous two sections both perform a limited form of community detection, the division of a network into exactly two communities, albeit of unspecified sizes. But "communities" are defined to be the natural groupings of vertices in networks and there is no reason to suppose that networks will in general have just two of them. They might have two, but they might have more than two, and we would like to be able to find them whatever their number. Moreover we don't, in general, want to have to specify the number of communities; that number should be fixed by the structure of the network and not by the experimenter.

In principle, the modularity maximization method can handle this problem perfectly well. Instead of maximizing modularity over divisions of a network into two groups, we should just maximize it over divisions into any number of groups. Modularity is supposed to be largest for the best division of the network, no matter how many groups that division possesses.

There are a number of community detection algorithms that take this "free maximization" approach to determining community number, and we discuss some of them in the following section. First, however, we discuss a simpler approach which is a natural extension of the methods of previous sections and of our graph partitioning algorithms, namely repeated bisection of a network. We start by dividing the network first into two parts and then we further subdivide those parts in to smaller ones, and so on.

One must be careful about how one does this, however. We cannot proceed as one can in the graph partitioning case and simply treat the communities found in the initial bisection of a network as smaller networks in their

---

[11]Note, however, that the vertex moving algorithm takes time $O(n^2)$ for each round of the algorithm, but we have not calculated, and do not in fact know, how many rounds are needed in general. As with the Kernighan–Lin algorithm, it is reasonable to suppose that the number of rounds needed might increase, at least slowly, with network size, which would make the time complexity of the vertex moving algorithm poorer than that of the spectral algorithm.

own right, applying our bisection algorithm to those smaller networks. The modularity of the complete network does not break up (as cut size does) into independent contributions from the separate communities and the individual maximization of the modularities of those communities treated as separate networks will not, in general, produce the maximum modularity for the network as a whole.

Instead, we must consider explicitly the change ΔQ in the modularity of the entire network upon further bisecting a community c of size n_c. That change is given by

$$
\begin{aligned}
\Delta Q &= \frac{1}{2m} \left[\frac{1}{2} \sum_{i,j \in c} B_{ij}(s_i s_j + 1) - \sum_{i,j \in c} B_{ij} \right] \\
&= \frac{1}{4m} \left[\sum_{i,j \in c} B_{ij} s_i s_j - \sum_{i,j \in c} B_{ij} \right] = \frac{1}{4m} \sum_{i,j \in c} \left[B_{ij} - \delta_{ij} \sum_{k \in c} B_{ik} \right] s_i s_j \\
&= \frac{1}{4m} \mathbf{s}^T \mathbf{B}^{(c)} \mathbf{s},
\end{aligned}
\tag{11.53}
$$

where we have made use of $s_i^2 = 1$, and $\mathbf{B}^{(c)}$ is the $n_c \times n_c$ matrix with elements

$$
B_{ij}^{(c)} = B_{ij} - \delta_{ij} \sum_{k \in c} B_{ik}.
\tag{11.54}
$$

Since Eq. (11.53) has the same general form as Eq. (11.45) we can now apply our spectral approach to this generalized modularity matrix, just as before, to maximize ΔQ, finding the leading eigenvector and dividing the network according to the signs of its elements.

In repeatedly subdividing a network in this way, an important question we need to address is at what point to halt the subdivision process. The answer is quite simple. Given that our goal is to maximize the modularity for the entire network, we should only go on subdividing groups so long as doing so results in an increase in the overall modularity. If we are unable to find any division of a community that results in a positive change ΔQ in the modularity, then we should simply leave that community undivided. The practical indicator of this situation is that our bisection algorithm will put all vertices in one of its two groups and none in the other, effectively refusing to subdivide the community rather than choose a division that actually decreases the modularity. When we have subdivided our network to the point where all communities are in this indivisible state, the algorithm is finished and we stop.

This repeated bisection method works well in many situations, but it is by no means perfect. A particular problem is that, as in the equivalent approach to graph partitioning, there is no guarantee that the best division of a network

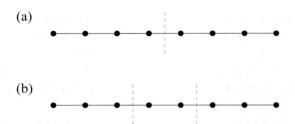

Figure 11.5: Division of a simple network by repeated maximization of the modularity. (a) The optimal bisection of this network of eight vertices and seven edges is straight down the middle. (b) The optimal division into an arbitrary number of groups is this division into three.

into, say, three parts, can be found by first finding the best division into two parts and then subdividing one of the two. Consider for instance the simple network shown in Fig. 11.5, which consists of eight vertices joined together in a line. The bisection of this network with highest modularity is the one shown in Fig. 11.5a, down the middle of the network, splitting it into two equally sized groups of four vertices each. The best modularity if the number of groups is unconstrained, however, is that shown in Fig. 11.5b, with three groups of sizes 3, 2, and 3, respectively. A repeated optimal bisection algorithm would never find the division in 11.5b because, having first made the bisection in 11.5a, there is no further bisection that will get us to 11.5b.

As mentioned above, an alternative method for dividing networks into more than two communities is to attempt to find directly the maximum modularity over divisions into any number of groups. This approach can, in principle, find better divisions than repeated bisection, but in practice is more complicated to implement and often runs slower. A number of promising methods have been developed, however, some of which are discussed in the next section.

11.10 OTHER MODULARITY MAXIMIZATION METHODS

There are a great variety of general algorithms for maximizing (or minimizing) functions over sets of states, and in theory any one of them could be brought to bear on the modularity maximization problem, thereby creating a new community detection algorithm. We describe briefly here three approaches that have met with some success. Each of these approaches attempts to maximize modularity over divisions into any number of communities of any sizes and thus to determine both the number and size of communities in the process.

One of the most widely used general optimization strategies is *simulated annealing*, which proceeds by analogy with the physics of slow cooling or "annealing" of solids. It is known that a hot system, such as a molten metal, will, if cooled sufficiently slowly to a low enough temperature, eventually find its *ground state*, that state of the system that has the lowest possible energy. Simulated annealing works by treating the quantity of interest—modularity in this case—as an energy and then simulating the cooling process until the system finds the state with the lowest energy. Since we are interested in finding the highest modularity, not the lowest, we equate energy in our case with *minus* the modularity, rather than with the modularity itself.

The details of the simulated annealing method are beyond the scope of this book, but the application to modularity maximization is a straightforward one and it appears to work very well [85, 150, 151, 215, 281]. For example, Danon *et al.* [85] performed an extensive test in which they compared the performance of a large number of different community detection algorithms on standardized tasks and found that the simulated annealing method gave the best results of any method tested. The main disadvantage of the approach is that it is slow, typically taking several times as long to reach an answer as competing methods do.

Another general optimization method is the genetic algorithm, a method inspired by the workings of biological evolution. Just as fitter biological species reproduce more and so pass on the genes that confer that fitness to future generations, so one can consider a population of different divisions of the same network and assign to each a "fitness" proportional to its modularity. Over a series of generations one simulates the preferential "reproduction" of high-modularity divisions, while those of low modularity die out. Small changes or mutations are introduced into the offspring divisions, allowing their modularity values either to improve or get worse and those that improve are more likely to survive in the next generation while those that get worse are more likely to be killed off. After many generations one has a population of divisions with good modularity and the best of these is the final division returned by the algorithm. Like simulated annealing the method appears to give results of high quality, but is slow, which restricts its use to networks of a few hundred vertices or fewer [295].

A third method makes use of a so-called *greedy algorithm*. In this very simple approach we start out with each vertex in our network in a one-vertex group of its own, and then successively amalgamate groups in pairs, choosing at each step the pair whose amalgamation gives the biggest increase in modularity, or the smallest decrease if no choice gives an increase. Eventually

381

all vertices are amalgamated into a single large community and the algorithm ends. Then we go back over the states through which the network passed during the course of the algorithm and select the one with the highest value of the modularity. A naive implementation of this idea runs in time $O(n^2)$, but by making use of suitable data structures the run time can be improved to $O(n \log^2 n)$ on a sparse graph [71, 319]. Overall the algorithm works only moderately well: it gives reasonable divisions of networks, but the modularity values achieved are in general somewhat lower than those found by the other methods described here. On the other hand, the running time of the method may be the best of any current algorithm, and this is one of the few algorithms fast enough to work on the very largest networks now being explored. Wakita and Tsurumi [319] have given one example of an application to a network of more than five million vertices, something of a record for studies of this kind.

11.11 OTHER ALGORITHMS FOR COMMUNITY DETECTION

As we have seen, the problem of detecting communities in networks is a less well-posed one than the problem of graph partitioning. In graph partitioning the goal is clear: to find the division of a network with the smallest possible cut size. There is, by contrast, no universally agreed upon definition of what constitutes a good division of a network into communities. In the previous sections we have looked at algorithms based one particular definition in terms of the modularity function, but there are a number of other definitions in common use that lead to different algorithms. In the following sections we look briefly at a few of these other algorithms.

11.11.1 BETWEENNESS-BASED METHODS

One alternative way of finding communities of vertices in a network is to look for the edges that lie between communities. If we can find and remove these edges, we will be left with just the isolated communities.

There is more than one way to quantify what we mean when we say an edge lies "between communities," but one common approach is to use betweenness centrality. As described in Section 7.7, the betweenness centrality of a vertex in a network is the number of geodesic (i.e., shortest) paths in the network that pass through that vertex. Similarly, we can define an *edge betweenness* that counts the number of geodesic paths that run along edges and, as shown in Fig. 11.6, edges that lie between communities can be expected to have high values of the edge betweenness.

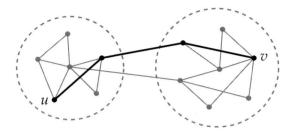

Figure 11.6: Identification of between-group edges. This simple example network is divided into two groups of vertices (denoted by the dotted lines), with only two edges connecting the groups. Any path joining vertices in different groups (such as vertices u and v) must necessarily pass along one of these two edges. Thus if we consider a set of paths between all pairs of vertices (such as geodesic paths, for instance), we expect the between-group edges to carry more paths than most. By counting the number of paths that pass along each edge we can in this way identify the between-group edges.

The calculation of edge betweenness is precisely analogous to the vertex case: we consider the geodesic path or paths between every pair of vertices in the network (except vertices in different components, for which no such path exists), and count how many such paths go along each edge. Edge betweenness can be calculated for all edges in time $O(n(m + n))$ using a slightly modified version of the algorithm described in Section 10.3.6 [250].

Our algorithm for detecting communities is then as follows. We calculate the betweenness scores of all edges in our network and then search through them for the edge with the highest score and remove it. In removing the edge we will change the betweenness scores of some edges, because any shortest paths that previously traversed the removed edge will now have to be rerouted another way. So we must recalculate the betweenness scores following the removal. Then we search again for the edge with the highest score and remove it, and so forth. As we remove one edge after another an initially connected network will eventually split into two pieces, and then into three, and so on.

The progress of the algorithm can be represented using a tree or *dendrogram* like that depicted in Fig. 11.7. At the bottom of the figure we have the "leaves" of the tree, which each represent one of the vertices of the network, and as we move up the tree, the leaves join together first in pairs and then in larger groups, until at the top of the tree all are joined together to form a single whole. Our algorithm in fact generates the dendrogram from the top, rather than the bottom, starting with a single connected network and splitting it repeatedly

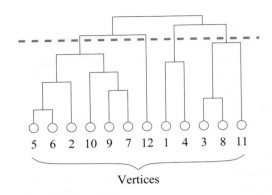

5   6   2   10   9   7   12   1   4   3   8   11

Vertices

Figure 11.7: A dendrogram. The results of the edge betweenness algorithm can be represented as a tree or "dendrogram" in which the vertices are depicted (conventionally) at the bottom of the tree and the "root" at the top represent the whole network. The progressive fragmentation of the network as edges are removed one by one is represented by the successive branching of the tree as we move down the figure and the identities of the vertices in a connected subset at any point in the procedure can be found by following the lines of the tree down to the bottom of the picture. Each intermediate division of the network through which the algorithm passes corresponds to a horizontal cut through the dendrogram. For instance, the cut denoted by the dotted line in this dendrogram splits the network into four groups of 6, 1, 2, and 3 vertices respectively.

until we get to the level of single vertices. Individual intermediate configurations of the network during the run of the algorithm correspond to horizontal cuts through the dendrogram, as indicated by the dotted line in the figure. Each branch of the tree that intersects this dotted line represents one group of vertices, whose membership we can determine by following the branch down to its leaves at the bottom of the figure. Thus the dendrogram captures in a single diagram the configuration of groups in the network at every stage from start to finish of the algorithm.

This algorithm is somewhat different from previous ones, therefore, in that it doesn't give a single decomposition of a network into communities, but a selection of different possibilities, ranging from coarse divisions into just a few large communities (at the top of the dendrogram) to fine divisions into many small communities (at the bottom). It is up to the user to decide which of the many divisions represented is most useful for their purposes. One could in principle use a measure such as modularity to quantify the quality of the different divisions and select the one with the highest quality in this sense. This, however, somewhat misses the point. If high modularity is what you care about, then you are better off simply using a modularity maximization algorithm in the first place. It is more appropriate simply to think of this betweenness-based algorithm as producing a different kind of output, one that has its own advantages and disadvantages but that can undoubtedly tell us interesting things about network structure.

The betweenness-based algorithm is, unfortunately, quite slow. As we have said the calculation of betweenness for all edges takes time of order $O(n(m + n))$ and we have to perform this calculation before the removal of each of the m edges, so the entire algorithm takes time $O(mn(m + n))$, or $O(n^3)$ on a sparse graph with $m \propto n$. This makes this algorithm one of the slower algorithms considered in this chapter. The algorithm gives quite good results in practice [138, 250], but has mostly been superseded by the faster modularity maximization methods of previous sections.

Nonetheless, the ability of the algorithm to return an entire dendrogram, rather than just a single division of a network, could be useful in some cases. The divisions represented in the dendrogram form a *hierarchical decomposition* in which the communities at one level are completely contained within the larger communities at all higher levels. There has been some interest in hierarchical structure in networks and hierarchical decompositions that might capture it. We look at another algorithm for hierarchical decomposition in Section 11.11.2.

An interesting variation on the betweenness algorithm has been proposed by Radicchi *et al.* [276]. Their idea revolves around the same basic principle of identifying the edges between communities and removing them, but the measure used to perform the identification is different. Radicchi *et al.* observe that the edges that fall between otherwise poorly connected communities are unlikely to belong to short loops of edges, since doing so would require that there be two nearby edges joining the same groups—see Fig. 11.8. Thus one way to identify the edges between communities would be to look for edges that belong to an unusually small number of short loops. Radicchi *et al.* found that loops of length three and four gave the best results. By repeatedly removing edges that belong to small numbers of such loops they were able to accurately uncover communities in a number of example networks.

An attractive feature of this method is its speed. The calculation of the number of short loops to which an edge belongs is a local calculation and can be performed for all edges in time that goes like the total size of the network. Thus, in the worst case, the running time of the algorithm will only go as $O(n^2)$ on a sparse graph, which is one order of system size faster than the betweenness-based algorithm and as fast as the earlier methods based on modularity maximization.

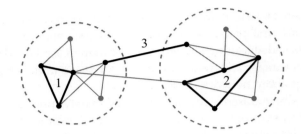

Figure 11.8: The algorithm of Radicchi *et al.* The algorithm of Radicchi *et al.* uses a different measure to identify between-group edges, looking for the edges that belong to the fewest short loops. In many networks, edges within groups typically belong to many short loops, such as the loops of length three and four labeled "1" and "2." But edges between groups, such as the edge labeled "3" here, often do not belong to such loops, because to do so would require there to be a return path along another between-group edge, of which there are, by definition, few.

On the other hand, the algorithm of Radicchi *et al.* has the disadvantage that it only works on networks that have a significant number of short loops in the first place. This restricts the method primarily to social networks, which indeed have large numbers of short loops (see Section 7.9). Other types of network, such as technological and biological networks, tend to have smaller numbers of short loops, and hence there is little to distinguish between-group edges from within-group ones.

11.11.2 HIERARCHICAL CLUSTERING

The algorithms of the previous section differ somewhat from the other community detection algorithms in this chapter in that they produce a hierarchical decomposition of a network into a set of nested communities, visualized in the form of a dendrogram as in Fig. 11.7, rather than just a single division into a unique set of communities. In this section we look at another algorithm that also produces a hierarchical decomposition, one of the oldest of community detection methods, the method of *hierarchical clustering*.[12]

Hierarchical clustering is not so much a single algorithm as an entire class

---

[12]The word "clustering" as used here just refers to community detection. We have mostly stayed away from using this word in this chapter, to avoid confusion with the other use of the word clustering introduced in Section 7.9 (see footnote 5 on page 354), but the name "hierarchical clustering" is a well established and traditional one, and we use it here in deference to convention.

of algorithms, with many variations and alternatives. Hierarchical clustering is an *agglomerative* technique in which we start with the individual vertices of a network and join them together to form groups. This contrasts with most of the other methods we have looked at for community detection and graph partitioning, which were *divisive* methods that took a complete network and split it apart. (One earlier algorithm, the greedy modularity maximization algorithm of Section 11.10, was an agglomerative method.)

The basic idea behind hierarchical clustering is to define a measure of similarity or connection strength between vertices, based on the network structure, and then join together the closest or most similar vertices to form groups. We discussed measures of vertex similarity in networks at some length in Section 7.12. Any of the measures of structural equivalence introduced there would be suitable as a starting point for hierarchical clustering, including cosine similarity (Section 7.12.1), correlation coefficients between rows of the adjacency matrix (Section 7.12.2), or the so-called Euclidean distance (Section 7.12.3). The regular equivalence measures of Section 7.12.4 might also be good choices, although the author is not aware of them having been used in this context.

That there are many choices for similarity measures is both a strength and a weakness of the hierarchical clustering method. It gives the method flexibility and allows it to be tailored to specific problems, but it also means that the method gives different answers depending on which measure we choose, and in many cases there is no way to know if one measure is more correct or will yield more useful information than another. Most often the choice of measure is determined more by experience or experiment than by argument from first principles.

Once a similarity measure is chosen we calculate it for all pairs of vertices in the network. Then we want to group together those vertices having the highest similarities. This, however, leads to a further problem: the similarities can give conflicting messages about which vertices should be grouped. Suppose vertices A and B have high similarity, as do vertices B and C. One might therefore argue that A, B, and C should all be in a group together. But suppose that A and C have *low* similarity. Now we are left with a dilemma. Should A and C be in the same group or not?

The basic strategy adopted by the hierarchical clustering method is to start by joining together those pairs of vertices with the highest similarities, forming a group or groups of size two. For these there is no ambiguity, since each pair only has one similarity value. Then we further join together the groups that are most similar to form larger groups, and so on. When viewed in terms of agglomeration of groups like this, the problem above can be stated in a new and

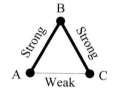

If the connections (A,B) and (B,C) are strong but (A,C) is weak, should A and C be in the same group or not?

useful way. Our process requires for its operation a measure of the similarity between *groups*, so that we can join the most similar ones together. But what we actually have is a measure of similarity between individual vertices, so we need to combine these vertex similarities somehow to create similarities for the groups. If we can do this, then the rest of the algorithm is straightforward and the ambiguity is resolved.

There are three common ways of combining vertex similarities to give similarity scores for groups. They are called single-, complete-, and average-linkage clustering. Consider two groups of vertices, group 1 and group 2, containing n_1 and n_2 vertices respectively. There are then $n_1 n_2$ pairs of vertices such that one vertex is in group 1 and the other in group 2. In the *single-linkage clustering* method, the similarity between the two groups is defined to be the similarity of the *most similar* of these $n_1 n_2$ pairs of vertices. Thus if the values of the similarities of the vertex pairs range from 1 to 100, the similarity of the two groups is 100. This is a very lenient definition of similarity: only a single vertex pair need have high similarity for the groups themselves to be considered similar. (This is the origin of the name "single-linkage clustering"—similarity between groups is a function of the similarity between only the single most similar pair of vertices.)

At the other extreme, *complete-linkage clustering* defines the similarity between two groups to be the similarity of the *least similar* pair of vertices. If the similarities range from 1 to 100 then the similarity of the groups is 1. By contrast with single-linkage clustering this is a very stringent definition of group similarity: every single vertex pair must have high similarity for the groups to have high similarity (hence the name "complete-linkage clustering").

In between these two extremes lies *average-linkage clustering*, in which the similarity of two groups is defined to be the mean similarity of all pairs of vertices. Average-linkage clustering is probably the most satisfactory choice of the three, being a moderate one—not extreme in either direction—and depending on the similarity of all vertex pairs and not just of the most or least similar pair. It is, however, relatively rarely used, for reasons that are not entirely clear.

The full hierarchical clustering method is as follows:

1. Choose a similarity measure and evaluate it for all vertex pairs.
2. Assign each vertex to a group of its own, consisting of just that one vertex. The initial similarities of the groups are simply the similarities of the vertices.
3. Find the pair of groups with the highest similarity and join them together into a single group.
4. Calculate the similarity between the new composite group and all others using one of the three methods above (single-, complete-, or average-

linkage clustering).

5. Repeat from step 3 until all vertices have been joined into a single group.

In practice, the calculation of the new similarities is relatively straightforward. Let us consider the three cases separately. For single-linkage clustering the similarity of two groups is equal to the similarity of their most similar pair of vertices. In this case, when we join groups 1 and 2 together, the similarity of the composite group to another group 3, is the greater of the similarities of 1 with 3 and 2 with 3, which can be found in $O(1)$ time.

For complete-linkage clustering the similarity of the composite group is the smaller of the similarities of 1 with 3 and 2 with 3, which can also be found in $O(1)$ time.

The average-linkage case is only slightly more complicated. Suppose as before that the groups 1 and 2 that are to be joined have n_1 and n_2 vertices respectively. Then if the similarities of 1 with 3 and 2 with 3 were previously σ_{13} and σ_{23}, the similarity of the composite group with another group 3 is given by the weighted average

$$\sigma_{12,3} = \frac{n_1\sigma_{13} + n_2\sigma_{23}}{n_1 + n_2}. \tag{11.55}$$

Again this can be calculated in $O(1)$ time.

On each step of the algorithm we have to calculate similarities in this way for the composite group with every other group, of which there are $O(n)$. Hence the recalculation of similarities will take $O(n)$ time on each step. A naive search through the similarities to find the greatest one, on the other hand, takes time $O(n^2)$, since there are $O(n^2)$ pairs of groups to check, so this will be the most time-consuming step in the algorithm. We can speed things up, however, by storing the similarities in a binary heap (see Section 9.7[13]), which allows us to add and remove entries in time $O(\log n)$ and find the greatest one in time $O(1)$. This slows the recalculation of the similarities to $O(n \log n)$ but speeds the search for the largest to $O(1)$.

Then the whole process of joining groups has to be repeated $n - 1$ times until all vertices have been joined into a single group. (To see this, simply consider that the number of groups goes down by one every time two groups are joined, so it takes $n - 1$ joins to go from n initial groups to just a single one

---

[13]The heap must be modified slightly from the one described in Section 9.7. First, the partial ordering must be inverted so that the largest, not the smallest, element of the heap is at its root. Second, we need to be able to remove arbitrary items from the heap, not just the root item, which we do by deleting the relevant item and then moving the last item in the heap to fill the vacated space. Then we have to sift the moved item both *up and* down the heap, since it might be either too large or too small for the position in which it finds itself.

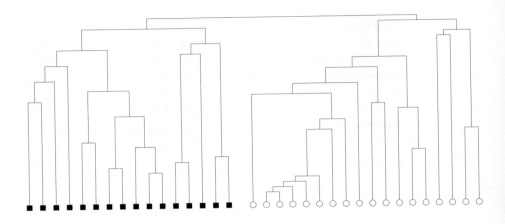

Figure 11.9: **Partitioning of the karate club network by average linkage hierarchical clustering.** This dendrogram is the result of applying the hierarchical clustering method described in the text to the karate club network of Fig. 11.4, using cosine similarity as our measure of vertex similarity. The shapes of the nodes represent the two known factions in the network, as in the two previous figures.

at the end.) Thus the total running time of the algorithm is $O(n^3)$ in the naive implementation or $O(n^2 \log n)$ if we use a heap.[14]

And how well does it work in practice? The answer depends on which similarity measure one chooses and which linkage method, but a typical application, again to the karate club network, is shown in Fig. 11.9. This figure shows what happens when we apply average-linkage clustering to the karate network using cosine similarity as our similarity measure. The figure shows the dendrogram that results from such a calculation and we see that there is a clear division of the dendrogram into two communities that correspond perfectly to the two known groups in the network.

Hierarchical clustering does not always work as well as this, however. In particular, though it is often good at picking out the cores of groups, where the vertices are strongly similar to one another, it tends to be less good at assigning peripheral vertices to appropriate groups. Such vertices may not be strongly similar to any others and so tend to get left out of the agglomerative

---

[14]For the special case of single-linkage clustering, there is a slightly faster way to implement the algorithm that makes use of a so-called union/find technique and runs in time $O(n^2)$. In practice the performance difference is not very large but the union/find method is considerably simpler to program. It is perhaps for this reason that single-linkage is more often used than complete- or average-linkage clustering.

clustering process until the very end. A common result of hierarchical cluster-ing is therefore a set of tightly knit cores surrounded by a loose collection of single vertices or smaller groups. Such a result may nonetheless contain a lot of valuable information about the underlying network structure.

Many other methods have been proposed for community detection and there is not room in this book to describe them all. For the reader interested in pursu-ing the topic further the review articles by Fortunato [124] and Schaeffer [291] provide useful overviews.

Problems

11.1 Show that the inverse of a symmetric matrix M is given by $M^{-1} = UDU^T$ where U is the orthogonal matrix whose columns are the normalized eigenvectors of M and D is the diagonal matrix whose elements are the reciprocals of the eigenvalues of M. Hence argue that the time complexity of the best algorithm for inverting a symmetric matrix can be no worse than the time complexity of finding all of its eigenvalues and eigenvectors. (In fact they are the same—both are $O(n)$ for an $n \times n$ matrix.)

11.2 Consider a general $n \times n$ matrix M with eigenvalues μ_i where $i = 1 \ldots n$.

a) Show that the matrix $M - aI$ has the same eigenvectors as M and eigenvalues $\mu_i - a$.

b) Suppose that the matrix's two eigenvalues of largest magnitude are both positive. Show that the time taken to find the leading eigenvector of the matrix using the power method of Section 11.1 can be improved by performing the calculation instead for the matrix $M - aI$, where a is positive.

c) What stops us from increasing the constant a arbitrarily until the calculation takes no time at all?

11.3 Consider a "line graph" consisting of n vertices in a line like this:

a) Show that if we divide the network into two parts by cutting any single edge, such that one part has r vertices and the other has $n - r$, the modularity, Eq. (7.76), takes the value

$$Q = \frac{3 - 4n + 4rn - 4r^2}{2(n-1)^2}.$$

b) Hence show that when n is even the optimal such division, in terms of modularity, is the division that splits the network exactly down the middle.

11.4 Using your favorite numerical software for finding eigenvectors of matrices, construct the Laplacian and the modularity matrix for this small network:

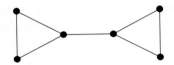

a) Find the eigenvector of the Laplacian corresponding to the second smallest eigenvalue and hence perform a spectral bisection of the network into two equally sized parts.

b) Find the eigenvector of the modularity matrix corresponding to the largest eigenvalue and hence divide the network into two communities.

You should find that the division of the network generated by the two methods is, in this case, the same.

11.5 Consider this small network with five vertices:

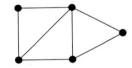

a) Calculate the cosine similarity for each of the $\binom{5}{2} = 10$ pairs of vertices.

b) Using the values of the ten similarities construct the dendrogram for the single-linkage hierarchical clustering of the network according to cosine similarity.

PART IV

NETWORK MODELS

CHAPTER 12

RANDOM GRAPHS

An introduction to the most basic of network models, the
random graph

SO FAR in this book we have looked at how we measure the structure of networks and at mathematical, statistical, and computational methods for making sense of the network data we get from our measurements. We have seen for instance how to measure the structure of the Internet, and once we have measured it how to determine its degree distribution, or the centrality of its vertices, or the best division of the network into groups or communities. An obvious next question to ask is, "If I know a network has some particular property, such as a particular degree distribution, what effect will that have on the wider behavior of the system?" It turns out that properties like degree distributions can in fact have huge effects on networked systems, which is one of the main reasons we are interested in them. And one of the best ways to understand and get a feel for these effects is to build mathematical models. The remainder of this book is devoted to the examination of some of the many network models in common use.

In Chapters 12 to 15 we consider models of the structure of networks, models that mimic the patterns of connections in real networks in an effort to understand the implications of those patterns. In Chapters 16 to 19 we consider models of processes taking place on networks, such as epidemics on social networks or search engines on the Web. In many cases these models of network processes are themselves built on top of our models of network structure, combining the two to shed light on the interplay between structure and dynamics in networked systems.

In Section 8.4, for instance, we noted that many networks have degree distributions that roughly follow a power law—the so-called scale-free networks.

A reasonable question would be to ask how the structure and behavior of such scale-free networks differs from that of their non-scale-free counterparts. A good way to address this question would be to create, on a computer for example, two artificial networks, one with a power-law degree distribution and one without, and explore their differences empirically. Better still, one could create a large number of networks in each of the two classes, to see what statistically significant features appear in one class and not in the other. This is precisely the rationale behind random graph models, which are the topic of this chapter and the following one. In random graph models, one creates networks that possess particular properties of interest, such as specified degree distributions, but which are otherwise random. Random graphs are interesting in their own right for the light they shed on the structural properties of networks, but have also been widely used as a substrate for models of dynamical processes *on* networks. In Chapter 17, for instance, we examine their use in epidemic modeling.

We also look at a number of other types of network model in succeeding chapters. In Chapter 14 we look at generative models of networks, models in which the network is "grown" according to a specified set of growth rules. Generative models are particularly useful for understanding how network structure arises in the first place. By growing networks according to a variety of different rules and comparing the results with real networks, we can get a feel for which growth processes are plausible and which can be ruled out. In Chapter 15 we look at "small-world models," which model the phenomenon of network transitivity or clustering (see Section 7.9), and at "exponential random graphs," which are particularly useful when we want to create model networks that match the properties of observed networks as closely as possible.

12.1 RANDOM GRAPHS

In general, a *random graph* is a model network in which some specific set of parameters take fixed values, but the network is random in other respects. One of the simplest examples of a random graph is the network in which we fix only the number of vertices n and the number of edges m. That is, we take n vertices and place m edges among them at random. More precisely, we choose m pairs of vertices uniformly at random from all possible pairs and connect them with an edge. Typically one stipulates that the network should be a simple graph, i.e., that it should have no multiedges or self-edges (see Section 6.1), in which case the position of each edge should be chosen among only those pairs that are distinct and not already connected.[1] This model is often referred to by its mathematical name $G(n, m)$.

Another entirely equivalent definition of the model is to say that the network is created by choosing uniformly at random among the set of all simple graphs with exactly n vertices and m edges.

Strictly, in fact, the random graph model is not defined in terms of a single randomly generated network, but as an *ensemble* of networks, i.e., a probability distribution over possible networks. Thus the model $G(n, m)$ is correctly defined as a probability distribution $P(G)$ over all graphs G in which $P(G) = 1/\Omega$ for simple graphs with n vertices and m edges and zero otherwise, where Ω is the total number of such simple graphs. We will see more complicated examples of random graph ensembles shortly.

When one talks about the properties of random graphs one typically means the average properties of the ensemble. For instance, the "diameter" of $G(n, m)$ would mean the diameter $\ell(G)$ of a graph G, averaged over the ensemble thus

$$\langle \ell \rangle = \sum_G P(G)\ell(G) = \frac{1}{\Omega} \sum_G \ell(G). \tag{12.1}$$

This is a useful definition for a several of reasons. First, it turns out to lend itself well to analytic calculations; many such average properties of random graphs can be calculated exactly, at least in the limit of large graph size. Second, it often reflects exactly the thing we want to get at in making our model network in the first place. Very often we are interested in the typical properties of networks. We might want to know, for instance, what the typical diameter is of a network with a given number of edges. Certainly there are special cases of such networks that have particularly large or small diameters, but these don't reflect the typical behavior. If it's typical behavior we are after, then the ensemble average of a property is often a good guide. Third, it can be shown that the distribution of values for many network measures is sharply peaked, becoming concentrated more and more narrowly around the ensemble average as the size of the network becomes large, so that in the large n limit essentially all values one is likely to encounter are very close to the mean.

Some properties of the random graph $G(n, m)$ are straightforward to calculate: obviously the average number of edges is m, for instance, and the average degree is $\langle k \rangle = 2m/n$. Unfortunately, other properties are not so easy to calculate, and most mathematical work has actually been conducted on a slightly different model that is somewhat easier to handle. This model is called

---

[1]It would in theory be perfectly possible, however, to create a variant of the model with multi-edges or self-edges, or both.

$G(n, p)$. In $G(n, p)$ we fix not the number but the *probability* of edges between vertices. Again we have n vertices, but now we place an edge between each distinct pair with independent probability p. In this network the number of edges is not fixed. Indeed it is possible that the network could have no edges at all, or could have edges between every distinct pair of vertices. (For most values of p these are not likely outcomes, but they could happen.)

Again, the technical definition of the random graph is not in terms of a single network, but in terms of an ensemble, a probability distribution over all possible networks. To be specific, $G(n, p)$ is the ensemble of networks with n vertices in which each simple graph G appears with probability

$$P(G) = p^m (1 - p)^{\binom{n}{2} - m}, \tag{12.2}$$

where m is the number of edges in the graph, and non-simple graphs have probability zero.

$G(n, p)$ was first studied, to this author's knowledge, by Solomonoff and Rapoport [303], but it is most closely associated with the names of Paul Erdős and Alfréd Rényi, who published a celebrated series of papers about the model in the late 1950s and early 1960s [105–107]. If you read scientific papers on this subject, you will sometimes find the model referred to as the "Erdős–Rényi model" or the "Erdős–Rényi random graph" in honor of their contribution. It is also sometimes called the "Poisson random graph" or the "Bernoulli random graph," names that refer to the distributions of degrees and edges in the model. And sometimes the model is referred to simply as "the" random graph—there are many random graph models, but $G(n, p)$ is the most fundamental and widely studied of them, so if someone is talking about a random graph but doesn't bother to mention which one, they are probably thinking of this one.

In this chapter we describe the basic mathematics of the random graph $G(n, p)$, focusing particularly on the degree distribution and component sizes, which are two of the model's most illuminating characteristics. The techniques we develop in this chapter will also prove useful for some of the more complex models examined later in the book.

12.2 Mean number of edges and mean degree

Let us start our study of the random graph $G(n, p)$ with a very simple calculation, the calculation of the expected number of edges in our model network. We have said that the number of edges in the model is not fixed, but we can calculate its mean or expectation value as follows. The number of graphs with exactly n vertices and m edges is equal to the number of ways of picking the positions of the edges from the $\binom{n}{2}$ distinct vertex pairs. Each of these graphs appears with the same probability $P(G)$, given by Eq. (12.2), and hence the total probability of drawing a graph with m edges from our ensemble is

$$P(m) = \binom{\binom{n}{2}}{m} p^m (1-p)^{\binom{n}{2}-m}, \qquad (12.3)$$

which is just the standard binomial distribution. Then the mean value of m is

$$\langle m \rangle = \sum_{m=0}^{\binom{n}{2}} m P(m) = \binom{n}{2} p. \qquad (12.4)$$

This result comes as no surprise. The expected number of edges between any individual pair of vertices is just equal to the probability p of an edge between the same vertices, and Eq. (12.4) thus says merely that the expected total number of edges in the network is equal to the expected number p between any pair of vertices, multiplied by the number of pairs.

We can use this result to calculate the mean degree of a vertex in the random graph. As pointed out in the previous section, the mean degree in a graph with exactly m edges is $\langle k \rangle = 2m/n$, and hence the mean degree in $G(n,p)$ is

$$\langle k \rangle = \sum_{m=0}^{\binom{n}{2}} \frac{2m}{n} P(m) = \frac{2}{n} \binom{n}{2} p = (n-1)p, \qquad (12.5)$$

where we have used Eq. (12.4) and the fact that n is constant. The mean degree of a random graph is often denoted c in the literature, and we will adopt this convention here also, writing

$$c = (n-1)p. \qquad (12.6)$$

This result is also unsurprising. It says that the expected number of edges connected to a vertex is equal to the expected number p between the vertex and any other vertex, multiplied by the number $n-1$ of other vertices.

12.3 DEGREE DISTRIBUTION

Only slightly more taxing is the calculation of the degree distribution of $G(n,p)$. A given vertex in the graph is connected with independent probability p to each of the $n-1$ other vertices. Thus the probability of being connected to a particular k other vertices and not to any of the others is $p^k(1-p)^{n-1-k}$. There are $\binom{n-1}{k}$ ways to choose those k other vertices, and hence the total probability of being connected to exactly k others is

$$p_k = \binom{n-1}{k} p^k (1-p)^{n-1-k}, \qquad (12.7)$$

which is a binomial distribution again. In other words, $G(n, p)$ has a binomial degree distribution.

In many cases we are interested in the properties of large networks, so that n can be assumed to be large. Furthermore, as discussed in Section 6.9, many networks have a mean degree that is approximately constant as the network size becomes large. (For instance, the typical number of friends a person has does not depend strongly on the total number of people in the world.) In such a case Eq. (12.7) simplifies as follows.

Equation (12.6) tells us that $p = c/(n-1)$ will become vanishingly small as $n \to \infty$, which allows us to write

$$\ln\left[(1-p)^{n-1-k}\right] = (n-1-k)\ln\left(1 - \frac{c}{n-1}\right)$$
$$\simeq -(n-1-k)\frac{c}{n-1} \simeq -c, \tag{12.8}$$

where we have expanded the logarithm as a Taylor series, and the equalities become exact as $n \to \infty$. Taking exponentials of both sizes, we thus find that $(1-p)^{n-1-k} = e^{-c}$ in the large-n limit. Also for large n we have

$$\binom{n-1}{k} = \frac{(n-1)!}{(n-1-k)!\,k!} \simeq \frac{(n-1)^k}{k!}, \tag{12.9}$$

and thus Eq. (12.7) becomes

$$p_k = \frac{(n-1)^k}{k!}p^k e^{-c} = \frac{(n-1)^k}{k!}\left(\frac{c}{n-1}\right)^k e^{-c} = e^{-c}\frac{c^k}{k!}, \tag{12.10}$$

in the limit of large n.

Equation (12.10) is the Poisson distribution: in the limit of large n, $G(n, p)$ has a Poisson degree distribution. This is the origin of the name *Poisson random graph*, which we will use occasionally to distinguish this model from some of the more sophisticated random graphs in the following chapter that don't in general have Poisson degree distributions.

12.4 CLUSTERING COEFFICIENT

A very simple quantity to calculate for the Poisson random graph is the clustering coefficient. Recall that the clustering coefficient C is a measure of the transitivity in a network (Section 7.9) and is defined as the probability that two network neighbors of a vertex are also neighbors of each other. In a random graph the probability that *any* two vertices are neighbors is exactly the same—all such probabilities are equal to $p = c/(n-1)$. Hence

$$C = \frac{c}{n-1}. \tag{12.11}$$

This is one of several respects in which the random graph differs sharply from most from real-world networks, many of which have quite high clustering coefficients—see Table 8.1—while Eq. (12.11) tends to zero in the limit $n \rightarrow \infty$ if the mean degree c stays fixed. This discrepancy is discussed further in Section 12.8.

12.5 GIANT COMPONENT

Consider the Poisson random graph $G(n,p)$ for $p = 0$. In this case there are no edges in the network at all and it is completely disconnected. Each vertex is an island on its own; the network has n separate components of exactly one vertex each.

In the opposite limit, when $p = 1$, every possible edge in the network is present and the network is an n-vertex clique in the technical sense of the word (see Section 7.8.1) meaning that every vertex is connected directly to every other. In this case, all the vertices are connected together in a single component that spans the entire network.

Now let us focus on the size of the largest component in the network in each of these cases. In the first case ($p = 0$) the largest component has size 1. In the second ($p = 1$) the largest component has size n. Apart from the second being much larger than the first, there is an important qualitative difference between these two cases: in the first case the size of the largest component is independent of the number of vertices n in the network; in the second it is proportional to n, or *extensive* in the jargon of theoretical physics. In the first case, the largest component will stay the same size if we make the network larger, but in the second it will grow with the network.

The distinction between these two cases is an important one. In many applications of networks it is crucial that there be a component that fills most of the network. For instance, in the Internet it is important that there be a path through the network from most computers to most others. If there were not, the network wouldn't be able to perform its intended role of providing computer-to-computer communications for its users. Moreover, as discussed in Section 8.1, most networks do in fact have a large component that fills most of the network. We can gain some useful insights about what is happening in such networks by considering how the components in our random graph behave. Although the random graph is a very simple network model and doesn't provide an accurate representation of the Internet or other real-world networks, we will see that when trying to understand the world it can be very helpful to study such simplified models.

So let us consider the largest component of our random graph, which, as we have said, has constant size 1 when $p = 0$ and extensive size n when $p = 1$. An interesting question to ask is how the transition between these two extremes occurs if we construct random graphs with gradually increasing values of p, starting at 0 and ending up at 1. We might guess, for instance, that the size of the largest component somehow increases gradually with p, becoming extensive only in the limit where $p = 1$. In reality, however, something much more interesting happens. As we will see, the size of the largest component undergoes a sudden change, or *phase transition*, from constant size to extensive size at one particular special value of p. Let us take a look at this transition.

A network component whose size grows in proportion to n we call a *giant component*. We can calculate the size of the giant component in the Poisson random graph exactly in the limit of large network size $n \to \infty$ as follows. We denote by u the average fraction of vertices in the random graph that do *not* belong to the giant component. Thus if there is no giant component in our graph, we will have $u = 1$, and if there is a giant component we will have $u < 1$. Alternatively, we can regard u as the probability that a randomly chosen vertex in the graph does not belong to the giant component.

For a vertex i not to belong to the giant component it must not be connected to the giant component via any other vertex. That means that for every other vertex j in the graph either (a) i is not connected to j by an edge, or (b) i is connected to j but j is itself not a member of the giant component. The probability of outcome (a) is simply $1 - p$, the probability of not having an edge between i and j, and the probability of outcome (b) is pu, where the factor of p is the probability of having an edge and the factor of u is the probability that vertex j doesn't belong to the giant component.[2] Thus the total probability of not being connected to the giant component via vertex j is $1 - p + pu$.

Then the total probability of not being connected to the giant component

---

[2]We need to be a little careful here: u here should really be the probability that j is not connected to the giant component via any of its connections other than the connection to i. However, it turns out that in the limit of large system size this probability is just equal to u. For large n the probability of not being connected to the giant component via any of the $n - 2$ vertices other than i is not significantly smaller than the probability for all $n - 1$ vertices.

via any of the $n-1$ other vertices in the network is

$$u = (1 - p + pu)^{n-1} = \left[1 - \frac{c}{n-1}(1-u)\right]^{n-1}, \qquad (12.12)$$

where we have used Eq. (12.6). Now we take logs of both sides thus:

$$\ln u = (n-1)\ln\left[1 - \frac{c}{n-1}(1-u)\right]$$
$$\simeq -(n-1)\frac{c}{n-1}(1-u) = -c(1-u), \qquad (12.13)$$

where the approximate equality becomes exact in the limit of large n. Taking exponentials of both sides, we then find that

$$u = e^{-c(1-u)}. \qquad (12.14)$$

But if u is the fraction of vertices not in the giant component, then the fraction of vertices that are in the giant component is $S = 1 - u$. Eliminating u in favor of S then gives us

$$S = 1 - e^{-cS}. \qquad (12.15)$$

This equation, which was first given by Erdős and Rényi in 1959 [105], tells us the size of the giant component as a fraction of the size of the network in the limit of large network size, for any given value of the mean degree c. Unfortunately, though the equation is very simple it doesn't have a simple solution for S in closed form.[3] We can however get a good feeling for its behavior from a graphical solution. Consider Fig. 12.1. The three curves show the function $y = 1 - e^{-cS}$ for different values of c. Note that S can take only values from zero to one, so only this part of the curve is shown. The dashed line in the

---

[3]One can write a closed-form solution in terms of the *Lambert W-function*, which is defined as the solution to the equation $W(z)e^{W(z)} = z$. In terms of this function the size of the giant component is

$$S = 1 + \frac{W(-ce^{-c})}{c},$$

where we take the principal branch of the W-function. This expression may have some utility for numerical calculations and series expansions, but it is not widely used. Alternatively, although we cannot write a simple solution for S as a function of c, we can write a solution for c as a function of S. Rearranging Eq. (12.15) for c gives

$$c = -\frac{\ln(1-S)}{S},$$

which can be useful, for instance, for plotting purposes. (We can make a plot of S as a function of c by first making a plot of c as a function of S and then swapping the axes.)

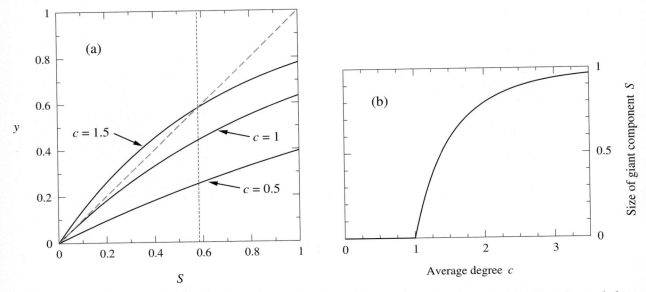

Figure 12.1: Graphical solution for the size of the giant component. (a) The three curves in the left panel show $y = 1 - e^{-cS}$ for values of c as marked, the diagonal dashed line shows $y = S$, and the intersection gives the solution to Eq. (12.15), $S = 1 - e^{-cS}$. For the bottom curve there is only one intersection, at $S = 0$, so there is no giant component, while for the top curve there is a solution at $S = 0.583\ldots$ (vertical dashed line). The middle curve is precisely at the threshold between the regime where a non-trivial solution for S exists and the regime where there is only the trivial solution $S = 0$. (b) The resulting solution for the size of the giant component as a function of c.

figure is the function $y = S$. Where line and curve cross we have $S = 1 - e^{-cS}$ and the corresponding value of S is a solution to Eq. (12.15).

As the figure shows, depending on the value of c there may be either one solution for S or two. For small c (bottom curve in the figure) there is just one solution at $S = 0$, which implies that there is no giant component in the network. (You can confirm for yourself that $S = 0$ is a solution directly from Eq. (12.15).) On the other hand, if c is large enough (top curve) then there are two solutions, one at $S = 0$ and one at $S > 0$. Only in this regime can there be a giant component.

The transition between the two regimes corresponds to the middle curve in the figure and falls at the point where the gradient of the curve and the gradient of the dashed line match at $S = 0$. That is, the transition takes place when

$$\frac{d}{dS}(1 - e^{-cS}) = 1, \qquad (12.16)$$

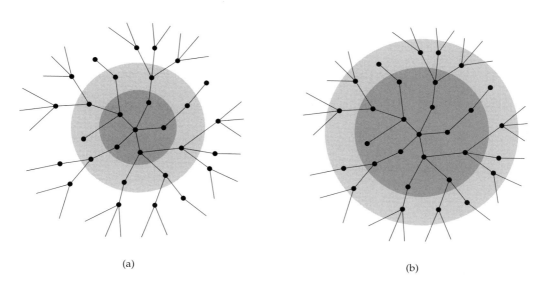

Figure 12.2: Growth of a vertex set in a random graph. (a) A set of vertices (inside the gray circles) consists of a core (dark gray) and a periphery (lighter). (b) If we grow the set by adding to it those vertices immediately adjacent to the periphery, then the periphery vertices become a part of the new core and a new periphery is added.

or

$$ce^{-cS} = 1. \tag{12.17}$$

Setting $S = 0$ we then deduce that the transition takes place at $c = 1$.

In other words, the random graph can have a giant component only if $c > 1$. At $c = 1$ and below we have $S = 0$ and there is no giant component.

This does not entirely solve the problem, however. Technically we have proved that there can be no giant component for $c \leq 1$, but not that there has to be a giant component at $c > 1$—in the latter regime there are two solutions for S, one of which is the solution $S = 0$ in which there is no giant component. So which of these solutions is the correct one that describes the true size of the giant component?

In answering this question, we will see another way to think about the formation of the giant component. Consider the following process. Let us find a small set of connected vertices somewhere in our network—say a dozen or so, as shown in Fig. 12.2a. In the limit of large $n \to \infty$ such a set is bound to exist somewhere in the network, so long as $c > 0$. We will divide the set into its *core* and its *periphery*. The core is the vertices that have connections only to other vertices in the set—the darker gray region in the figure. The *periphery* is the vertices that have at least one neighbor outside the set—the lighter gray.

Now imagine enlarging our set by adding to it all those vertices that are immediate neighbors, connected by at least one edge to the set—Fig. 12.2b. Now the old periphery is part of the core and there is a new periphery consisting of the vertices just added. How big is this new periphery? We don't know for certain, but we know that each vertex in the old periphery is connected with independent probability p to every other vertex. If there are s vertices in our set, then there are $n - s$ vertices outside the set, and the average number of connections a vertex in the periphery has to outside vertices is

$$p(n - s) = c\frac{n - s}{n - 1} \simeq c, \qquad (12.18)$$

where the equality becomes exact in the limit $n \to \infty$. This means that the average number of immediate neighbors of the set—the size of the new periphery when we grow the set—is c times the size of the old periphery.

We can repeat this argument, growing the set again and again, and each time the average size of the periphery will increase by another factor of c. Thus if $c > 1$ the average size of the periphery will grow exponentially. On the other hand, if $c < 1$ it will shrink exponentially and eventually dwindle to zero. Furthermore, if it grows exponentially our connected set of vertices will eventually form a component comparable in size to the whole network—a giant component—while if it dwindles the set will only ever have finite size and no giant component will form.

So we see that indeed we expect a giant component if (and only if) $c > 1$. And when there is a giant component the size of that giant component will be given by the larger solution to Eq. (12.15). This now allows us to calculate the size of the giant component for all values of c. (For $c > 1$ we have to solve for the larger solution of Eq. (12.15) numerically, since there is no exact solution, but this is easy enough to do.) The results are shown in Fig. 12.1. As the figure shows, the size of the giant component grows rapidly from zero as the value of c passes 1, and tends towards $S = 1$ as c becomes large.

12.6 SMALL COMPONENTS

In this section we look at the properties of random graphs from a different point of view, the point of view of the non-giant components. We have seen that in a random graph with $c > 1$ there exists a giant component that fills an extensive fraction of the network. That fraction is typically less than 100%, however. What is the structure of the remainder of the network? The answer is that it is made up of many small components whose average size is constant and doesn't increase with the size of the network.

The first step in demonstrating this result and shedding light on the structure of the small components is to show that there is only one giant component in a random graph, and hence that all other components are "non-giant" components. This is fairly easy to establish. Suppose that there were two or more giant components in a random graph. Take any two giant components, which have size $S_1 n$ and $S_2 n$, where S_1 and S_2 are the fractions of the network filled by each. The number of distinct pairs of vertices (i, j), where i is in the first giant component and j is in the second, is just $S_1 n \times S_2 n = S_1 S_2 n^2$. Each of these pairs is connected by an edge with probability p, or not with probability $1 - p$. For the two giant components to be separate components we require that there be zero edges connecting them together, which happens with probability q given by

$$q = (1 - p)^{S_1 S_2 n^2} = \left(1 - \frac{c}{n - 1}\right)^{S_1 S_2 n^2}, \tag{12.19}$$

where we have made use of Eq. (12.6).

Taking logs of both sides and going to the limit $n \to \infty$, we then find

$$\ln q = S_1 S_2 \lim_{n \to \infty} \left[n^2 \ln\left(1 - \frac{c}{n - 1}\right)\right] = S_1 S_2 \left[-c(n + 1) + \tfrac{1}{2} c^2\right]$$
$$= c S_1 S_2 \left[-n + \left(\tfrac{1}{2} c - 1\right)\right], \tag{12.20}$$

where we have dropped terms of order $1/n$. Taking the exponential again, we get

$$q = q_0 \, e^{-c S_1 S_2 n}, \tag{12.21}$$

where $q_0 = e^{c(c/2 - 1) S_1 S_2}$, which is independent of n if c is constant. Thus, for constant c, the probability that the two giant components are really separate components dwindles exponentially with increasing n, and in the limit of large n will vanish altogether. In a large random graph, therefore, there is only the very tiniest of probabilities that we will have two giant components, and for infinite n the probability is formally zero and it will never happen.

Given then that there is only one giant component in our random graph and that in most situations it does not fill the entire network, it follows that there must also be some non-giant components, i.e., components whose size does not increase in proportion to the size of the network. These are the *small components*.

12.6.1 SIZES OF THE SMALL COMPONENTS

The small components can, in general, come in various different sizes. We can calculate the distribution of these sizes as follows.

The basic quantity we focus on is the probability π_s that a randomly chosen vertex belongs to a small component of size exactly s vertices total. Note that if there is a giant component in our network then some vertices do not belong to a small component of any size and hence π_s is not normalized to unity. The sum of π_s over all sizes s is equal to the fraction of vertices that are not in the giant component. That is,

$$\sum_{s=0}^{\infty} \pi_s = 1 - S, \tag{12.22}$$

where S is, as before, the fraction of vertices in the giant component.

The crucial insight that allows us to calculate π_s is that the small components are trees, as we can see by the following argument. Consider a small component of s vertices that takes the form of a tree. A tree of s vertices contains $s - 1$ edges, as shown in Section 6.7, and this is the smallest number of edges that is needed to connect this many vertices together. If we add another edge to our component then we will create a loop, since we will be adding a new path between two vertices that are already connected (see figure). In a Poisson random graph the probability of such edge being present is the same as for any other edge, $p = c/(n-1)$. The total number of places where we could add such an extra edge to the component is given by the number of distinct pairs of vertices minus the number that are already connected by an edge, or

$$\binom{s}{2} - (s-1) = \tfrac{1}{2}(s-1)(s-2), \tag{12.23}$$

Recall that a tree is a graph or subgraph that has no loops—see Section 6.7.

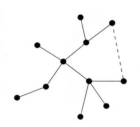

If we add an edge (dashed) to a tree we create a loop.

and the total number of extra edges in the component is $\tfrac{1}{2}(s-1)(s-2)c/(n-1)$. Assuming that s increases more slowly than $\sqrt{n}$ (and we will shortly see that it does), this probability tends to zero in the limit $n \to \infty$, and hence there are no loops in the component and the component is a tree.

We can use this observation to calculate the probability π_s as follows. Consider a vertex i in a small component of a random graph, as depicted in Fig. 12.3. Each of i's edges leads to a separate subgraph—the shaded regions in the figure—and because the whole component is a tree we know that these subgraphs are not connected to one another, other than via vertex i, since if they were there would be a loop in the component and it would not be a tree. Thus the size of the component to which i belongs is the sum of the sizes of the subgraphs reachable along each of its edges, plus 1 for vertex i itself. To put

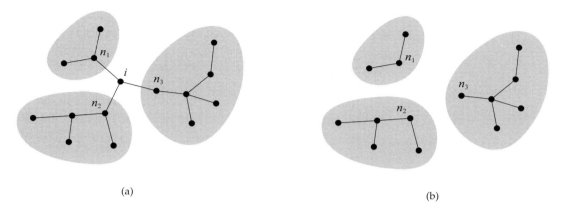

Figure 12.3: The size of one of the small components in a random graph. (a) The size of the component to which a vertex i belongs is the sum of the number of vertices in each of the subcomponents (shaded regions) reachable via i's neighbors n_1, n_2, n_3, plus one for i itself. (b) If vertex i is removed the subcomponents become components in their own right.

that another way, vertex i belongs to a component of size s if the sizes of the subgraphs to which its neighbors $n_1, n_2, \ldots$ belong sum to $s - 1$.

Bearing this in mind, consider now a slightly modified network, the network in which vertex i is completely removed, along with all its edges.[4] This network is still a random graph with the same value of p—each possible edge is still present with independent probability p—but the number of vertices has decreased by one, from n to $n - 1$. In the limit of large n, however, this decrease is negligible. The average properties, such as size of the giant component and size of the small components will be indistinguishable for random graphs with sizes n and $n - 1$, but the same p.

In this modified network, what were previously the subgraphs of our small component are now separate small components in their own right. And since the network has the same average properties as the original network for large n, that means that the probability that neighbor n_1 belongs to a small component of size s_1 (or a subgraph of size s_1 in the original network) is itself given by π_{s_1}. We can use this observation to develop a self-consistent expression for

---

[4]In the statistical physics literature, this trick of removing a vertex is called a *cavity method*. Cavity methods are used widely in the solution of all kinds of physics problems and are a powerful method for many calculations on lattices and in low-dimensional spaces as well as on networks [218].

the probability π_s.

Suppose that vertex i has degree k. As we have said, the probability that neighbor n_1 belongs to a small component of size s_1 when i is removed from the network is π_{s_1}. So the probability $P(s|k)$ that vertex i belongs to a small component of size s, given that its degree is k, is the probability that its k neighbors belong to small components of sizes $s_1, \ldots, s_k$—which is $\prod_{j=1}^{k} \pi_{s_j}$—and that those sizes add up to $s - 1$:

$$P(s|k) = \sum_{s_1=1}^{\infty} \cdots \sum_{s_k=1}^{\infty} \left[\prod_{j=1}^{k} \pi_{s_j} \right] \delta\big(s - 1, \textstyle\sum_j s_j\big), \tag{12.24}$$

where $\delta(m, n)$ is the Kronecker delta.

To get π_s, we now just average $P(s|k)$ over the distribution p_k of the degree thus:

$$\pi_s = \sum_{k=0}^{\infty} p_k P(s|k) = \sum_{k=0}^{\infty} p_k \sum_{s_1=1}^{\infty} \cdots \sum_{s_k=1}^{\infty} \left[\prod_{j=1}^{k} \pi_{s_j} \right] \delta\big(s - 1, \textstyle\sum_j s_j\big)$$

$$= e^{-c} \sum_{k=0}^{\infty} \frac{c^k}{k!} \sum_{s_1=1}^{\infty} \cdots \sum_{s_k=1}^{\infty} \left[\prod_{j=1}^{k} \pi_{s_j} \right] \delta\big(s - 1, \textstyle\sum_j s_j\big), \tag{12.25}$$

where we have made use of Eq. (12.10) for the degree distribution of the random graph.

This expression would be easy to evaluate if it were not for the delta function: one could separate the terms in the product, distribute them among the individual summations, and complete the sums in closed form. With the delta function, however, it is difficult to see how the sum can be completed.

Luckily there is a trick for problems like these, a trick that we will use many times in the rest of this book. We introduce a *generating function* or *z-transform*, defined by

$$h(z) = \pi_1 z + \pi_2 z^2 + \pi_3 z^3 + \ldots = \sum_{s=1}^{\infty} \pi_s z^s. \tag{12.26}$$

This generating function is a polynomial or series in z whose coefficients are the probabilities π_s. It encapsulates all of the information about the probability distribution in a single function. Given $h(z)$ we can recover the probabilities by differentiating:

$$\pi_s = \frac{1}{s!} \frac{d^s h}{dz^s} \bigg|_{z=0}. \tag{12.27}$$

Thus $h(z)$ is a complete representation of our probability distribution and if we can calculate it, then we can calculate π_s. We will look at generating functions in more detail in the next section, but for now let us complete the present calculation.

We can calculate $h(z)$ by substituting Eq. (12.25) into Eq. (12.26), which gives

$$
\begin{aligned}
h(z) &= \sum_{s=1}^{\infty} z^s e^{-c} \sum_{k=0}^{\infty} \frac{c^k}{k!} \sum_{s_1=1}^{\infty} \cdots \sum_{s_k=1}^{\infty} \left[\prod_{j=1}^{k} \pi_{s_j} \right] \delta\left(s-1, \textstyle\sum_j s_j\right) \\
&= e^{-c} \sum_{k=0}^{\infty} \frac{c^k}{k!} \sum_{s_1=1}^{\infty} \cdots \sum_{s_k=1}^{\infty} \left[\prod_{j=1}^{k} \pi_{s_j} \right] z^{1+\sum_j s_j} \\
&= z e^{-c} \sum_{k=0}^{\infty} \frac{c^k}{k!} \sum_{s_1=1}^{\infty} \cdots \sum_{s_k=1}^{\infty} \left[\prod_{j=1}^{k} \pi_{s_j} z^{s_j} \right] \\
&= z e^{-c} \sum_{k=0}^{\infty} \frac{c^k}{k!} \left[\sum_{s=1}^{\infty} \pi_s z^s \right]^k = z e^{-c} \sum_{k=0}^{\infty} \frac{c^k}{k!} \left[h(z) \right]^k \\
&= z \exp\left[c\left(h(z)-1\right) \right].
\end{aligned}
\tag{12.28}
$$

Thus we have a simple, self-consistent equation for $h(z)$ that eliminates the awkward delta function of (12.25).

Unfortunately, like the somewhat similar Eq. (12.15), this equation doesn't have a known closed-form solution for $h(z)$, but that doesn't mean the expression is useless. In fact we can calculate many useful things from it without solving for $h(z)$ explicitly. For example, we can calculate the mean size of the component to which a randomly chosen vertex belongs, which is given by

$$
\langle s \rangle = \frac{\sum_s s \pi_s}{\sum_s \pi_s} = \frac{h'(1)}{1-S},
\tag{12.29}
$$

where $h'(z)$ denotes the first derivative of $h(z)$ with respect to its argument and we have made use of Eqs. (12.22) and (12.26). (The denominator in this expression is necessary because π_s is not normalized to 1.)

From Eq. (12.28) we have

$$
\begin{aligned}
h'(z) &= \exp\left[c\left(h(z)-1\right)\right] + czh'(z)\exp\left[c\left(h(z)-1\right)\right] \\
&= \frac{h(z)}{z} + ch(z)h'(z),
\end{aligned}
\tag{12.30}
$$

or, rearranging,

$$
h'(z) = \frac{h(z)}{z[1-ch(z)]},
\tag{12.31}
$$

and thus

$$
h'(1) = \frac{h(1)}{1-ch(1)}.
\tag{12.32}
$$

413

But $h(1) = \sum_s \pi_s = 1 - S$, from Eqs. (12.22) and (12.26), so that

$$h'(1) = \frac{1 - S}{1 - c + cS}. \tag{12.33}$$

And so the average size $\langle s \rangle$ of Eq. (12.29) becomes

$$\langle s \rangle = \frac{1}{1 - c + cS}. \tag{12.34}$$

When $c < 1$ and there is no giant component, this gives simply $\langle s \rangle = 1/(1 - c)$. When there is a giant component, the behavior is more complicated, because we have to solve for S first before finding the value of $\langle s \rangle$, but the calculation can still be done. We first solve Eq. (12.15) for S and then substitute into Eq. (12.34).

It's interesting to note that Eq. (12.34) diverges when $c = 1$. (At this point $S = 0$, so the denominator vanishes.) Thus, if we slowly increase the mean degree c of our network from some small initial value less than 1, the average size of the component to which a vertex belongs gets bigger and bigger and finally becomes infinite exactly at the point where the giant component appears. For $c > 1$ Eq. (12.34) measures only the sizes of the non-giant components and the equation tells us that these get smaller again above $c = 1$. Thus the general picture we have is in one in which the small components get larger up to $c = 1$, where they diverge and the giant component appears, then smaller again as the giant component grows larger. Figure 12.4 shows a plot of $\langle s \rangle$ as a function of c with the divergence clearly visible.

Although the random graph is certainly not a realistic model of most networks, this general picture of the component structure of the network turns out to be a good guide to the behavior of networks in the real world. If a network has a low density of edges then typically it consists only of small components, but if the density is becomes enough then a single large component forms, usually accompanied by many separate small ones. Moreover, the small components tend on average to be smaller if the largest component is very large. This is a good example of the way in which simple models of networks can give us a feel for how more complicated real-world systems should behave in general.

12.6.2 AVERAGE SIZE OF A SMALL COMPONENT

A further important point to notice about Eq. (12.34) is that the average size of the small components does not grow with the number of vertices n. The typical size of the small components in a random graph remains constant as

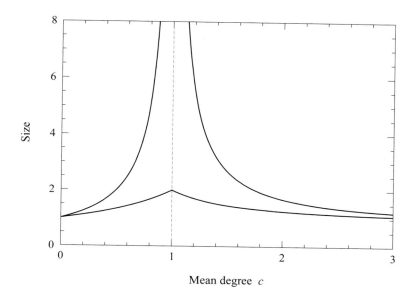

Figure 12.4: Average size of the small components in a random graph. The upper curve shows the average size $\langle s \rangle$ of the component to which a randomly chosen vertex belongs, calculated from Eq. (12.34). The lower curve shows the overall average size R of a component, calculated from Eq. (12.40). The dotted vertical line marks the point $c = 1$ at which the giant component appears. Note that, as discussed in the text, the upper curve diverges at this point but the lower one does not.

the graph gets larger. We must, however, be a little careful with these statements. Recall that π_s is the probability that a randomly chosen vertex belongs to a component of size s, and hence $\langle s \rangle$ as calculated here is not strictly the average size of a component, but the average size of the component to which a randomly chosen vertex belongs. Because larger components have more vertices in them, the chances of landing on them when we choose a random vertex is larger, in proportion to their size, and hence $\langle s \rangle$ is a biased estimate of the actual average component size. To get a correct figure for the average size of a component we need to make a slightly different calculation.

Let n_s be the actual number of components of size s in our random graph. Then the number of vertices that belong to components of size s is sn_s and hence the probability of a randomly chosen vertex belonging to such a component is

$$\pi_s = \frac{sn_s}{n}. \tag{12.35}$$

The average size of a component, which we will denote R, is

$$R = \frac{\sum_s s n_s}{\sum_s n_s} = \frac{n \sum_s \pi_s}{n \sum_s \pi_s/s} = \frac{1-S}{\sum_s \pi_s/s}, \tag{12.36}$$

where we have made use of Eq. (12.22). The remaining sum we can again evaluate using our generating function by noting that

$$\int_0^1 \frac{h(z)}{z}\, dz = \sum_{s=1}^{\infty} \pi_s \int_0^1 z^{s-1}\, dz = \sum_{s=1}^{\infty} \frac{\pi_s}{s}. \tag{12.37}$$

A useful expression for $h(z)/z$ can be obtained by rearranging Eq. (12.31) to yield

$$\frac{h(z)}{z} = \left[1 - ch(z)\right] \frac{dh}{dz}, \tag{12.38}$$

and hence we find that

$$\sum_{s=1}^{\infty} \frac{\pi_s}{s} = \int_0^1 \left[1 - ch(z)\right] \frac{dh}{dz} dz = \int_0^{1-S} (1 - ch)\, dh$$

$$= 1 - S - \tfrac{1}{2}c(1 - S)^2, \tag{12.39}$$

where we have used $h(1) = \sum_s \pi_s = 1 - S$ for the upper integration limit.

Substituting this result into Eq. (12.36), we find that the average component size is

$$R = \frac{2}{2 - c + cS}. \tag{12.40}$$

As with Eq. (12.34), this expression is independent of n, so the average size of a small component indeed does not grow as the graph becomes large.

On the other hand, R does not diverge at $c = 1$ as $\langle s \rangle$ does. At $c = 1$, with $S = 0$, Eq. (12.40) gives just $R = 2$. The reason for this is that, while the largest component in the network for $c = 1$ does become infinite in the limit of large n, so also does the total number of components. So the average size of a component is the ratio of two diverging quantities. Depending on the nature of the divergences, such a ratio could be infinite itself, or zero, or finite but non-zero in the special case where the two divergences have the same asymptotic form. In this instance the latter situation holds—both quantities are diverging linearly with n—and the average component size remains finite. A plot of R is included in Fig. 12.4 for comparison with $\langle s \rangle$.

12.6.3 THE COMPLETE DISTRIBUTION OF COMPONENT SIZES

So far we have calculated the average size of a small component in the random graph, but not the individual probabilities π_s that specify the complete distribution of sizes. In principle, we should be able to calculate the π_s by solving

Eq. (12.28) for the generating function $h(z)$ and then differentiating according to Eq. (12.27) to get π_s. Unfortunately we cannot follow this formula in practice because, as mentioned above, Eq. (12.28) does not have a known solution.

Remarkably, however, it turns out that we can still calculate the values of the individual π_s, by an alternative route. The calculations involve some more advanced mathematical techniques and if you are not particularly interested in the details it will do no harm to skip this section. If you're interested in this rather elegant development, however, read on.

To calculate an explicit expression for the probabilities π_s of the component sizes we make use of a beautiful result from the theory of complex variables, the *Lagrange inversion formula*. The Lagrange inversion formula is a formula that allows the explicit solution of equations of the form

$$f(z) = z\phi(f(z)) \tag{12.41}$$

for the unknown function $f(z)$, where $\phi(f)$ is a known function which at $f = 0$ is finite, non-zero, and differentiable.

Equation (12.41) has precisely the form of the equation for our generating function, Eq. (12.28). What's more, the Lagrange formula gives a solution for $f(z)$ in terms of the coefficients of the series expansion of $f(z)$ in powers of z, which is precisely what we want in the present case, since the coefficients are the probabilities π_s, which is what we want to calculate. The Lagrange formula is thus perfectly suited to the problem in hand. Here we first derive the general form of the formula then apply it to the current problem.[5]

Let us write the function $f(z)$ in Eq. (12.41) as a series expansion thus:

$$f(z) = \sum_{s=1}^{\infty} a_s z^s, \tag{12.42}$$

The coefficient a_s in this expansion is given explicitly by

$$a_s = \frac{1}{s!} \frac{d^s f}{dz^s}\bigg|_{z=0} = \frac{1}{s!} \left[\frac{d^{s-1}}{dz^{s-1}} \left(\frac{df}{dz} \right) \right]_{z=0}. \tag{12.43}$$

Cauchy's formula for the nth derivative of a function $g(z)$ at $z = z_0$ says that

$$\frac{d^n g}{dz^n}\bigg|_{z=z_0} = \frac{n!}{2\pi i} \oint \frac{g(z)}{(z - z_0)^{n+1}} \, dz, \tag{12.44}$$

---

[5]The formula derived here is not the *most* general form of the Lagrange inversion formula. It is adequate for the particular problem we are interested in solving, but the full Lagrange inversion formula is even more powerful, and can solve a broader range of problems. For details, see Wilf [329].

where the integral is around a contour that encloses z_0 in the complex plane but encloses no poles in $g(z)$. We will use an infinitesimal circle around z_0 as our contour.

Applying Cauchy's formula to (12.43) with $g(z) = f'(z)$, $z_0 = 0$, and $n = s - 1$, we get

$$a_s = \frac{1}{2\pi i s} \oint \frac{1}{z^s} \frac{df}{dz} \, dz = \frac{1}{2\pi i s} \oint \frac{df}{z^s}, \tag{12.45}$$

where the second integral is now around a contour in f rather than z. In this equation we are now thinking of z as being a function of f, $z = z(f)$, rather than the other way around. We are perfectly entitled to do this—knowing either quantity specifies the value of the other.[6]

It will be important later that the contour followed by f surrounds the origin, so let us pause for a moment to demonstrate that it does. Our choice of contour for z in the first integral of Eq. (12.45) is an infinitesimal circle around the origin. Expanding Eq. (12.41) to leading order around the origin, we find that

$$f(z) = z\phi(f(0)) + O(z^2) = z\phi(0) + O(z^2), \tag{12.46}$$

where we have made use of the fact that $f(0) = 0$, which is easily seen from Eq. (12.41) given that $\phi(f)$ is non-zero and finite at $f = 0$ by hypothesis. In the limit of small $|z|$ where the terms of order z^2 can be neglected, Eq. (12.46) implies that f traces a contour about the origin if z does, since the two are proportional to one another.

We now rearrange our original equation, Eq. (12.41), to give the value of z in terms of f thus

$$z(f) = \frac{f}{\phi(f)}, \tag{12.47}$$

and then substitute into Eq. (12.45) to get

$$a_s = \frac{1}{2\pi i s} \oint \frac{[\phi(f)]^s}{f^s} \, df. \tag{12.48}$$

Since, as we have said, the contour encloses the origin, this expression can be written in terms of a derivative evaluated at the origin by again making use of Cauchy's formula, Eq. (12.44):

$$a_s = \frac{1}{s!} \left[\frac{d^{s-1}}{df^{s-1}} [\phi(f)]^s \right]_{f=0}. \tag{12.49}$$

---

[6]The situation gets complicated if $z(f)$ is many-valued for some f, i.e., if $f(z)$ is non-monotonic. In our case, however, where the coefficients in the expansion of $f(z)$ are necessarily all non-negative because they are probabilities, $f(z)$ is monotonically increasing and no such problems arise.

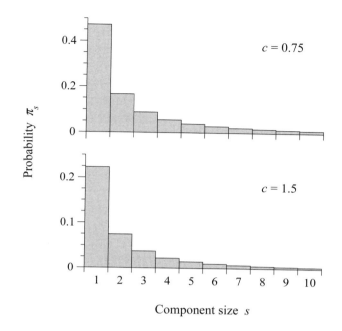

Figure 12.5: Sizes of small components in the random graph. This plot shows the probability π_s that a randomly chosen vertex belongs to a small component of size s in a Poisson random graph with $c = 0.75$ (top), which is in the regime where there is no giant component, and $c = 1.5$ (bottom), where there is a giant component.

This is the Lagrange inversion formula. This remarkably simple formula gives us, in effect, a complete series solution to Eq. (12.41).

To apply the formula to the current problem, of the component size distribution for the random graph, we set $f(z) \rightarrow h(z)$ and $\phi(f) \rightarrow e^{c(h-1)}$. Then the coefficients π_s of $h(z)$ are given by

$$
\pi_s = \frac{1}{s!} \left[\frac{d^{s-1}}{dh^{s-1}} e^{sc(h-1)} \right]_{h=0} = \frac{e^{-sc}(sc)^{s-1}}{s!}. \tag{12.50}
$$

These are the probabilities that a randomly chosen vertex belongs to a small component of size s in a random graph with mean degree c. Figure 12.5 shows the shape of π_s as a function of s for two different values of c. As the plot shows, the distribution is heavily skewed, with many components of small size and only a few larger ones.

12.7 PATH LENGTHS

In Sections 3.6 and 8.2 we discussed the small-world effect, the observation that the typical lengths of paths between vertices in networks tend to be short. Most people find the small-world effect surprising upon first learning about it.

See Section 6.10.1 for a discuss of geodesic distances and diameters.

We can use the random graph model to shed light on how the effect arises by examining the behavior of the network diameter in the model.

Recall that the diameter of a network is the longest geodesic distance between any two vertices in the same component of the network. As we now show, the diameter of a random graph varies with the number n of vertices as $\ln n$. Since $\ln n$ is typically a relatively small number even when n is large, this offers some explanation of the small-world effect, although it also leaves some questions open, as discussed further below.

The basic idea behind the estimation of the diameter of a random graph is simple. As discussed in Section 12.5, the average number of vertices s steps away from a randomly chosen vertex in a random graph is c^s. Since this number grows exponentially with s it doesn't take very many such steps before the number of vertices reached is equal to the total number of vertices in the whole network; this happens when $c^s \simeq n$ or equivalently $s \simeq \ln n / \ln c$. At this point, roughly speaking, every vertex is within s steps of our starting point, implying that the diameter of the network is approximately $\ln n / \ln c$.

Although the random graph is, as we have said, not an accurate model of most real-world networks, this is, nonetheless, believed to be the basic mechanism behind the small-world effect in most networks: the number of vertices within distance s of a particular starting point grows exponentially s and hence the diameter is logarithmic in n. We discuss the comparison with real-world networks in more detail below.

The argument above is only approximate. It's true that there are on average c^s vertices s steps away from any starting point so long as s is small. But once c^s becomes comparable with n the result has to break down since clearly the number of vertices at distance s cannot exceed the number of vertices in the whole graph. (Indeed it cannot exceed the number in the giant component.)

One way to deal with this problem is to consider two different starting vertices i and j. The average numbers of vertices s and t steps from them respectively will then be equal to c^s and c^t so long as we stay in the regime where both these numbers are much less than n. In the following calculation we consider only configurations in which both remain smaller than order n in the limit $n \rightarrow \infty$ so as to satisfy this condition.

The situation we consider is depicted in Fig. 12.6, with the two vertices i and j each surrounded by a "ball" or neighborhood consisting of all vertices with distances up to and including s and t respectively. If there is an edge between the "surface" (i.e., most distant vertices) of one neighborhood and the surface of the other, as depicted by the dashed line, then it is straightforward to show that there is also an edge between the surfaces of any pair of neighborhoods with larger s or t (or both). Turning that statement around, if there

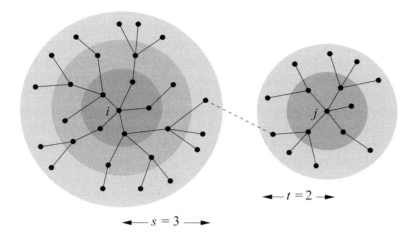

Figure 12.6: Neighborhoods of two vertices in a random graph. In the argument given in the text we consider the sets of vertices within distances s and t respectively of two randomly chosen vertices i and j. If there is an edge between any vertex on the surface of one neighborhood and any vertex on the surface of the other (dashed line), then there is a path between i and j of length $s + t + 1$.

is no edge between the surfaces of our neighborhoods, then there is also no edge between any smaller neighborhoods, which means that the shortest path between i and j must have length greater than $s + t + 1$. The reverse is also trivially true, that a shortest path longer than $s + t + 1$ implies there is no edge between our surfaces. Thus the absence of an edge between the surfaces is a necessary and sufficient condition for the distance d_{ij} between i and j to be greater than $s + t + 1$. This in turn implies that the probability $P(d_{ij} > s + t + 1)$ is equal to the probability that there is no edge between the two surfaces.

There are on average $c^s \times c^t$ pairs of vertices such that one lies on each surface, and each pair is connected with probability $p = c/(n - 1) \simeq c/n$ (assuming n to be large) or not with probability $1 - p$. Hence $P(d_{ij} > s + t + 1) = (1 - p)^{c^{s+t}}$. Defining for convenience $\ell = s + t + 1$, we can also write this as

$$P(d_{ij} > \ell) = (1 - p)^{c^{\ell-1}} = \left(1 - \frac{c}{n}\right)^{c^{\ell-1}}. \tag{12.51}$$

Taking logs of both sides, we find

$$\ln P(d_{ij} > \ell) = c^{\ell-1} \ln\left(1 - \frac{c}{n}\right) \simeq -\frac{c^\ell}{n}, \tag{12.52}$$

where the approximate inequality becomes exact as $n \to \infty$. Thus in this limit

$$P(d_{ij} > \ell) = \exp\left(-\frac{c^\ell}{n}\right).$$ (12.53)

The diameter of the network is the smallest value of ℓ such that $P(d_{ij} > \ell)$ is zero, i.e., the value such that no matter which pair of vertices we happen to pick there is zero chance that they will be separated by a greater distance. In the limit of large n, Eq. (12.53) will tend to zero only if c^ℓ grows faster than n, meaning that our smallest value of ℓ is the value such that $c^\ell = an^{1+\epsilon}$ with a constant and $\epsilon \to 0$ from above. Note that we can, as promised, achieve this while keeping both c^s and c^t smaller than order n, so that our argument remains valid.

Rearranging for ℓ, we now find our expression for the diameter:

$$\ell = \frac{\ln a}{\ln c} + \lim_{\epsilon \to 0} \frac{(1+\epsilon)\ln n}{\ln c} = A + \frac{\ln n}{\ln c},$$ (12.54)

where A is a constant.[7] Apart from the constant, this is the same result as we found previously using a rougher argument. The constant is known—it has a rather complicated value in terms of the Lambert W-function [114]—but for our purposes the important point is that it is (asymptotically) independent of n. Thus the diameter indeed increases only slowly with n, as $\ln n$, making it relatively small in large random graphs.

The logarithmic dependence of the diameter on n offers some explanation of the small-world effect of Section 3.6. Even in a network such as the acquaintance network of the entire world, with nearly seven billion inhabitants (at the time of writing), the value of $\ln n / \ln c$ can be quite small. Supposing each person to have about a thousand acquaintances,[8] we would get

$$\ell = \frac{\ln n}{\ln c} = \frac{\ln 6 \times 10^9}{\ln 1000} = 3.3\ldots,$$ (12.55)

which is easily small enough to account for the results of, for example, the small-world experiments of Milgram and others [93,219,311].

---

[7]There are still some holes in our argument. In particular, we have assumed that the product of the numbers of vertices on the surface of our two neighborhoods is c^{s+t} when in practice this is only the average value and there will in general be some variation. Also the calculation should really be confined to the giant component, since the longest path always falls in the giant component in the limit of large n. For a careful treatment of these issues see, for instance, Fernholz and Ramachandran [114].

[8]This appears to be a reasonable figure. Bernard et $al.$ [36] estimated the typical number of acquaintances for people in the United States to be about 2000—see Section 3.2.1.

On the other hand, although this calculation gives us some insight into the nature of the small-world effect, this cannot be the entire explanation. There are clearly many things wrong with the random graph as a model of real social networks, as we now discuss.

12.8 PROBLEMS WITH THE RANDOM GRAPH

The Poisson random graph is one of the best studied models of networks. In the half century since its first proposal it has given us a tremendous amount of insight into the expected structure of networks of all kinds, particularly with respect to component sizes and network diameters. The fact that it is both simple to describe and straightforward to study using analytic methods makes it an excellent tool for investigating all sorts of network phenomena. We will return to the random graph many times in the remainder of this book to help us understand the way networks behave.

The random graph does, however, have some severe shortcomings as a network model. There are many ways in which it is completely unlike the real-world networks we have seen in the previous chapters. One clear problem is that it shows essentially no transitivity or clustering. In Section 12.4 we saw at the clustering coefficient of a random graph is $C = c/(n - 1)$, which tends to zero in the limit of large n. And even for the finite values of n appropriate to real-world networks the value of C in the random graph is typically very small. For the acquaintance network of the human population of the world, with its $n \simeq 7$ billion people, each having about 1000 acquaintances [175], a random graph with the same n and c would have a clustering coefficient of

$$C \simeq \frac{1000}{7\,000\,000\,000} \simeq 10^{-7}. \qquad (12.56)$$

Whether the clustering coefficient of the real acquaintance network is 0.01 or 0.5 hardly matters. (It is probably somewhere in between.) Either way it is clear that the random graph and the true network are in strong disagreement.[9]

The random graph also differs from real-world networks in many other ways. For instance, there is no correlation between the degrees of adjacent vertices—necessarily so, since the edges are placed completely at random. The degrees in real networks, by contrast, are usually correlated, as discussed in Section 8.7. Many, perhaps most, real-world networks also show grouping of

---

[9]This disagreement, highlighted particularly by Watts and Strogatz [323], was one of the observations that prompted the current wave of interest in the properties of networks in the mathematical sciences, starting in the late 1990s.

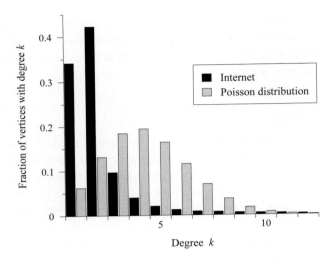

Figure 12.7: Degree distribution of the Internet and a Poisson random graph. The dark bars in this plot show the fraction of vertices with the given degrees in the network representation of the Internet at the level of autonomous systems. The lighter bars represent the same measure for a random graph with the same average degree as the Internet. Even though the two distributions have the same averages, it is clear that they are entirely different in shape.

their vertices into "communities," as discussed on Section 11.2.1, but random graphs have no such structure. And there are many other examples of interesting structure in real networks that is absent from the random graph.

However, perhaps the most significant respect in which the properties of random graphs diverge from those of real-world networks is the shape of their degree distribution. As discussed in Section 8.3, real networks typically have right-skewed degree distributions, with most vertices having low degree but with a small number of high-degree "hubs" in the tail of the distribution. The random graph on the other hand has a Poisson degree distribution, Eq. (12.10), which is not right-skewed to any significant extent. Consider Fig. 12.7, for example, which shows a histogram of the degree distribution of the Internet (darker bars), measured at the level of autonomous systems (Section 2.1.1). The right-skewed form is clearly visible in this example. On the same figure we show the Poisson degree distribution of a random graph (lighter bars) with the same average degree c as the Internet example. Despite having the same averages, the two distributions are clearly entirely different. It turns out that this difference has a profound effect on all sorts of properties of the network—we will see many examples in this book. This makes the Poisson random graph inadequate to explain many of the interesting phenomena we see in networks today, including resilience phenomena, epidemic spreading processes, percolation, and many others.

Luckily it turns out to be possible to generalize the random graph model to allow for non-Poisson degree distributions. This development, which leads to some of the most beautiful results in the mathematics of networks, is described in the next chapter.

_____

PROBLEMS

12.1 Consider the random graph $G(n, p)$ with mean degree c.

a) Show that in the limit of large n the expected number of triangles in the network is $\frac{1}{6}c^3$. This means that the number of triangles is constant, neither growing nor vanishing in the limit of large n.

b) Show that the expected number of connected triples in the network (as defined on page 200) is $\frac{1}{2}nc^2$.

c) Hence calculate the clustering coefficient C, as defined in Eq. (7.41), and confirm that it agrees for large n with the value given in Eq. (12.11).

12.2 Consider the random graph $G(n, p)$ with mean degree c.

a) Argue that the probability that a vertex of degree k belongs to a small component is $(1 - S)^k$, where S is the fraction of the network occupied by the giant component.

b) Thus, using Bayes' theorem (or otherwise) show that the fraction of vertices in small components that have degree k is $e^{-c}c^k(1 - S)^{k-1}/k!$.

12.3 Starting from the generating function $h(z)$ defined in Eq. (12.26), or otherwise, show that

a) the mean-square size of the component in a random graph to which a randomly chosen vertex belongs is $1/(1 - c)^3$ in the regime where there is no giant component;

b) the mean-square size of a randomly chosen component in the same regime is $1/[(1 - c)(1 - \frac{1}{2}c)]$.

Note that both quantities diverge at the phase transition where the giant component appears.

12.4 In Section 7.8.2 we introduced the idea of a bicomponent. A vertex in a random graph belongs to a bicomponent if two or more of its neighbors belong to the giant component of the network (since the giant component completes a loop between those neighbors forming a bicomponent). In principle, a vertex can also be in a bicomponent if two or more of its neighbors belong to the same small component, but in practice this never happens, since that would imply that the small component in question contained a loop and, as we have seen, the small components in a random graph are trees and so have no loops.

 a) Show that the fraction of vertices in a random graph that belong to a bicomponent is $S_2 = (1 - cu)(1 - u)$, where u is defined by Eq. (12.14).

 b) Show that this expression can be rewritten as $S_2 = S + (1 - S)\ln(1 - S)$, where S is the size of the giant component.

 c) Hence argue that the random graph contains a giant bicomponent whenever it contains an ordinary giant component.

12.5 The *cascade model* is a simple mathematical model of a directed acyclic graph, sometimes used to model food webs. We take n vertices labeled $i = 1 \ldots n$ and place an undirected edge between each distinct pair with independent probability p, just as in the ordinary random graph. Then we add directions to the edges such that each edge runs from the vertex with numerically higher label to the vertex with lower label. This ensures that all directed paths in the network run from higher to lower labels and hence that the network is acyclic, as discussed in Section 6.4.2.

 a) Show that the average in-degree of vertex i in the ensemble of the cascade model is $\langle k_i^{in} \rangle = (n - i)p$ and the average out-degree is $\langle k_i^{out} \rangle = (i - 1)p$.

 b) Show that the expected number of edges that connect to vertices i and lower from vertices above i is $(ni - i^2)p$.

 c) Assuming n is even, what are the largest and smallest values of this quantity and where do they occur?

In a food web this expected number of edges from high- to low-numbered vertices is a rough measure of energy flow and the cascade model predicts that energy flow will be largest in the middle portions of a food web and smallest at the top and bottom.

12.6 We can make a simple random graph model of a network with clustering or transitivity as follows. We take n vertices and go through each distinct trio of three vertices, of which there are $\binom{n}{3}$, and with independent probability p we connect the members of the trio together using three edges to form a triangle, where $p = c/\binom{n-1}{2}$ with c a constant.

 a) Show that the mean degree of a vertex in this model network is $2c$.

 b) Show that the degree distribution is

$$p_k = \begin{cases} e^{-c}c^{k/2}/(k/2)! & \text{if } k \text{ is even,} \\ 0 & \text{if } k \text{ is odd.} \end{cases}$$

 c) Show that the clustering coefficient, Eq. (7.41), is $C = 1/(2c + 1)$.

d) Show that when there is a giant component in the network its expected size S as a fraction of network size satisfies $S = 1 - e^{-cS(2-S)}$.

e) What is the value of the clustering coefficient when the giant component fills half of the network?

CHAPTER 13

RANDOM GRAPHS WITH GENERAL DEGREE DISTRIBUTIONS

*This chapter describes more sophisticated random graph
models that mimic networks with arbitrary degree
distributions*

I N THE previous chapter we looked at the classic random graph model, in which pairs of vertices are connected at random with uniform probabilities. Although this model has proved tremendously useful as a source of insight into the structure of networks, it also has, as described in Section 12.8, a number of serious shortcomings. Chief among these is its degree distribution, which follows the Poisson distribution and is quite different from the degree distributions seen in most real-world networks. In this chapter we show how we can create more sophisticated random graph models, which incorporate arbitrary degree distributions and yet are still exactly solvable for many of their properties in the limit of large network size.

The fundamental mathematical tool that we will use to derive the results of this chapter is the probability generating function. We have already seen in Section 12.6 one example of a generating function, which was useful in the calculation of the distribution of component sizes in the Poisson random graph. We begin this chapter with a more formal introduction to generating functions and to some of their properties which will be useful in later calculations. Readers interested in pursuing the mathematics of generating functions further may like to look at the book by Wilf [329].[1]

---

[1]Professor Wilf has generously made his book available for free in electronic form. You can download it from `www.math.upenn.edu/~wilf/DownldGF.html`.

13.1 GENERATING FUNCTIONS

Suppose we have a probability distribution for a non-negative integer variable, such that separate instances, occurrences, or draws of this variable are independent and have value k with probability p_k. A good example of such a distribution is the distribution of the degrees of randomly chosen vertices in a network. If the fraction of vertices in a network with degree k is p_k then p_k is also the probability that a randomly chosen vertex from the network will have degree k.

The *generating function* for the probability distribution p_k is the polynomial

$$g(z) = p_0 + p_1 z + p_2 z^2 + p_3 z^3 + \ldots = \sum_{k=0}^{\infty} p_k z^k. \tag{13.1}$$

Sometimes a function of this kind is called a *probability generating function* to distinguish it from another common type of function, the *exponential generating function*. We will not use exponential generating functions in this book, so for us all generating functions will be probability generating functions.

If we know the generating function for a probability distribution p_k then we can recover the values of p_k by differentiating:

$$p_k = \frac{1}{k!} \left. \frac{d^k g}{dz^k} \right|_{z=0}. \tag{13.2}$$

Thus the generating function gives us complete information about the probability distribution and vice versa. The distribution and the generating function are really just two different representations of the same thing. As we will see, it is easier in many cases to work with the generating function than with the probability distribution and doing so leads to many useful new results about networks.

13.1.1 EXAMPLES

Right away let us look at some examples of generating functions. Suppose our variable k takes only the values 0, 1, 2, and 3, with probabilities p_0, p_1, p_2, and p_3, respectively, and no other values. In that case the corresponding generating function would take the form of a cubic polynomial:

$$g(z) = p_0 + p_1 z + p_2 z^2 + p_3 z^3. \tag{13.3}$$

For instance, if we had a network in which vertices of degree 0, 1, 2, and 3 occupied 40%, 30%, 20%, and 10% of the network respectively then

$$g(z) = 0.4 + 0.3 z + 0.2 z^2 + 0.1 z^3. \tag{13.4}$$

As another example, suppose that k follows a Poisson distribution with mean c:

$$p_k = e^{-c} \frac{c^k}{k!}. \tag{13.5}$$

Then the corresponding generating function would be

$$g(z) = e^{-c} \sum_{k=0}^{\infty} \frac{(cz)^k}{k!} = e^{c(z-1)}. \tag{13.6}$$

Alternatively, suppose that k follows an exponential distribution of the form

$$p_k = C e^{-\lambda k}, \tag{13.7}$$

with $\lambda > 0$. The normalizing constant is fixed by the condition that $\sum_k p_k = 1$, which gives $C = 1 - e^{-\lambda}$ and hence

$$p_k = (1 - e^{-\lambda}) e^{-\lambda k}. \tag{13.8}$$

Then

$$g(z) = (1 - e^{-\lambda}) \sum_{k=0}^{\infty} (e^{-\lambda} z)^k = \frac{e^{\lambda} - 1}{e^{\lambda} - z}, \tag{13.9}$$

so long as $z < e^{\lambda}$. (If $z \geq e^{\lambda}$ the generating function diverges. Normally, however, we will be interested in generating functions only in the range $0 \leq z \leq 1$ so, given that $\lambda > 0$ and hence $e^{\lambda} > 1$, the divergence at e^{λ} will not be a problem.)

13.1.2 POWER-LAW DISTRIBUTIONS

One special case of particular interest in the study of networks is the power-law distribution. As we saw in Section 8.4, a number of networks, including the World Wide Web, the Internet, and citation networks, have degree distributions that follow power laws quite closely and this turns out to have interesting consequences that set these networks apart from others. To create and solve models of these networks it will be important for us to be able to write down generating functions for power-law distributions.

There are various forms that are used to represent power laws in practice but the simplest choice, which we will use in many of our calculations, is the "pure" power law

$$p_k = C k^{-\alpha}, \tag{13.10}$$

for constant $\alpha > 0$. This expression cannot apply all the way down to $k = 0$, however, or it would diverge. So commonly one stops at $k = 1$. The normalization constant C can then be calculated from the condition that $\sum_k p_k = 1$, which gives

$$C \sum_{k=1}^{\infty} k^{-\alpha} = 1. \tag{13.11}$$

The sum unfortunately cannot be performed in closed form. It is, however, a common enough sum that it has a name—it is called the *Riemann zeta function*, denoted $\zeta(\alpha)$:

$$\zeta(\alpha) = \sum_{k=1}^{\infty} k^{-\alpha}. \tag{13.12}$$

Thus we can write $C = 1/\zeta(\alpha)$ and

$$p_k = \begin{cases} 0 & \text{for } k = 0, \\ k^{-\alpha}/\zeta(\alpha) & \text{for } k \geq 1. \end{cases} \tag{13.13}$$

Although there is no closed-form expression for the zeta function, there exist good numerical methods for calculating its value accurately, and many programming languages and numerical software packages include functions to calculate it.

For this probability distribution the generating function is

$$g(z) = \frac{1}{\zeta(\alpha)} \sum_{k=1}^{\infty} k^{-\alpha} z^k. \tag{13.14}$$

Again the sum cannot be expressed in closed form, but again it has a name—it is called the *polylogarithm* of z and is denoted $\mathrm{Li}_\alpha(z)$:

$$\mathrm{Li}_\alpha(z) = \sum_{k=1}^{\infty} k^{-\alpha} z^k. \tag{13.15}$$

Thus we can write

$$g(z) = \frac{\mathrm{Li}_\alpha(z)}{\zeta(\alpha)}. \tag{13.16}$$

This is not completely satisfactory. We would certainly prefer a closed-form expression as in the case of the Poisson and exponential distributions of Eqs. (13.6) and (13.9). But we can live with it. Enough properties of the polylogarithm and zeta functions are known that we can carry out useful manipulations of the generating function. In particular, since derivatives of our generating functions will be important to us, we note the following useful relation:

$$\frac{\partial \mathrm{Li}_\alpha(z)}{\partial z} = \frac{\partial}{\partial z} \sum_{k=1}^{\infty} k^{-\alpha} z^k = \sum_{k=1}^{\infty} k^{-(\alpha-1)} z^{k-1} = \frac{\mathrm{Li}_{\alpha-1}(z)}{z}. \tag{13.17}$$

We should note also that in real-world networks the degree distribution does not usually follow a power law over its whole range—the distribution is not a "pure" power law in the sense above. Instead, it typically obeys a power law reasonably closely for values of k above some minimum value k_{min} but below that point it has some other behavior. In this case the generating function will take the form

$$g(z) = Q_{k_{min}-1}(z) + C \sum_{k=k_{min}}^{\infty} k^{-\alpha} z^k,$$ (13.18)

where $Q_n(z) = \sum_{k=0}^{n} p_k z^k$ is a polynomial in z of degree n and C is a normalizing constant. The sum in Eq. (13.18) also has its own name: it is called the *Lerch transcendent*.[2] In the calculations in this book we will stick to the pure power law, since it illustrates nicely the interesting properties of power-law degree distributions and is relatively simple to deal with, but for serious modeling one might sometimes have to use the cut-off form, Eq. (13.18).

13.1.3 NORMALIZATION AND MOMENTS

Let us now look briefly at some of the properties of generating functions that will be useful to us. First of all, note that if we set $z = 1$ in the definition of the generating function, $g(z) = \sum_k p_k z^k$ (Eq. (13.1)), we get

$$g(1) = \sum_{k=0}^{\infty} p_k.$$ (13.19)

If the probability distribution is normalized to unity, $\sum_k p_k = 1$, as are all the examples above, then this immediately implies that

$$g(1) = 1.$$ (13.20)

For most of the generating functions we will look at, this will be true, but not all. As a counter-example, consider the generating function for the sizes of the small components in the Poisson random graph defined in Eq. (12.26). The probabilities π_s appearing in this generating function were the probabilities that a randomly chosen vertex belongs to a small component of size s. If we are in the regime where there is a giant component in the network then not all vertices belong to a small component, and hence the probabilities π_s do not add up to one. In fact, their sum is equal to the fraction of vertices not in the giant component.

The derivative of the generating function $g(z)$ of Eq. (13.1) is

$$g'(z) = \sum_{k=0}^{\infty} k p_k z^{k-1}.$$ (13.21)

---

[2]No, really. I'm not making this up.

(We will use the primed notation $g'(z)$ for derivatives of generating functions extensively in this chapter, as it proves much less cumbersome than the more common notation dg/dz.)

If we set $z = 1$ in Eq. (13.21) we get

$$g'(1) = \sum_{k=0}^{\infty} k p_k = \langle k \rangle, \tag{13.22}$$

which is just the average value of k. Thus, for example, if p_k is a degree distribution, we can calculate the average degree directly from the generating function by differentiating. This is a very convenient trick. In many cases we will calculate a probability distribution of interest by calculating first its generating function. In principle, we can then extract the distribution itself by applying Eq. (13.2) and so derive any other quantities we want such as averages. But Eq. (13.22) shows us that we don't always have to do this. Some of the quantities we will be interested in can be calculated directly from the generating function without going through any intermediate steps.

In fact, this result generalizes to higher moments of the probability distribution as well. For instance, note that

$$z\frac{d}{dz}\left(z\frac{dg}{dz}\right) = \sum_{k=0}^{\infty} k^2 p_k z^k, \tag{13.23}$$

and hence, setting $z = 1$, we can write

$$\langle k^2 \rangle = \left[\left(z\frac{d}{dz}\right)^2 g(z)\right]_{z=1}. \tag{13.24}$$

It is not hard to show that this result generalizes to all higher moments as well:

$$\langle k^m \rangle = \left[\left(z\frac{d}{dz}\right)^m g(z)\right]_{z=1}. \tag{13.25}$$

This result can also be written as

$$\langle k^m \rangle = \frac{d^m g}{d(\ln z)^m}\bigg|_{z=1}. \tag{13.26}$$

13.1.4 POWERS OF GENERATING FUNCTIONS

Perhaps the most useful property of generating functions—and the one that makes them important for the study of networks—is the following. Suppose we are given a distribution p_k with generating function $g(z)$. And suppose we have m integers k_i, $i = 1 \ldots m$, which are independent random numbers drawn from this distribution. For instance, they could be the degrees of m randomly chosen vertices in a network with degree distribution p_k. Then the probability distribution of the sum $\sum_{i=1}^{m} k_i$ of those m integers has generating function $[g(z)]^m$. This is a very powerful result and it is worth taking a moment to see how it arises and what it means.

Given that our integers are independently drawn from the distribution p_k, the probability that they take a particular set of values $\{k_i\}$ is simply $\prod_i p_{k_i}$ and the probability π_s that the values drawn add up to a specific sum s is the sum of these probabilities over all sets $\{k_i\}$ that add up to s:

$$\pi_s = \sum_{k_1=0}^{\infty} \ldots \sum_{k_m=0}^{\infty} \delta\left(s, \sum_i k_i\right) \prod_{i=1}^{m} p_{k_i}, \tag{13.27}$$

where $\delta(a, b)$ is the Kronecker delta. Then the generating function $h(z)$ for the distribution π_s is

$$\begin{aligned}
h(z) &= \sum_{s=0}^{\infty} \pi_s z^s \\
&= \sum_{s=0}^{\infty} z^s \sum_{k_1=0}^{\infty} \ldots \sum_{k_m=0}^{\infty} \delta\left(s, \sum_i k_i\right) \prod_{i=1}^{m} p_{k_i} \\
&= \sum_{k_1=0}^{\infty} \ldots \sum_{k_m=0}^{\infty} z^{\sum_i k_i} \prod_{i=1}^{m} p_{k_i} \\
&= \sum_{k_1=0}^{\infty} \ldots \sum_{k_m=0}^{\infty} \prod_{i=1}^{m} p_{k_i} z^{k_i} = \left[\sum_{k=0}^{\infty} p_k z^k\right]^m \\
&= \left[g(z)\right]^m. \tag{13.28}
\end{aligned}$$

Thus, for example, if we know the degree distribution of a network, it is a straightforward matter to calculate the probability distribution of the sum of the degrees of m randomly chosen vertices from that network. This will turn out to be important in the developments that follow.

13.2 THE CONFIGURATION MODEL

Let us turn now to the main topic of this chapter, the development of the theory of random graphs with general degree distributions.

We can turn the random graph of Chapter 12 into a much more flexible model for networks by modifying it so that the degrees of its vertices are no longer restricted to having a Poisson distribution, and in fact it is possible to modify the model so as to give the network any degree distribution we please. Just as with the Poisson random graph, which can be defined in several slightly different ways, there is more than one way to define random graphs with general degree distributions. Here we describe two of them, which are roughly the equivalent of the $G(n, m)$ and $G(n, p)$ random graphs of Section 12.1.

The most widely studied of the generalized random graph models is the *configuration model*. The configuration model is actually a model of a random graph with a given degree *sequence*, rather than degree distribution. That is, the exact degree of each individual vertex in the network is fixed, rather than merely the probability distribution from which those degrees are chosen. This in turn fixes the number of edges in the network, since the number of edges is given by Eq. (6.21) to be $m = \frac{1}{2}\sum_i k_i$. Thus this model is in some ways analogous to $G(n, m)$, which also fixes the number of edges. (It is quite simple, however, to modify the model for cases where only the degree distribution is known and not the exact degree sequence. We describe how this is done at the end of this section.)

See Section 8.3 for a discussion of the distinction between degree sequences and degree distributions.

Suppose then that we specify the degree k_i that each vertex $i = 1\ldots n$ in our network is to take. We can create a random network with these degrees as follows. We give each vertex i a total of k_i "stubs" of edges as depicted in Fig. 13.1. There are $\sum_i k_i = 2m$ stubs in total, where m is the total number of edges. Then we choose two of the stubs uniformly at random and we create an edge by connecting them to one another, as indicated by the dashed line in the figure. Then we choose another pair from the remaining $2m - 2$ stubs, connect those, and so on until all the stubs are used up. The end result is a network in which every vertex has exactly the desired degree.

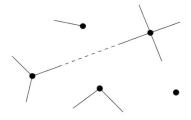

Figure 13.1: The configuration model. Each vertex is given a number of "stubs" of edges equal to its desired degree. Then pairs of stubs are chosen at random and connected together to form edges (dotted line).

More specifically the end result is a particular *matching* of the stubs, a particular set of pairings of stubs with other stubs. The process above generates each possible matching of stubs with equal probability. Technically the configuration model is defined as the ensemble in which each matching with the chosen degree sequence appears with the same probability (those with any other degree sequence having probability zero), and the process above is a process for drawing networks from the configuration model ensemble.

The uniform distribution over matchings in the configuration model has the important consequence that any stub in a configuration model network is equally likely to be connected to any other. This, as we will see, is the crucial property that makes the model solvable for many of its properties.

There are a couple of minor catches with the network generation process described here. First, there must be an even number of stubs overall if we want to end up with a network consisting only of vertices and edges, with no dangling stubs left over. This means that the sum $\sum_i k_i$ of the degrees must add up to an even number. We will assume that the degrees we have chosen satisfy this condition, otherwise it is clearly not possible to create a graph with the given degree sequence.

A second issue is that the network may contain self-edges or multiedges, or both. There is nothing in the network generation process that prevents us from creating an edge that connects a vertex to itself or that connects two vertices that are already connected by another edge. One might imagine that one could avoid this by rejecting the creation of any such edges during the process, but it turns out that this is not a good idea. A network so generated is no longer drawn uniformly from the set of possible matchings, which means that properties of the model can no longer be calculated analytically, at least by any means currently known. It can also mean that the network creation process breaks down completely. Suppose, for example, that we come to the end of the process, when there are just two stubs left to be joined, and find that those two both belong to the same vertex so that joining them would create a self-edge. Then either we create the self-edge or the network generation process fails.

In practice, therefore, it makes more sense to allow the creation of both multiedges and self-edges in our networks and the standard configuration model does so. Although some real-world networks have self-edges or multiedges in them, most do not, and to some extent this makes the configuration model less satisfactory as a network model. However, as shown below, the average number of self-edges and multiedges in the configuration model is a constant as the network becomes large, which means that the density of self-edges and multiedges tends to zero in this limit. This means, to all intents and purposes, that we can ignore the self-edges and multiedges in the large size limit.[3]

A further issue with the configuration model is that, while all matchings of stubs appear with equal probability in the model, that does not mean that all *networks* appear with equal probability because more than one matching can correspond to the same network, i.e., the same topological connections between vertices. If we label the stubs to keep track of which is which, then

---

[3]Even for finite-sized networks the difference between the properties of a configuration model network and a similar network without self-edges and multiedges would only result in a correction of order $1/n$ into our results. For the large networks that are the focus of most modern network studies this means that the error introduced by allowing self-edges and multiedges is small.

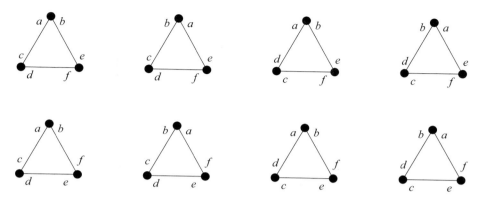

Figure 13.2: Eight stub matchings that all give the same network. This small network is composed of three vertices of degree two and hence having two stubs each. The stubs are lettered to identify them and there are two distinct permutations of the stubs at each vertex for a total of eight permutations overall. Each permutation gives rise to a different matching of stub to stub but all matchings correspond to the same topological configuration of edges, and hence there are eight ways in which this particular configuration can be generated by the stub matching process.

there are typically many different ways we can join up pairs of labeled stubs to create the same final configuration of edges. Figure 13.2 shows an example of a set of eight matchings that all correspond to the same three-vertex network.

In general, one can generate all the matchings that correspond to a given network by taking any one matching for that network and permuting the stubs at each vertex in every possible way. Since the number of permutations of the k_i stubs at a vertex i is $k_i!$, this implies that the number of matchings corresponding to each network is $N(\{k_i\}) = \prod_i k_i!$, which takes the same value for all networks, since the degrees are fixed. This implies that in fact networks occur with equal probability in the configuration model: if there are $\Omega(\{k_i\})$ matchings, each occurring with the same probability, then each *network* occurs with probability N/Ω.

However, this is not completely correct. If a network contains self-edges or multiedges then not all permutations of the stubs in the network result in a new matching of stubs. Consider Fig. 13.3. Panel (a) shows a network with the same degree sequence as those of Fig. 13.2, but a different matching of the stubs that creates a network with one self-edge and a multiedge consisting of two parallel single edges. In panel (b) we have permuted the stubs a and b at the ends of the self-edge but, as we can see, this has not resulted in a new matching of the stubs themselves. Stubs a and b are still connected to one another just as they were before. (The network is *drawn* differently now, but in terms of the matching and the topology of the edges nothing has changed

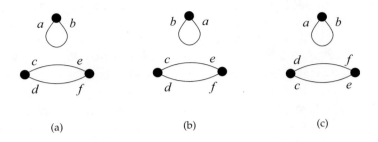

(a)   (b)   (c)

Figure 13.3: Permutations that do not produce new matchings. (a) The network shown here has the same degree sequence as those of Fig. 13.2 but a different configuration of edges, having one self-loop and a multiedge consisting of two parallel edges. (b) If we permute the stubs a and b of the self-edge we do not generate a new matching, because a is still matched with b, just as before. (c) If we permute the stubs at either end of a multiedge in exactly the same way we do not generate a new matching, since each stub at one end of the multiedge is still matched with the same stub at the other end.

from panel (a).) In panel (c) we have identically permuted the stubs at both ends of the multiedge. Again this has no effect on which stubs are matched with which others.

In general, for each multiedge in a network a permutation of the stubs at one end fails to generate a new matching if we simultaneously permute the stubs at the other end in the same way. This means that the total number of matchings is reduced by a factor of $A_{ij}!$, since A_{ij} is equal to the multiplicity of the edge between i and j. Indeed, this expression is correct even for vertex pairs not connected by a multiedge, if we adopt the convention that $0! = 1$. For self-edges there is a further factor of two because the interchange of the two ends of the edge does not generate a new matching. Combining these results, the number of matchings corresponding to a network turns out to be

$$N = \frac{\prod_i k_i!}{\prod_{i<j} A_{ij}! \prod_i A_{ii}!!},$$ (13.29)

where $n!! = n(n-2)(n-4)\ldots 2$ with n even is the so-called double factorial of n. Then the total probability of a particular network within the configuration model ensemble is N/Ω as before. Since the denominator in Eq. (13.29) depends not only on the degree sequence but also on the structure of the network itself, different networks do appear with different probabilities.

As we mentioned, however, the average densities of self-edges and multiedges in the configuration model vanish as n becomes large, so that the variation in probabilities is relatively small in the large-n limit, but it nonetheless does occasionally assume some importance and is therefore worth bearing in mind (see, for instance, Ref. [220]).

As discussed above, we are sometimes (indeed often) interested in the case where it is the degree distribution of the network that is specified rather than the degree sequence. That is, we specify the probability distribution p_k from which the degree sequence is drawn rather than the sequence itself. We can define an obvious extension of the configuration model to this case: we draw a degree sequence from the specified distribution and then generate a network with that degree sequence using the technique described above. More precisely, we define an ensemble in which each degree sequence $\{k_i\}$ appears with probability $\prod_i p_{k_i}$. Then if we can calculate an average value $X(\{k_i\})$ for some quantity of interest X in the standard configuration model, the average value in the extended model is given by

$$\langle X \rangle = \sum_{k_1=0}^{\infty} \cdots \sum_{k_n=0}^{\infty} X(\{k_i\}) \prod_{i=0}^{n} p_{k_i}. \tag{13.30}$$

In practice the difference between the two models is not actually very great. As we will see, the crucial parameter that enters into most of our configuration model calculations is the fraction of vertices that have each possible degree k. In the extended model above, this fraction is, by definition, equal to p_k in the limit of large n. If, on the other hand, the degree sequence is fixed then we simply calculate the fraction from the degree sequence and then use those numbers. In either the case the formulas for calculated quantities are the same.

13.2.1 EDGE PROBABILITY IN THE CONFIGURATION MODEL

A central property of the configuration model is the probability p_{ij} of the occurrence of an edge between two specified vertices, i and j. Obviously if either vertex i or vertex j has degree zero then the probability of an edge is zero, so let us assume that $k_i, k_j > 0$. Now consider any one of the stubs that emerges from vertex i. What is the probability that this stub is connected by an edge to any of the stubs of vertex j? There are $2m$ stubs in total, or $2m - 1$ excluding the one connected to i that we are currently looking at. Of those $2m - 1$, exactly k_j of them are attached to vertex j. So, given that any stub in the network is equally likely to be connected to any other, the probability that our particular stub is connected to any of those around vertex j is $k_j/(2m - 1)$. But there are k_i stubs around vertex i, so the total probability of a connection between i and j is

$$p_{ij} = \frac{k_i k_j}{2m - 1}. \tag{13.31}$$

Technically, since we have added the probabilities of independent events, this is really the average number of edges between i and j, rather than the probability of having an edge at all. But in the limit of large m, this number becomes small (for given k_i, k_j), and the average number of edges and the probability of an edge become equal. Also in the limit of large m we can ignore the -1 in the denominator and hence we can write

$$p_{ij} = \frac{k_i k_j}{2m}. \tag{13.32}$$

Note that, even though we assumed $k_i, k_j > 0$, this expression also gives the right result if either degree is zero, namely that in that case the probability of connection is zero.

We can use this result, for example, to calculate the probability of having two edges between the same pair of vertices. The probability of having one edge between vertices i and j is p_{ij} as above. Once we have one edge between the vertices the number of available stubs at each is reduced by one, and hence the probability of having a second edge is given by Eq. (13.32) but with k_i and k_j each reduced by one: $(k_i - 1)(k_j - 1)/2m$. Thus the probability of having (at least) two edges, i.e., of having a multiedge between i and j, is $k_i k_j (k_i - 1)(k_j - 1)/(2m)^2$ and, summing this probability over all vertices and dividing by two (to avoid double counting of vertex pairs), we find that the expected total number of multiedges in the network is

$$\frac{1}{2(2m)^2} \sum_{ij} k_i k_j (k_i - 1)(k_j - 1) = \frac{1}{2\langle k \rangle^2 n^2} \sum_i k_i(k_i - 1) \sum_j k_j(k_j - 1)$$

$$= \frac{1}{2} \left[\frac{\langle k^2 \rangle - \langle k \rangle}{\langle k \rangle} \right]^2, \tag{13.33}$$

where

$$\langle k \rangle = \frac{1}{n} \sum_i k_i, \qquad \langle k^2 \rangle = \frac{1}{n} \sum_i k_i^2, \tag{13.34}$$

and we have used $2m = \langle k \rangle n$ (see Eq. (6.23)). Thus the expected number of multiedges remains constant as the network grows larger, so long as $\langle k^2 \rangle$ is constant and finite, and the density of multiedges—the number per vertex—vanishes as $1/n$. We used this result in a number of our earlier arguments.[4]

Another way to derive the expression in Eq. (13.32) is to observe that there are $k_i k_j$ possible edges we could form between vertices i and j, while the total number of possible edges in the whole graph is the number of ways of

---

[4]For networks with power-law degree distributions $\langle k^2 \rangle$ diverges, as described in Section 8.4.2, and in that case the density of multiedges may not vanish or may do so more slowly than $1/n$.

choosing a pair of stubs from the $2m$ total stubs, or $\binom{2m}{2} = m(2m-1)$. The probability that any particular edge falls between i and j is thus given by the ratio $k_i k_j / m(2m-1)$, and if we make a total of m edges then the expected total number of edges between i and j is m times this quantity, which gives us Eq. (13.31) again.

The only case in which this derivation is not quite right is for self-edges. In that case the number of pairs of stubs is not $k_i k_j$ but instead is $\binom{k_i}{2} = \frac{1}{2}k_i(k_i-1)$ and hence the probability of a self-edge from vertex i to itself is

$$p_{ii} = \frac{k_i(k_i-1)}{4m}. \tag{13.35}$$

We can use this result to calculate the expected number of self-edges in the network, which is given by the sum over all vertices i:

$$\sum_i p_{ii} = \sum_i \frac{k_i(k_i-1)}{4m} = \frac{\langle k^2 \rangle - \langle k \rangle}{2\langle k \rangle}, \tag{13.36}$$

This expression remains constant as $n \to \infty$ provided $\langle k^2 \rangle$ remains constant, and hence, as with the multiedges, the density of self-edges in the network vanishes as $1/n$ in the limit of large network size.

We can use Eqs. (13.32) and (13.35) to calculate a number of other properties of vertices in the configuration model. For instance, we can calculate the expected number n_{ij} of common neighbors that vertices i and j share. The probability that i is connected to another vertex l is p_{il} and the probability that j is connected to the same vertex would likewise normally be p_{jl}. However, as with the calculation of multiedges above, if we already know that i is connected to l, then the number of available stubs at vertex l is reduced by one and, rather than being given by the normal expression (13.32), the probability of a connection between j and l is $k_j(k_l-1)/2m$. Multiplying the probabilities for the two edges and summing over l, we then get our expression for the expected number of common neighbors of i and j:

$$n_{ij} = \sum_l \frac{k_i k_l}{2m} \frac{k_j(k_l-1)}{2m} = \frac{k_i k_j}{2m} \frac{\sum_l k_l(k_l-1)}{n\langle k \rangle}$$

$$= p_{ij} \frac{\langle k^2 \rangle - \langle k \rangle}{\langle k \rangle}. \tag{13.37}$$

Thus the probability of sharing a common neighbor is equal to the probability $p_{ij} = k_i k_j / 2m$ of having a direct connection times a multiplicative factor that depends only on the mean and variance of the degree distribution but not on the properties of the vertices i and j themselves.

In this calculation we have ignored the fact that the probability of self-edges, Eq. (13.35), is different from the probability for other edges. As we have seen, however, the density of self-edges in the configuration model tends to zero as $n \to \infty$, so in that limit it is usually safe to make the approximation that Eq. (13.32) applies for all i and j.

13.2.2 RANDOM GRAPHS WITH GIVEN EXPECTED DEGREE

The configuration model of the previous section is, as we have said, similar in some ways to the standard random graph $G(n, m)$ described in Section 12.1, in which we distribute a fixed number m of edges at random between n vertices. In the configuration model the total number of edges is again fixed, having value $m = \frac{1}{2} \sum_i k_i$, but in addition we now also fix the individual degree of every vertex as well.

It is natural to ask whether there is also an equivalent of $G(n, p)$—the model in which only the probability of edges is fixed and not their number—and indeed there is. We simply place an edge between each pair of vertices i, j with independent probabilities taking the form of Eq. (13.32). We define a parameter c_i for each vertex and then place an edge between vertices i and j with probability $p_{ij} = c_i c_j / 2m$. As with the configuration model, we must allow self-edges if the model is to be tractable, and again self-edges have to be treated a little differently from ordinary edges. It turns out that the most satisfactory definition of the edge probability is[5]

$$p_{ij} = \begin{cases} c_i c_j / 2m & \text{for } i \neq j, \\ c_i^2 / 4m & \text{for } i = j, \end{cases} \tag{13.38}$$

where m is now defined by[6]

$$\sum_i c_i = 2m. \tag{13.39}$$

---

[5]As before, p_{ij} should really be regarded as the expected number of edges between i and j rather than the probability and in fact the proper formulation of the model is that we place a Poisson-distributed number of edges with mean p_{ij} between each pair of vertices i, j. Thus the model can in principle have multiedges as well as self-edges, just as in the configuration model. In the limit of large m and constant c_i, however, the probability and the expected number again become equal, and the density of multiedges tends to zero, so the distinction is unimportant.

[6]Another way of putting this is that the average value $\langle A_{ij} \rangle$ of an element of the adjacency matrix is simply $\langle A_{ij} \rangle = c_i c_j / 2m$ for all i, j—recall that the diagonal element A_{ii} of the adjacency matrix is defined to be twice the number of self-edges at vertex i, and this compensates for the extra factor of two in Eq. (13.38).

With this choice the average number of edges in the network is

$$\sum_{i\leq j} p_{ij} = \sum_{i<j} \frac{c_i c_j}{2m} + \sum_i \frac{c_i^2}{4m} = \sum_{ij} \frac{c_i c_j}{4m} = m, \tag{13.40}$$

as before. We can also calculate the average number of ends of edges connected to a vertex i, i.e., its average degree $\langle k_i \rangle$. Allowing for the fact that a self-edge contributes two ends of edges to the degree, we get

$$\langle k_i \rangle = 2p_{ii} + \sum_{j(\neq i)} p_{ij} = \frac{c_i^2}{2m} + \sum_{j(\neq i)} \frac{c_i c_j}{2m} = \sum_j \frac{c_i c_j}{2m} = c_i. \tag{13.41}$$

In other words the parameters c_i appearing in the definition of p_{ij}, Eq. (13.38), are the average or expected degrees in this model, just as the parameter c in $G(n, p)$ is the average degree of a vertex. The *actual* degree of a vertex could in principle take almost any value, depending on the luck of the draw about which edges happen to get randomly created and which do not. In fact one can show that the degree of vertex i will have a Poisson distribution with mean c_i, meaning that in practice it will be quite narrowly distributed about c_i, but there will certainly be some variation, unless c_i is zero.[7] Note that c_i does not have

---

[7]The probabilities of edges between vertex i and each other vertex are independent, which immediately implies that the degree has a Poisson distribution. This may be obvious to you— if you're a statistician, for example—but if not, here is a proof, which makes use of generating functions.

The probability that there are edges connecting vertex i to any specific set of vertices, including itself, is given by a product of factors p_{ij} for each edge present and $(1 - p_{ij})$ for each edge not present. This product can conveniently be written in the form

$$p_{ii}^{A_{ii}/2}(1 - p_{ii})^{1-A_{ii}/2} \prod_{j(\neq i)} p_{ij}^{A_{ij}}(1 - p_{ij})^{1-A_{ij}},$$

where A_{ij} is the standard adjacency matrix and we adopt the convention that $0^0 = 1$ for any cases where $p_{ij} = 0$. Note that it is important to separate out the term for p_{ii} as shown, since it takes a slightly different form from the others. Recall that a self-edge is represented by a diagonal element $A_{ii} = 2$ in the adjacency matrix (see Section 6.2) and we must allow for this with the factors of two above.

The probability $p_k^{(i)}$ that vertex i has degree exactly k is the sum of these probabilities over all cases where the ith row of the adjacency matrix adds up to k (including the 2s that appear for self-edges, since a self-edge contributes $+2$ to the degree). We can write this sum as

$$p_k^{(i)} = \sum_{A_{i1}=0,1} \cdots \sum_{A_{ii}=0,2} \cdots \sum_{A_{in}=0,1} \delta(k, \textstyle\sum_j A_{ij}) p_{ii}^{A_{ii}/2}(1 - p_{ii})^{1-A_{ii}/2} \prod_{j(\neq i)} p_{ij}^{A_{ij}}(1 - p_{ij})^{1-A_{ij}},$$

where $\delta(a, b)$ is the Kronecker delta. It is tricky to evaluate this sum directly because of the constraint imposed by the delta function, but we can do it using a generating function. Multiplying both sides of the equation by z^k, summing over all k, and defining the generating function $g_i(z) = \sum_k p_k^{(i)} z^k$, we get

443

to be an integer, unlike the degrees k_i appearing in the configuration model.

Thus in this model we specify the expected number of edges m and the expected degree sequence $\{c_i\}$ of the network but not the actual number of edges and actual degree sequence. This is again analogous to $G(n, p)$, in which we specify only the expected number of edges and not the actual number. Unfortunately, this means we usually cannot choose the degree *distribution* of our network, because the distribution of the actual degrees k_i is not the same as the distribution of the expected degrees c_i. This is a substantial disadvantage

$$
g_i(z) = \sum_{k=0}^{\infty} z^k \sum_{A_{i1}=0,1} \cdots \sum_{A_{ii}=0,2} \cdots \sum_{A_{in}=0,1} \delta\big(k, \textstyle\sum_j A_{ij}\big) p_{ii}^{A_{ii}/2}(1-p_{ii})^{1-A_{ii}/2} \prod_{j(\neq i)} p_{ij}^{A_{ij}}(1-p_{ij})^{1-A_{ij}}
$$

$$
= \sum_{A_{i1}=0,1} \cdots \sum_{A_{ii}=0,2} \cdots \sum_{A_{in}=0,1} z^{\sum_j A_{ij}} p_{ii}^{A_{ii}/2}(1-p_{ii})^{1-A_{ii}/2} \prod_{j(\neq i)} p_{ij}^{A_{ij}}(1-p_{ij})^{1-A_{ij}}
$$

$$
= \sum_{A_{i1}=0,1} \cdots \sum_{A_{ii}=0,2} \cdots \sum_{A_{in}=0,1} (p_{ii}z^2)^{A_{ii}/2}(1-p_{ii})^{1-A_{ii}/2} \prod_{j(\neq i)} (p_{ij}z)^{A_{ij}}(1-p_{ij})^{1-A_{ij}}
$$

$$
= (1 - p_{ii} + p_{ii}z^2) \prod_{j(\neq i)} (1 - p_{ij} + p_{ij}z)
$$

$$
= \left[1 + \frac{c_i^2}{4m}(z^2 - 1)\right] \prod_{j(\neq i)} \left[1 + \frac{c_i c_j}{2m}(z - 1)\right].
$$

Taking logs of both sides and going to the limit of large size, where $m \to \infty$ (with the c_i remaining finite), we then get

$$
\ln g_i(z) = \lim_{m\to\infty} \left\{ \ln\left[1 + \frac{c_i^2}{4m}(z^2 - 1)\right] + \sum_{j(\neq i)} \ln\left[1 + \frac{c_i c_j}{2m}(z - 1)\right] \right\}
$$

$$
= \frac{c_i^2}{4m}(z^2 - 1) + \sum_{j(\neq i)} \frac{c_i c_j}{2m}(z - 1)
$$

$$
= \frac{c_i^2}{4m}(z^2 - 1) - \frac{c_i^2}{2m}(z - 1) + \sum_{j=1}^{n} \frac{c_i c_j}{2m}(z - 1)
$$

$$
= \frac{c_i^2}{4m}(z^2 - 1) - \frac{c_i^2}{2m}(z - 1) + c_i(z - 1)
$$

$$
= c_i(z - 1)\left[1 + \frac{c_i(z - 1)}{4m}\right],
$$

where we have made use of Eq. (13.39) in the second-to-last line. For large m, the second term in the square brackets becomes negligible compared to the first and, taking exponentials again,

$$
g_i(z) = e^{c_i(z-1)}.
$$

Now we can derive the probability distribution of the degree of vertex i by differentiating:

$$
p_k^{(i)} = \frac{1}{k!} \frac{d^k g_i}{dz^k}\bigg|_{z=0} = e^{-c_i} \frac{c_i^k}{k!},
$$

which is indeed a Poisson distribution, with mean c_i, as promised.

of the model since the degree distribution is widely considered to be a crucial property of networks.[8]

This is unfortunate, because this model is in other respects a very nice one. It is straightforward to treat analytically and many of the derivations are substantially simpler for this model than for the configuration model. Nonetheless, because we place such a premium on being able to choose the degree distribution, this model is in fact hardly ever used in real calculations of the properties of networks. Instead, most calculations are made using the configuration model and this is the direction that we will take in this book as well. In the following sections, we describe how one can make use of the machinery of generating functions to calculate many of the properties of the configuration model exactly in the limit of large network size.

13.3 EXCESS DEGREE DISTRIBUTION

In the remainder of this chapter we describe the calculation of a variety of properties of the configuration model. We begin our discussion with some fundamental observations about the model—and networks in general—that will prove central to later developments.

Consider a configuration model with degree distribution p_k, meaning that a fraction p_k of the vertices have degree k. (We can consider either the standard version of the model in which the degree sequence is fixed, as in Section 13.2, or the version of Eq. (13.30) in which only the distribution is fixed but not the exact degree sequence.) The distribution p_k tells us the probability that a vertex chosen uniformly at random from our network has degree k. But suppose instead that we take a vertex (randomly chosen or not) and follow one of its edges (assuming it has at least one) to the vertex at the other end. What is the probability that this vertex will have degree k?

The answer cannot just be p_k. For instance, there is no way to reach a vertex with degree zero by following an edge in this way, because a vertex with degree zero has no edges. So the probability of finding a vertex of degree zero is itself zero, and not p_0.

In fact, the correct probability for general k is not hard to calculate. We know that an edge emerging from a vertex in a configuration model network has equal chance of terminating at any "stub" of an edge anywhere else in the network (see Section 13.2). Since there are $\sum_i k_i = 2m$ stubs in total, or $2m - 1$

---

[8]It is easy to see that there are some degree distributions that the model cannot reproduce at all—any distribution for which p_k is exactly zero for any k, for instance, since there is always a non-zero probability that any vertex can have any degree.

excluding the one at the beginning of our edge, and k of them are attached to any particular vertex with degree k, our edge has probability $k/(2m - 1)$ of ending at any particular vertex of degree k. In the limit of large network size, where m becomes large (assuming the degree distribution, and hence the average degree, remain constant), we can ignore the -1 and just write this as $k/2m$.

Given that p_k is the total fraction of vertices in the network with degree k, the total number of such vertices is np_k, and hence the probability of our edge attaching to *any* vertex with degree k is

$$\frac{k}{2m} \times np_k = \frac{kp_k}{\langle k \rangle},$$ (13.42)

where $\langle k \rangle$ is the average degree over the whole network and we have made use of the fact that $2m = n\langle k \rangle$, Eq. (6.23).

Thus the probability that we reach a vertex of degree k upon following an edge in this way is proportional not to p_k but to kp_k. To put that another way, the vertex you reach by following an edge is not a typical vertex in the network. It is more likely to have high degree than a typical vertex. Physically, the reasoning behind this observation is that a vertex with degree k has k edges attached to it, and you can reach that vertex by following any one of them. Thus if we choose an edge and follow it you have k times the chance of reaching a vertex with degree k that you have of reaching a vertex with degree 1.

It is important to recognize that this is a property specifically of the configuration model (or similar random graph models). In the real world, the degrees of adjacent vertices in networks are often correlated (see Section 7.13) and hence the probability of reaching a vertex of degree k when we follow an edge depends on what vertex we are coming from.[9] Nonetheless, it is found to apply approximately to many real-world networks, which is one of the reasons why insights gained from the configuration model are useful for understanding the world around us.

Equation (13.42) has some strange and counter-intuitive consequences. As an example, consider a randomly chosen vertex in the configuration model and let us calculate the average degree of a neighbor of that vertex. If we were using the configuration model to model a friendship network, for instance, the average degree of an individual's network neighbor would correspond to the average number of friends their friend has. This number is the average

---

[9]On the other hand, if we pick a *random* edge in a network and follow it to one of its ends, then the degree of the vertex we reach is distributed according to (13.42), regardless of whether degrees are correlated or not.

of the distribution in Eq. (13.42), which we get by multiplying by k and then summing over k thus:[10]

$$\text{average degree of a neighbor} = \sum_k k \frac{k p_k}{\langle k \rangle} = \frac{\langle k^2 \rangle}{\langle k \rangle}. \tag{13.43}$$

Note that the average degree of a neighbor is thus different from the average degree $\langle k \rangle$ of a typical vertex in the network. In fact, it is in general larger, as we can show by calculating the difference

$$\frac{\langle k^2 \rangle}{\langle k \rangle} - \langle k \rangle = \frac{1}{\langle k \rangle} \left(\langle k^2 \rangle - \langle k \rangle^2 \right) = \frac{\sigma_k^2}{\langle k \rangle}, \tag{13.44}$$

where $\sigma_k^2 = \langle k^2 \rangle - \langle k \rangle^2$ is the variance of the degree distribution. The variance, which is the square of the standard deviation, is necessarily non-negative and indeed is strictly positive unless every single vertex in the network has the same degree. Let us assume that there is some variation in the degrees so that σ_k^2 is greater than zero. The average degree $\langle k \rangle$ is also greater than zero, unless all vertices have degree zero. Thus Eq. (13.44) implies that $\langle k^2 \rangle / \langle k \rangle - \langle k \rangle > 0$, or

$$\frac{\langle k^2 \rangle}{\langle k \rangle} > \langle k \rangle. \tag{13.45}$$

In other words, the average degree of the neighbor of a vertex is greater than the average degree of a vertex. In colloquial terms, "Your friends have more friends than you do."

At first sight, this appears to be a very strange result. Certainly it seems likely that there will be some vertices in the network with higher degree than the average. But there will also be some who have lower degree and when you average over all neighbors of all vertices surely the two should cancel out. Surely the average degree of a neighbor should be the same as the average degree in the network as a whole. Yet Eq. (13.45) tells us that this is not so. And the equation really is correct. You can create a configuration model network on a computer and average the degrees of the neighbors of every vertex, and you'll find that the formula works to very high accuracy. Even more remarkably, as first shown by Feld [113], you can do the same thing with real networks and, although the configuration model formula doesn't apply exactly to these networks, the basic principle still seems to hold. Here, for instance, are some measurements for two academic collaboration networks, in which scientists

---

[10]The ratio $\langle k^2 \rangle / \langle k \rangle$ that appears in Eq. (13.43) crops up repeatedly in the study of networks. It appeared previously in Section 13.2.1 and it will come up in many later calculations.

are connected together by edges if they have coauthored scientific papers, and for a recent snapshot of the structure of the Internet at the autonomous system level:

Network	n	Average degree	Average neighbor degree	$\dfrac{\langle k^2 \rangle}{\langle k \rangle}$
Biologists	1 520 252	15.5	68.4	130.2
Mathematicians	253 339	3.9	9.5	13.2
Internet	22 963	4.2	224.3	261.5

According to these results a biologist's collaborators have, on average, more than four times as many collaborators as they do themselves. On the Internet, a node's neighbors have more than 50 times the average degree! Note that in each of the cases in the table the configuration model value of $\langle k^2 \rangle / \langle k \rangle$ overestimates the real average neighbor degree, in some cases by a substantial margin.[11] This is typical of calculations using simplified network models: they can give you a feel for the types of effect one might expect to see, or the general directions of changes in quantities. But they usually don't give quantitatively accurate predictions for the behavior of real networks.

The fundamental reason for the result, Eq. (13.45), is that when you go through the vertices of a network and average the degrees of the neighbors of each one, many of those neighbors appear in more than one average. In fact, a vertex with degree k will appear as one of the neighbors of exactly k other vertices, and hence appear in k of the averages. This means that high-degree vertices are over-represented in the calculations compared with low-degree ones and it is this bias that pushes up the overall average value.

In most of the calculations that follow, we will be interested not in the total degree of the vertex at the end of an edge but in the number of edges attached to that vertex *other* than the one we arrived along. For instance, if we want to calculate the size of the component to which a vertex i belongs then we will want to know first of all how many neighbors i has, and then how many neighbors those neighbors have, *other than i*, and so on.

The number of edges attached to a vertex other than the edge we arrived along is called the *excess degree* of the vertex and it is just one less than the total degree. Since the vertex at the end of an edge always has degree at least 1 (because of that edge) the minimum value of the excess degree is zero.

---

[11]There is no reason in principle why the configuration model should always overestimate the average degree of a neighbor. In some cases it could underestimate too.

We can calculate the probability distribution of the excess degree from Eq. (13.43). The probability q_k of having excess degree k is simply the probability of having total degree $k + 1$ and, putting $k \to k + 1$ in Eq. (13.43), we get

$$q_k = \frac{(k+1)p_{k+1}}{\langle k \rangle}. \tag{13.46}$$

(Note that the denominator is still just $\langle k \rangle$, and not $\langle k + 1 \rangle$, as you can verify for yourself by checking that Eq. (13.46) is correctly normalized so that $\sum_{k=0}^{\infty} q_k = 1$.)

The distribution q_k is called the *excess degree distribution* and it will come up repeatedly in the sections that follow. It is the probability distribution, for a vertex reached by following an edge, of the number of other edges attached to that vertex.

13.4 CLUSTERING COEFFICIENT

As a simple application of the excess degree distribution, let us calculate the clustering coefficient for the configuration model. Recall that the clustering coefficient is the average probability that two neighbors of a vertex are neighbors of each other.

Consider then a vertex v that has at least two neighbors, which we will denote i and j. Being neighbors of v, i and j are both at the ends of edges from v, and hence the number of other edges connected to them, k_i and k_j are distributed according to the excess degree distribution, Eq. (13.46). The probability of an edge between i and j is then $k_i k_j / 2m$ (see Eq. (13.32)) and, averaging both k_i and k_j over the distribution q_k, we get an expression for the clustering coefficient thus:

$$
\begin{aligned}
C &= \sum_{k_i,k_j=0}^{\infty} q_{k_i} q_{k_j} \frac{k_i k_j}{2m} = \frac{1}{2m} \left[\sum_{k=0}^{\infty} k q_k \right]^2 \\
&= \frac{1}{2m\langle k \rangle^2} \left[\sum_{k=0}^{\infty} k(k+1) p_{k+1} \right]^2 \\
&= \frac{1}{2m\langle k \rangle^2} \left[\sum_{k=0}^{\infty} (k-1) k p_k \right]^2 \\
&= \frac{1}{n} \frac{\left[\langle k^2 \rangle - \langle k \rangle \right]^2}{\langle k \rangle^3},
\end{aligned} \tag{13.47}
$$

where we have made use of $2m = n\langle k \rangle$, Eq. (6.23).

Like the clustering coefficient of the Poisson random graph, Eq. (12.11), this expression goes as n^{-1} for fixed degree distribution, and so vanishes in the limit of large system size. Hence, like the Poisson random graph, the configuration model appears to be an unpromising model for real-world networks with high clustering. Note, however, that Eq. (13.47) contains the second moment $\langle k^2 \rangle$ of the degree distribution in its numerator which can become large, for instance in networks with power-law degree distributions (see Section 8.4.2). This can result in surprisingly large values of C in the configuration model. For further discussion of this point see Section 8.6.

13.5 GENERATING FUNCTIONS FOR DEGREE DISTRIBUTIONS

In the calculations that follow, we will make heavy use of the generating functions for the degree distribution and the excess degree distribution of a network. We will denote these generating functions by $g_0(z)$ and $g_1(z)$ respectively. They are defined by

$$g_0(z) = \sum_{k=0}^{\infty} p_k z^k, \tag{13.48}$$

$$g_1(z) = \sum_{k=0}^{\infty} q_k z^k. \tag{13.49}$$

Although it will be convenient to have separate notations for these two commonly occurring functions, they are not really independent, since the excess degree distribution is itself defined in terms of the ordinary degree distribution via Eq. (13.46). Using Eq. (13.46) we can write $g_1(z)$ as

$$g_1(z) = \frac{1}{\langle k \rangle} \sum_{k=0}^{\infty} (k+1) p_{k+1} z^k = \frac{1}{\langle k \rangle} \sum_{k=0}^{\infty} k p_k z^{k-1}$$
$$= \frac{1}{\langle k \rangle} \frac{\mathrm{d}g_0}{\mathrm{d}z}. \tag{13.50}$$

But Eq. (13.22) tells us that the average vertex degree is $\langle k \rangle = g_0'(1)$, so

$$g_1(z) = \frac{g_0'(z)}{g_0'(1)}. \tag{13.51}$$

Thus if we can find $g_0(z)$, we can also find $g_1(z)$ directly from it, without the need to calculate the excess degree distribution explicitly.

For example, suppose our degree distribution is a Poisson distribution with mean c:

$$p_k = \mathrm{e}^{-c} \frac{c^k}{k!}. \tag{13.52}$$

Then its generating function is given by Eq. (13.6) to be

$$g_0(z) = e^{c(z-1)}.$$ (13.53)

Applying Eq. (13.51), we then find that

$$g_1(z) = e^{c(z-1)}.$$ (13.54)

In other words, $g_0(z)$ and $g_1(z)$ are identical in this case. (This is one reason why calculations are relatively straightforward for the Poisson random graph—there is no difference between the degree distribution and the excess degree distribution in that case, a fact you can easily demonstrate for yourself by substituting Eq. (13.52) directly into Eq. (13.46)).

A more complicated example is the power-law distribution, Eq. (13.10), which has a generating function given by Eq. (13.16) to be

$$g_0(z) = \frac{\mathrm{Li}_\alpha(z)}{\zeta(\alpha)},$$ (13.55)

where $\mathrm{Li}_\alpha(z)$ is the polylogarithm function and α is the exponent of the power law. Substituting this result into Eq. (13.51) and making use of Eq. (13.17) gives

$$g_1(z) = \frac{\mathrm{Li}_{\alpha-1}(z)}{z\,\mathrm{Li}_{\alpha-1}(1)} = \frac{\mathrm{Li}_{\alpha-1}(z)}{z\zeta(\alpha-1)},$$ (13.56)

where we have made use of the fact that $\mathrm{Li}_\alpha(1) = \zeta(\alpha)$ (see Eqs. (13.12) and (13.15)).

13.6 NUMBER OF SECOND NEIGHBORS OF A VERTEX

Armed with these results, we are now in a position to make some more detailed calculations of the properties of the configuration model. The first question we will address is a relatively simple one: what is the probability $p_k^{(2)}$ that a vertex has exactly k second neighbors in the network?

Let us break this probability down by writing it in the form

$$p_k^{(2)} = \sum_{m=0}^{\infty} p_m P^{(2)}(k|m),$$ (13.57)

where $P^{(2)}(k|m)$ is the probability of having k second neighbors given that we have m first neighbors and p_m is the ordinary degree distribution. Equation (13.57) says that the total probability of having k second neighbors is the probability of having k second neighbors given that we have m first neighbors,

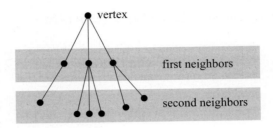

first neighbors

second neighbors

Figure 13.4: Calculation of the number of second neighbors of a vertex. The number of second neighbors of a vertex (top) is equal to the sum of the excess degrees of the first neighbors.

averaged over all possible values of m. We assume that we are given the degree distribution p_m; we need to find $P^{(2)}(k|m)$ and then complete the sum.

As illustrated in Fig. 13.4, the number of second neighbors of a vertex is equal to the sum of the excess degrees of the first neighbors. And as discussed in the previous section, the excess degrees are distributed according to the distribution q_k, Eq. (13.46), so that the probability that the excess degrees of our m first neighbors take the values $j_1 \ldots j_m$ is $\prod_{r=1}^{m} q_{j_r}$. Summing over all sets of values $j_1 \ldots j_m$, the probability that the excess degrees sum to k and hence that we have k second neighbors is

$$P^{(2)}(k|m) = \sum_{j_1=0}^{\infty} \cdots \sum_{j_m=0}^{\infty} \delta\big(k, \sum_{r=1}^{m} j_r\big) \prod_{r=1}^{m} q_{j_r}. \tag{13.58}$$

Substituting this expression into (13.57), we find that

$$p_k^{(2)} = \sum_{m=0}^{\infty} p_m \sum_{j_1=0}^{\infty} \cdots \sum_{j_m=0}^{\infty} \delta\big(k, \sum_{r=1}^{m} j_r\big) \prod_{r=1}^{m} q_{j_r}. \tag{13.59}$$

By now, you may be starting to find sums of this type familiar. We saw them previously in Eqs. (12.25) and (13.27), for example. We can handle this one by the same trick we used before: instead of trying to calculate $p_k^{(2)}$ directly, we calculate instead its generating function $g^{(2)}(z)$ thus:

$$g^{(2)}(z) = \sum_{k=0}^{\infty} p_k^{(2)} z^k$$

$$= \sum_{k=0}^{\infty} z^k \sum_{m=0}^{\infty} p_m \sum_{j_1=0}^{\infty} \cdots \sum_{j_m=0}^{\infty} \delta(k, \textstyle\sum_{r=1}^{m} j_r) \prod_{r=1}^{m} q_{j_r}$$

$$= \sum_{m=0}^{\infty} p_m \sum_{j_1=0}^{\infty} \cdots \sum_{j_m=0}^{\infty} z^{\sum_{r=1}^{m} j_r} \prod_{r=1}^{m} q_{j_r}$$

$$= \sum_{m=0}^{\infty} p_m \sum_{j_1=0}^{\infty} \cdots \sum_{j_m=0}^{\infty} \prod_{r=1}^{m} q_{j_r} z^{j_r}$$

$$= \sum_{m=0}^{\infty} p_m \left[\sum_{j=0}^{\infty} q_j z^j \right]^m. \tag{13.60}$$

But now we notice an interesting thing: the sum in square brackets in the last line is none other than the generating function $g_1(z)$ for the excess degree distribution, Eq. (13.49). Thus Eq. (13.60) can be written as

$$g^{(2)}(z) = \sum_{m=0}^{\infty} p_m [g_1(z)]^m = g_0(g_1(z)), \tag{13.61}$$

where $g_0(z)$ is the generating function for the ordinary degree distribution, defined in Eq. (13.48). So once we know the generating functions for our two basic degree distributions the generating function for the distribution of the second neighbors is very simple to calculate.

In fact, there was no need to go through this lengthy calculation to reach Eq. (13.61). We can derive the same result much more quickly by making use of the "powers" property of generating functions that we derived in Section 13.1.4. There we showed (Eq. (13.28)) that, given a quantity k distributed according to a distribution with generating function $g(z)$, m independent quantities drawn from the same distribution have a sum whose distribution is given by the generating function $[g(z)]^m$. We can apply this result here, by noting that the m excess degrees of the first neighbors of our vertex are just such a set of independent quantities. Given that $g_1(z)$ is the generating function for the distribution of a single one of them (Eq. (13.49)), the distribution $P^{(2)}(k|m)$ of their sum—which is the number of second neighbors—has generating function $[g_1(z)]^m$. That is,

$$\sum_{k=0}^{\infty} P^{(2)}(k|m) z^k = [g_1(z)]^m. \tag{13.62}$$

Now, using Eq. (13.57), the generating function for $p_k^{(2)}$ is

$$g^{(2)}(z) = \sum_{k=0}^{\infty} p_k^{(2)} z^k = \sum_{k=0}^{\infty} \sum_{m=0}^{\infty} p_m P^{(2)}(k|m) z^k$$

$$= \sum_{m=0}^{\infty} p_m \sum_{k=0}^{\infty} P^{(2)}(k|m) z^k = \sum_{m=0}^{\infty} p_m [g_1(z)]^m$$

$$= g_0(g_1(z)). \tag{13.63}$$

In future calculations, we will repeatedly make use of this shortcut to get our results, rather than taking the long route exemplified in Eq. (13.60).

We can also use similar methods to calculate the probability distribution of the number of third neighbors. The number of third neighbors is the sum of the excess degrees of each of the second neighbors. Thus, if there are m second neighbors, then the probability distribution $P^{(3)}(k|m)$ of the number of third neighbors has generating function $[g_1(z)]^m$ and the overall probability of having k third neighbors is exactly analogous to Eq. (13.63):

$$g^{(3)}(z) = \sum_{k=0}^{\infty} \sum_{m=0}^{\infty} p_m^{(2)} P^{(3)}(k|m) z^k = \sum_{m=0}^{\infty} p_m^{(2)} \sum_{k=0}^{\infty} P^{(3)}(k|m) z^k$$

$$= \sum_{m=0}^{\infty} p_m^{(2)} [g_1(z)]^m = g^{(2)}(g_1(z))$$

$$= g_0(g_1(g_1(z))). \tag{13.64}$$

Indeed, the generating function for the number of neighbors at any distance d can be expressed this way as

$$g^{(d)}(z) = \sum_{k=0}^{\infty} \sum_{m=0}^{\infty} p_m^{(d-1)} P^{(d)}(k|m) z^k$$

$$= \sum_{m=0}^{\infty} p_m^{(d-1)} \sum_{k=0}^{\infty} P^{(d)}(k|m) z^k = \sum_{m=0}^{\infty} p_m^{(d-1)} [g_1(z)]^m$$

$$= g^{(d-1)}(g_1(z)). \tag{13.65}$$

In other words $g^{(d)}(z) = g_0(g_1(\ldots g_1(z) \ldots))$, with $d-1$ copies of g_1 nested inside a single g_0. This expression is correct at arbitrary distances on an infinite network. On a finite network it will break down if d becomes large enough but will be accurate for small values of d.

These results are all very good, but what use are they? Even given the generating function $g^{(2)}(z)$ it is typically quite difficult to extract explicit probabilities for numbers of second neighbors in the network. For instance, if our degree distribution were Poisson with mean c then $g_0(z) = g_1(z) = e^{c(z-1)}$ as in Eqs. (13.53) and (13.54) and

$$g^{(2)}(z) = e^{c(e^{c(z-1)}-1)}. \tag{13.66}$$

But to find the actual probabilities we have to apply Eq. (13.2), which involves calculating derivatives of $g^{(2)}(z)$. One can, with a little work, calculate the first few derivatives, but finding a general formula for the nth derivative is hard.[12]

What we can do, however, is calculate the average number of neighbors at distance d. The average of a distribution is given by the first derivative of its generating function evaluated at $z = 1$ (see Eq. (13.22)) and the derivative of Eq. (13.63) is

$$\frac{dg^{(2)}}{dz} = g_0'(g_1(z)) g_1'(z). \tag{13.67}$$

Setting $z = 1$ and recalling that $g_1(1) = 1$ (Eq. (13.20)), we find that the average number c_2 of second neighbors is

$$c_2 = g_0'(1)g_1'(1). \tag{13.68}$$

But $g_0'(1) = \langle k \rangle$ and

$$g_1'(1) = \sum_{k=0}^{\infty} kq_k$$
$$= \frac{1}{\langle k \rangle} \sum_{k=0}^{\infty} k(k+1)p_{k+1} = \frac{1}{\langle k \rangle} \sum_{k=0}^{\infty} (k-1)kp_k$$
$$= \frac{1}{\langle k \rangle}(\langle k^2 \rangle - \langle k \rangle). \tag{13.69}$$

where we have used Eq. (13.46). Thus the mean number of second neighbors can also be written

$$c_2 = \langle k^2 \rangle - \langle k \rangle. \tag{13.70}$$

We can take this approach further and calculate the mean number c_d of neighbors at any distance d. Differentiating Eq. (13.65) we get

$$\frac{dg^{(d)}}{dz} = g^{(d-1)'}(g_1(z)) g_1'(z), \tag{13.71}$$

and setting $z = 1$ we get

$$c_d = g^{(d-1)'}(1) g_1'(1) = c_{d-1} g_1'(1). \tag{13.72}$$

---

[12]In fact, the general derivative in this case can be expressed in terms of the so-called Bell numbers. No closed-form solution exists for third-nearest neighbors or higher, however, nor for most other choices of degree distribution.

Making use of Eq. (13.68) to write $g_1'(1) = c_2/c_1$ where $c_1 = \langle k \rangle$, this can be expressed in the simple form

$$c_d = c_{d-1}\frac{c_2}{c_1},\tag{13.73}$$

which implies that

$$c_d = \left(\frac{c_2}{c_1}\right)^{d-1}c_1.\tag{13.74}$$

In other words, once we know the mean numbers of first and second neighbors, c_1 and c_2, we know everything. What's more, the average number of neighbors at distance d either grows or falls off exponentially, depending on whether c_2 is greater or less than c_1. This observation is strongly reminiscent of the argument we made in Section 12.5 for the appearance of a giant component in a random graph. There we argued that if the number of vertices you can reach within a certain distance is increasing with that distance (on average) then you must have a giant component in the network, while if it is decreasing there can be no giant component. Applying the same reasoning here, we conclude that the configuration model has a giant component if and only if we have

$$c_2 > c_1.\tag{13.75}$$

Using Eq. (13.70) for c_2 and putting $c_1 = \langle k \rangle$, we can also write this condition as $\langle k^2 \rangle - \langle k \rangle > \langle k \rangle$ or

$$\langle k^2 \rangle - 2\langle k \rangle > 0.\tag{13.76}$$

This condition for the existence of a giant component in the configuration model was first given by Molloy and Reed [224] in 1995.[13]

13.7 GENERATING FUNCTIONS FOR THE SMALL COMPONENTS

In this section and the following one we examine the sizes of components in the configuration model. As we will see, the situation is qualitatively similar

---

[13]This expression has an interesting history. In the 1940s Flory [123] considered a model of branching polymers in which elemental units with a fixed number of "legs"—vertices with uniform degree, in effect—joined together to form connected clumps. He showed that, if the system was restricted to forming only trees, then there was a transition at which the polymer "gelled" to create a clump of joined units which corresponds to our giant cluster and found the size of the gel. In effect, Flory's results were a special case of the solution given here for the uniform degree distribution, although they were not expressed in the language of networks. It was not until much later that Molloy and Reed, who were, as far as I know, unaware of Flory's work, gave the full solution for general degree distribution.

to that for the Poisson random graph in that a configuration model network generally has at most one giant component, plus a large number of small components. We will approach the calculation of component sizes by a route different from the one we took for the Poisson random graph and examine first the properties of the small components. We will see that it is possible to calculate the distribution of the sizes of the small components by a method similar to the one we used in the Poisson case. Then we can use these results to get at the properties of the giant component: once we have the sizes of the small components, we can subtract them from the size of the graph as a whole and whatever is left, if anything, must be the giant component.

Let π_s be the probability that a randomly chosen vertex belongs to a small (non-giant) component of size s. We will calculate π_s by first calculating its generating function

$$h_0(z) = \sum_{s=1}^{\infty} \pi_s z^s. \qquad (13.77)$$

Note that the minimum value of s is 1, since every vertex belongs to a component of size at least one (namely itself).

By an argument exactly analogous to that of Section 12.6.1 we can show that the small components in the configuration model are trees (in the limit of large n, provided the degree distribution is held constant as we go to the limit). We can use this fact to derive an expression for the distribution of small component sizes as follows.

Consider Fig. 13.5 (which is actually the same as the figure for the Poisson random graph in the previous chapter (Fig. 12.3), but it works just as well as an illustration of the configuration model). If vertex i is a member of a small component then that component is necessarily a tree. Just as in the Poisson case, this implies that the sets of vertices reachable along each of its edges (shaded areas in Fig. 13.5a) are not connected, other than via vertex i, since if they were connected there would be a loop in the component and hence it would not be a tree.

Now, taking a hint from our argument in the Poisson case, let us remove vertex i from the network along with all its edges—see Fig. 13.5b. The shaded areas in the figure are now not connected to one another at all and hence are each now separate components in their own right. And the size of the component to which vertex i belongs on the original network is equal to the sum of the sizes of these new components, plus one for vertex i itself.

A crucial point to notice, however, is that the neighbors $n_1, n_2, \ldots$ of vertex i are, by definition, reached by following an edge. Hence, as we have discussed, these are not typical network vertices, being more likely to have high degree

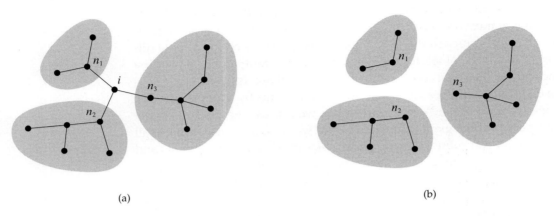

Figure 13.5: The size of one of the small components in the configuration model. (a) The size of the component to which a vertex i belongs is the sum of the number of vertices in each of the subcomponents (shaded regions) reachable via i's neighbors n_1, n_2, n_3, plus one for i itself. (b) If vertex i is removed the subcomponents become components in their own right.

than the typical vertex. Thus the components that they belong to in Fig. 13.5b—the shaded regions in the figure—are not distributed according to π_s. Instead they must have some other distribution. Let us denote this distribution by ρ_s. More specifically, let ρ_s be the probability that the vertex at the end of an edge belongs to a small component of size s after that edge is removed. Let us also define the generating function for this distribution to be

$$h_1(z) = \sum_{s=0}^{\infty} \rho_s z^s. \tag{13.78}$$

We don't yet know the value of ρ_s or its generating function and we will have to calculate them later, but for the moment let us proceed with the information we have.

Suppose that vertex i on the original network has degree k and let us denote by $P(s|k)$ the probability that, after i is removed, its k neighbors belong to small components of sizes summing to exactly s. Alternatively, $P(s-1|k)$ is the probability that i itself belongs to a small component of size s given that its degree is k. Then the total probability π_s that i belongs to a small component of size s is this probability averaged over k thus:

$$\pi_s = \sum_{k=0}^{\infty} p_k P(s-1|k). \tag{13.79}$$

Substituting this expression into Eq. (13.77) we then get an expression for the generating function for π_s as follows:

$$h_0(z) = \sum_{s=1}^{\infty} \sum_{k=0}^{\infty} p_k P(s-1|k) z^s = z \sum_{k=0}^{\infty} p_k \sum_{s=1}^{\infty} P(s-1|k) z^{s-1}$$

$$= z \sum_{k=0}^{\infty} p_k \sum_{s=0}^{\infty} P(s|k) z^s. \tag{13.80}$$

The final sum in this expression is the generating function for the probability that the k neighbors belong to small components whose size sums to s. But the sizes of the small components are independent of one another and hence we can use the "powers" property of generating functions (Section 13.1.4), which tells us that the generating function we want is just equal to the generating function for the size of the component any single neighbor belongs to—the function that we denoted $h_1(z)$ above—raised to the kth power. Thus

$$h_0(z) = z \sum_{k=0}^{\infty} p_k [h_1(z)]^k = z g_0(h_1(z)). \tag{13.81}$$

We still don't know the generating function $h_1(z)$ but we can derive it now quite easily. We consider the network in which vertex i is removed and ask what is the probability ρ_s that one of the neighbors of i belongs to a component of size s in this network. In the limit of large network size, the removal of the single vertex i will have no effect on the degree distribution, so the network still has the same distribution as before, which means that if the neighbor has degree k then its probability of belonging to a component of size s is $P(s-1|k)$, just as before. Note, however, that the degree k does not follow the ordinary degree distribution. Since the neighbor was reached by following an edge from i, its degree, discounting the edge to i that has been removed, follows the excess degree distribution q_k defined in Eq. (13.46), rather than the ordinary degree distribution. Thus

$$\rho_s = \sum_{k=0}^{\infty} q_k P(s-1|k), \tag{13.82}$$

and, substituting this expression into Eq. (13.78), we have

$$h_1(z) = \sum_{s=1}^{\infty} \sum_{k=0}^{\infty} q_k P(s-1|k) z^s = z \sum_{k=0}^{\infty} q_k \sum_{s=0}^{\infty} P(s|k) z^s. \tag{13.83}$$

As before, the last sum is the generating function for $P(s|k)$, which is equal to $[h_1(z)]^k$, and hence

$$h_1(z) = z \sum_{k=0}^{\infty} q_k [h_1(z)]^k = z g_1(h_1(z)). \tag{13.84}$$

Collecting together our results, the generating functions for π_s and ρ_s thus satisfy

$$h_0(z) = zg_0(h_1(z)), \tag{13.85}$$
$$h_1(z) = zg_1(h_1(z)). \tag{13.86}$$

If we can solve the second of these equations for $h_1(z)$ then we can substitute the result into the first equation and we have our answer for $h_0(z)$. In practice, it is often not easy to solve for $h_1(z)$, and, even if it is, extracting the actual component size distribution from the generating function can be difficult. But that does not mean that these results are useless. On the contrary, there are many useful things we can deduce from them. One important quantity we can calculate is the size of the giant component.

13.8 GIANT COMPONENT

Given the definition $h_0(z) = \sum_s \pi_s z^s$, where π_s is the probability that a randomly chosen vertex belongs to a small component of size s, we have $h_0(1) = \sum_s \pi_s$, which is the total probability that a randomly chosen vertex belongs to a small component. Unlike most generating functions, it is not necessarily the case that $h(1) = 1$ because there may be a giant component in the network. If there is a giant component then some of the vertices do not belong to any small component and $\sum_s \pi_s$ will be less than 1. In fact, $\sum_s \pi_s$ will be simply the fraction of vertices that belong to small components and hence the fraction S of vertices belonging to the giant component is

$$S = 1 - \sum_{s=0}^{\infty} \pi_s = 1 - h_0(1) = 1 - g_0(h_1(1)), \tag{13.87}$$

where we have used Eq. (13.85). The value of $h_1(1)$ we can get from Eq. (13.86):

$$h_1(1) = g_1(h_1(1)). \tag{13.88}$$

The quantity $h_1(1)$ will occur frequently in subsequent developments, so for convenience let us define the shorthand notation

$$u = h_1(1), \tag{13.89}$$

in which case Eqs. (13.87) and (13.88) can be written

$$S = 1 - g_0(u), \tag{13.90}$$
$$u = g_1(u). \tag{13.91}$$

In other words, u is a fixed point of the function $g_1(z)$—a point where the function is equal to its own argument—and if we can find this fixed point then we need only substitute the result into Eq. (13.90) and we have the size of the giant component.

Since $g_1(1) = 1$ (see Eq. (13.20) and the discussion that precedes it), there is always a fixed point of g_1 at $u = 1$, but this solution gives $S = 1 - g_0(1) = 0$ and hence no giant component. If there is to be a giant component there must be at least one other non-trivial solution to Eq. (13.91). We will see some examples of such solutions shortly.

The quantity $u = h_1(1)$ has a simple physical interpretation. Recall that $h_1(z) = \sum_s \rho_s z^s$ is the generating function for the probability ρ_s that the vertex reached by following an edge belongs to a small component of size s if that edge is removed. Thus $h_1(1) = \sum_s \rho_s$ is the total probability that such a vertex belongs to a small component of any size, or equivalently the probability that it doesn't belong to the giant component.

This observation suggests an alternative and simpler derivation of Eqs. (13.90) and (13.91) for the size of a giant component, as follows. To belong to the giant component, a vertex A must be connected to the giant component via at least one of its neighbors. Or equivalently, A does not belong to the giant component if (and only if) it is not connected to the giant component via any of its neighbors. Let us *define* u to be the average probability that a vertex is not connected to the giant component via its connection to some particular neighboring vertex. If vertex A has k neighbors, then the probability that it is not connected to the giant component via any of them is thus u^k. And the average of this probability over the whole network is $\sum_k p_k u^k = g_0(u)$, which is the average probability that a vertex is not in the giant component. But this probability is also, by definition, equal to $1 - S$, where S is the fraction of the graph occupied by the giant component and hence $1 - S = g_0(u)$ or

$$S = 1 - g_0(u), \tag{13.92}$$

which is Eq. (13.90) again.

Now let us ask what the value of u is. The probability that you are not connected to the giant component via a particular neighboring vertex is equal to the probability that *that* vertex is not connected to the giant component via any of its other neighbors. If there are k of those other neighbors, then that probability is again u^k. But because we are talking about a neighboring vertex, k is now distributed according to the excess degree distribution q_k, Eq. (13.46), and hence taking the average, we find that $u = \sum_k q_k u^k$ or

$$u = g_1(u), \tag{13.93}$$

which is Eq. (13.91) again. Thus we have rederived our two equations for the size of the giant component, but by a much shorter route. The main disadvantage of this method is that it only gives the size of the giant component and not the complete generating function for all the other components as well, and this is the reason why we took the time to go through the longer derivation. There are many further results we can derive by knowing the entire generating function, as we show in the next section.

13.8.1 EXAMPLE

Let's take a look at a concrete example and see how calculations for the configuration model work out in practice. Consider a network like that of the first example in Section 13.1.1 that has vertices only of degree 0, 1, 2 and 3, and no vertices of higher degree. Then the generating functions $g_0(z)$ and $g_1(z)$ take the form

$$g_0(z) = p_0 + p_1 z + p_2 z^2 + p_3 z^3, \tag{13.94}$$

$$g_1(z) = \frac{g_0'(z)}{g_0'(1)} = \frac{p_1 + 2p_2 z + 3p_3 z^2}{p_1 + 2p_2 + 3p_3}$$

$$= q_0 + q_1 z + q_2 z^2. \tag{13.95}$$

Equation (13.91) is thus quadratic in this case, $u = q_0 + q_1 u + q_2 u^2$, which has the solutions

$$u = \frac{1 - q_1 \pm \sqrt{(1 - q_1)^2 - 4q_0 q_2}}{2q_2}. \tag{13.96}$$

However, we know that $\sum_k q_k = 1$, and hence in this case $1 - q_1 = q_0 + q_2$. Using this result to eliminate q_1 we get

$$u = \frac{(q_0 + q_2) \pm \sqrt{(q_0 + q_2)^2 - 4q_0 q_2}}{2q_2}$$

$$= \frac{(q_0 + q_2) \pm (q_0 - q_2)}{2q_2}$$

$$= 1 \quad \text{or} \quad \frac{q_0}{q_2}. \tag{13.97}$$

Thus, as expected we have a solution $u = 1$, but we also have another non-trivial solution which *might* imply that we have a giant component.

If $q_2 < q_0$ then this non-trivial solution gives $u > 1$. Since u is a probability it cannot be greater than 1, so in this case we definitely do not have a giant component. On the other hand, if $q_2 > q_0$ we have a viable non-trivial solution

$u < 1$ equal to

$$u = \frac{q_0}{q_2} = \frac{p_1}{3p_3}, \tag{13.98}$$

where we have extracted values of q_0 and q_2 from Eq. (13.95). We can also write the condition $q_2 > q_0$ in terms of the p_k as

$$p_3 > \tfrac{1}{3}p_1. \tag{13.99}$$

In other words, there can be a giant component if the number of vertices of degree three exceeds one third the number of degree one. This is a remarkable result. It says that the number of vertices of degree zero and degree two don't matter at all (except to the extent that their absence makes room for more vertices of the other degrees). As we will see, this is actually a general result—the values of p_0 and p_2 never make any difference to the presence or absence of a giant component. On the other hand, the size of the giant component for the current example is given by Eq. (13.90) to be

$$S = 1 - g_0(u) = 1 - p_0 - \frac{p_1^2}{3p_3} - \frac{p_1^2 p_2}{9p_3^2} - \frac{p_1^3}{27p_3^2}. \tag{13.100}$$

Thus the size of the giant component does depend on p_0 and p_2, even though its presence or absence does not.

We have not, however, yet proved that a giant component actually does exist. In the regime where we have two solutions for u, one with $u = 1$ (no giant component) and one with $u < 1$ (there is a giant component) it is unclear which of these solutions we should believe. In Section 13.6, however, we showed that there is a giant component in the network when the degree sequence satisfies a specific condition, Eq. (13.76). In the next section, we show that in fact this condition is always satisfied whenever a non-trivial solution $u < 1$ exists, and hence that there is always a giant component when we have such a solution.

13.8.2 GRAPHICAL SOLUTIONS AND THE EXISTENCE OF THE GIANT COMPONENT

The example given in the last section is unusual in that we can solve the fixed-point equation (13.91) exactly for the crucial parameter u. In most other cases exact solutions are not possible, but we can nonetheless get a good idea of the behavior of u by graphical means. The derivatives of $g_1(z)$ are proportional to the probabilities ρ_s and hence are all non-negative. That means that for $z \geq 0$, $g_1(z)$ is in general positive, an increasing function of its argument, and upward concave. It also takes the value 1 when $z = 1$. Thus it must look qualitatively like one of the curves in Fig. 13.6. The solution of the fixed-point equation

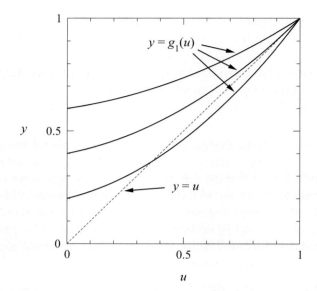

Figure 13.6: Graphical solution of Eq. (13.91). The solution of the equation $u = g_1(u)$ is given by the point at which the curve $y = g_1(u)$ intercepts the line $y = u$.

$u = g_1(u)$ is then given by the intercept of the curve $y = g_1(u)$ with the line $y = u$ (the dotted line in the figure).

As we already know, there is always a trivial solution at $u = 1$ (top right in the figure). But now we can see that there can be just one other solution with $u < 1$ and only if the curve takes the right form. In particular, we have a non-trivial solution at $u < 1$ if the slope $g_1'(1)$ of the curve at $u = 1$ is greater than the slope of the dotted line. That is, if

$$g_1'(1) > 1. \tag{13.101}$$

Using Eq. (13.49) for $g_1(z)$, we have

$$g_1'(1) = \sum_{k=0}^{\infty} k q_k = \frac{1}{\langle k \rangle} \sum_{k=0}^{\infty} k(k+1) p_k = \frac{1}{\langle k \rangle} \sum_{k=0}^{\infty} (k-1) k p_k$$

$$= \frac{\langle k^2 \rangle - \langle k \rangle}{\langle k \rangle}. \tag{13.102}$$

Thus our condition for the solution at $u < 1$ is

$$\frac{\langle k^2 \rangle - \langle k \rangle}{\langle k \rangle} > 1. \tag{13.103}$$

or equivalently,

$$\langle k^2 \rangle - 2\langle k \rangle > 0, \tag{13.104}$$

But this is none other than the condition for the existence of a giant component, Eq. (13.76). In other words, the conditions for the existence of a giant component and the existence of the non-trivial solution to Eq. (13.91) are exactly the same and and hence, as promised, there is always a giant component whenever a solution $u < 1$ exists for Eq. (13.91).

Writing $\langle k \rangle = n^{-1} \sum_i k_i$ and $\langle k^2 \rangle = n^{-1} \sum_i k_i^2$, we can also write Eq. (13.104) as

$$\sum_i k_i(k_i - 2) > 0. \tag{13.105}$$

But note now that, as before, vertices of degree zero and degree two make no contribution to the sum, since terms in which $k_i = 0$ or $k_i = 2$ vanish. Thus we can add as many vertices of degree zero or two to the network as we like (or take them away) and it will make no difference to the existence or not of a giant component. We noted a special case of this phenomenon in Section 13.8.1.

13.9 SIZE DISTRIBUTION FOR SMALL COMPONENTS

Having looked in some detail at the behavior of the giant component in the configuration model, let us return once more to the small components. In Eqs. (13.85) and (13.86) we have—in theory at least—the generating functions that give the entire distribution of sizes of the small components. Unfortunately, it is in most cases impossible to solve these equations exactly, but we can still extract plenty of useful information from them. For example, we can calculate the mean size of the component to which a randomly chosen vertex belongs, which is given by the equivalent of Eq. (12.29) thus:

$$\langle s \rangle = \frac{\sum_s s \pi_s}{\sum_s \pi_s} = \frac{h_0'(1)}{1 - S} = \frac{h_0'(1)}{g_0(u)}, \tag{13.106}$$

where we have used Eq. (13.90) in the final equality. Differentiating Eq. (13.85) we get

$$
\begin{aligned}
h_0'(z) &= g_0(h_1(z)) + z g_0'(h_1(z)) h_1'(z) \\
&= g_0(h_1(z)) + z g_0'(1) g_1(h_1(z)) h_1'(z) \\
&= \frac{h_0(z)}{z} + g_0'(1) h_1(z) h_1'(z),
\end{aligned} \tag{13.107}
$$

where we have used Eq. (13.51) in the second equality and Eqs. (13.85) and (13.86) in the third. Setting $z = 1$ we then get

$$h_0'(1) = h_0(1) + g_0'(1) h_1(1) h_1'(1) = 1 - S + g_0'(1) h_1'(1) u, \tag{13.108}$$

where we have used Eqs. (13.87) and (13.89). To calculate $h_1'(1)$ we differentiate Eq. (13.86) thus:

$$
\begin{aligned}
h_1'(z) &= g_1(h_1(z)) + zg_1'(h_1(z))h_1'(z) \\
&= \frac{h_1(z)}{z} + zg_1'(h_1(z))h_1'(z),
\end{aligned} \tag{13.109}
$$

or, rearranging,

$$
h_1'(z) = \frac{h_1(z)/z}{1 - zg_1'(h_1(z))}. \tag{13.110}
$$

Setting $z = 1$ in this expression gives

$$
h_1'(1) = \frac{u}{1 - g_1'(u)}. \tag{13.111}
$$

Combining Eqs. (13.106), (13.108), and (13.111), we then find that

$$
\langle s \rangle = 1 + \frac{g_0'(1)u^2}{g_0(u)[1 - g_1'(u)]}. \tag{13.112}
$$

Using values of S and u from Eqs. (13.90) and (13.91) we can then calculate $\langle s \rangle$ from this equation.

A simple case occurs when we are in the region where there is no giant component. In this region we have $S = 0$ and $u = 1$ by definition and hence

$$
\langle s \rangle = 1 + \frac{g_0'(1)}{1 - g_1'(1)}. \tag{13.113}
$$

Thus the average size of the component to which a vertex belongs diverges precisely at the point where $g_1'(1) = 1$, the point at which the curve in Fig. 13.6 is exactly tangent to the dotted line (the middle curve in the figure). This is, of course, also the point at which the giant component first appears.

Thus the picture we have is similar to that shown in Fig. 12.4 for the Poisson random graph, in which the typical size of the component to which a vertex belongs grows larger and larger until we reach the point, or *phase transition*, where the giant component appears, at which it diverges. Beyond this point the small components shrink in size again, although the overall mean component size, including the giant component, is infinite.

Equation (13.113) can also be expressed in a couple of other forms that may be useful in some circumstances. From Eq. (13.69) we know that $g_1'(1) = (\langle k^2 \rangle - \langle k \rangle)/\langle k \rangle$ and, putting $g_0'(1) = \langle k \rangle$ also, we find that

$$
\langle s \rangle = 1 + \frac{\langle k \rangle^2}{2\langle k \rangle - \langle k^2 \rangle}. \tag{13.114}
$$

This expression can be evaluated easily given only a knowledge of the degree sequence and avoids the need to calculate any generating functions. Using the notation introduced earlier in which c_1 and c_2 are the mean number of first and second neighbors of a vertex, with c_2 given by Eq. (13.70), we can also write (13.114) in the form

$$\langle s \rangle = 1 + \frac{c_1^2}{c_1 - c_2}, \tag{13.115}$$

so that the average size of the component a vertex belongs to is dictated entirely by the mean numbers of first and second neighbors.

13.9.1 AVERAGE SIZE OF A SMALL COMPONENT

As with the Poisson random graph, we must be careful about our claims in the previous section. We have calculated the average size $\langle s \rangle$ of the component to which a randomly chosen vertex belongs but this is not the same thing as the average size of a component, since more vertices belong to larger components, which biases the value of $\langle s \rangle$. If we want the true average size R of the small components, we must use Eq. (12.36), which we reproduce here for convenience:

$$R = \frac{1 - S}{\sum_s \pi_s / s}. \tag{13.116}$$

The sum can be calculated as before using the equivalent of Eq. (12.37):

$$\sum_{s=1}^{\infty} \frac{\pi_s}{s} = \int_0^1 \frac{h_0(z)}{z} \, dz. \tag{13.117}$$

Taking $h_0(z)/z$ from Eq. (13.107), we get

$$\begin{aligned}
\sum_{s=1}^{\infty} \frac{\pi_s}{s} &= \int_0^1 \frac{dh_0}{dz} \, dz - g_0'(1) \int_0^1 h_1(z) \frac{dh_1}{dz} \, dz \\
&= \int_0^{1-S} dh_0 - \langle k \rangle \int_0^u h_1 \, dh_1 \\
&= 1 - S - \tfrac{1}{2} \langle k \rangle u^2.
\end{aligned} \tag{13.118}$$

Then

$$R = \frac{2}{2 - \langle k \rangle u^2 / (1 - S)}. \tag{13.119}$$

Note that the value of this average at the transition point where $S = 0$ and $u = 1$ is just $2/(2 - \langle k \rangle)$, which is normally perfectly finite.[14] Thus the average component size does not normally diverge at the transition (unlike $\langle s \rangle$).

---

[14]The only exception is when $\langle k \rangle = 2$ at the transition point.

13.9.2 COMPLETE DISTRIBUTION OF SMALL COMPONENT SIZES

One of the most surprising results concerning the configuration model is that it is possible to derive an expression not just for the average size of the component to which a vertex belongs, but for the exact probability that it belongs to a component of any specific size—the probability that it belongs to a component of size ten, or a hundred, or a million. The derivation of this result is similar to the derivation given in Section 12.6.3 for the corresponding quantity for the Poisson random graph.

Since a component cannot have size zero, the generating function for the probabilities π_s has the form

$$h_0(z) = \sum_{s=1}^{\infty} \pi_s z^s, \tag{13.120}$$

with the sum starting at 1. Dividing by z and differentiating $s - 1$ times, we then find that

$$\pi_s = \frac{1}{(s-1)!} \left[\frac{d^{s-1}}{dz^{s-1}} \left(\frac{h_0(z)}{z} \right) \right]_{z=0}, \tag{13.121}$$

(which is just a minor variation on the standard formula, Eq. (13.2)). Using Eq. (13.85), this can also be written

$$\begin{aligned} \pi_s &= \frac{1}{(s-1)!} \left[\frac{d^{s-1}}{dz^{s-1}} g_0(h_1(z)) \right]_{z=0} \\ &= \frac{1}{(s-1)!} \left[\frac{d^{s-2}}{dz^{s-2}} [g_0'(h_1(z)) h_1'(z)] \right]_{z=0}. \end{aligned} \tag{13.122}$$

Now we make use of the Cauchy formula for the n derivative of a function, which says that

$$\left. \frac{d^n f}{dz^n} \right|_{z=z_0} = \frac{n!}{2\pi i} \oint \frac{f(z)}{(z-z_0)^{n+1}} dz, \tag{13.123}$$

where the integral is around a contour that encloses z_0 in the complex plane but encloses no poles in $f(z)$. Applying this formula to Eq. (13.122) with $z_0 = 0$ we get

$$\pi_s = \frac{1}{2\pi i(s-1)} \oint \frac{g_0'(h_1(z))}{z^{s-1}} \frac{dh_1}{dz} dz. \tag{13.124}$$

For our contour, we choose an infinitesimal circle around the origin.

Changing the integration variable to h_1, we can also write this as

$$\pi_s = \frac{1}{2\pi i(s-1)} \oint \frac{g_0'(h_1)}{z^{s-1}} dh_1. \tag{13.125}$$

Here we are regarding z now as a function of h_1, rather than the other way around. Furthermore, since $h_1(z)$ goes to zero as $z \to 0$, the contour in h_1 surrounds the origin too. (The proof is the same as for Eq. (12.46).)

Now we make use of Eq. (13.86) to eliminate z and write

$$
\begin{aligned}
\pi_s &= \frac{1}{2\pi i(s-1)} \oint \frac{[g_1(h_1)]^{s-1} g_0'(h_1)}{h_1^{s-1}} \, dh_1 \\
&= \frac{g_0'(1)}{2\pi i(s-1)} \oint \frac{[g_1(h_1)]^{s}}{h_1^{s-1}} \, dh_1,
\end{aligned}
\tag{13.126}
$$

where we have made use of Eq. (13.51) in the second line. Given that the contour surrounds the origin, this integral is now in the form of Eq. (13.123) again, and hence

$$
\pi_s = \frac{\langle k \rangle}{(s-1)!} \left[\frac{d^{s-2}}{dz^{s-2}} [g_1(z)]^{s} \right]_{z=0},
\tag{13.127}
$$

where we have written $g_0'(1) = \langle k \rangle$.

The only exception to this formula is for the case $s = 1$, for which Eq. (13.124) gives $0/0$ and is therefore clearly incorrect. However, since the only way to belong to a component of size 1 is to have no connections to any other vertices, the probability π_1 is trivially equal to the probability of having degree zero:

$$
\pi_1 = p_0.
\tag{13.128}
$$

Equations (13.127) and (13.128) give the probability that a randomly chosen vertex belongs to a component of size s in terms of the degree distribution. In principle if we know p_k we can calculate π_s. It is not always easy to perform the derivatives in practice and in some cases we may not even know the generating function $g_1(z)$ in closed form, but at least in some cases the calculations are possible. As an example, consider a network with the exponential degree distribution

$$
p_k = (1 - e^{-\lambda}) e^{-\lambda k},
\tag{13.129}
$$

with exponential parameter $\lambda > 0$. From Eqs. (13.9) and (13.51) the generating functions $g_0(z)$ and $g_1(z)$ are given by

$$
g_0(z) = \frac{e^{\lambda} - 1}{e^{\lambda} - z}, \qquad g_1(z) = \left(\frac{e^{\lambda} - 1}{e^{\lambda} - z} \right)^{2}.
\tag{13.130}
$$

Then it is not hard to show that

$$
\frac{d^n}{dz^n} [g_1(z)]^{s} = \frac{(2s - 1 + n)!}{(2s - 1)!} \frac{[g_1(z)]^{s}}{(e^{\lambda} - z)^{n}},
\tag{13.131}
$$

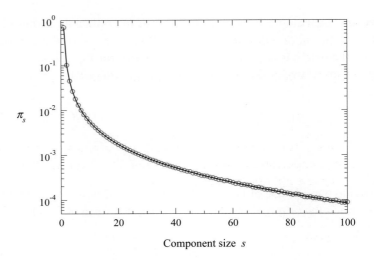

Figure 13.7: The distribution of component sizes in a configuration model. The probability π_s that a vertex belongs to a component of size s for the configuration model with an exponential degree distribution of the form (13.129) for $\lambda = 1.2$. The solid lines represent the exact formula, Eq. (13.132), for the $n \to \infty$ limit and the points are measurements of π_s averaged over 100 computer-generated networks with $n = 10^7$ vertices each.

and hence

$$\pi_s = \frac{(3s-3)!}{(s-1)!(2s-1)!}\, e^{-\lambda(s-1)}\left(1 - e^{-\lambda}\right)^{2s-1}. \qquad (13.132)$$

Figure 13.7 shows a comparison of this formula with the results of numerical simulations for $\lambda = 1.2$ and, as we can see, the agreement between formula and simulations is good—our calculations seem to describe the simulated random graph well even though the graph is necessarily finite in size while the calculations are performed in the limit of large n.

13.10   POWER-LAW DEGREE DISTRIBUTIONS

As we saw in Section 8.4, a number of networks have degree distributions that approximately obey a power law. As an example of the application of the machinery developed in this chapter, let us look at the properties of a random graph with a power-law degree distribution.

Suppose we have a network with a "pure" power-law degree distribution of the form

$$p_k = \begin{cases} 0 & \text{for } k = 0, \\ k^{-\alpha}/\zeta(\alpha) & \text{for } k \geq 1. \end{cases} \qquad (13.133)$$

(See Eq. (13.13).) Here $\alpha > 0$ is a constant exponent and $\zeta(\alpha)$ is the Riemann zeta function:

$$\zeta(\alpha) = \sum_{k=1}^{\infty} k^{-\alpha}. \qquad (13.134)$$

Using the results of the previous sections we can, for instance, say whether there is a giant component in this network or not. Equation (13.76) tells us that there will be a giant component if and only if

$$\langle k^2 \rangle - 2\langle k \rangle > 0. \qquad (13.135)$$

In the present case

$$\langle k \rangle = \sum_{k=0}^{\infty} k p_k = \frac{1}{\zeta(\alpha)} \sum_{k=1}^{\infty} k^{-\alpha+1} = \frac{\zeta(\alpha - 1)}{\zeta(\alpha)}, \qquad (13.136)$$

and

$$\langle k^2 \rangle = \sum_{k=0}^{\infty} k^2 p_k = \frac{1}{\zeta(\alpha)} \sum_{k=1}^{\infty} k^{-\alpha+2} = \frac{\zeta(\alpha - 2)}{\zeta(\alpha)}. \qquad (13.137)$$

Thus there is a giant component if

$$\zeta(\alpha - 2) > 2\zeta(\alpha - 1). \qquad (13.138)$$

Figure 13.8 shows this inequality in graphical form. The two curves in the figure show the values of $\zeta(\alpha - 2)$ and $2\zeta(\alpha - 1)$ as functions of α and, as we can see, the inequality (13.138) is satisfied only for sufficiently low values of α, below the dotted line in the figure. In fact a numerical solution of the equation $\zeta(\alpha - 2) = 2\zeta(\alpha - 1)$ indicates that the network will have a giant component only for $\alpha < 3.4788\ldots$, a result first given by Aiello *et al.* [9] in 2000.

In practice this result is of only limited utility because it applies only for the pure power law. In general, other distributions with power-law tails but different behavior for low k will have different thresholds at which the giant component appears. There is however a general result we can derive that applies to all distributions with power-law tails. In Section 8.4.2 we noted that the

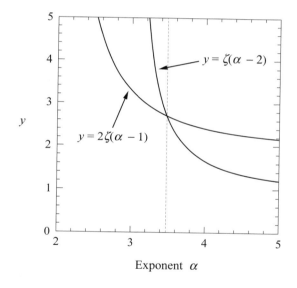

Figure 13.8: Graphical solution of Eq. (13.138). The configuration model with a pure power-law degree distribution (Eq. (13.133)) has a giant component if $\zeta(\alpha - 2) > 2\zeta(\alpha - 1)$. This happens for values of α below the crossing point of the two curves.

second moment $\langle k^2 \rangle$ diverges for any distribution with a power-law tail with exponent $\alpha \leq 3$, while the first moment $\langle k \rangle$ remains finite so long as $\alpha > 2$. This means that Eq. (13.135) is always satisfied for any configuration model with a power-law tail to its degree distribution so long as α lies in the range $2 < \alpha \leq 3$, and hence there will always be a giant component no matter what else the distribution does. For $\alpha > 3$, on the other hand, there may or may not be a giant component, depending on the precise functional form of the degree distribution. (For $\alpha \leq 2$ it turns out that there is always a giant component, although more work is needed to demonstrate this.) Note that, as discussed in Section 8.4, most observed values of α for real-world networks lie in the range $2 < \alpha \leq 3$ and hence we tentatively expect such networks to have a giant component, although we must also bear in mind that the configuration model is a simplified model of a network and is not necessarily a good representation of any specific real-world network.

Returning to the pure power law let us calculate the size S of the giant component, when there is one. The fundamental generating functions $g_0(z)$ and $g_1(z)$ for the power-law distribution are given by Eqs. (13.55) and (13.56), which we repeat here for convenience:

$$g_0(z) = \frac{\mathrm{Li}_\alpha(z)}{\zeta(\alpha)}, \qquad g_1(z) = \frac{\mathrm{Li}_{\alpha-1}(z)}{z\zeta(\alpha-1)}. \tag{13.139}$$

Here $\zeta(\alpha)$ is the Riemann zeta function again and $\mathrm{Li}_\alpha(z)$ is the polylogarithm

$$\mathrm{Li}_\alpha(z) = \sum_{k=1}^{\infty} k^{-\alpha} z^k. \tag{13.140}$$

(See Eq. (13.15).) Now the crucial equation (13.91) for the probability $u = h_1(1)$ reads

$$u = \frac{\mathrm{Li}_{\alpha-1}(u)}{u\zeta(\alpha-1)} = \frac{\sum_{k=1}^{\infty} k^{-\alpha+1} u^k}{u\zeta(\alpha-1)} = \frac{\sum_{k=0}^{\infty}(k+1)^{-\alpha+1} u^k}{\zeta(\alpha-1)}, \tag{13.141}$$

where we have used the explicit definition of the polylogarithm for clarity.

In general there is no closed-form solution for this equation, but we do notice some interesting points. In particular, note that the sum in the numerator is strictly positive for $u \geq 0$, which means that if $\zeta(\alpha-1)$ diverges we will get a solution $u = 0$. And indeed $\zeta(\alpha-1)$ does diverge. It diverges at $\alpha = 2$ and all values below, as one can readily verify from the definition, Eq. (13.134).[15]

---

[15]Traditionally $\zeta(x)$ is actually defined to have finite values below $x = 1$ by analytic continuation. But in our case we are really interested in the value of the sum $\sum_{k=1}^{\infty} k^{-x}$, which diverges for all $x \leq 1$.

Thus for $\alpha \leq 2$ we have $u = 0$ and Eq. (13.90) then tells us that the giant component has size $S = 1 - g_0(0) = 1 - p_0$. However, for our particular choice of degree distribution, Eq. (13.133), there are no vertices with degree zero, and hence $p_0 = 0$ and $S = 1$. That is, the giant component fills the entire network and there are no small components at all!

Technically, this statement is not quite correct. There is always some chance that, for instance, a vertex of degree 1 will connect to another vertex of degree 1, forming a small component. What we have shown is that the probability that a randomly chosen vertex belongs to a small component is zero in the limit of large n, i.e., that what small components there are fill a fraction of the network that vanishes as $n \rightarrow \infty$. In the language used by mathematicians, a randomly chosen vertex "almost surely" belongs to the giant component, meaning it is technically possible to observe another outcome, but the probability is vanishingly small.

Thus our picture of the pure power-law configuration model is one in which there is a giant component for values of $\alpha < 3.4788\ldots$ and that giant component fills essentially the entire network when $\alpha \leq 2$. In the region between $\alpha = 2$ and $\alpha = 3.4788$ there is a giant component but it does not fill the whole network and some portion of the network consists of small component. If $\alpha > 3.4788\ldots$ there are only small components. As a confirmation of this picture, Fig. 13.9 shows the size of the giant component extracted from a numerical solution of Eq. (13.141).[16] As we can see it fits nicely with the picture described above.

We could in principle take our calculations further, calculating, for instance, the mean size of the small components in the region $\alpha > 2$ using Eq. (13.112), or the entire distribution of their sizes using Eq. (13.127).

13.11 Directed random graphs

In this chapter we have studied random graph models that go a step beyond the Poisson random graph of Chapter 12 by allowing us to choose the degree distribution of our model network. This introduces an additional level of realism to the model that makes it substantially more informative. It is, however, only a first step. There are many other features we can add to the model to make it more realistic still. We can for instance create random graph models of networks with assortative (or disassortative) mixing [237], bipartite struc-

---

[16]The numerical solution is simple: we just choose a suitable starting value ($u = \frac{1}{2}$ works fine) and iterate Eq. (13.141) until it converges. Fifty iterations are easily enough to give a highly accurate result.

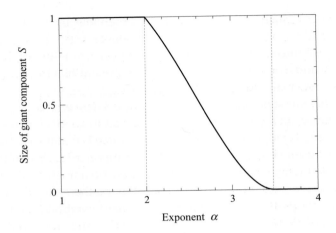

Figure 13.9: Size of the giant component for the configuration model with a power-law degree distribution. This plot shows the fraction of the network filled by the giant component as a function of the exponent α of the power law, calculated by numerical solution of Eqs. (13.91) and (13.141). The dotted lines mark the value $\alpha = 2$ below which the giant component has size 1 and the value $\alpha = 3.4788$ above which there is no giant component.

ture [253], or clustering [247]. All of these models are still exactly solvable in the limit of large system size, although the solutions are more complicated than for the models we have seen in this chapter. For instance, in the case of the random graph with assortative mixing the fundamental generating function $g_1(z)$ becomes a vector, the corresponding equation (13.86) for the distribution of component sizes becomes a vector equation, and the condition for the existence of a giant component, Eq. (13.76), becomes a condition on the determinant of a matrix.

We will not go into detail on all of the many random graph models that have been proposed and studied, but in this section we take a look at one case, that of the directed random graph, as an example of the types of calculation that are possible.

13.11.1 GENERATING FUNCTIONS FOR DIRECTED GRAPHS

As discussed in Section 6.4, many networks, including the World Wide Web, metabolic networks, food webs, and others, are directed. The configuration model can be generalized to directed networks in a straightforward fashion, although the generalization displays some new behaviors not seen in the undirected case. Our presentation follows that of Refs. [100] and [253].

To create a directed equivalent of the configuration model, we must specify a double degree sequence, consisting of an in-degree j_i and an out-degree k_i for each vertex i. We can think of these as specifying the numbers of ingoing and outgoing stubs of edges at each vertex. Then we create a network by repeatedly choosing pairs of stubs—one ingoing and one outgoing—uniformly at random and connecting them to make directed edges, until no unused stubs remain. The result is a matching of the stubs drawn uniformly at random from the set of all possible matchings, just as in the configuration model, and the model itself is defined to be the ensemble of such directed networks in which each matching appears with equal probability. (The only small catch is that we must make sure that the total number of ingoing and outgoing stubs is the same, so that none are left over at the end of the process. We will assume this to be the case in the following developments.)

The probability that a particular outgoing stub at vertex w attaches to one of the j_v ingoing stubs at vertex v is

$$\frac{j_v}{\sum_i j_i} = \frac{j_v}{m}, \tag{13.142}$$

where m is the total number of edges and we have made use of Eq. (6.26). Since the total number of outgoing stubs at w is k_w, the total expected number of directed edges from vertex w to vertex v is then $j_v k_w / m$, which is also the probability of an edge from w to v in the limit of large network size, provided the network is sparse. This is similar to the corresponding result, Eq. (13.32), for the undirected configuration model, but not identical—notice that there is no factor of two now in the denominator.

As in the undirected case we can, if we prefer, work with the degree distribution, rather than the degree sequence. As discussed in Section 8.3, the most correct way to describe the degree distribution of a directed network is by a joint distribution: we define p_{jk} to be the fraction of vertices in the network that have in-degree j and out-degree k. This allows for the possibility that the in- and out-degrees of vertices are correlated. For instance, it would allow us to represent a network in which the in- and out-degrees of each vertex were exactly equal to one another.[17] (This is rather an extreme example, but it demonstrates the point.)

The joint degree distribution can be captured in generating function form

---

[17]Given the joint degree distribution we can still, if we wish, calculate the distributions of in- or

by defining a *double generating function* $g_{00}(x, y)$ thus:

$$g_{00}(x, y) = \sum_{j,k=0}^{\infty} p_{jk} x^j y^k. \qquad (13.143)$$

(The two subscript zeros are the equivalent for the double generating function of the subscript zero in our previous generating function $g_0(z)$ for the undirected network.) As in the undirected case, the generating function $g_{00}(x, y)$ captures all the information contained in the degree distribution. Given the generating function we can reconstruct the degree distribution by differentiating:

$$p_{jk} = \frac{1}{j!\, k!} \frac{\partial^j \partial^k g_{00}}{\partial x^j \partial y^k} \bigg|_{x,y=0}. \qquad (13.144)$$

This is the equivalent for the directed case of Eq. (13.2) in the undirected case.

Just as in the undirected case the generating function satisfies certain conditions. First, since the degree distribution must be normalized according to $\sum_{jk} p_{jk} = 1$, the generating function satisfies

$$g_{00}(1, 1) = 1. \qquad (13.145)$$

Second, the average in- and out-degrees are given by

$$\langle j \rangle = \sum_{j,k=0}^{\infty} j p_{jk} = \frac{\partial g_{00}}{\partial x} \bigg|_{x,y=1}, \qquad (13.146)$$

$$\langle k \rangle = \sum_{j,k=0}^{\infty} k p_{jk} = \frac{\partial g_{00}}{\partial y} \bigg|_{x,y=1}. \qquad (13.147)$$

In a directed graph, however, the average in- and out-degrees are equal—see Eq. (6.27)—so $\langle j \rangle = \langle k \rangle$ and

$$\frac{\partial g_{00}}{\partial x} \bigg|_{x,y=1} = \frac{\partial g_{00}}{\partial y} \bigg|_{x,y=1}. \qquad (13.148)$$

For convenience we will denote the average in-degree and out-degree by c in the equations that follow. Thus $\langle j \rangle = \langle k \rangle = c$.

---

out-degrees alone. They are given by

$$p_j^{(in)} = \sum_{k=0}^{\infty} p_{jk}, \qquad p_k^{(out)} = \sum_{j=0}^{\infty} p_{jk}.$$

We can also write down generating functions for the excess degree distribution of vertices reached by following an edge in the network. There are two different ways of following a directed edge—either forward or backward. Consider first the forward case. If we follow an edge forward to the vertex it points to, then the probability of reaching a particular vertex will be proportional to the number of edges pointing to that vertex, i.e., to its in-degree. Thus the joint degree distribution of such a vertex is proportional not to p_{jk} but to jp_{jk}. As before, we will be interested primarily in the number of edges entering and leaving a vertex other than the one we arrived along. If j and k denote these numbers then the total in-degree is $j + 1$ and the total out-degree is just k, so the distribution we want is proportional to $(j + 1)p_{j+1,k}$ or, correctly normalized, $(j + 1)p_{j+1,k}/c$. The double generating function for this excess degree distribution is then

$$g_{10}(x, y) = \frac{\sum_{jk}(j + 1)p_{j+1,k}x^j y^k}{\sum_{jk}(j + 1)p_{j+1,k}} = \frac{\sum_{jk}jp_{jk}x^{j-1}y^k}{\sum_{jk}jp_{jk}}$$
$$= \frac{1}{c}\frac{\partial g_{00}}{\partial x}. \qquad (13.149)$$

The backward case is similar. The appropriate excess degree distribution for the vertex from which an edge originates is $(k + 1)p_{j,k+1}/c$ and has generating function

$$g_{01}(x, y) = \frac{\sum_{jk}(k + 1)p_{j,k+1}x^j y^k}{\sum_{jk}(k + 1)p_{j,k+1}} = \frac{1}{c}\frac{\partial g_{00}}{\partial y}. \qquad (13.150)$$

13.11.2 Giant components

A directed graph has various different types of component, as discussed in Sections 6.11.1 and 8.1.1, including strongly and weakly connected components, in-components, and out-components. (Take a look at the "bow tie" diagram, Fig. 8.2 on page 240, for a reminder of the definitions of the components.) In general there can be both small and giant components of each of these types. Let us look at the giant components.

A strongly connected component is a set of vertices in which every vertex is reachable by a directed path from every other in the set. To put that a different way, for a vertex to belong to a strongly connected component at least one of its outgoing edges must lead to another vertex from which there is a path to the strongly connected component, and at least one of its ingoing edges must lead from a vertex to which there is path from the strongly connected component (see figure).

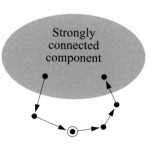

A vertex belongs to a strongly connected component if it has a directed path to the component and another from the component.

Let v be the probability that the vertex to which a randomly chosen edge in our graph leads has no directed path to the giant strongly connected component. For this to happen, it must be that none of the other outgoing edges from that vertex themselves have such a path. If the vertex has out-degree k, this happens with probability v^k. But j and k are distributed according to the excess degree distribution $(j+1)p_{j+1,k}/c$ and hence, averaging over both, we find that

$$v = \frac{1}{c} \sum_{j,k=0}^{\infty} (j+1)p_{j+1,k}v^k = g_{10}(1,v).$$ (13.151)

Similarly, consider the vertex from which a randomly chosen edge originates and let u be the probability that there is no path from the giant strongly connected component to that vertex. Then u is the solution to

$$u = g_{01}(u,1).$$ (13.152)

Now consider a vertex with in-degree j and out-degree k. The probability that there is no path to the vertex from the giant strongly connected component via any of the vertex's j ingoing edges is u^j and the probability that there *is* such a path is $1 - u^j$. Similarly the probability that there is a path from the vertex to the giant strongly connected component is $1 - v^k$. And the probability that there are both—and hence that the vertex itself belongs to the giant strongly connected component—is the product of these two, or $(1 - u^j)(1 - v^k)$. Averaging this expression over the joint distribution of j and k we then find that the average probability S_s that a vertex lies in the giant strongly connected component, which is also the size of the giant strongly connected component measured as a fraction of the network size, is

$$S_s = \sum_{j,k=0}^{\infty} p_{jk}(1 - u^j)(1 - v^k)$$

$$= \sum_{j,k=0}^{\infty} p_{jk} - \sum_{j,k=0}^{\infty} p_{jk}u^j - \sum_{j,k=0}^{\infty} p_{jk}v^k + \sum_{j,k=0}^{\infty} p_{jk}u^j v^k$$

$$= 1 - g_{00}(u,1) - g_{00}(1,v) + g_{00}(u,v),$$ (13.153)

with u and v given by Eqs. (13.151) and (13.152).

As discussed in Section 6.11, each strongly connected component in a network also has an in-component and an out-component associated with it—the sets of vertices from which it can be reached, and which can be reached from it. The in- and out-components of the giant strongly connected component are usually called the giant in- and out-components. By their definition, both are supersets of the giant strongly connected component itself, and we can calculate the size of both for our directed random graph. In fact, we have performed most of the calculation already.

A vertex with out-degree k fails to belong to the giant in-component only if none of its outgoing edges leads to a vertex that has a path to the strongly connected component. This happens with probability v^k, where v is as above. Averaging over j and k, we find the probability S_i that the vertex does belong to the giant in-component (which is also the size of the giant in-component) to be

$$S_i = 1 - \sum_{j,k=0}^{\infty} p_{jk} v^k = 1 - g_{00}(1, v). \tag{13.154}$$

Similarly the size of the giant out-component is given by

$$S_o = 1 - g_{00}(u, 1). \tag{13.155}$$

Using these results we can also write an expression for the combined size of the giant strongly connected component and its in- and out-components— the entire "bow tie" in Fig. 8.2. Since the giant in- and out-components both include the giant strongly connected component as a subset, their sum is equal to the size of the whole bow tie except that it counts the strongly connected part twice. Subtracting S_s to allow for this overcounting we then find the size of the bow tie to be

$$S_i + S_o - S_s = 1 - g_{00}(u, v), \tag{13.156}$$

and the fraction of the network not in the bow tie is just $g_{00}(u, v)$. (We could have derived this result by more direct means, just by noting that a vertex not in the bow tie has a path neither to nor from the giant strongly connected component.)

And what about the giant weakly connected component? A weakly connected component in a directed graph is a normal graph component of connected vertices in which we ignore the directions of all the edges. At first glance one might imagine that the size of the giant weakly connected component was just equal to the combined size $S_i + S_o - S_s$ of the in-, out-, and strongly connected components calculated above. This, however, is not correct because the definition of the giant weakly connected component includes some vertices that are not in the in-, out-, or strongly connected components. An example would be any vertex that is reachable from the giant in-component but that does not itself have a path to the strongly connected component and hence is not in the giant in-component. Thus the size of the giant weakly connected component is, in general, larger than $S_i + S_o - S_s$. Nonetheless, we can still calculate the size of the giant weakly connected component by an argument quite similar to the ones we have already seen.

A vertex belongs to the giant weakly connected component if any of its edges, ingoing or outgoing, are connected to a vertex in that component. Let u now be the probability that a vertex is not connected to the giant weakly connected component via the vertex at the other end of one of its ingoing edges and let v be the equivalent probability for an outgoing edge. Then the probability that a vertex with in-degree j and out-degree k is not in the giant weakly connected component is $u^j v^k$ and the probability that it is in the giant weakly connected component is $1 - u^j v^k$. Averaging over the joint distribution p_{jk} of the two degrees we then find that the size S_w of the giant weakly connected component is

$$S_w = \sum_{jk} p_{jk} - \sum_{jk} p_{jk} u^j v^k = 1 - g_{00}(u,v). \tag{13.157}$$

We can derive the value of u by noting that the vertex at the end of an ingoing edge is not in the giant weakly connected component with probability $u^j v^k$ again, but with j and k being the numbers of edges excluding the edge we followed to reach the vertex. These numbers are distributed according to the appropriate excess degree distribution and, performing the average, we find that

$$u = g_{01}(u,v). \tag{13.158}$$

Similarly we can show that

$$v = g_{10}(u,v), \tag{13.159}$$

and Eqs. (13.157) to (13.159) between them give us our solution for the size of the giant weakly connected component.

13.11.3 THE APPEARANCE OF THE GIANT COMPONENTS

As in the undirected random graph, there may or may not be giant components in the direct random graph, depending on the degree distribution. We can derive conditions for the existence of the giant components using the machinery developed above. The calculation is easiest for the giant in- and out-components. Their size is given by Eqs. (13.154) and (13.155). Given that $g_{00}(1,1) = 1$ (Eq. (13.145)), these equations give a non-zero size only if u or v is less than 1. Looking at Eq. (13.151) for the value of v we see a similar situation to that depicted in Fig. 13.6: we can have a solution with $v < 1$, and hence a giant in-component, only if

$$\left. \frac{\partial g_{10}}{\partial y} \right|_{x,y=1} > 1, \tag{13.160}$$

or equivalently if

$$\frac{\partial^2 g_{00}}{\partial x \partial y}\bigg|_{x,y=1} > c, \qquad (13.161)$$

where we have made use of Eqs. (13.148) and (13.149). Similarly we can have a giant out-component only if

$$\frac{\partial g_{01}}{\partial x}\bigg|_{x,y=1} > 1, \qquad (13.162)$$

or equivalently,

$$\frac{\partial^2 g_{00}}{\partial x \partial y}\bigg|_{x,y=1} > c. \qquad (13.163)$$

Interestingly, Eqs. (13.161) and (13.163) are identical, meaning that the conditions for the giant in- and out-components to appear are the same. If there is a phase transition at which one appears, the other also appears at the exact same moment.[18]

We can express (13.161) directly in terms of the degree distribution if we want. Substituting from Eq. (13.143) we find that

$$\frac{\partial^2 g_{00}}{\partial x \partial y}\bigg|_{x,y=1} = \sum_{j,k=0}^{\infty} jk p_{jk} \qquad (13.164)$$

and hence the giant in- and out-components appear if

$$\sum_{j,k=0}^{\infty} jk p_{jk} > c. \qquad (13.165)$$

If we prefer we can write $c = \sum_j j p_{jk} = \sum_j k p_{jk}$ to give the alternative form

$$\sum_{j,k=0}^{\infty} (2jk - j - k) p_{jk} > 0. \qquad (13.166)$$

This result is the equivalent for a directed network of Eq. (13.76) for the undirected case.

The calculation for the giant strongly connected component is similar. From Eq. (13.153) we see that $S_s = 0$ unless at least one of u and v is non-zero,

---

[18]Strictly we have not shown that a giant component actually appears if the condition (13.161) is satisfied. We have only shown that (13.161) is a necessary condition for a giant component to appear. However, we can go through an argument similar to the one we made for the undirected case in Section 13.6 to convince ourselves that there must be a giant component when the condition is satisfied.

so the condition for the existence of a giant strongly connected component is the same as for the in- and out-components, Eq. (13.166). In other words, the giant in-, out-, and strongly connected components all appear or disappear simultaneously.

The giant weakly connected component, however, is different. It is possible for there to be a giant weakly connected component in a network but no giant strongly connected component and hence the condition for the existence of a giant weakly connected component must be different from that for the other giant components. For instance, a network in which all vertices have either only ingoing edges or only outgoing edges can have a giant weakly connected component but trivially has no strongly connected component of size greater than one, since there are only paths to or from each vertex, but not both. Weakly connected components, however, are generally of less interest than strongly connected ones, and the calculation of the condition for the existence of the giant weakly connected component is non-trivial, so we leave it as an exercise for the motivated reader and move on to other things.

13.11.4 SMALL COMPONENTS

We can also calculate the distribution of small components in a directed random graph. In fact the distribution of small strongly connected components is trivial: there aren't any. Or more properly the probability that a randomly chosen vertex belongs to a strongly connected component of size greater than one other than the giant strongly connected component is zero in the limit of large network size. To see this, recall that the small components in the undirected configuration model take the form of trees (see Section 13.7). If we consider a small strongly connected component in a directed network and ignore the directions of its edges, then the same argument we used before indicates that the resulting subgraph will also be a tree. But a tree has no loops in it, which leads to a contradiction because a strongly connected component must have loops—the paths in either direction between any pair of vertices form a loop. Thus, we conclude, there cannot be any small strongly connected components of size greater than one in the network.

In fact, this is not precisely true. In a random network there is always some chance that, for example, two vertices will each have a directed edge to the other, forming a strongly connected component of two vertices. In the limit of large n, however, the probability that a randomly chosen vertex belongs to such a component tends to zero. A detailed calculation shows that on average there is only a constant number of short loops in the network and their density vanishes as $1/n$ in the limit of large network size.

There can however be small in- and out-components. In a directed network, each strongly connected component has its own in- and out-components. In the present model, as we have said, we have no small strongly connected components, other than single vertices, so the component structure consists of the giant in- and out-components and then a large number of small in- and out-components for single vertices. Let us ask what the probability is that a randomly chosen vertex has a small out-component of size s, i.e., that there are s vertices including itself that can be reached by directed paths starting from the vertex. We can calculate the distribution of sizes by the same method we used for the undirected case. We define a generating function $h_1(y)$ for the distribution of the size of the out-component of a vertex reached by following an edge in the forward direction, which then satisfies an equation of the form of Eq. (13.86), except that the generating function g_1 for the excess degree distribution is replaced by the corresponding generating function for the directed network, Eq. (13.149), giving

$$h_1(y) = yg_{10}(1, h_1(y)).$$
(13.167)

And the generating function $h_0(y)$ for the size of the out-component to which a randomly chosen vertex belongs is then

$$h_0(y) = yg_{00}(1, h_1(y)).$$
(13.168)

We can write similar equations for in-components too and, armed with these equations, we can find the average size of the in- or out-component to which a vertex belongs, or even find the entire distribution of component sizes using the equivalent of Eq. (13.127).

PROBLEMS

13.1 Consider the binomial probability distribution $p_k = \binom{n}{k} p^k (1-p)^{n-k}$.

a) Show that the distribution has probability generating function $g(z) = (pz + 1 - p)^n$.

b) Find the first and second moments of the distribution from Eq. (13.25) and hence show that the variance of the distribution is $\sigma^2 = np(1-p)$.

c) Show that the sum of two numbers drawn independently from the same binomial distribution is distributed according to $\binom{2n}{k} p^k (1-p)^{2n-k}$.

13.2 Consider a configuration model in which every vertex has the same degree k.
   a) What is the degree distribution p_k? What are the generating functions g_0 and g_1 for the degree distribution and the excess degree distribution?
   b) Show that the giant component fills the whole network for all $k \geq 3$.
   c) What happens when $k = 1$?
   d) When $k = 2$ show that in the limit of large n the probability π_s that a vertex belongs to a component of size s is given by $\pi_s = 1/\left[2\sqrt{n(n-s)}\right]$.

13.3 Consider the configuration model with exponential degree distribution $p_k = (1 - e^{-\lambda})e^{-\lambda k}$ with $\lambda > 0$, so that the generating functions $g_0(z)$ and $g_1(z)$ are given by Eq. (13.130).
   a) Show that the probability u of Eq. (13.91) satisfies the cubic equation

$$u^3 - 2e^{\lambda}u^2 + e^{2\lambda}u - (e^{\lambda} - 1)^2 = 0.$$

   b) Noting that $u = 1$ is always a trivial solution of this equation, show that the non-trivial solution corresponding to the existence of a giant component satisfies the quadratic equation $u^2 - (2e^{\lambda} - 1)u + (e^{\lambda} - 1)^2 = 0$, and hence that the size of the giant component, if there is one, is

$$S = \tfrac{3}{2} - \sqrt{e^{\lambda} - \tfrac{3}{4}}.$$

   c) Show that the giant component exists only if $\lambda < \ln 3$.

13.4 Equation (13.74) tells us on average how many vertices are a distance d away from a given vertex.
   a) Assuming that this expression works for all values of d (which is only a rough approximation to the truth), at what value of d is this average number of vertices equal to the number n in the whole network?
   b) Hence derive a rough expression for the diameter of the network in terms of c_1 and c_2, and so argue that configuration model networks display the small-world effect in the sense that typical geodesic distances between vertices are $O(\log n)$.

13.5 Consider a network model in which edges are placed independently between each pair of vertices i, j with probability $p_{ij} = Kf_if_j$, where K is a constant and f_i is a number assigned to vertex i. Show that the expected degree c_i of vertex i within the model is proportional to f_i, and hence that the only possible choice of probability with this form is $p_{ij} = c_ic_j/2m$, as in the model of Section 13.2.2.

13.6 As described in Section 13.2, the configuration model can be thought of as the ensemble of all possible matchings of edge stubs, where vertex i has k_i stubs. Show that for a given degree sequence the number Ω of matchings is

$$\Omega = \frac{(2m)!}{2^m m!},$$

which is independent of the degree sequence.

13.7 Consider the example model discussed in Section 13.8.1, a configuration model with vertices of degree three and less only and generating functions given by Eqs. (13.94) and (13.95).

a) In the regime in which there is no giant component, show that the average size of the component to which a randomly chosen vertex belongs is

$$\langle s \rangle = 1 + \frac{(p_1 + 2p_2 + 3p_3)^2}{p_1 - 3p_3}.$$

b) In the same regime find the probability that such a vertex belongs to components of size 1, 2, and 3.

13.8 Consider a directed random graph of the kind discussed in Section 13.11.

a) If the in- and out-degrees of vertices are uncorrelated, i.e., if the joint in/out-degree distribution p_{jk} is a product of separate functions of j and k, show that a giant strongly connected component exists in the graph if and only if $c(c-1) > 0$, where c is the mean degree, either in or out.

b) In real directed graphs the degrees are usually correlated (or anti-correlated). The correlation can be quantified by the covariance ρ of in- and out-degrees. Show that in the presence of correlations, the condition above for the existence of a giant strongly connected component generalizes to $c(c-1) + \rho > 0$.

c) In the World Wide Web the in- and out-degrees of the vertices have a measured covariance of about $\rho = 180$. The mean degree is around $c = 4.6$. On the basis of these numbers, do we expect the Web to have a giant strongly connected component?

CHAPTER 14

MODELS OF NETWORK FORMATION

A discussion of models of the formation of networks,
particularly models of networks that grow by addition of
vertices, such as the World Wide Web or citation
networks

T HE MODELS described in Chapters 12 and 13 provide an excellent tool for studying the structural features of networks, such as giant and small components, degree distributions, the lengths of paths in networks, and so forth. Moreover, as we will see in later chapters, they can also serve as a convenient basis for further modeling work, such as the modeling of network resilience or of the spread of diseases over contact networks.

But there is another important class of network model that has an entirely different purpose. In the models we have seen so far, the parameters of the network, such as the number of vertices and edges or the degree distribution, are fixed from the outset—chosen by the modeler to have some desired values. For instance, if we are interested in networks with power-law degree distributions, we make a random graph model with a power-law degree distribution as in Section 13.10 and then explore its structure analytically or computationally. But models of this kind offer no explanation of *why* the network should have a power-law degree distribution in the first place. In this chapter we describe models of a different kind that offer such an explanation.

The models in this chapter are *generative network models*. That is, they model the mechanisms by which networks are created. The idea behind models such as these is to explore hypothesized generative mechanisms to see what structures they produce. If the structures are similar to those of networks we observe in the real world, it suggests—though does not prove—that similar generative mechanisms may be at work in the real networks. The best-known example of a generative network model, and the one that we study first in this chapter, is the "preferential attachment" model for the growth of networks with power-law degree distributions. Later in the chapter we examine a number of other models, including generalizations of preferential attachment models, vertex copying models, and models based on optimization.

14.1 PREFERENTIAL ATTACHMENT

As discussed in Section 8.4, many networks are observed to have degree distributions that approximately follow power laws, at least in the tail of the distribution. Examples include the Internet, the World Wide Web, citation networks, and some social networks and biological networks. The power law is a somewhat unusual distribution and its occurrence in empirical data is often considered a potential indicator of interesting underlying processes.[1] A natural question to ask therefore is how might a network come to have such a distribution? This question was first directly considered in the 1970s by Price [275], who proposed a simple and elegant model of network formation that gives rise to power-law degree distributions.

Price was interested in, among other things, the citation networks of scientific papers, having authored an important early paper on the topic in the 1960s in which he pointed out the power-law degree distribution seen in these networks [274]. In considering the possible origins of the power law, Price was inspired by the work of economist Herbert Simon [299], who noted the occurrence of power laws in a variety of (non-network) economic data, such as the distribution of people's personal wealth. Simon proposed an explanation for the wealth distribution based on the idea that people who have money already gain more at a rate proportional to how much they already have. This seems a reasonable supposition. Wealthy individuals make money by investing the money they have, and the return on their investment is essentially proportional to the amount invested. Simon was able to show mathematically that this "rich-get-richer" effect can give rise to a power-law distribution and Price adapted Simon's methods, with relatively little change, to the network context. Price gave a name to Simon's mechanism:[2] he called it *cumulative advantage*, although it is more often known today by the name *preferential attachment*, which was coined in 1999 by Barabási and Albert [27]. In this book we use principally the latter term, which has become the accepted name in recent years.

See Section 4.2 for a discussion of citation networks.

---

[1]For a discussion see, for instance, Refs. [222] and [244].

[2]Simon himself called the mechanism the *Yule process*, in recognition of the statistician Udny Yule, who had studied a simple version many years earlier [333].

Price's model of a citation network is as follows. We assume that papers are published continually (though they do not have to be published at a constant rate) and that newly appearing papers cite previously existing ones. As discussed in Section 4.2, the papers and citations form a directed citation network, the papers being the vertices and the citations being the directed edges between them. Since no paper ever disappears after it is published, vertices in this network are created but never destroyed.

Let the average number of papers cited by a newly appearing paper be c. In the language of graph theory, c is the average out-degree of the network; in the language of publishing, c is the average size of the bibliography of a paper. The model allows for the actual sizes of the bibliographies to fluctuate about c. So long as the distribution of sizes satisfies a few basic sanity conditions,[3] only the average value is important for the behavior of the model in the limit of large network size. In real citation networks the sizes of bibliographies also vary from one field to another and depend on when papers were published, the average bibliography having grown larger over the years in most fields, but these effects are neglected in the model.

The crucial central assumption of Price's model is that a newly appearing paper cites previous ones chosen at random with probability proportional to the *number of citations those previous papers already have*. In this most basic of models there is no question of which papers are most relevant topically or which papers are most original or best written or the difference between research articles and reviews, or any of the many other factors that certainly affect real citation patterns. The model is thus very much a simplified representation of the citation process. As we have seen with the random graphs of previous chapters, however, even simple models can lead to real insights. We certainly need to remember that the model only represents one aspect of the citation process—and a hypothetical one at that—but with this in mind let us press on and see what we can discover.

As with personal wealth, it is not implausible that the number of citations a paper receives could increase with the number it already has. When one reads papers, one often looks up the other works that those papers cite and reads some of them too. If a work is cited often, then, all other things being equal, we are more likely to come across it than a less cited work. And if we

---

[3]The main condition on the distribution is that it should have finite variance. This rules out, for example, cases in which bibliographies have a power-law distribution of sizes with exponent less than three. Empirical evidence suggests that real bibliographies have an unexceptionable distribution of sizes with a modest and finite variance, so the assumptions of Price's model are met.

read it and like it, then perhaps we will cite it ourselves if we write a paper on the same topic. This does not mean that the probability of a paper receiving a citation is precisely proportional to the number of citations the paper has already, but it does at least give some justification for why the rich should get richer in this paper citation context.

In fact, upon further thought, it's clear that the probability of receiving a new citation cannot be *precisely* proportional to the number of citations a paper already has. Except under unusual circumstances, papers start out life with zero citations, which, with a strict proportionality rule, would mean that their probability of getting new citations would also be zero and so they would have zero citations for ever afterwards. To get around this hitch, Price proposed that in fact the probability that a paper receives a new citation should be proportional to the number that it already has plus a positive constant a. (In fact, Price only considered one special case $a = 1$ in his original model, but there seems to be no particular reason to limit ourselves to this case, so we will treat the case of general $a > 0$.)

The constant a in effect gives each paper a number of "free" citations to get it started in the race—each paper acts as though it started off with a citations instead of none. An alternative interpretation is that a certain fraction of citations go to papers chosen uniformly at random without regard for how many citations they current have, while the rest go to papers chosen in proportion to current citation count.[4] This gives all papers a chance to accrue citations, even if they currently have none. (We discuss this interpretation in more detail in Section 14.1.1, where we use it to construct a fast algorithm for simulating Price's model.)

We also need to specify what the starting state of the network is, how we initialize the model to begin with. It turns out in fact that in the limit of large network size the predictions of the model don't depend on the initial conditions, but we could, for instance, start the network out with a small set of initial papers having zero citations each.

Thus, in summary, Price's model consists of a growing network of papers and their citations in which vertices (papers) are continually added but none are ever taken away, each paper cites on average c others (so that the mean out-degree is c), and the cited papers are chosen at random[5] with probability

---

[4]The two fractions are in fact $a/(c + a)$ and $c/(c + a)$, respectively, as shown in Section 14.1.1.

[5]There is nothing in the definition of Price's model to prevent a paper from listing the same other paper twice in its bibliography, something that doesn't happen in real citation networks. In the language of graph theory, such double citations would correspond to directed multiedges in the citation network (see Section 6.1) while true citation networks are simple graphs having no

See Section 6.4.2 for a discussion of acyclic networks.

proportional to their in-degree plus a constant a.

One important property of Price's model is immediately apparent: it generates purely acyclic networks, since every edge points from a more recently added vertex to a less recently added one, i.e., backward in time. Thus all directed paths in the network point backward in time and hence there can be no closed loops, because to close a loop we would need edges pointing forward in time as well. This fits well with the original goal of the model as a model of citation, since citation networks are acyclic, or very nearly so (see Section 4.2). On the other hand it fits poorly with some other directed networks such as the World Wide Web, although the model is still sometimes used as a model for power-law distributions in the Web.

Armed with our definition of Price's model, we will now write down equations governing the distribution of the in-degrees of vertices, i.e., the numbers of citations received by papers in terms of the parameters c and a, and hence solve for the degree distribution and various other quantities, at least in the limit of large network size. We will discuss models of both directed and undirected graphs in this chapter, so we will need to be careful to distinguish in-degree in the directed case from ordinary undirected degree in the undirected case. Previously in this book we have done this by denoting the in-degree of a vertex i by k_i^{in} (see Section 6.9), but this notation can make our equations quite difficult to read, so in the interests of clarity we will in this chapter adopt instead the notation introduced by Dorogovtsev *et al.* [99] in which the in-degree of vertex i is denoted q_i. Degrees in undirected graphs will still be denoted k_i just as before.

So consider Price's model of a growing network and let $p_q(n)$ be the fraction of vertices in the network that have in-degree q when the network contains n vertices—this is the in-degree distribution of the network—and let us examine what happens when we add a single new vertex to the network.

Consider one of the citations made by this new vertex. In the model, the probability that the citation is to a particular other vertex i is proportional to $q_i + a$, where a is a positive constant. Since the citation in question has to be to *some* paper, this probability must be normalized such that its sum over all i is 1. In other words the correctly normalized probability must be

$$\frac{q_i + a}{\sum_i (q_i + a)} = \frac{q_i + a}{n\langle q \rangle + na} = \frac{q_i + a}{n(c + a)}, \tag{14.1}$$

---

multiedges. However, as with random graph models, the probability of generating a multiedge vanishes in the limit of large network size and so the predictions of the model in this limit are not altered by allowing them, and doing so makes the mathematical treatment of the model much simpler.

where we have written the average in-degree as $\langle q \rangle = n^{-1} \sum_i q_i$. In the second equality we have made use of the fact that the average out-degree of the network is c by definition, and that the average in- and out-degrees of a directed network are equal (see Eq. (6.27)) so that $\langle q \rangle = c$.

Each newly appearing paper cites c others on average, so the expected number of new citations to vertex i upon appearance of our new paper is c times Eq. (14.1). And there are $np_q(n)$ vertices with degree q in our network and hence the expected number of new citations to all vertices with degree q is

$$np_q(n) \times c \times \frac{q+a}{n(c+a)} = \frac{c(q+a)}{c+a} p_q(n). \tag{14.2}$$

Now we can write down a so-called *master equation* for the evolution of the in-degree distribution as follows. When we add a single new vertex to our network of n vertices, the number of vertices in the network with in-degree q increases by one for every vertex previously of in-degree $q-1$ that receives a new citation,[6] thereby becoming a vertex of in-degree q. From Eq. (14.2) we know that the expected number of such vertices is

$$\frac{c(q-1+a)}{c+a} p_{q-1}(n). \tag{14.3}$$

Similarly, we lose one vertex of in-degree q every time such a vertex receives a new citation, thereby becoming a vertex of in-degree $q+1$. The expected number of such vertices receiving citations is

$$\frac{c(q+a)}{c+a} p_q(n). \tag{14.4}$$

The number of vertices with in-degree q in the network after the addition of a single new vertex is $(n+1)p_q(n+1)$ which, putting together the results above, is given by

$$(n+1)p_q(n+1) = np_q(n) + \frac{c(q-1+a)}{c+a} p_{q-1}(n) - \frac{c(q+a)}{c+a} p_q(n). \tag{14.5}$$

The first term on the right-hand side here represents the number of vertices previously of in-degree q, the second term represents the vertices gained, and the third term the vertices lost.

---

[6]In theory, it also increases by one if a vertex of in-degree $q-2$ receives two new citations, and similarly for larger numbers of citations. This, however, would create a multiedge, and multiedges, as we have said, are vanishingly improbable in the limit of large network size, so we can ignore this possibility.

Equation (14.5) applies for all values of q except $q = 0$. When $q = 0$ there are no vertices of lower degree that can gain an edge to become vertices of degree zero, and hence the second term in Eq. (14.5) doesn't appear. On the other hand, we gain a vertex of degree zero whenever a new vertex is added to the network, since by hypothesis papers have no citations when they are first published. Since exactly one vertex is added in going from a network of n vertices to a network of $n + 1$, the appropriate equation for $q = 0$ is:

$$(n+1)p_0(n+1) = np_0(n) + 1 - \frac{ca}{c+a}p_0(n). \tag{14.6}$$

Now let us consider the limit of large network size $n \to \infty$ and calculate the asymptotic form of the degree distribution in this limit.[7] Taking the limit $n \to \infty$ and using the shorthand $p_q = p_q(\infty)$, Eqs. (14.5) and (14.6) become

$$p_q = \frac{c}{c+a}\left[(q-1+a)p_{q-1} - (q+a)p_q\right] \qquad \text{for } q \geq 1, \tag{14.7}$$

$$p_0 = 1 - \frac{ca}{c+a}p_0 \qquad \text{for } q = 0. \tag{14.8}$$

The second of these equations we can easily rearrange to give an explicit expression for the fraction p_0 of degree-zero vertices:

$$p_0 = \frac{1+a/c}{a+1+a/c}. \tag{14.9}$$

The solution for $q \geq 1$ is a little more complicated, though only a little. Rearranging Eq. (14.7) for p_q we find that

$$p_q = \frac{q+a-1}{q+a+1+a/c}p_{q-1}. \tag{14.10}$$

We can use this equation to calculate p_q iteratively for all values of q starting from our solution for p_0, Eq. (14.9). First, we set $q = 1$ in Eq. (14.10) to get

$$p_1 = \frac{a}{a+2+a/c}p_0 = \frac{a}{(a+2+a/c)}\frac{(1+a/c)}{(a+1+a/c)}. \tag{14.11}$$

Now we can use this result to calculate p_2:

$$p_2 = \frac{a+1}{a+3+a/c}p_1 = \frac{(a+1)a}{(a+3+a/c)(a+2+a/c)}\frac{(1+a/c)}{(a+1+a/c)}, \tag{14.12}$$

---

[7]Strictly we should first prove that the degree distribution *has* an asymptotic form in the limit of large n and doesn't go on changing forever, but for the purposes of the present discussion let us assume that there is an asymptotic form.

and

$$p_3 = \frac{a+2}{a+4+a/c}\, p_2$$

$$= \frac{(a+2)(a+1)a}{(a+4+a/c)(a+3+a/c)(a+2+a/c)}\, \frac{(1+a/c)}{(a+1+a/c)}, \qquad (14.13)$$

and so forth. It's easy to see that for general q the correct expression must be

$$p_q = \frac{(q+a-1)(q+a-2)\ldots a}{(q+a+1+a/c)\ldots(a+2+a/c)}\, \frac{(1+a/c)}{(a+1+a/c)}. \qquad (14.14)$$

This is effectively a complete solution for the degree distribution of Price's model, but there is a little more we can do to write it in a useful form. We make use of the gamma function,

$$\Gamma(x) = \int_0^\infty t^{x-1} e^{-t}\, dt, \qquad (14.15)$$

which has the useful property that[8]

$$\Gamma(x+1) = x\Gamma(x) \qquad (14.16)$$

for all $x > 0$. Iterating this formula, we see that

$$\frac{\Gamma(x+n)}{\Gamma(x)} = (x+n-1)(x+n-2)\ldots x. \qquad (14.17)$$

Using this result in Eq. (14.14) we can write

$$p_q = (1+a/c)\frac{\Gamma(q+a)\Gamma(a+1+a/c)}{\Gamma(a)\Gamma(q+a+2+a/c)}. \qquad (14.18)$$

This expression can be simplified further by writing it in terms of Euler's beta function, which is defined by

$$B(x,y) = \frac{\Gamma(x)\Gamma(y)}{\Gamma(x+y)}. \qquad (14.19)$$

If we multiply both the numerator and the denominator of Eq. (14.18) by $\Gamma(2+a/c) = (1+a/c)\Gamma(1+a/c)$, we find that

$$p_q = \frac{\Gamma(q+a)\Gamma(2+a/c)}{\Gamma(q+a+2+a/c)} \times \frac{\Gamma(a+1+a/c)}{\Gamma(a)\Gamma(1+a/c)}, \qquad (14.20)$$

---

[8]The proof of this result is simple, making use of integration by parts:

$$\Gamma(x+1) = \int_0^\infty t^x e^{-t}\, dt = -[t^x e^{-t}]_0^\infty + x\int_0^\infty t^{x-1} e^{-t}\, dt = x\Gamma(x),$$

where the boundary term $[\ldots]$ disappears at both limits.

or

$$p_q = \frac{B(q + a, 2 + a/c)}{B(a, 1 + a/c)}. \tag{14.21}$$

Note that this expression is not only correct for $q \geq 1$ but also gives the correct value when $q = 0$.

One of the nice things about Eq. (14.21) is that it depends on q only via the first argument of the upper beta function. Thus if we want to understand the shape of the degree distribution we need only to understand the behavior of this one function.

In particular, let us examine the behavior for large q and fixed a and c. For large values of its first argument, we can rewrite the beta function using Stirling's approximation for the gamma function [2]

$$\Gamma(x) \simeq \sqrt{2\pi}\, e^{-x} x^{x - \frac{1}{2}}, \tag{14.22}$$

which means that

$$B(x, y) = \frac{\Gamma(x)\Gamma(y)}{\Gamma(x + y)} \simeq \frac{e^{-x} x^{x - \frac{1}{2}}}{e^{-(x+y)}(x + y)^{x+y-\frac{1}{2}}} \Gamma(y). \tag{14.23}$$

But

$$(x + y)^{x+y-\frac{1}{2}} = x^{x+y-\frac{1}{2}} \left[1 + \frac{y}{x}\right]^{x+y-\frac{1}{2}} \simeq x^{x+y-\frac{1}{2}} e^{y}, \tag{14.24}$$

where the last equality becomes exact in the limit of large x. Then

$$B(x, y) \simeq \frac{e^{-x} x^{x - \frac{1}{2}}}{e^{-(x+y)} x^{x+y-\frac{1}{2}} e^{y}} \Gamma(y) = x^{-y} \Gamma(y). \tag{14.25}$$

In other words, the beta function $B(x, y)$ falls off as a power law for large values of x, with exponent y.

Applying this finding to Eq. (14.21) we then discover that for large values of q the degree distribution of our network goes as $p_q \sim (q + a)^{-\alpha}$, or simply

$$p_q \sim q^{-\alpha} \tag{14.26}$$

when $q \gg a$, where the exponent α is

$$\alpha = 2 + \frac{a}{c}. \tag{14.27}$$

Thus Price's model for a citation network gives rise to a degree distribution with a power-law tail. This is very much in keeping with the degree distributions of real citation networks, which, as we saw in Fig. 8.8, appear to have clear power-law tails.

Note that the exponent $\alpha = 2 + a/c$ is strictly greater than two (since a and c are both strictly positive). Most measurements put the exponent of the power law for citation networks around $\alpha = 3$ (see Table 8.1), which is easily achieved in the model by setting the constants a and c equal. In a typical experimental situation the exponent α and the parameter c, the mean size of a paper's bibliography, are easily measured, but the parameter a, which represents the number of "free" effective citations a paper receives upon publication, is not. Typically therefore the value of a is extracted by rearranging Eq. (14.27) to give $a = c(\alpha - 2)$.

While it is delightful that Price's simple model generates a power-law degree distribution similar to that seen in real networks, we should not take the details of the model too seriously, nor the exact relation between the parameters and the exponent of the power law. As we noted at the start of this section, the model is highly simplified and substantially incomplete as a model of the citation process, omitting many factors that are undoubtedly important for real citations, including the quality and relevance of papers, developments and fashions in the field of study, the reputation of the publishing journal and of the author, and many others besides. Still, Price's model is striking in its ability to reproduce one of the most interesting features of citation networks using only a small number of reasonable assumptions, and many scholars believe that it may capture the fundamental mechanism behind the observed power-law degree distribution.

14.1.1 COMPUTER SIMULATION OF PRICE'S MODEL

When Price proposed his model in the 1970s, analytic treatments like the one above were essentially the only tool available for understanding the behavior of such models. Today, however, we can go further and study the operation of the model explicitly by performing computer simulations following the rules Price laid down. In addition to providing a useful check on our solution for the degree distribution, such simulations also allow us to generate real examples of networks on our computer. We can then measure these networks to determine the values, within the model, of any network quantities we like— path lengths, correlations, clustering coefficients, and so forth—including ones for which we do not at present have an analytic solution. Researchers have also made use of simulated networks as a convenient but still relatively realistic substrate for other kinds of calculation, including solutions of dynamical models, percolation processes, opinion formation models, and others.

In principle, simulation of Price's model appears straightforward. Typically one simulates the model with the out-degrees of vertices fixed to be exactly equal to c, where c is restricted to integer values. (In the original model and our analysis above, c was only the average out-degree—actually out-degree could fluctuate about the average.) Then the only complicated part of the simulation is the selection of the vertices that receive new edges, which has to be done in a random but non-uniform way as a function of the vertices' current in-degree. There are standard techniques for simulating such non-uniform random processes and one can without too much labor create a simple program that carry out the steps of the model. This, however, is not usually the best way to proceed. A naive direct simulation of this kind becomes slow when the network gets large, which limits the size of the network that can be generated. Luckily, there is a much faster way to perform the simulation that allows larger networks to be generated in shorter times, while still being simple to program on a computer. This method, first proposed by Krapivsky and Redner [187], works as follows.

When we create a new edge in Price's model we attach it to a vertex chosen in proportion to in-degree plus a constant a. Let us denote by θ_i the probability that an edge attaches to vertex i, which from Eq. (14.1) is given by

$$\theta_i = \frac{q_i + a}{n(c + a)}. \tag{14.28}$$

Now consider an alternative process in which upon creating a new edge we do one of two things. With some probability ϕ we attach the edge to a vertex chosen strictly in proportion to its current in-degree, i.e., with probability

$$\frac{q_i}{\sum_j q_j} = \frac{q_i}{nc}. \tag{14.29}$$

Alternatively, with probability $1 - \phi$, we attach to a vertex chosen uniformly at random from all n possibilities, i.e., with probability $1/n$. Then the total probability θ_i' of attaching to vertex i in this process is

$$\theta_i' = \phi \frac{q_i}{nc} + (1 - \phi)\frac{1}{n}. \tag{14.30}$$

Now let us make the choice $\phi = c/(c + a)$, so that

$$\theta_i' = \frac{c}{c + a}\frac{q_i}{nc} + \left(1 - \frac{c}{c + a}\right)\frac{1}{n} = \frac{q_i + a}{n(c + a)}. \tag{14.31}$$

This, however, is precisely equal to the probability θ_i, Eq. (14.28), of selecting a vertex in the Price model and the two processes thus choose vertices with the exact same probabilities.

So an alternative way of performing a step of Price's model is the following:

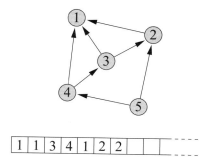

Figure 14.1: The vertex label list used in the simulation of Price's model. The list (bottom) contains one entry for the target of each edge in the network (top). In this example, there are three edges that point to vertex 1 and hence there are three elements containing the number 1 in the list. Similarly there are two containing the number 2, because vertex 2 is the target of two edges. And so forth.

> With probability $c/(c+a)$ choose a vertex in strict proportion to in-degree. Otherwise choose a vertex uniformly at random from the set of all vertices.

The choice between the two parts can be achieved, for example, by generating a random number r in the range $0 \leq r < 1$. If $r < c/(c+a)$ then we choose a vertex in proportion to in-degree. Otherwise we choose uniformly.

Choosing a vertex uniformly is easily accomplished. Choosing a vertex in proportion to in-degree is only slightly harder. It can be done rapidly by noting that choosing in proportion to in-degree is equivalent to picking an edge in the network uniformly at random and choosing the vertex which that edge points to. By definition this makes a vertex with in-degree q exactly q times as likely to be chosen as a vertex with in-degree 1, since it has q opportunities to be chosen, one for each of the edges that point to it.

To turn this observation into a computer algorithm we make a list, stored for instance in an ordinary array, of the target of each directed edge in the network. That is, the list's elements contain the vertex labels i of the vertices to which each edge points. Figure 14.1 shows an example for a small network. Note that the edges do not have to be in any particular order. Any order will do. Nor does the size of the array used to store the list have to match the length of the list exactly; it can contain empty elements at the end as shown in the figure. Indeed, since making already existing arrays larger is difficult in most computer languages, it makes sense to initially create an array that is large enough to hold the longest list we will need. (This means that it should have length nc if the out-degree of vertices is constant. If out-degree is allowed to fluctuate then the longest list might be a bit larger or smaller than nc, in which case one might create an array of size nc plus a few percent, to be on the safe side.)

Once we have our list, choosing a vertex in proportion to its in-degree becomes a trivial operation: we simply choose an element uniformly at random from the list and our vertex is identified by the contents of that element. When a new edge is added to the network, we must also update the list by adding the target of that edge to the end of the list.

Thus our algorithm for creating a new edge is the following:

1. Generate a random number r in the range $0 \le r < 1$.
2. If $r < c/(c+a)$, choose an element uniformly at random from the list of targets.
3. Otherwise choose a vertex uniformly at random from the set of all vertices.
4. Create an edge linking to the vertex thus selected, and add that vertex to the end of the list of targets.

Each step in this process can be accomplished in constant time and hence the growth of a network of n vertices can be accomplished in time $O(n)$ (provided other parts of the program are implemented efficiently so that they also take constant time per step).

Figure 14.2a shows the degree distribution of a 100-million-node network generated computationally in this fashion, and the power-law form in the tail of the distribution is clearly visible. A practical problem, however, is the noise in the tail of the histogram, which makes the exact form of the distribution hard to gauge. This is exactly the same problem as we encountered for real-world data in Section 8.4.1: the bins in the tail of the histogram have relatively few samples in them and so the statistical fluctuations are large as a fraction of the number of samples. Indeed, in many respects simulation data often behave in similar ways to experimental data and they can often be treated using the same techniques. In this case we can take a hint from Section 8.4.1 and plot a cumulative distribution function instead of a histogram. To recap, the cumulative distribution function P_q is

$$P_q = \sum_{q'=q}^{\infty} p_{q'}, \tag{14.32}$$

(see Eq. (8.4)) and is expected to have a power-law tail with an exponent $\alpha - 1 = 1 + a/c$, one less than the exponent of the degree distribution itself. Figure 14.2b shows the cumulative distribution of degrees for our simulation and we now see a much cleaner power-law behavior over several decades in q.

For comparison we can also calculate the cumulative distribution function analytically from our solution of the model. To do this, we make use of the standard integral form for the beta function:[9]

$$B(x, y) = \int_0^1 u^{x-1}(1-u)^{y-1}\, du. \tag{14.33}$$

Using this expression we find that

$$
\begin{aligned}
P_q = \sum_{q'=q}^{\infty} p_q &= \frac{1}{B(a, 1+a/c)} \sum_{q'=q}^{\infty} \int_0^1 u^{q'+a-1}(1-u)^{1+a/c}\, du \\
&= \frac{1}{B(a, 1+a/c)} \int_0^1 u^{a-1} \sum_{q'=q}^{\infty} u^{q'}(1-u)^{1+a/c}\, du \\
&= \frac{1}{B(a, 1+a/c)} \int_0^1 u^{q+a-1}(1-u)^{a/c}\, du \\
&= \frac{B(q+a, 1+a/c)}{B(a, 1+a/c)}.
\end{aligned}
\tag{14.34}
$$

Given that $B(x, y)$ goes as x^{-y} for large x (Eq. (14.25)), this implies that indeed the cumulative distribution function has a power-law tail with exponent $1 + a/c$.

In Fig. 14.2b we show Eq. (14.34) along with the simulation data, and the simulation and analytic solution agree well, as we would hope.

---

[9]The integral form can be derived by making use of the definition of $B(x, y)$ in terms of gamma functions and the integral form of the gamma function, Eq. (14.15):

$$B(x, y) = \frac{\Gamma(x)\Gamma(y)}{\Gamma(x+y)} = \frac{1}{\Gamma(x+y)} \int_0^\infty s^{x-1}e^{-s}\, ds \int_0^\infty t^{y-1}e^{-t}\, dt.$$

We change variables to $u = s/(s+t)$, $v = s+t$ (which implies $s = uv$, $t = (1-u)v$ and a Jacobian of v), giving

$$
\begin{aligned}
B(x, y) &= \frac{1}{\Gamma(x+y)} \int_0^1 du \int_0^\infty v\, dv\, (uv)^{x-1}e^{-uv}\left[(1-u)v\right]^{y-1}e^{-(1-u)v} \\
&= \frac{1}{\Gamma(x+y)} \int_0^\infty v^{x+y-1}e^{-v}\, dv \int_0^1 u^{x-1}(1-u)^{y-1}\, du.
\end{aligned}
$$

The first integral, however, is equal to $\Gamma(x+y)$ by Eq. (14.15) and hence we recover Eq. (14.33).

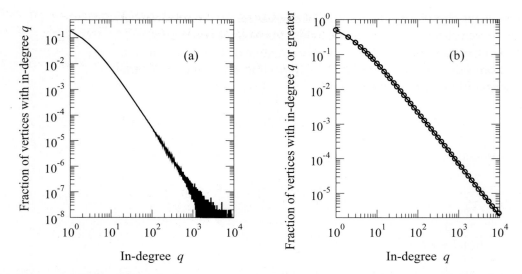

Figure 14.2: Degree distribution in Price's model of a growing network. (a) A histogram of the in-degree distribution for a computer-generated network with $c = 3$ and $a = 1.5$ which was grown until it had $n = 10^8$ vertices. The simulation took about 80 seconds on the author's computer using the fast algorithm described in the text. (b) The cumulative distribution function for the same network. The points are the results from the simulation and the solid line is the analytic solution, Eq. (14.34).

14.2 THE MODEL OF BARABÁSI AND ALBERT

Price's model of a growing network is an elegant one and the existence of an exact solution showing that its degree distribution has a power-law tail makes a persuasive case for preferential attachment as a possible origin for power-law behavior. At least until recently, however, Price's work in this area was not well known outside of the information science community. Preferential attachment did not become widely accepted as a mechanism for generating power laws in networks until much later, in the 1990s, when it was independently discovered by Barabási and Albert [27], who proposed their own model of a growing network (along with the name "preferential attachment"). The Barabási–Albert model, which is certainly the best known generative network model in use today, is similar to Price's, though not identical, being a model of an undirected rather than a directed network.

In the model of Barabási and Albert, vertices are again added one by one to a growing network and each vertex connects to a suitably chosen set of previously existing vertices. The connections, however, are now undirected and the number of connections made by each vertex is exactly c (unlike Price's model, where the number of connections was required only to take an average value of c but might vary from step to step). Note that this implies that c must be an integer, since a vertex cannot have non-integer degree. Connections are made to vertices with probability precisely proportional to the vertices' current degree. Notice that there is no in- or out-degree now because the network is undirected. Connections are made simply in proportion to the (undirected) degree. We will denote the degree of vertex i by k_i to distinguish it from the directed in-degree q_i of the last section. As before, vertices and edges are only ever added to the network and never taken way, which means, among other things, that there are no vertices with degree $k < c$. The smallest degree in the network is always $k = c$.

One can write down a solution for the model of Barabási and Albert using a master equation method similar to that of Section 14.1,[10] but in fact there is no need, because it is straightforward to show that the model is equivalent to a special case of Price's model. Imagine that, purely for the purposes of our discussion, we give each edge added to the network a direction, running from the vertex just added to the previously existing vertex that the edge connects to. That is each edge runs from the more recent of the two vertices it connects to to the less recent. In this way we convert our network into a directed network in which each vertex has out-degree exactly c (since this is the number of outgoing edges a vertex starts with and it never gains any more). And the total degree k_i of a vertex in the sense of the original undirected network is the sum of the vertex's in-degree and out-degree, which is $k_i = q_i + c$ where q_i is the in-degree as before.

But given that the probability of an edge attaching to a vertex is simply proportional to k_i, it is thus also proportional to $q_i + c$, which is the same as in Price's model if we make the particular choice $a = c$. Thus the distribution of in-degrees in this directed network is the same as for Price's model with $a = c$, which we find from Eq. (14.21) to be

$$p_q = \frac{B(q + c, 3)}{B(c, 2)}.$$ (14.35)

To get the distribution of the total degree we then simply replace $q + c$ by k to get

$$p_k = \begin{cases} \dfrac{B(k, 3)}{B(c, 2)} & \text{for } k \geq c, \\ 0 & \text{for } k < c. \end{cases}$$ (14.36)

---

[10]See for instance [99] or [189].

This expression can be simplified further by making use of Eq. (14.17) to write

$$B(k,3) = \frac{\Gamma(k)\Gamma(3)}{\Gamma(k+3)} = \frac{\Gamma(3)}{k(k+1)(k+2)}, \tag{14.37}$$

and similarly

$$B(c,2) = \frac{\Gamma(2)}{c(c+1)}, \tag{14.38}$$

so that

$$p_k = \frac{\Gamma(3)}{\Gamma(2)} \frac{c(c+1)}{k(k+1)(k+2)} = \frac{2c(c+1)}{k(k+1)(k+2)} \tag{14.39}$$

for $k \geq c$, where we have used (14.17) again to get rid of the remaining gamma functions. In the limit where k becomes large, this gives

$$p_k \sim k^{-3}, \tag{14.40}$$

and hence the Barabási–Albert model generates a degree distribution with a power-law tail that always has an exponent $\alpha = 3$.

Equation (14.39) was first derived by Krapivsky et al. [189] and independently by Dorogovtsev et al. [99]. A more detailed treatment was later given by Bollobás et al. [48], which clarifies precisely the domain of validity of the solution and the possible deviations from the expected value of p_k.

The model of Barabási and Albert can be simulated efficiently on a computer by exploiting the same mapping to Price's model and the simulation method described in Section 14.1.1. Again we regard the network as a directed one and maintain a list of targets of every directed edge, i.e., the vertices that the edges point to. Then, setting $a = c$, the algorithm of Section 14.1.1 becomes particularly simple: with probability $\frac{1}{2}$ we choose an element from our list uniformly at random and take the contents of that element as our target vertex. Otherwise we choose a target uniformly at random from the set of all vertices currently in existence. Then we create a new edge from the vertex just added to the target we have selected and also add that target to the end of our list.

The Barabási–Albert model is attractive for its simplicity—it doesn't require the offset parameter a of Price's model and hence has one less parameter to worry about. It is also satisfying that one can write the degree distribution without using special functions such as the beta and gamma functions that appear in the solution of Price's model. The price one pays for this simplicity is that the model can no longer match the exponents observed in real networks, being restricted to just a single exponent value $\alpha = 3$.

14.3 FURTHER PROPERTIES OF PREFERENTIAL ATTACHMENT MODELS

The models of Price and of Barabási and Albert were proposed as explanations for the observed power-law degree distributions in networks such as the Web, citation networks, and others, and for this reason the degree distribution is the property of these models that has attracted most interest. However, it is not the only property that can be calculated. The master equation method can be extended to the calculation of a number of other properties, many of which are interesting in their own right. We describe a few of these calculations in this section.

14.3.1 DEGREE DISTRIBUTION AS A FUNCTION OF TIME OF CREATION

Consider a network grown according to Price's model of Section 14.1. Older vertices in the network—those added earlier in the growth process—have more time to acquire links from other vertices and hence we might expect that they would on average have higher in-degree. This indeed turns out to be the case, as we can show by calculating the degree distribution as a function of the time at which vertices are created.

Let $p_q(t, n)$ be the average fraction of vertices in our directed network that were created at time t and have in-degree q when the network has n vertices total. The time of creation is measured in terms of the number of vertices, the first vertex having $t = 1$ and the last having $t = n$. Alternatively, you can just think of t as counting the vertices from 1 to n, recording the order in which they were added. Strictly t need not reflect actual time, because the vertices need not have been added at a constant rate, but if we know the real times at which vertices were added we can easily convert between our timescale and real time.

We can write down a master equation for the evolution of $p_q(t, n)$ as follows. Upon the addition of a new vertex to the network, the expected number of new edges acquired by previously existing vertices with in-degree q is independent of time of those vertices creation and, following Eq. (14.2), is given by

$$np_q(t, n) \times c \times \frac{q + a}{n(c + a)} = \frac{c(q + a)}{c + a} p_q(t, n), \qquad (14.41)$$

with the parameters c and a defined as in Section 14.1. Then the master equation takes the form

$$(n + 1)p_q(t, n + 1) = np_q(t, n) + \frac{c}{c + a} \left[(q - 1 + a)p_{q-1}(t, n) - (q + a)p_q(t, n) \right].$$

$$(14.42)$$

The only exception to this equation is, as before, for the case of $q = 0$, where we get

$$(n+1)p_0(t, n+1) = np_0(t, n) + \delta_{tn} - \frac{ca}{c+a}p_0(t, n). \tag{14.43}$$

Notice the Kronecker delta, which adds a single vertex of in-degree zero if $t = n$, but none otherwise.

These equations, though correct, don't make much sense in the limit of large n, since the fraction of vertices created at time t goes to zero in this limit because only one vertex is created at any particular t. So instead we change variables to a rescaled time

$$\tau = \frac{t}{n}, \tag{14.44}$$

which takes values between zero (oldest vertices) and one (youngest vertices). At the same time we also change from $p_q(t, n)$ to a density function $\pi_q(\tau, n)$ such that $\pi_q(\tau, n)\, d\tau$ is the fraction of vertices that have in-degree q and fall in the interval from τ to $\tau + d\tau$. The number of vertices in the interval $d\tau$ is $n\, d\tau$, which implies that $\pi_q\, d\tau = p_q \times n\, d\tau$ and hence

$$\pi_q(\tau, n) = np_q(t, n). \tag{14.45}$$

Being a density function, π_q does not vanish as $n \to \infty$.

The downside of this variable change is that τ is no longer constant for a given vertex. A vertex created at time t has rescaled time t/n when there are n vertices in the network but $t/(n+1)$ when there are $n+1$. Thus, in terms of τ and π_q, Eq. (14.42) becomes

$$\pi_q\left(\frac{n}{n+1}\tau, n+1\right) = \pi_q(\tau, n)$$
$$+ \frac{c}{c+a}\left[(q-1+a)\frac{\pi_{q-1}(t, n)}{n} - (q+a)\frac{\pi_q(t, n)}{n}\right]. \tag{14.46}$$

Now we consider the limit where $n \to \infty$. If we define the shorthand notation $\pi_q(t) = \pi_q(t, \infty)$ and the small quantity $\epsilon = 1/n$, Eq. (14.42) becomes

$$\frac{\pi_q(\tau) - \pi_q(\tau - \epsilon\tau)}{\epsilon} + \frac{c}{c+a}\left[(q-1+a)\pi_{q-1}(\tau) - (q+a)\pi_q(\tau)\right] = 0, \tag{14.47}$$

where we have dropped terms of order ϵ^2.

As $n \to \infty$, we have $\epsilon \to 0$ and the first two terms become a derivative thus:

$$\lim_{\epsilon \to 0} \frac{\pi_q(\tau) - \pi_q(\tau - \epsilon\tau)}{\epsilon} = \tau\frac{d\pi_q}{d\tau}, \tag{14.48}$$

and so our master equation becomes a differential equation in this case:

$$\tau \frac{d\pi_q}{d\tau} + \frac{c}{c+a}\left[(q-1+a)\pi_{q-1}(\tau) - (q+a)\pi_q(\tau)\right] = 0. \qquad (14.49)$$

The corresponding equation for $q = 0$ is

$$\tau \frac{d\pi_0}{d\tau} - \frac{ca}{c+a}\pi_0(\tau) = 0, \qquad (14.50)$$

so long as $\tau < 1$. For the special case $\tau = 1$ the δ_{tn} in Eq. (14.43) presents a problem, but on the other hand for $\tau = 1$ we know what the answer is anyway: there is always exactly one vertex created at time $t = n$, which has in-degree zero, so $\pi_0(1) = 1$ in the language of our rescaled variables.[11] In effect, this just provides a boundary condition on $\pi_0(\tau)$. The corresponding boundary condition for $q \geq 1$ is $\pi_q(1) = 0$, since there are no vertices with $t = n$ and $q \geq 1$.

We can solve Eqs. (14.49) and (14.50) by starting with a solution for $q = 0$ and working up through increasing values of q. This is similar to our solution for the degree distribution, Eq. (14.21), except that the equations we are solving are now differential equations.

The solution for the $q = 0$ case is straightforward—Eq. (14.50) is homogeneous in π_0 and can be solved by standard methods. You can easily verify that the solution is $\pi_0(\tau) = A\tau^{ca/(c+a)}$ where A is an integration constant. The constant is fixed by the boundary condition $\pi_0(1) = 1$, which implies that $A = 1$ and hence

$$\pi_0(\tau) = \tau^{ca/(c+a)}. \qquad (14.51)$$

As a check, we can integrate over τ to get the total fraction of vertices with in-degree zero:

$$\int_0^1 \tau^{ca/(c+a)}\, d\tau = \frac{1 + a/c}{a + 1 + a/c}, \qquad (14.52)$$

which agrees nicely with our previous result for the same quantity, Eq. (14.9).

Now we can use this solution to find $\pi_1(\tau)$. Equation (14.49) tells us that

$$\tau \frac{d\pi_1}{d\tau} - \frac{c}{c+a}(a+1)\pi_1(\tau) = -\frac{c}{c+a}a\pi_0(\tau) = -\frac{ca}{c+a}\tau^{ca/(c+a)}. \qquad (14.53)$$

This is again just an ordinary first-order differential equation, although an inhomogeneous one this time (i.e., it has a driving term on the right-hand side).

---

[11]It may not be immediately obvious that $\pi_0(1)$ must equal 1. There is one vertex in the time interval between $\tau = 1$ and $\tau = 1 - 1/n$ and it always has in-degree zero. One vertex is a fraction $1/n$ of the whole network, so there is $1/n$ of the network in an interval of width $1/n$, which corresponds to a density $\pi_0(1) = 1$.

We tackle it in standard fashion. First we find the general solution for the homogeneous equation in which the right-hand side is set to zero, which is $B\tau^{c(a+1)/(c+a)}$ where B is an integration constant. Then we find any (non-general) solution to the full equation with the driving term included—the obvious one is $a\tau^{ca/(c+a)}$—and sum the two. The constant is fixed by the boundary condition $\pi_1(1) = 0$, which implies that $B = -a$, and we get

$$\pi_1(\tau) = a\tau^{ca/(c+a)}\left(1 - \tau^{c/(c+a)}\right). \tag{14.54}$$

Now, by a similar method, we can use *this* solution to solve for $\pi_2(\tau)$, and so forth to higher and higher values of q. The algebra is tedious, but with persistence you can show that the next two results are

$$\pi_2(\tau) = \tfrac{1}{2}a(a+1)\tau^{ca/(c+a)}\left(1 - \tau^{c/(c+a)}\right)^2, \tag{14.55}$$

$$\pi_3(\tau) = \tfrac{1}{6}a(a+1)(a+2)\tau^{ca/(c+a)}\left(1 - \tau^{c/(c+a)}\right)^3. \tag{14.56}$$

These results suggest the general solution (first given by Dorogovtsev *et al.* [99])

$$\pi_q(\tau) = \frac{1}{q!}\left[a(a+1)\ldots(a+q-1)\right]\tau^{ca/(c+a)}\left(1 - \tau^{c/(c+a)}\right)^q$$

$$= \frac{\Gamma(q+a)}{\Gamma(q+1)\Gamma(a)}\tau^{ca/(c+a)}\left(1 - \tau^{c/(c+a)}\right)^q, \tag{14.57}$$

where we have made use again of the convenient property of the gamma function derived in Eq. (14.17) as well as the result that $\Gamma(n+1) = n!$ when n is a positive integer.[12] With a little work you can verify that this is indeed a complete solution of Eq. (14.49) for all q. As a check we can also integrate over τ to find the total fraction of vertices with in-degree q and confirm that the result agrees with Eq. (14.21). We leave this calculation as an exercise for the reader.[13]

Let us take a moment to examine the structure of our solution for $\pi_q(\tau)$ and see what it tells us about the network. The general shape of the solution is shown in Fig. 14.3. Panel (a) shows the distribution of creation times τ for vertices of given in-degree q for various values of q and for each value there is a clear peak in the distribution, indicating that vertices of a given degree are concentrated around a particular era in the growth of the graph. As degree increases, that era gets earlier, so that the times of creation of vertices that ultimately achieve high in-degree are strongly concentrated around the beginning of the growth process.

---

[12]To prove this we set $x = 1$ in Eq. (14.17) to get $\Gamma(n+1)/\Gamma(1) = n(n-1)\ldots 1 = n!$ and from Eq. (14.15) we have $\Gamma(1) = \int_0^\infty e^{-t}\,dt = 1$.

[13]Hint: you will probably need Eq. (14.33).

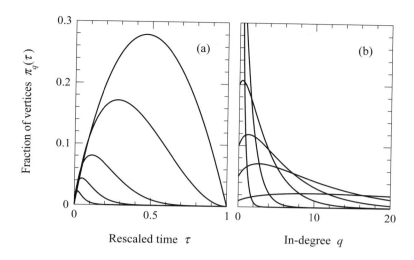

Figure 14.3: Distribution of vertices in Price's model as a function of in-degree and time of creation. The two panels show the distribution $\pi_q(\tau)$, Eq. (14.57), for $c = 3$ and $a = 1.5$ as (a) a function of τ for (top to bottom) $q = 1, 2, 5, 10$, and 20, and (b) a function of q for $\tau = 0.01$ (flattest curve), 0.05, 0.1, 0.5, and 0.9 (steepest curve).

Panel (b) of Fig. 14.3 shows the distribution of in-degrees for vertices created at a selection of different times τ. This distribution also has a peak, then falls off sharply as q becomes large.[14] Indeed the distribution falls off roughly exponentially as q becomes large, as we can see from Eq. (14.57) by writing

$$\frac{\Gamma(q+a)}{\Gamma(q+1)\Gamma(a)} = \frac{\Gamma(q+a)}{q\Gamma(q)\Gamma(a)} = \frac{1}{qB(q,a)}, \tag{14.58}$$

where we have used Euler's beta function again, Eq. (14.19). As shown in Section 14.1, the beta function has a power-law tail $B(x,y) \sim x^{-y}$ for large x (Eq. (14.25)) so π_q goes with q as

$$\pi_q(\tau) \sim q^{a-1}\left(1 - \tau^{c/(c+a)}\right)^q. \tag{14.59}$$

In other words it decays exponentially except for a leading algebraic factor. Thus the degree distribution for vertices with specific values of τ does not follow a power law. The power-law behavior seen in the full degree distribution

---

[14]Actually, the peak only exists for small values of τ and disappears once τ becomes large enough. There are no peaks in the degree distribution for the $\tau = 0.5$ and $\tau = 0.9$ curves in Fig. 14.3b.

of the model, Eq. (14.21), only appears when we integrate over all times τ. However, the decay of the exponential in Eq. (14.59) is slower for smaller τ, so older vertices are more likely to have high in-degree than younger ones, as we saw in Fig. 14.3.

To investigate this point further, we can calculate the mean in-degree $\gamma(\tau)$ for a vertex created at time τ thus:

$$\gamma(\tau) = \sum_{q=0}^{\infty} q\pi_q(\tau) = a(\tau^{-c/(c+a)} - 1). \tag{14.60}$$

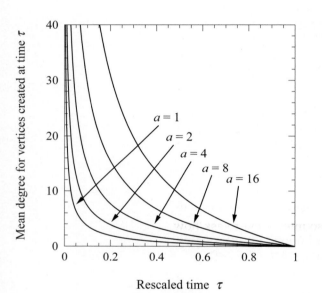

Figure 14.4: **Average in-degree of vertices as a function of their time of creation.** The average in-degree of vertices in Price's network model as a function of the rescaled time $\tau = t/n$ at which they were added to the network, in the limit of large n for various values of the parameter a. The out-degree parameter c was in each case $c = 2a$, so that the exponent of the power-law degree distribution $\alpha = 2 + a/c$ (Eq. (14.27)) is 2.5 for all curves, which is a typical value for real-world networks.

Figure 14.4 shows the shape of $\gamma(\tau)$ for a variety of choices of the parameters and, as we expect, the mean value of the in-degree increases with decreasing τ and eventually diverges as τ approaches zero. Notice, though, that no vertex ever actually has $\tau = 0$. The first vertex added to the network has $t = 1$, so the smallest value of τ is $1/n$. Nonetheless, we see that vertices added to the network early have an enormous advantage in terms of in-degree over those added even a little later. For a citation network, for instance, this suggests that the early papers in a field will receive substantially more citations than later ones, purely because they were published first.

Indeed, this is a pattern seen in many different areas, not just in networks. In any situation where success begets more success, first movers are expected to have a large advantage over others. Any small lead gained early in the process is quickly amplified by the preferential attachment process into a bigger lead and soon the lucky first movers find themselves racing ahead of the pack. Those who enter the game later may experience chance fluctuations that give them a small boost, but since there are probably many others already ahead of them, that boost is not amplified significantly because most of the wealth is already going to the leaders under the preferential attachment rule.

A nice demonstration of this process, although not in the field of networks, has been given by Salganik *et al.* [288], who examined the behavior of a group of people downloading popular music online. Salganik *et al.* created a website on which participants could download and listen to songs by little-known artists for free. Participants were told how many times each song had previously been downloaded and Salganik and co-workers found that there was a clear preferential attachment effect: songs with many previous downloads were downloaded far more than those with few. As a result there was a strong first-mover advantage, with songs that took an early lead benefiting from the preferential attachment and turning that lead into a much larger one, resulting in a roughly power-law distribution in the numbers of downloads.

To test the theory that they were seeing a preferential attachment process rather than actual differences in song quality leading to different download rates, Salganik *et al.* then changed the download numbers reported for each song, deliberately misrepresenting the number of times each had been downloaded. They discovered when they did this that the songs with the highest *reported* numbers of downloads were still downloaded most often, even though the reported numbers no longer corresponded to true popularity.[15] These results strongly suggest that success is, at least in this context and at least in part, a result of previous success and that a good way to be successful is to get in at the beginning and get an early lead. Of course, that may be easier said than done. Many people would like to get in at the beginning of a new field of scientific research or a new business opportunity, but it's not always clear how one should do it.

Returning to our network growth model, it is also interesting to ask how the expected in-degree of a vertex varies with its age after it enters the network. This differs from the expected degree for a particular τ calculated above because a given vertex does not have a fixed value of τ. The value of $\tau = t/n$ for a vertex decreases as time passes because n is increasing. For this reason the behavior of individual vertices is more easily understood in terms of our original non-rescaled time t, which does remain constant.

So let t again be the time at which a vertex is added to the network and let s be the subsequent elapsed time, i.e., the age of the vertex. Necessarily we have $s + t = n$ and hence

$$\tau = \frac{t}{n} = \frac{t}{t+s}. \tag{14.61}$$

Substituting this expression into Eq. (14.60), we then find the expected in-

---

[15]Salganik *et al.* did find a weak effect of song quality—songs that had proved popular when the download numbers were reported faithfully continued to do better than expected even when the download numbers were misreported.

degree $\gamma_t(s)$ of the vertex added at time t, as a function of its age s, to be

$$\gamma_t(s) = a\left[\left(1 + \frac{s}{t}\right)^{c/(c+a)} - 1\right].\qquad(14.62)$$

When a vertex is first added to the network and $s \ll t$, we can expand in the small quantity s/t to give

$$\gamma_t(s) \simeq \frac{ca}{c+a}\left(\frac{s}{t}\right).\qquad(14.63)$$

In other words, the in-degree of a vertex initially grows linearly with the age of the vertex, on average, but with a constant of proportionality that is smaller the later the vertex entered the network—again we see that there is a substantial advantage for vertices that enter early.

As the vertex ages, there is a crossover to another regime around the point $s = t$, i.e., at the point where the vertex switches from being in the younger half of the population to being in the older. For $s \gg t$,

$$\gamma_t(s) \simeq a\left(\frac{s}{t}\right)^{c/(c+a)},\qquad(14.64)$$

which has a similar form to Eq. (14.63) but with a different exponent, $c/(c+a)$, which is always less than 1, so the growth is slower than linear but still favors vertices that appear early. Figure 14.5 shows the behavior of $\gamma_t(s)$ with time for vertices created at a selection of different times t.

All of these results can be applied to the Barabási–Albert model as well by setting $a = c$ with c an integer and writing the formulas in terms of total degree $k = q + c$ rather than in-degree. For instance, the joint degree/time distribution, Eq. (14.57), becomes

$$\pi_k(\tau) = \binom{k-1}{k-c}\sqrt{\tau}(1 - \sqrt{\tau})^{k-c}\qquad(14.65)$$

for $k \geq c$ and $\pi_k(\tau) = 0$ for $k < c$. This result was first given by Krapivsky and Redner [187] for the case $c = 1$.

14.3.2 SIZES OF IN-COMPONENTS

The in-components of vertices in our growing networks have some interesting properties. Recall that the in-component of vertex i is the set of vertices from which i can be reached by following a directed path through the network (see Section 6.11). In a citation network, for instance, the in-component of paper A

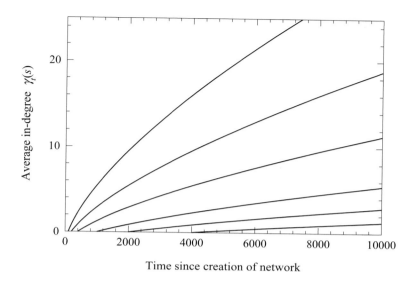

Figure 14.5: Average in-degree of vertices created at different times. The curves show the average in-degrees in Price's model of vertices created at times (top to bottom) $t =$ 100, 200, 400, 1000, 2000, and 4000 as a function of time since the creation of the network. The model parameters were $c = 3$ and $a = 1.5$.

is the set of all papers from which A can be reached by following some trail of successive citations. The reader reading any paper in the in-component can look up other papers in its bibliography, find those papers and look up further ones in *their* bibliographies and so forth, and ultimately reach paper A. One can think of the in-component as representing the set of all papers that "indirectly cite" paper A and the size of the in-component can be considered a measure of the total impact of paper A.

We can study the distribution of in-component sizes by a method similar to the one we used for the degree distribution. Consider Fig. 14.6, which shows a sketch of the in-component of a single vertex A. We have drawn the in-component as a tree, which is accurate so long as the size s of the component is small, $s \ll n$. Just as in the random graph models of Chapters 12 and 13, the probability of a small component having an extra edge that destroys its tree structure vanishes in the limit of large n (see Section 12.6.1). We must be careful however. As we will shortly see, it is possible for the sizes of in-components to become comparable with n in preferential attachment models, in which case the arguments below break down. For the moment, however, let us proceed under the assumption that our component is a tree.

Our in-component will grow in size as vertices and edges are added to the network. Specifically, it will grow larger every time a newly added edge links to any of its members. The probability of a new edge linking to vertex i is given by Eq. (14.1) and summing this probability over all vertices in the in-component $\mathscr{C}$ gives a total probability of

$$\sum_{i \in \mathscr{C}} \frac{q_i + a}{n(c + a)} = \frac{1}{n(c + a)} \left[as + \sum_{i \in \mathscr{C}} q_i \right], \tag{14.66}$$

where s is the number of vertices in the in-component.

Considering Fig. 14.6, we see that every incoming edge in the in-component is necessarily also an outgoing edge from another vertex in the in-component, and moreover that there is exactly one such outgoing edge from each vertex in the in-component, except for A itself. (There are also other outgoing edges from vertices in the in-component, as shown in gray in the figure, but these connect to vertices outside the component itself and play no part in our calculation.) Thus the total number $\sum_{i \in \mathscr{C}} q_i$ of incoming edges is equal to the number of vertices in the in-component minus one, which means our probability of connection, Eq. (14.66), can also be written as

$$\frac{1}{n(c + a)} (as + s - 1) = \frac{(1 + a)s - 1}{n(c + a)}. \tag{14.67}$$

Let us define $p_s(n)$ to be the probability that a randomly chosen vertex has an in-component of size s when the network has n vertices (still assuming $s \ll n$). Then $np_s(n)$ is the number of in-components of size s and, given that each new vertex arrives with c outgoing links, the total number of in-components of size s receiving a new link upon the addition of a new vertex to the network is

$$np_s(n) \times c \times \frac{(1 + a)s - 1}{n(c + a)} = \frac{c}{c + a} [(1 + a)s - 1] p_s(n). \tag{14.68}$$

Now, by an argument similar to the one used to derive Eq. (14.5), we can show that $p_s(n)$ satisfies the master equation

$$(n + 1)p_s(n + 1) = np_s(n)$$
$$+ \frac{c + ca}{c + a} \left[\left(s - 1 - \frac{1}{1 + a} \right) p_{s-1}(n) - \left(s - \frac{1}{1 + a} \right) p_s(n) \right]. \tag{14.69}$$

The only exception is for in-components of size 1, the smallest size possible, for which

$$(n + 1)p_1(n + 1) = np_1(n) + 1 - \frac{ca}{c + a} p_1(n). \tag{14.70}$$

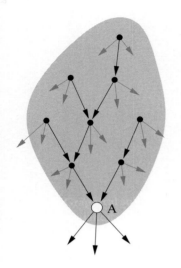

Figure 14.6: The in-component of a vertex A. The total number of incoming links attached to vertices in an in-component is equal to the number of vertices in the component minus 1. Note that there are, in general, many edges outgoing from vertices in the in-component (shown in gray) which connect to vertices not in the in-component and which can thus be ignored for the purposes of our calculation.

The $+1$ here represents the fact that there is one new in-component of size 1 created for each vertex added.

We now take the limit $n \to \infty$ and write $p_s = p_s(\infty)$ to get

$$p_s = \frac{c+ca}{c+a}\left[\left(s - 1 - \frac{1}{1+a}\right)p_{s-1} - \left(s - \frac{1}{1+a}\right)p_s\right] \qquad (14.71)$$

for $s \geq 2$ and

$$p_1 = 1 - \frac{ca}{c+a}p_1. \qquad (14.72)$$

The solution of these equations follows exactly the same lines as the solution for degree distribution. The final result is

$$p_s = \frac{B\left(s - 1/(1+a), \beta\right)}{B\left(1 - 1/(1+a), \beta - 1\right)}, \qquad (14.73)$$

where $B(x, y)$ is the Euler beta function, Eq. (14.19), and

$$\beta = 1 + \frac{1 + a/c}{1 + a}. \qquad (14.74)$$

As we have seen previously, the beta function has a power-law tail $B(x, y) \sim x^{-y}$ (see Eq. (14.25)), so the in-component size distribution also has a power-law tail:

$$p_s \sim s^{-\beta}, \qquad (14.75)$$

although with an exponent β that is in general different from that of the degree distribution (see Eq. (14.27)). Indeed, note that for the normal situation where $c \geq 1$ we have $1 < \beta \leq 2$, which is puzzling: power laws with $\beta \leq 2$ have no finite mean, but for any finite value of n our network must certainly have a finite average component size. The solution to this conundrum is relatively simple, however. As we pointed out above, our calculations are only valid for component sizes $s \ll n$. For larger sizes the method will break down and the power-law behavior will be lost. In physical terms, in-components clearly cannot be larger than the size of the whole network, and so we must expect finite-size effects that cut off the size distribution as s approaches n. In mathematical terms, the components stop being trees as their size becomes comparable with n and hence Eq. (14.67) ceases to be correct.

It's also possible to derive solutions for the component size distribution as a function of the time of creation of a vertex or the age of a vertex, just as we did for the degree distribution in the first part of this section. Indeed there are many more properties of these models that can be calculated using the master equation approach, which is an immensely useful technique for problems such as these. Life, however, is short, and there are many other interesting matters to look into, so we will move on to other things.

See Section 8.4.2 for a discussion of the mean and other moments of power-law distributions.

513

14.4 EXTENSIONS OF PREFERENTIAL ATTACHMENT MODELS

Many extensions and generalizations of preferential attachment models have been suggested, typically addressing questions about what happens when we vary the details of the model definition or attempt to make the model more faithful to the way real networks grow. For instance:

1. By contrast with citations, links in the Web are not permanent. They can and frequently do disappear as well as appear, and links can be added between vertices not just at the moment a vertex is created but at any later time too.

2. Entire web pages also disappear as well as appear.

3. There is no obvious reason why the preferential attachment process has to be linear in the degree. What happens if it is non-linear?

4. Not all vertices are created equal. Some papers or websites might be intrinsically more interesting or important by virtue of their content and hence attract more links. Can this process be incorporated into the model?

In this section we describe modifications of the preferential attachment process that address each of these questions. In the interests of simplicity, we describe the developments in the context of the Barabási–Albert model, rather than the more general Price model. Generalizations of Price's model are certainly possible but the algebra is in many cases unwieldy, and the main conclusions are easier to understand in the context of the simpler model.

14.4.1 ADDITION OF EXTRA EDGES

Price proposed his model of a growing network with citation networks in mind. Since the bibliography of a paper cannot be changed after the paper is published, the edges in a citation network are effectively frozen in place from the moment they are first created, and Price's model mimics this behavior with edges being added only at the moment a vertex is created and never moved or removed thereafter.

This is not true of all networks, however. The World Wide Web, for example, is constantly changing. Links between web pages can, and often are, added or removed after the pages are created. This state of flux is not captured by Price's model or by the Barabási–Albert model of Section 14.2. Yet the Web still has a power-law degree distribution. This leads us to wonder whether it is possible to create a generalized model that includes the addition and removal of edges after vertices are created but still generates power-law distributions. It turns out that we can indeed, as we now describe.

We first consider the relatively simple case in which edges are added to our network but never taken away, which has been studied by a number of authors [11, 96, 190]. The case of edge removal is more complex and is considered in the following section. The model we consider is a generalization of the Barabási–Albert model in which vertices are added to the network one by one as before and each starts out with c undirected edges which attach to other vertices with probability proportional to degree k. But we now include a second process in the model as well: at each step some number w of extra edges are added to the network with *both* ends attaching to vertices chosen in proportional to degree. Thus when the network has n vertices it will have a total of $n(c + w)$ edges. (In fact, it is only necessary that an *average* number w of extra edges be added at each step. The actual number can fluctuate around this figure, provided the fluctuations satisfy some modest constraints on their size, and the net result, in the limit of large network size, will be the same. This allows us to give w a non-integer value if we wish.)

This model turns out to be quite easy to solve given the results of previous sections. The only difference between it and the standard Barabási–Albert model is that, instead of c new ends of edges attaching to old vertices for every new vertex added, we now have $c + 2w$ new ends of edges—two extra for each of the w extra edges. The probability of attachment of any one of those edges to a particular vertex i is $k_i / \sum_i k_i$ as before. The sum in the denominator is equal to twice the number of edges in the network (see Eq. (6.20)), or $\sum_i k_i = 2n(c + w)$.

Then, if $p_k(n)$ denotes the fraction of vertices with degree k when the network has n vertices in total, the number of vertices of degree k receiving a new edge, per vertex added, is

$$np_k(n) \times (c + 2w) \times \frac{k}{2n(c + w)} = \frac{c + 2w}{2(c + w)} kp_k(n). \qquad (14.76)$$

We can use this result to write a master equation for $p_k(n)$ thus:

$$(n + 1)p_k(n + 1) = np_k(n) + \frac{c + 2w}{2(c + w)} \left[(k - 1)p_{k-1}(n) - kp_k(n) \right], \qquad (14.77)$$

for $k > c$ and

$$(n + 1)p_c(n + 1) = np_c(n) + 1 - \frac{c + 2w}{2(c + w)} cp_c(n), \qquad (14.78)$$

for $k = c$. (There are, as before, no vertices of degree less than c.) Taking the limit of large n and writing $p_k = p_k(\infty)$, these equations simplify to

$$p_k = \frac{c + 2w}{2(c + w)} \left[(k - 1)p_{k-1} - kp_k \right] \qquad \text{for } k > c, \qquad (14.79)$$

$$p_c = 1 - \frac{c + 2w}{2(c + w)} cp_c \qquad \text{for } k = c. \qquad (14.80)$$

Rearranging these equations along the lines of Eqs. (14.9) to (14.21), we then find that

$$p_k = \frac{B(k, \alpha)}{B(c, \alpha - 1)},$$

(14.81)

where $B(x, y)$ is the Euler beta function again, Eq. (14.19), and

$$\alpha = 2 + \frac{c}{c + 2w}.$$

(14.82)

Since $B(x, y)$ goes as x^{-y} for large x (Eq. (14.25)), our degree distribution has a power-law tail with exponent α. For the special case of $w = 0$, in which no additional edges are added to the network, we recover the standard result $\alpha = 3$ for the Barabási–Albert model; for $w > 0$ we get exponents in the range $2 < \alpha < 3$, which agrees nicely with the values typically observed for degree distributions on the Web (see Table 8.1). Bear in mind though that the Web is a directed network while the model described here is undirected. If we want to build a model of a directed network we would need to start with something like the Price model of Section 14.1. Generalizations of Price's model that include addition of extra edges as above are certainly possible—see for example Krapivsky *et al.* [190].

14.4.2 REMOVAL OF EDGES

Now consider the case of a network in which edges can be removed. To keep things simple let us first consider the case where edges can be removed at any time but are only added at the initial creation of a vertex, as in the standard Barabási–Albert model. (In a moment we will consider the general case of addition and removal at any time.)

There are many ways in which edges could be removed from a network, but let us consider the most basic case in which they are simply deleted uniformly at random. What then is the probability that a particular vertex i loses an edge when a single edge is removed from the network? When an edge is deleted both of its two ends vanish. Given that the deletion is uniformly random, the probability that one of those two ends is attached to vertex i is simply proportional to the total number of ends attached to i, which is equal to the degree k_i. Properly normalized, the probability that vertex i loses an edge is thus $2k_i / \sum_i k_i$, the factor of two coming from the two ends of the edge. In other words, the random deletion of edges is like a type of preferential attachment in reverse: the higher the degree of the vertex, the more likely it is to lose an edge.

So consider the undirected network model in which vertices with degree c are added to the network following the normal preferential attachment scheme and an average of v edges are deleted at random for each vertex added. (As with the model of Section 14.4.1 the actual number of edges deleted can fluctuate about the mean and v can take a non-integer value if we wish.) In order that the number of edges in the network grow, rather than shrinking to zero and vanishing, we require that the net number of edges added per vertex $c - v$ be positive, i.e., $v < c$. Then when the network has n vertices the number of edges will be $n(c - v)$.

In writing down a master equation for this model there are several processes we need to consider. As before, the number of vertices with degree k increases whenever a vertex of degree $k - 1$ gains a new edge and decreases when a vertex of degree k gains a new edge. By an argument analogous to the one leading to Eq. (14.76), the number of vertices of degree k gaining an edge per vertex added to the network is

$$np_k(n) \times c \times \frac{k}{2n(c - v)} = \frac{c}{2(c - v)} k p_k(n). \quad (14.83)$$

But we also now have a new process in which a vertex can lose an edge, which means that the number of vertices of degree k also increases when a vertex of degree $k + 1$ loses an edge and decreases when a vertex of degree k loses an edge. The number of vertices of degree k losing an edge per vertex added is given by

$$np_k(n) \times 2v \times \frac{k}{2n(c - v)} = \frac{v}{c - v} k p_k(n), \quad (14.84)$$

with the factor of $2v$ reflecting the fact that each of the v edges removed has two ends.

Another important thing to notice is that, by contrast with the original Barabási–Albert model, vertices can now have any degree $k \geq 0$—vertices can lose any or all of their edges, right down to the last one, so there is no restriction $k \geq c$ on the degree as before.

Our master equation now takes the form

$$(n + 1)p_k(n + 1) = np_k(n) + \frac{c}{2(c - v)}(k - 1)p_{k-1}(n)$$
$$+ \frac{v}{c - v}(k + 1)p_{k+1}(n) - \frac{c + 2v}{2(c - v)} k p_k(n) \quad (14.85)$$

for $k \neq c$ and

$$(n + 1)p_c(n + 1) = np_c(n) + 1 + \frac{c}{2(c - v)}(c - 1)p_{c-1}(n)$$
$$+ \frac{v}{c - v}(c + 1)p_{c+1}(n) - \frac{c + 2v}{2(c - v)} c p_c(n) \quad (14.86)$$

for $k = c$. These two equations can conveniently be combined by writing

$$(n+1)p_k(n+1) = np_k(n) + \delta_{kc} + \frac{c}{2(c-v)}(k-1)p_{k-1}(n)$$
$$+ \frac{2v}{2(c-v)}(k+1)p_{k+1}(n) - \frac{c+2v}{2(c-v)}kp_k(n),$$

(14.87)

where δ_{kc} is the Kronecker delta, which is 1 if $k = c$ and 0 otherwise.

The only exception to this master equation is for the case $k = 0$, where the term proportional to $k - 1$ vanishes because there are no vertices of degree -1. A simple way of enforcing this exception is to define $p_{-1}(n) = 0$ for all n, in which case Eq. (14.87) then applies for all $k \geq 0$. We will adopt this convention henceforth.

The model as we have described it so far incorporates the processes of vertex addition and edge removal, but, given Eq. (14.87), it is only a small extra step to incorporate the edge addition process of Section 14.4.1 as well. If as before we add w extra edges per vertex added, then $c + w - v$ edges are added net per vertex, and our master equation becomes

$$(n+1)p_k(n+1) = np_k(n) + \delta_{kc} + \frac{c+2w}{2(c+w-v)}(k-1)p_{k-1}(n)$$
$$+ \frac{v}{c+w-v}(k+1)p_{k+1}(n) - \frac{c+2w+2v}{2(c+w-v)}kp_k(n).$$

(14.88)

The equation for edge removal only, Eq. (14.87), can then be considered a special case of this equation with $w = 0$. As before, we require that the net number of edges added per vertex be positive, or $v < c + w$.

Now taking the limit as $n \to \infty$ and writing $p_k = p_k(\infty)$ we find that

$$p_k = \delta_{kc} + \frac{c+2w}{2(c+w-v)}(k-1)p_{k-1}$$
$$+ \frac{v}{c+w-v}(k+1)p_{k+1} - \frac{c+2w+2v}{2(c+w-v)}kp_k.$$

(14.89)

This equation differs in a crucial way from the master equations we have encountered previously, such as Eq. (14.7), because the right-hand side contains terms for vertices of three different degrees ($k - 1$, k, and $k + 1$) rather than just two. This makes the equation substantially more difficult to solve. We can no longer simply rearrange to derive an expression for p_k in terms of p_{k-1} and then apply that expression repeatedly to itself. A solution is still possible, but it's not simple. Here we give just an outline of the method. The gory details, for those interested in them, are spelled out by Moore *et al.* [226].[16]

The basic strategy for solving Eq. (14.89) is to use a generating function of the kind we introduced in Section 13.1. We define

$$g(z) = \sum_{k=0}^{\infty} p_k z^k. \tag{14.90}$$

Substituting for p_k from Eq. (14.89) we get

$$g(z) = \sum_{k=0}^{\infty} \delta_{kc} z^k + \frac{c + 2w}{2(c + w - v)} \sum_{k=0}^{\infty} (k - 1) p_{k-1} z^k$$

$$+ \frac{2v}{2(c + w - v)} \sum_{k=0}^{\infty} (k + 1) p_{k+1} z^k - \frac{c + 2w + 2v}{2(c + w - v)} \sum_{k=0}^{\infty} k p_k z^k. \tag{14.91}$$

The first term on the right is simple—it is equal to z^c. The others require a little more care. Consider the second term, for example. Note that the first term in the sum, the term for $k = 0$, is necessarily zero because, as we have said, $p_{-1} = 0$. Hence we can write

$$\sum_{k=0}^{\infty} (k - 1) p_{k-1} z^k = \sum_{k=1}^{\infty} (k - 1) p_{k-1} z^k = \sum_{k=0}^{\infty} k p_k z^{k+1}$$

$$= z^2 \sum_{k=0}^{\infty} k p_k z^{k-1} = z^2 \frac{d}{dz} \sum_{k=0}^{\infty} p_k z^k$$

$$= z^2 \frac{dg}{dz}, \tag{14.92}$$

where in the first line we have made the substitution $k - 1 \to k$ and in the second line we have made use of the fact that the $k = 0$ term is again zero (because of the factor of k).

For the third and fourth terms in (14.91) we can similarly write

$$\sum_{k=0}^{\infty} (k + 1) p_{k+1} z^k = \sum_{k=1}^{\infty} k p_k z^{k-1} = \sum_{k=0}^{\infty} k p_k z^{k-1} = \frac{dg}{dz}, \tag{14.93}$$

and

$$\sum_{k=0}^{\infty} k p_k z^k = z \sum_{k=0}^{\infty} k p_k z^{k-1} = z \frac{dg}{dz}. \tag{14.94}$$

---

[16]In fact, Moore *et al.* give a solution for a model in which vertices rather than edges are deleted, but the two can be treated by virtually the same means. The calculation given here is adapted from their work with only minor changes.

Combining Eqs. (14.91) to (14.94) and rearranging, we then get

$$\frac{(c+2w)z - 2v}{2(c+w-v)}(1-z)\frac{dg}{dz} + g(z) = z^c. \tag{14.95}$$

This is a first-order linear differential equation and is solvable by standard—if tedious—methods. To cut a long story short, one can find an integrating factor for the left-hand side and hence express the solution in terms of an integral that, provided $v < \frac{1}{2}c + w$, can be reduced by repeated integration by parts to give

$$p_k = Ak^{-\alpha} \int_0^k \frac{(1-x/k)^k}{(1-\gamma x/k)^k} x^{\alpha-2}\,dx, \tag{14.96}$$

for $k \geq c$, where A is a k-independent constant and

$$\alpha = 2 + \frac{v-w}{c+2w-2v}, \tag{14.97}$$

$$\gamma = \frac{2v}{c+2w}. \tag{14.98}$$

The remaining integral can be written in terms of hypergeometric functions, but we can find the asymptotic behavior of the degree distribution for large k more directly by noticing that as k becomes large

$$\left[1 - \frac{x}{k}\right]^k \to e^x, \qquad \left[1 - \frac{\gamma x}{k}\right]^k \to e^{\gamma x}, \tag{14.99}$$

so that

$$p_k \sim k^{-\alpha} \int_0^\infty e^{-(1-\gamma)x} x^{\alpha-2}\,dx = \frac{\Gamma(\alpha-1)}{(1-\gamma)^{\alpha-1}} k^{-\alpha}. \tag{14.100}$$

Thus we once again find that our degree distribution has a power-law tail, with an exponent given this time by Eq. (14.97). Note that this exponent can take values both greater than and less than two. What's more for the case where $v = \frac{1}{2}c + w$ it actually becomes infinite. Moore *et al.* [226] show that at this point we lose the power-law behavior and the distribution becomes instead a stretched exponential. Up until this point, however, the distribution still follows a power law, albeit with a very large exponent as v grows larger. For values of $v > \frac{1}{2}c + w$, the solution becomes nonsensical, with a negative value of α, and one must return to the original differential equation (14.95) to find the solution for this case. We leave the developments, however, as an exercise for the especially avid reader.

Before we leave this topic, however, let us point out that the methods used to solve Eq. (14.89) can also be used to calculate what happens when we remove not edges but vertices from our network. Loss of vertices does occur in some networks, such as the World Wide Web, so it is potentially of interest to ask what effect it has on the degree distribution. In fact, the solution for this case is very similar to the solution for loss of edges, with a power-law distribution and an exponent that depends on the vertex loss rate, diverging as the rate of loss approaches the rate at which vertices are added. The details can be found in [226].

14.4.3 NON-LINEAR PREFERENTIAL ATTACHMENT

In the models we have considered so far, the probability that a new edge attaches to a vertex is linear in the degree of the vertex. Although this is a reasonable first guess about the way things might be, it's certainly also possible that attachment processes might not be linear. Indeed, there is some empirical evidence that this is the case. For instance, Jeong *et al.* [165] looked at the growth of several real-world networks, measuring the rate at which nodes acquired new edges. To avoid problems associated with the fact that the rate can depend not only on degree but also on the total size n of the network (see Eq. (14.1)), they restricted their observations to relatively short intervals of time. The measured rates, plotted as a function of vertex degree, showed that for some networks there was a roughly linear preferential attachment effect, but for others attachment appeared to be non-linear, going as some power γ of the degree with γ being significantly different from 1. (The values they observed were around $\gamma = 0.8$.)

What effect would non-linear preferential attachment have on the degree distribution of our network? Should we still expect to see power-law behavior in the non-linear case? The answers to these questions depend on the particular functional form of the attachment probability and there are an infinite variety of functional forms. We will look at some specific examples shortly, but for the moment let us keep the discussion completely general. Following an approach introduced by Krapivsky *et al.* [189], we define an *attachment kernel*, denoted a_k, which specifies the functional form of the attachment probability. For the model of Barabási and Albert, where attachment is simply proportional to degree, the attachment kernel would be $a_k = k$. For the non-linear attachment observed by Jeong *et al.* and discussed above, it would be $a_k = k^\gamma$. Note that the attachment kernel is not a probability, merely a functional form. The correctly normalized probability that a newly added edge attaches to a specific vertex i having degree k_i is $a_k / \sum_i a_{k_i}$.

So consider again a growing undirected network of the type discussed in previous sections and let $p_k(n)$ be the fraction of vertices with degree k when the network has n vertices. As before, an average of c new edges are added to the network with each new vertex, but preferential attachment is now non-linear, governed by the attachment kernel a_k, which means that, by analogy with Eq. (14.2), the expected number of vertices of degree k receiving a new connection when a single new vertex is added to the network is

$$n p_k(n) \times c \times \frac{a_k}{\sum_i a_{k_i}} = \frac{c}{\mu(n)} a_k p_k(n), \tag{14.101}$$

where

$$\mu(n) = \frac{1}{n} \sum_{i=1}^{n} a_{k_i} = \sum_k a_k p_k(n). \tag{14.102}$$

Now the master equation for $p_k(n)$ is

$$(n+1) p_k(n+1) = n p_k(n) + \frac{c}{\mu(n)} \left[a_{k-1} p_{k-1}(n) - a_k p_k(n) \right]. \tag{14.103}$$

As before the term in $p_{k-1}(n)$ represents new vertices of degree k created when vertices of degree $k-1$ receive new edges and the last term in $p_k(n)$ represents vertices of degree k lost when they gain new edges to become vertices of degree $k+1$.

The only exception to this equation is for vertices of degree c, for which

$$(n+1) p_c(n+1) = n p_c(n) + 1 - \frac{c}{\mu(n)} a_c p_c(n). \tag{14.104}$$

(And there are no vertices of degree less than c, since all vertices are created with degree c initially and edges are never removed.)

Taking the limit as $n \to \infty$ and writing $p_k = p_k(\infty)$ and $\mu = \mu(\infty)$, these equations become

$$p_k = \frac{c}{\mu} \left[a_{k-1} p_{k-1} - a_k p_k \right] \tag{14.105}$$

for $k > c$ and

$$p_c = 1 - \frac{c a_c}{\mu} p_c. \tag{14.106}$$

Note that μ depends via Eq. (14.102) on the degree distribution, which we don't yet know. For now, however, it will be enough that μ is independent of k; we will derive an expression for its exact value in a moment.

Equations (14.105) and (14.106) can be rearranged to give

$$p_c = \frac{\mu/c}{a_c + \mu/c} \tag{14.107}$$

and

$$p_k = \frac{a_{k-1}}{a_k + \mu/c} p_{k-1}. \tag{14.108}$$

Applying the latter repeatedly we get

$$
\begin{aligned}
p_k &= \frac{a_{k-1}a_{k-2}\ldots a_c}{(a_k + \mu/c)\ldots(a_{c+1} + \mu/c)}\, p_c \\
&= \frac{\mu}{ca_k}\, \frac{a_k \ldots a_c}{(a_k + \mu/c)\ldots(a_c + \mu/c)} \\
&= \frac{\mu}{ca_k} \prod_{r=c}^{k}\left[1 + \frac{\mu}{ca_r}\right]^{-1}.
\end{aligned}
\tag{14.109}
$$

All we need to complete our solution is the value of μ. Taking Eq. (14.102) and letting $n \to \infty$, we get

$$
\mu = \sum_{k=c}^{\infty} a_k p_k = \frac{\mu}{c} \sum_{k=c}^{\infty} \prod_{r=c}^{k}\left[1 + \frac{\mu}{ca_r}\right]^{-1}.
\tag{14.110}
$$

Canceling μ from both sides we arrive at the equation

$$
\sum_{k=c}^{\infty} \prod_{r=c}^{k}\left[1 + \frac{\mu}{ca_r}\right]^{-1} = c.
\tag{14.111}
$$

In principle we should be able to solve this equation for μ and substitute the result into Eq. (14.109) to get the complete degree distribution. In practice, unfortunately, the equation is not solvable in closed form for most choices of the attachment kernel a_k, although an approximate value for μ can usually be calculated numerically on a computer. Even without knowing μ, however, we can still find the overall functional form of p_k, which is enough to answer many of the questions we are interested in.

As an example, consider a network of the type observed by Jeong *et al.* [165] and discussed above in which attachment goes as k^γ for some positive constant γ, and let us assume that (as found by Jeong *et al.*) we have $\gamma < 1$. The solution for this particular choice was given by Krapivsky *et al.* [189] and shows a number of interesting features.

Putting $a_k = k^\gamma$ in Eq. (14.109) gives

$$
p_k = \frac{\mu}{ck^\gamma} \prod_{r=c}^{k}\left[1 + \frac{\mu}{cr^\gamma}\right]^{-1}.
\tag{14.112}
$$

This degree distribution turns out *not* have a power-law tail, by contrast with the case of linear preferential attachment. In other words the power-law form is sensitive to the precise shape of the attachment kernel. We can see this by writing

$$
\prod_{r=c}^{k}\left[1 + \frac{\mu}{cr^\gamma}\right]^{-1} = \exp\left[-\sum_{r=c}^{k}\ln\left(1 + \frac{\mu}{cr^\gamma}\right)\right]
\tag{14.113}
$$

and then expanding the logarithm as a Taylor series in μ/cr^γ:

$$\sum_{r=c}^{k} \ln\left(1 + \frac{\mu}{cr^\gamma}\right) = -\sum_{r=c}^{k} \sum_{s=1}^{\infty} \frac{(-1)^s}{s} \left(\frac{\mu}{c}\right)^s r^{-s\gamma}$$

$$= -\sum_{s=1}^{\infty} \frac{(-1)^s}{s} \left(\frac{\mu}{c}\right)^s \sum_{r=c}^{k} r^{-s\gamma}. \qquad (14.114)$$

The sum over r cannot be expressed in closed form, but we can approximate it using the *trapezoidal rule*,[17] which says that for any function $f(r)$:

$$\sum_{r=a}^{b} f(r) = \int_a^b f(r)\, dr + \tfrac{1}{2}\left[f(a) + f(b)\right] + O\big(f'(b) - f'(a)\big). \qquad (14.115)$$

(For those not familiar with it, the derivation of the trapezoidal rule is illustrated in Fig. 14.7.[18])

In our case $f(r) = r^{-s\gamma}$ and Eq. (14.115) gives

$$\sum_{r=c}^{k} r^{-s\gamma} = A_s + \frac{k^{1-s\gamma}}{1 - s\gamma} + \tfrac{1}{2} k^{-s\gamma} + O\big(k^{-(s\gamma+1)}\big), \qquad (14.116)$$

where A_s is a constant depending on s (and on c) but not on k.

Consider now what happens when k becomes large. Since $\gamma > 0$, the term in $k^{-s\gamma}$ and all subsequent terms vanish as $k \to \infty$ and Eq. (14.114) becomes

$$\sum_{r=c}^{k} \ln\left(1 + \frac{\mu}{cr^\gamma}\right) \simeq A - \sum_{s=1}^{\infty} \frac{(-1)^s}{s} \left(\frac{\mu}{c}\right)^s \frac{k^{1-s\gamma}}{1 - s\gamma}, \qquad (14.117)$$

where A is a k-independent constant equal to $\sum_{s=1}^{\infty} A_s (-\mu/c)^s/s$.

This expression can be simplified still further by noting that, in the limit $k \to \infty$, all terms in $k^{1-s\gamma}$ where $1 - s\gamma < 0$ also vanish. Thus for any given value of γ we need keep terms in k up to only a certain value of s. The simplest case is when $\tfrac{1}{2} < \gamma < 1$. In this case only the term for $s = 1$ grows as k increases, all others vanishing, and

$$\sum_{r=c}^{k} \ln\left(1 + \frac{\mu}{cr^\gamma}\right) \simeq A + \frac{\mu k^{1-\gamma}}{c(1 - \gamma)} \qquad (14.118)$$

as $k \to \infty$.

---

[17] Also called the trapezium rule in British English.

[18] Explicit expressions are known for the correction terms (the terms in $f'(a)$ and $f'(b)$)—they are given in terms of the Bernoulli numbers by the so-called Euler–Maclaurin formula [2]—but they're not necessary in our application because the correction terms vanish anyway.

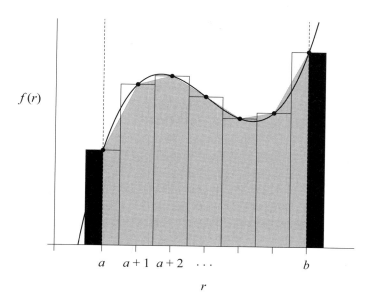

$f(r)$

$a \quad a+1 \; a+2 \quad \cdots \quad\quad\quad\quad b$

r

Figure 14.7: The trapezoidal rule. The trapezoidal rule approximates a sum by an integral (or vice versa). The sum of the function $f(r)$ from $r = a$ to $r = b$ (dotted lines) is equal to the sum of the areas of the rectangular bars, which is also equal to the area shaded in gray. This shaded area can be approximated by the integral of $f(r)$ between a and b (smooth curve) plus the two extra rectangular sections at either end (hatched), which have area $\frac{1}{2}f(a)$ and $\frac{1}{2}f(b)$ respectively. Add everything up and we get Eq. (14.115). The error in the approximation is equal to the sum of the relatively small regions between the curve and the shaded area.

Now, combining Eqs. (14.112), (14.113), and (14.118), we find that the asymptotic form of p_k is

$$p_k \sim k^{-\gamma} \exp\left(-\frac{\mu k^{1-\gamma}}{c(1-\gamma)}\right), \tag{14.119}$$

for $\frac{1}{2} < \gamma < 1$.

Distributions of this general form, in which the dominant contribution to the probability falls off as the exponential of a power of k, are called *stretched exponentials*. Since the exponent $1 - \gamma$ is less than one, the distribution falls off more slowly than an ordinary exponential in k, which is why we called it "stretched."[19] On the other hand, the distribution still falls off a good deal

---

[19]Although, confusingly, people often still call it a stretched exponential even when the expo-

faster than the power law that we found in the case of linear preferential attachment, and this is really the important point here. This calculation reveals that the power-law distribution in the Barabási–Albert model is a special feature of the linear attachment process assumed by that model. (Note that this observation is valid even though we haven't calculated the value of the constant μ. The general functional form of the degree distribution doesn't depend on the value of the constant.)

For other values of γ the calculation is similar but involves more terms in Eq. (14.117). For instance, if $\frac{1}{3} < \gamma < \frac{1}{2}$ then the terms in $k^{1-s\gamma}$ for $s = 1$ and 2 both grow as k becomes large while all others vanish, and we find that

$$\sum_{r=k_0+1}^{k} \ln\left(1 + \frac{\mu}{cr^\gamma}\right) \simeq A + \frac{\mu k^{1-\gamma}}{c(1-\gamma)} - \frac{\mu^2 k^{1-2\gamma}}{2c^2(1-2\gamma)}, \tag{14.120}$$

which gives

$$p_k \sim k^{-\gamma} \exp\left(-\frac{\mu k^{1-\gamma}}{c(1-\gamma)} + \frac{\mu^2 k^{1-2\gamma}}{2c^2(1-2\gamma)}\right). \tag{14.121}$$

In between the solutions (14.119) and (14.121) there is a special case solution when γ is exactly equal to $\frac{1}{2}$. For $\gamma = \frac{1}{2}$ and $s = 2$ the integral in Eq. (14.115) gives rise not to a power of k but to a log and Eq. (14.114) becomes

$$\sum_{r=c}^{k} \ln\left(1 + \frac{\mu}{cr^\gamma}\right) \simeq A + \frac{2\mu}{c}\sqrt{k} - \frac{\mu^2}{2c^2} \ln k, \tag{14.122}$$

all other terms vanishing in the limit of large k. Substituting this expression into Eq. (14.113), we then arrive at

$$p_k \sim \left(\sqrt{k}\right)^{\mu^2/c^2 - 1} \exp\left(-\frac{2\mu}{c}\sqrt{k}\right), \tag{14.123}$$

for $\gamma = \frac{1}{2}$.

We can continue in this vein ad infinitum. There are distinct solution forms for $\frac{1}{4} < \gamma < \frac{1}{3}$ and $\frac{1}{5} < \gamma < \frac{1}{4}$ and so forth, as well as special case solutions for $\gamma = \frac{1}{3}, \frac{1}{4}, \frac{1}{5}$, and so forth. Figure 14.8 shows the degree distribution for the case $\gamma = 0.8$, along with the asymptotic form (14.119). Note the convex form of the curve on the semilogarithmic scales, which indicates a function decaying slower than an exponential.

One can also calculate the degree distribution for superlinear preferential attachment, i.e., for values of γ greater than one. This case also shows some

---

nent is greater than one. This case should really be called a "squeezed exponential."

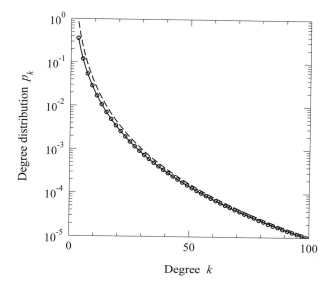

Figure 14.8: Degree distribution for sublinear preferential attachment. This plot shows the fraction p_k of vertices with degree k in a growing network with attachment kernel k^γ as described in the text. In this case $\gamma = 0.8$ and $c = 3$. The points are results from computer simulations, averaged over 100 networks of (final) size 10^7 vertices each. The solid line is the exact solution, Eq. (14.112), evaluated numerically. The dashed line is the asymptotic form, Eq. (14.119), with the overall constant of proportionality chosen to coincide with the exact solution for large values of k.

interesting behaviors: it turns out that for $\gamma > 1$ the typical behavior is for one vertex to emerge as a "leader" in the network, gaining a non-zero fraction of all edges, with the rest of the vertices having small degree (almost all having degree less than some fixed constant). Readers interested in these developments can find them described in detail in Ref. [189].

14.4.4 VERTICES OF VARYING QUALITY OR ATTRACTIVENESS

The models of growing networks we have examined so far assume that all vertices of a given degree are equally likely to gain a new edge. In these models, for example, all papers that have never been cited before are equally likely to get new citations. All websites that no one has linked to yet are equally likely to receive links.

In the real world, of course, nothing could be farther from the truth. There are huge differences in the perceived importance and quality of scientific papers or websites that mean some are far more likely to gain edges than others. A website, for instance, that provides a useful service, such as a directory or an encyclopedia, will almost certainly receive new links at a higher rate than most people's personal home pages. Indeed, search engines use the numbers of links web pages receive precisely as a measure of which pages people find most useful. Similarly, people look at the numbers of citations a paper receives to try to gauge how influential that paper has been. These approaches would

See Sections 7.4 and 19.1 for a discussion of the operation of search engines.

527

not work unless there were some correlation between the degree and the perceived quality of a vertex.

If one allows for variations in the intrinsic quality or attractiveness of vertices, then, presumably it will have an effect on the degree distribution. It seems entirely possible that, with such effects at work, the power laws generated by preferential attachment models might completely disappear, leaving us at a loss to explain how power laws might arise in real-world networks. In this section we study a model of the growth of a network proposed by Bianconi and Barabási [42,43] that includes effects of varying node quality—or *fitness* as they call it. As we will see, the power-law behavior of traditional models disappears once vertex fitness enters the picture, although the distribution for vertices of a given fitness still follows a power law.

The model of Bianconi and Barabási is defined as follows. Vertices are added one by one with each attaching by undirected edges to c prior vertices, just as before. Now, however, each vertex i has a fitness η_i that is assigned at the moment of the vertex's creation and never changed thereafter. The fitnesses are real numbers with values drawn from some distribution $\rho(\eta)$, so that the probability of a value falling between η and $\eta + d\eta$ is $\rho(\eta)\, d\eta$. Each of the c new edges added with each new vertex attaches to a previously existing vertex with probability proportional to an attachment kernel $a_k(\eta)$ that depends now on both the degree k of the target vertex and its fitness η. (In fact, Bianconi and Barabási examined only the special case $a_k(\eta) = \eta k$. The general model considered here was proposed and solved subsequently by Krapivsky and Redner [188].)

This model can be solved by the same method as the model of Section 14.4.3. We define $p_k(\eta, n)\, d\eta$ to be the fraction of vertices with degree k and fitness in the interval η to $\eta + d\eta$ when the network has n vertices. Writing down a master equation as before and taking the limit $n \to \infty$ we arrive at equations that read

$$p_k(\eta) = \frac{c}{\mu}\left[a_{k-1}(\eta)p_{k-1}(\eta) - a_k(\eta)p_k(\eta)\right] \tag{14.124}$$

for $k > c$ and

$$p_c(\eta) = \rho(\eta) - \frac{ca_c(\eta)}{\mu}p_c(\eta), \tag{14.125}$$

where $p_k(\eta) = p_k(\eta, \infty)$ and μ is again the appropriate normalizing factor

$$\mu = \frac{1}{n}\sum_{i=1}^{n} a_{k_i}(\eta_i) = \sum_{k=c}^{\infty}\int_{-\infty}^{\infty} a_k(\eta)p_k(\eta)\, d\eta. \tag{14.126}$$

(See Eq. (14.110).) Note that the $+1$ of Eq. (14.106) has been replaced by $\rho(\eta)$ in Eq. (14.125), because the average number of new vertices of degree c added to the network with fitness in the interval η to $\eta + d\eta$ is not 1 but $\rho(\eta)$.

Following the same steps that led to Eq. (14.109), we can solve the master equation to show that

$$p_k(\eta) = \rho(\eta) \frac{\mu}{ca_k(\eta)} \prod_{r=c}^{k} \left[1 + \frac{\mu}{ca_r(\eta)} \right]^{-1}, \qquad (14.127)$$

and the value of μ can be determined by substituting this result back into Eq. (14.126) (although usually an analytic solution is not possible and the equations must be solved numerically).

As an example consider the case where the attachment kernel is linear in the degree, $a_k(\eta) = \eta k$, which was studied by Bianconi and Barabási.[20] Since $a_k(\eta)$ is (proportional to) a probability it cannot be negative, so we must restrict η to non-negative values. Then the product in Eq. (14.127) becomes

$$\prod_{r=c}^{k} \left[1 + \frac{\mu}{c\eta k} \right]^{-1} = \prod_{r=c}^{k} \frac{k}{k + \mu/c\eta}$$
$$= \frac{\Gamma(k+1)\Gamma(c + \mu/c\eta)}{\Gamma(c)\Gamma(k + 1 + \mu/c\eta)}$$
$$= \frac{c\eta k}{\mu} \frac{B(k, 1 + \mu/c\eta)}{B(c, \mu/c\eta)}, \qquad (14.128)$$

where we have made use of Eq. (14.17) and $B(x, y)$ is Euler's beta function, Eq. (14.19), again. Substituting into Eq. (14.127), we then find that

$$p_k(\eta) = \rho(\eta) \frac{B(k, 1 + \mu/c\eta)}{B(c, \mu/c\eta)}. \qquad (14.129)$$

We showed previously that the beta function goes as a power law $B(x, y) \sim x^{-y}$ for large values of its first argument (Eq. (14.25)) so Eq. (14.129) implies that the distribution of the degrees of vertices with a particular value of the fitness η has a power-law tail with exponent

$$\alpha(\eta) = 1 + \frac{\mu}{c\eta}. \qquad (14.130)$$

However, the overall degree distribution for the entire network may or may not have a power-law tail, depending on the distribution $\rho(\eta)$. It is clear that it is a power law for some choices of $\rho(\eta)$—for example the trivial choice where

---

[20]The most general form of linear kernel would be $a_k(\eta) = f(\eta) k$ where $f(\eta)$ is an increasing function of η. However, this form can be turned into the one above by a simple change of variables to $\eta' = f(\eta)$, so in fact we are not losing any generality by assuming $a_k(\eta) = \eta k$.

all vertices have the same η, which just reduces to the original Barabási–Albert model. If η is broadly distributed, however, the degree distribution will be a sum over power laws with a wide range of different exponents, which will not in general yield another power law.

The solution above does not tell the whole story. There are some interesting features of this model that are missing from Eq. (14.129). To see this, let us calculate the average degree of a vertex in our network. This might seem like a pointless exercise—the average degree must take the value $2c$ since exactly c edges are added for every vertex—but in fact the calculation is quite revealing. The average degree is given by

$$\langle k \rangle = \sum_{k=c}^{\infty} \int_0^{\infty} k p_k(\eta) \, d\eta = \sum_{k=c}^{\infty} \int_0^{\infty} k \rho(\eta) \, \frac{B(k, 1 + \mu/c\eta)}{B(c, \mu/c\eta)} \, d\eta$$

$$= \int_0^{\infty} \frac{\rho(\eta)}{B(c, \mu/c\eta)} \sum_{k=c}^{\infty} k \, B(k, 1 + \mu/c\eta) \, d\eta. \qquad (14.131)$$

The sum can be performed by making use of the integral form of the beta function, Eq. (14.33), and gives[21]

$$\sum_{k=c}^{\infty} k \, B(k, 1 + \mu/c\eta) = \frac{c}{1 - c\eta/\mu} \, B(c, \mu/c\eta). \qquad (14.132)$$

An important point to notice, however, is that this result only works if $\eta < \mu/c$. If $\eta \geq \mu/c$ the sum diverges making the average degree in the network infinite, which cannot be the case since, as we have said, the average degree is always $2c$. To avoid the divergence we will impose the restriction that $\rho(\eta) = 0$ for all $\eta \geq \eta_0$, where η_0 is a constant in the range $0 \leq \eta_0 < \mu/c$. (The interesting question of what happens to the network if we choose a $\rho(\eta)$ that violates this condition is dealt with below.)

Combining Eqs. (14.131) and (14.132), we then find that

$$\langle k \rangle = c \int_0^{\eta_0} \frac{\rho(\eta) \, d\eta}{1 - c\eta/\mu}. \qquad (14.133)$$

And since $\langle k \rangle = 2c$ this immediately implies that

$$\int_0^{\eta_0} \frac{\rho(\eta) \, d\eta}{1 - c\eta/\mu} = 2. \qquad (14.134)$$

We haven't yet calculated a value for the constant μ, but even without it this equation tells us something interesting. The integral is a monotonically

---

[21] The calculation is essentially the same as the one leading to Eq. (14.34).

decreasing function of μ: it takes its smallest value of 1 when $\mu \to \infty$, and its largest value when $\mu \to c\eta_0$. (Recall that $\eta_0 < \mu/c$ so μ can get no smaller than $c\eta_0$.) But if this largest value is still less than two then there is no way to satisfy Eq. (14.134) and no value of μ such that Eq. (14.133) gives the correct answer for $\langle k \rangle$. In the limit $\mu \to c\eta_0$ the denominator of the integrand equals $1 - \eta/\eta_0$, which tends to zero at the upper limit of the integral, but provided $\rho(\eta)$ also tends to zero in this limit the integral can take a finite value and, as we will see in a moment, this value can certainly be less than two for some choices of $\rho(\eta)$.

How can it be that our solution does not give the correct value for the average degree? Have we made a mistake somewhere? More importantly, what does the network actually do in this regime?

The answer to this conundrum turns out to be a subtle and interesting one. The important point is that there are some behaviors of the vertex degrees in a growing network that cannot be captured by a simple probability distribution p_k. In particular, if there are a fixed, finite number of vertices in the network with degrees that scale in proportion to the size n of the entire network, those vertices do not appear in the degree distribution: because there are only a fixed number of them they constitute a fraction $1/n$ of the network and hence contribute zero to the degree distribution as $n \to \infty$. Nonetheless, they make a non-zero contribution to the average degree of the network in the limit of large n and hence must be taken into account in the calculation of $\langle k \rangle$.

Bianconi and Barabási referred to the appearance of such vertices in the network as "condensation" by analogy with similar behaviors seen in low-temperature physics,[22] and to the vertices themselves as a *condensate*. For some choices of $\rho(\eta)$ this kind of condensation does indeed occur and a condensate of "superhubs" with very high degree forms in the network.

Suppose we are in such a regime and let us write the sum of the degrees of the vertices in the condensate as K. Then the full expression for the average degree, including the condensate, becomes

$$\langle k \rangle = \frac{K}{n} + \sum_{k=c}^{\infty} \int_0^{\infty} k p_k(\eta) \, d\eta = \frac{K}{n} + c \int_0^{\eta_0} \frac{\rho(\eta) \, d\eta}{1 - c\eta/\mu}. \tag{14.135}$$

Clearly, no matter what the value of the integral, it is now always possible to

---

[22]Bianconi and Barabási, who are physicists, solved their model initially by showing that it can be mapped onto the standard physics problem of "Bose–Einstein condensation" in an ensemble of non-interacting bosons. For a physicist already familiar with Bose–Einstein condensation, this provides a quick and elegant way of deriving a solution. For non-physicists, on the other hand, the mapping is probably not very illuminating.

achieve $\langle k \rangle = 2c$ by making K sufficiently large. The appropriate value of K is given by setting $\langle k \rangle = 2c$ and rearranging thus:

$$K = nc \left[2 - \int_0^{\eta_0} \frac{\rho(\eta)\, d\eta}{1 - c\eta/\mu} \right]. \tag{14.136}$$

Again, the largest value of the integral is achieved when $\mu \to c\eta_0$, which means that

$$K > nc \left[2 - \int_0^{\eta_0} \frac{\rho(\eta)\, d\eta}{1 - \eta/\eta_0} \right]. \tag{14.137}$$

So if the value of this integral is less than two, then in order to get the correct value for the average degree we require that K scales with the size n of the network just as we hypothesized.

As an example, suppose the distribution of fitnesses takes the form $\rho(\eta) = A(\eta_0 - \eta)^\tau$, where τ is a positive exponent and A is a normalization constant whose value is easily shown to be $A = (\tau + 1)/\eta_0^{\tau+1}$. Then the integral in Eq. (14.137) is

$$\int_0^{\eta_0} \frac{\rho(\eta)\, d\eta}{1 - \eta/\eta_0} = \frac{\tau + 1}{\eta_0^\tau} \int_0^{\eta_0} (\eta_0 - \eta)^{\tau-1}\, d\eta = 1 + \frac{1}{\tau}, \tag{14.138}$$

and

$$K > nc \left(1 - \frac{1}{\tau} \right). \tag{14.139}$$

If $\tau < 1$ then this result tells us only that $K \geq 0$, which is trivially true. But if $\tau > 1$ then the sum of the degrees in the condensate must vary in proportion to the network size n.

Thus, depending on our choice for the distribution of fitnesses, the network can show two different behaviors. In one case, the distributions of the degrees of vertices with any given fitness follow a power law with a fitness-dependent exponent, but no vertices in the network are special or distinguished by any particular behavior. In the other case, a condensate forms consisting of one or more "superhubs," which connect to a non-zero fraction of all other vertices. None of the remaining non-condensate vertices show any special behavior, however; they still have power-law degree distributions for each value of fitness, with the same exponents as before. Some authors have likened the condensation phase of the model to the monopolistic dominance of a market by a single vendor or a small number of vendors—once one vertex (vendor) gets a non-zero fraction of all edges (business), preferential attachment guarantees that it will go on doing so thereafter.

We still do not have a complete solution of the model, because we are missing the value of μ which is required to evaluate Eq. (14.136). Unfortunately, to calculate μ we need to know the exact form of the condensate. In most discussions of the model in the literature it is assumed that the condensate consists of just a single vertex of degree K at or close to the maximum fitness η_0, in which case we can evaluate μ using Eq. (14.126). Including the contribution from the condensate vertex, this equation becomes

$$\mu = \frac{a_K(\eta_0)}{n} + \sum_{k=c}^{\infty} \int_{-\infty}^{\infty} a_k(\eta) p_k(\eta) \, d\eta, \tag{14.140}$$

and, setting $a_k(\eta) = \eta k$ again and making use of Eqs. (14.129) and (14.132), we then find

$$\frac{\eta_0 K}{n\mu} + \int_0^{\eta_0} \frac{\rho(\eta) \, d\eta}{\mu/c\eta - 1} = 1. \tag{14.141}$$

Equations (14.136) and (14.141) together now give us two equations in the two unknowns K and μ, which we can, at least in theory, solve for both given $\rho(\eta)$, although in practice closed-form solutions are rare because the integrals are non-trivial.

There is, however, no reason why the condensate must consist of just a single vertex. It could in principle consist of more than one. It could even consist not of a fixed number of vertices but of a growing number provided the number grows slower than linearly with the size of the network, so that again the condensate makes no contribution to the degree distribution p_k in the limit of large n. In the latter case, the superhubs that make up the condensate would have degrees that themselves scaled sublinearly with system size, but would still become arbitrarily large as $n \rightarrow \infty$. To the best of the author's knowledge, it is not known how to predict which of these behaviors will happen for a given choice of $\rho(\eta)$. Indeed, an exact prediction may not even be possible: computer simulations of the model appear to indicate that the exact nature of the condensate—how many vertices it contains and how their degrees grow with system size—is not deterministic but depends on the details of fluctuations taking place in the early growth of the network. If one performs repeated computer simulations with the same choice of $\rho(\eta)$, the macroscopic behavior of the condensate varies from one run of the program to another.[23]

---

[23]The model is, in this respect, reminiscent of the much simpler and older model of a growing system called *Pólya's urn*. In this model an urn (i.e., a large pot) initially contains two balls, one green and one red. Repeatedly we draw one ball at random from the urn and replace it with two of the same color. In the limit where the number of balls becomes large, the fraction of green (or red) balls tends to a constant, but the value of that constant is entirely unpredictable—it depends on the details of the fluctuations in the numbers of balls at the early stages of the growth process and all values of the constant are equally likely in the $n \rightarrow \infty$ limit.

We have also assumed in our discussion that the distribution of the fitness is bounded, that there is a maximum value η_0 that the fitness can take. What happens if this assumption is violated? In this case there will of course still be a fittest vertex in the network and the network cannot "tell" whether the fitness distribution is bounded above that point or extends to $\eta = \infty$, and hence the behavior of the model will be the essentially the same as in the bounded case. The main difference from the bounded case is that the value of the highest fitness may change from time to time, which also changes the value of μ via Eq. (14.126). However, the changes in the highest fitness become rarer and rarer as time goes by[24] so that asymptotically the behavior of the system is the same in the bounded and unbounded cases for arbitrarily long periods of time.

Many other extensions and variations of preferential attachment models have been studied in addition to the ones described in this chapter. If you're interested in learning more, there are a number of review articles that go into the subject in some detail—see Refs. [12], [46], and [98]. The rest of this chapter is devoted to the discussion of other models of network formation and growth that don't rely on preferential attachment.

14.5 VERTEX COPYING MODELS

Preferential attachment models offer a plausible, if simplified, explanation for power-law degree distributions in networks such as citation networks and the World Wide Web. Preferential attachment, however, is by no means the only mechanism by which a network can grow, nor even the only mechanism known to generate power laws. In the remainder of this chapter we look at a number of other models and mechanisms for the formation of networks, starting in this section with models based on vertex copying.

In Section 14.1 we introduced the preferential attachment mechanism and suggested a possible explanation of its origin in citation networks, that a reader perusing the literature in a given academic field would encounter citations to frequently cited papers more often than citations to less cited ones, and hence would be more likely to cite those frequently cited papers themselves. Another way of saying this is that, in effect, researchers are copying citations from the

---

[24]The statistics of these leader changes are themselves non-trivial. They obey a so-called *record dynamics*, an interesting non-stationary process that has been studied in its own right, for example by Sibani and Littlewood [296].

bibliographies of papers they read.[25]

Kleinberg *et al.* [180] have proposed an alternative mechanism for network formation that takes this idea one step further. What if people simply copied the entire bibliography of a single paper to create the new bibliography of their own paper? This would then create a new vertex in the network with the same pattern of outgoing edges as the vertex they copied from.

As we will see, this process, with slight modifications, can give rise to a power-law degree distribution. First, however, we note that the process as stated has some problems. To begin with, it's clearly rather far-fetched. Authors of papers do take note of who other authors have cited, but it seems unlikely that an author would copy the entire bibliography from someone else's paper. Moreover, if they did just copy the entire bibliography then previously cited papers would get new citations as a result, but there would be no way for papers to receive citations if they had never been cited before.

Both of these problems can be solved by changing the model a little. Instead of assuming that the bibliography of the new paper is copied wholesale from the bibliography of an older one, let us assume that only some fraction of the entries in the old bibliography are copied. Then the remainder of the new bibliography is filled out with references to other papers. These other papers could be selected in a variety of way, but a simple choice would be to select them uniformly at random from the entire network.

These modifications insure that bibliographies are now no longer copied in their entirety and papers with no previous citations have a chance of being cited. The model is, however, still not a very plausible model of a real citation network. But, like Price's preferential attachment model (which is also not very realistic), it can be regarded as a simplified and tractable version of the vertex copying mechanism that allows us to investigate quantitatively the consequences of that mechanism.[26] The precise definition of the model is as

---

[25] We use the word "copying" figuratively here, but in fact there is evidence to suggest that some people really do just copy citations from other papers, possibly without even looking at the cited paper. Simkin and Roychowdhury [297,298] have noted that there is a statistically surprising regularity to the typographical errors people make in citing papers. For instance, many different authors will use the same wrong page number in citing a particular paper, which suggests that rather than copying the citation from the paper itself, they have copied it from an erroneous entry in another bibliography. This does not prove that they did not read the paper in question, but it makes it more likely—if they had actually looked up the paper, there is a good chance they would have noticed that they had the page number wrong.

[26] Kleinberg *et al.* themselves proposed a different model of the copying process in their paper, but their model is quite complex and doesn't lend itself easily to analysis. The model described here is a simplified realization that possesses the important features of the process while remaining relatively tractable. We note also that Kleinberg *et al.* were not in fact concerned with citation

follows.

Let us suppose for simplicity that each new vertex added to our network has the same out-degree c. In the language of citations, the bibliographies are all the same size. For each vertex added we choose uniformly at random a previous vertex and go one by one through the c entries in the bibliography of that previous vertex. For each entry we either (a) with probability $\gamma < 1$ copy that entry to the bibliography of the new vertex or (b) with probability $1 - \gamma$ add to the bibliography of the new vertex a citation to another vertex chosen uniformly at random from the entire network. The end result is a bibliography for the new vertex in which, on average, γc of the entries are copied from the old vertex and the remainder are chosen at random. In effect, we have made an imperfect copy of the old vertex in which the destinations of some fraction of the outgoing edges have been randomly reassigned.

We also need to specify the starting state of the network, but, as with our preferential attachment models, it turns out that the asymptotic properties of the network do not depend on the state we choose. Thus the choice is not particularly important, but we could, for instance, specify a starting network consisting of some number $n_0 > c$ vertices in which each points randomly to c of the others.

We can solve for the degree distribution of the network generated by this model as follows. Let us ask what the probability is that vertex i receives a new incoming edge upon the addition of a new vertex to our network. For i to receive a new edge, one of two things has to happen. Either the newly added vertex happens to copy connections from a vertex that already points to vertex i, in which case with probability γ the connection to i will itself get copied, or i could be one of the vertices chosen at random to receive a new edge. Let us treat these two processes separately.

Suppose that a particular existing vertex happens to have a link to our vertex i. The probability that a newly added vertex will choose to copy its own links from this existing vertex is simply $1/n$, since the source for the copies is chosen uniformly at random from the whole network. Thus if i has in-degree q_i, the chance that any one of the q_i vertices that point to it gets chosen is q_i/n. And the chance that the link from that vertex to i gets copied is γ, for a total probability of $\gamma q_i/n$.

The average number of random links that a newly added vertex makes—ones not copied from a previous vertex—is $1 - \gamma$ for each of its c outgoing

---

networks in their paper. Their focus was the World Wide Web. We use the language of citation networks here to emphasize the parallels with Price's model, but the discussion could equally have been framed in the language of the Web.

edges, or $(1 - \gamma)c$ overall. And the probability that our vertex i happens to be the target of one of these random links is $1/n$, for an overall probability of $(1 - \gamma)c/n$.

Putting everything together, the total probability that vertex i gets a new link is[27]

$$\frac{\gamma q_i}{n} + \frac{(1 - \gamma)c}{n} = \frac{\gamma q_i + (1 - \gamma)c}{n}. \tag{14.142}$$

Defining $p_q(n)$ as before to be the fraction of vertices with in-degree q when the network has n vertices, the total expected number of vertices of in-degree q receiving a new edge is

$$n p_q(n) \times \frac{\gamma q + (1 - \gamma)c}{n} = \left[\gamma q + (1 - \gamma)c\right] p_q(n). \tag{14.143}$$

But now we notice a remarkable fact. If we define a new constant a by

$$a = c\left(\frac{1}{\gamma} - 1\right), \tag{14.144}$$

then

$$\gamma = \frac{c}{c + a} \tag{14.145}$$

and Eq. (14.143) becomes

$$\left[\gamma q + (1 - \gamma)c\right] p_q(n) = \frac{c(q + a)}{c + a} p_q(n), \tag{14.146}$$

which is exactly the same as the probability (14.2) for the equivalent quantity in Price's model.

We can now use this probability to write down a master equation for the evolution of the degree distribution p_q, which will be precisely the same as the master equation (14.5) for Price's model and all subsequent developments follow through just as in Section 14.1. The end result is that our vertex copying model behaves precisely as the Price model does, but with a value of a specified now in terms of the parameter γ by Eq. (14.144). Thus, for example, the degree distribution in the limit of large n will obey Eq. (14.21) and hence will asymptotically follow a power law with exponent α given by Eq. (14.27) to be

$$\alpha = 2 + \frac{a}{c} = 1 + \frac{1}{\gamma} \tag{14.147}$$

---

[27]In simply adding together our probabilities we are technically writing down an expression for the expected number of new edges the vertex receives, rather than the probability of receiving a new edge. However, in the limit of large n the two become the same.

This gives exponents in the range from 2 to ∞, with the value depending on how faithfully vertices are copied. Faithful copies (γ close to one) give exponents close to two, while sloppy copies give exponents that can be arbitrarily large. Other properties of Price's model carry over as well, such as the distribution of in-degree as a function of age given in Eq. (14.57).

This is not to say, however, that vertex copying generates networks identical in every respect to preferential attachment. With vertex copying, for instance, many of the links that a newly appearing vertex makes are typically copied from the same other vertex and hence most vertices in the network will have connections that are similar to those of at least one other vertex. In preferential attachment models, on the other hand, there is no such correlation between the connections of different vertices—each link is chosen independently from the available possibilities at the time it is created and not copied from anywhere else. The two networks therefore, while they may have the same degree distribution, are different in the details of their structure.

In addition to being interesting in its own right, the vertex copying model serves as a useful cautionary tale concerning the mechanisms of network formation. We have seen that many real networks have degree distributions that follow a power law, at least approximately, and that preferential attachment models can generate such degree distributions. A natural conclusion is that real networks are the product of preferential attachment processes, and this may indeed be correct. We should be careful, however, not to jump immediately to conclusions because, as we have now seen, there exists at least one other mechanism—vertex copying—that produces precisely the same degree distribution. Without further information we have no way of telling which of these mechanisms is the correct one, or whether some other third mechanism that we have not yet thought of is at work.

One could in principle examine details of the structure of specific real-world networks in an attempt to tell which, if either, of our two mechanisms is the better model for their creation. For instance, one might examine a network to see if there appear to be pairs of vertices whose outgoing connections are approximate copies of one another. In fact, in real citation networks it turns out that there are many such pairs, an observation that appears to lend weight to the vertex copying scenario. However, we must remember that both of our models are much simplified and it's likely that neither of them is an accurate representation of the way real networks are created. A simple explanation for vertices in citation networks with similar patterns of links is that they correspond to papers on similar topics and so tend to cite the same literature; there is no need to assume that one of them copied from the other. As a result, it may not be possible to distinguish firmly between preferential attachment and

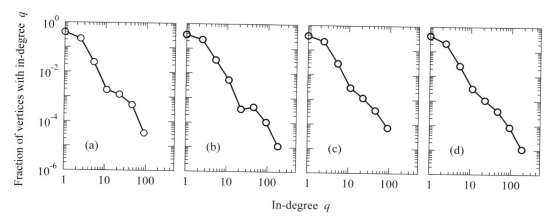

Figure 14.9: Distribution of in-degrees in the metabolic networks of various organisms. Jeong *et al.* [166] examined the degree distributions of the known portions of the metabolic networks of 43 organisms, finding some of them to follow power laws, at least approximately. Show here are the in-degree distributions for (a) the archaeon *A. fulgidus*, (b) the bacterium *E. coli*, (c) the worm *C. elegans* (a eukaryote), and (d) the aggregated in-degree distribution for all 43 organisms. After Jeong *et al.* [166].

vertex copying in many cases.

There are, however, some cases where preferential attachment appears to be an implausible candidate to explain the structure of a network, and in some of these cases vertex copying is the most promising remaining option. A good example comes from the realm of biology, where vertex copying is considered a strong candidate for explaining the structure of metabolic networks and protein–protein interaction networks. As discussed in Chapter 5, these are networks of chemical and physical interactions between molecules in the cell and, although our knowledge of their structure is currently quite incomplete, there is at least tentative evidence to suggest that they have power-law degree distributions—see Fig. 14.9 and Refs. [164] and [166]. It seems unlikely, however, that preferential attachment is the cause of these power laws: there is no obvious mechanism by which preferential attachment could take place in this context. Vertex copying, on the other hand, may be a reasonable candidate.

Consider for example a protein interaction network. As described in Section 5.1.3, proteins in the cell are created by the processes of molecular transcription and translation from codes stored in the cell's DNA. The section of code that defines a single protein is called a gene and it turns out that genes are sometimes inadvertently copied when cells reproduce.

When a cell splits in two to reproduce, its DNA is copied so that each half of the split cell will have a complete copy. The cellular machinery responsible for the copying is highly reliable, but not perfect. Very occasionally, a section of DNA will be copied twice, giving rise to a repeated section, which can mean that the new cell has two copies of a certain gene or genes where the old cell had only one. Many examples of such repeated sections are known in the human genome and the genomes of other animals and plants.

Another common type of copying error is the *point mutation*, whereby individual nucleotides—letters in the DNA code—are copied incorrectly. Over the course of many cell divisions, point mutations can accumulate, and as a result two initially identical versions of the same gene can become no longer identical, with some fraction of their bases changed to new and (roughly speaking) random values. These processes typically happen slowly over the course of evolutionary time, taking thousands or even millions of years. The end result, however, is that a gene is copied and then mutated to be slightly different from the original.

And these processes are reflected in the network of protein interactions. Typically both copies of a duplicated gene in a genome can generate the corresponding protein; the subsequent mutation of one or both copies can result in the two producing similar but slightly different versions of the protein, different enough in some cases to also have slightly different sets of interactions in the network. Some interactions may be common to both proteins but, just as in our vertex copying model, some may also be different.

This picture is made more plausible by the fact that changes in genes are not purely random but are subject to Darwinian selection under which some gene mutations are more advantageous than others. A cell with two copies of a particular protein may gain a selective advantage if those copies do slightly different things, rather than needlessly duplicating functionality that a single copy alone could achieve. Thus it seems possible that nature may actually favor duplicated proteins that have slightly different functions and hence different sets of network connections. Moreover an examination of the data for real-world protein–protein interaction networks turns up many examples of pairs of proteins that are similar but not identical in their patterns of interactions, and gene duplication is widely, if not universally, believed to be the cause.

Several models of vertex copying and mutation in biological networks have been proposed and studied. The model proposed by Solé *et al.* [302], for example, is very similar to the model described above, the main difference being that it is a model of an undirected network rather than a directed one. Another model, put forward by Vázquez *et al.* [317], is also similar but includes a mechanism whereby the connections of the *copied* vertex can be changed as well as those of the copying vertex. Although the latter mechanism would make little sense in a model of a citation network (the bibliography of a paper never changes after publication), it is appropriate in the biological context, where all genes are potentially mutating all the time.

14.6 NETWORK OPTIMIZATION MODELS

In the models we've looked at so far in this chapter, network structure is determined by the way in which the network grows—how newly added vertices connect to others, where newly added edges get placed, and so forth. Furthermore, the structure of these networks is for the most part a result of a succession of random processes, often decentralized and quite blind to the large-scale structure they are creating.

An alternative network formation mechanism, important in certain types of network, is structural optimization. In some cases, such as transportation networks (Section 2.4) or distribution networks (Section 2.5), a network has been specifically designed to achieve a particular goal or goals, such as the delivery of mail or packages around the country or the transportation of airline passengers to their destinations, and the structure of the network can heavily influence the efficiency with which that goal is accomplished. Networks of airline routes, for example, are typically based on a hub-and-spoke arrangement with a small number of busy airport hubs and a large number of minor destinations.[28] (Package delivery companies also use a similar scheme.) The reason is that it makes little sense to fly airplanes directly between minor destinations—there will typically be very few passengers interested in the service and the planes will be half empty. By ensuring that the only flights in and out of minor destinations are to and from major hubs, one concentrates the passengers on those routes, ensuring fuller planes while still giving the passengers a reasonably short journey.

In other words, the hub-and-spoke design of the airline networks *optimizes* the network, making it more efficient, and hence more profitable, for the airline. In such cases, the structure of the network is explained not by a growth mechanism but by the fact that the network has been designed to optimize certain characteristics. In this section we look briefly at some models of network optimization.

---

[28]This is a relatively recent development, at least in the United States, where industry regulations made the hub-and-spoke system impractical until 1978. After regulations were lifted the hub-and-spoke system was rapidly adopted by most of the major airlines. Hub-and-spoke systems were also adopted by the package delivery industry around the same time.

14.6.1 TRADE-OFFS BETWEEN TRAVEL TIME AND COST

The example given above of an airline network is a good place for us to start. Airline networks are, in fact, highly optimized: the airline industry operates on very small (sometimes even negative) profit margins, and optimization of operations to trim even a tiny percentage off their enormous costs can make a substantial difference to the bottom line. Airlines employ large staffs of researchers whose sole task is to find new ways to optimize aspects of their business, including particularly their network of routes. At the same time, airlines need to keep their customers happy if they are to avoid losing market share to their competitors. This means, for instance, that they need to provide short, quick routes between as many pairs of destinations as possible—travelers are strongly averse to long journeys that wear them out or waste their time. The twin goals of cost-efficient operation and short routes are to some extent at odds with one another. The quickest way to get passengers from any place to any other, for example, would be to fly separate planes between every pair of airports in the country, but this would be immensely costly. The observed structure of real airline networks is a compromise response to the conflicting needs of the company and its passengers.

The optimization problems faced by real airlines are, inevitably, hugely complex, involving as they do organizations with thousands of employees, billions of dollars worth of material resources, and rapidly changing parameters such as fuel costs, consumer demand, and the nature of the competition. Nonetheless, there is insight to be gained by creating and studying simplified models of the optimization process in the same way that simple models of, for example, citation networks can grant us insight despite the many features of real citation processes that they omit.

One of the simplest models of network optimization is that proposed by Ferrer i Cancho and Solé [117], which balances two elements of exactly the types discussed above. In this model the cost of maintaining and operating the network is represented by the number of edges m in the network. This would be equivalent to saying that the cost of running an airline is proportional to the number of routes it operates. Obviously this is a vast simplification of the real situation, but let us accept it for the moment and see where it leads. The customer satisfaction half of the equation is represented by the mean geodesic distance ℓ between vertex pairs. In our airline example ℓ would be the average number of legs required to journey from one point to another, which is certainly one element of customer satisfaction, though not the only one. Technically, ℓ is a *dissatisfaction* measure, since large values correspond to disgruntled customers.

We would like to design a network that minimizes both m and ℓ but this is in general not possible: the minimum value of ℓ is achieved by placing an edge between every pair of vertices, but this maximizes the value of m. Thus our two goals are, as discussed above, at odds with one another and the best we can hope for is a reasonable compromise between them. In search of such a compromise, Ferrer i Cancho and Solé studied the quality function

$$E(m, \ell) = \lambda m + (1 - \lambda)\ell, \tag{14.148}$$

where λ is a parameter in the range $0 \leq \lambda \leq 1$. For any given network and a given value of λ we can calculate $E(m, \ell)$; the value of ℓ for instance can be computed using the breadth-first search algorithm of Section 10.3. Ferrer i Cancho and Solé considered networks of a given number of vertices n and then asked what happens when we try to minimize $E(m, \ell)$ by varying the position of the edges in that network to find the smallest value possible. If $\lambda = 1$, then $E = m$ and this process is equivalent to just minimizing the number of edges without regard for path lengths. If $\lambda = 0$ then $E = \ell$ and we are minimizing only average path length without regard for m. For values in between, we are striking a balance between number of edges and path length, with the precise weight of each term controlled by our choice of λ.

At some level, this model is a trivial one. Observe that the value of ℓ becomes formally infinite if there is any pair of vertices in the network that is not connected by a path—i.e., if the network has more than one component—since the distance between such pairs is by convention considered infinite (see Section 6.10.1). Thus the minimum value of E must be for a connected network, a network with just one component. Observe also that the minimum value of m for a connected network is $m = n - 1$, where n is the number of edges. This is the value for a tree, which is the connected network with the smallest number of edges (see Section 6.7).

Provided λ is reasonably large, so that we place a moderate amount of weight on minimizing m, the network with the best value of $E(m, \ell)$ is then found by giving m its minimum value of $n - 1$, which constrains the network to be a tree, and searching through the set of possible trees to find the one that minimizes ℓ. In fact, the latter task has a simple, known solution: the minimum value of ℓ among trees with n vertices is obtained by the *star graph*, the network in which there is a single central hub connected by a single edge to each of the $n - 1$ remaining vertices. By definition there are always exactly m pairs of vertices with geodesic distance one in any network—the pairs that are directly connected by an edge—which means that in a tree there are $n - 1$ such pairs. Among the set of all trees, therefore, the value of the mean distance ℓ is governed by the numbers of pairs with distances of two or more, since the

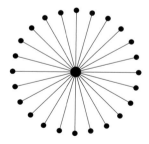

A star graph of 25 vertices.

number with distance one is fixed. But in the star graph all other pairs have distance exactly two—the shortest (indeed only) path from any (non-hub) vertex to any other is the path of length two via the hub. Thus there can be no other tree with a smaller value of ℓ.

Thus, for sufficiently large λ, the optimum network under the quality function (14.148) is always the star graph. This is satisfying to some extent: it offers a simple explanation of why the hub-and-spoke system is so efficient. It offers short journeys while still being economic in terms of the number of different routes the airline has to operate. But it is also, as we have said, somewhat trivial. The model shows essentially only the one behavior. For smaller values of λ other behaviors are possible, but it turns out that the value of λ has to be *really* small: non-star-graph solutions only appear when[29]

$$\lambda < \frac{2}{n^2 + 2}. \tag{14.149}$$

Since the expression on the right-hand side dwindles rapidly as n becomes

---

[29]The derivation of this result is as follows. If λ is sufficiently large then, as we have shown, the optimal network is the star graph. If we now reduce λ slowly then at some point we enter a regime in which the cost of adding an edge is sufficiently offset by the corresponding reduction in the mean geodesic distance that it becomes worthwhile to add edges between the "spoke" vertices in the star graph. To calculate the point at which such additions become beneficial let us take our star graph and add to it some number r of extra edges. Necessarily these edges fall between the spoke vertices, since there is nowhere else for them fall, and in doing so they form paths of length one between pairs of vertices whose previous shortest path was of length two. The shortest paths between no other vertices are affected by the addition. Thus the total number of vertex pairs connected by paths of length 1 is $n - 1 + r$ and all the rest have paths of length two. Then the mean geodesic distance, as defined in Eq. (7.31) is

$$\ell = \frac{1}{n^2} \sum_{ij} d_{ij} = 2\frac{(n-1+r) + 2[\frac{1}{2}n(n-1) - (n-1+r)]}{n^2} = 2\frac{(n-1)^2 - r}{n^2}.$$

(The leading factor of two comes from the fact that the sum over i, j counts each pair of vertices twice.)

Substituting this expression, along with $m = n - 1 + r$, into Eq. (14.148) then gives

$$E = \lambda(n - 1 + r) + 2(1 - \lambda)\frac{(n-1)^2 - r}{n^2} = \text{constant} + \left[\lambda + \frac{2(\lambda - 1)}{n^2}\right]r.$$

This will decrease with growing r only if the quantity in square brackets [...] is negative, i.e., if

$$\lambda < \frac{2}{n^2 + 2}.$$

If this condition is satisfied then it becomes advantageous to add edges between the spoke vertices, and to keep on doing so until the network becomes a complete graph, with every vertex connected to every other. Thus there is a discontinuous transition between two behaviors—the star graph and the complete graph—at the point $\lambda = 2/(n^2 + 2)$. Real distribution and transportation networks appear to be in the star-graph regime.

large, the optimal network is the star graph for almost all values of λ, even for networks of quite modest size.

In their paper, however, Ferrer i Cancho and Solé did not perform precisely the calculation we have done here. Instead, they took a different and interesting approach, in which they looked for *local minima* of $E(m, \ell)$, rather than the global minimum. They did this numerically, starting with a random network, repeatedly choosing a pair of vertices at random, and either connecting them by an edge if they were not already connected or deleting the edge between them if they were. Then they compared the value of E before and after the change. If E decreased or stayed the same, they kept the change. If not, they reverted back to the state of the network before the change. The whole procedure was then repeated until the value of E stopped improving, meaning in practice that a long string of attempted changes were rejected because they increased E.

An algorithm of this kind is called a *random hill climber* or *greedy algorithm*. The networks it finds are networks for which the value of E cannot be reduced any further by the addition or removal of any single edge. This does not mean, however, that no lower values of E exist: there may be states of the network that differ by more than one edge—the addition and deletion of whole regions of the network—that have better values of E. But if so, the algorithm will not find them. It comes to a halt at a local minimum where no single-edge change will improve the value of E.

When studied in this way, the model shows an interesting behavior. For large values of λ, where the addition of an edge costs a great deal in terms of the value of E, the algorithm rapidly runs into trouble and cannot find a way to improve the network, long before it gets anywhere close to the optimum hub-and-spoke arrangement. When λ is small, on the other hand, the algorithm typically manages to find the star graph solution. The result is a spectrum of networks that range from a random-looking tree to a star-graph, as shown in Fig. 14.10.

What's more, Ferrer i Cancho and Solé found that the degree distributions of their networks show interesting behavior, passing from an exponential distribution for large λ, though a transition point with a power-law degree distribution, to approximately star-like graphs for small λ in which one vertex gets a finite fraction of all the edges and the remaining vertices have low degree. This spectrum is reminiscent of the behavior of continuous phase transitions such as the transition at which a giant component appears in a random graph (see Section 12.5), in which an initially exponential distribution of component sizes passes through a transition to a regime in which one component gets a finite fraction of all vertices and the rest are small.

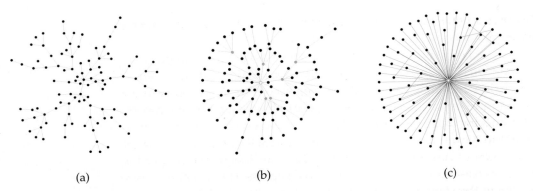

(a)   (b)   (c)

Figure 14.10: Networks generated by the optimization model of Ferrer i Cancho and Solé. The optimization model described in the text generates a range of networks, all trees or approximate trees, running from (a) distributed networks with exponential degree distributions, through (b) power-law degree distributions, to (c) star graphs in which there is just one major hub. Figure adapted from [117]. Original figure Copyright 2003 Springer-Verlag Berlin Heidelberg. Reproduced with kind permission of Springer Science and Business Media.

Sadly, this observation does not go any further than an intriguing hint. The work of Ferrer i Cancho and Solé is entirely numerical and they do not give any analytic treatment of the model. In addition there are some other problems with the model. In particular, it is not clear why one should look at local minima of E rather than global ones: the researchers who work for real airlines are certainly capable of realizing when they are stuck in a local optimum and better profits are available by changing the network in some substantial way that moves them to a different and better optimum. It seems likely therefore that, to the extent that real networks show interesting structural behavior of the type observed here, it is not a result of getting stuck in local minima and hence that a model with a different approach is needed.

One such model, proposed by Gastner and Newman [137], generalizes that of Ferrer i Cancho and Solé by considering not only number of legs in a journey but also the geographic distance traveled. Suppose that airline travelers are principally concerned not with the number of legs in their journey but with the total time it takes them to travel from origin to destination. Number of legs can be regarded as a simple proxy for travel time, but a better proxy would be to take the length of those legs into account as well as their number. The travel time contributed to a journey by one leg is composed of the time spent in the airport (checking in, waiting, embarking, taxiing, disembarking, etc.) plus the time spent in the air. A simple formula would be to assume that the former is roughly constant, regardless of the distance being traveled, while the latter is roughly proportional to the distance traveled. Thus, the time taken by a leg from vertex i to vertex j in our network would be

$$t_{ij} = \mu + \nu r_{ij}, \qquad (14.150)$$

where μ and ν are constants and r_{ij} is the distance flown from i to j. By varying the values of μ and ν, we can place more or less emphasis on the fixed "airport" time cost versus the time spent in the air.

Gastner and Newman used this expression for travel time in place of the simple hop-count of the model of Ferrer i Cancho and Solé, redefining ℓ to be the average shortest-path distance between pairs of vertices when distances are measured in terms of travel time. The quality function E is defined just as before, Eq. (14.148), but using this new definition of ℓ.

Despite the superficial similarity between this model and that of Ferrer i Cancho and Solé there is a crucial difference between the two: the model of Gastner and Newman depends on actual spatial distances between airports and hence requires that the vertices of the network be placed at some set of positions on a map. The model of Ferrer i Cancho and Solé by contrast depends only on the network topology and has no spatial element. There are various ways in which the vertices can be positioned on the map. Gastner and Newman, for instance, specifically considered the map of the United States and took the real US population distribution into account, placing vertices with greater density in areas with greater populations. While this adds a level of realism to the calculations, the interesting behavior of their model can be seen without going so far. In the examples given here we consider a fictional map in which vertices are just placed uniformly at random in a square with periodic boundary conditions.

Another important difference between the two models is that Gastner and Newman considered the global optimum of the quality function rather than local optima as Ferrer i Cancho and Solé did. In practice, unfortunately, the global optimum is hard to find, so one often has to make do with approximate optima. Gastner and Newman used the numerical optimization technique called simulated annealing to find good approximations to the global optimum, but we should bear in mind that they are only approximations.

Figure 14.11 shows optimal or approximately optimal networks for various values of the parameters μ and ν. The leftmost frames of the figure correspond to small μ and large ν, meaning that the cost to the traveler of a trip is roughly proportional to the total mileage traveled and the number of legs has little effect. In this case, the best networks are ones that allow travelers to travel in roughly straight lines from any origin to any destination. As the figure shows, the networks are roughly planar in appearance. They look reminiscent of road

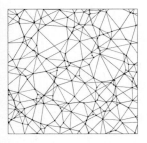 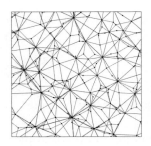 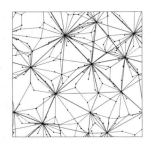 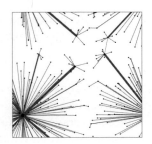

Figure 14.11: Networks generated by the spatial network model of Gastner and Newman. The four frames show networks that optimize or nearly optimize the quality function, Eq. (14.148), with ℓ defined according to the prescription of Gastner and Newman [137] in which the lengths of edges in the network are chosen to represent the approximate travel time to traverse the edge. Travel time has two components, a fixed cost per edge and a cost that increases with the Euclidean length of an edge. The frames show the resulting networks as the relative weight of these two components is varied between the extremes represented by the network on the left, for which all of the weight is on the Euclidean length, and by the network on the right, for which cost is the same for all edges. The resulting structures range from road-like in the former case, to airline-like in the latter. Adapted from Gastner [135].

networks, rather than airline routes, and this is no coincidence. Travel times for road travelers are indeed dominated by total mileage: there is almost no "per leg" cost associated with road travel, since it takes only a few seconds to turn from one road onto another. It is satisfying to see therefore that the simple model of Gastner and Newman generates networks that look rather like real road maps in this limit.

The rightmost frames in the figure show optimal networks for large μ and small ν—the case where it is mostly the number of legs that matters and the length of those legs is relatively unimportant. As we saw for the model of Ferrer i Cancho and Solé, the best networks in this case are star-like hub-and-spoke networks, and this is what we see in the present model too.

Thus this model interpolates between road-like and airline-like networks as the parameters are varied from one extreme to the other. Note that the parameter λ governing the cost of building or maintaining the network is held constant in Fig. 14.11. In principle, we could vary this parameter too, which would affect the total number of edges in the network. For higher λ sparser networks with fewer edges would be favored, while for lower λ we would see denser networks.

The work of Gastner and Newman still suffers from the drawback that the results are numerical only. More recently, however, some results for the model have been derived analytically by Aldous [15]. The interested reader is encouraged to consult his paper.

PROBLEMS

14.1 Consider the growing network model of Price, as described in Section 14.1.

a) From the results given in this chapter write down an expression in terms of the parameters a and c for the expected in-degree of the ith vertex added to the network just before the jth vertex is added, where $i < j$.

b) Hence show that the average probability of a directed edge from j to i in a network with n vertices, where $n \geq j$, is

$$P_{ij} = \frac{ca}{c+a} \, i^{-c/(c+a)} (j-1)^{-a/(c+a)}.$$

14.2 Consider Price's model as a model of a citation network, applied to publications in a single field, a field that is currently, say, ten years old.

a) Suppose that you are the author of the tenth paper published in the field. How long will it be from now before the expected number of citations your paper has within the field is equal to the expected number that the first paper published currently has?

b) Derive an expression for the average number of citations per paper to papers published between times τ_1 and τ_2, where time is defined as in Eq. (14.44).

c) Reasonable values of the model parameters for real citation networks are $c = 20$ and $a = 5$. For these parameter choices, what is the average number of citations to a paper in the first 10% of those published? And what is the average number for a paper in the last 10%?

These perhaps surprising numbers are examples of the first-mover advantage discussed in Section 14.3.1—the substantial bias of citation numbers in favor of the first papers published in a field.

14.3 Consider a model of a growing directed network similar to Price's model described in Section 14.1, but without preferential attachment. That is, vertices are added one by one to the growing network and each has c outgoing edges, but those edges now attach to existing vertices uniformly at random, without regard for degrees or any other vertex properties.

a) Derive master equations, the equivalent of Eqs. (14.7) and (14.8), that govern the distribution of in-degrees q in the limit of large network size.

b) Hence show that in the limit of large size the in-degrees have an exponential distribution $p_q = Ce^{-\lambda q}$, where C is a normalization constant and $\lambda = \ln(1 + 1/c)$.

14.4 Consider a model network similar to the model of Barabási and Albert described in Section 14.2, in which undirected edges are added between vertices according to a preferential attachment rule, but suppose now that the network does not grow—it starts off with a given number n of vertices and neither gains nor loses any vertices thereafter. In this model, starting with an initial network of n vertices and some specified arrangement of edges, we add at each step one undirected edge between two vertices, both of which are chosen at random in direct proportion to degree k. Let $p_k(m)$ be the fraction of vertices with degree k when the network has m edges in total.

a) Show that when the network has m edges, the probability that vertex i will get a new edge upon the addition of the next edge is k_i/m.

b) Write down a master equation giving $p_k(m+1)$ in terms of $p_{k-1}(m)$ and $p_k(m)$. Be sure to give the equation for the special case of $k = 0$ also.

c) Eliminate m from the master equation in favor of the mean degree $c = 2m/n$ and take the limit $n \to \infty$ with c held constant to show that $p_k(c)$ satisfies the differential equation

$$c\frac{dp_k}{dc} = (k-1)p_{k-1} - kp_k.$$

d) Define a generating function $g(c,z) = \sum_{k=0}^{\infty} p_k(c) z^k$ and show that it satisfies the partial differential equation

$$c\frac{\partial g}{\partial c} + z(1-z)\frac{\partial g}{\partial z} = 0.$$

e) Show that $g(c,z) = f(c - c/z)$ is a solution of this differential equation, where $f(x)$ is any differentiable function of x.

f) The particular choice of f depends on the initial conditions on the network. Suppose the network starts off in a state where every vertex has degree one, which means $c = 1$ and $g(1,z) = z$. Find the function f that corresponds to this initial condition and hence find $g(c,z)$ for all values of c and z.

g) Show that, for this solution, the degree distribution as a function of c takes the form

$$p_k(c) = \frac{(c-1)^{k-1}}{c^k},$$

except for $k = 0$, for which $p_0(c) = 0$ for all c.

Note that this distribution decays exponentially in k, implying that preferential attachment does not, in general, generate a power-law degree distribution if the network is not also growing.

14.5 Consider a model of a growing network similar to Price's model described in Section 14.1, but in which the parameter a, which governs the rate at which vertices receive new incoming links when their current in-degree is zero, varies from vertex to vertex. That is the probability of a new edge attaching to vertex i is proportional to $q_i + a_i$, where q_i is the current in-degree and a_i is a specified parameter. In the context of citation networks, for example, a_i could be considered a measure of the intrinsic merit of a paper, controlling as it does the rate at which the paper gets citations immediately after first publication, when $q_i = 0$. (This differs from the model discussed in Section 14.4.4, where the preferential attachment term was multiplied by a varying factor to represent variations in the merit or fitness of vertices.)

a) Suppose that a_i is drawn at random from some stationary distribution with a well-defined mean. Show that, in the limit of large n, the probability that the $(n+1)$th vertex added to the network attaches to a previous vertex i with in-degree q_i is $c(q_i + a_i)/n(c + \bar{a})$, where $\bar{a}$ is the average value of a_i.

b) Hence show that the in-degree distribution of the network satisfies the same master equations, (14.7) and (14.8), as Price's model, but with a replaced by $\bar{a}$.

(It immediately follows that the degree distribution of the network is also the same as for Price's model with the same substitution.)

14.6   Consider the following simple model of a growing network. Vertices are added to a network at a rate of one per unit time. Edges are added at a mean rate of β per unit time, where β can be anywhere between zero and ∞. (That is, in any small interval δt of time, the probability of an edge being added is $\beta\,\delta t$.) Edges are placed uniformly at random between any pair of vertices that exist at that time. They are never moved after they are first placed.

We are interested in the component structure of this model, which we will tackle using a master equation method. Let $a_k(n)$ be the fraction of vertices that belong to components of size k when there are n vertices in the network. Equivalently, if we choose a vertex at random from the n vertices currently in the network, $a_k(n)$ is the probability the vertex will belong to a component of size k.

a) What is the probability that a newly appearing edge will fall between a component of size r and another of size s? (You can assume that n is large and the probability of both ends of an edge falling in the same component is small.) Hence, what is the probability that a newly appearing edge will join together two pre-existing components to form a new one of size k?

b) What is the probability that a newly appearing edge joins a component of size k to a component of *any* other size, thereby creating a new component of size larger than k?

c) Thus write down a master equation that gives the fraction of vertices $a_k(n+1)$ in components of size k when there are $n+1$ vertices in total.

d) The only exception to the previous result is that components of size 1 appear at a rate of one per unit time. Write a separate master equation for $a_1(n+1)$.

e) If a steady-state solution exists for the component size distribution, show that it satisfies the equations

$$(1+2\beta)a_1 = 1, \qquad (1+2\beta k)a_k = \beta k \sum_{j=1}^{k-1} a_j a_{k-j}.$$

f) Multiply by z^k and sum over k from 1 to ∞ and hence show that the generating function $g(z) = \sum_k a_k z^k$ satisfies the ordinary differential equation

$$2\beta\,\frac{dg}{dz} = \frac{1-g/z}{1-g}.$$

Unfortunately, the solution to this equation is not known, so for the moment at least we do not have a complete solution for the component sizes in the model.

CHAPTER 15

OTHER NETWORK MODELS

A brief introduction to two specialized network models,
the small-world model and the exponential random graph

T HE RANDOM graph and preferential attachment models of previous chapters are the most widely studied of network models, but they are not the
only ones. Many other models have been proposed, either as a way of shedding light on specific observed features of networks or as tools to help in the
analysis of network data. In this chapter we describe briefly two of the best-
known additional types of network models, the small-world model and exponential random graphs.

15.1 THE SMALL-WORLD MODEL

One of the least well-understood features of real-world networks is transitivity, the propensity for two neighbors of a vertex also to be neighbors of one
another. (See Section 7.9 for an introduction to the phenomenon of transitivity.) Neither the random graph models of Chapters 12 and 13 nor the models
of network growth discussed in Chapter 14 generate networks with any significant level of transitivity, as quantified by the clustering coefficient, Eq. (7.41).
The Poisson random graph of Chapter 12, for instance, has a clustering coefficient $c/(n-1)$, where c is the mean degree of a vertex (see Eq. (12.11)). Thus
the clustering coefficient vanishes as n becomes large for constant c. In practice, as discussed in Section 7.9, this often results in values of the clustering
coefficient that are orders of magnitude smaller than those observed in real
networks.

It is not that difficult to come up with a network model that does have a
high clustering coefficient. For example, a simple triangular lattice, as shown

in Fig. 15.1, has significant transitivity.

There are twice as many triangles in such a lattice as there are vertices, for a total of $2n$ triangles in a network of n vertices. At the same time there are $\binom{6}{2} = 15$ connected triples for each vertex, so, following Eq. (7.41), the clustering coefficient is

$$C = \frac{(\text{number of triangles}) \times 3}{(\text{number of connected triples})} = \frac{2n \times 3}{15n} = \frac{2}{5} = 0.4. \quad (15.1)$$

A value of 0.4 is comparable with the clustering coefficients measured for many social networks (see Section 7.9 again). Moreover, this value does not depend on the size of the network, as the value for the random graph (and many other models) does, so it remains large even as the network size diverges.

Another simple model network with high transitivity is depicted in Fig. 15.2a. Unlike the triangular lattice, this model allows the value of the clustering coefficient to be varied. In this model the vertices are arranged on a one-dimensional line, and each vertex is connected by an edge to the c vertices nearest to it, where for consistency c should be an even number. To make analytic treatment easier, we can apply periodic boundary conditions to the line, effectively bending it around into a circle, as in Fig. 15.2b.

To calculate the number of triangles in such a network, we observe that a trip around any triangle must consist of two steps in the same direction around

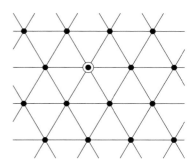

Figure 15.1: A triangular lattice. Any vertex in a triangular lattice, such as the one highlighted here, has six neighbors and hence $\binom{6}{2} = 15$ pairs of neighbors, of which six are connected by edges, giving a clustering coefficient of $\frac{6}{15} = 0.4$ for the whole network, regardless of size.

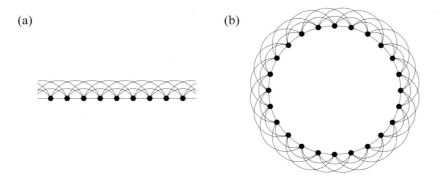

(a)   (b)

Figure 15.2: A simple one-dimensional network model. (a) Vertices are arranged on a line and each is connected to its c nearest neighbors, where $c = 6$ in this example. (b) The same network with periodic boundary conditions applied, making the line into a circle.

the circle—say clockwise—followed by one step back to close the triangle. The number of triangles per vertex in the whole network is then equal to the number of such triangles that start from any given point.

Traversing a "triangle" in our circle model means taking two steps forward around the circle and one step back.

Note, however, that the third and final step in the triangle can go at most $\frac{1}{2}c$ units or lattice spacings around the circle, since this is the length of the longest link in the network. And the number of ways to choose the two steps forward is simply the number of distinct ways of choosing the target vertices for those steps from the $\frac{1}{2}c$ possibilities, which is $\binom{c/2}{2} = \frac{1}{4}c(\frac{1}{2}c - 1)$. Thus the total number of triangles is $\frac{1}{4}nc(\frac{1}{2}c - 1)$.

The number of connected triples centered on each vertex is just $\binom{c}{2} = \frac{1}{2}c(c - 1)$ and hence the total number of connected triples is $\frac{1}{2}nc(c - 1)$.

Putting these results together, the clustering coefficient for the complete network is

$$C = \frac{\frac{1}{4}nc(\frac{1}{2}c - 1) \times 3}{\frac{1}{2}nc(c - 1)} = \frac{3(c - 2)}{4(c - 1)}. \tag{15.2}$$

As c is varied, this clustering coefficient ranges from zero for $c = 2$ up to a maximum of $\frac{3}{4}$ when $c \to \infty$. And, as with the triangular lattice, the value does not fall off with increasing network size, since Eq. (15.2) is independent of n.

While this simple "circle model" and the triangular lattice both give large values of the clustering coefficient, they are clearly unsatisfactory in other respects as models of networks. One obvious problem is the degree distribution. The circle model, for instance, gives all vertices the same degree c. In the language of graph theory the model generates a regular graph, which is entirely unlike most real-world networks with their broad distributions of vertex degree. This problem however could quite easily be solved by making a circle of vertices with varying degrees instead of constant ones.

A more serious problem with models of this type is that they are "large worlds"—they don't display the small-world effect characteristic of essentially every observed network in the real world and discussed previously in Sections 3.6 and 8.2. The small-world effect is the observation that the geodesic or shortest-path distance between most pairs of vertices in a network is small—typically just a few steps even in networks with billions of vertices such as the acquaintance network of the entire world population.

The shortest distance between two vertices in the circle model above is straightforward to calculate: the farthest one can move around the ring in a single step is $\frac{1}{2}c$ lattice spacings, so two vertices m lattice spacings apart are connected by a shortest path of $2m/c$ steps.[1] Averaging over the complete range of m from 0 to $\frac{1}{2}n$ then gives a mean shortest path of $n/2c$. In a network such as the acquaintance network of the world, with $n = O(10^9)$ people each acquainted with, say, $c = O(10^3)$ others, this expression yields an average shortest path length on the order of a million steps, which is wildly off the mark—a more realistic figure would be six or maybe ten, but not a million.

By contrast, the random graph studied in Chapter 12 *does* capture the small-world effect rather well (as indeed do most of the other network models discussed in previous chapters). As shown in Section 12.7, the typical shortest path between connected vertices in a random graph has length about $\ln n / \ln c$ which has a value on the order of $\frac{9}{3} = 3$ for the acquaintance network above. On the other hand, as we have said, the random graph has an unrealistically low clustering coefficient.

Thus we have two models, our simple circle model and the random graph, that between them each capture one property of real networks—high transitivity and short path lengths—but neither captures both. This leads us to ask whether it is possible to create a hybrid of the two that, like real-world networks, displays both high transitivity and short path lengths simultaneously. The *small-world model*, proposed in 1998 by Watts and Strogatz [323], does exactly this.

The small-world model, in its original form, interpolates between our circle model and the random graph by moving or *rewiring* edges from the circle to random positions. The detailed structure of the model is shown in Fig. 15.3a. Starting with a circle model of n vertices in which every vertex has degree c, we go through each of the edges in turn and with some probability p we remove that edge and replace it with one that joins two vertices chosen uniformly at random.[2] The randomly placed edges are commonly referred to as *shortcuts* because, as shown in Fig. 15.3a, they create shortcuts from one part of the circle to another.

---

[1] Strictly it's $\lceil 2m/c \rceil$, where $\lceil x \rceil$ is the smallest integer not less than x.

[2] In fact, in the original small-world model, as defined by Watts and Strogatz, only one end of each edge—say the more clockwise end—was rewired and the other left where it was. This, however, results in a model that never becomes a true random graph even when all edges are rewired, as one can easily see, since each vertex is still attached to half of its original edges and hence would have degree at least $\frac{1}{2}c$. In a true random graph there is no such constraint on degrees; vertices can have degrees of any value between zero and $n-1$. The original model also imposed some other constraints, such as the constraint that no two edges may connect the same vertex pair. This constraint could be imposed in the version we discuss here, although it makes little difference in practice, since the number of such multiedges is of order $1/n$ in the limit of large n and therefore the multiedges make a small contribution to any results if the network is large.

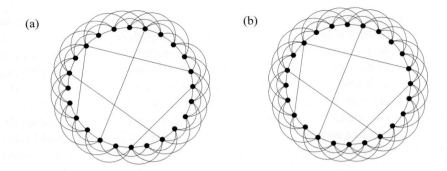

Figure 15.3: Two versions of the small-world model. (a) In the original version of the small-world model, edges are with independent probability p removed from the circle and placed between two vertices chosen uniformly at random, creating shortcuts across the circle as shown. In this example $n = 24$, $c = 6$, and $p = 0.07$, so that 5 out of 72 edges are "rewired" in this fashion. (b) In the second version of the model only the shortcuts are added and no edges are removed from the circle.

The parameter p in the small-world model controls the interpolation between the circle model and the random graph. When $p = 0$ no edges are rewired and we retain the original circle. When $p = 1$ all edges are rewired to random positions and we have a random graph. For intermediate values of p we generate networks that lie somewhere in between. Thus for $p = 0$ the small-world model shows clustering (so long as $c > 2$—see Eq. (15.2)) but no small-world effect. For $p = 1$ it does the reverse. The crucial point about the model is that as p is increased from zero the clustering is maintained up to quite large values of p while the small-world behavior, meaning short average path lengths, already appears for quite modest values of p. As a result there is a substantial range of intermediate values for which the model shows both effects simultaneously, thereby demonstrating that the two are in fact entirely compatible and not exclusive at all.

Unfortunately, it is hard to demonstrate this result rigorously because the small-world model as defined above is difficult to treat by analytic means. For this reason we will in this chapter study a slight variant of the model, which is easier to treat [254]. In this variant, shown in Fig. 15.3b, edges are added between randomly chosen vertex pairs just as before, but no edges are removed from the original circle. This leaves the circle intact, which makes our calculations much simpler. For ease of comparison with the original small-world model, the definition of the parameter p is kept the same: for every edge in the circle we add with independent probability p an additional shortcut between two vertices chosen uniformly at random.[3]

A downside of this version of the model is that it no longer becomes a random graph in the limit $p = 1$. Instead it becomes a random graph plus the original circle. This, however, turns out not to be a significant problem, since most of the interest in the model lies in the regime where p is small and in this regime the two models differ hardly at all; the only difference is the presence in the second variant of a small number of edges around the circle that would be absent in the first, having been rewired. Henceforth, we will study the variant model in which no edges are removed and we will refer to it, as others have, as the small-world model, although the reader should bear in mind that there are two slightly different models that carry this name.

15.1.1 DEGREE DISTRIBUTION

In the circle model described in the last section every vertex has the same degree c—the network is a regular graph. Once we add shortcuts to the circle to make the small-world model, the degree of a vertex is c plus the number of shortcut edges attached to it. The definition of the small-world model says that for each of the non-shortcut edges around the circle, of which there are $\frac{1}{2}nc$, we add a shortcut with probability p at a random location, so that there are $\frac{1}{2}ncp$ shortcuts on average and ncp ends of shortcuts. This means that cp shortcuts on average end at any particular vertex. And the specific number s of shortcuts attached to any one vertex is Poisson distributed with mean cp thus:

$$p_s = e^{-cp} \frac{(cp)^s}{s!}. \tag{15.3}$$

The total degree of a vertex is $k = s + c$. Putting $s = k - c$ into Eq. (15.3) then gives us the degree distribution of the small-world model:

$$p_k = e^{-cp} \frac{(cp)^{k-c}}{(k-c)!} \tag{15.4}$$

for $k \geq c$ and $p_k = 0$ if $k < c$.

Figure 15.4 shows the form of this distribution for $c = 6$, $p = \frac{1}{2}$. As we can see, the distribution has an unusual peaked shape with a lower cut-off, quite unlike the degree distributions we saw for real networks in Section 8.3. In this respect, therefore, the small-world model does not mimic well the structure of

---

[3]Equivalently, one could just say that the number of shortcuts added is drawn from a Poisson distribution with mean $\frac{1}{2}ncp$.

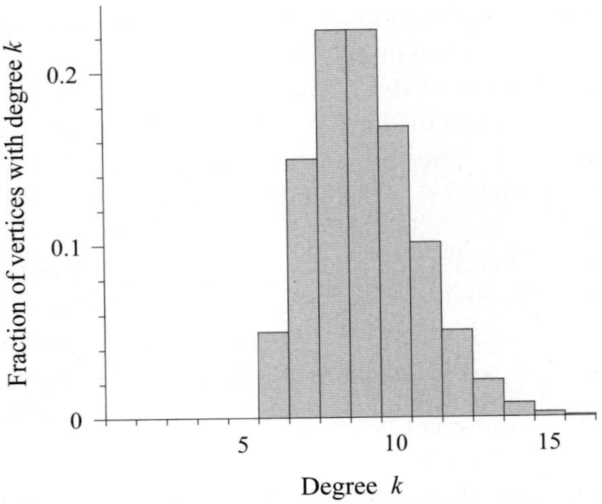

Figure 15.4: The degree distribution of the small-world model. The frequency distribution of vertex degrees in a small-world model with parameters $c = 6$ and $p = \frac{1}{2}$.

networks in the real world. On the other hand, the model was never intended to mimic real-world degree distributions. What it does do well is mimic the clustering and short path lengths seen in real networks.

15.1.2 CLUSTERING COEFFICIENT

The clustering coefficient C is defined by Eq. (7.41), which we reproduce here:

$$C = \frac{(\text{number of triangles}) \times 3}{(\text{number of connected triples})}. \tag{15.5}$$

To evaluate C for the small-world model we need to calculate the numbers of triangles and connected triples in the network. Let us start with the former.

Since the underlying circle in the model is unchanged by the addition of shortcuts, every triangle in that circle, of which there are, as before, $\frac{1}{4}nc(\frac{1}{2}c - 1)$, is still present. Some new triangles are also introduced by the shortcuts. For example, vertex pairs $\frac{1}{2}c + 1$ to c steps apart on the circle are connected by one or more paths of length two, and if the same vertices are also connected by a shortcut those paths are turned into triangles.

The number of such paths of length two is clearly proportional to n—if we double the length of the circle we double the number of paths. The average number of shortcuts in the small-world model is, as we have said, $\frac{1}{2}ncp$ and there are $\binom{n}{2}$ places they can fall, meaning that any particular pair of vertices is connected with probability

$$\frac{\frac{1}{2}ncp}{\frac{1}{2}n(n-1)} = \frac{cp}{n-1},$$ (15.6)

or just cp/n in the limit of large n. The number of paths of length two that are completed by shortcuts to form triangles is thus proportional to $n \times cp/n = cp$, which is a constant. This means that in the limit of large network size we can safely ignore these triangles, because they will be negligible compared to the $O(n)$ triangles in the main circle.

Triangles can also be formed from two or three shortcuts, but these also turn out to be negligible in number. Thus, to leading order in n, the number of triangles in the small-world model is simply equal to the number in the circle, which is $\frac{1}{4}nc(\frac{1}{2}c-1)$.

And what about the number of connected triples? Once again, all connected triples in the circle model are still present in the small-world model. As shown in Section 15.1, there are $\frac{1}{2}nc(c-1)$ such triples. There are, however, also triples created by a shortcut combining with an edge in the circle. There are $\frac{1}{2}ncp$ shortcuts and c edges that they can form a triple with at each of their two ends, for a total of $\frac{1}{2}ncp \times c \times 2 = nc^2p$ connected triples.

There are also triples created by pairs of shortcuts. If a vertex is connected to m shortcuts then there are $\binom{m}{2}$ triples made of two shortcuts centered on that vertex and, averaging over the Poisson distribution of m, with mean cp, the expected number of connected triples centered at a vertex is $\frac{1}{2}c^2p^2$, for a total of $\frac{1}{2}nc^2p^2$ triples over all vertices.

Thus the expected total number of connected triples of all types in the whole network is $\frac{1}{2}nc(c-1) + nc^2p + \frac{1}{2}nc^2p^2$. Substituting the numbers of triangles and triples into Eq. (15.5), we then find that

$$C = \frac{\frac{1}{4}nc(\frac{1}{2}c-1) \times 3}{\frac{1}{2}nc(c-1) + nc^2p + \frac{1}{2}nc^2p^2}$$

$$= \frac{3(c-2)}{4(c-1) + 8cp + 4cp^2}.$$ (15.7)

Note that this becomes the same as Eq. (15.2), as it should, when $p = 0$. And as p grows it becomes smaller, with a minimum value of $C = \frac{3}{4}(c-2)/(4c-1)$ when $p = 1$. For instance when $c = 6$, the minimum value of the clustering coefficient is $\frac{3}{23} = 0.130\ldots$ (This behavior contrasts with that of the original Watts–Strogatz version of the small-world model in which edges are removed from the circle. In that version the clustering coefficient tends to zero as $n \to \infty$ when $p = 1$, since the network becomes a random graph at $p = 1$.)

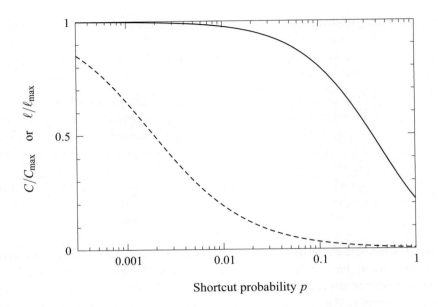

Figure 15.5: Clustering coefficient and average path length in the small-world model. The solid line shows the clustering coefficient, Eq. (15.7), for a small-world model with $c = 6$ and $n = 600$, as a fraction of its maximum value $C_{max} = \frac{3}{4}(c-2)/(c-1) = 0.6$, plotted as a function of the parameter p. The dashed line shows the average geodesic distance between vertices for the same model as a fraction of its maximum value $\ell_{max} = n/2c = 50$, calculated from the mean-field solution, Eq. (15.14). Note that the horizontal axis is logarithmic.

Figure 15.5 shows a plot of the clustering coefficient as a function of p for a small-world network with $c = 6$.

15.1.3 AVERAGE PATH LENGTHS

See Section 6.10.1 for a discussion of geodesic distances in networks.

Calculating the average path length in the small-world model, i.e., the mean geodesic or shortest-path distance between pairs of vertices, is harder than calculating the degree distribution or clustering coefficient. Indeed, no exact expression for mean distance has yet been found, though some approximate expressions are known and have been found in simulations of the model to be reasonably accurate.

One thing that is known about path lengths in the model is how they *scale* with the model parameters. Consider the simple case of a small-world model with $c = 2$, so that around the circle each vertex is connected only to its immediate neighbors, and consider the following dimensional argument. We define a length measure in our network by saying that the distance covered by an edge in the network is one length unit—a meter say, or a foot.[4] Then we can ask what other quantities in the model have the dimensions of length. One candidate is the distance around the whole circle, which is just n.

But there is another length in the model also, which is the mean distance between the ends of shortcuts around the circle. Suppose there are s shortcuts in our network, which means there $2s$ ends of shortcuts. (We know in fact that $s = \frac{1}{2}ncp$, but the point of this argument will be clearer if we stick with the simple notation s for the moment.) Then the average distance ξ between ends around the circle is

$$\xi = \frac{n}{2s}. \tag{15.8}$$

Once we specify the two distances n and ξ, we have specified the entire model, because once we have n the value of ξ fixes s, which fixes p, which is the only free parameter in the model given that $c = 2$.

Now consider the ratio of the length of the average shortest path in the network, which we will denote ℓ, to the length of the path around the entire circle, which is n. This ratio can, by definition, be written as a function of n and ξ, since n and ξ specify the entire model. However, it is also the ratio of two distances, meaning that it is dimensionless, and hence can be a function of only of dimensionless combinations of n and ξ. But there is only one such dimensionless combination, the ratio n/ξ. Thus it must be the case that

$$\frac{\ell}{n} = F(n/\xi) = F(2s), \tag{15.9}$$

where $F(x)$ is some function that doesn't depend on any of the parameters, a *universal function* in the language of scaling theory.

In other words, the mean geodesic distance ℓ between vertices in the small-world model with $c = 2$ is simply equal to the number of vertices n times some function of the number of shortcuts:

$$\ell = nF(2s). \tag{15.10}$$

And what happens for larger values of c? When we increase c the lengths of the shortest paths between vertices decrease. If we keep everything the same

---

[4]We consider all edges, including the shortcuts, to be the same length, even though the shortcuts are drawn as being longer in figures like Fig. 15.3. We are regarding the network as a purely topological object, not a spatial one.

in our model—number of vertices, number of shortcuts—but increase c from two to four, then we will roughly halve the shortest path between any pair of vertices. This is because we now have edges connecting next-nearest neighbor vertices around the circle as well as nearest neighbors, which means that we can traverse a given distance around the circle in half as many hops as we could previously. If the path incorporates any shortcuts then that part of the distance doesn't change—the shortcuts are as long as they ever were. However, if the density of shortcuts is low then most of the hops in most paths will be around the circle rather than along shortcuts and to a good approximation we can say that the length of the paths has simply halved. Similarly, for general values of c the length of the paths is decreased by a factor of $\frac{1}{2}c$ over its value for the $c = 2$ case.

Thus, provided the density of shortcuts is low, the equation corresponding to Eq. (15.10) for general values of c is:

$$\ell = \frac{2n}{c}F(2s).\tag{15.11}$$

We can derive an alternative form by making use of the fact that the number of shortcuts is $s = \frac{1}{2}ncp$, which gives us $\ell = 2(n/c)F(ncp)$. In fact, conventionally we absorb the leading factor of two into the definition of F, defining a new universal function $f(x) = 2F(x)$, so that

$$\ell = \frac{n}{c}f(ncp).\tag{15.12}$$

This *scaling form*, first proposed by Barthélémy and Amaral [31], tells us how the average path length in the small-world model depends on the model parameters n, c, and p when the density of shortcuts is low.

The catch is that we don't know the form of the function $f(x)$. We can, however, get an idea of its shape by numerical simulation of the model. We can generate random small-world networks and measure the mean distance ℓ between their vertices using breadth-first search (Section 10.3). Equation (15.12) tells us that if we perform such measurements for many different networks with many different values of the parameters we should find that the combination $c\ell/n$ is equal to the same function of ncp in all of them:

$$\frac{c\ell}{n} = f(ncp).\tag{15.13}$$

Figure 15.6 shows the results of such simulations for many different networks, and indeed we see that all of the points in the figure follow, roughly speaking, a single curve. This is the curve of $f(x)$.

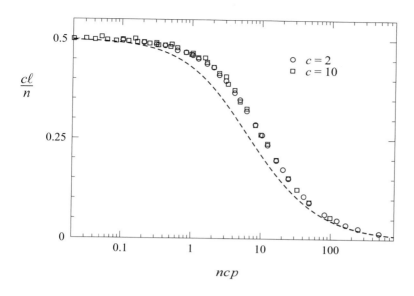

Figure 15.6: Scaling function for the small-world model. The points show numerical results for $c\ell/n$ as a function of ncp for the small-world model with a range of parameter values $n = 128$ to $32\,768$ and $p = 1 \times 10^{-6}$ to 3×10^{-2}, and two different values of c as marked. Each point is averaged over 1000 networks with the same parameter values. The points collapse, to a reasonable approximation, onto a single scaling function $f(ncp)$ in agreement with Eq. (15.13). The dashed curve is the mean-field approximation to the scaling function given in Eq. (15.14).

Another approach is to try to calculate $f(x)$ approximately in some fashion. Various approaches have been tried, including series approximations, distributional approximations, and mean-field methods. A mean-field approximation, for example, gives the result [251]

$$f(x) = \frac{2}{\sqrt{x^2 + 4x}} \tanh^{-1} \sqrt{\frac{x}{x + 4}}. \tag{15.14}$$

The methods used to derive this form become exact in the limit of either very small or very large numbers of shortcuts in the network,[5] but in between around $x = 1$ they are only approximate. The form of Eq. (15.14) is shown as the dashed line in Fig. 15.6 and indeed we see that it agrees well with the

---

[5]Note that the number of shortcuts can be large even when the density of shortcuts remains small, as it must for the scaling form (15.12) to be valid at all.

numerical results at the ends of the range but less well in the middle.

This, however, is enough for us to prove that the small-world model is indeed a "small world." Consider Eq. (15.14) for large values of x. Making use of the standard identity

$$\tanh^{-1} u = \tfrac{1}{2} \ln \frac{1+u}{1-u}, \tag{15.15}$$

we can write $f(x)$ as

$$f(x) = \frac{1}{\sqrt{x^2+4x}} \ln \frac{\sqrt{1+4/x}+1}{\sqrt{1+4/x}-1}, \tag{15.16}$$

and then taking the limit of large x we find

$$f(x) = \frac{\ln x}{x} \tag{15.17}$$

for $x \gg 1$. Substituting this into Eq. (15.12), we then have

$$\ell = \frac{\ln(ncp)}{c^2 p} \tag{15.18}$$

for $ncp \gg 1$. Recalling that ncp is simply twice the number of shortcuts in the network, this implies that, provided the number of shortcuts in the network is significantly greater than 1, the average distance between vertices will increase logarithmically with n, i.e., very slowly, for fixed c and p. Thus the number of vertices in the network can become very large and the value of ℓ will remain small, which is precisely the phenomenon we call the small-world effect.

Moreover, since only the *number* of shortcuts, and not the number per vertex, has to be large, the model tells us that the addition of only a small density of random shortcuts to a large network can produce small-world behavior. This helps explain why most real-world networks show the small-world effect. Most networks contain long-range connections and have at least some randomness in them—very few are perfectly regular or have only short-range connections—so we should not be surprised to see small-world behavior in almost all cases.

It is important to notice that the small-world model not only shows the small-world effect, but that it does so at the same time as displaying clustering. Since the number of shortcuts in the network is $\tfrac{1}{2}ncp$, we can always make it much larger than one simply by increasing the size n of the network, while keeping c and p constant. At the same time, the clustering coefficient, Eq. (15.7), is independent of n and hence retains its (non-zero) value as $n \to \infty$. In this limit, therefore, we simultaneously have non-zero clustering and the small-world effect, demonstrating conclusively that the two are not at odds with one another—it is perfectly possible to have both in the same network at the same time.

Figure 15.5 shows a plot of the approximate value of ℓ as a function of p from Eqs. (15.12) and (15.14) for a small-world model with $n = 600$ vertices and $c = 6$, along with the curve for the clustering coefficient of the same model that we plotted earlier and, as we can see, there is a substantial range of values of p in which the value of ℓ is low while the value of C is high.

Many other properties and quantities can be calculated for the small-world model, either analytically or numerically. For a short review of results concerning the model see Ref. [232].

15.2 EXPONENTIAL RANDOM GRAPHS

Many of the networks we observe in the real world exist in only one instantiation, one example that we can study. There's only one Internet, for instance, and only one World Wide Web. But is the precise structure of such a network—the precise pattern of connections in the Internet, say—the only possible structure the network could have? Common sense suggests that it is not. For a start, the Internet evolves in time, so we see different structures if we look at different times and all of them are by definition plausible structures for the network. More importantly, it's clear that, had circumstances been slightly different, the Internet could easily have evolved to have a different topology, but one that in practical terms would probably have worked about as well as the present one.

On the other hand, we can say that the structure of such an alternate Internet would probably have been "similar" to the real Internet, in some sense. That is, all reasonable choices for the structure of the Internet have some basic features in common, even if they differ in smaller details. Similar considerations also apply to other types of network, including social networks, biological networks, and information networks.

In some cases the questions we want to answer about a networked system can be tackled by studying the structure of only a single observed example—the real Internet for instance. But there are other cases where we would like to know about the entire set of possible networks that *could* represent a system. If we are studying some social process in a social network, for instance, such as opinion formation or the spread of a disease, we can measure a social network and then calculate or simulate the effects of the process of interest on that network. More often, however, we would like to know how the process behaves on social networks generally, rather than on the one particular network we have measured.

Considerations of this kind lead us to consider ensemble models of networks, an *ensemble*, in this context, meaning a set of possible networks plus a probability distribution over them. We have seen some examples of ensemble models in previous chapters, such as the random graphs of Chapters 12 and 13. In this section we introduce a beautiful and general formalism for ensemble network models called the *exponential random graph*, which includes random graphs as special cases but also extends to many other network ensembles that describe all sorts of network phenomena.[6]

Elegant though this formalism is, however, it also has some serious drawbacks. For reasons that are still not entirely understood, exponential random graphs fail as models of some common network phenomena such as transitivity (see Section 7.9). We will examine the nature of some of these failures towards the end of the chapter.

15.2.1 DEFINITION OF THE EXPONENTIAL RANDOM GRAPH

Suppose we want to create an ensemble of networks with a given set of properties, such as a given number of edges or a given value of the clustering coefficient. We can do that, as ordinary random graph models do, by fixing absolutely the values of the quantity or quantities of interest and then drawing uniformly from the set of all networks with the desired values. For instance, if we draw uniformly from the set of all graphs with a given number of edges we have the $G(n, m)$ random graph model of Section 12.1.

In many cases, however, this approach is not exactly what we want. If we observe that a social network, for example, has a given number of edges, it does not necessarily mean that every possible social network for the given community would have exactly that many edges. Had the world evolved slightly differently, the number of edges might well have turned out differently as well.

Often, therefore, a better approach is to fix the *average* value of the property or properties of interest. We might fix the average number of edges, for instance, so that some networks in our ensemble have more than the average and some have less, but over the whole ensemble we get the right average value. Moreover, we can arrange that networks with numbers of edges close to the desired value have higher probabilities in the ensemble than networks further away, so that the ensemble is dominated by networks with properties close to the desired ones. The exponential random graph provides an elegant

---

[6]In the sociology literature exponential random graphs are also called *p-star models* (sometimes written "p^*").

way of achieving these goals.

Suppose, therefore, that we have some set of network measures whose numerical values we want to fix. Examples might include number of edges or mean degree of a vertex, degrees of individual vertices, number of triangles or clustering coefficient, and so forth. Let us denote these measures by $x_1, x_2, \ldots$

Now consider the set $\mathcal{G}$ of all simple graphs[7] with n vertices and let us define an ensemble by giving each graph G in the set a probability $P(G)$, normalized so that

$$\sum_{G \in \mathcal{G}} P(G) = 1. \tag{15.19}$$

Recall that a simple graph is a graph with no multi-edges and no self-edges—see Section 6.1.

The mean or expectation value $\langle x_i \rangle$ of a network measure x_i within this ensemble is given by

$$\langle x_i \rangle = \sum_{G \in \mathcal{G}} P(G) \, x_i(G), \tag{15.20}$$

where $x_i(G)$ is the value of x_i measured on the graph G (e.g., number of edges in graph G, number of triangles, etc.).

Now, following the prescription outlined above, let us fix the mean value of each of our measures within our ensemble. If we do this, then Eq. (15.20) is turned around and becomes a constraint on the probability distribution over graphs:

$$\sum_{G \in \mathcal{G}} P(G) \, x_i(G) = \langle x_i \rangle, \tag{15.21}$$

where $\langle x_i \rangle$ is now a specified number. We have one such constraint for each network measure.

The number of measures, however, is typically quite small—maybe only one or two, maybe hundreds or even thousands, but usually nowhere near the number of graphs in our ensemble. The number of simple graphs of n vertices is $2^{n(n-1)/2}$, which becomes very large even for relatively modest values of n. This means that the constraints in Eqs. (15.19) and (15.21) do not specify the probability distribution $P(G)$ completely. Indeed, they leave an enormous amount of flexibility about the values of $P(G)$. There are many more unknowns $P(G)$ than there are constraints in our equations and hence a wide range of choices of $P(G)$ that will satisfy the constraints. How do we choose between them?

This question, of making the best choice of a probability distribution given only a relatively small number of constraints on that distribution, is one that

---

[7]One can also define exponential random graphs models for sets that include non-simple graphs, but the case considered here of simple graphs is the most commonly studied one.

is familiar to physicists and statisticians, having been studied for over a hundred years since the pioneering work of Willard Gibbs in the latter part of the nineteenth century. The solution is remarkably simple, although deriving it is not. It can be shown that the best choice of probability distribution is the one that maximizes the *Gibbs entropy*

$$S = -\sum_{g \in \mathcal{G}} P(G) \ln P(G), \tag{15.22}$$

subject to the known constraints.

One may well ask what we mean by "best choice" in this context. The maximum entropy choice is best in the sense that it makes the minimum assumptions about the distribution other than those imposed upon us by the constraints. There are choices of distribution we could make that would satisfy the constraints but would effectively make additional assumptions. For instance, some choices might make a particular graph or graphs highly probable while other graphs, only slightly different, are given far lower probabilities. These would be considered "bad" choices in the sense that they assume things about the ensemble for which we have no supporting evidence. The Gibbs entropy is precisely a measure of the amount of "assumption" that goes into a particular choice of distribution $P(G)$, or more precisely it is the amount of "anti-assumption" or ignorance, and by maximizing it we minimize unjustified assumptions as much as possible. The derivation of the formula, Eq. (15.22), would take us some way away from our central topic of networks, so we will not go through it here, but the interested reader is encouraged to look for example at the books by Grandy [142] and Cover and Thomas [82].

The maximization of the entropy, subject to the constraints of Eqs. (15.19) and (15.21), can be achieved by the method of Lagrange multipliers. The optimum is the set of values of the $P(G)$ that maximizes the quantity

$$-\sum_{G \in \mathcal{G}} P(G) \ln p(G) - \alpha \left[1 - \sum_{G \in \mathcal{G}} P(G)\right] - \sum_i \beta_i \left[\langle x_i \rangle - \sum_{G \in \mathcal{G}} P(G)\, x_i(G)\right], \tag{15.23}$$

where α and β_i are Lagrange multipliers whose values will be determined shortly. Differentiating with respect to the probability $P(G)$ of a particular graph G and setting the result to zero, we then find that

$$-\ln P(G) - 1 + \alpha + \sum_i \beta_i x_i(G) = 0, \tag{15.24}$$

which implies

$$P(G) = \exp\left[\alpha - 1 + \sum_i \beta_i x_i(G)\right], \tag{15.25}$$

or

$$P(G) = \frac{e^{H(G)}}{Z},\qquad(15.26)$$

where $Z = e^{1-\alpha}$ is called the *partition function* and

$$H(G) = \sum_i \beta_i x_i(G)\qquad(15.27)$$

is the *graph Hamiltonian*.[8]

It remains to fix the values of Z and β_i (for all i). Z is fixed by the normalization condition, Eq. (15.19), which requires that

$$\sum_{G\in\mathcal{G}} P(G) = \frac{1}{Z}\sum_{G\in\mathcal{G}} e^{H(G)} = 1,\qquad(15.28)$$

and hence

$$Z = \sum_{G\in\mathcal{G}} e^{H(G)}.\qquad(15.29)$$

There is no equivalent general formula for the values of the β_i. They are calculated by substituting Eq. (15.26) into Eq. (15.21) and solving the resulting set of non-linear simultaneous equations, but the particular solution depends on the form of the Hamiltonian. We will see some examples of the process shortly.

There are some cases in which we are interested in an exponential random graph only as a class of models. That is, we are concerned not as much with the model's properties for a particular set of values $\{\beta_i\}$ as with the behavior of the model in general. In such cases we can regard the β_i as free parameters controlling the structure of the network, much as the edge probability p controls the structure of the network in a Poisson random graph.

15.2.2 EXPECTATION VALUES

Once we have determined the probability distribution $P(G)$ over graphs, we can use it to calculate estimates of quantities of interest within the ensemble. The most common objects of interest are expectation values (i.e., averages) of quantities, the expectation value of a quantity y in the ensemble being given by

$$\langle y \rangle = \sum_{G\in\mathcal{G}} P(G)\, y(G) = \frac{1}{Z}\sum_{G\in\mathcal{G}} e^{H(G)}\, y(G).\qquad(15.30)$$

---

[8]Sometimes the graph Hamiltonian is defined to be minus this quantity and a corresponding minus sign is introduced in Eq. (15.26). This is by analogy with similar quantities in statistical physics, where the Hamiltonian is an energy function and lower energies correspond to higher probabilities. In studies of networks, however, the definitions are most commonly as given here.

In effect, this calculation gives us a "best estimate" of the value of y. That is, given a certain set of observations or constraints on our network, embodied in Eq. (15.21), but no other information about the network structure, we can calculate a best-guess ensemble of networks subject to those constraints and then use that ensemble to calculate the expectation value of the quantity y, giving us a best guess at the value of that quantity given only the constraints. Thus the exponential random graph model enables us to answer questions of the type, "If I know certain things, A, B, and C, about a network, what is my best estimate of some other thing D?" For instance, if I know the average degree of a vertex in a network, what is my best estimate of the degree distribution? Or the clustering coefficient? The exponential random graph gives a rigorous and principled answer to questions of this kind.

An interesting special case arises when the quantity y that we want to estimate is itself one of the set of network measures x_i that we used to specify our ensemble in the first place. You might ask why we would want to do this, given that, by hypothesis, we already know the expectation values of these quantities—they are precisely the quantities that we used as inputs to our model in the first place. The answer is that we still need to fix the parameters β_i and we do this by calculating the expectation values $\langle x_i \rangle$ for given β_i and then varying the β_i until the $\langle x_i \rangle$ take the desired values.

The value of $\langle x_i \rangle$ within the ensemble is given by

$$\langle x_i \rangle = \frac{1}{Z} \sum_{G \in \mathcal{G}} e^{\sum_i \beta_i x_i(G)} x_i(G) = \frac{1}{Z} \frac{\partial}{\partial \beta_i} \sum_{G \in \mathcal{G}} e^{\sum_i \beta_i x_i(G)}$$

$$= \frac{1}{Z} \frac{\partial Z}{\partial \beta_i} = \frac{\partial \ln Z}{\partial \beta_i}, \tag{15.31}$$

where we have made use of Eq. (15.29). The quantity

$$F = \ln Z \tag{15.32}$$

is called the *free energy* of the ensemble and Eq. (15.31) can be written simply as

$$\langle x_i \rangle = \frac{\partial F}{\partial \beta_i}. \tag{15.33}$$

To calculate $\langle x_i \rangle$, therefore, all we need to do is calculate the partition function Z, from it evaluate the free energy, and then differentiate.

Calculating expectation values for other quantities is harder, and indeed this is one of the main practical problems with exponential random graphs: the actual calculations of quantities of interest can be very difficult and in many cases can only be performed using numerical methods. If we are clever, however, we can still use the machinery embodied in Eq. (15.33) in some cases. The trick is to introduce an extra term involving y into our Hamiltonian thus:

$$H(G) = \sum_i \beta_i x_i(G) + \mu y(G). \tag{15.34}$$

If we set the parameter μ to zero, then the answers we get out of our calculations will be unchanged from before and hence will still be correct. However, we can now differentiate with respect to μ (at the point $\mu = 0$) to calculate the expectation value of y:

$$\langle y \rangle = \left. \frac{\partial F}{\partial \mu} \right|_{\mu=0}. \tag{15.35}$$

This allows us again to calculate just the one sum, the partition function Z, and from it calculate the free energy and thus the average $\langle y \rangle$. The catch is that we have to calculate Z for general (non-zero) values of μ so that we can perform the derivative—we only set μ to zero at the end of the calculation. In many cases it can be quite difficult to calculate Z in this way, which makes the exponential random graph, though elegant, technically tricky.

15.2.3 SIMPLE EXAMPLES

Probably the simplest example of an exponential random graph model is the model in which we fix the expected number of edges in an undirected network and nothing else. Following the formalism above, this gives us a graph Hamiltonian, Eq. (15.27), of $H = \beta m$, where m is the number of edges. Then individual graphs appear in the ensemble with probability

$$P(G) = \frac{e^{\beta m}}{Z}, \tag{15.36}$$

where

$$Z = \sum_G e^{\beta m}. \tag{15.37}$$

Thus higher values of β in this model correspond to denser networks, those with more edges.

To make further progress with this model we need a way to perform the sum over graphs G in Eq. (15.37). The standard way to achieve this is to sum over possible values of the elements A_{ij} of the adjacency matrix. In this case we are considering undirected graphs, so we need to specify only the matrix elements above the diagonal or those below it, but not both, since the matrix is symmetric. And since we are restricting ourselves to simple graphs the only allowed values of A_{ij} are 0 and 1 if $i \neq j$ and $A_{ii} = 0$.

We can write the number of edges m in terms of the adjacency matrix thus:

$$m = \sum_{i<j} A_{ij}, \tag{15.38}$$

and hence the partition function is

$$
\begin{aligned}
Z &= \sum_{\{A_{ij}\}} \exp\left(\beta \sum_{i<j} A_{ij}\right) \\
&= \sum_{\{A_{ij}\}} \prod_{i<j} e^{\beta A_{ij}} = \prod_{i<j} \sum_{A_{ij}=0,1} e^{\beta A_{ij}} = \prod_{i<j} (1 + e^{\beta}) \\
&= (1 + e^{\beta})^{\binom{n}{2}}, \tag{15.39}
\end{aligned}
$$

where the notation $\{A_{ij}\}$ indicates summation over all allowed values of the adjacency matrix.

From this expression we can calculate the free energy:[9]

$$F = \ln Z = \binom{n}{2} \ln(1 + e^{\beta}), \tag{15.40}$$

and thus, using Eq. (15.33), the average number of edges in the model is

$$\langle m \rangle = \frac{\partial F}{\partial \beta} = \binom{n}{2} \frac{1}{1 + e^{-\beta}}. \tag{15.41}$$

If we have a particular desired value that $\langle m \rangle$ should take, we can now achieve it by rearranging this expression to find the appropriate value for the Lagrange multiplier β thus:

$$\beta = \ln \frac{\langle m \rangle}{\binom{n}{2} - \langle m \rangle}. \tag{15.42}$$

We can also calculate, for example, the probability p_{vw} that there will be an edge between a particular pair of vertices v, w, which is given by the average of the corresponding element A_{vw} of the adjacency matrix. From Eq. (15.30) we

---

[9]Those familiar with free energy in its original thermodynamic context may find this expression odd because it varies with network size as n^2 to leading order. In thermodynamic systems, by contrast, free energy is always directly proportional to system size. However the degrees of freedom or "particles" in our network are really the edges (or absence of edges) between vertex pairs, not the vertices themselves, and there are $\binom{n}{2}$ vertex pairs, which is why Eq. (15.40) is proportional to $\binom{n}{2}$.

have

$$p_{vw} = \langle A_{vw} \rangle = \frac{1}{Z} \sum_{\{A_{ij}\}} A_{vw} \exp\left(\beta \sum_{i<j} A_{ij} \right) = \frac{\sum_{A_{vw}=0,1} A_{vw} e^{\beta A_{vw}}}{\sum_{A_{vw}=0,1} e^{\beta A_{vw}}}$$

$$= \frac{1}{1+e^{-\beta}} = \frac{\langle m \rangle}{\binom{n}{2}}. \tag{15.43}$$

Thus the probability of an edge between a given pair of vertices is the same in this model for every pair. In other words, this model is just the ordinary Poisson random graph of Chapter 12 with $p = \langle m \rangle / \binom{n}{2}$. The random graph can thus be regarded as a special case of the more general exponential random graph model.

The random graph, as we saw in Chapter 12, is in many respects a poor model of real-world networks. In particular, its degree distribution is Poissonian and hence very different from the highly right-skewed degree distributions in most observed networks. It is natural to ask, therefore, whether we can make an exponential random graph model that has a more realistic degree distribution. There are a number of ways of doing this, but one of the simplest is to create a model in which we specify the expected degree of each vertex within the ensemble. That is, we create an exponential random graph model with the graph Hamiltonian

$$H = \sum_i \beta_i k_i, \tag{15.44}$$

where k_i is the degree of vertex i. Note that we do not also need a term that fixes the average number of edges in this model, since fixing the average degree of each vertex already fixes the average number of edges (see Eq. (6.20)).

We can write the degrees in terms of the adjacency matrix as

$$k_i = \sum_j A_{ij}, \tag{15.45}$$

and hence write the Hamiltonian as

$$H = \sum_{ij} \beta_i A_{ij}$$
$$= \sum_{i<j} \beta_i A_{ij} + \sum_{i>j} \beta_i A_{ij} = \sum_{i<j} \beta_i A_{ij} + \sum_{i<j} \beta_j A_{ji}$$
$$= \sum_{i<j} (\beta_i + \beta_j) A_{ij}, \tag{15.46}$$

where in the second line we have interchanged the dummy variables i and j and in the third line we have made use of $A_{ji} = A_{ij}$. We have also again assumed that there are no self-edges, so that $A_{ii} = 0$ for all i.

Now we can write the partition function as

$$Z = \sum_{\{A_{ij}\}} \exp\left(\sum_{i<j}(\beta_i + \beta_j)A_{ij}\right) = \prod_{i<j}\sum_{A_{ij}=0,1} e^{(\beta_i+\beta_j)A_{ij}}$$
$$= \prod_{i<j}\left[1 + e^{\beta_i+\beta_j}\right], \tag{15.47}$$

and the probability of an edge between vertices u and v is

$$p_{vw} = \langle A_{vw}\rangle = \frac{1}{Z}\sum_{\{A_{ij}\}} A_{vw}\exp\left(\sum_{i<j}(\beta_i + \beta_j)A_{ij}\right)$$
$$= \frac{\sum_{A_{vw}=0,1} A_{vw}\, e^{(\beta_v+\beta_w)A_{vw}}}{\sum_{A_{vw}=0,1} e^{(\beta_v+\beta_w)A_{vw}}}$$
$$= \frac{1}{1 + e^{-(\beta_v+\beta_w)}}. \tag{15.48}$$

Thus edges in this model now have different probabilities. Of particular interest is the case of a sparse network, one in which the probability of any individual edge is small, $p_{vw} \ll 1$. (As we have seen throughout this book, most real-world networks are very sparse.) To achieve this, we need $e^{-(\beta_v+\beta_w)} \gg 1$ in Eq. (15.48), which means that

$$p_{vw} \simeq e^{\beta_v}e^{\beta_w}. \tag{15.49}$$

In other words, in a sparse network the probability of an edge is simply a product of two terms, one for each of the vertices at either end of the edge. Moreover, it turns out that these terms are simply related to the expected degrees of the vertices. The expected degree of vertex v, for instance, is just the sum of the expected number p_{vw} of edges between it and every other vertex:

$$\langle k_v\rangle = \sum_w p_{vw} = e^{\beta_v}\sum_w e^{\beta_w}, \tag{15.50}$$

so that

$$e^{\beta_v} = C\langle k_v\rangle, \tag{15.51}$$

where $C = 1/\sum_w e^{\beta_w}$.

Thus $p_{vw} = C^2\langle k_v\rangle\langle k_w\rangle$ in this model, and since we require that $\sum_{vw} p_{vw} = \sum_v \langle k_v\rangle = 2\langle m\rangle$ (see Eqs. (6.19) and (6.20)), it's then straightforward to show that

$$p_{vw} = \frac{\langle k_v\rangle\langle k_w\rangle}{2\langle m\rangle}. \tag{15.52}$$

Once again, this is a model we have seen before. It is the random graph model that we studied in Section 13.2.2 in which we specify the expected degrees of vertices (rather than their exact degrees, as in the more common configuration model).

We can also create exponential random graph models of directed networks. For instance, we can make a model in which the constrained quantities are the expected values of the in- and out-degrees of a directed network by using a Hamiltonian of the form

$$H = \sum_i \beta_i^{\text{in}} k_i^{\text{in}} + \sum_j \beta_j^{\text{out}} k_j^{\text{out}}. \tag{15.53}$$

Writing $k_i^{\text{in}} = \sum_{j(\neq i)} A_{ij}$ and $k_j^{\text{out}} = \sum_{i(\neq j)} A_{ij}$, we have

$$H = \sum_{i \neq j} (\beta_i^{\text{in}} + \beta_j^{\text{out}}) A_{ij}. \tag{15.54}$$

The ensemble is now a distribution over (simple) directed graphs, which means that the adjacency matrix is in general asymmetric and each element A_{ij} can take its own value. Thus the partition function is

$$Z = \sum_{\{A_{ij}\}} \exp\left[\sum_{i \neq j} (\beta_i^{\text{in}} + \beta_j^{\text{out}}) A_{ij}\right] = \prod_{i \neq j} \sum_{A_{ij}=0,1} e^{(\beta_i^{\text{in}} + \beta_j^{\text{out}}) A_{ij}}$$

$$= \prod_{i \neq j} \left[1 + e^{\beta_i^{\text{in}} + \beta_j^{\text{out}}}\right], \tag{15.55}$$

and the probability of an edge from w to v is

$$p_{vw} = \langle A_{vw} \rangle = \frac{1}{Z} \sum_{\{A_{ij}\}} A_{vw} \exp\left(\sum_{i \neq j} (\beta_i^{\text{in}} + \beta_j^{\text{out}}) A_{ij}\right)$$

$$= \frac{\sum_{A_{vw}=0,1} A_{vw} \, e^{(\beta_v^{\text{in}} + \beta_w^{\text{out}}) A_{vw}}}{\sum_{A_{vw}=0,1} e^{(\beta_v^{\text{in}} + \beta_w^{\text{out}}) A_{vw}}}$$

$$= \frac{1}{1 + e^{-(\beta_v^{\text{in}} + \beta_w^{\text{out}})}}. \tag{15.56}$$

In the case of a sparse network this becomes

$$p_{vw} \simeq e^{\beta_v^{\text{in}}} e^{\beta_w^{\text{out}}} = \frac{\langle k_v^{\text{in}} \rangle \langle k_w^{\text{out}} \rangle}{\langle m \rangle}, \tag{15.57}$$

by an argument similar to the one leading to Eq. (15.52). This expression is similar to that for the corresponding quantity in the directed version of the configuration model (see page 475), and indeed the model above is the equivalent for the directed case of the random graph in which we specify the expected degrees of the vertices rather than the exact degrees.

15.2.4   RECIPROCITY MODEL

We now turn to some more complex examples of exponential random graphs, ones that are not equivalent to models we have already seen. The first example we look at is the "reciprocity model" proposed by Holland and Leinhardt [157].

As discussed in Section 7.10, many directed networks exhibit the phenomenon of reciprocity, whereby edges between vertices tend to be reciprocated. If I say that you are my friend, for example, then it is likely that you will also say that I am your friend. We can create an exponential random graph model of reciprocity by fixing the expected number of reciprocated edges in the network. The number of reciprocated edges, m_r is given by $m_r = \sum_{i \neq j} A_{ij} A_{ji}$, so we need to introduce a term proportional to this into our graph Hamiltonian. We can also introduce other terms, such as terms to fix the expected degrees of vertices as in the previous section. Here let us look the simple case where we fix only the number of edges as we did with the Poisson random graph. The number of edges in a simple directed network is given by $m = \sum_{i \neq j} A_{ij}$ and hence our Hamiltonian takes the form

$$H = \beta \sum_{i \neq j} A_{ij} + \gamma \sum_{i \neq j} A_{ij} A_{ji}$$
$$= \sum_{i<j} \left[\beta \left(A_{ij} + A_{ji} \right) + 2\gamma A_{ij} A_{ji} \right], \tag{15.58}$$

where β and γ are free parameters that can be varied to create the desired numbers of edges and reciprocated edges. This is actually a simplified version of the model proposed by Holland and Leinhardt, but it will serve our purpose nicely, and it is easy to solve.

The partition function for this model is

$$Z = \sum_{\{A_{ij}\}} \exp\left(\sum_{i<j} \left[\beta \left(A_{ij} + A_{ji} \right) + 2\gamma A_{ij} A_{ji} \right] \right)$$
$$= \prod_{i<j} \sum_{A_{ij}=0,1} \sum_{A_{ji}=0,1} e^{\beta(A_{ij}+A_{ji})+2\gamma A_{ij} A_{ji}} = \prod_{i<j} \left[1 + 2e^{\beta} + e^{2(\beta+\gamma)} \right]$$
$$= \left[1 + 2e^{\beta} + e^{2(\beta+\gamma)} \right]^{\binom{n}{2}}. \tag{15.59}$$

The free energy for the network is then

$$F = \binom{n}{2} \ln\left(1 + 2e^{\beta} + e^{2(\beta+\gamma)} \right), \tag{15.60}$$

and, applying Eq. (15.33), we find that the expected numbers of edges and reciprocated edges are

$$\langle m \rangle = \frac{\partial F}{\partial \beta} = n(n-1)\frac{e^{\beta} + e^{2(\beta+\gamma)}}{1 + 2e^{\beta} + e^{2(\beta+\gamma)}},$$

$$\langle m_r \rangle = \frac{\partial F}{\partial \gamma} = n(n-1)\frac{e^{2(\beta+\gamma)}}{1 + 2e^{\beta} + e^{2(\beta+\gamma)}}. \tag{15.61}$$

In Section 7.10 we defined the reciprocity r of a directed network to be the fraction of edges that are reciprocated, which in our model is given by the ratio

$$r = \frac{\langle m_r \rangle}{\langle m \rangle} = \frac{1}{1 + e^{-(\beta+2\gamma)}}. \tag{15.62}$$

Thus we can control both the number of edges and the level of reciprocity in the network by suitable choices of β and γ.

15.2.5 TWO-STAR MODEL

After ordinary random graphs, probably the simplest undirected exponential random graph is the so-called *two-star model*. In this model one specifies the expected number $\langle m \rangle$ of edges in the network and the expected number $\langle m_2 \rangle$ of *two-stars*, meaning a vertex connected by edges to two others (which we called a "connected triple" in other circumstances—see Eq. (7.41) on page 200). Varying the number of two-stars allows us to control the extent to which edges in the network "stick together," meaning they share common vertices. If we fix only the number of edges in a network, then those edges may stick together or they may not, but if we also give the network a lot of two-stars, then the edges have to stick together to make the required number of two-stars. Thus the two-star model allows us to control the "clumpiness" of the network, the extent to which the edges gather together in clumps or are distributed more randomly.

A two-star is a vertex connected by edges to two other vertices.

The number of two-stars in a network is

$$m_2 = \sum_i \sum_{j(\neq i)} \sum_{k(\neq i,j)} A_{ij}A_{ik} = \tfrac{1}{2}\sum_{i\neq j} A_{ij} \sum_{k(\neq i,j)} (A_{ik} + A_{jk}), \tag{15.63}$$

and the number of edges is, as before, $m = \sum_{i<j} A_{ij} = \tfrac{1}{2}\sum_{i\neq j} A_{ij}$. Thus the Hamiltonian is

$$H = \tfrac{1}{2}\beta \sum_{i\neq j} A_{ij} + \tfrac{1}{2}\gamma \sum_{i\neq j} A_{ij} \sum_{k(\neq i,j)} (A_{ik} + A_{jk})$$

$$= \tfrac{1}{2}\sum_{i\neq j} A_{ij}\left[\beta + \gamma \sum_{k(\neq i,j)} (A_{ik} + A_{jk})\right], \tag{15.64}$$

We encountered mean-field theory briefly earlier in the chapter, in our study of the small-world model, though we did not elaborate on it there. See Eq. (15.14) and the associated discussion.

where β and γ are our two parameters.

We can solve this model using *mean-field theory*, a technique borrowed from statistical physics. We note that the term $\sum_{k(\neq i,j)} A_{ik}$ is simply the number of edges attached to vertex i, excluding any edge between i and j. All vertex pairs are equivalent in this model—vertices have no individual properties to distinguish them—so the mean probability $\langle A_{ij} \rangle$ of an edge between any pair is the same. If we denote this probability by p then the expected value of the term above is just

$$\left\langle \sum_{k(\neq i,j)} A_{ik} \right\rangle = \sum_{k(\neq i,j)} \langle A_{ik} \rangle = \sum_{k(\neq i,j)} p = (n-2)p. \qquad (15.65)$$

But, assuming that the network is large, this is, to a good approximation, just np, which is the mean degree of a vertex.

The mean-field approach consists of replacing the actual term in the Hamiltonian with the expected value np. We also make the same replacement for the term $\sum_{k(\neq i,j)} A_{jk}$. These replacements are a good approximation so long as $np \gg 1$ since for large values of np the statistical variation from vertex to vertex around the expected value becomes negligible. If the value of p is kept fixed as we make our network larger then np will always be large in the limit $n \to \infty$. Thus, in the limit of large network size, this mean-field approximation is a good one.

In this large-n regime, making the replacement described above, we have

$$H = \tfrac{1}{2}(\beta + 2\gamma np) \sum_{i \neq j} A_{ij} = (\beta + 2\gamma np)m, \qquad (15.66)$$

where m is the number of edges as before.

Now, however, this is the same as the Hamiltonian for the ordinary Poisson random graph in Section 15.2.3, except for the replacement $\beta \to \beta + 2\gamma np$, so we can immediately write down the partition function and other quantities using the results of that section. In particular, Eq. (15.41) tells us that the average number of edges in the network will be

$$\langle m \rangle = \binom{n}{2} \frac{1}{1 + e^{-(\beta + 2\gamma np)}}. \qquad (15.67)$$

But the average number of edges is related to the mean probability of an edge by $\langle m \rangle = \binom{n}{2}p$ and hence

$$p = \frac{\langle m \rangle}{\binom{n}{2}} = \frac{1}{1 + e^{-(\beta + 2\gamma np)}} = \tfrac{1}{2}\left[\tanh(\tfrac{1}{2}\beta + \gamma np) + 1\right]. \qquad (15.68)$$

This gives us a self-consistent equation that we can solve to find p as a function of the parameters β and γ, and once we have p we can solve for other properties of the network by treating it as a normal Poisson random graph.

For convenience in solving for p, let us define $B = \frac{1}{2}\beta$ and $C = \frac{1}{2}\gamma n$ so that Eq. (15.68) becomes

$$p = \frac{1}{2}[\tanh(B + 2Cp) + 1]. \tag{15.69}$$

There is no known closed-form solution for this equation in general, but we can visualize the solution easily enough using a graphical method. If we make plots of the lines $y = p$ and $y = \frac{1}{2}[\tanh(B + 2Cp) + 1]$ as functions of p on the same axes, they will intersect at the solution (or solutions) of Eq. (15.69). Three such plots are shown in Fig. 15.7 for different choices of the parameters.

Consider first panel (a), which shows the curve of $y = \frac{1}{2}[\tanh(B + 2Cp) + 1]$ for $C = \frac{1}{2}$ and three different values of B (solid lines). Varying B merely shifts the entire curve horizontally without changing its overall shape. For each curve there is a single point of intersection with the line $y = p$, indicated by a small circle. As B is varied this intersection point moves smoothly between high and low values of p. Thus in this regime we can tune the density of the network to any desired value by varying the parameter B (or equivalently the parameter $\beta = 2B$).

Now take a look at the last panel in Fig. 15.7, panel (c), which shows curves for $C = \frac{3}{2}$ and again three difference values of B. Again varying B shifts the curve horizontally, but now there is an important difference. Because of the higher value of C, the shape of the curve has changed. It is steeper in the middle than it was previously and as a result it is now possible at suitable values of B for the curve to intersect with the line $y = p$ not just in one place but in three different places. In this regime there are three different possible solutions for p for the same values of the parameters. In fact it turns out that the middle solution is unphysical and only the two outer solutions are realized in practice. These two, however, correspond to very different networks. One has very high density with many edges while the other is very sparse with few edges. Yet both solutions are real. If one were to simulate the two-star model on a computer, generating networks at random according to the model prescription, one would in this regime sometimes find a high-density network and sometimes a low-density one for the same parameter values, and one would not be able to predict in advance which would occur.

This peculiar behavior is called *spontaneous symmetry breaking*. It is a behavior well known to physicists, who study it in condensed matter physics, where it gives rise to the phenomenon of ferromagnetism, and in particle physics, where it gives rise to the phenomenon of particle mass. In network models,

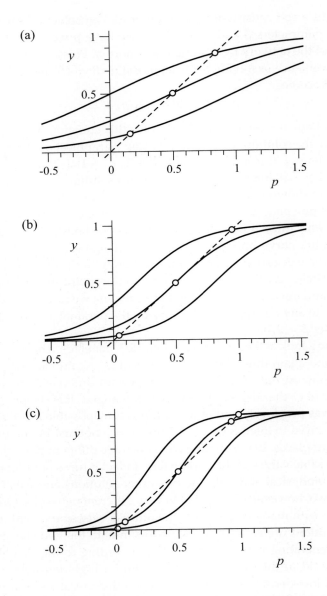

Figure 15.7: Graphical solutions of the properties of the two-star model. Curves for $y = \frac{1}{2}[\tanh(B + 2Cp) + 1]$ for varying values of B and (a) $C = \frac{1}{2}$, (b) $C = 1$, and (c) $C = \frac{3}{2}$. The points where the curves intersect the line $y = p$ (dotted line in each panel) are solutions of Eq. (15.69).

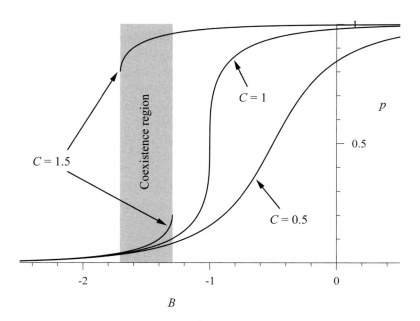

Figure 15.8: Edge probability in the two-star model. Plot of solutions of Eq. (15.69) for the edge probability p as a function of B for the same three values of C as were used in the three panels of Fig. 15.7. Note that there are two possible solutions within the coexistence region for the case $C = \frac{3}{2}$, and more importantly that for this case there is no value of B that gives any intermediate value of p. For $C = \frac{3}{2}$ the only possible values of p lie either above about 0.8 or below about 0.2.

however, it is primarily an annoyance, and sometimes a grave weakness. A model that can produce two radically different classes of network for the same values of the model parameters is, at the least, troubling. But worse, for values of C as in Fig. 15.7c there some values of p that are simply impossible to reach.

Figure 15.8 shows the values of the solutions for p for the cases depicted in Fig. 15.7 as a function of B and, as we have said, p is a smooth function of B for the $C = \frac{1}{2}$ case, so that any value of p is reachable. For the case of $C = \frac{3}{2}$, however, there are only very high and very low values of p. There is no value of B that produces intermediate values of p and hence no way in this model to generate graphs with such intermediate values if $C = \frac{3}{2}$. If we wanted to generate a graph with $p = \frac{1}{2}$, for instance, there is simply no way to do it in the two-star model when $C = \frac{3}{2}$.

This is a fundamental problem with the two-star model and with many other exponential random graphs. We will see in the following section an example of an exponential random graph where this kind of behavior renders the model essentially useless as a model of a network.

Panel (b) of Fig. 15.7 shows the borderline case that falls between panels (a) and (c). When the parameter C is such that the curve of $y = \frac{1}{2}[\tanh(B + 2Cp) + 1]$ has gradient exactly one at its steepest point then we are right on the boundary between the two different types of behavior. In the present case, this happens at $C = 1$. If C is increased any further beyond this point, spontaneous symmetry breaking occurs. Below it, there is no symmetry breaking. In the physics jargon this transition is called a *continuous phase transition* and the point at which it occurs is called a *critical point*.[10]

Note that, even when the value of C is greater than 1 and we are above the critical point, spontaneous symmetry breaking still only occurs within a certain range of values of B, as Fig. 15.7c shows. If B is either too small or too large then there is only one solution to Eq. (15.69) (the two outer curves in Fig. 15.7c). The portion of parameter space where there are two solutions is called the *coexistence region*. The boundaries of the coexistence region correspond to the values of B such that the curve is tangent to the line $y = p$, as shown in Fig. 15.9. Put another way, we are on the boundary when the point at which the curve has gradient one falls on the line $y = p$. The gradient of $y = \frac{1}{2}[\tanh(B + 2Cp) + 1]$ is given by

$$\frac{dy}{dp} = C \operatorname{sech}^2(B + 2Cp), \qquad (15.70)$$

and setting this equal to one and making use of $\operatorname{sech}^2 x = 1 - \tanh^2 x$, we have

$$1 - \tanh^2(B + 2Cp) = \frac{1}{C}. \qquad (15.71)$$

But p is also a solution of Eq. (15.69), so $\tanh(B + 2Cp) = 2p - 1$ and Eq. (15.71) becomes $1 - (2p - 1)^2 = 1/C$, or

$$p^2 - p + \frac{1}{4C} = 0, \qquad (15.72)$$

---

[10]We encountered continuous phase transitions previously in Chapters 12 and 13—the point at which the giant component first appears in a random graph is a continuous phase transition, although admittedly it is not obvious that there is a connection between the behavior of the giant component and that seen in Fig. 15.7. The study of phase transitions is an intriguing and beautiful branch of physics that has important implications in areas as diverse as superconductivity, elementary particles, and the origin of the universe. Readers interested in learning more are encouraged to consult, for example, the book by Yeomans [331].

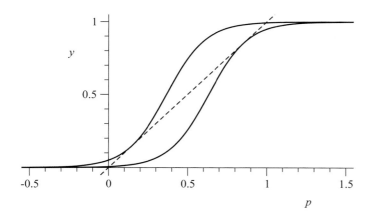

Figure 15.9: The boundaries of the coexistence region in the two-star model. The ends of the coexistence region for a given value of C correspond to those values of B that place the curve $y = \frac{1}{2}[\tanh(B + 2CP) + 1]$ precisely tangent to the line $y = p$.

which has solutions

$$p = \tfrac{1}{2}\left[1 \pm \sqrt{1 - 1/C}\right].\qquad(15.73)$$

Rearranging Eq. (15.69) for B and substituting for p we then find that

$$B = \tanh^{-1}(2p - 1) - 2Cp$$
$$= \pm\tanh^{-1}\sqrt{1 - 1/C} - C\left[1 \pm \sqrt{1 - 1/C}\right],\qquad(15.74)$$

where we either take both the plus signs or both the minus signs.

Figure 15.10 shows a plot of this result in the form of a *phase diagram* of the two-star model showing the different regimes or "phases" of the model as a function of its two parameters. The two lines corresponding to the solutions in Eq. (15.74) form the boundaries of the coexistence region. Inside this region there are values of p than cannot be reached for any choice of parameters. Outside it, we can generate networks with any value of p.

15.2.6 STRAUSS'S MODEL OF TRANSITIVE NETWORKS

As our last example in this chapter, we look at another exponential random graph model that shows spontaneous symmetry breaking, the transitive network model of Strauss [306]. Where the two-star model is something of a toy model—useful for demonstrating the mathematics, but not especially important in practice—the model of this section is one of some importance, and the

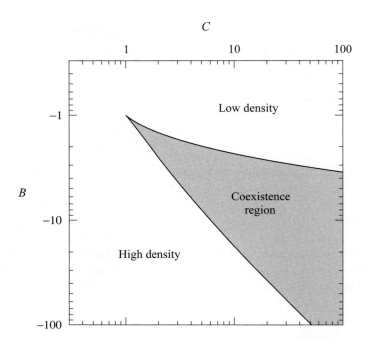

Figure 15.10: Phase diagram of the two-star model. The phases of the two-star model as a function of the parameters B and C. Density generally increases as B becomes more negative and for $C > 1$ there is a coexistence region at intermediate values of B in which spontaneous symmetry breaking occurs. Notice that the scales are logarithmic and that $B < 0$. (There are no other phases for positive B or negative C, so these values are not shown.) Adapted from Park and Newman [260]. Original figure Copyright 2009 American Physical Society. Reproduced with permission.

fact that it shows pathological behavior with the variation of its parameters is a puzzle and a significant hindrance to progress, one that has not, at least at the time of writing, been fully resolved.

Strauss's model is a model of a simple undirected network that shows clustering or transitivity, the propensity for triangles to form in the network, which, as discussed in Section 7.9, is a common phenomenon, particularly in social networks. In this model one specifies the expected number of edges $\langle m \rangle$ in the network and also the expected number of triangles $\langle m_3 \rangle$. The number of triangles can be expressed in terms of the elements of the adjacency matrix as

$$m_3 = \tfrac{1}{3} \sum_{ijk} A_{ij} A_{jk} A_{ki}, \qquad (15.75)$$

where the factor of $\frac{1}{3}$ accounts for the fact that each triangle in the network appears three times in the sum. The number of edges is just $m = \frac{1}{2}\sum_{ij} A_{ij}$. (For simplicity of notation we have included the diagonal terms in these sums. They are zero since the network is simple, so it makes no difference whether we include them or not.) Thus the graph Hamiltonian is

$$H = \tfrac{1}{2}\beta \sum_{ij} A_{ij} + \tfrac{1}{3}\gamma \sum_{ijk} A_{ij} A_{jk} A_{ki}. \qquad (15.76)$$

This model, like the two-star model, can be solved exactly in the limit of large network size using a mean-field technique. The details of the calculation are more complicated than for the two-star model. As well as replacing sums of the form $\sum_k A_{ik}$ by their average value, we also make a similar replacement for sums of the form $\sum_k A_{jk} A_{ki}$, and the values of these two quantities are expressed self-consistently in terms of each other. We will not go into the details of the calculation here—the interested reader is invited to consult Ref. [261]. The end result, however, is similar to that for the two-star model: there is a phase transition in the model beyond which the system develops a coexistence region where there are two distinct solutions to the equations, both of which are realized in simulations of the model. One solution corresponds to a network of high density and the other a network of low density but, as in the two-star case, there is in this regime no choice of model parameters that will give networks of medium density and as a result there is a wide range of networks that simply cannot be generated by this model. If one were to observe a network in the real world whose properties fell within this unattainable range, then Strauss's model could not be used to mimic its properties.

This is a fundamental problem with Strauss's model and with many similar exponential random graphs. The entire point of a model such as this one is to create model networks with properties similar to those seen in real networks. Moreover, this model in particular and exponential random graphs in general seem at first sight to be a very logical approach to the creation of such networks: from a statistical point of view the construction of the model using a maximum entropy ensemble is natural and should, one might imagine, give sensible answers. The fact that it does not is a disturbing finding that is still not properly understood. That there are ranges of network properties that simply cannot be created using the model, while at the same time real-world networks can and do display properties in these ranges, indicates that there is a fundamental flaw or gap in our reasoning, or perhaps in our understand of the nature of networks themselves. Strauss himself was already aware of these issues when he proposed his model in the 1980s, and the fact that they are still unresolved indicates that there are some difficult issues here.

PROBLEMS

15.1 Consider the following variation on the small-world model. Again we have a ring of n vertices in which each is connected to its c nearest neighbors, where c is even. And again a shortcut is added to the network with probability p for each edge around the ring, but now instead of connecting random vertex pairs, each shortcut connects a random vertex to the same single hub vertex in the center of the network:

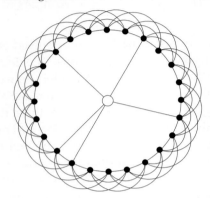

This model could be, for example, a model of a (one-dimensional) world connected together by a bus or train (the central vertex) whose stops are represented by the shortcuts.

Show that the mean distance between two vertices in this network in the limit of large n is $\ell = 2(c^2p+1)/c^2p$ (which is a constant, independent of n).

15.2 One of the difficulties with the original small-world model depicted in Fig. 15.3a is that vertices can become disconnected from the rest of the network by the rewiring process. For instance, a single vertex can become disconnected if all of its incident edges around the ring are rewired and it has no shortcuts.

a) Show that the probability of this happening to any given vertex is $[pe^{-p}]^c$.

b) Hence, how large must the network be before we expect that one vertex will be disconnected, if $c = 6$ and $p = 0.01$?

15.3 Consider an undirected exponential random graph model in which the Hamiltonian takes the form $H = \sum_{i<j} \Theta_{ij} A_{ij}$, where the Θ_{ij} are parameters we control.

a) Derive an expression for the free energy.

b) Hence show that the probability of an edge between vertices i and j is $1/(e^{\Theta_{ij}}+1)$.

15.4 Consider the mean-field solution of the two-star model, as described in Section 15.2.5 for the case $\beta = -\gamma n$, or equivalently $B = -C$ in the notation of Eq. (15.69). Let us define an *order parameter* $x = 2p - 1$.

a) Show that the order parameter obeys the equation $x = \tanh Cx$.

b) Sketch the solutions to this equation as a function of C. Argue that the order parameter must be zero on one side of the phase transition at $C = 1$ but takes non-zero values on the other.

PART V

PROCESSES ON NETWORKS

CHAPTER 16

PERCOLATION AND NETWORK RESILIENCE

A discussion of one of the simplest of processes taking place on networks, percolation, and its use as a model of network resilience

T HE ULTIMATE goal in studying networks is to better understand the behavior of the systems networks represent. For instance, we study the structure of the Internet to understand better how Internet traffic flows or why communications protocols function the way they do or how we could change or rearrange the network to make it perform better. We study biochemical networks like metabolic networks because we hope they will lead to an understanding of the complex chemical processes taking place in the cell or perhaps to algorithmic tools that can help us extract biological insights from the large volumes of data generated by modern laboratory techniques.

Studies of the *structure* of networks, such as those discussed in the previous chapters of this book, are only one step towards this kind of understanding. Another important step is to make the connection between network structure and function: once we have measured and quantified the structure of a network, how do we turn the results into predictions or conclusions about how the overall system will behave? Unfortunately, progress in this area has been far slower than progress on characterizing structure, which is why a majority of this book is devoted to the discussion of structure. Nonetheless, there are some areas in which substantial progress has been made and illuminating theories and models developed. Among these are studies of network failure and resilience, of dynamical systems on networks, and of epidemic and other spreading processes. The remaining chapters of this book are devoted to a description of our current understanding of these and similar network processes. We begin in this chapter with a study of one of the simplest of network processes, percolation, which leads to an elegant theory of the robustness of

591

networked systems to the failure of their components.

16.1   PERCOLATION

Imagine taking a network and removing some fraction of its vertices, along with the edges connected to those vertices—see Fig. 16.1. This process is called *percolation* (or, more precisely, *site percolation*—see below), and can be used as a model of a variety of real-world phenomena. The failure of routers on the Internet, for instance, can be formally represented by removing the corresponding vertices and their attached edges from a network representation of the Internet. In fact, about 3% of the routers on the Internet are non-functional for one reason or another at any one time, and it is a question of some practical interest what effect this will have on the performance of the network. The theory of percolation processes can help us answer this question.

Another example of a percolation phenomenon is the vaccination or immunization of individuals against the spread of disease. As discussed in Chapter 1, and at greater length in Chapter 17, diseases spread through populations over the networks of contacts between individuals. But if an individual is vaccinated against a disease and therefore cannot catch it, then that individual does not contribute to the spread of the disease. Of course, the individual is still present in the network, but, from the point of view of the spread of the disease, might as well be absent, and hence the vaccination process can again be formally represented by removing vertices.

One can see immediately that percolation processes can give rise to some interesting behaviors. The vaccination of an individual in a population, for example, not only prevents that individual from becoming infected but also prevents them from infecting others, and so has a "knock-on" effect in which the benefit of vaccinating one individual is felt by more than one. As we will show, this knock-on effect means that in some cases the vaccination of a relatively small fraction of the population can effectively prevent the spread of disease to anyone, an outcome known as *herd immunity*.

Similar effects crop up in our Internet example, although in that case they are usually undesirable. The removal or failure of a single router on the Internet prevents that router from receiving data, but also prevents data from reaching others via the failed one, forcing traffic to take another route—possibly longer or more congested—or even cutting off some portions of the network altogether. One of the goals of percolation theory on networks is to understand how the knock-on effects of vertex removal or failure affect the network as a whole.

Sometimes it is not the vertices in the network that fail but the edges. For

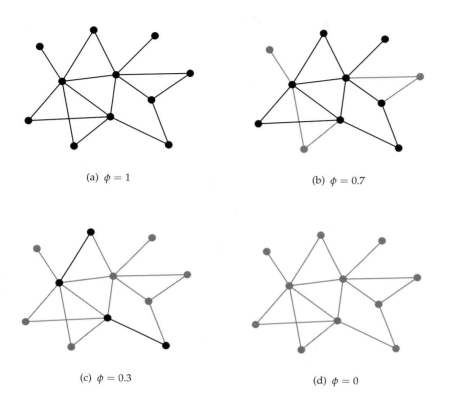

(a) $\phi = 1$ (b) $\phi = 0.7$

(c) $\phi = 0.3$ (d) $\phi = 0$

Figure 16.1: Percolation. A depiction of the site percolation process on a small network for various values of the occupation probability ϕ. Gray denotes vertices that have been removed, along with their associated edges, and black denotes those that are still present. The networks in panels (a) and (b) are above the percolation threshold while those in panels (c) and (d) are below it.

instance, communication lines on the Internet can fail, disconnecting routers from one another, even though the routers themselves are still functioning perfectly. Phenomena like this can be modeled using a slightly different percolation process in which edges rather than vertices are removed from the appropriate network. If we need to distinguish between the two types of percolation process we could refer to them as vertex percolation on the one hand and edge percolation on the other, but in fact they are more commonly called *site percolation* and *bond percolation*, a nomenclature that derives from studies of percolation on low-dimensional lattices in physics and mathematics.[1] In this chapter we will focus principally on site percolation (i.e., removal of vertices) but bond percolation (removal of edges) will become important in Chapter 17 when we look at epidemic processes.

There is more than one way in which vertices can be removed from a network. In the simplest case they could be removed purely at random: we could for example take away some specified fraction of the vertices chosen uniformly at random from the entire network. This is the most commonly studied form of site percolation, and indeed for many people the word "percolation" refers specifically to this particular process. But there are many other ways in which vertices could be removed and "percolation" as used in this chapter is considered to include all of them. One popular alternative removal scheme is to remove vertices according to their degree in some fashion. For instance, we could remove vertices in order of degree from highest to lowest, an approach that turns out to make an effective vaccination strategy for the control of disease. Other approaches have also been considered occasionally, such as removing vertices with high betweenness centrality. Let us begin, however, by examining the simplest case of uniformly random removal.

16.2 UNIFORM RANDOM REMOVAL OF VERTICES

Consider a network in which some fraction of the vertices, selected uniformly at random, are removed. As discussed above, in many real-world situations "removal" does not imply actual physical removal of the vertices, but only that they are non-functional in some way, such as routers that have failed on the Internet, or vaccinated individuals in a network of disease-causing contacts.

Traditionally the percolation process is parametrized by a probability ϕ, which is probability that a vertex is present or functioning in the network. In the parlance of percolation theory, one says that the functional vertices are *occupied* and ϕ is called the *occupation probability*. Thus $\phi = 1$ indicates that all vertices in the network are occupied (i.e., no vertices have been removed) and $\phi = 0$ indicates that no vertices are occupied (i.e., all of them have been removed).[2]

---

[1]If you're interested in the study of percolation in physics the book by Stauffer and Aharony [304] contains a lot of interesting material on the subject, although most of it is not directly relevant to percolation on networks.

[2]In most of the physics literature on percolation the occupation probability is denoted p, but we use ϕ because the letter p is used for many other things in the theory of networks and could cause confusion.

Now look again at Fig. 16.1 and consider panel (a), in which $\phi = 1$, all vertices are present or occupied, and all vertices are connected together into a single component. (The network could have more the one component, but in this example it has only one.) Now look at the other panels. In panel (b) a few vertices have been removed, but those that remain are all still connected together by the remaining edges. In panel (c) still more vertices have been removed, and now so many are gone that the remaining vertices are no longer all connected together, having split into two small components. In the final panel, panel (d), all vertices have been removed and there is no network left at all.

The behavior we see in this small example is typical of percolation processes. When ϕ is large the vertices tend to be connected together, forming a giant component that fills most of the network (although there may be small components also). But as ϕ is decreased there comes a point where the giant component breaks apart and we are left only with small components. Conversely, if we increase ϕ from zero we first form small components, which then grow in size and eventually coalesce to form a giant component that fills a large fraction the network.

The formation or dissolution of a giant component in this fashion is called a *percolation transition*. When the network contains a giant component we say that it *percolates* and the point at which the percolation transition occurs is called the *percolation threshold*.

The percolation transition is similar in many ways to the phase transition in the Poisson random graph at which a giant component forms (see Section 12.5). In the random graph we vary not the fraction of occupied vertices but the probability of connection between those vertices. In both cases, however, when enough of the network is removed the giant component is destroyed and we are left with only small components.

In studies of percolation the "components" that remain after vertices have been removed are in fact usually called *clusters*, another term inherited from the physics and mathematics literature and one that we will use here—it will be useful to distinguish between the "components" of the underlying network and the "clusters" of the percolation process. That is, we will use "component" to refer to connected groups of vertices on the original network before any vertices have been removed and "cluster" to refer to those after removal. The giant component of the percolation process, if there is one, is thus properly called the *giant cluster*.[3]

---

[3]In most of the literature on percolation theory, the giant cluster is called the *spanning cluster*. The reason is that most work on percolation has considered low-dimensional lattices such as the

The percolation transition plays a central role in our interpretation of percolation phenomena. In a network like the Internet, for example, there has to be a giant cluster if the network is to perform its intended function as a communications network. If the network has only small clusters, as in Fig. 16.1c, then every vertex has a connection to, at most, a handful of others and is cut off from everyone else. If there is a giant cluster, on the other hand, then the members of that giant cluster, who are a finite fraction of all vertices in the network, are connected and can communicate with one another, although the remainder of the network is still cut off. Thus the presence of a giant cluster is an indicator of a network that is at least partly performing its intended function, while the size of the giant cluster tells us exactly how much of the network is working.

16.2.1 UNIFORM REMOVAL IN THE CONFIGURATION MODEL

To gain some understanding of the percolation transition and the giant cluster, let us consider the behavior of the site percolation process on networks generated using the configuration model of Chapter 13, a simple but useful model of a network with a specified degree distribution. We can calculate the properties of the giant percolation cluster in the configuration model by a method similar to the one we used for the giant component of configuration model in Section 13.8.

Consider a configuration model network with degree distribution p_k and a percolation process on that network in which vertices are present or occupied with occupation probability ϕ as above. Now consider one of the vertices that is present in the network (i.e., one that has not been removed). If that vertex is to belong to the giant cluster it must be connected to it via at least one of its neighbors. Equivalently, it is not a member of the giant cluster if and only if it is not connected to the giant cluster via any of its neighbors. Following the notation of Section 13.8, let us define u to be the average probability that a vertex is not connected to the giant cluster via a particular neighbor. Then if the vertex in question has degree k, the total probability of its not belonging to the giant cluster is u^k. And if we then average over the probability distribution p_k of the degree we find that the average probability of not being in the giant

---

square lattice. On such lattices the giant cluster is distinguished by being the only cluster that spans the lattice from one side to the other in the limit of large n. There is no equivalent phenomenon for percolation on general networks, however, since networks don't have "sides," so the concept of spanning is not a useful one.

cluster is $\sum_k p_k u^k = g_0(u)$, where

$$g_0(z) = \sum_{k=0}^{\infty} p_k z^k \qquad (16.1)$$

is the generating function for the degree distribution, as defined previously in Eq. (13.48). Then the average probability that a vertex *does* belong to the giant cluster is $1 - g_0(u)$.

Bear in mind, however, that this is for a vertex that is itself assumed not to have been removed from the network. Vertices that have been removed are obviously not members of the giant cluster either. Thus out of all the original vertices in the network the total fraction S that are in the giant cluster is equal to the fraction ϕ that have not been removed times the probability $1 - g_0(u)$ that they are in the giant cluster:

$$S = \phi[1 - g_0(u)]. \qquad (16.2)$$

We still need to calculate the value of u, which is the average probability that a vertex is not connected to the giant cluster via a particular neighboring vertex. There are two ways to not be connected to the giant cluster via a neighbor: either the neighbor in question—let us call it vertex A—has been removed, which happens with probability $1 - \phi$, or it is present (probability ϕ) but it is not itself a member of the giant cluster. The latter happens if A is not connected to the giant cluster via any of its other neighbors. Suppose there are k of these. Then the probability that none of them connects us to the giant cluster is u^k. Adding everything together, the total probability that we are not connected to the giant cluster via A is $1 - \phi + \phi u^k$.

Since A is reached by following an edge, the value of k in this case is distributed according to the excess degree distribution

$$q_k = \frac{(k+1)p_{k+1}}{\langle k \rangle}, \qquad (16.3)$$

(see Section 13.3) where $\langle k \rangle$ is the average degree in the network. Averaging over this distribution, we then arrive at an expression for the average probability u thus:

$$u = \sum_{k=0}^{\infty} q_k(1 - \phi + \phi u^k) = 1 - \phi + \phi \sum_{k=0}^{\infty} q_k u^k$$
$$= 1 - \phi + \phi g_1(u), \qquad (16.4)$$

where

$$g_1(z) = \sum_{k=0}^{\infty} q_k z^k \qquad (16.5)$$

is the generating function for the excess degree distribution, defined previously in Eq. (13.49), and we have made use of the normalization condition $\sum_k q_k = 1$.

Equations (16.2) and (16.4) give us a complete solution for the size of the giant cluster in our network.[4] In practice it is often not possible to solve Eq. (16.4) in closed form, but there is an elegant graphical representation of the solution as follows.

Consider Fig. 16.2a, which gives a sketch of the form of the function $g_1(u)$. The exact form of the curve will depend on the degree distribution, but we know the general shape: g_1 is a polynomial with all coefficients non-negative (because they are probabilities), so it must have a non-negative value and all derivatives non-negative for $u \geq 0$. Thus in general it is an increasing function of u and curves upward as shown in the figure.

To get the function $1 - \phi + \phi g_1(u)$ that appears on the right-hand side of Eq. (16.4) we first multiply $g_1(u)$ by ϕ then add $1 - \phi$. Graphically that is equivalent to compressing the unit square of Fig. 16.2a (along with the curve it contains) until it has height ϕ and then shifting it upward a distance $1 - \phi$ as shown in Fig. 16.2b. The point or points at which the resulting curve crosses the line $y = u$ (dotted line in Fig. 16.2b) are then the solutions to Eq. (16.4).

In Fig. 16.2b there are two such solutions. One is a trivial solution at $u = 1$. This solution always exists because $g_1(1) = 1$ for any correctly normalized excess degree distribution q_k. But there is also a non-trivial solution with $u < 1$, indicated by the dot in the figure. Only if we have such a non-trivial solution can there be a giant cluster in the network and the value of u for this solution gives us the size of the giant cluster via Eq. (16.2). (The $u = 1$ solution gives $S = 0$ in Eq. (16.2) and so doesn't give us a giant cluster.)

Now consider Fig. 16.2d, which shows the equivalent graphical solution of Eq. (16.4) for a smaller value of ϕ. Now the curve of the generating function has been compressed more and the result is that the non-trivial solution for u has vanished. Only the trivial solution at $u = 1$ remains and so in this regime there can be no giant cluster.

Figure 16.2c shows the borderline case between cases (b) and (d). The non-trivial solution for u vanishes at the point shown, where the curve just meets the dotted line. Mathematically this is the point at which the curve is tangent

---

[4]This solution of the percolation problem has a history stretching back some years. In 1961, Fisher and Essam [120] derived a solution for percolation on regular trees (called Cayley trees or Bethe lattices in physics), which is equivalent to the solution given here for the case where every vertex has the same degree. The developments for general degree distributions, however, were not given till some decades later [62, 74].

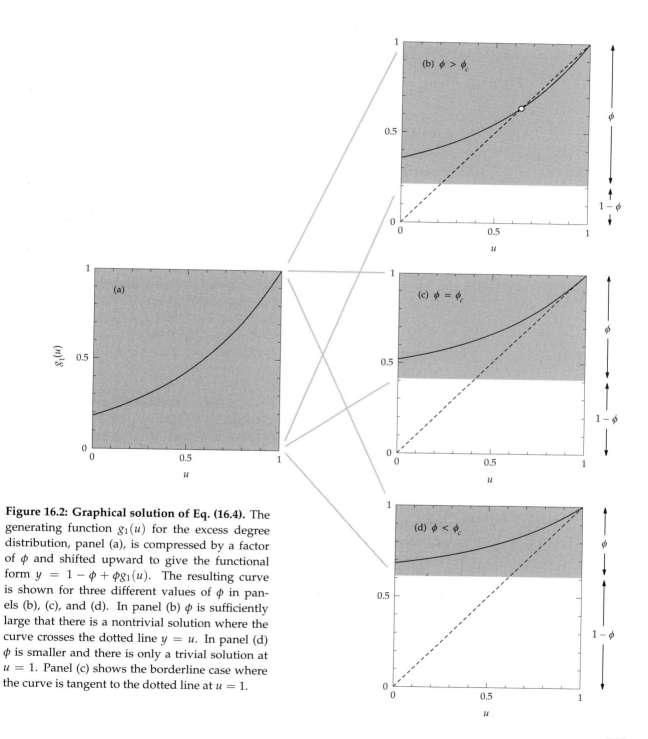

Figure 16.2: Graphical solution of Eq. (16.4). The generating function $g_1(u)$ for the excess degree distribution, panel (a), is compressed by a factor of ϕ and shifted upward to give the functional form $y = 1 - \phi + \phi g_1(u)$. The resulting curve is shown for three different values of ϕ in panels (b), (c), and (d). In panel (b) ϕ is sufficiently large that there is a nontrivial solution where the curve crosses the dotted line $y = u$. In panel (d) ϕ is smaller and there is only a trivial solution at $u = 1$. Panel (c) shows the borderline case where the curve is tangent to the dotted line at $u = 1$.

to the dotted line at $u = 1$, i.e., the point where its gradient at $u = 1$ is 1. In other words the percolation threshold occurs when

$$\left[\frac{d}{du} \left(1 - \phi + \phi g_1(u) \right) \right]_{u=1} = 1. \tag{16.6}$$

Performing the derivative we then find that the value of ϕ at the transition, which we call the *critical value*, denoted ϕ_c, is

$$\phi_c = \frac{1}{g_1'(1)}. \tag{16.7}$$

We can express the critical value more directly in terms of the degree distribution by making use of the definitions of the generating function g_1 and the excess degree distribution, Eqs. (16.3) and (16.5). Substituting one into the other and differentiating, we find that

$$g_1'(1) = \frac{1}{\langle k \rangle} \sum_{k=0}^{\infty} k(k+1) p_{k+1} = \frac{1}{\langle k \rangle} \sum_{k=0}^{\infty} k(k-1) p_k$$
$$= \frac{\langle k^2 \rangle - \langle k \rangle}{\langle k \rangle}, \tag{16.8}$$

and hence the critical occupation probability ϕ_c is given by

$$\phi_c = \frac{\langle k \rangle}{\langle k^2 \rangle - \langle k \rangle}, \tag{16.9}$$

an expression first given by Cohen *et al.* [74].

This equation tells us the minimum fraction of vertices that must be present or occupied in our configuration model network for a giant cluster to exist. Thus, for instance, if we were to consider the configuration model as a simple model of the Internet, we would want to make ϕ_c low, so that the network will have a giant cluster even when some fraction of vertices are non-functional, and hence go on functioning as a communication network. We can arrange this by making sure that $\langle k^2 \rangle \gg \langle k \rangle$ for the network. If, for instance, the network had a Poisson degree distribution,

$$p_k = e^{-c} \frac{c^k}{k!}, \tag{16.10}$$

where c is the mean degree, then $\langle k \rangle = c$ and $\langle k^2 \rangle = c(c+1)$, so

$$\phi_c = \frac{1}{c}. \tag{16.11}$$

Then if we can make c large we will have a network that can withstand the loss of many of its vertices. For $c = 4$, for example, we would have $\phi_c = \frac{1}{4}$, meaning that $\frac{3}{4}$ of the vertices would have to fail before the giant cluster is destroyed. A network that can tolerate the loss of a large fraction of its vertices in this way is said to be *robust* against random failure.

The degree distribution of the Internet, however, is not Poissonian. In fact, as discussed in Section 8.4, the Internet's degree distribution appears roughly to follow a power law with an exponent $\alpha \simeq 2.5$ (see Table 8.1). As we showed in Section 8.4.2, power laws with exponents in the range $2 < \alpha < 3$, which includes most real-world examples, have a finite mean $\langle k \rangle$, but their second moment $\langle k^2 \rangle$ diverges. In this case Eq. (16.9) implies that $\phi_c = 0$. In other words, no matter how many vertices we remove from the network there will *always* be a giant cluster. Scale-free networks—those with power-law degree distributions—are thus highly robust networks that can survive the failure of any number of their vertices, a point first highlighted in the work of Albert *et al.* [14].

In practice, as discussed in Section 8.4.2, the second moment of the degree distribution is never actually infinite in any finite network. Even for finite n though it can still become very large, which can result in non-zero but very small values of ϕ_c, so that the network is still highly robust.

The structure of the real Internet is not the same as that of a configuration model with the same degree distribution. It has all sorts of layers and levels of structure engineered into it, as discussed in Section 2.1. Nonetheless, it does appear to be quite robust to random removal of its vertices. For instance, Albert *et al.* [14] simulated the behavior of the Internet as vertices were randomly removed from its structure and found that performance is hardly affected at all by the removal of even a significant fraction of vertices. (Performance is of course completely destroyed for the vertices that are themselves removed, but for the remaining ones the effects are relatively minor.) These and related results are discussed further in Section 16.3.

Network robustness also plays an important role in the vaccination example mentioned at the start of the chapter. A disease spreading over a contact network between individuals can only reach a significant fraction of the population if there is a giant cluster in the network. If the network contains only small clusters then an outbreak of the disease will be hemmed in by vaccinated individuals and unable to spread further than the small cluster in which it starts. Thus one does not have to vaccinate the entire population to prevent disease spread. One need only vaccinate enough of them to bring the network below its percolation threshold. This is the herd immunity effect mentioned earlier.

In this example, network robustness is a bad thing. The fewer individuals we have to vaccinate to destroy the giant cluster the better. Thus small values of ϕ_c are bad in this case and large values are good. Unfortunately, we usually don't have much control over the degree distributions of contact networks, so we may be stuck with a low value of ϕ_c whether we like it or not. In particular, if the network in question has a power-law (or approximately power-law) degree distribution, then ϕ_c may be very small, implying that almost all vertices have to be vaccinated to wipe out the disease. Some contact networks do indeed appear to have roughly power-law degree distributions [167, 197, 198] and it may be very difficult to eradicate some diseases as a result [264].

It is interesting to ask how the special behavior of power-law networks shows up in the graphical solution of Fig. 16.2. The answer is that, since $g_1'(1)$ is infinite in the power-law case (because $\langle k^2 \rangle$ diverges in Eq. (16.8) while $\langle k \rangle$ remains finite), the curve of $g_1(u)$ has infinite slope at $u = 1$. Thus $g_1(u)$ must look something like Fig. 16.3. Because of the infinite slope, it makes no difference how much we compress the function (as in Fig. 16.2)—the curve will always drop below the line of $y = u$ before coming back up again and crossing it to give a non-trivial solution for u.

The position of the percolation threshold is not the only quantity important in assessing the robustness of a network. The size of the giant cluster also plays a role because it tells us what fraction of the network will be connected and functional. To find the size of the giant cluster we need to solve Eq. (16.4) for u and then substitute the result back into Eq. (16.2). In many cases, as we have said, we cannot solve for u exactly, but in some cases we can. Consider, for example, a network with an exponential degree distribution given by

$$p_k = (1 - e^{-\lambda}) e^{-\lambda k}, \qquad (16.12)$$

where $\lambda > 0$ and the leading factor of $1 - e^{-\lambda}$ insures that the distribution is properly normalized. Then, as shown in Section 13.9.2, we have

$$g_0(z) = \frac{e^{\lambda} - 1}{e^{\lambda} - z}, \qquad g_1(z) = \left(\frac{e^{\lambda} - 1}{e^{\lambda} - z} \right)^2, \qquad (16.13)$$

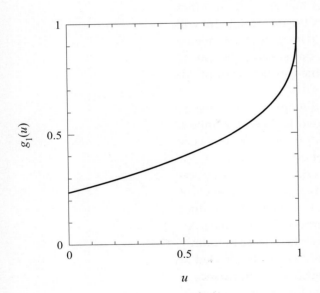

Figure 16.3: Generating function for the excess degree distribution in a scale-free network. The generating function $g_1(u)$ for a network with a power-law degree distribution has a derivative that diverges as $u \to 1$, though the value of the generating function remains finite and tends to 1 in this limit. Thus the function looks generically like the curve sketched here.

and Eq. (16.4) becomes

$$u(e^\lambda - u)^2 - (1 - \phi)(e^\lambda - u)^2 - \phi(e^\lambda - 1)^2 = 0. \tag{16.14}$$

This is a cubic equation, which is ugly (though not impossible) to solve. In this case, however, we don't have to solve it directly. We observe instead that $u = 1$ is always a solution of Eq. (16.4) and hence that our cubic equation must contain a factor of $u - 1$. A few moments work reveals that indeed this is the case. Equation (16.14) factorizes as

$$(u - 1)\left[u^2 + (\phi - 2e^\lambda)u + \phi - 2\phi e^\lambda + e^{2\lambda}\right] = 0. \tag{16.15}$$

Thus the two other solutions for u satisfy the quadratic equation

$$u^2 + (\phi - 2e^\lambda)u + \phi - 2\phi e^\lambda + e^{2\lambda} = 0. \tag{16.16}$$

Of these two solutions one is greater than one for $\lambda > 0$ and so cannot be our probability u. The other is

$$u = e^\lambda - \tfrac{1}{2}\phi - \sqrt{\tfrac{1}{4}\phi^2 + \phi(e^\lambda - 1)}. \tag{16.17}$$

Now we can plug this value back into Eq. (16.2) to get an expression for the size of the giant cluster as a fraction of the whole network:

$$\begin{aligned}
S &= \phi\left[1 - \frac{2(e^\lambda - 1)}{\phi + \sqrt{\phi^2 + 4\phi(e^\lambda - 1)}}\right] \\
&= \phi\left[1 - 2(e^\lambda - 1)\frac{\phi - \sqrt{\phi^2 + 4\phi(e^\lambda - 1)}}{\phi^2 - (\phi^2 + 4\phi(e^\lambda - 1))}\right] \\
&= \tfrac{3}{2}\phi - \sqrt{\tfrac{1}{4}\phi^2 + \phi(e^\lambda - 1)}.
\end{aligned} \tag{16.18}$$

Notice that the solution for u, Eq. (16.17), can become greater than 1 for sufficiently small ϕ, which is unphysical. In this regime the only acceptable solution is the trivial $u = 1$ solution, which gives $S = 0$ and so there is no giant cluster when this happens. This gives us an alternative way to derive the position of the percolation transition. The transition takes place at the point where Eq. (16.17) equals one, i.e., when

$$e^\lambda - 1 - \tfrac{1}{2}\phi = \sqrt{\tfrac{1}{4}\phi^2 + \phi(e^\lambda - 1)}. \tag{16.19}$$

Squaring both sides and rearranging for ϕ we find that the percolation threshold falls at

$$\phi_c = \tfrac{1}{2}(e^\lambda - 1). \tag{16.20}$$

It is left as an exercise to demonstrate that this is the same result we get if we apply the general formula, Eq. (16.7).

Note also that if λ becomes sufficiently large then the value of ϕ_c given by Eq. (16.20) can become greater than one. For values of λ this large there is no percolation transition and the system never percolates because ϕ can never be greater than ϕ_c. The value of λ at which we enter this regime is the value at which $\frac{1}{2}(e^\lambda - 1) = 1$, which gives $\lambda = \ln 3$. Upon closer inspection, it turns out that this is precisely the point at which the network itself loses its giant *component*,[5] which explains why percolation is not possible beyond this point. For $\lambda > \ln 3$ the network has no giant component, and hence it is not possible to have a giant cluster even if every vertex in the network is present. (A similar result of course applies to all networks—a giant percolation cluster is never possible in a network without a giant component.)

Figure 16.4 shows a plot of the value of S for our exponential network with $\lambda = \frac{1}{2}$ as a function of ϕ. For small ϕ there is a region in which there are only small clusters and no giant cluster. When we pass through the percolation transition, marked by the dotted line in the figure, a giant cluster appears and grows smoothly from zero as ϕ increases. This is an example of what a physicist would call a continuous phase transition.[6] We saw other examples in Sections 13.9 and 15.2.5.

The overall behavior shown in Fig. 16.4 is typical of percolation in networks. For most degree distributions we expect S to take a similar form with a continuous phase transition, as we can demonstrate by the following argument. Suppose the generating function $g_1(u)$ is well-behaved near $u = 1$, having all its derivatives finite,[7] then we can expand it about this point as

$$g_1(u) = g_1(1) + (u-1)g_1'(1) + \tfrac{1}{2}(u-1)^2 g_1''(1) + \mathrm{O}(u-1)^3$$

$$= 1 + \frac{u-1}{\phi_c} + \tfrac{1}{2}(u-1)^2 g_1''(1) + \mathrm{O}(u-1)^3, \tag{16.21}$$

where we have made use of $g_1(1) = 1$ (see Eq. (13.20)) and Eq. (16.7). Substi-

---

[5]From Eq. (13.101) we know that a configuration model network has a giant component if and only if $g_1'(1) > 1$, and thus loses its giant component at the point where $g_1'(1) = 1$. Substituting from Eq. (16.13), our network loses its giant component when $2/(e^\lambda - 1) = 1$, i.e., when $\lambda = \ln 3$. See also Problem 13.3 on page 484 for another derivation of this result.

[6]A phase transition is *continuous* if the quantity of interest, also called the *order parameter* (S in this case), is zero on one side of the transition and non-zero on the other, but its value is continuous at the transition itself. The alternative to a continuous phase transition is a *first-order phase transition*, in which the order parameter jumps discontinuously as it crosses the transition point.

[7]This excludes the power-law case shown in Fig. 16.3, which is discussed separately below.

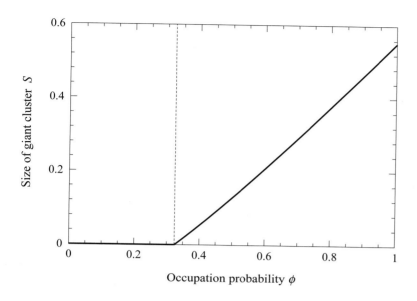

Figure 16.4: Size of the giant cluster for site percolation in the configuration model.
The curve indicates the size of the giant cluster for a configuration model with an exponential degree distribution of the form (16.12) with $\lambda = \frac{1}{2}$, as given by Eq. (16.18). The dotted line indicates the position of the percolation transition, Eq. (16.20).

tuting into Eq. (16.4), we then find that

$$u = 1 + \frac{\phi}{\phi_c}(u - 1) + \frac{1}{2}\phi(u - 1)^2 g_1''(1) + O(u - 1)^3 \qquad (16.22)$$

or

$$u - 1 = \frac{2}{g_1''(1)}\frac{\phi_c - \phi}{\phi_c\phi} + O(u - 1)^2. \qquad (16.23)$$

We can similarly expand $g_0(u)$ as

$$g_0(u) = g_0(1) + (u - 1)g_0'(1) + O(u - 1)^2$$
$$= 1 + \frac{2\langle k \rangle}{g_1''(1)}\frac{\phi_c - \phi}{\phi_c\phi} + O(\phi - \phi_c)^2, \qquad (16.24)$$

where we have used $g_0(1) = 1$ and Eqs. (13.22) and (16.23). Substituting into Eq. (16.2) then gives us

$$S = \frac{2\langle k \rangle}{g_1''(1)}\frac{\phi - \phi_c}{\phi_c} + O(\phi - \phi_c)^2, \qquad (16.25)$$

In other words, S varies linearly with $\phi - \phi_c$ just above the percolation transition, going to zero continuously as we approach the transition from above. Thus we would expect the percolation transition for essentially all degree distributions to look generically like the curve in Fig. 16.4, with a continuous phase transition as we pass the percolation threshold.[8]

This result is important, because it implies that the giant cluster becomes very small as we approach the percolation transition from above. In other words, the network may be "functional" in the sense of having a giant cluster, but the functional portion of the network is vanishingly small. If the network is a communication network, for example, then a finite fraction of all the vertices in the network can communicate with one another so long as there is a giant cluster, but that fraction becomes very small as we approach the percolation threshold, meaning that in practice most vertices are cut off. Thus one could argue that it is misleading to interpret the percolation threshold as the point where the network stops functioning: in effect most of it has stopped functioning before we reach this point. To fully describe the functional state of the network one should specify not only whether it contains a giant cluster but also what the size of that cluster is.

It is also important to note that the sharp percolation transition of Fig. 16.4 is only truly seen in an infinite network. For networks of finite size—which is all real networks, of course—the transition gets rounded off. To see this, consider the behavior of the giant cluster in a finite-sized network. Technically, in fact, there *is* no giant cluster for an individual finite network. The proper definition of the giant cluster, like the giant component in a random graph, is as a cluster whose size scales in proportion to the size of the network (see Section 12.5). But it makes no sense to talk about the scaling of a cluster with network size when the size of the network is fixed. In practice, therefore, we normally consider instead the largest cluster, which is a reasonable proxy for the giant cluster in a finite-size network. Its size as a fraction of the size of the network should be a reasonable approximation to the size of the giant cluster given by our theory when we are above the percolation transition.

Below the transition the largest cluster will be small in size, but not zero, and hence fills a small but non-zero fraction of the network, in rough but not perfect agreement with the theoretical prediction $S = 0$. Furthermore, this non-zero value grows as we approach the transition point because small clusters in general, including the largest one, grow as the occupation probability ϕ increases. The net result is a slight rounding of the sharp transition predicted

---

[8]To be more precise, the transition is a *second-order* transition—one where the order parameter is continuous at the transition but its derivative is not.

by the theory, which is often visible, for example, in computer simulations of percolation on smaller networks. Effects such as this that show up only in finite-sized systems are known as *finite size effects*.

Even in the limit of large network size there are exceptions to the behavior of Fig. 16.4 and Eq. (16.25). Consider a network with a power-law degree distribution with exponent $2 < \alpha < 3$, as discussed above. In this case our assumption that the derivatives of g_1 are finite does not hold (see Fig. 16.3 and the accompanying discussion), so the argument above breaks down. Not only does the percolation threshold fall at $\phi_c = 0$ for power-law networks, but the giant cluster does not grow linearly as ϕ increases. In general it will grow slower than linearly, the exact functional form depending on the shape of $g_1(u)$ near $u = 1$. For example, a typical form is

$$1 - g_1(u) = c(1 - u)^\beta, \tag{16.26}$$

near $u = 1$ with c and β positive constants. Provided $\beta < 1$ this makes the gradient of $g_1(u)$ (and all higher derivatives) infinite at $u = 1$ while still ensuring that $g_1(1) = 1$. With this form for $g_1(u)$, Eq. (16.4) implies

$$1 - u = (c\phi)^{1/(1-\beta)}. \tag{16.27}$$

Then[9]

$$g_0(u) \simeq g_0(1) + g_0'(1)(u - 1) = 1 + \langle k \rangle (u - 1), \tag{16.28}$$

close to $u = 1$, with $\langle k \rangle$ finite so long as the power-law exponent $\alpha > 2$, and hence the giant cluster has size

$$S = \phi[1 - g_0(u)] \simeq \phi \langle k \rangle (1 - u) \sim \phi^{(2-\beta)/(1-\beta)}, \tag{16.29}$$

which goes to zero faster than linearly[10] as $\phi \to 0$ since $(2 - \beta)/(1 - \beta) > 1$ if $\beta < 1$.

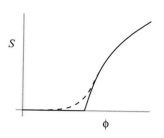

The phase transition at which the giant cluster appears is only sharp in an infinite system (solid line). In a finite sized system it gets rounded off (dashed line).

---

[9]If we want to be more careful and keep track of the correction terms we can make use of Eq. (13.51) and integrate Eq. (16.26) to show that $g_0(u) = 1 - \langle k \rangle (1 - u) + c(1 - u)^{\beta+1}/(\beta + 1)$. The last term vanishes faster than those before it as $u \to 1$ because $\beta > 0$ and hence $g_0(u) \simeq 1 - \langle k \rangle (1 - u)$. This is at first slightly surprising—one would imagine that the correction term ought to be $O(1 - u)^2$—but this type of behavior is common with power-law distributions.

[10]To the extent that one can regard a power-law network as having a percolation transition at $\phi = 0$ it is interesting to ask what the order of this transition is. The answer is unclear since Eq. (16.29) doesn't perfectly fit the standard forms for continuous phase transitions. If we define a transition to be second-order if the order parameter is continuous at the transition and third-order if its derivative is continuous, then the transition is third-order in this case. But one could also argue that the transition is of fractional order between two and three since it varies from zero as a fractional power of the occupation probability ϕ.

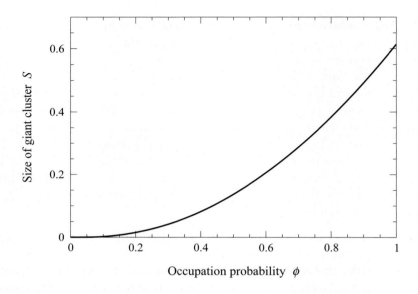

Figure 16.5: Size of the giant cluster for a network with power-law degree distribution. The size of the giant cluster for a scale-free configuration model network with exponent $\alpha = 2.5$, a typical value for real-world networks. Note the non-linear form of the curve near $\phi = 0$, which means that S, while technically non-zero, becomes very small in this regime. Contrast this figure with Fig. 16.4 for the giant cluster size in a network with an exponential degree distribution.

Thus we expect the giant cluster to become very small as $\phi \rightarrow 0$. Figure 16.5 shows the equivalent of Fig. 16.4 for a scale-free network with exponent $\alpha = 2.5$, derived from numerical solutions of Eqs. (16.2) and (16.4) and the non-linear form of S close to $\phi = 0$ is clear.

This result mitigates somewhat our earlier statement that scale-free networks are highly robust because $\phi_c = 0$. It is true that the percolation threshold is zero in these networks and hence that there is a giant cluster for any positive ϕ, but that giant cluster can become exceedingly small. A communication network with a power-law degree distribution, for instance, might be formally functional for very small values of ϕ, but in practice the fraction of vertices that could communicate with one another would be so small that the network would probably not be of much use.

16.3 NON-UNIFORM REMOVAL OF VERTICES

In the first part of this chapter we have considered percolation phenomena in the case where vertices are removed from a network uniformly at random. This is the classical form of percolation long studied by physicists and mathematicians. When discussing networks, however, it is interesting also to consider other ways in which vertices might be removed. In Section 16.1, for example, we mentioned the possibility of removing vertices in order of their degrees, starting with the highest degrees and working down. This might be effective, for example, as a vaccination strategy for preventing the spread of disease: should they become infected, the high degree vertices in the network clearly present a disease risk to their many neighbors, so perhaps vaccinating them first would be a sensible approach.

Let us consider a generalization of our percolation process in which the occupation probability of a vertex can now depend on its degree. We define ϕ_k to be the probability that a vertex with degree k is present or occupied in our network. If ϕ_k is a constant, independent of k, then we recover the uniform scenario of previous sections. On the other hand, if $\phi_k = 1$ for all vertices with degree $k < k_0$ for some constant k_0, and $\phi_k = 0$ for all vertices with $k \geq k_0$, then we effectively remove from the network all vertices with degree k_0 or greater. A host of other choices are also possible, resulting in more complex removal patterns.

Let us again look at percolation on configuration model networks and as before define u to be the average probability a vertex is not connected to the giant cluster via one of its neighbors. If the vertex has degree k then the probability that it is not connected to the giant cluster via any of its neighbors is u^k and the probability that it *is* connected to the giant cluster is $1 - u^k$. But in order to belong to the giant cluster, the vertex itself must also be present, which happens with probability ϕ_k, so the probability of it being a member of the giant cluster is $\phi_k(1 - u^k)$.

Now we average over the probability distribution p_k of the degree to find the average probability of being in the giant cluster and get

$$S = \sum_{k=0}^{\infty} p_k \phi_k (1 - u^k) = \sum_{k=0}^{\infty} p_k \phi_k - \sum_{k=0}^{\infty} p_k \phi_k u^k$$
$$= f_0(1) - f_0(u), \tag{16.30}$$

where

$$f_0(z) = \sum_{k=0}^{\infty} p_k \phi_k z^k. \tag{16.31}$$

Notice that this new generating function is not normalized in the conventional fashion—the value $f_0(1)$ that appears in Eq. (16.30) is not in general equal to one. Instead it is given by

$$f_0(1) = \sum_{k=0}^{\infty} p_k \phi_k = \overline{\phi}, \qquad (16.32)$$

which is the average probability that a vertex is occupied.

We can calculate the value of u using an approach similar to that for the uniform percolation scenario. The value of u is the probability that you are not connected to the giant cluster via your neighbor, which happens if either the neighbor is not occupied or if it is occupied but it is not connected to the giant cluster via any of its other neighbors. Let k now be the excess degree of the neighboring vertex. Then the probability that the neighbor is not occupied is $1 - \phi_{k+1}$. Notice that the index is $k + 1$ because ϕ_k is defined in terms of the total degree of a vertex, which is one greater than the excess degree (see Section 13.3). The probability that the neighbor *is* occupied but is itself not connected to the giant cluster is $\phi_{k+1} u^k$. Adding up the terms and averaging over the distribution q_k of the excess degree, we then find that

$$u = \sum_{k=0}^{\infty} q_k(1 - \phi_{k+1} + \phi_{k+1} u^k) = 1 - f_1(1) + f_1(u), \qquad (16.33)$$

where

$$f_1(z) = \sum_{k=0}^{\infty} q_k \phi_{k+1} z^k \qquad (16.34)$$

and we have used $\sum_k q_k = 1$.

Like $f_0(z)$, the function $f_1(z)$ is not normalized to unity. The definition of $f_1(z)$ looks slightly odd because of the subscript $k + 1$. If we prefer we can write it using the full expression for the excess degree distribution, Eq. (16.3), which gives

$$f_1(z) = \frac{1}{\langle k \rangle} \sum_{k=0}^{\infty} (k+1) p_{k+1} \phi_{k+1} z^k$$

$$= \frac{1}{\langle k \rangle} \sum_{k=1}^{\infty} k p_k \phi_k z^{k-1}, \qquad (16.35)$$

which has a more symmetric look about it. Note also that

$$f_1(z) = \frac{f_0'(z)}{g_0'(1)}, \qquad (16.36)$$

where $g_0(z)$ is defined as before. This expression can be useful for calculating $f_1(z)$ once $f_0(z)$ has been found.

Equations (16.30) and (16.33), which were first given by Callaway *et al.* [62], give us a complete solution for the size of the giant cluster for our generalized percolation process.

As an example of their use, consider again a network with exponential degree distribution given by Eq. (16.12) and suppose we remove all vertices that have degree k_0 or greater. That is, we choose

$$\phi_k = \begin{cases} 1 & \text{if } k < k_0, \\ 0 & \text{otherwise.} \end{cases} \qquad (16.37)$$

Then we have

$$f_0(z) = (1 - e^{-\lambda}) \sum_{k=0}^{k_0-1} e^{-\lambda k} z^k = (1 - e^{-\lambda k_0} z^{k_0}) \frac{e^\lambda - 1}{e^\lambda - z}, \qquad (16.38)$$

and

$$f_1(z) = \frac{f_0'(z)}{g_0'(1)}$$

$$= \left[(1 - e^{-\lambda k_0} z^{k_0}) - k_0 e^{-\lambda(k_0-1)} z^{k_0-1} (1 - e^{-\lambda} z) \right] \left(\frac{e^\lambda - 1}{e^\lambda - z} \right)^2. \qquad (16.39)$$

For this choice Eq. (16.33) becomes a polynomial equation of order k_0 and unfortunately such equations are not solvable exactly for their roots (unless $k_0 \leq 4$). It is, however, fairly easy to find the roots numerically, especially given that we know that the root of interest in this case lies in the range between zero and one, and then we can calculate the size of the giant cluster from (16.30).

Figure 16.6a shows the results of such a calculation, plotted as a function of k_0. Looking at this figure, consider what happens as we lower k_0 from an initial high value, effectively removing more and more of the high-degree vertices in our network. As the figure shows, the size of the giant cluster decreases only slowly at first. This is because there are not many vertices of very high degree in the network, so very few have been removed. Once k_0 passes a value around 10, however, our attack on the network starts to become evident in a shrinking of the giant cluster, which becomes progressively more rapid until the size of the cluster reaches zero around $k_0 = 5$.

One might be forgiven for thinking that Fig. 16.6a portrays a network quite resilient to the removal of even its highest-degree vertices: it appears that we have to remove vertices all the way down to degree five in order to break up

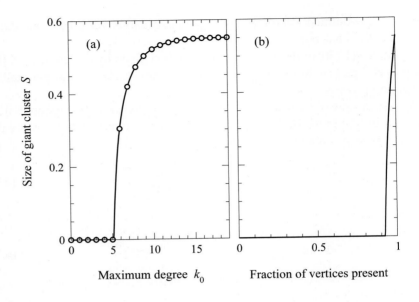

Figure 16.6: Size of the giant percolation cluster as the highest degree vertices in a network are removed. (a) The size of the giant cluster in a network with an exponential degree distribution $p_k \sim e^{-\lambda k}$ with $\lambda = \frac{1}{2}$ as vertices are removed in order of degree, starting from those with the highest degree. The curve is shown as a function of the degree k_0 of the highest-degree vertex remaining in the network. Technically, since k_0 must be an integer, the plot is only valid at the integer points marked by the circles; the curves are just an aid to the eye. (b) The same data plotted now as a function of the fraction $\overline{\phi}$ of vertices remaining in the network.

the giant cluster. This impression is misleading, however, because it fails to take account of the fact that the vast majority of vertices in the network are of very low degree, so that even when we have removed all vertices with degree greater than five, we have still removed only a small fraction of all vertices.

Perhaps a more useful representation of the solution is to plot it as a function of the fraction $\overline{\phi}$ of occupied vertices in the network, which is

$$\overline{\phi} = f_0(1) = 1 - e^{-\lambda k_0}. \tag{16.40}$$

Figure 16.6b shows the result replotted in this way and reveals that the giant cluster in fact disappears completely when only about 8% of the highest-degree vertices in the network have been removed. By contrast, when we removed vertices uniformly at random, as shown in Fig. 16.4, we had to remove nearly 70% of the vertices to destroy the giant cluster. Though the difference

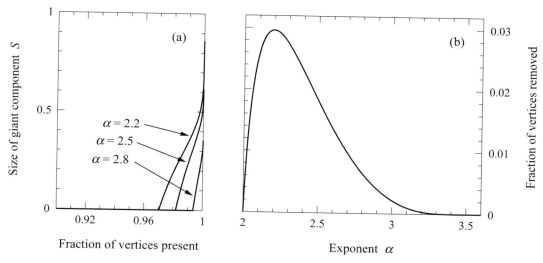

Figure 16.7: Removal of the highest-degree vertices in a scale-free network. (a) The size of the giant cluster in a configuration model network with a power-law degree distribution as vertices are removed in order of their degree, starting with the highest-degree vertices. Only a small fraction of the vertices need be removed to destroy the giant cluster completely. (b) The fraction of vertices that must be removed to destroy the giant cluster as a function of the exponent α of the power-law distribution. For no value of α does the fraction required exceed 3%.

is startling, however, it is also intuitively reasonable. The high-degree vertices have a lot of connections, all of which are lost if we remove those vertices.

These results suggest, for example, that were we able to find the highest degree vertices in a network of disease-causing contacts and vaccinate them to effectively remove them from the network, it would be a much more efficient strategy for disease control than simply vaccinating at random.

A particularly striking example of the effect described here arises in networks with power-law degree distributions. In these networks, as we have seen, uniform removal of vertices *never* destroys the giant cluster, provided the exponent of the power-law lies between two and three. By contrast, removal of the highest-degree vertices in these networks has a devastating effect. Once again we cannot solve for S in closed form in the power-law case but it is reasonably straightforward to perform a numerical solution. Figure 16.7a shows the equivalent of Fig. 16.6b for the power-law case, and as we can see the giant cluster disappears extraordinarily rapidly as the high-degree vertices are removed. Only a few percent of the vertices need be removed to completely destroy the giant cluster, the exact value depending on the exponent of the power law.

Indeed, if we want to calculate only the fraction that need be removed to destroy the giant cluster, we can do so by observing once again that the phase transition at which the giant cluster appears or disappears falls at the point where the non-trivial solution of Eq. (16.33) appears or disappears, which is the point at which the right-hand side of the equation is tangent to the line $y = u$ at $u = 1$. That is, the general criterion for the transition point is

$$f_1'(1) = 1. \tag{16.41}$$

(Alternatively, we could say that the giant cluster exists if and only if $f_1'(1) > 1$.) Again, exact solutions are often not possible but we can solve numerically. Doing this for the power-law case we find the results shown in Figure 16.7b, which plots the fraction of vertices that need be removed to destroy the giant cluster as a function of the exponent α. As we can see, the curve peaks around $\alpha = 2.2$ at a value just below 3%. Thus in no case need we remove more than 3% to destroy the connectivity in the network.

Scale-free networks are thus paradoxically both robust and fragile, a point first emphasized by Albert *et al.* [14]. On the one hand, they are remarkably robust to the random failure of their vertices, with the giant cluster persisting no matter how many vertices we remove. (Although one should bear in mind the *proviso* of Section 16.2.1 that the size of the giant cluster matters also, and this becomes very small when the fraction ϕ of occupied vertices tends to zero.) On the other hand, scale-free networks are very fragile to attacks targeted specifically at their highest-degree vertices. We need remove only the tiniest fraction of the high-degree hubs in such a network to entirely destroy the giant cluster.

The fragility of scale-free network to such targeted attack is both bad news and good news. Some networks we wish to defend against possible attack. The Internet is an example: a communication network that can easily be brought down by a malicious adversary targeting just a few of its most crucial vertices may be a disaster waiting to happen.

On the other hand, results like these could also be exploited to help eradicate or reduce disease by targeting vaccination efforts at network hubs. It is worth noting, however, that it's not necessarily easy to find the hubs in a network, so that implementation of a targeted vaccination strategy may be difficult. In most cases one does not know the entire network and so cannot simply pick out the high-degree vertices from a list.

One intriguing way of getting around this problem has been put forward by Cohen *et al.* [76], who suggest that we make use of the structure of the network itself to find the high-degree vertices. In their scheme, which they call "acquaintance immunization," they propose that one choose members of the population at random and then get each of them to nominate an acquaintance. Then that acquaintance receives a vaccination against the disease under consideration. The acquaintance in this scenario is a "vertex at the end of an edge," so in the configuration model it would have degree distributed according to the excess degree distribution, Eq. (13.46), rather than the original degree distribution of the network. But the excess degree distribution, as discussed in Section 13.3, is biased towards high-degree vertices since there are more edges that end at a high-degree vertex than at a low-degree one. Thus the selection of individuals in the scheme of Cohen *et al.* is also biased towards those with high degree. The selected individuals are not guaranteed to be the highest-degree vertices in the network, but we are a lot more likely to find the hubs this way than if we just choose vertices at random and in simulations the acquaintance immunization scheme appears to work quite well.

The acquaintance immunization scheme does have some drawbacks. First, contact networks in the real world are of course not configuration models and it is unclear how accurately the theoretical results describe real situations. Second, real contact networks mostly don't have power-law degree distributions, instead having somewhat shorter tails than the typical power law, which will reduce the effectiveness of the scheme, or indeed of any scheme based on targeting the highly connected vertices. Another issue is that, in asking people to name their acquaintances, the acquaintance immunization scheme necessarily probes the network of who is acquainted with whom, which is in general not the same as the network of disease transmission, since people who are acquainted don't necessarily have regular physical contact of the type necessary to spread disease and because diseases can be and often are transmitted between people who don't know one another. We can do our best to make the networks similar, asking participants to name only acquaintances whom they have seen recently and in person, rather than those they might not have seen for a while or might only have to talked to on the phone. Still, the differences between the two networks means that the scheme might end up focusing vaccination efforts on the wrong set of people.

16.4 PERCOLATION IN REAL-WORLD NETWORKS

Having seen how percolation plays out in model networks, let us now take a look at some real ones. If we have data on the structure of a network then we can simulate the percolation process on a computer, removing vertices one by one and examining the resulting clusters. Although this is straightforward in theory, it requires some care to get good results in practice. The main issue is that the percolation process is normally a random one: the vertices are removed in random order, which means that the cluster sizes can vary depending on the precise order we choose. Even in the case where vertices are removed in decreasing order of their degree the process is still random to some extent since there can be many vertices with a given degree, among which we must choose somehow. To avoid possible biases, we usually choose among them at random.

This randomness can easily be simulated on a computer using standard random number generators, but the results of the simulation will then vary from one run of our simulation to another depending on the output of the generator. To get a reliable picture of how percolation affects a network we must perform the entire calculation many times, removing the vertices in different random orders each time, so that we can see what the typical behavior is, as well as the range of variation around that typical behavior. And this in turn means that we need to be able to perform the percolation calculation quickly. In a typical situation we might want to repeat the percolation calculation a thousand times with different random orders of removal and even if each calculation took just one minute of computer time, all thousand runs would still take a day.

If we are crafty, however, we can do much better than this and get an answer in just a few seconds for networks of the typical sizes we have been considering in this book.

16.5   COMPUTER ALGORITHMS FOR PERCOLATION

The simplest way to simulate the percolation process on a computer is to make use of the breadth-first search algorithm of Section 10.3.4, which can find all components in a network in time $O(m + n)$, where m is the total number of edges in the network and n is the total number of vertices, or just $O(n)$ for a sparse network in which $m \propto n$. If we remove a certain randomly chosen set of vertices from a network, along with the edges attached to them, then the resulting percolation clusters are by definition the components of the network that remains, and hence we can use the component-finding algorithm to find the clusters. Then we can, for example, look through those clusters until we find the largest one.

In the case of uniformly random removal of vertices, for instance, we would go through each vertex in turn, removing it (and its edges) from the network with probability $1 - \phi$, finding the clusters, and (say) measuring the size of the largest one. Then we repeat the entire calculation, starting with the complete network again, removing a different set of vertices, and finding the clusters. Repeating the calculation a large number of times, we can calculate a mean value $S(\phi)$ for the size of the largest component when vertices are present or functioning with probability ϕ.

If we are interested in only a single value of ϕ, this is, in fact, the best algorithm to use and the fastest known way of getting an answer. Usually, however, we are interested, as in previous sections, in the behavior of the system over the whole range of ϕ from zero to one, or at least some portion of that range. In that case, we would have to repeat the whole calculation above for many values of ϕ in the range of interest and this process is time-consuming and is not the best way to approach the problem.

Consider instead the following alternative approach, which appears at first to be only a slight variation on the previous one, but leads, as we will see, to much more efficient algorithms. Instead of making each vertex in the network occupied with independent probability ϕ, let us make a fixed number r of vertices occupied, repeating the calculation many times for a given value of r and averaging to get a figure S_r for the size of the largest component (or any other quantity of interest) as a function of r.

The calculation doesn't directly give us the result we want: S_r is not the same as $S(\phi)$ and it is the latter we are interested in. If, however, we know the value of S_r for every allowed value of the integer r, i.e., from 0 to n, then we can calculate $S(\phi)$ as follows. If each vertex in the network is occupied with probability ϕ, then the probability that there are exactly r vertices occupied is given by the binomial distribution

$$p_r = \binom{n}{r} \phi^r (1 - \phi)^{n-r}. \tag{16.42}$$

Averaging over this distribution, the average size of the largest component as a function of ϕ is then

$$S(\phi) = \sum_{r=0}^{n} p_r S_r = \sum_{r=0}^{n} \binom{n}{r} \phi^r (1 - \phi)^{n-r} S_r. \tag{16.43}$$

At first sight, this appears to be a less promising approach for calculating $S(\phi)$ than the previous approach. To make use of Eq. (16.43) we need to know S_r for all r and it takes time $O(m + n)$ to calculate S_r for one value of r using breadth-first search, so it is going to take $O(n(m + n))$ to calculate for all n values, or $O(n^2)$ on a sparse network. Given that we also need to perform each calculation of S_r many times to average over the randomness, the entire process could take a very long time to complete.

There is however a faster way to calculate S_r for all r, inspired by the simple observation that if we have already found all the clusters in a network with r vertices present, then we can find the clusters with $r + 1$ vertices simply by adding one more vertex. Most of the clusters do not change very much when

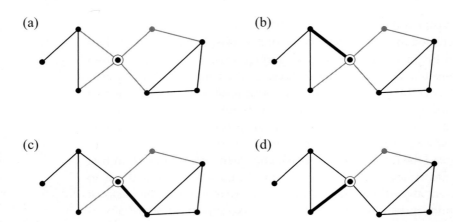

Figure 16.8: Percolation algorithm. In the percolation algorithm described in the text we add vertices to our network one by one, rather than taking them away. Each addition consists of several steps. (a) We add the vertex itself but none of its accompanying edges yet. At this stage the vertex constitutes a new cluster in its own right. (b) We start adding the accompanying edges (if any) in any order we like. Only edges that connect to other vertices already present in the network are added. The first edge added (if any are added) will thus, by definition, always join the new vertex to one of the previously existing clusters. Or to put it another way, it will join two clusters together, one of the old clusters and the new cluster that consists of just the single added vertex. (c) In this example the next edge added also joins two clusters together. (d) The final edge added joins two vertices that are already members of the same cluster, so the cluster structure of the network does not change.

we add just one vertex, and if we can find only the clusters that change upon adding a vertex, then we can save ourselves the work of performing an entire new breadth-first search, and hence save ourselves a lot of computer time. A simple algorithm for doing this works as follows.

Rather than removing vertices from the complete network, our algorithm works by building the network up from an initial state in which no vertices are occupied and switching on vertices one by one until we recover the entire network. As we add each vertex to the network we also add the accompanying edges that join it to other vertices. Only connections to other vertices that are already present need be added.

For the purposes of our algorithm, let us break down this process as shown in Fig. 16.8. Each new vertex is first added with, initially, no accompanying edges (panel (a) in the figure). In this state it forms a cluster all on its own. Then, one by one, we add its edges, those that connect it to other vertices already present. If there are no edges attached to the vertex or none connect to vertices already present, then our new vertex remains a cluster on its own. If

(a)
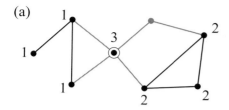

(b)
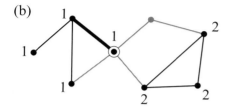

Figure 16.9: Using labels to keep track of clusters. In the algorithm described in the text, each vertex is given a label, typically an integer, to denote which cluster it belongs to. In this example there are initially two clusters, labeled 1 and 2. Then a new vertex is added between them. (a) The new vertex is added initially without its accompanying edges and is labeled as a new cluster, cluster 3. (b) An edge is added that connects cluster 3 to cluster 1, so we relabel one cluster to give it the same label as the other. In the algorithm described in the text we always relabel the smaller of the two clusters, which is cluster 3 in this case. (c) The next edge added joins clusters 1 and 2 and we relabel cluster 2 since it is smaller.

(c)
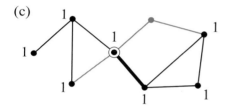

there are edges, however, then the first edge we add joins our vertex to an adjacent cluster—see Fig. 16.8b. Subsequent edges are more complicated. They can do one of two things. An edge can connect our vertex to another, different cluster, in which case in the process it joins two clusters together making them into a single cluster—see Fig. 16.8c. Alternatively, it could join our vertex to another member of the same cluster that it already belongs to, as in Fig. 16.8d. In this case, no clusters are joined together, and in terms of the size and identity of the clusters the added edge has no effect.

To keep track of the clusters in the network, therefore, our algorithm needs to do two things. First, when an edge is added it needs to identify the clusters to which the vertices at either end belong. Second, if the clusters are different, it needs to join them together into a single cluster. (If they are the same nothing need be done.)

There are various ways of achieving this but a simple one is just to put a label, such as an integer, on each vertex denoting the cluster to which it belongs—see Fig. 16.9a. Then it is a simple matter to determine if two vertices belong to the same cluster (they do if their labels are the same), and joining two clusters together is just a matter of relabeling all the vertices in one of the clusters to match the label of the other cluster. This process is illustrated in Fig. 16.9.

Then our algorithm is as follows:

1. Start with an empty network with no occupied vertices. Let $c = 0$ be the number of clusters in the network initially. Choose at random an order in which the vertices will be added to the network.

2. Add the next vertex in the chosen order, initially with no edges. This vertex is a cluster in its own right, so increase c by one and label the vertex with label c to indicate which cluster it belongs to. Also make a note that cluster c has size 1.

3. Go through the edges attached to this vertex one by one. For each edge determine whether the vertex at the other end has already been added to the network. If it has, add the edge to the network.

4. As each edge is added, examine the cluster labels of the vertices at either end. If they are the same, do nothing. If they are different, choose one of the clusters and relabel all its vertices to have the same label as the other cluster. Update the record of the size of the cluster to be equal to the sum of the sizes of the two clusters from which it was formed.

5. Repeat from step 2 until all vertices have been added.

At the end of this process, we have gone from an entirely empty network to the complete final network with all vertices and edges present and in between we have passed through a state with every possible intermediate number r of vertices. Moreover, in each of those states we had a complete record of the identities and sizes of all the clusters which we can use, for instance, to find the size S_r of the largest cluster. Then we can feed the results into Eq. (16.43) to get $S(\phi)$ for any ϕ. As before, we will typically want to average the results over many runs of the algorithm to allow for random variations from one run to another, which arise from variations in the order in which the vertices are added. This, however, is no longer a serious impediment to finishing the calculation because, if implemented appropriately, the algorithm can be made to run very quickly.

The most time-consuming part of the algorithm is the relabeling of clusters when they are joined together. Note however that when an edge joins two different clusters we are free to choose which of the two we relabel. It turns out that the speed of the algorithm can be improved greatly if we choose always to relabel the smaller one. (If the two clusters have the same size, it does not matter which we choose to relabel.) To see this, consider the following argument.

If we always relabel the smaller of two clusters, then the relabeled cluster must have been joined with one at least as large as itself and hence it is now a part of a cluster at least twice its size. Thus every time a vertex is relabeled the cluster it belongs to at least doubles in size. Given that each vertex starts off as a cluster in its own right of size 1, the size of the cluster to which it belongs after k relabelings is thus at least 2^k. Since no vertex can belong to a cluster of size greater than the size n of the whole network, the maximum number of relabelings a vertex can experience during the entire algorithm is given by $2^k = n$ or $k = \log_2 n$, and the maximum number of relabeling operations on all n vertices is thus $n \log_2 n$. Thus the total time to perform the relabeling part of the algorithm is $O(n \log n)$.

The other parts of the algorithm are typically faster than this. The adding of the vertices takes $O(n)$ time and the adding of the edges takes $O(m)$ time, which is the same as $O(n)$ on a sparse network with $m \propto n$. So the overall running time of the algorithm to leading order is $O(m + n \log n)$, or $O(n \log n)$ on a sparse network, which is much better than our first estimate of $O(n(m + n))$ above.

Essentially the same algorithm can also be used when vertices are added or removed with probabilities other than the uniformly random ones considered here. For instance, if vertices are to be removed in decreasing order of their degrees we simply reverse that process and add vertices to an initially empty network in increasing order of degrees. The details of the algorithm itself are unchanged—only the order of the vertices changes.

This algorithm works well in practice for almost all calculations. It is not, however, the very fastest algorithm for the percolation problem. There exists an even faster one, which runs in $O(m + n)$ time (or $O(n)$ for a sparse network) and is also considerably simpler to program, although its outward simplicity hides some subtleties. The reader interested in learning more about this approach is encouraged to look at Ref. [255].

16.5.1 RESULTS

Figure 16.10 shows results for four different networks as a function of the fraction of occupied vertices. In this case, the occupied vertices are chosen uniformly at random. The figure shows in each case the size S of the largest cluster as a fraction of system size, plotted as a function of ϕ. As described in Section 16.2.1, the largest cluster acts as a proxy for the giant cluster in numerical calculations on fixed networks for which the idea of a giant cluster, as a cluster that scales with system size, is meaningless.

The top two networks in the figure, a power grid and a road network, are

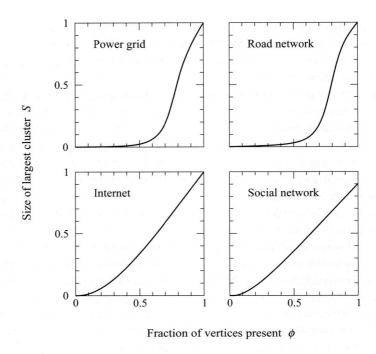

Figure 16.10: Size of the largest percolation cluster as a function of occupation probability for four networks. The four frames of this figure show the size of the largest cluster, measured as a fraction of network size, for random removal of vertices from four real-world networks: the western United States power grid, the network formed by the US Interstate highways, the Internet at the level of autonomous systems, and a social network of professional collaborations between physicists. Each curve is averaged over 1000 random repetitions of the calculation, which is why the curves appear smooth.

both networks with non-power-law degree distributions—the power grid has a roughly exponential distribution while the road network has only vertices of degree one to four and nothing else. For these cases, we expect to see behavior of the generic type described in Section 16.2.1: a continuous percolation transition at a non-zero value of ϕ from a regime in which $S \simeq 0$ to a regime of non-zero S. Because the networks are relatively small, however (4941 vertices for the power grid, 935 for the road network), we also expect to see some rounding of the transition (see Section 16.2.1).

And this is in fact what we do see. In each of these two cases S is close to zero below a certain value of ϕ, then grows rapidly but with a certain amount of rounding near the transition. Overall, other than the rounding, the shape of the curves is qualitatively similar to that of Fig. 16.4. One could even tentatively make an estimate of the percolation transition, which appears to fall around $\phi = 0.6$ or 0.7 in both networks.

The bottom two frames in the figure tell a different story. These show results for percolation on the Internet and a social network. Both of these networks have approximately power-law degree distributions and thus, based on the insights of Section 16.2.1, might be expected to show no percolation transition (or a transition at $\phi = 0$ if you prefer) and non-linear growth of the largest cluster with growing ϕ. Again our expectations seem to be borne out, at least qualitatively, by the numerical results. In both networks the value of S appears to take non-zero values for all $\phi > 0$ and the initial growth for small ϕ shows some curvature, indicating non-linear behavior.

Thus our percolation theory for random graphs seems in this case to provide a good general guide to the robustness of networks. The power-law networks are robust against random removal of vertices, in the sense that a fraction of the vertices that haven't been removed remain connected in a large cluster even when most vertices have been removed. The non-power-law networks, by contrast, become essentially disconnected after relatively few vertices have been removed—just about 40% in this case.

Figure 16.11 shows results for the same four networks when vertices are removed in order of degree, highest degrees first. As we can see, this "attack" on the network is more effective at reducing the size of the largest component than is random removal for all four networks. However, the difference between Figs. 16.10 and 16.11 is not so great for the first two networks, the power grid and the road network. The giant component in both of these networks survives nearly as long under the targeted attack as under random removal. This is as we would expect, since neither has a significant number of very high-degree vertices (the road network, with maximum degree four, has none at all), so that removal of the highest-degree vertices is not so very different from the removal of vertices of average degree.

For the second two networks, however, the Internet and the collaboration network, which both have roughly power-law degree distributions, the effect is far larger. Where these networks were more resilient to random removal than they others, they are clearly less resilient, at least by this measure, to targeted attack. The Internet in particular has a largest cluster size that falls essentially to zero when only about 5% of its highest-degree vertices have been removed, a behavior similar again to our theoretical calculations (see Fig. 16.7 on page 613). Thus the real Internet appears to show the mix of robust and fragile behavior that we saw in our calculations for the configuration model,

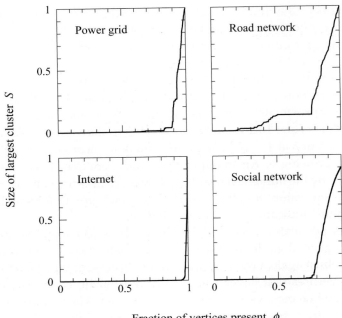

Figure 16.11: Size of the largest percolation cluster as a function of occupation probability for targeted attacks on four networks. The four frames in this figure show the size of the largest cluster, measured as a fraction of network size, for the same four networks as Fig. 16.10, when vertices are removed in degree order, highest-degree vertices first. Since this is mostly a deterministic process and not a random one (except for random choices between vertices of the same degree) the curves cannot be averaged as in Fig. 16.10 and so are relatively jagged.

being remarkably resilient to the random removal of vertices but far more susceptible to targeted attacks.

Overall, therefore, the percolation theory seems to be successful as a qualitative guide to the resilience of networks. Certainly it does not perfectly predict the exact behavior of individual networks, but it gives a good feel for the behavior we expect of networks as vertices fail or are removed, as a function of their degree distribution.

In the next chapter we will see another application of percolation, to the spread of diseases in networks.

PROBLEMS

16.1 Consider the problem of bond percolation on a square lattice and consider the following construction:

Here we have taken a bond percolation system (in black) and constructed another one interlocking it (in gray), such that the bonds of the new system are occupied if and only if the intersecting bond on the old system was not. Such an interlocking system is called a *dual lattice*.

a) If the fraction of occupied bonds on the original lattice is ϕ, what is the fraction of occupied bonds on the dual lattice?

b) Show that there is a path from top to bottom of the dual lattice if and only if there is no path from side to side of the original lattice.

c) Hence show that the percolation transition for the square lattice occurs at $\phi = \frac{1}{2}$.

16.2 Consider the site percolation problem with occupation probability ϕ on a Poisson random graph with mean degree c. Let π_s be the probability that a vertex belongs to an non-giant percolation cluster of s vertices and define a generating function $h(z) = \sum_{s=0}^{\infty} \pi_s z^s$.

a) Show that $h(z) = 1 - \phi + \phi z e^{c[h(z)-1]}$.

b) Hence show that the mean size of a small cluster in the non-percolating regime (no giant cluster) is

$$\langle s \rangle = \frac{\phi}{1 - c\phi}.$$

c) Define $f(z) = [h(z) - 1 + \phi]/\phi$. Using the Lagrange inversion formula, Eq. (12.49), solve for the coefficients in the series expansion of $f(z)$ and hence show that

$$\pi_s = \begin{cases} 1 - \phi & \text{if } s = 0, \\ \phi e^{-sc\phi}(sc\phi)^{s-1}/s! & \text{if } s > 0. \end{cases}$$

16.3 Consider a configuration model network that has vertices of degree 1, 2, and 3 only, in fractions p_1, p_2, and p_3, respectively.

a) Find the value of the critical vertex occupation probability ϕ_c at which site percolation takes place on the network.

b) Show that there is no giant cluster for any value of the occupation probability ϕ if $p_1 > 3p_3$. Why does this result not depend on p_2?

c) Find the size of the giant cluster as a function of ϕ. (Hint: you may find it useful to remember that $u = 1$ is always a solution of the equation $u = 1 - \phi + \phi g_1(u)$.)

16.4 In Section 16.3 we examined what happens when the highest-degree vertices are removed from a configuration model network with a power-law degree distribution $p_k = k^{-\alpha}/\zeta(\alpha)$ for $k \geq 1$ and $p_0 = 0$.

a) Show that in this case the phase transition at which the giant cluster disappears occurs when all vertices with degree $k > k_0$ have been removed, where the cut-off parameter k_0 satisfies

$$\sum_{k=1}^{k_0} (k^{-\alpha+2} - k^{-\alpha+1}) = \zeta(\alpha - 1).$$

b) Using the fact that $\sum_1^{k_0} k^{-x} + \sum_{k_0+1}^{\infty} k^{-x} = \zeta(x)$, and making use of the trapezoidal rule (Eq. (14.115) on page 524) for large values of k, show that

$$\sum_1^{k_0} k^{-x} \simeq \zeta(x) - \tfrac{1}{2}(k_0 + 1)^{-x} - \frac{(k_0 + 1)^{-x+1}}{x - 1}.$$

c) Keeping leading-order terms in k_0 only, show that the giant cluster disappears approximately when

$$(k_0 + 1)^{-\alpha+3} = (\alpha - 3)\left[\zeta(\alpha - 2) - 2\zeta(\alpha - 1)\right].$$

d) Find the approximate value of k_0 at the point where the giant cluster disappears for $\alpha = 2.5$.

16.5 Consider the computer algorithm for percolation described in Section 16.5, but suppose that upon the addition of an edge between two clusters we relabel not the smaller of the two clusters but one or the other chosen at random. Show by an argument analogous to the one in Section 16.5 that the worst-case running time of this algorithm is $O(n^2)$, which is substantially worse than the $O(n \log n)$ of the algorithm that always relabels the smaller cluster.

CHAPTER 17

EPIDEMICS ON NETWORKS

An introduction to the theory of the epidemic processes
by which diseases spread over networks of contact
between humans, animals, plants, and even computers

O NE OF the reasons for the large investment the scientific community has made in the study of social networks is their connection with the spread of disease. Diseases spread over networks of contacts between individuals: airborne diseases like influenza or tuberculosis are communicated when two people breathe the air in the same room; contagious diseases and parasites can be communicated when people touch; HIV and other sexually transmitted diseases are communicated when people have sex. The patterns of such contacts can be represented as networks and a good deal of effort has been devoted to empirical studies of these networks' structure. We have already discussed some network aspects of epidemiology in the previous chapter when we considered site percolation as a model for the effects of vaccination. In this chapter we look in more detail at the connections between network structure and disease dynamics and at mathematical theories that allow us to understand and predict the outcomes of epidemics.

On a related topic, recent years have seen the emergence of a new type of infection, the computer virus, a self-reproducing computer program that spreads from computer to computer in a manner similar to the spread of pathogenic infections between humans or animals. Many of the ideas described in this chapter can be applied not only to human diseases but also to computer viruses.

17.1 MODELS OF THE SPREAD OF DISEASE

The biology of what happens when an individual (a "host" in the epidemiology jargon) catches an infection is complicated. The pathogen responsible for the infection typically multiplies in the body while the immune system attempts to beat it back, often causing symptoms in the process. One or the other usually wins in the end, though sometimes neither, with the final result being the individual's recovery, their death, or a chronic disease state of permanent infection. In theory if we want to understand fully how diseases spread through populations we need to take all of this biology into account, but in practice that's usually a dauntingly large job and it is rarely, if ever, attempted. Luckily there are more tractable approaches based on simplified models of disease spread that give a good guide to disease behavior in many cases and it is on these that we focus in this chapter.

17.2   THE SI MODEL

In the typical mathematical representation of an epidemic the within-host dynamics of the disease is reduced to changes between a few basic disease states. In the simplest version there are just two states, *susceptible* and *infected*. An individual in the susceptible state is someone who does not have the disease yet but could catch it if they come into contact with someone who does. An individual in the infected state is someone who has the disease and can, potentially, pass it on if they come into contact with a susceptible individual.[1] Although this two-state classification sweeps a lot of biological details under the rug, it captures some of the gross features of disease dynamics and is a useful simplification in the case where, as here, we are focused more on what's happening at the level of networks and populations than on what's happening within the bodies of the individual population members.

Mathematical modeling of epidemics predates the study of networks by many years, stretching back at least as far as the pioneering work of Anderson McKendrick, a doctor and amateur mathematician who made foundational contributions to the field early in the twentieth century. The theories that he and others developed form the core of traditional mathematical epidemiology, which is an extensive and heavily researched field. Classic introductions to the

---

[1]If you look at the epidemiology literature you will sometimes see the infected state referred to as "infective." There's no difference between the two terms; they are synonymous. You may also see the word "infectious" used, but this may mean something slightly different. As discussed later in the chapter, more sophisticated models of disease distinguish between a state in which an individual has a disease but it has not yet developed to the point where the individual can pass it on, and a state where they can pass it on. This latter stage is sometimes called the "infectious" stage, a name chosen to emphasize that the disease can be communicated. (The former state is usually called the "exposed" state.) In the present simple two-state model, however, there is no difference between infected and infectious; all individuals who are one are also the other.

subject include the highly theoretical 1975 book by Bailey [25] and the more recent and practically oriented book by Anderson and May [17]. The review article by Hethcote is also a good resource [156].

The traditional approach avoids discussing contact networks at all by making use of a *fully mixed* or *mass-action approximation*, in which it is assumed that every individual has an equal chance, per unit time, of coming into contact with every other—people mingle and meet completely at random in this approach. This is, of course, not a realistic representation of the way the world is. In the real world, people have contact with only a small fraction of the population of the world, and that fraction is not chosen at random, which is precisely why networks play an important role in the spread of disease. Nonetheless, a familiarity with the traditional approaches will be useful to us in our study of network epidemiology, so we will spend a little time looking at its basic principles.

Consider a disease spreading through a population of individuals. Let $S(t)$ be the number of individuals who are susceptible at time t and let $X(t)$ be the number who are infected.[2] Technically, since the disease-spreading process is a random one, these numbers are not uniquely determined—if the disease were to spread through the same population more than once, even under very similar conditions, the numbers would probably be different each time. To get around this problem let us define S and X more carefully to be the average or expected numbers of susceptible and infected individuals, i.e., the numbers we would get if we ran the process many times under identical conditions and then averaged the results.[3]

The number of infected individuals goes up when susceptible individuals contract the disease from infected ones. Suppose that people meet and make contacts sufficient to result in the spread of disease entirely at random with a per-individual rate β, meaning that each individual has, on average, β contacts with randomly chosen others per unit time.

The disease is transmitted only when an infected person has contact with a susceptible one. If the total population consists of n people, then the average probability of a person you meet at random being susceptible is S/n, and hence an infected person has contact with an average of $\beta S/n$ susceptible people per unit time. Since there are on average X infected individuals in total that means the overall average rate of new infections will be $\beta SX/n$ and we can write a

The allowed transitions between states can be represented by flow charts like this simple one for the SI model.

---

[2]It might be more logical to use $I(t)$ for the number infected, and many authors do so, but we use X instead to avoid later confusion with the index i used to label vertices.

[3]For convenience we will usually drop the explicit t-dependence of $S(t)$ and $X(t)$ and, as here, just write S and X.

differential equation for the rate of change of X thus:

$$\frac{dX}{dt} = \beta \frac{SX}{n}.$$ (17.1)

At the same time the number of susceptible individuals goes down at the same rate:

$$\frac{dS}{dt} = -\beta \frac{SX}{n}.$$ (17.2)

This simple mathematical model for the spread of a disease is called the *fully mixed susceptible–infected model*, or *SI model* for short.

It is often convenient to define variables representing the fractions of susceptible and infected individuals thus:

$$s = \frac{S}{n}, \qquad x = \frac{X}{n},$$ (17.3)

in terms of which Eqs. (17.1) and (17.2) can be written

$$\frac{ds}{dt} = -\beta s x,$$ (17.4a)

$$\frac{dx}{dt} = \beta s x.$$ (17.4b)

In fact, we don't really need both of these equations, since it is also true that $S + X = n$ or equivalently $s + x = 1$ because every individual must be either susceptible or infected. With this condition it is easy to show that Eqs. (17.1) and (17.2) are really the same equation. Alternatively, we can eliminate s from the equations altogether by writing $s = 1 - x$, which gives

$$\frac{dx}{dt} = \beta(1 - x)x.$$ (17.5)

This equation, which occurs in many places in biology, physics, and elsewhere, is called the *logistic growth equation*. It can be solved using standard methods to give

$$x(t) = \frac{x_0 e^{\beta t}}{1 - x_0 + x_0 e^{\beta t}}$$ (17.6)

where x_0 is the value of x at $t = 0$. Generically this produces an S-shaped "logistic growth curve" for the fraction of infected individuals, as shown in Fig. 17.1. The curve increases exponentially for short time, corresponding to the initial phase of the disease in which most of the population is susceptible, and then saturates as the number of susceptibles dwindles and the disease has a harder and harder time finding new victims.[4]

---

[4]There aren't many diseases that really saturate their population like this. Most real diseases

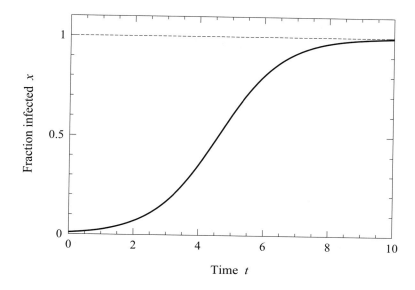

Figure 17.1: The classic logistic growth curve of the SI epidemic model. A small initial number of infected individuals in an SI model (1% in this example) will at first grow exponentially as they infect others, but growth eventually saturates as the supply of susceptible individuals is exhausted, and the curve levels off at $x = 1$.

17.3 THE SIR MODEL

The SI model is the simplest possible model of infection. There are many ways in which it can be extended to make it more realistic or more appropriate as a model of specific diseases. One common extension deals with recovery from disease.

In the SI model individuals, once infected, are infected (and infectious) forever. For many real diseases, however, people recover from infection after a certain time because their immune system fights off the agent causing the disease. Furthermore, people often retain their immunity to the disease after such a recovery so that they cannot catch it again. To represent this behavior in our

---

that don't kill their victims are eventually defeated by the immune system. In addition, for many diseases some fraction of the population has a natural immunity that prevents them from being infected (meaning that when exposed to the pathogen their immune system sees it off so quickly that they never become infectious). And some diseases spread so slowly that a large fraction of the population never catches them because they die of other causes first. None of these phenomena is represented in this model.

model we need a new third disease state, usually denoted R for *recovered*. The corresponding three-state model is called the *susceptible–infected–recovered* or *SIR model*.

With some other diseases people do not recover, but instead die after some interval. Although this is the complete opposite of recovery in human terms, it is essentially the same thing in epidemiological terms: it makes little difference to the disease whether a person is immune or dead—either way they are effectively removed from the pool of potential hosts for the disease.[5] Both recovery and death can be represented by the R state in our model. Diseases with mixed outcomes where people sometimes recover and sometimes die can also be modeled in this way—from a mathematical point of view we don't care whether the individuals in the R state are recovered or dead. For this reason some people say that the R stands for *removed* rather than recovered, so as to encompass both possibilities, and they refer to the corresponding model as the *susceptible–infected–removed* model.

The dynamics of the fully mixed SIR model has two stages. In the first stage, susceptible individuals become infected when they have contact with infected individuals. Contacts between individuals are assumed to happen at an average rate β per person as before. In the second stage, infected individuals recover (or die) at some constant average rate γ.

Given the value of γ we can calculate the length of time τ that an infected individual is likely to remain infected before they recover. The probability of recovering in any time interval $\delta\tau$ is $\gamma\,\delta\tau$ and the probability of not doing so is $1 - \gamma\,\delta\tau$. Thus the probability that the individual is still infected after a total time τ is given by

$$\lim_{\delta\tau\to 0}(1 - \gamma\,\delta\tau)^{\tau/\delta\tau} = e^{-\gamma\tau}, \tag{17.7}$$

and the probability $p(\tau)\,d\tau$ that the individual remains infected this long and then recovers in the interval between τ and $\tau + d\tau$ is this quantity times $\gamma\,d\tau$:

$$p(\tau)\,d\tau = \gamma e^{-\gamma\tau}\,d\tau, \tag{17.8}$$

The flow chart for the SIR model.

---

[5]This is only approximately true. If people really do have a certain average number of contacts per unit time and assuming those contacts are with living people, then the presence of living but recovered people in the population reduces the number of contacts between infected and susceptible individuals. If, on the other hand, people die rather than recover from the disease then only susceptible and infected individuals are alive and the number of contacts between them will be correspondingly greater. In effect, a person whose acquaintance dies from the disease will (on average) gain one new acquaintance from among the living to replace them, and that new acquaintance might be infected, or might become infected, thereby increasing the chance of transmission of the disease. This effect can easily be incorporated into the model, but we don't do so here.

which is a standard exponential distribution. Thus an infected person is most likely to recover just after becoming infected, but might in theory remain in the infected state for quite a long time—many times the mean infectious time (which is just $1/\gamma$).

Neither of these behaviors is very realistic for most real diseases. With real diseases, most victims remain infected for about the same length of time, such as a week, say, or a month. Few stay in the infected state for much longer or shorter than the average (see figure). Nonetheless, we will for the moment stick with this model because it makes the mathematics simple. This is one thing that will improve when we come to look at network models of epidemics.

In terms of the fractions s, x, and r of individuals in the three states, the equations for the SIR model are

$$\frac{ds}{dt} = -\beta s x, \tag{17.9a}$$

$$\frac{dx}{dt} = \beta s x - \gamma x, \tag{17.9b}$$

$$\frac{dr}{dt} = \gamma x, \tag{17.9c}$$

and in addition the three variables necessarily satisfy

$$s + x + r = 1. \tag{17.10}$$

To solve these equations we eliminate x between Eqs. (17.9a) and (17.9c), giving

$$\frac{1}{s}\frac{ds}{dt} = -\frac{\beta}{\gamma}\frac{dr}{dt}, \tag{17.11}$$

and then integrate both sides with respect to t to get:

$$s = s_0 e^{-\beta r/\gamma}, \tag{17.12}$$

where s_0 is the value of s at $t = 0$ and we have chosen the constant of integration so that there are no individuals in the recovered state at $t = 0$. (Other choices are possible but we'll use this one for now.)

Now we put $x = 1 - s - r$ in Eq. (17.9c) and use Eq. (17.12) to get

$$\frac{dr}{dt} = \gamma\left(1 - r - s_0 e^{-\beta r/\gamma}\right). \tag{17.13}$$

If we can solve this equation for r then we can find s from Eq. (17.12) and x from Eq. (17.10).

The solution is easy to write down in principle. It is given by

$$t = \frac{1}{\gamma}\int_0^r \frac{du}{1 - u - s_0 e^{-\beta u/\gamma}}. \tag{17.14}$$

The distribution of times for which an individual remains infected is typically narrowly peaked around some average value for real diseases, quite unlike the exponential distribution assumed by the SIR model.

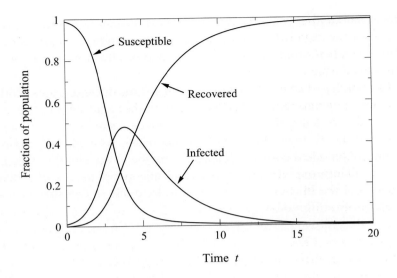

Figure 17.2: Time evolution of the SIR model. The three curves in this figure show the fractions of the population in the susceptible, infected, and recovered states as a function of time. The parameters are $\beta = 1$, $\gamma = 0.4$, $s_0 = 0.99$, $x_0 = 0.01$, and $r_0 = 0$.

Unfortunately, in practice we can't evaluate the integral in closed form. We can however evaluate it numerically. An example is shown in Fig. 17.2.

There are a number of notable things about this figure. The fraction of susceptibles in the population decreases monotonically as susceptibles are infected and the fraction of recovered individuals increases monotonically. The fraction infected, however, goes up at first as people get infected, then down again as they recover, and eventually goes to zero as $t \to \infty$.

Note however that the number of susceptibles does not go to zero; a close inspection shows that the curve for $s(t)$ ends a little above the axis. This is because when $x \to 0$ there are no infected individuals left to infect the remaining susceptibles. Any individuals who survive to late enough times without being infected will probably never get the disease at all. They are the lucky ones who made it through the outbreak and out the other side. Similarly the fraction of recovered individuals does not quite reach one as $t \to \infty$.

The asymptotic value of r has an important practical interpretation: it is the total number of individuals who ever catch the disease during the entire course of the epidemic—the total size of the outbreak. It can be calculated from Eq. (17.13) as the value at which $dr/dt = 0$, which gives $r = 1 - s_0 e^{-\beta r/\gamma}$.

The initial conditions for the model can be chosen in a variety of ways, but the most common is to assume that the disease starts with either a single infected individual or a small number c of individuals and everyone else in the susceptible state. In other words, the initial values of the variables are $s_0 = 1 - c/n$, $x_0 = c/n$, and $r_0 = 0$. In the limit of large population size $n \to \infty$, we can then write $s_0 \simeq 1$, and our final value of r satisfies

$$r = 1 - e^{-\beta r/\gamma}. \tag{17.15}$$

Interestingly, this is the same as the equation we derived in Section 12.5 for the size S of the giant component of a Poisson random graph, Eq. (12.15), provided we equate β/γ with the mean degree of the random graph, and this correspondence allows us immediately to say several useful things. First, we know what the size of the epidemic must look like (in the limit of large n) as a function of the parameters β and γ: it will look like the plot of giant component size shown in the right-hand panel of Fig. 12.1 on page 406, with $c = \beta/\gamma$. Second, it tells us that the size of the epidemic goes continuously to zero as β/γ approaches one from above and for $\beta/\gamma \leq 1$, or equivalently $\beta \leq \gamma$, there is no epidemic at all. The simple explanation for this result is that if $\beta \leq \gamma$ then infected individuals recover faster than susceptible individuals become infected, so the disease cannot get a toehold in the population. The number of infected individuals, which starts small, goes down, not up, and the disease dies out instead of spreading.

The transition between the epidemic and non-epidemic regimes happens at the point $\beta = \gamma$ and is called the *epidemic transition*. Note that there was no epidemic transition in the simpler SI model: in that model the disease always spreads because individuals once infected never recover and hence the number of infected individuals cannot decrease. (One can think of the SI model as the special case of the SIR model in which $\gamma = 0$, so that β can never be less than γ.)

An important quantity in the study of epidemics is the *basic reproduction number*, denoted R_0, which is defined as follows. Consider the spread of a disease when it is just starting out, when there are only a few cases of the disease and the rest of the population is susceptible—what is called a *naive population* in the epidemiology jargon—and consider a susceptible who catches the disease in this early stage of the outbreak. The basic reproduction number is defined to be the average number of additional people that such a person passes the disease onto before they recover. For instance, if each person catching the disease passes it onto two others on average, then $R_0 = 2$. If half of them pass it on to just one person and the rest to none at all, then $R_0 = \frac{1}{2}$, and so forth.

If we had $R_0 = 2$ then each person catching the disease would pass it on to two others on average, each of them would pass it on to two more, and so forth, so that the number of new cases of the disease would double at each round, thus growing exponentially. Conversely if $R_0 = \frac{1}{2}$ the disease would die out exponentially. The point $R_0 = 1$ separates the growing and shrinking behaviors and thus marks the *epidemic threshold* between regimes in which the disease either multiplies or dies out.

We can calculate R_0 straightforwardly for our model. If an individual remains infectious for a time τ then the expected number of others they will have contact with during that time is $\beta\tau$. The definition of R_0 is specifically for a naive population, and in a naive population all of the people with whom one has contact will be susceptible, and hence $\beta\tau$ is also the total number of people our infected individual will infect. Then we average over the distribution of τ, Eq. (17.8), to get the average number R_0:

$$R_0 = \beta\gamma \int_0^\infty \tau e^{-\gamma\tau} \, d\tau = \frac{\beta}{\gamma}. \tag{17.16}$$

This gives us an alternative way of deriving the epidemic threshold in the SIR: the epidemic threshold falls at $R_0 = 1$, which corresponds in this model to the point $\beta = \gamma$, the same result as we found above by considering the long-time behavior.[6]

17.4 THE SIS MODEL

A different extension of the SI model is one that allows for *reinfection*, i.e., for diseases that don't confer immunity on their victims after recovery, or confer only limited immunity, so that individuals can be infected more than once. The simplest such model is the *SIS model*, in which there are just two states, susceptible and infected, and infected individuals move back into the susceptible state upon recovery. The differential equations for this model are

$$\frac{ds}{dt} = \gamma x - \beta s x, \tag{17.17a}$$

$$\frac{dx}{dt} = \beta s x - \gamma x, \tag{17.17b}$$

with

$$s + x = 1. \tag{17.18}$$

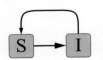

Flow chart for the SIS model.

---

[6]Note that when $\gamma = 0$, as in the SI model, Eq. (17.16) implies that $R_0 \to \infty$. This is because an infected individual remains infected indefinitely in the SI model and hence can infect an arbitrary number of others, so that R_0 is formally infinite. In any population of finite size, however, the empirical value of R_0 will be finite.

Putting $s = 1 - x$ in Eq. (17.17b) gives

$$\frac{dx}{dt} = (\beta - \gamma - \beta x)x, \tag{17.19}$$

which has the solution

$$x(t) = (1 - \gamma/\beta)\frac{Ce^{(\beta-\gamma)t}}{1 + Ce^{(\beta-\gamma)t}}, \tag{17.20}$$

where the integration constant C is fixed by the initial value of x to be

$$C = \frac{\beta x_0}{\beta - \gamma - \beta x_0}. \tag{17.21}$$

In the case of a large population and a small number of initial carriers of the disease we have $x_0 \to 0$ and $C = \beta x_0/(\beta - \gamma)$, which gives us the simpler solution

$$x(t) = x_0\frac{(\beta - \gamma)e^{(\beta-\gamma)t}}{\beta - \gamma + \beta x_0 e^{(\beta-\gamma)t}}. \tag{17.22}$$

If $\beta > \gamma$ this produces a logistic growth curve similar to that of the basic SI model—see Fig. 17.3—but differing in one important respect: we never have the whole population infected with the disease. In the limit of long time the system finds a stable state where the rates at which individuals are infected and recover from infection are exactly equal and a steady fraction of the population—but not all of them—is always infected with the disease. (Which particular individuals are infected changes over time, however, as some recover and others are infected.) The fraction of infected individuals can be found from Eq. (17.22), or more directly from Eq. (17.19) by setting $dx/dt = 0$ to give $x = (\beta - \gamma)/\beta$. In the epidemiology jargon the steady state is called an *endemic disease state*.

Note that the fraction infected in the endemic state goes to zero as β approaches γ, and if $\beta < \gamma$ then Eq. (17.22) predicts that the disease will die out exponentially. Thus, as in the SIR model, the point $\beta = \gamma$ marks an epidemic transition between a state in which the disease spreads and one in which it doesn't. As before, we can calculate a basic reproduction number R_0, which again takes the value $R_0 = \beta/\gamma$, giving us an alternative derivation of the position of the transition as the point at which $R_0 = 1$.

17.5 THE SIRS MODEL

We will look at one more epidemic model before we turn to the properties of these models on networks. This is the *SIRS model*, another model incorporating reinfection. In this model individuals recover from infection and gain

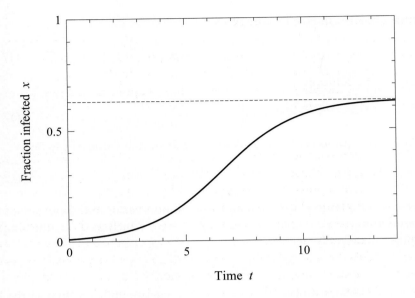

Figure 17.3: Fraction of infected individuals in the SIS model. The fraction of infected individuals in the SIS model grows with time following a logistic curve, as in the SI model. Unlike the SI model, however, the fraction infected never reaches unity, tending instead to an intermediate value at which the rates of infection and recovery are balanced. (Compare this figure with Fig. 17.1 for the SI model.)

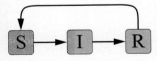

Flow chart for the SIRS model

immunity as in the SIR model, but that immunity is only temporary, and after a certain period of time individuals lose it and become susceptible again. We introduce a new parameter δ to represent the average rate at which individuals lose immunity. Then the equations for this model are

$$\frac{ds}{dt} = \delta r - \beta s x, \tag{17.23a}$$

$$\frac{dx}{dt} = \beta s x - \gamma x, \tag{17.23b}$$

$$\frac{dr}{dt} = \gamma x - \delta r, \tag{17.23c}$$

and

$$s + x + r = 1. \tag{17.24}$$

The SIRS model cannot be solved analytically, although it can be treated using linear stability analysis and other tricks from the non-linear dynamics toolbox. A more straightforward approach is numerical integration of the differential equations, which reveals that the SIRS model has a rich palette of behaviors depending on the values of the three parameters, including behaviors where the disease persists in an endemic state, where it dies out, and where it oscillates between outbreaks and periods of remission. We will not delve into the behavior of the SIRS model further in this chapter; the interested reader can find more details in Ref. [156].

Many other epidemic models have also been proposed to model the spread of particular types of diseases. Extra states can be introduced such as an "exposed" state that represents people who have caught a disease but whose infection has not yet developed to the point where they can pass it on to others; or an initial immune state coming before the susceptible state, often used to represent the maternally derived immunity that newborn babies possess. There are also models that allow for new individuals to enter the population, by being born or immigrating, and models that distinguish between people who recover fully from disease and those who recover but remain carriers who can pass the disease to others. Those interested in pursuing the subject further are encouraged to take a look at the references given at the beginning of the chapter. For our purposes, however, the models we have seen so far will be enough. Let's look at how these models behave when we include network structure in our calculations.

17.6 EPIDEMIC MODELS ON NETWORKS

As discussed in Section 17.2, the standard approach to epidemic modeling described in the first part of this chapter assumes "full mixing" of the population, meaning that each individual can potentially have contact with any other, those contacts being realized, at a level sufficient to transmit the disease, with probability β per unit time.

In the real world, however, it is not a good assumption to say that any two people could potentially have contact with one another. The chance of a meeting between two people chosen at random from the population of the entire world is probably small enough to be negligible. Most people have a set of regular acquaintances, neighbors, coworkers, and so forth whom they meet with some regularity and most other members of the world population can safely be ignored. The set of a person's potential contacts can be represented as a network and the structure of that network can have a strong effect on the way a disease spreads through the population.

Network models of disease typically work in the same way as the fully mixed models we have already seen but make use of this network of potential contacts instead of assuming that contact is possible with the entire population. Let us define the *transmission rate* or *infection rate* for our network disease process to be the probability per unit time that infection will be transmitted between two individuals, one susceptible and one infected, who are connected by an edge in the appropriate network. Alternatively it is the rate at which contact sufficient to spread the disease occurs between any two individuals connected by an edge. The transmission rate is commonly denoted β by analogy with the quantity appearing in the fully mixed models, and we will adopt that notation here, although you should note that the two parameters are not exactly equivalent since β in the fully mixed case is the rate of contacts between an infected individual and all others in the population, whereas in the network case it is the rate of contacts with just one other.

The transmission rate is a property of the disease. Some diseases are transmitted more easily than others and so have higher transmission rates. But transmission rate is also a property of the social and behavioral parameters of the population. In some countries, for example, it is common etiquette for people with minor respiratory infections such as colds to wear surgical face masks to prevent the spread of disease. Such conventions are absent in other countries, and the difference in conventions could produce a difference in transmission rate.

17.7   Late-time properties of epidemics on networks

Given a value for the transmission rate one can define models for the spread of disease over a network. Each of the models introduced in the first part of the chapter can be generalized to the network case. Consider the SI model, for instance. In the network version of this model we have n individuals represented by the vertices of our network, with most of them in the susceptible state at time $t = 0$ and just a small fraction x_0, or maybe even just a single vertex, in the infected state. With probability β per unit time, infected nodes spread the disease to their susceptible neighbors and over time the disease spreads across the network.

It is difficult to solve a model such as this for a general network, and in many cases the best we can do is to simulate it on a computer. There is, however, one respect in which the model is straightforward, and that is its late-time properties. It is clear that as $t \to \infty$ in this model every individual who *can* be infected by the disease is infected: since infected individuals remain infectious forever, their susceptible neighbors will always, in the end, also become infected, no matter how small the transmission rate, so long as it is not zero. The only condition for being infected therefore is that a vertex must be connected to at least one infected individual by at least one path through the network, so that the disease can reach them.

Thus in the limit of long times the disease will spread from every initial carrier to infect all reachable vertices, meaning all vertices in the component to which the carrier belongs. In the simplest case, where the disease starts out with a single infected carrier, just one component will be infected.

As we have seen, however, most networks have a one large component that contains a significant fraction of all vertices in the network, plus, typically, a selection of smaller components. If we have this kind of structure then an interesting behavior emerges. If we start with a single infected individual, and if that individual turns out to belong to the large component, then the disease will infect the large component and we will have a large outbreak. If the individual belongs to one of the small components, however, the disease will only infect the few members of that small component and then die out. If the initial carrier of the disease is chosen uniformly at random from the network, the probability that it will fall in the large component and we will have a large outbreak is simply equal to S, the fraction of the network occupied by the large component, and the size of the outbreak as a fraction of the network will also be S. Conversely, with probability $1 - S$ the initial carrier will fall in one of the small components and the outbreak will be small. In the latter case the size of the outbreak will be given by the size of the appropriate small component. If we can calculate the distribution of sizes of the small components, either analytically or numerically, for the network of interest, then we also know the distribution of possible sizes of these small outbreaks, although unless we know exactly which component the disease will start in we cannot predict its size exactly.

This constitutes a new type of behavior not seen in fully mixed models. In fully mixed models the possible behaviors are also either a run-away epidemic that affects a large fraction of the population, or an outbreak that affects only a few then dies out. But the choice between these outcomes was uniquely determined by the choice of model and the model parameters. For a given model and parameter values the disease always either did one thing or the other. In our network model, however, the behavior depends on the network structure and on the position in the network of the first infected individual. Thus there is a new stochastic element in the process: with identical model parameters and an identical network the disease sometimes takes off and sometimes dies out.

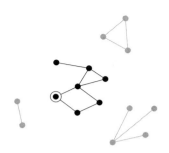

An outbreak starting with a single infected individual (circled) will eventually affect all those in the same component of the network, but leave other components untouched.

17.8 LATE-TIME PROPERTIES OF THE SIR MODEL

The situation becomes more interesting still when we look at the SIR model. In the SIR model individuals remain infectious for only a finite amount of time and then they recover, so it is in general no longer true (as in the SI model) that the susceptible neighbor of an infected individual will always get infected in the end. If they are lucky, such neighbors may never catch the disease. The probability of this happening can be calculated in a manner similar to the calculation of Eq. (17.7), and is equal to $e^{-\beta\tau}$, where β is again the transmission rate and τ is the amount of time for which the infected individual remains infected. Thus the probability that the disease *is* transmitted is

$$\phi = 1 - e^{-\beta\tau}. \tag{17.25}$$

For simplicity, let us suppose that every infected individual remains infectious for the same length of time. This differs from the fully mixed version of the model, where τ was distributed according to an exponential distribution (see Eq. (17.8)), but in many cases is actually more realistic. As mentioned in Section 17.3, observed values of τ for many diseases are narrowly concentrated about a mean value, and their distribution is far from being exponential.

With this assumption, the probability of transmission ϕ is a constant across the whole network. Every susceptible individual has equal probability ϕ of catching the disease from their infected neighbor. (Of course, if they have more than one infected neighbor the total probability is higher.)

Now here is a nice trick, developed originally by Mollison [223] and Grassberger [144]. Let us take our network and "color in" or "occupy" each edge with probability ϕ, or not with probability $1 - \phi$. This is just the ordinary bond percolation process introduced in Section 16.1, where a fraction ϕ of edges are occupied uniformly at random. The occupied edges represent those along which disease will be transmitted if it reaches either of the vertices at the ends of the edge. That is, the occupied edges represent contacts sufficient to spread the disease, but not necessarily actual disease transmission: if the disease doesn't reach either end of an occupied edge then disease will not be transmitted along that edge, so edge occupation only represents the potential for transmission if the disease reaches an edge.

With this in mind consider now the spread of a disease that starts at a randomly chosen vertex. We can immediately see that the set of vertices to which the disease will ultimately spread is precisely the set connected to the initial vertex by any path of occupied edges—the disease simply passes from one vertex to another by traversing occupied edges until all reachable vertices have been infected. The end result is that the disease infects all members of the bond

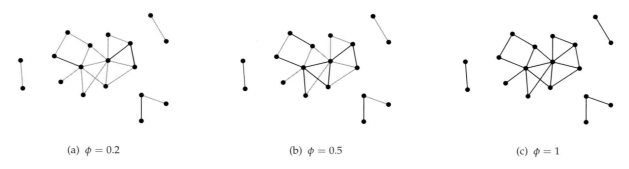

(a) $\phi = 0.2$ (b) $\phi = 0.5$ (c) $\phi = 1$

Figure 17.4: Bond percolation. In bond percolation, a fraction ϕ of the edges in a network are filled in or "occupied" at random to create connected clusters of vertices. (a) For small occupation probability ϕ the clusters are small. (b) Above the percolation threshold a large cluster forms, though there are usually still some small clusters as well. (c) When $\phi = 1$ all edges are occupied but the large cluster may still not fill the whole network: at $\phi = 1$ the largest cluster corresponds to the largest component of the network, which is often just a subset of the whole network.

percolation cluster to which the initial carrier belongs.

It is important to appreciate that, as with our treatment of the network SI model in the previous section, this process does not give us any information about the temporal evolution of the disease outbreak. Individual infection events are stochastic and a calculation of the curve of infections as a function of time requires a more complicated analysis that takes their randomness into account. However, if we want to know only about long-time behavior, about the overall total number of individuals infected by the disease, then all we need do is count the vertices in the appropriate percolation cluster.

Bond percolation is in many ways similar to the site percolation processes we studied in Chapter 16. Consider Fig. 17.4. For low edge occupation probability ϕ there are just a few occupied bonds which group into small disconnected clusters. But as ϕ increases there comes a point, the percolation transition, where the disconnected clusters grow large enough to join together and form a giant cluster, although usually there exist other small clusters as well that are not joined to the giant cluster. As ϕ increases still further, the giant cluster grows, reaching its maximum size when $\phi = 1$. Notice, however, that this maximum size is not generally equal to the size of the whole network. Even when every edge in the network is occupied, the size of the largest cluster is still limited to the size of the largest component on the network, which is usually smaller than the whole network.

Translating these ideas into the language of epidemiology, we see that for small values of ϕ the cluster to which the initial carrier of a disease belongs must be small, since all clusters are small. Thus in this regime we will have only a small disease outbreak and most members of the population will be uninfected. Once we reach the percolation transition, however, and a giant cluster forms, then a large outbreak of the disease—an epidemic—becomes possible, although not guaranteed. If the giant cluster of the percolation process occupies a fraction S of the entire network, then our randomly chosen initial vertex will fall within it with probability S, and if it does then the disease will spread to infect the whole giant cluster, creating an epidemic reaching a fraction of the population also equal to S. With probability $1 - S$, on the other hand, the initial vertex will fall in one of the small clusters and we will have only a small outbreak of the disease. As ϕ increases, S also increases and hence both the probability and the size of an epidemic increase with ϕ.

Thus the percolation transition for bond percolation on our network corresponds precisely to the epidemic threshold for a disease on the same network, where the edge occupation probability ϕ is given in terms of the transmission rate β and recovery time τ for the disease by Eq. (17.25), and the sizes of outbreaks are given by the sizes of the bond percolation clusters. This mapping between percolation and epidemics is a powerful one that allows us to make a whole range of calculations of the effects of network structure on the spread of disease.

It is important to note that even when ϕ is above the epidemic threshold we are not guaranteed that there will be an epidemic. This is similar to the situation we saw in the simpler SI model, but different from the situation in the fully mixed SIR model of Section 17.3, where an epidemic always takes place if we are above the epidemic threshold. In many ways the behavior of our network model is more realistic than that of the fully mixed model. For many diseases it is true that outbreaks do not always result in epidemics. Sometimes a disease dies out because, just by chance, its earliest victims happen not to pass the disease on to others. Our theory tells us that the probability of this happening is $1 - S$, where S is the size of the giant cluster, which is also the size of the epidemic if it does happen. The value of $1 - S$ is usually small when we are well above the epidemic threshold, but can be quite large if we are only a little above threshold, meaning that the probability of the disease dying out can be quite large in this regime.

It is also important to bear in mind that percolation is a stochastic process. We occupy edges at random on our network to represent the random nature of the contacts that result in transmission of the disease. Two outbreaks happening under the same conditions on the same networks would not necessarily travel along the same edges and the shapes of the percolation clusters would not necessarily be the same. Thus a vertex that happens to belong to the giant cluster on one occasion might not belong to it on another and our theory cannot make exact predictions about disease outcomes. The best we can do is calculate probabilities or average behaviors. We could for instance calculate the expected number of people who would be affected by an outbreak, but we cannot predict the exact number for any given outbreak.

17.8.1 SIR MODEL AND THE CONFIGURATION MODEL

In Section 16.2.1 we showed that it is possible to calculate exactly the average behavior of a site percolation process on configuration model networks. With only slight modification the same approach can also be used for bond percolation and hence we can make predictions about the size distribution of epidemics and the position of the epidemic threshold in such networks.

Consider an SIR epidemic process of the kind discussed in the previous section, taking place on a configuration model network with degree distribution p_k. Let u be the average probability that a vertex is not connected to the giant cluster via a specific one of its edges. There are two ways this can happen: either the edge in question can be unoccupied (with probability $1 - \phi$), or it is occupied (probability ϕ) but the vertex at the other end of the edge is itself not a member of the giant cluster. The latter happens only if that vertex is not connected to the giant cluster via any of its other edges, which happens with probability u^k if there are k such edges. Thus the total probability is $1 - \phi + \phi u^k$.

The value of k is distributed according to the excess degree distribution

$$q_k = \frac{(k+1)p_{k+1}}{\langle k \rangle} \tag{17.26}$$

(see Eq. (16.3)). Averaging over k we then arrive at a self-consistent expression for u thus:

$$u = 1 - \phi + \phi \sum_{k=0}^{\infty} q_k u^k = 1 - \phi + \phi g_1(u), \tag{17.27}$$

where g_1 is the probability generating function for the excess degree distribution, defined in Eq. (13.49). Equation (17.27) is the same as the corresponding equation for the site percolation case, Eq. (16.4), and has the same solutions.

The probability that a vertex of total degree k does not belong to the giant cluster is now simply u^k, and the average such probability over the whole network, which is equal to $1 - S$, is calculated by averaging u^k over the degree distribution p_k giving

$$S = 1 - \sum_{k=0}^{\infty} p_k u^k = 1 - g_0(u). \tag{17.28}$$

This equation differs from the corresponding equation in the site percolation case, Eq. (16.2), by an overall factor of ϕ, but is otherwise the same. Thus the shape of the curve for S as a function of ϕ will be different from the site percolation case, but the position ϕ_c of the percolation transition, which is dictated by the solution of Eq. (17.27), will be the same. The solution of Eq. (17.27) was shown graphically in Fig. 16.2 and the position of the transition is given by Eq. (16.7) to be

$$\phi_c = \frac{1}{g_1'(1)} = \frac{\langle k \rangle}{\langle k^2 \rangle - \langle k \rangle}. \tag{17.29}$$

This equation thus also gives us the position of the epidemic threshold in terms of the probability ϕ. If we prefer our solution in terms of the more fundamental parameters β and τ we can rearrange Eq. (17.25) to give

$$\beta\tau = -\ln(1 - \phi_c) = \ln\frac{\langle k^2 \rangle - \langle k \rangle}{\langle k^2 \rangle - 2\langle k \rangle}. \tag{17.30}$$

If $\beta\tau$ exceeds this value then there is the possibility of an epidemic, though not the certainty, since the initial carrier or carriers of the disease could by chance fall outside the giant cluster. If $\beta\tau$ is smaller than this value then an epidemic is impossible, no matter where the initial carrier falls. The probability of the epidemic, if one is possible, is given by S, Eq. (17.28), as is the size of the epidemic if and when one occurs.

Since the epidemic behavior of the model is controlled by the combination of parameters $\beta\tau$, the epidemic transition can be driven either by an increase in the infectiousness time τ, which is a property of the particular disease under study, or by an increase in the transmission rate β, which is a property both of the disease and of the behavior of members of the population. At the same time, the precise position of the transition in terms of these variables, as well as the probability and size of any epidemic that occurs, depend on the structure of the network via the moments $\langle k \rangle$ and $\langle k^2 \rangle$ of the degree distribution. This contrasts with the fully mixed model of Section 17.3, which incorporated no network effects.

Because of the close similarity between the site and bond percolation problems, we can easily translate a number of the results of Section 16.2.1 into the language of epidemics. For instance, a random graph with a Poisson degree distribution with mean c, which has $g_0(z) = g_1(z) = e^{c(z-1)}$, has an epidemic threshold falls at $\phi_c = 1/c$ (Eq. (16.11)), or

$$\beta\tau = \ln\frac{c}{c-1}, \tag{17.31}$$

and the size of the epidemic, when there is one, is given by the solution to the equations

$$u = 1 - \phi + \phi e^{c(u-1)}, \qquad S = 1 - e^{c(u-1)}. \qquad (17.32)$$

The first of these equations can be rearranged to read $1 - u = \phi(1 - e^{c(u-1)}) = \phi S$ and substituting into the second then gives

$$S = 1 - e^{-\phi c S}, \qquad (17.33)$$

which has no simple closed-form solution,[7] but can easily be solved numerically by making an initial guess at the solution ($S = \frac{1}{2}$ seems to work well) and then iterating the equation to convergence.

Note that this equation is similar to Eq. (17.15) for the fully mixed model, but with different parameters. The similarity is not coincidental. In the fully mixed model an infected individual infects others chosen uniformly at random from the population, and in the Poisson random graph the network neighbors of any individual are also chosen uniformly at random. It is possible to show that there is a direct correspondence between the traditional fully mixed model and the network model on a random graph [30].[8]

Another important case is the scale-free network with its power-law degree distribution. As we saw in Section 16.2.1, if the exponent α of the power law in such a network lies in the usual range $2 < \alpha < 3$ then $\phi_c = 0$, because $\langle k^2 \rangle$ diverges while $\langle k \rangle$ remains constant and hence Eq. (17.29) goes to zero. Thus in the power-law case there is always an epidemic, no matter how small the probability of transmission of the disease, at least in the limit of infinite network size. (For finite networks, $\langle k^2 \rangle$ is not infinite, but very large, and ϕ_c is correspondingly very small, but not precisely zero.)

---

[7]The solution can be written in closed form using the *Lambert W-function*, which is defined to be the solution of the equation $W(z)e^{W(z)} = z$. In terms of this function, the size of the epidemic is given by

$$S = 1 + \frac{W(-\phi c e^{-\phi c})}{\phi c}.$$

Alternatively, we can rearrange Eq. (17.33) to give ϕ as a function of S rather than the other way around:

$$\phi = -\frac{\ln(1-S)}{cS}.$$

This expression can be useful for making plots of S.

[8]The differences in parameters arise because we are considering a slightly different disease process (one in which each individual is infectious for the same amount of time, rather than the exponential distribution used in the fully mixed model), and also because in the network model β is the transmission rate per edge, rather than the rate for the whole network—this is what gives us the factor of c in the exponent of Eq. (17.33).

This statement is, however, slightly misleading since, as we saw in the previous chapter, the size of the giant cluster in a scale-free network becomes very small as we approach $\phi = 0$; it generally decays faster than linearly with ϕ. Thus although technically there may be an epidemic for all positive values of ϕ, it can be very small in practice, affecting only the tiniest fraction of the population. (On the other hand, the difference between non-epidemic behavior and epidemic behavior, even with a tiny value of S, will become very important when we look at models such as the SIS model that incorporate reinfection. In such models the epidemic threshold separates the regime in which the disease persists and the regime in which it becomes extinct, an important distinction even if the number of individuals infected is small.)

17.9 TIME-DEPENDENT PROPERTIES OF EPIDEMICS ON NETWORKS

The techniques of the previous section can tell us about the late-time properties of epidemics on networks, such as how many people will eventually be affected in an outbreak of a disease. If we want to know about the detailed progression of an outbreak as a function of time, however, then we need another approach that takes dynamics into account. Moreover, the techniques we have used so far cannot tell us about even the late-time behavior of models with reinfection, such as the SIS and SIRS models of Sections 17.4 and 17.5. For these models the equivalence between epidemics and percolation that we used above does not hold, and to understand their behavior, including at long times, we need to address the dynamics of the epidemic.

A number of approaches have been proposed for tackling the dynamics of epidemics on networks, some exact and some approximate. Of course, given a specific network, one can always perform computer simulations of epidemics and get numerical answers for typical disease outbreaks. Analytic approaches, however, offer more insight and some results are known, as discussed below, but they are mostly confined to specific classes of model network, such as random graphs and their generalizations. In the following sections we will look at some of the most straightforward and general approaches to epidemic dynamics on networks, starting with the simple SI model and progressing to the more complex (and interesting) models in later sections.

17.10 TIME-DEPENDENT PROPERTIES OF THE SI MODEL

The analytic treatment of the time-dependent properties of epidemic models revolves around the time evolution of the probabilities for vertices to be in specific disease states. One can imagine having repeated outbreaks of the same disease on the same network, starting from the same initial conditions, and calculating for example the average probabilities $s_i(t)$ and $x_i(t)$ that vertex i is susceptible or infective at time t. Given the adjacency matrix of a network one can write down equations for the evolution of such quantities in a straightforward manner. Consider for instance the SI model.

An SI outbreak starting with a single randomly chosen vertex somewhere eventually spreads, as we have seen, to all members of the component containing that vertex. Our main interest is in epidemics occurring in the giant component of the network, since all other outbreaks will only affect a small component and then die out, so let us focus on the giant component case.

Consider a vertex i. If the vertex is not a member of the giant component then by hypothesis $s_i = 0$ at all times, since we are assuming the epidemic to take place in the giant component. For i in the giant component we can write down a differential equation for s_i by considering the probability that i becomes infected between times t and $t + dt$. To become infected an individual must catch the disease from a neighboring individual j, meaning j must already be infected, which happens with probability $x_j = 1 - s_j$, and must transmit the disease during the given time interval, which happens with probability $\beta\, dt$. In addition we also require that i be susceptible in the first place, which happens with probability s_i. Multiplying these probabilities and then summing over all neighbors of i, the total probability of i becoming infected is $\beta s_i \sum_j A_{ij} x_j$, where A_{ij} is an element of the adjacency matrix. Thus s_i obeys the coupled set of n non-linear differential equations:

$$\frac{ds_i}{dt} = -\beta s_i \sum_j A_{ij} x_j = -\beta s_i \sum_j A_{ij}(1 - s_j). \tag{17.34}$$

Note the leading minus sign on the right-hand side—the probability of being susceptible goes down when vertices become infected.

Similarly we can write an equation for x_i thus:

$$\frac{dx_i}{dt} = \beta s_i \sum_j A_{ij} x_j = \beta(1 - x_i) \sum_j A_{ij} x_j, \tag{17.35}$$

although the two equations are really the same equation, related to one another by $s_i + x_i = 1$.

We will use the same initial conditions as we did in the fully mixed case, assuming that the disease starts with either a single infected vertex or a small number c of vertices, chosen uniformly at random, so that $x_i = c/n$ and $s_i = 1 - c/n$ for all i. In the limit of large system size n, these become $x_i = 0$, $s_i = 1$, and we will use this large-n limit to simplify some of the expression derived in this and the following sections.

Equation (17.34) is not solvable in closed form for general A_{ij} but we can calculate some features of its behavior by considering suitable limits. Consider for example the behavior of the system at early times. For large n, and assuming initial conditions as above, x_i will be small in this regime. Working with Eq. (17.35) and ignoring terms of quadratic order in small quantities, we have

$$\frac{dx_i}{dt} = \beta \sum_j A_{ij} x_j, \tag{17.36}$$

or in matrix form

$$\frac{d\mathbf{x}}{dt} = \beta \mathbf{A} \mathbf{x}, \tag{17.37}$$

where $\mathbf{x}$ is the vector with elements x_i.

Now let us write $\mathbf{x}$ as a linear combination of the eigenvectors of the adjacency matrix:

$$\mathbf{x}(t) = \sum_{r=1}^n a_r(t) \mathbf{v}_r, \tag{17.38}$$

where $\mathbf{v}_r$ is the eigenvector with eigenvalue κ_r. Then

$$\frac{d\mathbf{x}}{dt} = \sum_{r=1}^n \frac{da_r}{dt} \mathbf{v}_r = \beta \mathbf{A} \sum_{r=1}^n a_r(t) \mathbf{v}_r = \beta \sum_{r=1}^n \kappa_r a_r(t) \mathbf{v}_r. \tag{17.39}$$

Then, comparing terms in $\mathbf{v}_r$, we get

$$\frac{da_r}{dt} = \beta \kappa_r a_r, \tag{17.40}$$

which has the solution

$$a_r(t) = a_r(0) \, e^{\beta \kappa_r t}. \tag{17.41}$$

Substituting this expression back into Eq. (17.38), we then have

$$\mathbf{x}(t) = \sum_{r=1}^n a_r(0) \, e^{\beta \kappa_r t} \, \mathbf{v}_r. \tag{17.42}$$

The fastest growing term in this expression is the term corresponding to the largest eigenvalue κ_1. Assuming this term dominates over the others we will get

$$\mathbf{x}(t) \sim e^{\beta \kappa_1 t} \mathbf{v}_1. \tag{17.43}$$

So we expect the number of infected individuals to grow exponentially, just as it does in the fully mixed version of the SI model, but now with an exponential constant that depends not just on β but also on the leading eigenvalue of the adjacency matrix.

Moreover, the probability of infection in this early period varies from vertex to vertex roughly as the corresponding element of the leading eigenvector $\mathbf{v}_1$. The elements of the leading eigenvector of the adjacency matrix are the same quantities that in other circumstances we called the eigenvector centrality—see Section 7.2. Thus eigenvector centrality is a crude measure of the probability of early infection of a vertex in an SI epidemic.

At long times in the SI model the probability of infection of a vertex in the giant component tends to one (again assuming the epidemic takes place in the giant component). Thus overall we expect the SI epidemic to have a similar form to that seen in the fully mixed version of the model, producing curves qualitatively like that in Fig. 17.1 but with vertices of higher eigenvector centrality becoming infected faster than those of lower.

Reasonable though this approach appears to be, it is not precisely correct, as we can see by integrating Eq. (17.35) numerically. Figure 17.5a shows the results of such a numerical integration (the curve labeled "first-order") on a network generated using the configuration model (Section 13.2), compared against an average over a large number of simulated epidemics with the same β spreading on the same network (the circular dots). As the figure shows, the agreement between the two is good, but definitely not perfect.

The reason for this disagreement is an interesting one. Equation (17.34) may appear to be a straightforward generalization of the equivalent equation for the fully mixed SI model, Eq. (17.4), but there are some subtleties involved. The right-hand side of the equation contains two average quantities, s_i and x_j, and in multiplying these quantities we are implicitly assuming that the product of the averages is equal to the average of their product. In the fully mixed model this is true (for large n) because of the mixing itself, but in the present case it is, in general, not, because the probabilities are not independent. The quantity s_i measures a vertex's probability of being susceptible and x_j measures the probability of its neighbor being infected. It should come as no surprise that in general these quantities will be correlated between neighboring vertices. Correlations of this type can be incorporated into our calculations, at least approximately, by using a so-called pair approximation or moment closure method, as described in the following section.

17.10.1 PAIR APPROXIMATIONS

Correlations between the disease states of different vertices can be handled by augmenting our theory to take account of the joint probabilities for pairs of vertices to have given pairs of states. To handle such joint probabilities we will need to make our notation a little more sophisticated. Let us denote

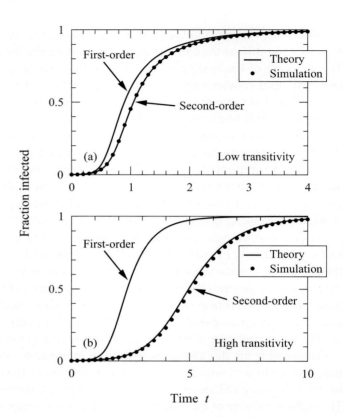

Figure 17.5: Comparison of theory and simulation for the SI model on two different networks. (a) The fraction of infected individuals as a function of time on the giant component of a network with low transitivity (i.e., low clustering coefficient), calculated by numerical solution of the differential equations for the first- and second-order moment closure methods, and by direct simulation. (b) The same comparison for a network with high transitivity. The networks have one million vertices each and the transmission rate is $\beta = 1$ in all cases. Simulation results were averaged over 500 runs.

by $\langle s_i \rangle$ the average probability that vertex i is susceptible. This is the same quantity that we previously called s_i, but, as we will see, it will be useful to indicate the average explicitly with the angle brackets $\langle \ldots \rangle$. If you like, you can think of $s_i(t)$ as now being a variable with value one if i is susceptible at time t and zero otherwise and $\langle s_i \rangle$ as being the average of this quantity over many different instances of disease outbreaks on the same network. Similarly $\langle x_i \rangle$ will be the average probability that i is infected. And $\langle s_i x_j \rangle$ indicates the average probability that i is susceptible *and* j is infected at the same time.

In this notation it is now straightforward to write down a truly exact version of Eq. (17.34), taking correlations into account. It is

$$\frac{\mathrm{d}\langle s_i \rangle}{\mathrm{d}t} = -\beta \sum_j A_{ij} \langle s_i x_j \rangle. \tag{17.44}$$

Equation (17.34) is an approximation to this true equation in which we assume that $\langle s_i x_j \rangle \simeq \langle s_i \rangle \langle x_j \rangle$.

The trouble with Eq. (17.44) is that we cannot solve it directly because it contains the unknown quantity $\langle s_i x_j \rangle$ on the right-hand side. To find this quantity, we need another equation for $\langle s_i x_j \rangle$, which we can deduce as follows. To reach the state in which i is susceptible and j is infected in an SI model it must be the case that both i and j are susceptible to begin with and then j becomes infected. Even though i and j are neighbors j cannot be infected by i, since i is not infected, so j must be infected by some other neighboring vertex k, which itself must therefore be infected. In our new notation, the probability for the configuration in which i and j are susceptible and k is infected is $\langle s_i s_j x_k \rangle$. If we have this configuration, then j will become infected via k with rate β. Summing over all neighbors k except for i, the total rate at which j becomes infected is then $\beta \sum_{k(\neq i)} A_{jk} \langle s_i s_j x_k \rangle$.

Unfortunately, this is not the end of the story because $\langle s_i x_j \rangle$ can also *decrease*—it decreases if i becomes infected. This can happen in two different ways. Either i can be infected by its infected neighbor j, which happens with rate $\beta \langle s_i x_j \rangle$, or it can be infected by another neighbor $l \neq j$ that happens to be infected, which happens with rate $\beta \langle x_l s_i x_j \rangle$. Summing the latter expression over all neighbors l other than j gives a total rate of $\beta \sum_{l(\neq j)} A_{il} \langle x_l s_i x_j \rangle$.

Putting all of these terms together, with minus signs for those that decrease the probability, we get a final equation for $\langle s_i x_j \rangle$ thus:

$$\frac{d\langle s_i x_j \rangle}{dt} = \beta \sum_{k(\neq i)} A_{jk} \langle s_i s_j x_k \rangle - \beta \sum_{l(\neq j)} A_{il} \langle x_l s_i x_j \rangle - \beta \langle s_i x_j \rangle. \qquad (17.45)$$

In theory this equation will now allow us to calculate $\langle s_i x_j \rangle$. In practice, however, it involves yet more terms that we don't know on the right-hand side, the three-variable averages $\langle s_i s_j x_k \rangle$ and $\langle x_l s_i x_j \rangle$. We can write down further equations for these averages but, as you can no doubt guess, those equations involve still higher-order (four-variable) terms, and so forth. The succession of equations will never end—in the jargon of mathematics, it doesn't *close*—and so looks as though it will be of no use to us.[9]

In fact, however, we can still make progress by approximating our three-variable averages with appropriate combinations of one- and two-variable averages, which allows us to close the equations and get a set we can actually solve. This process is called *moment closure* and the method described in this section is called a *moment closure method*. The moment closure method at the

---

[9]On a finite network with n vertices the equations will in fact close once we get all the way up to combinations of n variables, but this limit is not useful in practice as the equations will become unmanageably numerous and complicated long before we reach it.

level of two-variable averages that we discuss here is also called a *pair approximation method*.

In fact, our first attempt at writing equations for the SI model on a network, Eq. (17.34), was itself a simple moment-closure method. We approximated the true equation, Eq. (17.44), by writing $\langle s_i x_j \rangle \simeq \langle s_i \rangle \langle x_j \rangle$, closing the equations at the level of one-variable averages. By going a step further and closing at the pair approximation level of two-variable averages, we can make our equations more precise because we will be taking two-variable correlations into account. In fact, as we will see, this "second-order" moment closure approach is exact for some networks, although only approximate for others. Even in the latter case, however, the method gives a remarkably good approximation. The approximation can be further improved by going to third order, but the equations rapidly become complicated and researchers have rarely used moment closure methods beyond the second-order, pair approximation level.

The pair approximation is relatively straightforward however. Starting with Eq. (17.45) our goal is to approximate the three-variable averages on the right-hand side with lower-order ones. We do this by making use of Bayes theorem for probabilities thus:

$$\langle s_i s_j x_k \rangle = P(i, j \in S, k \in I) = P(i, j \in S)P(k \in I | i, j \in S), \qquad (17.46)$$

where $P(i \in S)$ means the probability that vertex i is in the set S of susceptible vertices. We know that i and j are neighbors in the network and that j and k are neighbors, and our approximation involves assuming that the disease state of k doesn't depend on the disease state of i. This is a good approximation—indeed not an approximation at all—if the only path in the network from i to k is through j. In that case, given that we know j to be susceptible, there is no way that the disease state of i can affect that of k because there is no other path by which the disease could spread from i to k. On the other hand, if there is another path from i to k that avoids vertex j then the disease can spread along that path, which will introduce correlations between i and k and in that case our approximation is just that—an approximation—although as we will see it may be a very good one.

Assuming the state of k to be independent of the state of i, we have

$$P(k \in I | i, j \in S) = P(k \in I | j \in S) = \frac{P(j \in S, k \in I)}{P(j \in S)} = \frac{\langle s_j x_k \rangle}{\langle s_j \rangle}, \qquad (17.47)$$

where we have used Bayes theorem again in the second equality. Putting Eqs. (17.46) and (17.47) together, we then have

$$\langle s_i s_j x_k \rangle = \frac{\langle s_i s_j \rangle \langle s_j x_k \rangle}{\langle s_j \rangle}. \qquad (17.48)$$

We can write a similar expression for the other three-variable average appearing in Eq. (17.45):

$$\langle x_l s_i x_j \rangle = \frac{\langle x_l s_i \rangle \langle s_i x_j \rangle}{\langle s_i \rangle}, \tag{17.49}$$

and, substituting both into Eq. (17.45), we then get the pair approximation equation

$$\frac{d\langle s_i x_j \rangle}{dt} = \beta \frac{\langle s_i s_j \rangle}{\langle s_j \rangle} \sum_{k(\neq i)} A_{jk} \langle s_j x_k \rangle - \beta \frac{\langle s_i x_j \rangle}{\langle s_i \rangle} \sum_{l(\neq j)} A_{il} \langle s_i x_l \rangle - \beta \langle s_i x_j \rangle. \tag{17.50}$$

This equation now contains only averages over two variables at a time. It does also contain a new average $\langle s_i s_j \rangle$ that we have not encountered before, but this can easily be rewritten as $\langle s_i s_j \rangle = \langle s_i(1 - x_j) \rangle = \langle s_i \rangle - \langle s_i x_j \rangle$ and so our equation becomes

$$\frac{d\langle s_i x_j \rangle}{dt} = \beta \frac{\langle s_i \rangle - \langle s_i x_j \rangle}{\langle s_j \rangle} \sum_{k(\neq i)} A_{jk} \langle s_j x_k \rangle - \beta \frac{\langle s_i x_j \rangle}{\langle s_i \rangle} \sum_{l(\neq j)} A_{il} \langle s_i x_l \rangle - \beta \langle s_i x_j \rangle. \tag{17.51}$$

This equation is more complex than Eq. (17.34) but it can be simplified by rewriting it as follows. We define p_{ij} to be the conditional probability that j is infected given that i is not:

$$p_{ij} = P(j \in I | i \in S) = \frac{P(i \in S, j \in I)}{P(i \in S)} = \frac{\langle s_i x_j \rangle}{\langle s_i \rangle}. \tag{17.52}$$

Then the time evolution of p_{ij} is given by

$$
\begin{aligned}
\frac{dp_{ij}}{dt} &= \frac{d}{dt} \left(\frac{\langle s_i x_j \rangle}{\langle s_i \rangle} \right) \\
&= \frac{1}{\langle s_i \rangle} \frac{d\langle s_i x_j \rangle}{dt} - \frac{\langle s_i x_j \rangle}{\langle s_i \rangle^2} \frac{d\langle s_i \rangle}{dt} \\
&= \beta \left(1 - \frac{\langle s_i x_j \rangle}{\langle s_i \rangle} \right) \sum_{k(\neq i)} A_{jk} \frac{\langle s_j x_k \rangle}{\langle s_j \rangle} - \beta \frac{\langle s_i x_j \rangle}{\langle s_i \rangle} \sum_{l(\neq j)} A_{il} \frac{\langle s_i x_l \rangle}{\langle s_i \rangle} \\
&\quad - \beta \frac{\langle s_i x_j \rangle}{\langle s_i \rangle} + \beta \frac{\langle s_i x_j \rangle}{\langle s_i \rangle} \sum_l A_{il} \frac{\langle s_i x_l \rangle}{\langle s_i \rangle} \\
&= \beta(1 - p_{ij}) \sum_{k(\neq i)} A_{jk} p_{jk} - \beta p_{ij} \sum_{l(\neq j)} A_{il} p_{il} - \beta p_{ij} + \beta p_{ij} \sum_l A_{il} p_{il}, \tag{17.53}
\end{aligned}
$$

where we have used Eqs. (17.44) and (17.51) in the third line. All but one of the terms in the two sums over l now cancel out, leaving us with the relatively simple equation

$$\frac{dp_{ij}}{dt} = \beta(1 - p_{ij}) \left[-p_{ij} + \sum_{k(\neq i)} A_{jk} p_{jk} \right], \tag{17.54}$$

where we have used the fact that $A_{ij} = 1$ (since i and j are neighbors). We can also rewrite Eq. (17.44) in terms of p_{ij} thus:

$$\frac{\mathrm{d}\langle s_i \rangle}{\mathrm{d}t} = -\beta \langle s_i \rangle \sum_j A_{ij} p_{ij}, \qquad (17.55)$$

which has the solution

$$\langle s_i(t) \rangle = \langle s_i(0) \rangle \exp\left(-\beta \sum_j A_{ij} \int_0^t p_{ij}(t')\,\mathrm{d}t' \right). \qquad (17.56)$$

Between them, Eqs. (17.54) and (17.56) now give us our solution for the evolution of the epidemic. Note that there are two equations of the form (17.54) for each edge in the network, since p_{ij} is not symmetric in i and j.

Figure 17.5a shows results from a numerical solution of these equations (the curve marked "second-order"), again on a configuration model network and, as the figure shows, the calculation now agrees very well with the simulation results represented by the dots in the figure. By accounting for correlations between adjacent vertices we have created a much more accurate theory.

This near-perfect agreement, however, is something of a special case. Configuration model networks are locally tree-like, meaning they have no short loops, and, as discussed above, our second-order moment closure approximation is exact when non-adjacent vertices i and k have only a single path between them through some intermediate j. When there are no short loops in our network this is true to an excellent approximation—the only other way to get from i to k in such a network is by going around a long loop and the length of such loops dilutes any resulting correlations between the states of i and k, often to the point where they can be ignored. The network used in the simulations for Fig. 17.5a was sufficiently large (a million vertices) and the resulting loops sufficiently long that the pair approximation equations are an excellent approximation, which is why the agreement is so good in the figure.

Unfortunately, as we saw in Section 7.9, most real social networks have a lot of short loops, which raises the question of how well our method does on such networks. Figure 17.5b shows a comparison between the predictions of our equations and direct simulations for a network with many short loops,[10] for both the simple first-order moment closure, Eq. (17.34), and for our more sophisticated second-order approach. As the plot shows, the first-order calculation agrees quite poorly with the simulations, its predictions being inaccurate enough to be of little use in this case. The second-order equations, however,

---

[10]The network was generated using the clustered network model of Ref. [240].

still do remarkably well. Their predictions are not in perfect agreement with the simulations, but they are close.

Thus the pair approximation method offers a significant improvement on networks both with and without short loops, providing a usefully accurate approximation in the former case and being essentially exact in the latter.

17.10.2 DEGREE-BASED APPROXIMATION FOR THE SI MODEL

The analysis of the previous section gives exact equations for the dynamics of the SI model on a network with few short loops and an excellent approximation in other cases. Unfortunately those equations cannot in general be solved analytically, even for simple networks such as those of the configuration model. The solutions presented in Fig. 17.5 were derived by integrating the equations numerically.

In this section we describe an alternative approximate approach that gives good, though not perfect, results in practice and produces equations that can be solved analytically. Moreover, the method can, as we will see, be generalized to other epidemic models such as the SIR model. The method was pioneered by Pastor-Satorras and coworkers [32, 33, 263, 264], though it has precursors in earlier work by May and others [199, 212]. It takes its simplest form when applied to networks drawn from the configuration model and so it is on this model that we focus here, although in principle the method can be extended to other networks.

Consider a disease propagating on a configuration model network, i.e., a random graph with a given degree distribution p_k, as discussed in Chapter 13. As before we focus on outbreaks taking place in the giant component of the network, this being the case of most interest—outbreaks in small components by definition die out quickly and do not give rise to epidemics.

An important point to notice is that the degree distribution of vertices in the giant component of a configuration model network is not the same as the degree distribution of vertices in the network as a whole. As shown in Section 13.8, the probability of a vertex of degree k belonging to the giant component goes up with vertex degree. This means that the degree distribution of vertices in the giant component is skewed towards higher degrees. (For a start, notice that there are trivially no vertices of degree zero in the giant component, since by definition such vertices are not attached to any others.) We will, as before, denote the degree distribution and the excess degree distribution in our calculations by p_k and q_k, but bear in mind that these are for vertices in the giant component, which means they are not the same as the distributions for the network as a whole.

The approximation introduced by Pastor-Satorras *et al.* was to assume that all vertices of the same degree have the same probability of infection at any given time. Certainly this *is* an approximation. The probability of infection of a vertex of degree, say, five situated in the middle of the dense core of a network will presumably be larger than the probability for a vertex of degree five that is out on the periphery. Nonetheless, if the distribution of probabilities for vertices of given degree is relatively narrow it may be a good approximation to set them all equal to the same value. And in practice, as we have said, the approximation appears to work very well.

Returning, for the sake of simplicity, to our earlier notation style, let us define $s_k(t)$ and $x_k(t)$ to be the probabilities that a vertex with degree k is susceptible or infected, respectively, at time t. Now consider a susceptible vertex A. To become infected, A has to contract the infection from one of its network neighbors. The probability that a particular neighbor B is infected depends on the neighbor's degree, but we must be careful. By hypothesis vertex A is not infected and so B cannot have caught the disease from A. If B is infected it must have caught the disease from one of its remaining neighbors. In effect this reduces the degree of B by one—B will have the same probability of being infected at the current time as the average vertex with degree one less. To put that another way, B's probability of infection depends upon its excess degree, the number of edges it has other than the edge we followed from A to reach it. B's probability of infection is thus x_k, but where k indicates the excess degree, not the total degree.

The advantage of the degree-based approach now becomes clear: the probability of B being infected depends, in this approach, only on B's excess degree and not on A's degree. By contrast, the conditional probability p_{ij} in our earlier formalism was a function of two indices, making the equations more complicated. To derive the equations for the degree-based approximation, consider the probability that vertex A becomes infected between times t and $t + dt$. To become infected it must catch the disease from one of its neighbors, meaning that neighbor must be infected. The probability of a neighbor being infected is x_k where k is the excess degree of the neighbor, and the excess degree is distributed according to the distribution q_k of Eq. (13.46), which means that the average probability that the neighbor is infected is

$$v(t) = \sum_{k=0}^{\infty} q_k x_k(t). \tag{17.57}$$

If the neighbor is infected then the probability that the disease will be transmitted to vertex A in the given time interval is $\beta\, dt$. Then the total probability of transmission from a single neighbor during the time interval is $\beta v(t)\, dt$ and the probability of transmission from any neighbor is $\beta k v(t)\,dt$, where k is now the number of A's neighbors. In addition we also require that A itself be susceptible, which happens with probability $s_k(t)$, so our final probability that A becomes infected is $\beta k v s_k\,dt$. Thus the rate of change of s_k is given by

$$\frac{ds_k}{dt} = -\beta k v s_k. \tag{17.58}$$

This equation can be solved exactly. We can formally integrate it thus:

$$s_k(t) = s_0 \exp\left(-\beta k \int_0^t v(t')\,dt'\right), \tag{17.59}$$

where we have fixed the integration constant so that all vertices have probability s_0 of being susceptible at $t = 0$. Although we don't yet know the form of the function $v(t)$ this expression tells us that s_k depends on k as a simple power of some universal k-independent function $u(t)$:

$$s_k(t) = s_0[u(t)]^k, \tag{17.60}$$

where in this case

$$u(t) = \exp\left(-\beta \int_0^t v(t')\,dt'\right). \tag{17.61}$$

Writing $x_k = 1 - s_k$ and substituting into Eq. (17.57) we then get

$$v(t) = \sum_{k=0}^{\infty} q_k(1 - s_k) = \sum_{k=0}^{\infty} q_k(1 - s_0 u^k) = 1 - s_0 g_1(u), \tag{17.62}$$

where $g_1(u)$ is the generating function for q_k and we have made use of $\sum_k q_k = 1$. Substituting Eq. (17.60) into Eq. (17.58) then gives us

$$\frac{du}{dt} = -\beta u v = -\beta u[1 - s_0 g_1(u)]. \tag{17.63}$$

This is a straightforward linear differential equation for u that, given the degree distribution, can be solved by direct integration.

Finally, to calculate the total fraction $x(t)$ of infected individuals in the network we average over k thus:

$$x(t) = \sum_{k=1}^{\infty} p_k x_k(t) = \sum_{k=1}^{\infty} p_k(1 - s_0 u^k) = 1 - s_0 g_0(u). \tag{17.64}$$

Notice that the sums here start at $k = 1$ because there are no vertices of degree zero in the giant component.

Equations (17.63) and (17.64) between them give us an approximate solution for the SI model on the giant component of a configuration model network with any degree distribution.

Although the solution is elegant in principle, in most practical cases we cannot integrate Eq. (17.63) in closed form. Even without completing the integral, however, we can already see the basic form of the solution. First of all, at time $t = 0$ we have $u = 1$ by Eq. (17.61). Since $v(t)$ is, by definition, positive and non-decreasing with time, the same equation also implies that $u(t)$ always decreases and tends to zero as $t \rightarrow \infty$. This implies that at long times Eq. (17.63) becomes

$$\frac{\mathrm{d}u}{\mathrm{d}t} = -\beta u \left[1 - s_0 g_1(0)\right] = -\beta u (1 - s_0 p_1 / \langle k \rangle), \tag{17.65}$$

and hence $u(t)$ decays exponentially as $\mathrm{e}^{-\beta(1 - s_0 p_1 / \langle k \rangle)t}$. Assuming the infection starts with only one or a handful of cases, so that $s_0 = 1 - c/n$ for some constant c, we have $s_0 \rightarrow 1$ in the limit of large n and

$$u(t) \sim \mathrm{e}^{-\beta(1 - p_1 / \langle k \rangle)t}. \tag{17.66}$$

Note that the long-time behavior is dictated by the fraction p_1 of vertices with total degree one. This is because these are the last vertices to be infected—individuals with only one contact are best protected from infection, although even they are guaranteed to become infected in the end. In networks where the fraction p_1 is zero or very small we have $u(t) \sim \mathrm{e}^{-\beta t}$ and the functional form of the long-time behavior depends only on the infection rate and not on the network structure.

At short times we can write $u = 1 - \epsilon$ and to leading order in ϵ Eq. (17.63) becomes

$$\frac{\mathrm{d}\epsilon}{\mathrm{d}t} = \beta \left[x_0 + (g_1'(1) - 1)\epsilon\right], \tag{17.67}$$

where $x_0 = 1 - s_0$ is the initial value of x_k. This has solution

$$\epsilon(t) = \frac{\beta x_0}{g_1'(1) - 1} \left[\mathrm{e}^{\beta(g_1'(1) - 1)t} - 1\right], \tag{17.68}$$

where we have made use of the initial condition $\epsilon = 0$. Equivalently we can write[11]

$$u(t) = 1 - \epsilon = 1 - \frac{\beta x_0}{g_1'(1) - 1} \left[\mathrm{e}^{\beta(g_1'(1) - 1)t} - 1\right]. \tag{17.69}$$

---

[11]This equation will diverge if $g_1'(1) = 1$. However, since we are performing the calculation on the giant component of the network, and since the giant component only exists if $g_1'(1) > 1$—see Section 13.8—we can safely rule out this possibility.

Given the short- and long-time behavior and the fact that $u(t)$ is monotonically decreasing, we can now guess that $u(t)$ has a form something like Fig. 17.6. Then, since g_0 is a monotonically increasing function of its argument, $x(t)$ in Eq. (17.64) has a similar shape but turned upside down, so that it looks qualitatively similar to the curve for the fully mixed version of the model shown in Fig. 17.1, although quantitatively it may be different.

The initial growth of $x(t)$ can be calculated by putting $u = 1 - \epsilon$ in Eq. (17.64) to give $g_0(1 - \epsilon) \simeq 1 - g_0'(1)\epsilon$ and

$$x(t) = 1 - s_0 + s_0 g_0'(1)\epsilon$$
$$= x_0 \left[1 + \frac{\beta g_0'(1)}{g_1'(1) - 1} \left[e^{\beta(g_1'(1)-1)t} - 1 \right] \right], \quad (17.70)$$

where we have again set $s_0 = 1$. Thus, as we would expect, the initial growth of infection is roughly exponential.

The appearance of $g_1'(1)$ in Eq. (17.70) is of interest. As we saw in Eq. (13.68), $g_1'(1)$ is equal to the ratio c_2/c_1 of the average number of second neighbors to first neighbors of a vertex and hence is a measure of how fast the network branches as we move away from the vertex where the disease first starts. It should be not surprising therefore (though it's still satisfying) to see that this same quantity—along with the transmission rate β—controls the rate at which the disease spreads in our SI model.

Another interesting feature of the model is the behavior of the quantities $s_k(t)$ that measure the probability that a vertex of a given degree is susceptible. Since these quantities are all proportional to powers of $u(t)$—see Eq. (17.60)—they form a family of curves as shown in Fig. 17.7. Thus, as we might expect, the vertices with highest degree are the ones that become infected first, on average, while those with low degree hold out longer.

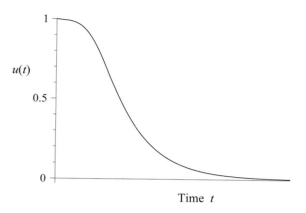

Figure 17.6: The function $u(t)$ in the solution of the SI model. Generically we expect $u(t)$ to have the form sketched here: it is monotonically decreasing from an initial value of 1 and has an exponential tail at long times.

17.11 Time-dependent properties of the SIR model

It is relatively straightforward to extend the techniques of Section 17.10 to the more complex (and interesting) SIR model. Again we concentrate on outbreaks taking place in the giant component of the network and we define s_i, x_i, and r_i to be the probabilities that vertex i is susceptible, infected, or recovered respectively. The evolution of s_i is (approximately) governed by the same equation

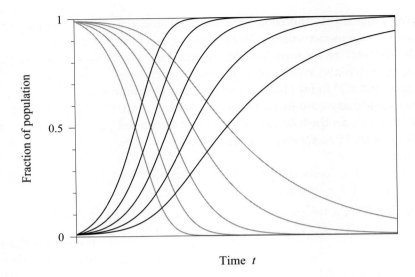

Figure 17.7: Fractions of susceptible and infected vertices of various degrees in the SI model. The various curves show the fraction of vertices of degree k that are susceptible (gray) and infected (black) as a function of time for $k = 1, 2, 4, 8$, and 16. The highest values of k give the fastest changing (leftmost) curves and the lowest values the slowest changing. The curves were calculated by integrating Eq. (17.63) numerically with $\beta = 1$ and a Poisson degree distribution with mean degree four.

as before:

$$\frac{\mathrm{d}s_i}{\mathrm{d}t} = -\beta s_i \sum_j A_{ij} x_j, \tag{17.71}$$

while x_i and r_i obey

$$\frac{\mathrm{d}x_i}{\mathrm{d}t} = \beta s_i \sum_j A_{ij} x_j - \gamma x_i, \tag{17.72}$$

$$\frac{\mathrm{d}r_i}{\mathrm{d}t} = \gamma x_i, \tag{17.73}$$

where, as previously, γ is the recovery rate, i.e., the probability per unit time that an infected individual will recover.[12]

---

[12]This contrasts with the approach we took in Section 17.8 where all vertices remained infected for the same amount of time and then recovered. Thus the model studied in this section is not exactly the same as that of Section 17.8, being more similar to the traditional SIR model of Section 17.3. We will see some minor consequences of this difference shortly.

We can choose the initial conditions in various ways, but let us here make the same assumption as we did for the SI model, that at $t = 0$ we have a small number c of infected individuals and everyone else is susceptible, so that $s_i(0) = 1 - c/n$, $x_i(0) = c/n$, and $r_i(0) = 0$.

As with the SI model we cannot solve these equations exactly, but we can extract some useful results by examining their behavior at early times. In the limit $t \to 0$, x_i is small and $s_i = 1 - c/n$, which tends to 1 as n becomes large, so Eq. (17.72) can be approximated as

$$\frac{\mathrm{d}x_i}{\mathrm{d}t} = \beta \sum_j A_{ij} x_j - \gamma x_i = \sum_j (\beta A_{ij} - \gamma \delta_{ij}) x_j, \qquad (17.74)$$

where δ_{ij} is the Kronecker delta. This can be written in matrix form as

$$\frac{\mathrm{d}\mathbf{x}}{\mathrm{d}t} = \beta \mathbf{M} \mathbf{x}, \qquad (17.75)$$

where $\mathbf{M}$ is the $n \times n$ symmetric matrix

$$\mathbf{M} = \mathbf{A} - \frac{\gamma}{\beta} \mathbf{I}. \qquad (17.76)$$

As before we can write $\mathbf{x}$ as a linear combination of eigenvectors, though they are now eigenvectors of $\mathbf{M}$ rather than of the simple adjacency matrix as in the case of the SI model. But now we notice a useful thing: since $\mathbf{M}$ differs from the adjacency matrix only by a multiple of the identity matrix, it has the same eigenvectors $\mathbf{v}_r$ as the adjacency matrix:

$$\mathbf{M} \mathbf{v}_r = \mathbf{A} \mathbf{v}_r - \frac{\gamma}{\beta} \mathbf{I} \mathbf{v}_r = \left(\kappa - \frac{\gamma}{\beta} \right) \mathbf{v}_r. \qquad (17.77)$$

Only the eigenvalue has been shifted downward by γ/β.

The equivalent of Eq. (17.42) is now

$$\mathbf{x}(t) = \sum_{r=1}^{n} a_r(0) \mathbf{v}_r \mathrm{e}^{(\beta \kappa_r - \gamma)t}. \qquad (17.78)$$

Note that the exponential constant now depends on $\beta \kappa_r - \gamma$ and so is a function not only of the adjacency matrix and the infection rate but also of the recovery rate, as we would expect—the faster people recover from infection the less chance they have to spread the disease and the slower it will spread.

Again the fastest growing term is that corresponding to the most positive eigenvalue κ_1 of the adjacency matrix and individuals having the highest eigenvector centrality get infected first. Note, however, that it is now possible for γ to be sufficiently large that the exponential constant in the leading term becomes negative, meaning that the term decays exponentially rather than grows. And if the leading term decays, so necessarily do all other terms, and so the total number of infected individuals will decay over time and the disease will die out without causing an epidemic.

The point at which this happens is the epidemic threshold for our model and it occurs at $\beta \kappa_1 - \gamma = 0$, or equivalently

$$\frac{\beta}{\gamma} = \frac{1}{\kappa_1}. \tag{17.79}$$

Thus the position of the epidemic threshold depends on the leading eigenvalue of the adjacency matrix. If the leading eigenvalue is small, then the probability of infection β must be large, or the recovery rate γ small, for the disease to spread. In other words a small value of κ_1 makes it harder for the disease to spread and a large value easier. This makes intuitive sense, since large values of κ_1 correspond to denser adjacency matrices and smaller values to sparser ones.

As in the case of the SI model, Eqs. (17.71–17.73) are only approximate, because they neglect correlations between the states of adjacent vertices. And as before we can allow for these correlations by using a pair approximation, but here we take a different approach and consider instead the equivalent of the methods of Section 17.10.2 for the SIR model.[13]

17.11.1 DEGREE-BASED APPROXIMATION FOR THE SIR MODEL

As with the SI model, let us make the approximation that all vertices with the same degree behave in the same way. Again we concentrate on the example of the configuration model [229] and on outbreaks taking place in the giant component of the network. We define $s_k(t)$, $x_k(t)$, and $r_k(t)$ to be the probabilities that a vertex with degree k is susceptible, infected, or recovered, respectively, at time t. Then we consider the state of a vertex B that is the neighbor of a susceptible vertex A. For such a vertex to be infected it must have contracted the disease from one of its neighbors other than A, since A is susceptible. That means, as before, that B's probability of being infected is given by x_k, but with k equal to the excess degree, which is one less than the total degree. And the

---

[13]We can see that the approach of this section cannot be exactly correct from the behavior of Eq. (17.79) on very sparse networks. On a vanishingly sparse network, with only a very few edges and no giant component, κ_1 becomes very small, though still non-zero. On such a network Eq. (17.79) implies that we could, nonetheless, have an epidemic if β is very large or γ very small. Clearly this is nonsense—there can be no epidemic in a network with no giant component. Thus the equation cannot be exactly correct.

probability that B is recovered depends only on the probability that it was previously infected, which is given by r_k where k is the excess degree, and the probability s_k of being susceptible can be derived from $s_k + x_k + r_k = 1$.

Armed with these observations, we can now write down an appropriate set of equations for the epidemic. The rate at which the probability of being susceptible decreases is given by the same equation as before, Eq. (17.58):

$$\frac{ds_k}{dt} = -\beta k v s_k, \tag{17.80}$$

where $v(t)$ is the average probability that a neighbor is infected:

$$v(t) = \sum_{k=0}^{\infty} q_k x_k(t), \tag{17.81}$$

and the equations for x_k and r_k are

$$\frac{dx_k}{dt} = \beta k v s_k - \gamma x_k, \tag{17.82}$$

$$\frac{dr_k}{dt} = \gamma x_k. \tag{17.83}$$

We can solve these equations exactly by a combination of the methods of Sections 17.3 and 17.10. We define the average probability that a neighbor is recovered thus:

$$w(t) = \sum_{k=0}^{\infty} q_k r_k(t). \tag{17.84}$$

Then, using Eqs. (17.81) and (17.83), we find

$$\frac{dw}{dt} = \sum_{k=0}^{\infty} q_k \frac{dr_k}{dt} = \gamma \sum_{k=0}^{\infty} q_k x_k = \gamma v, \tag{17.85}$$

which we use to eliminate v from Eq. (17.80), giving

$$\frac{ds_k}{dt} = -\frac{\beta}{\gamma} k \frac{dw}{dt} s_k. \tag{17.86}$$

This equation can be integrated to give

$$s_k = s_0 \exp\left(-\frac{\beta}{\gamma} k w\right), \tag{17.87}$$

where we have fixed the constant of integration so that at $t = 0$ all vertices have the same probability s_0 of being susceptible and there are no recovered vertices ($w = 0$).

Equation (17.87) implies that s_k is again proportional to a power of a universal function:

$$s_k(t) = s_0 \big[u(t) \big]^k, \tag{17.88}$$

where in this case

$$u(t) = e^{-\beta w / \gamma}. \tag{17.89}$$

Then, using Eq. (17.87), we find

$$v(t) = \sum_k q_k x_k = \sum_k q_k (1 - r_k - s_k) = 1 - w(t) - s_0 \sum_k q_k u^k$$

$$= 1 + \frac{\gamma}{\beta} \ln u - s_0 g_1(u), \tag{17.90}$$

and Eq. (17.85) becomes

$$\frac{du}{dt} = -\beta u \left[1 + \frac{\gamma}{\beta} \ln u - s_0 g_1(u) \right]. \tag{17.91}$$

This is the equivalent for the SIR model of Eq. (17.63), and indeed differs from that equation only by the new term in $\ln u$ on the right-hand side.

As before, Eq. (17.91) is a first-order linear differential equation in u and hence can, in principle, be solved by direct integration, although for any given degree distribution the integral may not have a closed-form solution. Once we have $u(t)$ the probability s_k of a vertex being susceptible is given by Eq. (17.88), or we can write the total fraction of susceptibles as

$$s(t) = \sum_k p_k s_k = s_0 \sum_k p_k u^k = s_0 g_0(u). \tag{17.92}$$

Solving for x_k and r_k requires a little further work but with perseverance it can be achieved.[14] Figure 17.8 shows the equivalent of Fig. 17.7 for vertices of a range of degrees. As we can see, the solution has the expected form, with the number of infected individuals rising, peaking, then dropping off as the system evolves to a final state in which some fraction of the population is recovered from the disease and some fraction has never caught it (and never will). Among vertices of different degrees the number infected goes up sharply with degree, as we would expect.

Even in cases where the integral in Eq. (17.91) cannot be performed, our solution can still shed light on features of the epidemic. Consider for example

---

[14]We observe that

$$\frac{d}{dt} \left(e^{\gamma t} x_k \right) = e^{\gamma t} \left(\frac{dx_k}{dt} + \gamma x_k \right) = e^{\gamma t} \beta k v s_k,$$

where we've used Eq. (17.82) in the second equality. Integrating and using Eqs. (17.81) and (17.88),

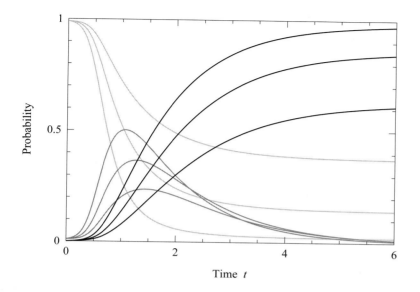

Figure 17.8: Fractions of susceptible, infected, and recovered vertices of various degrees in the SI model. The fraction of vertices of degree k that are susceptible (light gray), infected (darker gray), and recovered (black) as a function of time for $k = 1, 2, 4$ and $\beta = \gamma = 1$ on a network with an exponential degree distribution (Eq. (13.129)) with $\lambda = 0.2$. The highest values of k give the fastest growing numbers of infected and recovered vertices and the lowest values the slowest growing.

the long-time behavior. In the limit of long time we expect that the number of infected individuals will vanish leaving some individuals recovered and some who have never caught the disease. At $t = \infty$ the total fraction $r(t)$ of recovered individuals measures the overall size of the outbreak of the disease and is given by

$$r(\infty) = 1 - s(\infty) = 1 - g_0(u(\infty)).\qquad(17.93)$$

where we have set $s_0 = 1$ as before on the assumption that the system is large and the number of initially infected individuals small.

We can find the stationary value of u by setting $du/dt = 0$ in Eq. (17.91) to

---

we then have

$$x_k(t) = e^{-\gamma t}\left[x_0 + \beta k s_0 \int_0^t e^{\gamma t'} [u(t')]^k \left(1 + \frac{\gamma}{\beta}\ln u(t') - s_0 g_1(u(t'))\right) dt'\right],$$

and $r_k = 1 - s_k - x_k$.

give

$$1 + \frac{\gamma}{\beta} \ln u - g_1(u) = 0. \tag{17.94}$$

In the special case where the outbreak is small, so that the final value of u is close to 1, we can expand $\ln u = \ln[1 + (u - 1)] \simeq u - 1$ and Eq. (17.94) becomes

$$u \simeq 1 - \frac{\beta}{\gamma} + \frac{\beta}{\gamma} g_1(u). \tag{17.95}$$

Equations (17.93) and (17.95) are similar in form to Eqs. (17.27) and (17.28) which give the final size of the outbreak in our treatment of the SIR model using percolation theory. The reason why Eq. (17.95) is only approximate in the present case where Eq. (17.28) was exact is that the model treated in this section is slightly different from the one treated earlier, having (as discussed in footnote 12 on page 662) a constant probability γ per unit time of recovery from disease for each infected individual as opposed to a fixed infection time for the model of Section 17.8.1.

We can also examine the early-time behavior of the outbreak by looking at the behavior of Eq. (17.91) close to $u = 1$. Writing $u = 1 - \epsilon$ and keeping terms to leading order in ϵ we get

$$\frac{d\epsilon}{dt} = [\beta g_1'(1) - \gamma]\epsilon, \tag{17.96}$$

assuming $s_0 = 1$ again, which means that

$$u(t) = 1 - \epsilon(t) = 1 - e^{(\beta g_1'(1) - \gamma)t}. \tag{17.97}$$

This is similar to Eq. (17.69) for the SI model, except for the inclusion of the term in γ. The fraction of susceptible degree-k vertices is given by

$$s_k(t) = u^k = \left[1 - e^{(\beta g_1'(1) - \gamma)t}\right]^k \simeq 1 - e^{k(\beta g_1'(1) - \gamma)t}, \tag{17.98}$$

and total cases of the disease, infected and recovered, which is just $1 - s_k$, grows exponentially as $e^{k[\beta g_1'(1) - \gamma]t}$.

The epidemic threshold for the model is the line that separates an initially growing number of cases of the disease from an initially decreasing one and is given in this case by the point at which the exponential constant in Eq. (17.98) equals zero, which gives

$$\frac{\beta}{\gamma} = \frac{1}{g_1'(1)}. \tag{17.99}$$

This result is similar in form to Eq. (17.79) for the epidemic threshold on a general network,[15] but with the leading eigenvalue of the adjacency matrix κ_1 replaced with $g_1'(1)$. It also looks similar to Eq. (17.29) for the percolation threshold for bond percolation, but this similarity is somewhat deceptive. In fact, the result most nearly corresponding to this one in the percolation treatment is Eq. (17.30). If we equate our recovery rate γ with the reciprocal of the infectiousness time τ in that previous treatment, then the two are roughly equivalent when the epidemic threshold is low, meaning either that β is small or that γ is large. If the threshold is higher then the match between the two models is poorer, which is again a result of the fact that the models are defined in slightly different ways.

17.12 TIME-DEPENDENT PROPERTIES OF THE SIS MODEL

It is straightforward to extend our methods to the SIS model also. By analogy with Eqs. (17.71–17.73) we have

$$\frac{ds_i}{dt} = -\beta s_i \sum_j A_{ij} x_j + \gamma x_i, \qquad (17.100a)$$

$$\frac{dx_i}{dt} = \beta s_i \sum_j A_{ij} x_j - \gamma x_i \qquad (17.100b)$$

for the SIS model. Caveats similar to those for previous models apply here: these equations ignore correlations between the states of adjacent vertices and hence are only an approximation.

Equations (17.100a) and (17.100b) are not independent since $s_i + x_i = 1$, so only one is needed to form a solution. Taking the second and eliminating s_i we get

$$\frac{dx_i}{dt} = \beta(1 - x_i) \sum_j A_{ij} x_j - \gamma x_i. \qquad (17.101)$$

At early times, assuming as before that $x_i(0) = x_0 = 1 - c/n$ for all i and constant c, we can drop terms at quadratic order in small quantities to get

$$\frac{dx_i}{dt} = \beta \sum_j A_{ij} x_j - \gamma x_i, \qquad (17.102)$$

which is identical to Eq. (17.74) for the SIR model at early times. Hence we can immediately conclude that the early-time behavior of the model is the same,

---

[15]And like Eq. (17.79) it is also clearly wrong on sparse networks for the same reasons—see footnote 13 on page 664.

with initially exponential growth and an epidemic threshold given by

$$\frac{\beta}{\gamma} = \frac{1}{\kappa_1}. \tag{17.103}$$

(See Eq. (17.79).) Also as in the SIR model the probability of infection of a given vertex at early times will be proportional to the vertex's eigenvector centrality.

At late times we expect the probability of infection to settle to a constant endemic level, which we can calculate by setting $dx_i/dt = 0$ in Eq. (17.101) and rearranging, to give

$$x_i = \frac{\sum_j A_{ij}x_j}{\gamma/\beta + \sum_j A_{ij}x_j}. \tag{17.104}$$

Typically we cannot derive a closed-form solution for x_i from this expression, but we can solve it numerically by iteration starting from a random initial guess. We can also see the general form the solution will take by considering limiting cases. If β/γ is large, meaning that we are well above the epidemic threshold given in Eq. (17.103), then we can ignore the term γ/β in the denominator and $x_i \simeq 1$ for all i, meaning that essentially all vertices will be infected all the time. This makes good sense since if β/γ is large then the rate of infection is very high while the rate of recovery is negligible.

Conversely, if β/γ is only just above the epidemic threshold level set by Eq. (17.103) then x_i will be small—the disease only just manages to stay alive—and we can ignore the sum in the denominator of Eq. (17.104) so that

$$x_i \simeq \frac{\beta}{\gamma} \sum_j A_{ij}x_j, \tag{17.105}$$

or

$$\kappa_1 x_i \simeq \sum_j A_{ij}x_j, \tag{17.106}$$

where we have used Eq. (17.103). This implies that x_i is proportional to the leading eigenvector of the adjacency matrix or, equivalently, proportional to the eigenvector centrality. (Note that this is at late times so this result is distinct from the finding above that x_i is proportional to eigenvector centrality at early times.)

Thus the long-time endemic disease behavior of the SIS model varies from a regime just above the epidemic threshold in which the probability of a vertex being infected is proportional to its eigenvector centrality, to a regime well above the threshold in which essentially every vertex is infected at all times.

17.12.1 DEGREE-BASED APPROXIMATION FOR THE SIS MODEL

We can also write down approximate equations for the evolution of the SIS model in which, as in Sections 17.10.2 and 17.11.1, we assume that the probability of infection is the same for all vertices with a given degree. Focusing once again on configuration model networks, the equivalent of Eqs. (17.80–17.82) is

$$\frac{ds_k}{dt} = -\beta kv s_k + \gamma x_k, \tag{17.107a}$$

$$\frac{dx_k}{dt} = \beta kv s_k - \gamma x_k, \tag{17.107b}$$

where the variables s_k and x_k are as before, and again

$$v(t) = \sum_{k=0}^{\infty} q_k x_k(t). \tag{17.108}$$

As before Eqs. (17.107a) and (17.107b) are not independent and only one is need to form a solution. Let us take the second and rewrite it using $s_k = 1 - x_k$ to give

$$\frac{dx_k}{dt} = \beta kv(1 - x_k) - \gamma x_k. \tag{17.109}$$

Unfortunately, there is no known complete solution to this equation but we can once again find its behavior at early and late times.

Assuming, as previously, that our epidemic starts off with only a single case or a small number of cases, the probability x_k of being infected at early times is c/n for constant c and hence small in the limit of large n. Dropping terms of second order in small quantities then gives us the linear equation

$$\frac{dx_k}{dt} = \beta kv - \gamma x_k, \tag{17.110}$$

which can be rewritten using an integrating factor to read

$$\frac{d}{dt}\left(e^{\gamma t} x_k\right) = e^{\gamma t}\frac{dx_k}{dt} + \gamma e^{\gamma t} x_k = \beta k e^{\gamma t} v, \tag{17.111}$$

and hence integrated to give

$$x_k(y) = \beta k e^{-\gamma t} \int_0^t e^{\gamma t'} v(t')\, dt'. \tag{17.112}$$

Thus $x_k(t)$ for short times takes the form

$$x_k = ku(t), \tag{17.113}$$

671

where $u(t)$ is some universal, k-independent function. Substituting into Eqs. (17.108) and (17.110), we then have

$$v(t) = u(t) \sum_{k=0} kq_k = g'_1(1)u(t),$$ (17.114)

and

$$\frac{du}{dt} = [\beta g'_1(1) - \gamma]u(t).$$ (17.115)

Thus we have exponential growth or decay of the epidemic at early times, with the epidemic threshold separating the two falling at the point where $\beta g'_1(1) - \gamma = 0$, or

$$\frac{\beta}{\gamma} = \frac{1}{g'_1(1)},$$ (17.116)

just as for the SIR model (see Eq. (17.99)).

At late times the disease to settles down into an endemic state in which some constant fraction of the population is infected. We can solve for this endemic state by setting $dx_k/dt = 0$ for all k in Eq. (17.109) to give

$$x_k = \frac{kv}{kv + \gamma/\beta}.$$ (17.117)

Substituting this expression into Eq. (17.108), we then find that

$$\sum_{k=0}^{\infty} \frac{kq_k}{kv + \gamma/\beta} = 1.$$ (17.118)

In general there is no closed-form solution to this implicit equation for v, although it can typically be solved numerically for any given q_k, and given the value we can then get x_k from Eq. (17.117).

What we can tell from Eq. (17.118) is that, given the degree distribution, v at late times is a function solely of β/γ (or γ/β if you prefer) and hence x_k is solely a function of β/γ and k. Moreover, in order for Eq. (17.118) to be satisfied v must be an increasing function of β/γ—as β gets larger or γ smaller, v must increase in order to keep the sum in the equation equal to one. This means that x_k will also be an increasing function of β/γ. (Equation (17.117) implies that it is an increasing function of k as well.) Thus the equations give us a qualitative picture of the behavior of the SIS model, although quantitative details require a numerical solution.

We have in this chapter only brushed the surface of what is possible in the modeling of epidemics spreading across networks. We can extend our studies to more complicated network structures, such as networks with degree corre-lations, networks with transitivity, networks with community structure, and even epidemics on empirically observed networks. More complicated mod-els of the spread of infection are also possible, such as the SIRS model men-tioned in Section 17.5, as well as models that incorporate birth, death, or ge-ographic movement of individuals [17, 156]. In recent years, scientists have developed extremely sophisticated computer models of disease spread using complex simulations of the behavior patterns of human populations, including models of entire cities down to the level of individual people, cars, and build-ings [110], and models of the international spread of disease that incorporate detailed data on the flight patterns and timetables of international airlines [79]. These developments, however, are beyond the scope of our necessarily brief treatment in this chapter.

PROBLEMS

17.1 Consider an SIR epidemic on a configuration model network with exponential degree distribution $p_k = (1 - e^{-\lambda})e^{-\lambda k}$.

a) Using the results of Section 16.2.1 write down an expression for the probability u appearing in Eq. (17.27) in terms of ϕ and λ.

b) Hence find an expression for the probability that a vertex is infected by the disease if it has degree k.

c) Evaluate this probability for the case $\lambda = 1$ and $\phi = 0.9$, for $k = 0, 1$, and 10.

17.2 Consider the spread of an SIR-type disease on a network in which some fraction of the individuals have been vaccinated against the disease. We can model this situation using a joint site/bond percolation model in which a fraction ϕ_s of the vertices are occupied, to represent the vertices not vaccinated, and a fraction ϕ_b of the edges are occupied to represent the edges along which contact takes place.

a) Show that the fraction S of individuals infected in the limit of long time is given by the solution of the equations

$$S = \phi_s[1 - g_0(u)], \qquad u = 1 - \phi_s\phi_b + \phi_s\phi_b g_1(u),$$

where $g_0(z)$ and $g_1(z)$ are the generating functions for the degree distribution and excess degree distribution, as usual.

b) Show that for a given probability of contact ϕ_b the fraction of individuals that need to be vaccinated to prevent spread of the disease is $1 - 1/[\phi_b g_1'(1)]$.

17.3 We have been concerned in this chapter primarily with *epidemic* disease out-breaks, meaning outbreaks that affect a finite fraction of all individuals in a network. Consider, by contrast, a small SIR outbreak—an outbreak that corresponds to one of the non-giant percolation clusters in the bond percolation approach of Section 17.8—occurring on a configuration model network with degree distribution p_k.

a) What is the probability of such an outbreak occurring if the disease starts at a vertex chosen uniformly at random from the whole network (including vertices both within and outside the giant component)?

b) Show that if the probability of transmission along an edge is ϕ then the generating function $h_0(z)$ for the probability π_s that the outbreak has size s is given by the equations

$$h_0(z) = zg_0(h_1(z)), \qquad h_1(z) = 1 - \phi + \phi z g_1(h_1(z)),$$

where $g_0(z)$ and $g_1(z)$ are the generating functions for the degree distribution and excess degree distribution respectively.

c) What is the mean size of such an outbreak?

17.4 Consider an SI-type epidemic spreading on the giant component of a k-regular random graph, i.e., a configuration model network in which all vertices have the same degree k. Assume that some number c of vertices, chosen at random, are infected at time $t = 0$.

a) Show using the results of Section 17.10 that the probability of infection of every vertex increases at short times as $e^{\beta kt}$.

b) Show that within the first-order moment closure approximation of Eq. (17.35) the average probability of infection x of every vertex is the same and give the differential equation it satisfies.

c) Hence show that

$$x(t) = \frac{ce^{\beta kt}}{n - c + ce^{\beta kt}}.$$

d) Find the time at which the "inflection point" of the epidemic occurs, the point at which the rate of appearance of new disease cases stops increasing and starts decreasing.

17.5 Consider a configuration model network containing vertices of degrees 1, 2, and 3 only, such that the fractions of vertices of each degree in the giant component are $p_1 = 0.3$, $p_2 = 0.3$, and $p_3 = 0.4$.

a) Find an expression for the excess-degree generating function $g_1(z)$ appearing in Eq. (17.63).

b) Hence, by solving Eq. (17.63), find an expression for t as a function of u for an SI epidemic on the giant component of the network, assuming that $s_0 \simeq 1$, and with initial condition $u(0) = 1 - \epsilon$, where ϵ is small.

c) Show that in the limit of long times the number of susceptibles falls off in proportion to $e^{-21\beta t/2}$.

17.6 Consider the spread of an SIR-type disease in a network in which some fraction of the individuals have been vaccinated against the disease. We can model this situation using a joint site/bond percolation model in which a fraction ϕ_s of the vertices are occupied, to represent the vertices not vaccinated, and a fraction ϕ_b of the edges are occupied to represent the edges along which contact takes place.'

a) Show that the fraction S of individuals infected in the limit of long time is given by the solution of the equations

$$S = \phi_s[1 - g_0(u)], \qquad u = 1 - \phi_s\phi_b + \phi_s\phi_b g_1(u),$$

where $g_0(z)$ and $g_1(z)$ are the generating functions for the degree distribution and excess degree distribution, as usual.

b) Show that for a given probability of contact ϕ_b the fraction of individuals that need to be vaccinated to prevent spread of the disease is $1 - 1/[\phi_b g_1'(1)]$.

CHAPTER 18

DYNAMICAL SYSTEMS ON NETWORKS

*A discussion of dynamical systems on networks, a
subject area that is in its infancy but about which we
nonetheless have some interesting results*

T HE epidemic models of Chapter 17 are a particular example of the more
general concept of dynamical systems on networks. A dynamical system
is any system whose state, as represented by some set of quantitative variables, changes over time according to some given rules or equations. Dynamical systems come in continuous- and discrete-time varieties and can be either
deterministic or stochastic. The epidemic models we looked at, for instance,
were continuous-time dynamical systems because their equations described
the continuous-time variation of the variables. They were also deterministic
because the equations we wrote down exactly determine the values of all variables for all time: there was no random or external element affecting the evolution whose value was not known in advance. On the other hand, an explicit
computer simulation of, say, an SI epidemic model on a network would be
a stochastic dynamical system and might use either continuous- or discrete-
time. The stochastic element in this case corresponds to the chance infection of
a susceptible individual by an infectious neighbor. And time might be represented in discrete time-steps, although it might not, depending on the decision
of the researcher.

Many other real-world processes—or simplified models of real-world processes—can be represented as dynamical systems on networks. The spread
of news or information between friends, the movement of money through an
economy, the flow of traffic on roads, data over the Internet, or electricity over
the grid, the evolution of populations in an ecosystem, the changing concentrations of metabolites in a cell, and many other systems of scientific interest

676

are best thought of as dynamical processes of one kind or another taking place on an appropriate network.

In other, non-network contexts, the theory of dynamical systems is a well-developed branch of mathematics and physics. (See, for example, the book by Strogatz [307].) In this chapter we delve into some of this theory and show how it can be applied to dynamical systems on networks. Necessarily our introduction only skims the surface of what could be said; dynamical systems is a topic of entire books in its own right. But the material covered here gives a flavor of the kinds of calculation that are possible.

18.1 DYNAMICAL SYSTEMS

Our discussion in this chapter will concentrate principally on deterministic systems of continuous real-valued variables evolving in continuous time t. We begin by introducing some of the basic ideas in a non-network context, then we extend these ideas to networks.

A simple (non-network) example of a continuous dynamical system is a system described by a single real variable $x(t)$ that evolves according to a first-order differential equation

$$\frac{dx}{dt} = f(x), \tag{18.1}$$

where $f(x)$ is some specified function of x. Typically we will also give an initial condition that specifies the value x_0 taken by x at some initial time t_0.

The fully-mixed SI model of Section 17.2 is an example of a dynamical system of this kind, having a single variable x representing the fraction of infected individuals in the system, obeying the equation

$$\frac{dx}{dt} = \beta x(1 - x). \tag{18.2}$$

(See Eq. (17.5).) Thus in this case we have $f(x) = \beta x(1 - x)$.

One can also have dynamical systems of two variables:

$$\frac{dx}{dt} = f(x, y), \qquad \frac{dy}{dt} = g(x, y), \tag{18.3}$$

and the approach can be extended to larger numbers of variables as well. When we come to consider systems on networks we will put separate variables on each vertex of the network.

One could also imagine making the functions on the right-hand sides of our equations depend explicitly on time t:

$$\frac{dx}{dt} = f(x, t). \tag{18.4}$$

This, however, can be regarded as merely a special case of Eq. (18.3). If we write

$$\frac{dx}{dt} = f(x, y), \qquad \frac{dy}{dt} = 1, \tag{18.5}$$

with initial condition $y(0) = 0$, then we have $y = t$ for all times and $dx/dt = f(x, t)$ as required. By this trick it is always possible to turn equations with explicit dependence on t into equations without explicit dependence on t but with one extra variable. For this reason we will confine ourselves in this chapter to systems with no explicit dependence on t.

Another possible generalization would be to consider systems governed by equations containing higher derivatives, such as second derivatives. But these can also be reduced to simpler cases by introducing extra variables. For instance the equation

$$\frac{d^2x}{dt^2} + \left(\frac{dx}{dt}\right)^2 - \frac{dx}{dt} = f(x), \tag{18.6}$$

can be transformed by introducing a new variable $y = dx/dt$ so that we have

$$\frac{dx}{dt} = y, \qquad \frac{dy}{dt} = f(x) - y^2 + y, \tag{18.7}$$

which is a special case of Eq. (18.3) again.

Thus the study of systems of equations like (18.1) and (18.3) covers a broad range of situations of scientific interest. Let us look at some of the techniques used to analyze such equations.

18.1.1 FIXED POINTS AND LINEARIZATION

Equation (18.1), which involves only the one variable x, can, at least in principle, always be solved by simply rearranging and integrating:

$$\int_{x_0}^{x} \frac{dx'}{f(x')} = t - t_0, \tag{18.8}$$

although in practice the integral may not be known in closed form. For cases with two or more variables, on the other hand, it is not in general possible to find a solution. And for the network examples that we will be studying shortly the number of variables is typically very large, so that, unless we are lucky (as we were with some of the epidemiological models of the previous chapter), full analytic solutions are unlikely to be forthcoming.

We can of course integrate the equations numerically and in some cases this can give useful insight. But let's not give up on analytic approaches yet. There is in fact a well-developed set of techniques for understanding how dynamical systems work without first solving their equations exactly. Most of those techniques focus on the properties of fixed points.

A *fixed point* is a steady state of the system—any value of the variable or variables for which the system is stationary and doesn't change over time. In the one-variable system, Eq. (18.1), for example, a fixed point is any point $x = x^*$ for which the function on the right-hand side of the equation is zero

$$f(x^*) = 0, \qquad (18.9)$$

so that $dx/dt = 0$ and x doesn't move. If, in the evolution of the system, x ever reaches a fixed point then it will remain there forever. The fixed points of a one-variable system can be found simply by solving Eq. (18.9) for x.

In a two-variable system like Eq. (18.3) a fixed point is a pair of values (x^*, y^*) such that $f(x^*, y^*) = 0$ and $g(x^*, y^*) = 0$, making $dx/dt = dy/dt = 0$ so that both variables stand still at this point.

Consider the SI model of Eq. (18.2). Putting $f(x) = 0$ in this model gives us $\beta x(1 - x) = 0$, which has solutions $x = 1$ and $x = 0$ for the fixed points. We can see immediately what these fixed points mean in epidemiological terms. The first at $x = 1$ represents the steady state in which everyone in the system is infected. Clearly once everyone is infected the system doesn't change any more, because there is no one else to infect and because in the SI model no one recovers either. The second fixed point $x = 0$ corresponds to the state of the system where no one is infected. In this state no one will ever become infected, since there is no one to catch the disease from, so again we have a steady state.

The importance of fixed points in the study of dynamical systems derives from two key features of these points: first, they are relatively easy to find, and second, it is straightforward to determine the dynamics of the system when it is close to, but not exactly at, a fixed point. The dynamics close to a fixed point is found by expanding about the point as follows.

Consider first a simple one-variable system obeying Eq. (18.1). We represent the value of x close to a fixed point at x^* by writing $x = x^* + \epsilon$ where ϵ, which represents our distance from the fixed point, is small. Then

$$\frac{dx}{dt} = \frac{d\epsilon}{dt} = f(x^* + \epsilon). \qquad (18.10)$$

Now we perform a Taylor expansion of the right-hand side about the point $x = x^*$ to get

$$\frac{d\epsilon}{dt} = f(x^*) + \epsilon f'(x^*) + O(\epsilon^2), \qquad (18.11)$$

where f' represents the derivative of f with respect to its argument. Neglecting terms of order ϵ^2 and smaller and noting that $f(x^*) = 0$ (see Eq. (18.9)), we then have

$$\frac{\mathrm{d}\epsilon}{\mathrm{d}t} = \epsilon f'(x^*). \tag{18.12}$$

This is a linear first-order differential equation with solution

$$\epsilon(t) = \epsilon(0)\,\mathrm{e}^{\lambda t}, \tag{18.13}$$

where

$$\lambda = f'(x^*). \tag{18.14}$$

Note that λ is just a simple number, which we can calculate provided we know the position x^* of the fixed point and the function $f(x)$. Depending on the sign of λ, Eq. (18.13) tells us that our distance ϵ from the fixed point will either grow or decay exponentially in time. Thus this analysis allows us to classify our fixed points into two types. An *attracting fixed point* is one with $\lambda < 0$, for which points close by are attracted towards the fixed point and eventually flow into it. A *repelling fixed point* is one with $\lambda > 0$, for which points close by are repelled away. In between these two types there is a special case when $\lambda = 0$ exactly. Fixed points with $\lambda = 0$ are usually still either attracting or repelling,[1] but one cannot tell which is which from the analysis here; one must retain some of the higher-order terms that we dropped in Eq. (18.11) to determine what happens.

Analysis of the kind represented by Eq. (18.12) is known as *linear stability analysis*. It can be applied to systems with two or more variables as well. Consider, for instance, a dynamical system governed by equations of the form of Eq. (18.3), with a fixed point at (x^*, y^*), meaning that

$$f(x^*, y^*) = 0, \qquad g(x^*, y^*) = 0. \tag{18.15}$$

We represent a point close to the fixed point in the two-dimensional x, y space by $x = x^* + \epsilon_x$ and $y = y^* + \epsilon_y$, where ϵ_x and ϵ_y are both assumed small. As before we expand about the fixed point, performing now a double Taylor expansion:

$$\begin{aligned}
\frac{\mathrm{d}x}{\mathrm{d}t} = \frac{\mathrm{d}\epsilon_x}{\mathrm{d}t} &= f(x^* + \epsilon_x, y^* + \epsilon_y) \\
&= f(x^*, y^*) + \epsilon_x f^{(x)}(x^*, y^*) + \epsilon_y f^{(y)}(x^*, y^*) + \ldots,
\end{aligned} \tag{18.16}$$

---

[1] There are also a couple of other rarer possibilities. A fixed point with $\lambda = 0$ can be *neutral*, meaning it neither attracts nor repels. Points near a neutral fixed point stay exactly where they are, meaning that they are fixed points too. For example, the choice $f(x) = 0$ for all x has a neutral fixed point at every value of x. Another less trivial possibility is that a fixed point with $\lambda = 0$ may be of *mixed* type, meaning that it attracts on one side and repels on the other. An example is $f(x) = x^2$ which has a fixed point at $x = 0$ that is attracting for $x < 0$ and repelling for $x > 0$.

where $f^{(x)}$ and $f^{(y)}$ indicate the derivatives of f with respect to x and y. Making use of Eq. (18.15) and neglecting all higher-order terms in the expansion, we can simplify this expression to

$$\frac{d\epsilon_x}{dt} = \epsilon_x f^{(x)}(x^*, y^*) + \epsilon_y f^{(y)}(x^*, y^*). \tag{18.17}$$

Similarly

$$\frac{d\epsilon_y}{dt} = \epsilon_x g^{(x)}(x^*, y^*) + \epsilon_y g^{(y)}(x^*, y^*). \tag{18.18}$$

We can combine Eqs. (18.17) and (18.18) and write them in matrix form as

$$\frac{d\epsilon}{dt} = \mathbf{J}\epsilon, \tag{18.19}$$

where ϵ is the two-component vector (ϵ_x, ϵ_y) and $\mathbf{J}$ is the Jacobian matrix

$$\mathbf{J} = \begin{pmatrix} \dfrac{\partial f}{\partial x} & \dfrac{\partial f}{\partial y} \\[2mm] \dfrac{\partial g}{\partial x} & \dfrac{\partial g}{\partial y} \end{pmatrix}, \tag{18.20}$$

where the derivatives are all evaluated at the fixed point.

For systems of three or more variables we an employ the same approach and again arrive at Eq. (18.19), but with the rank of the vectors and matrices increasing with increasing number of variables.

Equation (18.19) is again a linear first-order differential equation but its solution is more complicated than for the one-variable equivalent. Let us begin with a particular simple case, the case where the Jacobian matrix is diagonal:

$$\begin{pmatrix} \dfrac{d\epsilon_x}{dt} \\[2mm] \dfrac{d\epsilon_y}{dt} \end{pmatrix} = \begin{pmatrix} \lambda_1 & 0 \\ 0 & \lambda_2 \end{pmatrix} \begin{pmatrix} \epsilon_x \\ \epsilon_y \end{pmatrix}, \tag{18.21}$$

where λ_1 and λ_2 are real numbers. In this case, the equations for ϵ_x and ϵ_y separate from one another thus:

$$\frac{d\epsilon_x}{dt} = \lambda_1 \epsilon_x, \qquad \frac{d\epsilon_y}{dt} = \lambda_2 \epsilon_y, \tag{18.22}$$

and we can solve them separately to get

$$\epsilon_x(t) = \epsilon_x(0)\, e^{\lambda_1 t}, \qquad \epsilon_y(t) = \epsilon_y(0)\, e^{\lambda_2 t}, \tag{18.23}$$

or equivalently

$$x(t) = x^* + \epsilon_x(0)\, e^{\lambda_1 t}, \qquad y(t) = y^* + \epsilon_y(0)\, e^{\lambda_2 t}, \qquad (18.24)$$

so that x and y are independently either attracted or repelled from the fixed point over time, depending on the signs of the two quantities

$$\lambda_1 = \left(\frac{\partial f}{\partial x}\right)_{\substack{x=x^* \\ y=y^*}}, \qquad \lambda_2 = \left(\frac{\partial g}{\partial y}\right)_{\substack{x=x^* \\ y=y^*}}. \qquad (18.25)$$

These results give rise to a variety of possible behaviors of the system near the fixed point, as shown in Fig. 18.1. If λ_1 and λ_2 are both negative, for instance, then the fixed point will be attracting, while if they are both positive it will be repelling. If they are of opposite signs then we have a new type of point called a *saddle point* that attracts along one axis and repels along the other. In some respects a saddle point is perhaps best thought of as a form of repelling fixed point, since a system that starts near a saddle point will not stay near it, the dynamics being repelled along the unstable direction.

Unless we are very lucky, however, the Jacobian matrix is unlikely to be diagonal. In the general case it will have off-diagonal as well as diagonal elements and the solution above will not be correct. With a little more work, however, we can make progress in this case too. The trick is to find combinations of the variables x and y that move independently as x and y alone do above.

Consider the combinations of variables

$$\xi_1 = a\epsilon_x + b\epsilon_y, \qquad \xi_2 = c\epsilon_x + d\epsilon_y. \qquad (18.26)$$

In matrix form we can write these as

$$\begin{pmatrix} \xi_1 \\ \xi_2 \end{pmatrix} = \begin{pmatrix} a & b \\ c & d \end{pmatrix} \begin{pmatrix} \epsilon_x \\ \epsilon_y \end{pmatrix}, \qquad (18.27)$$

or simply

$$\boldsymbol{\xi} = \mathbf{Q}\boldsymbol{\epsilon}, \qquad (18.28)$$

where $\mathbf{Q}$ is the matrix of the coefficients a, b, c, d.

The time evolution of $\boldsymbol{\xi}$ close to the fixed point is given by

$$\frac{d\boldsymbol{\xi}}{dt} = \mathbf{Q}\frac{d\boldsymbol{\epsilon}}{dt} = \mathbf{Q}\mathbf{J}\boldsymbol{\epsilon} = \mathbf{Q}\mathbf{J}\mathbf{Q}^{-1}\boldsymbol{\xi} \qquad (18.29)$$

where we have used Eqs. (18.19) and (18.28). If ξ_1 and ξ_2 are to evolve independently, then we require that the matrix $\mathbf{Q}\mathbf{J}\mathbf{Q}^{-1}$ be diagonal, just as $\mathbf{J}$ itself

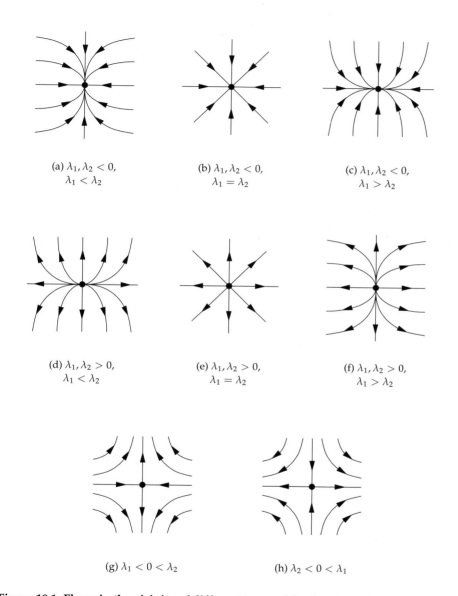

(a) $\lambda_1, \lambda_2 < 0$,
$\lambda_1 < \lambda_2$

(b) $\lambda_1, \lambda_2 < 0$,
$\lambda_1 = \lambda_2$

(c) $\lambda_1, \lambda_2 < 0$,
$\lambda_1 > \lambda_2$

(d) $\lambda_1, \lambda_2 > 0$,
$\lambda_1 < \lambda_2$

(e) $\lambda_1, \lambda_2 > 0$,
$\lambda_1 = \lambda_2$

(f) $\lambda_1, \lambda_2 > 0$,
$\lambda_1 > \lambda_2$

(g) $\lambda_1 < 0 < \lambda_2$

(h) $\lambda_2 < 0 < \lambda_1$

Figure 18.1: Flows in the vicinity of different types of fixed points. The flows around a fixed point in a two-variable dynamical system with a diagonal Jacobian matrix, as described in the text, can take a variety of different forms as shown. (a), (b), and (c) are all attracting fixed points, (d), (e), and (f) are repelling, and (g) and (h) are saddle points.

was in the simple case we studied above. Linear algebra then tells us that $\mathbf{Q}$ must be the matrix of eigenvectors of $\mathbf{J}$. More specifically, since $\mathbf{J}$ is in general asymmetric, $\mathbf{Q}$ is the matrix whose rows are the left eigenvectors of $\mathbf{J}$ and $\mathbf{Q}^{-1}$ is the inverse of that matrix, which is the matrix whose columns are the right eigenvectors of $\mathbf{J}$ (since the left and right eigenvectors of a matrix are mutually orthogonal).

Thus, provided we can find the eigenvectors of $\mathbf{J}$ we can also find the combinations ξ_1 and ξ_2 that move independently of one another near the fixed point. These combinations satisfy the equations

$$\frac{\mathrm{d}\xi_1}{\mathrm{d}t} = \lambda_1\xi_1, \qquad \frac{\mathrm{d}\xi_2}{\mathrm{d}t} = \lambda_2\xi_2, \tag{18.30}$$

where λ_1 and λ_2 are the elements of our diagonal matrix, which are also the eigenvalues of $\mathbf{J}$ corresponding to the two eigenvectors. Equation (18.30) has the obvious solution

$$\xi_1(t) = \xi_1(0)\,\mathrm{e}^{\lambda_1 t}, \qquad \xi_2(t) = \xi_2(0)\,\mathrm{e}^{\lambda_2 t}. \tag{18.31}$$

The lines $\xi_1 = 0$ and $\xi_2 = 0$ play the role of the axes in Fig. 18.1—they are lines along which we move either directly away from or directly towards the fixed point—and Eq. (18.31) indicates that our distance from the fixed point along these lines will either grow or decay exponentially according to the signs of the two eigenvalues. Since the eigenvectors of an asymmetric matrix are not in general orthogonal to one another, these lines are not in general at right angles, so the flows around the fixed point will look similar to those of Fig. 18.1 but squashed, as shown in Fig. 18.2. Nonetheless, we can still classify our fixed points as attracting, repelling, or saddle points as shown in the figure. Similar analyses can be performed for systems with larger numbers of variables and the basic results are the same: by finding the eigenvectors of the Jacobian matrix we can determine the combinations of variables that move independently and hence solve the evolution of the system in the vicinity of the fixed point.

There is another subtlety that arises for systems of two or more variables that is not found in the one-variable case. The eigenvalues of an asymmetric matrix need not be real. Even if the elements of the matrix itself are real, the eigenvalues can be imaginary or complex. What does it mean if the eigenvalues of the Jacobian matrix in our derivation are complex? Putting such eigenvalues into Eq. (18.31) gives us a solution that *oscillates* around the fixed point, rather than simply growing or decaying. Indeed, the substitution actually gives us a value for ξ_1 and ξ_2 that itself is complex, which looks like it might be a problem, since the coordinates are supposed to be real. However,

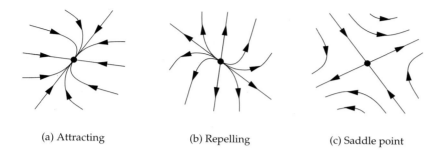

(a) Attracting　　　　　(b) Repelling　　　　　(c) Saddle point

Figure 18.2: Examples of flows around general fixed points. When the Jacobian matrix is not diagonal the flows around a fixed point look like squashed or stretched versions of those in Fig. 18.1.

our equations are linear, so the real part of that solution is also a solution, as is the imaginary part, or any combination of the two.

If $\lambda_1 = \alpha + i\omega$, for example, where α and ω are real numbers, then the general real solution for ξ_1 is

$$\xi_1(t) = \mathrm{Re}\left[C\,e^{(\alpha+i\omega)t}\right] = e^{\alpha t}\left(A\cos\omega t + B\sin\omega t\right), \qquad (18.32)$$

where A and B are real constants and C is a complex constant. Thus the solution is the product of a part that oscillates and a part that either grows or decays exponentially. For the case of two variables, it turns out that the eigenvalues are always either both real or both complex, and if both are complex then they are complex conjugates of one another. In the latter case, both ξ_1 and ξ_2 then have this combined behavior of oscillation with exponential growth or decay, with the same frequency ω of oscillation and the same rate of growth or decay. The net result is a trajectory that describes a spiral around the fixed point. Depending on whether α is positive or negative the spiral either moves outward from the fixed point or inward. If it moves inward, i.e., if $\alpha < 0$. then the fixed point is a stable one; otherwise, of $\alpha > 0$, it is unstable. Thus stability is in this case determined solely by the real part of the eigenvalues. (In the special case where $\alpha = 0$ we must, as before, look at higher-order terms in the expansion around the fixed point to determine the nature of the point.)

When there are more than two variables, the eigenvalues must either be real or they appear in complex conjugate pairs. Thus again we have eigendirections that simply grow or decay, or that spiral in or out.

We are, however, not done yet. There is a further interesting behavior aris-

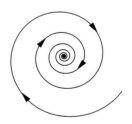

The flows around a fixed point whose Jacobian matrix has complex eigenvalues describe a spiral.

ing in systems with two or more variables that will be important when we come to study networked systems. In addition to fixed points, one also finds in some systems *limit cycles*. A limit cycle is a closed loop in the dynamics such that a system finding itself on such a loop remains there indefinitely, circulating around and returning repeatedly to its starting point. Limit cycles can be treated in many ways rather like fixed points: we can study the dynamics close to the limit cycle by expanding in a small displacement coordinate. Like fixed points, limit cycles tend to be either attracting or repelling, meaning that points close to them either spiral inwards toward the limit cycle or outwards away from it.

Physically, limit cycles represent stable oscillatory behaviors in systems. We mentioned one such behavior in Section 17.5 in our brief discussion of the SIRS model. In certain parameter regimes, the SIRS model can show "waves" of infection—oscillatory behaviors under which a disease infects a large fraction of the population, who then recover and gain immunity, reducing substantially the number of victims available to the disease and therefore causing the number of cases to drop dramatically. When the first wave of individuals later loses their immunity they move back into the susceptible state, become infected again, and another wave starts. Another example of oscillation in a dynamical system is the oscillation of the numbers of predators and prey in a two-species ecosystem represented, for example, by the Lotka–Volterra predator-prey equations [307]. Such oscillations have been famously implicated in the mysterious periodic variation in the populations of hares and lynx recorded by the Hudson Bay Company in Canada during the nineteenth century. A further discussion of this and other aspects of limit cycles can be found in Ref. [307].

18.2   DYNAMICS ON NETWORKS

Let us now apply some of the ideas of the previous section to dynamical systems on networks. First, we need to be clear what we mean by such systems. Typically, we mean that we have independent dynamical variables $x_i, y_i, \ldots$ on each vertex i of our network and that they are coupled together only along the edges of the network. That is, when we write our equation for the time evolution of a variable x_i, the individual terms appearing in that equation each involve only x_i, other variables on vertex i, or one or more variables on a vertex adjacent to i in the network. There are no terms involving variables on non-adjacent vertices and no terms involving variables on more than one adjacent vertex.

An example of a dynamical system of this type is our equation (17.35) for the probability of infection of a vertex in the network version of the SI epidemic model:

$$\frac{dx_i}{dt} = \beta(1 - x_i) \sum_j A_{ij} x_j. \tag{18.33}$$

This equation only has terms involving pairs of variables that are connected by edges since these are the only pairs for which A_{ij} is non-zero.

For a system with a single variable on each vertex we can write a general first-order equation

$$\frac{dx_i}{dt} = f_i(x_i) + \sum_j A_{ij} g_{ij}(x_i, x_j), \tag{18.34}$$

where we have separated terms that involve variables on adjacent vertices from those that do not. You can think of f_i as specifying the intrinsic dynamics of a vertex—it specifies how the variable x_i would evolve in the absence of any connections between vertices, i.e., if $A_{ij} = 0$ for all i, j. Conversely, g_{ij} describes the contribution from the connections themselves; it represents the coupling between variables on different vertices.

Notice that we have specified different functions f_i and g_{ij} for each vertex or pair of vertices, so the dynamics obeyed by each vertex can be different. In many cases, however, when each of the vertices represents a similar thing—such as a person in the case of an epidemic model—the dynamics for each vertex may be the same, or at least similar enough that we can ignore any differences. In such cases, the functions in Eq. (18.34) are the same for all vertices and the equation becomes

$$\frac{dx_i}{dt} = f(x_i) + \sum_j A_{ij} g(x_i, x_j). \tag{18.35}$$

In the examples in this chapter we will assume that this is the case. We will also assume that the network is undirected so that A_{ij} is symmetric—if x_i is affected by x_j then x_j is similarly affected by x_i. (Note, however, that we do not assume that the function g is symmetric in its arguments: $g(u, v) \neq g(v, u)$.) Again, the SI model of Eq. (18.33) is an example of a system of this kind, one in which $f(x) = 0$ and $g(x_i, x_j) = \beta(1 - x_i)x_j$.

18.2.1 LINEAR STABILITY ANALYSIS

Let us try applying the tools of linear stability analysis to Eq. (18.35). Suppose we are able to find a fixed point $\{x_i^*\}$ of Eq. (18.35) by solving the simultaneous equations

$$f(x_i^*) + \sum_j A_{ij} g(x_i^*, x_j^*) = 0 \tag{18.36}$$

for all i. Note that finding a fixed point in this case means finding a value $x_i = x_i^*$ for every vertex i—the fixed point is the complete set $\{x_i^*\}$. Note also that in general the position of the fixed point depends both on the particular dynamical process taking place on the network (via the functions f and g) and on the structure of the network (via the adjacency matrix). If either is changed then the position of the fixed point will also change.

Now we can linearize about this fixed point in the usual way by writing $x_i = x_i^* + \epsilon_i$, performing a multiple Taylor expansion in all variables simultaneously, and dropping terms at second order in small quantities and higher:

$$
\frac{dx_i}{dt} = \frac{d\epsilon_i}{dt} = f(x_i^* + \epsilon_i) + \sum_j A_{ij} g(x_i^* + \epsilon_i, x_j^* + \epsilon_j)
$$

$$
= f(x_i^*) + \epsilon_i \frac{df}{dx}\bigg|_{x=x_i^*} + \sum_j A_{ij} g(x_i^*, x_j^*)
$$

$$
+ \epsilon_i \sum_j A_{ij} \frac{\partial g(u,v)}{\partial u}\bigg|_{u=x_i^*, v=x_j^*} + \sum_j A_{ij} \epsilon_j \frac{\partial g(u,v)}{\partial v}\bigg|_{u=x_i^*, v=x_j^*} + \ldots
$$

$$
= \epsilon_i \frac{df}{dx}\bigg|_{x=x_i^*} + \epsilon_i \sum_j A_{ij} \frac{\partial g(u,v)}{\partial u}\bigg|_{u=x_i^*, v=x_j^*} + \sum_j A_{ij} \epsilon_j \frac{\partial g(u,v)}{\partial v}\bigg|_{u=x_i^*, v=x_j^*} + \ldots,
$$

$$(18.37)$$

where we have used Eq. (18.36).

If we know the position of the fixed point, then the derivatives in these expressions are simply numbers. For convenience, let us write

$$
\alpha_i = \frac{\partial f}{\partial x}\bigg|_{x=x_i^*}, \tag{18.38a}
$$

$$
\beta_{ij} = \frac{\partial g(u,v)}{\partial u}\bigg|_{u=x_i^*, v=x_j^*}, \tag{18.38b}
$$

$$
\gamma_{ij} = \frac{\partial g(u,v)}{\partial v}\bigg|_{u=x_i^*, v=x_j^*}. \tag{18.38c}
$$

Then

$$
\frac{d\epsilon_i}{dt} = \left[\alpha_i + \sum_j \beta_{ij} A_{ij}\right] \epsilon_i + \sum_j \gamma_{ij} A_{ij} \epsilon_j, \tag{18.39}
$$

which we can write in matrix form as

$$
\frac{d\boldsymbol{\epsilon}}{dt} = \mathbf{M}\boldsymbol{\epsilon}, \tag{18.40}
$$

where **M** is the matrix with elements

$$M_{ij} = \delta_{ij}\left[\alpha_i + \sum_k \beta_{ik}A_{ik}\right] + \gamma_{ij}A_{ij}, \tag{18.41}$$

and δ_{ij} is the Kronecker delta.

We can solve Eq. (18.40) by writing ϵ as a linear combination of the eigenvectors of **M**, specifically the *right* eigenvectors, since **M** is in general not symmetric:

$$\epsilon(t) = \sum_r c_r(t)\mathbf{v}_r, \tag{18.42}$$

so that Eq. (18.40) becomes

$$\sum_r \frac{dc_r}{dt}\mathbf{v}_r = \mathbf{M}\sum_r c_r(t)\mathbf{v}_r = \sum_r \mu_r c_r(t)\mathbf{v}_r, \tag{18.43}$$

where μ_r is the eigenvalue corresponding to the eigenvector $\mathbf{v}_r$. Comparing terms in each eigenvector we then have

$$\frac{dc_r}{dt} = \mu_r c_r(t), \tag{18.44}$$

which implies that

$$c_r(t) = c_r(0)\,e^{\mu_r t}. \tag{18.45}$$

Immediately we see that if the real parts of all of the eigenvalues μ_r are negative, then $c_r(t)$—and hence ϵ—is decaying in time for all r and our fixed point will be attracting. If the real parts are all positive the fixed point will be repelling. And if some are positive and some are negative then the fixed point is a saddle, although, as before, this is perhaps best looked at as a form of repelling fixed point: the flows near a saddle have at least one repelling direction, which means that a system starting in the vicinity of such a point will not in general stay near it, regardless of whether the other directions are attracting or not.

18.2.2 SPECIAL CASES

Let us look at some common special cases of the general formalism above. A particularly simple case is when the fixed point is *symmetric*, meaning that x_i^* has the same value for every i: $x_i^* = x^*$. This occurs in the SI model for instance—there is a fixed point at $x_i^* = 1$ for all i.

For a symmetric fixed point, the fixed point equation, Eq. (18.36), becomes

$$f(x^*) + \sum_j A_{ij}g(x^*, x^*) = f(x^*) + k_i g(x^*, x^*) = 0, \tag{18.46}$$

where k_i is the degree of vertex i and we have made use of $k_i = \sum_j A_{ij}$ (see Eq. (6.19)). Given the appearance of k_i here, there are only two ways this equation can be satisfied for all i: either all vertices must have the same degree or $g(x^*, x^*) = 0$. Since the former is not really realistic—few networks of interest have all degrees the same—let us concentrate on the latter and assume that

$$g(x^*, x^*) = 0. \tag{18.47}$$

Again the SI model provides an example of this type of behavior. The coupling function g is of the form $\beta x(1 - x)$ in that model, which is zero at the two fixed points at $x = 0, 1$.

Equations (18.46) and (18.47) together imply also that $f(x^*) = 0$ and hence the fixed point x^* is the same in this case as the fixed point for the "intrinsic" dynamics of a vertex: it falls at the same place as it would if there were no connections between vertices at all. The position of the fixed point is also independent of the network structure in this case, a point that will shortly be important.

For a symmetric fixed point, the quantities α_i, β_{ij}, and γ_{ij} defined in Eq. (18.38) become

$$\alpha_i = \alpha = \left.\frac{\partial f}{\partial x}\right|_{x=x^*}, \tag{18.48a}$$

$$\beta_{ij} = \beta = \left.\frac{\partial g(u, v)}{\partial u}\right|_{u,v=x^*}, \tag{18.48b}$$

$$\gamma_{ij} = \gamma = \left.\frac{\partial g(u, v)}{\partial v}\right|_{u,v=x^*}. \tag{18.48c}$$

Then Eq. (18.39) becomes

$$\frac{d\epsilon_i}{dt} = (\alpha + \beta k_i)\epsilon_i + \gamma \sum_j A_{ij}\epsilon_j. \tag{18.49}$$

The situation simplifies further if the coupling function $g(x_i, x_j)$ depends only on x_j and not on x_i, i.e., if x_i obeys an equation of the form $dx_i/dt = f(x_i) + \sum_j A_{ij}g(x_j)$. Then $\beta = 0$ and

$$\frac{d\epsilon_i}{dt} = \alpha\epsilon_i + \gamma \sum_j A_{ij}\epsilon_j, \tag{18.50}$$

which we can write in matrix form as

$$\frac{d\epsilon}{dt} = (\alpha \mathbf{I} + \gamma \mathbf{A})\epsilon. \tag{18.51}$$

As in the general case, the fixed point will be stable if and only if all of the eigenvalues of the matrix $\alpha\mathbf{I} + \gamma\mathbf{A}$ are negative. Let $\mathbf{v}_r$ be the eigenvector of the adjacency matrix with eigenvalue κ_r. Then

$$(\alpha\mathbf{I} + \gamma\mathbf{A})\mathbf{v}_r = \alpha\mathbf{I}\mathbf{v}_r + \gamma\mathbf{A}\mathbf{v}_r = \alpha\mathbf{v}_r + \gamma\kappa_r\mathbf{v}_r = (\alpha + \gamma\kappa_r)\mathbf{v}_r. \qquad (18.52)$$

Hence $\mathbf{v}_r$ is also an eigenvector of $\alpha\mathbf{I} + \gamma\mathbf{A}$, but with eigenvalue $\alpha + \gamma\kappa_r$. Now if all eigenvalues are to be negative, we require that

$$\alpha + \gamma\kappa_r < 0 \qquad (18.53)$$

for all r and from this we can deduce a number of things. First of all it implies that $\alpha < -\gamma\kappa_r$ for all r. The adjacency matrix always has both positive and negative eigenvalues (a result that we will prove in Section 18.3.2), which means that for this inequality to be satisfied for all r we must have $\alpha < 0$. If $\alpha > 0$ then the fixed point is never stable.

Second, we can rearrange Eq. (18.53) to give

$$\kappa_r < -\alpha/\gamma \qquad \text{if } \gamma > 0, \qquad (18.54a)$$
$$\kappa_r > -\alpha/\gamma \qquad \text{if } \gamma < 0, \qquad (18.54b)$$

for all r. Note, however, that if Eq. (18.54a) is satisfied for the largest (most positive) eigenvalue κ_1 of the adjacency matrix, then it is necessarily satisfied by all the other eigenvalues as well. Similarly if Eq. (18.54b) is satisfied for the most negative eigenvalue κ_n then it is satisfied by all others. Thus the conditions above can be simplified to a single condition each:

$$\kappa_1 < -\alpha/\gamma \qquad \text{if } \gamma > 0, \qquad (18.55a)$$
$$\kappa_n > -\alpha/\gamma \qquad \text{if } \gamma < 0. \qquad (18.55b)$$

Alternatively, we can take reciprocals of these conditions and combine them into a single statement:

$$\frac{1}{\kappa_n} < -\frac{\gamma}{\alpha} < \frac{1}{\kappa_1}. \qquad (18.56)$$

If we want we can fill in the explicit values of α and γ thus:

$$\frac{1}{\kappa_n} < -\left[\frac{\mathrm{d}g}{\mathrm{d}x}\Big/\frac{\mathrm{d}f}{\mathrm{d}x}\right]_{x=x^*} < \frac{1}{\kappa_1}, \qquad (18.57)$$

where we have written g as a function of a single variable since, by hypothesis, it only depends on one argument in this case.

Equation (18.57) is sometimes called a *master stability condition*. It has a special form: note that κ_1 and κ_n depend only on the structure of the network and not on anything about the dynamics, while α and γ depend only the nature of the dynamics and not on the network structure. Thus Eq. (18.57) effectively gives us a single condition that must be satisfied by any type of dynamics and its associated fixed point if that dynamics is to be stable on our network. Or conversely, it gives a condition on the network structure, via the largest and smallest eigenvalues, that guarantees stability of a given fixed point for a given type of dynamics.

Another case where we can derive a master stability condition is the case in which the coupling function g depends on its two arguments according to $g(x_i, x_j) = g(x_i) - g(x_j)$. A physicist might think of this as a "spring-like" interaction—if $g(x)$ were a simple linear function of its argument then x_i and x_j would act upon one another like two masses coupled by a spring, exerting forces that depend on the difference of their positions. More generally, $g(x)$ is non-linear and we have a non-linear spring.

For this choice of coupling, and still assuming a symmetric fixed point, we have $g(x^*, x^*) = 0$ as before and hence also $f(x^*) = 0$, and the quantities defined in Eq. (18.38) become

$$\alpha_i = \alpha = \left.\frac{df}{dx}\right|_{x=x^*}, \tag{18.58a}$$

$$\beta_{ij} = \beta = \left.\frac{dg}{dx}\right|_{x=x^*}, \tag{18.58b}$$

$$\gamma_{ij} = -\beta. \tag{18.58c}$$

Then Eq. (18.39) becomes

$$\frac{d\epsilon_i}{dt} = (\alpha + \beta k_i)\epsilon_i - \beta \sum_j A_{ij}\epsilon_j$$

$$= \alpha\epsilon_i + \beta \sum_j (k_i\delta_{ij} - A_{ij})\epsilon_j, \tag{18.59}$$

or in matrix form

$$\frac{d\epsilon}{dt} = (\alpha\mathbf{I} + \beta\mathbf{L})\epsilon, \tag{18.60}$$

where $\mathbf{L}$ is the matrix with elements

$$L_{ij} = k_i\delta_{ij} - A_{ij}. \tag{18.61}$$

We have encountered this matrix before. It is the *graph Laplacian*—see Eq. (6.43).

Equation (18.60) is of the same form as Eq. (18.51), with the adjacency matrix replaced by the graph Laplacian. Thus we can immediately see that the fixed point will be stable if and only if the eigenvalues λ_r of the Laplacian satisfy

$$\alpha + \beta\lambda_r < 0 \tag{18.62}$$

for all r.

As shown in Section 6.13.2, the smallest eigenvalue of the Laplacian matrix is always zero, and hence Eq. (18.62), when applied to the smallest eigenvalue, implies again that $\alpha < 0$ is a necessary (but not sufficient) condition for the fixed point to be stable, or equivalently

$$\left.\frac{df}{dx}\right|_{x=x^*} < 0. \tag{18.63}$$

Assuming this condition is satisfied then, since all eigenvalues of the Laplacian are non-negative it follows that $1/\lambda_r > -\beta/\alpha$ for stability, regardless of the sign of β. Furthermore, if this condition is true for the largest eigenvalue, traditionally denoted λ_n, then it is true for all smaller eigenvalues as well, so the requirement for stability can be reduced to the requirement that $1/\lambda_n > -\beta/\alpha$, or

$$\frac{1}{\lambda_n} > -\left[\frac{dg}{dx}\bigg/\frac{df}{dx}\right]_{x=x^*}, \tag{18.64}$$

along with the condition in Eq. (18.63).

Again, Eq. (18.64) neatly separates questions of dynamics from questions of network structure. The structure appears only on the left of the inequality, via the eigenvalues of the graph Laplacian, and the dynamics appears only on the right, via derivatives of the functions f and g.

Apart from establishing a condition for the stability of a fixed point, the master stability condition is of particular interest in the study of bifurcations— situations in which a fixed point loses stability as the parameters of a system change. If we vary parameters appearing in the definitions of f and g, for example, then we can cause a fixed point that initially satisfies a condition like (18.64) to stop satisfying it and so become unstable. In practice, this means that the system will suddenly change its behavior as it passes through the point where $1/\lambda_n = -\beta/\alpha$. At one moment it will be sitting happily at its stable fixed point, going nowhere, and at the next, as that point becomes unstable, it will start moving, gathering speed exponentially, and quite likely wind up in some completely different state far from where it started, as it falls into the basin of attraction of a different stable fixed point or limit cycle. We will see some examples of behavior of this kind shortly.

18.2.3 AN EXAMPLE

As an example, consider the following simple model of "gossip," or diffusion of an idea or fad across a social network. Suppose some new idea is circulating through a community and x_i represents the amount person i is talking about it, which will be governed by an equation of the form (18.35). We will put

$$f(x) = a(1 - x) \tag{18.65}$$

with $a > 0$, which means that the intrinsic dynamics of a single vertex has a stable fixed point at $x^* = 1$—each person has an intrinsic tendency to talk this much about the latest craze, regardless of whether their friends want to hear about it or not. For the interaction term we will assume that people tend to copy their friends: they increase the amount they are talking about whatever it is if their friends are talking about it more than they are, and decrease if their friends are talking about it less. We represent this by putting $g(x_i, x_j) = g(x_j) - g(x_i)$ with

$$g(x) = \frac{bx}{1 + x} \tag{18.66}$$

and $b > 0$. This is an increasing function of its argument, as it should be, but saturates when $x \gg 1$—beyond some point, it makes no difference if your friends shout louder.

Now we can apply the general formalism developed above. The symmetric fixed point for the model is at $x_i = 1$ for all i. At this point everyone is talking about the topic *du jour* with equal enthusiasm. This fixed point, however, is stable only provided Eqs. (18.63) and (18.64) are satisfied. Equation (18.63) is always satisfied, given that $a > 0$. Equation 18.64 implies that $1/\lambda_n > b/4a$, or equivalently

$$\lambda_n < \frac{4a}{b}. \tag{18.67}$$

Thus we can make the fixed point unstable, for example, by increasing b to the point where the right-hand side of this inequality falls below the largest eigenvalue λ_n of the Laplacian for the particular network we are looking at. Increasing b in this case corresponds to increasing the amount of influence your friends have on you.

And what happens when the fixed point becomes unstable? There are no other symmetric fixed points for this particular system, since there are no other values that give $f(x) = 0$ (which is a requirement for our symmetric fixed point). So the system cannot switch to another symmetric fixed point. One possibility is that the variables might diverge to $\pm\infty$, and this happens in some systems, but not in this one, where the form of $f(x)$ prevents it. Another possibility is that the system might begin to oscillate, or even enter a chaotic regime in which it meanders around in pseudorandom fashion indefinitely. In the present case, however, it does something simpler. It moves to a *non-symmetric* fixed point, one in which the fixed-point values of the variables x_i are not all equal. This is an interesting and perhaps unexpected behavior. Our calculations are telling us when the influence between neighboring individuals in the network becomes very strong that instead of driving everyone to behave in the same way, as one might expect, it actually causes behaviors to differ. People spontaneously develop idiosyncrasies and start doing things their own way.

18.3 DYNAMICS WITH MORE THAN ONE VARIABLE PER VERTEX

Our developments so far have assumed that there is only a single variable x_i on each vertex i of the network. Many systems, however, have more than one variable per vertex. The epidemiological examples of Chapter 17, mostly have several—s, x, r, and so forth.

Consider a system with an arbitrary number of variables $x_1^i, x_2^i, \ldots$ on each vertex i, but let us assume that we have the same number of variables on each vertex and that, as before, they obey equations of the same form. For convenience let us write the set of variables on a single vertex as a vector $\mathbf{x}^i = (x_1^i, x_2^i, \ldots)$ and then write the equations governing their time evolution as

$$\frac{d\mathbf{x}^i}{dt} = \mathbf{f}(\mathbf{x}^i) + \sum_j A_{ij}\mathbf{g}(\mathbf{x}^i, \mathbf{x}^j). \qquad (18.68)$$

Note that the functions f and g, representing the intrinsic dynamics and the coupling, have now become vector functions $\mathbf{f}$ and $\mathbf{g}$ of vector arguments, with the same rank as $\mathbf{x}$.

Following the same line of reasoning as before, we can study the stability of a symmetric fixed point $\mathbf{x}^i = \mathbf{x}^*$ by writing $\mathbf{x}^i = \mathbf{x}^* + \boldsymbol{\epsilon}^i$ and performing a Taylor expansion. The resulting linearized equation for the evolution of the μth component of $\boldsymbol{\epsilon}^i$ is then

$$\frac{d\epsilon_\mu^i}{dt} = \left[\epsilon_1^i \frac{\partial f_\mu(\mathbf{x})}{\partial x_1} + \epsilon_2^i \frac{\partial f_\mu(\mathbf{x})}{\partial x_2} + \ldots\right]_{\mathbf{x}=\mathbf{x}^*}$$

$$+ \sum_j A_{ij}\left[\epsilon_1^i \frac{\partial g_\mu(\mathbf{u}, \mathbf{v})}{\partial u_1} + \epsilon_2^i \frac{\partial g_\mu(\mathbf{u}, \mathbf{v})}{\partial u_2} + \ldots + \epsilon_1^j \frac{\partial g_\mu(\mathbf{u}, \mathbf{v})}{\partial v_1} + \epsilon_2^j \frac{\partial g_\mu(\mathbf{u}, \mathbf{v})}{\partial v_2} + \ldots\right]_{\mathbf{u},\mathbf{v}=\mathbf{x}^*}$$

$$= \sum_v \left[\epsilon_v^i \frac{\partial f_\mu(\mathbf{x})}{\partial x_v}\bigg|_{\mathbf{x}=\mathbf{x}^*} + k_i \epsilon_v^i \frac{\partial g_\mu(\mathbf{u}, \mathbf{v})}{\partial u_v}\bigg|_{\mathbf{u},\mathbf{v}=\mathbf{x}^*} + \sum_j A_{ij}\epsilon_v^j \frac{\partial g_\mu(\mathbf{u}, \mathbf{v})}{\partial v_v}\bigg|_{\mathbf{u},\mathbf{v}=\mathbf{x}^*}\right], \qquad (18.69)$$

where f_μ and g_μ represent the μth components of $\mathbf{f}$ and $\mathbf{g}$.

As before, the derivatives in this expression are simply constants, and for convenience let us define

$$\alpha_{\mu\nu} = \left. \frac{\partial f_\mu(\mathbf{x})}{\partial x_\nu} \right|_{\mathbf{x}=\mathbf{x}^*}, \tag{18.70a}$$

$$\beta_{\mu\nu} = \left. \frac{\partial g_\mu(\mathbf{u}, \mathbf{v})}{\partial u_\nu} \right|_{\mathbf{u},\mathbf{v}=\mathbf{x}^*}, \tag{18.70b}$$

$$\gamma_{\mu\nu} = \left. \frac{\partial g_\mu(\mathbf{u}, \mathbf{v})}{\partial v_\nu} \right|_{\mathbf{u},\mathbf{v}=\mathbf{x}^*}, \tag{18.70c}$$

so that

$$\frac{d\epsilon_\mu^i}{dt} = \sum_\nu \left[(\alpha_{\mu\nu} + k_i \beta_{\mu\nu}) \epsilon_\nu^i + \sum_j A_{ij} \gamma_{\mu\nu} \epsilon_\nu^j \right]$$

$$= \sum_{j\nu} \left[\delta_{ij} (\alpha_{\mu\nu} + k_i \beta_{\mu\nu}) + A_{ij} \gamma_{\mu\nu} \right] \epsilon_\nu^j \tag{18.71}$$

where δ_{ij} is the Kronecker delta again.

We can write this equation in the matrix form

$$\frac{d\epsilon}{dt} = \mathbf{M}\epsilon, \tag{18.72}$$

where $\mathbf{M}$ is a matrix whose rows (and columns) are labeled by a double pair of indices (i, μ) and whose elements are

$$M_{i\mu,j\nu} = \delta_{ij}\alpha_{\mu\nu} + \delta_{ij}k_i\beta_{\mu\nu} + A_{ij}\gamma_{\mu\nu}. \tag{18.73}$$

In principle, we can now determine whether the fixed point is stable by examining the eigenvalues of this new matrix. If the real parts of the eigenvalues are all negative then the fixed point is stable, otherwise it is not. In practice this can be a difficult thing to do in general but, as before, there are some common special cases where the calculation simplifies, yielding a master stability condition.

18.3.1 SPECIAL CASES

As before we consider the case where $\mathbf{g}(\mathbf{x}^i, \mathbf{x}^j)$ depends only on its second argument and not on its first. In this case $\beta_{\mu\nu} = 0$ for all μ, ν and Eq. (18.71) becomes

$$\frac{d\epsilon_\mu^i}{dt} = \sum_{j\nu} \left[\delta_{ij}\alpha_{\mu\nu} + A_{ij}\gamma_{\mu\nu} \right] \epsilon_\nu^j. \tag{18.74}$$

Now let v_r^i be the ith component of the eigenvector $\mathbf{v}_r$ of the adjacency matrix corresponding to eigenvalue κ_r. Let us write

$$\epsilon_\mu^i(t) = \sum_r c_\mu^r(t) v_r^i. \tag{18.75}$$

This equation expresses the vector of elements ϵ_μ^i as a linear combination of eigenvectors in the usual way, but with a separate set of coefficients c_μ^r for each dynamical variable μ. Substituting into Eq. (18.74), we get

$$\sum_r \frac{dc_\mu^r}{dt} v_r^i = \sum_r \sum_{jv} \left[\delta_{ij} \alpha_{\mu v} + A_{ij} \gamma_{\mu v} \right] c_v^r(t) v_r^j$$

$$= \sum_{rv} \left[\alpha_{\mu v} + \kappa_r \gamma_{\mu v} \right] c_v^r(t) v_r^i. \tag{18.76}$$

Equating terms in the individual eigenvectors on both sides of the equation, we thus conclude that

$$\frac{dc_\mu^r}{dt} = \sum_v \left[\alpha_{\mu v} + \kappa_r \gamma_{\mu v} \right] c_v^r(t). \tag{18.77}$$

We can think of this as itself a matrix equation for a vector $\mathbf{c}^r = (c_1^r, c_2^r, \ldots)$ thus:

$$\frac{d\mathbf{c}^r}{dt} = [\boldsymbol{\alpha} + \kappa_r \boldsymbol{\gamma}] \mathbf{c}^r(t), \tag{18.78}$$

where $\boldsymbol{\alpha}$ and $\boldsymbol{\gamma}$ are matrices with elements $\alpha_{\mu v}$ and $\gamma_{\mu v}$, respectively.

This equation expresses the dynamics of the system close to the fixed point as a decoupled set of n separate systems, one for each eigenvalue κ_r of the adjacency matrix. If the fixed point of the system as a whole is to be stable, then each of these individual systems also needs to be stable, meaning that *their* eigenvalues need to be negative, or, more simply, the largest (i.e., most positive) eigenvalue of $\boldsymbol{\alpha} + \kappa_r \boldsymbol{\gamma}$ needs to be negative for every r.

Let us define the function $\sigma(\kappa)$ to be equal to the most positive eigenvalue of the matrix $\boldsymbol{\alpha} + \kappa \boldsymbol{\gamma}$, or the most positive real part in the case where the eigenvalues are complex. Typically this is an easy function to evaluate numerically. Notice that $\boldsymbol{\alpha} + \kappa \boldsymbol{\gamma}$ has only as many rows and columns as there are variables on each vertex of the network. If we have three variables on each vertex, for instance, the matrix has size 3×3, which is easily diagonalized.

The function $\sigma(\kappa)$ is called a *master stability function*. If our system is to be stable, the master stability function evaluated at the eigenvalue κ_r should be negative for all r:

$$\sigma(\kappa_r) < 0. \tag{18.79}$$

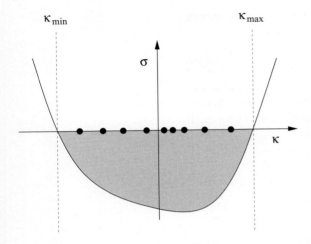

Figure 18.3: A sketch of a master stability function. One possible form for the master stability function $\sigma(\kappa)$ might be as shown here (solid curve), with positive values for large and small κ but negative values in the intermediate range between κ_{min} and κ_{max}. If all the eigenvalues of the adjacency matrix (represented by the dots) fall in this intermediate range, then the system is stable.

One possible form for the master stability function is shown in Fig. 18.3—it becomes large and positive for κ sufficiently small or sufficiently big, but is negative in some intermediate range $\kappa_{min} < \kappa < \kappa_{max}$. In that case, the system is stable provided all eigenvalues κ_r of the adjacency matrix fall in this range. Again this gives us a master stability condition that separates network structure from dynamics. The eigenvalues κ_r are properties solely of the structure, being derived from the adjacency matrix alone, while the limits κ_{min} and κ_{max} are properties solely of the dynamics, being derived from the matrices α and γ, which are determined by the derivatives of the functions f and g.

We can similarly write down the generalization of Eq. (18.58) to the case of many variables per vertex. If the interaction between vertices takes the form $\mathbf{g}(\mathbf{x}^i, \mathbf{x}^j) = \mathbf{g}(\mathbf{x}^i) - \mathbf{g}(\mathbf{x}^j)$, then $\gamma_{\mu\nu} = -\beta_{\mu\nu}$ and

$$\frac{d\epsilon_\mu^i}{dt} = \sum_{j\nu} [\delta_{ij}\alpha_{\mu\nu} + L_{ij}\beta_{\mu\nu}]\epsilon_\nu^j, \qquad (18.80)$$

where $L_{ij} = \delta_{ij} - A_{ij}$ is an element of the Laplacian. Then the equivalent of Eq. (18.78) is

$$\frac{d\mathbf{c}^r}{dt} = [\alpha + \lambda_r\beta]\mathbf{c}^r(t), \qquad (18.81)$$

where λ_r is an eigenvalue of the Laplacian and β is the matrix with elements $\beta_{\mu\nu}$. Again we can define a master stability function $\sigma(\lambda)$ equal to the most positive eigenvalue of $\alpha + \lambda\beta$ (or the most positive real part for complex eigenvalues) and for overall stability of the system this function must be negative when $\lambda = \lambda_r$ for all r:

$$\sigma(\lambda_r) < 0. \qquad (18.82)$$

And once again, for suitable forms of the master stability function, this allows us to develop a stability criterion that separates structure from dynamics.

18.3.2 SPECTRA OF COMPLEX NETWORKS

The formalism of the previous section turns questions about the stability of dynamical systems on networks into questions about the eigenvalue spectra of matrices. Given the definition of the dynamics taking place on the vertices of a network we calculate the master stability function and then the stability or not of the system depends on whether the master stability function is negative when evaluated at each of the eigenvalues of the appropriate matrix, such as the adjacency matrix or graph Laplacian. In particular, when the master stability function takes a relatively simple form like that sketched in Fig. 18.3, so that stability requires only that the eigenvalues fall in some specified range, then it is enough to know the smallest (most negative) and largest (most positive) eigenvalues of the matrix to insure stability—if the smallest and largest fall in the given range then necessarily all the others do too.

A number of results are known about the spectra of networks, and in particular about the smallest and largest eigenvalues, which allow us to make quite general theoretical statements about stability. For the adjacency matrix, for example, we can derive limits on the eigenvalues as follows.

Let x be an arbitrary real vector of n elements, which we will write as a linear combination of the eigenvectors v_r of the adjacency matrix A thus:

$$x = \sum_r c_r v_r. \tag{18.83}$$

Then

$$\frac{x^T A x}{x^T x} = \frac{\sum_s c_s v_s^T A \sum_r c_r v_r}{\sum_s c_s v_s^T \sum_r c_r v_r} = \frac{\sum_{rs} c_s c_r \kappa_r v_s^T v_r}{\sum_{rs} c_s c_r v_s^T v_r} = \frac{\sum_r c_r^2 \kappa_r}{\sum_r c_r^2} \le \frac{\sum_r c_r^2 \kappa_1}{\sum_r c_r^2} = \kappa_1, \tag{18.84}$$

where, as before, κ_1 is the largest eigenvalue and we have made use of the fact that $v_s^T v_r = \delta_{rs}$. This inequality is correct for any choice of x. Thus, for instance, if $x = 1 = (1, 1, 1, \ldots)$ then

$$\kappa_1 \ge \frac{1^T A 1}{1^T 1} = \frac{2m}{n} = \langle k \rangle. \tag{18.85}$$

So the largest eigenvalue of the adjacency matrix is never less than the average degree of the network.

Alternatively, suppose that vertex v is the highest-degree vertex in the network, with degree k_{max}, and let us choose the elements of x thus:

$$x_i = \begin{cases} \sqrt{k_{max}} & \text{if } i = v, \\ 1 & \text{if } A_{iv} = 1, \\ 0 & \text{otherwise.} \end{cases} \tag{18.86}$$

Then

$$\sum_j A_{ij} x_j \ge \begin{cases} k_{max} & \text{if } i = v \\ \sqrt{k_{max}} & \text{if } A_{iv} = 1 \\ 0 & \text{otherwise} \end{cases} = \sqrt{k_{max}} \, x_i. \tag{18.87}$$

(This result is non-trivial and you may find it helpful to work through each of the three cases to convince yourself that it is indeed correct.)

Multiplying both sides of Eq. (18.87) by x_i and summing over i we now get $\mathbf{x}^T \mathbf{A} \mathbf{x} \geq \sqrt{k_{max}}\, \mathbf{x}^T \mathbf{x}$ or, using Eq. (18.84),

$$\kappa_1 \geq \frac{\mathbf{x}^T \mathbf{A} \mathbf{x}}{\mathbf{x}^T \mathbf{x}} \geq \sqrt{k_{max}}. \tag{18.88}$$

Thus the largest eigenvalue of the adjacency matrix is never less than the square root of the largest degree.

Equations (18.85) and (18.88) imply that if we increase either the average or the maximum degree in our network, we will eventually increase the maximum eigenvalue also. In a system with a master stability function like that depicted in Fig. 18.3, this will in the end cause the system to become unstable.

We can also derive similar results for the lowest (most negative) eigenvalue κ_n of the adjacency matrix. We have

$$\frac{\mathbf{x}^T \mathbf{A} \mathbf{x}}{\mathbf{x}^T \mathbf{x}} = \frac{\sum_r c_r^2 \kappa_r}{\sum_r c_r^2} \geq \frac{\sum_r c_r^2 \kappa_n}{\sum_r c_r^2} = \kappa_n \tag{18.89}$$

for any real vector $\mathbf{x}$. So, for instance, if vertex v is again the highest-degree vertex in the network and we make the choice

$$x_i = \begin{cases} \sqrt{k_{max}} & \text{if } i = v, \\ -1 & \text{if } A_{iv} = 1, \\ 0 & \text{otherwise,} \end{cases} \tag{18.90}$$

then, following the same approach as before, we find that

$$\kappa_n \leq -\sqrt{k_{max}}. \tag{18.91}$$

Thus increasing the highest degree in the network can also make the system unstable by the alternative route of decreasing the lowest eigenvalue. Whichever eigenvalue passes out of the region of stability first will be the one that makes the system unstable.

(We note in passing that Eqs. (18.88) and (18.91) together also tell us that the adjacency matrix of an undirected network always has both positive and negative eigenvalues, unless the network has no edges in it at all, in which case all eigenvalues are zero. We used this result previously in Section 18.2.2.)

Other results for the eigenvalues of the adjacency matrix can be derived for specific models of networks. For example, Chung *et al.* [68] have shown for the configuration model that the expected value of the largest eigenvalue in the limit of large network size is

$$\kappa_1 = \frac{\langle k^2 \rangle}{\langle k \rangle}. \tag{18.92}$$

In many cases this gives values of κ_1 considerably above the limits set by Eqs. (18.85) and (18.88). On configuration model networks with power-law degree distributions, for instance, where $\langle k^2 \rangle$ formally diverges in the limit of large n, we expect that κ_1 will similarly diverge.

One can also derive results for eigenvalues of the Laplacian. The smallest eigenvalue of the Laplacian is simple—it is always zero. For large networks the largest eigenvalue λ_n can be shown to lie in the range [18]

$$k_{\max} \leq \lambda_n \leq 2k_{\max}, \qquad (18.93)$$

which appears to be a relatively large range but in fact tells us a lot, ensuring again that the largest eigenvalue is guaranteed to increase if the highest degree in the network increases sufficiently.

18.4 SYNCHRONIZATION

A topic closely related to the study of dynamical stability is the study of synchronization. Many systems of scientific interest can be modeled as oscillators of one sort or another. The flashing of fireflies, the ticking of clocks, the synchronized clapping of a large audience, and the pathologically synchronized firing of brain cells during an epileptic attack can all be modeled as networks of oscillators coupled in such a way that the coupling causes the oscillators to synchronize.

The periodic, synchronized oscillations of such an oscillator network correspond, in dynamical systems terms, to a limit cycle of the overall dynamics (see Section 18.1.1). Like fixed points, limit cycles can be stable or unstable, attracting or repelling, depending on whether small perturbations away from the periodic behavior tend to grow or decay over time. The mathematics of whether synchronized states are stable is very similar to that for fixed points. Again one starts with a system of equations of the form of Eq. (18.68) but now assumes a periodic limit-cycle solution, $\mathbf{x}_i(t) = \mathbf{s}(t)$ for all i. Perturbing around this solution one can linearize the equations and, depending on the particular form of the interaction between vertices, expand the linearized solution as a combination of the eigenvectors of an appropriate matrix, such as the adjacency matrix or Laplacian. The result is a set of n decoupled systems, each oscillating independently and each of which must be stable if the system as a whole is to be stable. One can define a master stability function $\sigma(\lambda)$ again, corresponding to the growth rate of perturbations away from the periodic solution, which in this context is known as a *Lyapunov exponent*, although it plays exactly the same role as the leading eigenvector in our earlier analysis. Once again this master stability function must be negative when evaluated at each of the eigenvalues λ of the appropriate matrix and this gives us a condition for stability of the synchronized state.

Many details of the network synchronization process and many special cases have been studied. For a comprehensive discussion, the interested reader is encouraged to consult the review by Arenas *et al.* [23].

PROBLEMS

18.1 Consider a dynamical system on a k-regular network (i.e., one in which every vertex has the same degree k) satisfying

$$\frac{dx_i}{dt} = f(x_i) + \sum_j A_{ij}g(x_i, x_j),$$

and in which the initial condition is uniform over vertices, so that $x_i(0) = x_0$ for all i.

a) Show that $x_i(t) = x(t)$ for all i where

$$\frac{dx}{dt} = f(x) + kg(x, x),$$

and hence that one has to solve only one equation to solve the dynamics.

b) Show that for stability around a fixed point at $x_i = x^*$ for all i we require that

$$k < -\frac{1}{f'(x^*)}\left[\left(\frac{\partial}{\partial u} + \frac{\partial}{\partial v}\right)g(u, v)\right]_{u=v=x^*}.$$

18.2 Consider a dynamical system on an undirected network, with one variable per vertex obeying

$$\frac{dx_i}{dt} = f(x_i) + \sum_j A_{ij}[g(x_i) - g(x_j)],$$

as in Section 18.2.2. Suppose that the system has a symmetric fixed point at $x_i = x^*$ for all i.

a) Show, using results given in this chapter, that the fixed point is always stable if the largest degree k_{max} in the network satisfies

$$\frac{1}{k_{max}} > -2\left[\frac{dg}{dx}\bigg/\frac{df}{dx}\right]_{x=x*}.$$

b) Suppose that $f(x) = rx(1 - x)$ and $g(x) = ax^2$. Show that there are two symmetric fixed points for this system, but that one if them is always unstable.

c) Give a condition on the maximum degree in the network that will ensure the stability of the other fixed point.

18.3 The dynamical systems we have considered in this chapter have all been on undirected networks, but systems on directed networks are possible too. Consider a dynamical system on a directed network in which the sign of the interaction along an edge attached to a vertex depends on the direction of the edge, ingoing edges having positive sign and outgoing edges having negative sign. An example of such a system is a food web of predator–prey interactions, in which an ingoing edge indicates in-flow of energy to a predator from its prey and an outgoing edge indicates out-flow from a prey to its predator. Such a system can be represented by a dynamics of the form

$$\frac{\mathrm{d}x_i}{\mathrm{d}t} = f(x_i) + \sum_j (A_{ij} - A_{ji})g(x_i, x_j),$$

where g is a symmetric function of its arguments: $g(u, v) = g(v, u)$.

a) Consider a system of this form in which the in- and out-degrees of every vertex are equal to the same constant k. Show that such a system has a symmetric fixed point $x_i^* = x^*$ for all i satisfying $f(x^*) = 0$.

b) Writing $x_i = x^* + \epsilon_i$ linearize around this fixed point to show that in the vicinity of the fixed point the vector $\epsilon = (\epsilon_1, \epsilon_2, \dots)$ satisfies

$$\frac{\mathrm{d}\epsilon}{\mathrm{d}t} = (\alpha \mathbf{I} + \beta \mathbf{M})\epsilon,$$

where $\mathbf{M} = \mathbf{A} - \mathbf{A}^T$. Determine the values of the constants α and β.

c) Show that the matrix $\mathbf{M}$ has the property $\mathbf{M}^T = -\mathbf{M}$. Matrices with this property are called *skew-symmetric matrices*.

d) If $\mathbf{v}$ is a right eigenvector of a skew-symmetric matrix $\mathbf{M}$ with eigenvalue μ, show that $\mathbf{v}^T$ is a left eigenvector with eigenvalue $-\mu$. Hence by considering the equality

$$\mu = \frac{\mathbf{v}^{*T}\mu\mathbf{v}}{\mathbf{v}^{*T}\mathbf{v}} = \frac{\mathbf{v}^{*T}\mathbf{M}\mathbf{v}}{\mathbf{v}^{*T}\mathbf{v}}$$

show that the complex conjugate of the eigenvalue is $\mu^* = -\mu$ and hence that all eigenvalues of a skew-symmetric matrix are imaginary.

e) Show that the dynamical system is stable if $\mathrm{Re}(\alpha + \beta\mu_r) < 0$ for all eigenvalues μ_r of the matrix $\mathbf{M}$, and hence that the condition for stability is simply $\alpha < 0$.

The last result means that if the individual vertices are stable in the absence of interaction with other vertices, then the coupled dynamical system is also stable at the symmetric fixed point.

18.4 Following the arguments of Section 18.2.2 the stability of a fixed point in certain dynamical systems on networks depends on the spectrum of eigenvalues of the adjacency matrix. Suppose we have a dynamical system on a network that takes the form of an $L \times L$ square lattice with periodic (toroidal) boundary conditions along its edges, and suppose we label each vertex of the lattice by its position vector $\mathbf{r} = (i, j)$ where $i, j = 1 \dots L$ are the row and column indices of the vertex.

a) Consider the vector $\mathbf{v}$ with one element for each vertex such that $v_{\mathbf{r}} = \exp(i\mathbf{k}^T\mathbf{r})$. Show that this vector is an eigenvector of the adjacency matrix provided

$$\mathbf{k} = \frac{2\pi}{L}\begin{pmatrix} n_1 \\ n_2 \end{pmatrix},$$

where n_1 and n_2 are integers.

b) What range of values is permitted for the integers n_1 and n_2? Hence find the largest and smallest eigenvalues.

18.5 Consider a network with an oscillator on every vertex. The state of the oscillator on vertex i is represented by a phase angle θ_i and the system is governed by dynamical equations of the form

$$\frac{d\theta_i}{dt} = \omega + \sum_j A_{ij}g(\theta_i - \theta_j),$$

where ω is a constant and the function $g(x)$ respects the rotational symmetry of the phases, meaning that $g(x + 2\pi) = g(x)$ for all x.

a) Show that the synchronized state $\theta_i = \theta^* = \omega t$ for all i is a solution of the dynamics.

b) Consider a small perturbation away from the synchronized state $\theta_i = \theta^* + \epsilon_i$ and show that the vector $\epsilon = (\epsilon_1, \epsilon_2, \ldots)$ satisfies

$$\frac{d\epsilon}{dt} = g'(0)\,\mathbf{L}\epsilon,$$

where $\mathbf{L}$ is the graph Laplacian.

c) Hence show that the synchronized state is stable against small perturbations if and only if $g'(0) < 0$.

CHAPTER 19

NETWORK SEARCH

*A discussion of methods for searching networks for
particular vertices or items, a process important for web
search and peer-to-peer networks, and for our
understanding of the workings of social networks*

I N CHAPTER 4 we saw a number of examples of networks that have information stored at their vertices: the World Wide Web, citation networks,
peer-to-peer networks, and so forth. These networks can store large amounts
of data but those data would be virtually useless without some way of searching through them for particular items. So important is it to be able to perform
fast and accurate searches that the companies that provide the most popular
search services are now some of the largest in their respective industries—
Google, Thomson Reuters, LexisNexis—and constitute multibillion dollar international operations. In this chapter we examine some of the network issues
involved in efficient searching and some implications of search ideas for the
structure and behavior of networks.

19.1   WEB SEARCH

We have already discussed briefly some aspects of how web search engines
work in Sections 4.1, 7.4, and 7.5. In this section we discuss the issue in more
detail.

Traditional, or offline, web search is a multistage process. It involves first
"crawling" the Web to find web pages and recording their contents, then creating an annotated index of those contents, including lists of words and estimates of the importance of pages based on a variety of criteria. And then there
is the search process itself, in which a user submits a text query to a search

engine and the search engine extracts a list of pages matching that query from the index.

The process of web crawling by which web pages are discovered is interesting in itself and exploits the network structure of the Web directly. The crawler follows hyperlinks between web pages in a manner similar to the breadth-first search algorithm for finding components described in Section 10.3. The basic process is described in Section 4.1. Practical web crawlers for big search operations employ many elaborations of this process, including:

- Searching in parallel at many locations on the Web simultaneously using many different computers,
- Placing the computers at distributed locations around the world to speed access times to pages coming from different places,
- Repeatedly crawling the same web pages at intervals of a few days or weeks to check for changes in page contents or pages that appear or disappear,
- Checking on pages more often if their contents have historically changed more often,
- Checking on pages more often if they are popular with users of the search engine,
- Heuristics to spot dynamically generated pages that can lead a crawler into an infinite loop or tree of pages and waste time,
- Targeted crawling that probes more promising avenues in the network first, and
- Altered behavior depending on requests from owners of specific sites, who often allow only certain crawlers to crawl their pages, or allow crawlers to crawl only certain pages, in order to reduce the load on their servers.

At their heart, however, most web crawlers are still dumb animals, following links and recording what they see for later processing.

The processing of the raw crawler output also has interesting network-related elements. Early search engines simply compiled indexes of words or phrases occurring in web pages, so one could look up a word and get a list of pages containing it. Pages containing combinations of words could also be found by taking the sets of pages containing each individual word in the combination of interest and forming the intersection of those sets. Indexes can be extended by adding annotations indicating, for example, how often a word appears on a page or whether it appears in the page title or in a section heading, which might indicate a stronger connection between that word and the subject matter of the page. Such annotations allow the search engine to make choices about which are the pages most relevant to a given query. Even so, search engines based solely on indexes and textual criteria of this sort do not return very good results and have been superseded by more sophisticated technology.

Modern search engines do still use indexes in their search process, but only as a first step. A typical modern search engine will use an index to find a set of candidate pages that might be relevant to the given query and then narrow that set down using other criteria, some of which may be network-based. The initial set is usually chosen deliberately to be quite broad. It will typically include pages on which the words of the query appear, but also pages on which they don't appear but that link to, or are linked to by, pages that do contain the query words. The net result is a set of pages that probably includes most of those that might be of interest to the user submitting the query, but also many irrelevant pages as well. The strength of the search engine, its ability to produce useful results, therefore rests primarily on the criteria it uses to narrow the search within this broad set.

The classic example of a criterion for narrowing web searches comes from the Google search engine, which makes use of the eigenvector centrality measure known as PageRank, discussed in Section 7.4. PageRank accords pages a high score if they receive hyperlinks from many other pages, but does so in a way such that the credit received for a link is higher if it comes from a page that is itself highly ranked. PageRank, however, is only one of many elements that go into the formula Google uses to rank web pages. Others include traditional measures such as frequency of occurrence of query words in the page text and position of occurrence (near the top or bottom, in titles and headings, etc.), as well as occurrence of query words in "anchor text" (the highlighted text that denotes a hyperlink in a referring page) and previous user interest in a particular page (whether people selected this page from the list of search results on other occasions when the same text query, or a similar one, was entered).

Google gives each web page in the initial set a score that is a weighted combination of these elements and others. The particular formula used is a closely guarded secret and is moreover constantly changing, partly just to try and improve results, but also to confound the efforts of web page creators, who try to increase their pages' ranking by working out what particular elements carry high weight in Google's formula and incorporating those elements into their pages.

An important point to appreciate is that some parts of the score a page receives depend on the particular search query entered by the user—frequency of occurrence of query words, for instance—but others, such as PageRank, do not. This allows Google's computers (or their counterparts in other search companies) to calculate the latter parts "offline," meaning they are calculated ahead of time and not at the time of the query itself. This has some advantages. PageRank, for instance, is computationally intensive to calculate and it saves a lot of time if you only have to calculate it once. But there are disadvantages too. PageRank measures the extent to which people link to a given web page, but people may link to a page for many reasons. Thus a page may have a high PageRank for a reason unrelated to the current search query. A page whose text makes mention of two or more different topics (and many do) may be a crucial authority on one topic but essentially irrelevant on another, and PageRank cannot distinguish between the two.

One could imagine a version of PageRank that was specific to each individual query. One could calculate a PageRank score within just the subnetwork formed by the set of pages initially selected from the index to match the query. But this would be computationally expensive and it's not what Google does. As a result it is not uncommon for a page to be ranked highly in a particular search even though a casual human observer could quickly see that it was irrelevant to the search topic. In fact, a large fraction of "bad" search results returned by search engines probably fall in this category: they are pages that are important in some context, but not in the context of the specific search conducted.

The overall process behind searches on Google and similar large search engines is thus as follows [55]. First the Web is crawled to find web pages. The text of those web pages is processed to created an annotated index, and the link structure of the hyperlinks between them is used to calculate a centrality score or scores for each page, such as PageRank in Google's case or (presumably) some similar measure for other search engines. When a user enters a query the search engine extracts a deliberately broad set of matching pages from the index, scores them according to various query-specific measures such as frequency of occurrence of the query words, then combines those scores with the pre-computed centrality measure and possibly other pre-computed quantities, to give each page in the set an overall score. Then the pages are sorted in order of their scores and the ones with the highest scores are transmitted to the user. Typically only a small number of the highest-scoring pages are transmitted— say the first ten—but with an option to see lower-scoring pages if necessary.

Despite the reservations mentioned above, this system works well in practice, far better than early web search engines based on textual content alone, and provides useful search results for millions of computer users every day.

19.2 SEARCHING DISTRIBUTED DATABASES

Some information networks form distributed databases. A typical example is a peer-to-peer file-sharing network, in which individual computers in the network each store a subset of the data stored in the network as a whole. The form and function of peer-to-peer networks were described in Section 4.3.1.

The "network" in a peer-to-peer network is typically a virtual one, in which individual computers maintain contacts with a subset of others, which are not necessarily those with which they have direct physical data connections. In this respect peer-to-peer networks are somewhat similar to the World Wide Web, in which the hyperlinks between websites are virtual links chosen by a page's creator and their topology need have nothing to do with the topology of the underlying physical Internet. Indeed, the World Wide Web is itself, in a sense, a distributed database, storing information in the pages at its vertices, but web search works in a fundamentally different way from search in other distributed databases, so we treat the two separately.

Search is a fundamental problem in peer-to-peer networks and similar systems: how do we find specific items among those stored at the many vertices of the network? One way would be to copy the web search approach of Section 19.1 and construct a comprehensive index of all items at some central location and then search that index for items of interest. For a variety of reasons, however, most peer-to-peer networks don't go this route, but instead make use of distributed search techniques in which the search task is shared among the computers in the network via messages passed along network edges. Indeed the performance of such distributed searches is the primary reason for linking the vertices into a network in the first place and there are some interesting principles relating the structure of the network to the efficiency with which searches can be performed.

Suppose that we have a peer-to-peer network composed of n individual computers and each computer is linked by virtual connections to a selection of the others, where "linked" in this context merely means that these others are the ones with which a computer has agreed to communicate directly in the course of performing searches. There is no reason in principle why a computer could not communicate with *all* others if it wanted to, but in practice this would demand too much effort or data bandwidth, and limiting the number of network neighbors a computer has brings the resources required within reasonable bounds.

The simplest form of distributed search, used in some of the earliest peer-to-peer networks, is a version of the breadth-first search algorithm described in Section 10.3 (where it was used for finding network components and shortest paths). Under this approach, a user gives the computer a search term, such as the name of a computer file, and the computer sends a query to each of its neighbors in the peer-to-peer network, asking if they have the file in question. If they do, they send the file to the first computer and the search is complete. If they don't, then they send a further query to each of *their* neighbors asking for the file. Any neighbor that has seen the query before, such as the computer that originated it in the first place, ignores it. All others check to see if the have the requested file and send it back to the originating computer if they do. If not, they pass the query on to their neighbors, and so on.

This simple strategy certainly works and it has some advantages. For instance, assuming that the network displays the small-world effect (Section 8.2), the number of steps we will have to take in our breadth-first search will be small even when the network is large (typically increasing only logarithmically with n—see Section 12.7). This means that most searches will take only a short amount of time to find the desired file.

But there are some serious disadvantages with the approach as well. First, as we have described it the search doesn't actually stop when the target file is found. There is no mechanism to inform computers that the file has been found and that they don't need to pass the query on to anyone further. This problem can be fixed relatively easily, however, for example by requiring each computer receiving the query to check with the originating computer to see if the file has been found before they do anything else.

A more serious problem is that the messages transmitted in the process of spreading a query across the network quickly add up to a huge amount of data and can easily overwhelm the capacity of the computers involved. Assuming a worst-case scenario in which a desired file exists on only a single computer in the network, we will, on average, have to pass our query to half of all computers before we find the file. That means the number of messages sent in the course of one query is $O(n)$. Suppose that users perform queries at some constant average rate r, so that the overall rate of queries is $rn = O(n)$. Then the total number of messages sent per unit time is $O(n) \times O(n) = O(n^2)$ and the number of messages per computer per unit time is, on average, $O(n^2)/n = O(n)$, which goes up linearly with the size of the network. This means that, no matter how much bandwidth our computers have to send and receive data, it will in the end always become swamped if the network becomes large enough. And peer-to-peer networks can become extremely large. Some of the largest have millions of users.

Luckily this worst-case scenario is not usually realized. It is in fact rarely the case that an item of interest exists on only one computer in a network. Most items in typical peer-to-peer networks exist in many places. Indeed, assuming that some fraction of the user population likes or needs each item, it is more reasonable that any given item appears on some fixed fraction c of the vertices in the network, so that the total number of copies cn goes up as the size of the network increases. If this is the case, and assuming for the moment that the value of c is the same for every item, then one will have to search on average only $1/c$ vertices before finding a copy of an item. This means that the total number of query messages sent over the network per unit time is $O(n/c)$ and the number per computer per unit time is $O(1/c)$, which is just a constant and does not increase with increasing network size.

A more realistic calculation allows for the fact that some items are more popular than others. Suppose that the factors c, which are proportional to popularity, have a distribution $p(c)$, meaning that the probability of falling in the interval c to $c + dc$ is $p(c) \, dc$. Also important to note is that not all items are searched for with equal frequency. Indeed a more reasonable assumption is that they are searched for with frequency proportional to their popularity, i.e., that the probability of a search query asking for an item with popularity in the interval c to $c + dc$ is $cp(c) \, dc / \langle c \rangle$, where the factor of $\langle c \rangle = \int cp(c) \, dc$ insures that the distribution is properly normalized. Then the average number of vertices we have to examine before we find the item corresponding to a typical query is

$$\int_0^1 \frac{1}{c} \frac{cp(c) \, dc}{\langle c \rangle} = \frac{1}{\langle c \rangle}, \tag{19.1}$$

and hence the number of query messages sent or received per computer per unit time is $O(1/\langle c \rangle)$, which is again a constant as network size becomes large.

In principle, therefore, if a node can handle messages at the rate given by Eq. (19.1) then the network should go on functioning just fine as its size becomes large. In practice, however, there are still problems. The main difficulty is that vertices in the network vary enormously in their bandwidth capabilities. Most vertices have relatively slow communications with the network, i.e., low bandwidth, while a few have much better, higher-bandwidth connections. This means that even if bandwidth requirements per vertex are reduced to a constant as above, the network will still run at a speed dictated by the majority slow vertices, making queries slow and possibly overwhelming the capacity of some vertices.

To get around this problem, most modern peer-to-peer networks make use of *supernodes* (also called *superpeers*). Supernodes are high-bandwidth nodes chosen from the larger population in the network and connected to one another to form a supernode network over which searches can be performed quickly—see Fig. 19.1.

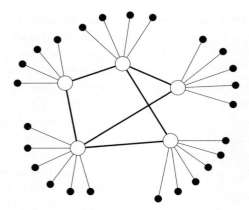

Figure 19.1: The structure of a peer-to-peer network with supernodes. Client nodes (filled circles) are connected to a network of supernodes (open circles) that have above-average network bandwidth and hence can conduct searches quickly.

A supernode acts a little like a local exchange in a telephone network (see Section 2.2). Each normal user, or *client*, in the network attaches to a super-node (or sometimes to more than one) that acts as their link to the rest of the network. Each supernode has a number of such clients and the clients communicate to the supernode a list of the files or other data items they possess so that the supernode can respond appropriately to search queries from other supernodes. An individual client wanting to perform a search then sends their search query to the local supernode, which conducts a breadth-first search interrogation of the network of supernodes to find the desired item. Since the supernodes possess records of all the items that the clients have, the entire search can be performed on the network of supernodes alone and no client resources are used at all. And since the supernodes are deliberately selected to have fast network connections, the search runs at the speed of the quickest vertices in the network.

In practice schemes like this work quite well—well enough to be in wide use in peer-to-peer networks of millions of users. More sophisticated schemes have been devised that in theory could work better still—an example is the "Chord" system proposed by Stoica *et al.* [305]—but such systems have yet to find widespread adoption since the more traditional supernode approach appears to work well enough for practical purposes.

19.3 MESSAGE PASSING

A different variation of the distributed search problem is the problem of getting a message to a particular node in a network. The classic example of this problem is Stanley Milgram's "small-world" experiment, described in Section 3.6. In this experiment participants were asked to get a message to a specific target individual by passing it from acquaintance to acquaintance through the social network. Milgram famously found that messages that arrived at the destination passed through only about six people on their way, which is the origin of the popular concept of the "six degrees of separation." As discussed in Section 3.6, however, there is another perhaps more surprising implication of the experiment, first pointed out by Kleinberg [177], which is that short paths not only exist in the network but that people are remarkably good at finding them. Of course if one knows the structure of an entire network then one can find short paths directly using, for example, the breadth-first search method of Section 10.3.5. Participants in Milgram's experiment, however, did not know the whole network and probably only knew a very small part of it, and yet they were still able to get a message rapidly to the desired target.

This observation raises a number of interesting questions. How, in practice, did people find these short paths to the target? Can we come up with an algorithm that will do the job efficiently? How does the performance of that algorithm depend on the structure of the network? In the following sections we consider two different models of the message passing process that address these questions. As we will see, these models suggest that social (or other) networks must have a very particular type of structure if one wants to be able to find short paths easily without a global knowledge of the network.

19.3.1 KLEINBERG'S MODEL

The instructions to the participants in Milgram's experiment were that upon receiving the message (actually a small booklet or "passport" sent through the mail), they were to forward it to an acquaintance who they believed to be closer to the target than they were. The definition of "closer" was left vague, however, and one of the first things we need to do if we want to model the mechanics of the experiment is decide on a practical definition.

An illuminating attempt at modeling Milgram's experiment was made by Kleinberg [177, 178], who employed a variant of the small-world model of Section 15.1, as shown in Fig. 19.2. As in the standard small-world model, it has a ring of vertices around the edge plus a number of "shortcut" edges that connect vertex pairs at random points around the ring. In Kleinberg's model all

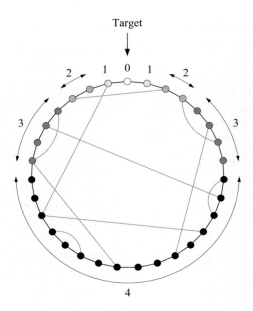

Figure 19.2: The variant small-world model used to model message passing. In the variant of the small-world model used here, vertices are connected around a ring and shortcuts added between them as in the normal small-world model. However, the shortcuts are now biased so that there are more of them connecting nearby vertices than distant vertices. The strength of the bias is controlled by the parameter α. In the proof given in the text, the vertices are divided into numbered classes, class 0 consisting of just the target vertex and higher classes radiating out from the target, each successive class containing twice as many vertices as the previous one.

vertices are connected to their two immediate neighbors around the ring—$c = 2$ in the notation of Chapter 15—and Kleinberg made use of the connections in the ring to define the "closeness" of vertices for the purposes of the message-passing experiment. He proposed that individuals in the network, represented by vertices, are aware of the distance around the ring to other individuals, and hence can say when one of their acquaintances is "closer" to the target vertex than they are in this sense.

In his calculations Kleinberg considered a *greedy algorithm* for message passing in which each individual receiving a message passes it on to the one of their neighbors who is closest to the target in the sense above. This algorithm is guaranteed always to get the message to the target eventually. Every individual has at least one neighbor who is closer to the target in the Kleinberg sense than they are—their neighbor around the ring in the direction towards the target. Thus on each step of the message passing process the message is guaranteed to get at least one step closer to the target around the ring and hence it must eventually get to the target. In the worst case individuals simply pass the message around the ring until it reaches its destination but generally we can expect to do better than this because of the shortcuts. The question is how much better. Kleinberg showed that it is possible for the greedy algorithm to find the target vertex in $O(\log^2 n)$ steps, but that it can do so only for particular choices of the arrangement of the shortcuts.

Kleinberg considered a one-parameter family of models that generalizes the standard small-world model by allowing for different arrangements of the shortcuts.[1] Instead of assuming that shortcuts are placed uniformly at random, we assume (not unreasonably) that people have more acquaintances among those close to them (in the sense defined above) than among those far away. By analogy with the standard small-world model let us place shortcuts around the ring equal in number to p times the number of edges in the ring itself, which in this case is just n. Since each shortcut has two ends this means that the average number of shortcuts attached to each vertex will be $2p$ (and the actual number will be Poisson distributed with mean $2p$). Where we differ from the standard small-world model is in how these shortcuts are placed. Shortcuts are still placed at random, but they are chosen so that the probability of a particular shortcut covering a distance r around the ring is $Kr^{-\alpha}$, where α is a non-negative constant and K is a normalizing constant. That is, for each shortcut we choose first its length r from this distribution, then we place the shortcut, spanning exactly r vertices, at a position around the ring chosen uniformly at random. If $\alpha = 0$ then we recover the standard small-world model of Section 15.1, but more generally, for $\alpha > 0$, the model has a preference for connections between nearby vertices.

Note that the probability that a particular shortcut connects a specific pair of vertices a distance r apart is equal to $Kr^{-\alpha}/n$, which is the probability $Kr^{-\alpha}$ that the shortcut has length r multiplied by the probability $1/n$ that out of the n possible choices it falls in the specific position around the ring that connects the two vertices in question. Given that there are np shortcuts in the whole network, this means that the total probability of having a shortcut between

---

[1]The model we use is a somewhat simplified version of Kleinberg's. His model, for instance, used a two-dimensional lattice instead of a one-dimensional ring as the underlying structure on which the model was built. The calculations, however, work just as well in either case. Our model also places shortcuts at random, where Kleinberg's fixed the number attached to each vertex to be constant and also made them directed rather than undirected.

a given pair of vertices is $np \times Kr^{-\alpha}/n = pKr^{-\alpha}$. (More correctly, this is the expected number of such shortcuts, but so long as the number is small, the difference is negligible.)

The normalizing constant K is fixed by the condition that every shortcut must have *some* length, and that all lengths lie between 1 and $\frac{1}{2}(n-1)$, so that[2]

$$K \sum_{r=1}^{\frac{1}{2}(n-1)} r^{-\alpha} = 1. \tag{19.2}$$

We can approximate the sum by an integral using the trapezoidal rule of Eq. (14.115) thus:

$$\sum_{r=1}^{\frac{1}{2}(n-1)} r^{-\alpha} \simeq \int_1^{\frac{1}{2}(n-1)} r^{-\alpha}\,\mathrm{d}r + \tfrac{1}{2} + \tfrac{1}{2}\left[\tfrac{1}{2}(n-1)\right]^{-\alpha}$$

$$= \frac{\left[\tfrac{1}{2}(n-1)\right]^{1-\alpha} - 1}{1-\alpha} + \tfrac{1}{2} + \tfrac{1}{2}\left[\tfrac{1}{2}(n-1)\right]^{-\alpha}, \tag{19.3}$$

which gives

$$K \simeq \begin{cases} (1-\alpha)(\tfrac{1}{2}n)^{\alpha-1} & \text{for } \alpha < 1, \\ 1/\ln \tfrac{1}{2}n & \text{for } \alpha = 1, \\ 2(\alpha-1)/(\alpha+1) & \text{for } \alpha > 1, \end{cases} \tag{19.4}$$

as n becomes large.[3]

We can now show that, for suitable choice of α, the greedy algorithm on this network can indeed find a given target vertex quickly. The proof is as follows. Suppose, without loss of generality, that the target vertex is at the top of the ring, as depicted in Fig. 19.2, and let us divide up the other vertices into classes according to their distance from the target. Class 0 consists of just the target itself. Class 1 consists of all vertices distance $d = 1$ from the target around the ring, of which there are two. Class 2 consists of vertices with distances in the range $2 \leq d < 4$, class 3 of vertices $4 \leq d < 8$, and so forth. Each class is double the size of the previous one. In general, class k consists of vertices at distances $2^{k-1} \leq d < 2^k$ and contains $n_k = 2^k$ vertices. (For simplicity, let us assume that the total number n of vertices is a power of two, minus one, so that everything works out neatly.)

Now consider a message being passed through the network according to the greedy algorithm described above and suppose that at a particular step of

---

[2]The maximum length of a shortcut is $\frac{1}{2}(n-1)$ if n is odd and $\frac{1}{2}n$ if n is even. We will assume that n is odd in this case, which avoids some small annoyances in the derivations.

[3]Note that both the numerator and denominator of the fraction in Eq. (19.3) vanish at $\alpha = 1$, so one must use l'Hopital's rule to extract the limiting value. The same goes for Eq. (19.8).

the process the message is at a vertex of class k. How many more steps will it take before the message leaves class k and passes into a lower class? The total number of vertices in lower classes is

$$\sum_{m=0}^{k-1} n_m = \sum_{m=0}^{k-1} 2^m = 2^k - 1 > 2^{k-1}, \tag{19.5}$$

and from Fig. 19.2 we can see that all of these are, at most, a distance $3 \times 2^k - 2 < 2^{k+2}$ from the vertex in class k that currently holds the message. Thus the probability of the vertex with the message having a shortcut to a particular one of these vertices in lower classes is at least $pK\,2^{-(k+2)\alpha}$, and the probability of having a shortcut to any of them is at least $pK\,2^{k-1-(k+2)\alpha}$.

If our vertex has no shortcut that takes the message out of class k then, in the worst case, it simply passes the message to another vertex in class k that is closer to the target, either via a shortcut or by passing around the ring. Using the probability above, the expected number of such moves made before we find a shortcut that takes us out of class k is at most

$$\frac{1}{pK\,2^{k-1-(k+2)\alpha}} = \frac{1}{pK}\,2^{2\alpha+1}2^{(\alpha-1)k}. \tag{19.6}$$

Finally, again in the worst case, the message will pass through each of the classes before reaching the target. There are $\log_2(n+1)$ classes in total and summing over them we find that an upper bound on the expected number of steps ℓ needed to reach the target is

$$\ell \leq \frac{1}{pK}\,2^{2\alpha+1}\sum_{k=0}^{\log_2(n+1)} 2^{(\alpha-1)k} = \frac{1}{pK}\,2^{2\alpha+1}\frac{2^{(\alpha-1)[1+\log_2(n+1)]} - 1}{2^{\alpha-1} - 1}$$

$$= \frac{1}{pK}\,2^{2\alpha+1}\frac{[2(n+1)]^{\alpha-1} - 1}{2^{\alpha-1} - 1}. \tag{19.7}$$

Making use of Eq. (19.4) for the constant K and taking the limit of large n we find that asymptotically

$$\ell \leq \begin{cases} A\,n^{1-\alpha} & \text{if } \alpha < 1, \\ B\log^2 n & \text{if } \alpha = 1, \\ C\,n^{\alpha-1} & \text{if } \alpha > 1, \end{cases} \tag{19.8}$$

where A, B, and C are constants depending on α and p, but not n, whose rather complicated values we can work out from Eqs. (19.4) and (19.7) if we want.

Since Eq. (19.8) gives an upper bound on ℓ, this result guarantees that for the particular case $\alpha = 1$ we will be able to find the target vertex in a time that increases as $\log^2 n$ with the size of the network. This is not quite as good as $\log n$, which is the actual length of the shortest path in a typical network, but it is still a slowly growing function of n and it would be fair to claim that the small-world experiment would succeed in finding short paths in a network that had $\alpha = 1$. Thus it is possible, provided the network has the correct structure, for a simple strategy like the greedy algorithm, in which vertices have knowledge only of their immediate network neighborhood, to produce results similar to those observed by Milgram in his experiment.

On the other hand, if $\alpha \neq 1$ then Eq. (19.8) increases as a power of n, suggesting that it would take much longer in such networks to find the target vertex. In particular, for the original small-world model of Section 15.1, which corresponds to $\alpha = 0$, Eq. (19.8) grows linearly with n, suggesting that the Milgram experiment could take millions of steps to find a target in a social network of millions of people. Equation (19.8) is only an upper bound on the time taken, so if one is lucky one may be able to find the target faster. For instance, if the message starts at a vertex that happens to have a shortcut directly to the target vertex then one can find the target in a single step. However, Kleinberg [178] was also able to prove that the *average* time it takes to find the target increases at least as fast as a power of n except in the special case $\alpha = 1$, so in general the greedy algorithm for $\alpha \neq 1$ will not work well.[4]

These results tell us two things. First, they tell us that it is indeed possible for the small-world experiment to work as observed even if the participants don't know the details of the whole network. Second, they tell us that, at least within the context of the admittedly non-realistic model used here, the experiment only works for certain very special values of the parameters of the network. Thus the success of Milgram's experiment suggests not only that, as Milgram concluded, there are short paths in social networks, but also that they have a particular structure that makes path finding possible.

19.3.2   A HIERARCHICAL MODEL OF MESSAGE PASSING

While interesting, the results of the previous section are not wholly convincing because the model is clearly not a realistic one. People don't live around a circle with just a few shortcuts to others, and message passing doesn't work because people know where others live on the circle.

---

[4]In fact, since Kleinberg was studying a two-dimensional version of the small-world model, his result was for $\alpha = 2$, not $\alpha = 1$. In general, on a small-world network built on a d-dimensional lattice, the greedy algorithm succeeds in finding the target in time $O(\log^2 n)$ only when $\alpha = d$ and for all other values takes time increasing at least as a power of n.

So can we derive similar results for a more realistic network model? To answer this question let us first ask how message passing *does* work. We can get a hint from the "reverse small-world" experiments of Killworth and Bernard [39, 174] discussed in Section 3.6. Recall that in these experiments researchers asked subjects to imagine that they were participating in Milgram's small-world experiment and then asked them what information they would want to know about the target in order to make a decision about who to pass their message on to. Killworth and Bernard found that three pieces of information were sought more often than any others, and by almost all subjects: the name, occupation, and geographic location of the target.

The target's name is an obvious requirement in the small-world experiment, since it's needed to recognize the target when you find him or her. Beyond that, however, it probably doesn't play much role in the message passing, except perhaps in cultures where names can give a clue as to the location or social status of an individual. Occupation and geographic location, on the other hand, are of great use in deciding how to forward a message, and these appear to be the primary pieces of information participants in the experiment use.

Take geographic location as an example. How would one use information on geography to route a message? Presumably, one would attempt to pass the message to someone closer geographically to the target than oneself. Suppose for instance that the target lives, as Milgram's did, in a suburb of the city of Boston, Massachusetts, in the United States. A participant in, say, England, attempting to get a message to this target, would perhaps first forward it to someone they knew in the US. That person might forward it in turn to someone they knew in the state of Massachusetts, who would forward it to someone in Boston, who would forward it to the target's specific suburb, and so forth. At each step in the process, the participants narrow down the search to a smaller and smaller geographic area until, with luck, the area is so small that someone there knows the target individual directly.

In a sense, this is what happens in Kleinberg's model. In Section 19.3.1 we divided Kleinberg's circle into zones or classes that get ever smaller as they close in on the target and showed that under suitable circumstances it takes only a small number of steps of the message-passing process to find a connection from one class to the next smaller one. Since the number of classes is logarithmic in the size of the network, this means that it also takes only a small number of steps overall to home in on the target. Kleinberg's network structure was unrealistic, but the basic idea of progressively narrowing the field is a good one and we would like to find a more realistic network model to which the same type of argument can be applied.

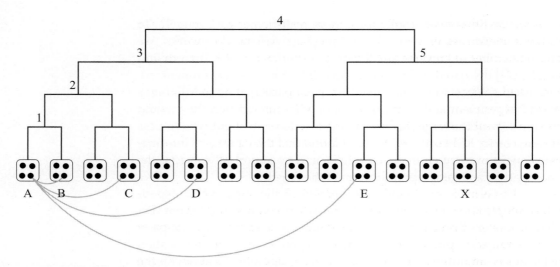

Figure 19.3: The hierarchical model of Watts *et al.* Small groups of individuals (boxes) are divided up in a hierarchical structure represented by a binary tree, which might, for instance, correspond to the hierarchical division of geographic space into countries, regions, towns, and so forth. The hierarchy dictates which social connections (indicated by curves) are most likely. A vertex in group A, for instance, is most likely to be connected to others close to it in the tree (B, C) and less likely to be connected to those further away (D, E).

Such a model is the hierarchical model of Watts *et al.* [322], in which the interplay of social structure and geographic or other dimensions is represented by a tree or dendrogram.[5] In the context of geography, for example, the world would be divided into countries, the countries into regions, states, or provinces, the regions into cities and towns, and so forth. The division ends when we reach units so small that it can reasonably be assumed that everyone knows everyone else—a single family, for instance.

The divisions can be represented by a tree structure like that shown in Fig. 19.3. The tree depicted is a binary tree in this case. Each branch splits in two, then in two again, and so forth. In the real world branches might easily split into more than two parts. There are more than two countries in the world after all. However, the binary tree is the simplest case to study (and the one studied by Watts *et al.*), and the analysis given here for the binary case can be generalized to other cases quite easily.

Let us also assume that the groups at the bottom of the tree all have the

---

[5]A similar model was also proposed independently by Kleinberg [179].

same size g. Again this is a simplification, but a useful one that does not have a major effect on the results. If the total number of individuals in the network is n then the number of groups is n/g, and the number of levels in the tree is $\log_2(n/g)$.

The model of Watts *et al.* makes two other important assumptions. First, it assumes that people measure distance to a target individual in terms of the tree, and more specifically in terms of the lowest common ancestor in the tree that they share with the target. That is, people are able to tell when someone lives in the same country as themselves, or the same region or town, but do not have any detailed information beyond that. This is a more conservative assumption than is made by Kleinberg's model. In Kleinberg's model it is assumed that people know their exact geometric distance to the target, no matter where in the network the target falls. In the present model people have a much more coarse-grained impression of how close they are to the target.

The second assumption in the model of Watts *et al.* is that the social network itself is correlated with the hierarchical tree structure so that people who are closer together in the tree, in the sense of sharing a lower common ancestor, are also more likely to be acquainted. Thus people are more likely to know others in their own country than in other countries, more likely to know others in their own town than in other towns, and so forth. A few sample acquaintances are represented by the curves at the bottom of the figure.

Thus there are really two networks present in this model. There is the "real" network of actual acquaintances represented by the curves, and a "shadow" network, the hierarchical tree, which is not a network of actual acquaintances but which influences the acquaintance network and of which individuals are somewhat aware, in the sense that they know how close they are to others in the tree.

An important point to note about this model is that although an individual is less likely to know others far away in the tree, there are also more such far-away individuals than there are ones close by, and the two effects cancel out to some extent so that it is quite possible for a given individual to know others who are both near and far. The people who live on your street, for instance, are close by, so you are likely to know some of them, but they are also few in number. By contrast, India may be far away for you (depending on where you live) but there are a lot of people there, so even though you are not very likely to know any particular inhabitant, it is nonetheless quite likely that you know at least one out of the whole population. This behavior is crucial to making the message-passing experiment work on this network.

Consider an individual in group A in Fig. 19.3. Let us suppose that, because of the effect above, this individual has at least one acquaintance at every "distance" in the tree, i.e., one acquaintance in every subtree of the hierarchy with which they share a common ancestor. That is, they know one of the individuals in group B, the one group with which they share ancestor 1, and they also know (say) someone in group C, one of the two with whom they share ancestor 2, and so on through groups D and E as shown. And suppose that a similar pattern of acquaintances holds for every individual in the network: everyone knows at least one person in every subtree with whom they share a common ancestor.

Now consider a greedy algorithm for message passing on this network. Suppose the message starts at a vertex in group A and, as before, the holder of the message at each step passes it to an acquaintance closer to the target than they are, distance now being measured in the sense of the hierarchical tree as described above.

Suppose the target vertex is in group X, which shares a common ancestor with A only at the highest and coarsest level marked 4 in the figure. That is, the target is in the opposite subtree of ancestor 4 from A. By hypothesis, the individual holding the message knows this and hence knows that in order to get the message closer to the target they must pass it to someone in that opposite subtree. Luckily, under the assumption above they always have such an acquaintance, in this case in group E. So they pass the message to their friend in group E. The friend now notes that the target X is in the subtree with whom they share the common ancestor marked 5 and hence knows that they must pass the message to a neighbor in that subtree to get it closer to the target. Again, by definition, they have at least one such neighbor, to whom they pass the message. And so the process proceeds. At each step we narrow down our search to a smaller subtree of the overall network, or equivalently we move to a lower level in the hierarchy, pivoting about a lower common ancestor. But the total number of levels in the hierarchy is $\log_2(n/g)$ and hence this is the maximum number of steps that the process will take to reach the target. In this model, therefore, the message always reaches its target in a logarithmic number of steps.

It's not, however, very realistic to assume that each individual in the network knows at least one person at each distance. Watts *et al.* considered a more realistic probabilistic model in which there is a probability p_m of two individuals knowing one another when their lowest common ancestor is at level m in the tree. The level is defined to be $m = 0$ for groups that are immediately adjacent, as A and B are in Fig. 19.3, and to increase by one for each higher level up to a maximum of $m = \log_2(n/g) - 1$ at the top of the tree.

Watts *et al.* considered the particular choice

$$p_m = C2^{-\beta m}, \tag{19.9}$$

where C and β are constants.[6] So long as β is positive this choice gives, as desired, a lower probability of acquaintance with more distant individuals, the exact rate of variation being controlled by the value of β. The parameter C controls the overall number of acquaintances that each individual has.

The number of vertices with which any given vertex shares its ancestor at level m is just $2^m g$ and hence the expected number of such vertices that it will be connected to is

$$2^m g p_m = Cg\,2^{(1-\beta)m} \tag{19.10}$$

with the choice above for p_m. Summing over all levels the total expected number of acquaintances an individual has, their average degree in the network, is

$$\langle k \rangle = Cg \sum_{m=0}^{\log_2(n/g)-1} 2^{(1-\beta)m} = Cg\frac{2^{(1-\beta)\log_2(n/g)} - 1}{2^{1-\beta} - 1} = Cg\frac{(n/g)^{1-\beta} - 1}{2^{1-\beta} - 1}. \tag{19.11}$$

Thus the constant C is given by

$$C = \frac{\langle k \rangle}{g}\frac{2^{1-\beta} - 1}{(n/g)^{1-\beta} - 1}. \tag{19.12}$$

In the limit of large n this simplifies to

$$C = \begin{cases} (\langle k \rangle/g)(2^{1-\beta} - 1)(n/g)^{1-\beta} & \text{for } \beta < 1, \\ (\langle k \rangle/g)/\log_2(n/g) & \text{for } \beta = 1, \\ (\langle k \rangle/g)(1 - 2^{1-\beta}) & \text{for } \beta > 1. \end{cases} \tag{19.13}$$

Now if a particular vertex receives a message and wants to pass it to a member of the opposite subtree at level m, it can do so provided it has a suitable acquaintance. If (19.10) is small, however, then most likely it will not, in which case the best it can do is to pass the message to someone else in the subtree it is already in, who can then repeat the process. The expected number of times this will happen before one person does have a neighbor in the opposite subtree is given by the reciprocal of (19.10), which is $2^{(\beta-1)m}/Cg$. Then, summing this over all levels, the total expected number of steps to reach the target

---

[6]Watts *et al.* actually wrote the expression as $Ce^{-\beta m}$, but the difference is only in the value of β and we find the definition (19.9) more convenient.

is

$$\ell = \frac{1}{Cg} \sum_{m=0}^{\log_2(n/g)-1} 2^{(\beta-1)m} = \frac{1}{Cg} \frac{2^{(\beta-1)\log_2(n/g)} - 1}{2^{\beta-1} - 1}$$

$$= \frac{1}{Cg} \frac{(n/g)^{\beta-1} - 1}{2^{\beta-1} - 1}. \tag{19.14}$$

It is also possible that the vertex holding the message will not have a neighbor either in the opposite subtree or in its own subtree. If this happens then the vertex has only neighbors further from the target than it is and none nearer. In this case the Milgram experiment fails—recall that participants were asked to pass the message to someone closer to the target. This, however, is not necessarily unrealistic. As Watts *et al.* pointed out, this presumably does happen in the real experiment sometimes, and moreover it is well documented that many messages, a majority in fact, get lost and never reach their target. For messages that do get through, however, Eq. (19.14) gives an estimate of the number of steps they take to arrive.

Equation (19.14) is rather similar to the corresponding expression for the model of Kleinberg, Eq. (19.7), which is not a coincidence since the mechanisms by which the message-passing process proceeds are similar in the two cases. Taking the limit of large n and making use of Eq. (19.13), we find that

$$\ell = \begin{cases} D(n/g)^{1-\beta} & \text{for } \beta < 1, \\ E\log^2(n/g) & \text{for } \beta = 1, \\ F(n/g)^{\beta-1} & \text{for } \beta > 1, \end{cases} \tag{19.15}$$

where D, E, and F are constants.

These results have the same functional form as those of Eq. (19.8) for Kleinberg's model and tell us that it is indeed possible for Milgram's experiment to succeed in networks of this type, but only for the special parameter value $\beta = 1$. For all other values, the number of steps ℓ taken to reach the target increases as a power of n.

Thus the model of Watts *et al.* confirms Kleinberg's results in the context of a more realistic network. The results are, however, somewhat mysterious in a way. The idea that the network must be tuned to a special point in order for the Milgram experiment to succeed is surprising. The Milgram experiment does appear to succeed when conducted on real-world social networks, but on the face of it there is no clear reason why real-world networks should fall at this special point. Is it really true that if the world happened to be a little different from the way it is, Milgram's experiment would fail? This is a point that is not yet fully understood. It is possible that our model misses some important feature of the network structure that makes message passing more robust in the real world and less dependent on the precise tuning of the network, or that people are using a different scheme for passing messages that works substantially better than our greedy algorithm. On the other hand, it is also possible that our model is basically correct but that the world is in fact only rather loosely tuned to the special point $\beta = 1$ at which message passing succeeds in finding short paths. For values of β close to 1 the power of n in Eq. (19.14) is small and hence ℓ still grows quite slowly. Indeed it is in general difficult to distinguish experimentally between low powers and logarithms, so any value of β in the rough vicinity of $\beta = 1$ could result in good apparent performance in the message passing experiment.

PROBLEMS

19.1 Suppose that we use a web crawler to crawl a small portion of the Web, starting from a randomly chosen web page somewhere in the large in-component. Let us model the crawl as a breadth-first search starting from the given vertex and proceeding for r "waves" of search, i.e., until it reaches vertices that are r steps away from the start. Let S_i be the size of the large in-component.

a) What is the probability that a given web page has been crawled at the "zeroth" wave of the algorithm, i.e., when only the one starting page has been crawled?

b) Argue that the probability p_i that a page is first reached by the crawl on the rth wave is given approximately by $\mathbf{p}(r) = \mathbf{A}\mathbf{p}(r-1)$, where $\mathbf{p} = (p_1, p_2, \ldots)$. Why is this relation only approximate in general?

c) Hence argue that the probability of a page being found in a small crawl is roughly proportional to the eigenvector centrality of the page. Recall that the eigenvector centrality is zero for vertices in the in-component that don't also belong to the strongly connected component (see Section 7.2). Explain why this makes sense in the present context.

19.2 Suppose that a search is performed on a peer-to-peer network using the following algorithm. Each vertex on the network maintains a record of the items held by each of its neighbors. The vertex originating a search queries one of its neighbors, chosen uniformly at random, for a desired item and the neighbor responds either that it or one of its neighbors has the item, in which case the search ends, or that they do not. In the latter case, the neighboring vertex then passes the query on to one of *its* neighbors, chosen at random, and the process repeats until the item is found. Effectively, therefore, the search query makes a random walk on the network.

a) Argue that, in the limit of a large number of steps, the probability that the query encounters a vertex i on any particular step is $k_i/2m$, where k_i is the degree as usual and m is the total number of edges in the network.

b) Upon arriving at a vertex of degree k, the search learns (at most) about the items held by all of that vertex's k neighbors except for the one the query is coming from, for a total of $k - 1$ vertices. Show that on average at each step the search learns about the contents of approximately $\langle k^2 \rangle / \langle k \rangle - 1$ vertices and hence that, for a target item that can be found at a fraction c of the vertices in the network, the expected number of copies of the item found on a given step is $c(\langle k^2 \rangle / \langle k \rangle - 1)$.

c) Argue that the probability of not finding the target item on any particular step is approximately $q = \exp[c(1 - \langle k^2 \rangle / \langle k \rangle)]$ and that average number of steps it takes to find a copy of the item is $1/(1 - q)$.

d) On a network with a power-law degree distribution with exponent less than 3, so that $\langle k^2 \rangle \to \infty$, this last result implies that in the limit of large network size the search should end after only one step. Is this really true? If not, explain why not.

Although the random walk is not a realistic model for actual network search it is nonetheless useful: presumably more intelligent search strategies will find results quicker than a mindless random walk and hence the random walk provides an upper bound on the length of search needed to find an item. In particular, if the random walk works well, as in the example above, then it suggests that more intelligent forms of search will also work well.

19.3 The network navigation model of Kleinberg described in Section 19.3.1 is a one-dimensional version of what was, originally, a two-dimensional model. In Kleinberg's original version, the model was built on a two-dimensional square lattice with vertices connected by shortcuts with probability proportional to $r^{-\alpha}$ where r is the "Manhattan" distance between the vertices, i.e., the geodesic network distance in terms of number of edges traversed (rather than the Euclidean distance). Following the outline of Section 19.3.1, sketch an argument to show for this variant of the model that it is possible to find a target vertex in $O(\log^2 n)$ steps, but only if $\alpha = 2$.

19.4 Show that the ability to find short paths (of order $\log^2 n$) in the hierarchical model of Section 19.3.2 coincides with the state of the network in which a vertex has equal numbers of neighbors on average at each possible distance, where "distance" is defined by the lowest common ancestor two vertices share.

REFERENCES

[1] Abello, J., Buchsbaum, A., and Westbrook, J., A functional approach to external graph algorithms, in *Proceedings of the 6th European Symposium on Algorithms*, Springer, Berlin (1998).

[2] Abramowitz, M. and Stegun, I. A., eds., *Handbook of Mathematical Functions*, Dover Publishing, New York (1974).

[3] Achlioptas, D., Clauset, A., Kempe, D., and Moore, C., On the bias of traceroute sampling: or, Power-law degree distributions in regular graphs, *J. ACM* **56**(4), 21 (2009).

[4] Adamic, L. A. and Glance, N., The political blogosphere and the 2004 US election, in *Proceedings of the WWW-2005 Workshop on the Weblogging Ecosystem* (2005).

[5] Adamic, L. A. and Huberman, B. A., The nature of markets in the World Wide Web, *Q. J. Electronic Commerce* **1**, 512 (2000).

[6] Adamic, L. A., Lukose, R. M., Puniyani, A. R., and Huberman, B. A., Search in power-law networks, *Phys. Rev. E* **64**, 046135 (2001).

[7] Adleman, L. M., Molecular computation of solutions to combinatorial problems, *Science* **266**, 1021–1024 (1994).

[8] Ahuja, R. K., Magnanti, T. L., and Orlin, J. B., *Network Flows: Theory, Algorithms, and Applications*, Prentice Hall, Upper Saddle River, NJ (1993).

[9] Aiello, W., Chung, F., and Lu, L., A random graph model for massive graphs, in *Proceedings of the 32nd Annual ACM Symposium on Theory of Computing*, pp. 171–180, Association of Computing Machinery, New York (2000).

[10] Aiello, W., Chung, F., and Lu, L., Random evolution of massive graphs, in J. Abello, P. M. Pardalos, and M. G. C. Resende, eds., *Handbook of Massive Data Sets*, pp. 97–122, Kluwer, Dordrecht (2002).

[11] Albert, R. and Barabási, A.-L., Topology of evolving networks: Local events and universality, *Phys. Rev. Lett.* **85**, 5234–5237 (2000).

[12] Albert, R. and Barabási, A.-L., Statistical mechanics of complex networks, *Rev. Mod. Phys.* **74**, 47–97 (2002).

[13] Albert, R., Jeong, H., and Barabási, A.-L., Diameter of the world-wide web, *Nature* **401**, 130–131 (1999).

[14] Albert, R., Jeong, H., and Barabási, A.-L., Attack and error tolerance of complex networks, *Nature* **406**, 378–382 (2000).

[15] Aldous, D., Spatial transportation networks with transfer costs: Asymptotic optimality of hub and spoke models, *Mathematical Proceedings of the Cambridge Philosophical Society* **145**, 471–487 (2008).

[16] Amaral, L. A. N., Scala, A., Barthélémy, M., and Stanley, H. E., Classes of small-world networks, *Proc. Natl. Acad. Sci. USA* **97**, 11149–11152 (2000).

[17] Anderson, R. M. and May, R. M., *Infectious Diseases of Humans*, Oxford University Press, Oxford (1991).

[18] Anderson, W. N. and Morley, T. D., Eigenvalues of the Laplacian of a graph, *Linear and Multilinear Algebra* **18**, 141–145 (1985).

[19] Anthonisse, J. M., The rush in a directed graph, Technical Report BN 9/71, Stichting Mathematisch Centrum, Amsterdam (1971).

[20] Appel, K. and Haken, W., Every planar map is four colorable. II: Reducibility, *Illinois J. Math.* **21**, 491–567 (1977).

[21] Appel, K. and Haken, W., The solution of the four-color map problem, *Sci. Am.* **237**, 108–121 (1977).

[22] Appel, K., Haken, W., and Koch, J., Every planar map is four colorable. I: Discharging, *Illinois J. Math.* **21**, 429–490 (1977).

[23] Arenas, A., Díaz-Guilera, A., Kurths, J., Moreno, Y., and Zhou, C., Synchronization in complex networks, *Phys. Rep.* **469**, 93–153 (2008).

[24] Auerbach, F., Das Gesetz der Bevölkerungskonzentration, *Petermanns Geographische Mitteilungen* **59**, 74–76 (1913).

[25] Bailey, N. T. J., *The Mathematical Theory of Infectious Diseases and Its Applications*, Hafner Press, New York (1975).

[26] Banavar, J. R., Maritan, A., and Rinaldo, A., Size and form in efficient transportation networks, *Nature* **399**, 130–132 (1999).

[27] Barabási, A.-L. and Albert, R., Emergence of scaling in random networks, *Science* **286**, 509–512 (1999).

[28] Barabási, A.-L., Albert, R., and Jeong, H., Scale-free characteristics of random networks: The topology of the World Wide Web, *Physica A* **281**, 69–77 (2000).

[29] Barabási, A.-L., Jeong, H., Ravasz, E., Néda, Z., Schuberts, A., and Vicsek, T., Evolution of the social network of scientific collaborations, *Physica A* **311**, 590–614 (2002).

[30] Barbour, A. and Mollison, D., Epidemics and random graphs, in J. P. Gabriel, C. Lefevre, and P. Picard, eds., *Stochastic Processes in Epidemic Theory*, pp. 86–89, Springer, New York (1990).

[31] Barthélemy, M. and Amaral, L. A. N., Small-world networks: Evidence for a crossover picture, *Phys. Rev. Lett.* **82**, 3180–3183 (1999).

[32] Barthélemy, M., Barrat, A., Pastor-Satorras, R., and Vespignani, A., Velocity and hierarchical spread of epidemic outbreaks in scale-free networks, *Phys. Rev. Lett.* **92**, 178701 (2004).

[33] Barthélemy, M., Barrat, A., Pastor-Satorras, R., and Vespignani, A., Dynamical patterns of epidemic outbreaks in complex heterogeneous networks, *J. Theor. Bio.* **235**, 275–288 (2005).

[34] Bearman, P. S., Moody, J., and Stovel, K., Chains of affection: The structure of adolescent romantic and sexual networks, *Am. J. Sociol.* **110**, 44–91 (2004).

[35] Bern, M., Eppstein, D., and Gilbert, J., Provably good mesh generation, in *Proceedings of the 31st Annual IEEE Symposium on the Foundations of Computer Science*, pp. 231–241, Institute of Electrical and Electronics Engineers, New York (1990).

[36] Bernard, H. R., Johnsen, E. C., Killworth, P. D., and Robinson, S., Estimating the size of an average personal network and of an event population, in M. Kochen, ed., *The Small World*, pp. 159–175, Ablex Publishing, Norwood, NJ (1989).

[37] Bernard, H. R., Johnsen, E. C., Killworth, P. D., and Robinson, S., Estimating the size of an average personal network and of an event population: Some empirical results, *Social Science Research* **20**, 109–121 (1991).

[38] Bernard, H. R. and Killworth, P. D., Informant accuracy in social network data II, *Hum. Commun. Res.* **4**, 3–18 (1977).

[39] Bernard, H. R., Killworth, P. D., Evans, M. J., McCarty, C., and Shelley, G. A., Studying social relations cross-culturally, *Ethnology* **2**, 155–179 (1988).

[40] Bernard, H. R., Killworth, P. D., and Sailer, L., Informant accuracy in social network data IV: A comparison of clique-level structure in behavioral and cognitive network data, *Soc. Networks* **2**, 191–218 (1980).

[41] Bernard, H. R., Killworth, P. D., and Sailer, L., Informant accuracy in social network data V: An experimental attempt to predict actual communication from recall data, *Soc. Sci. Res.* **11**, 30–66 (1982).

[42] Bianconi, G. and Barabási, A.-L., Bose–Einstein condensation in complex networks, *Phys. Rev. Lett.* **86**, 5632–5635 (2001).

[43] Bianconi, G. and Barabási, A.-L., Competition and multiscaling in evolving networks, *Europhys. Lett.* **54**, 436–442 (2001).

[44] Bianconi, G. and Capocci, A., Number of loops of size h in growing scale-free networks, *Phys. Rev. Lett.* **90**, 078701 (2003).

[45] Blondel, V. D., Gajardo, A., Heymans, M., Senellart, P., and Dooren, P. V., A measure of similarity between graph vertices: Applications to synonym extraction and web searching, *SIAM Review* **46**, 647–666 (2004).

[46] Boccaletti, S., Latora, V., Moreno, Y., Chavez, M., and Hwang, D.-U., Complex networks: Structure and dynamics, *Phys. Rep.* **424**, 175–308 (2006).

[47] Boguñá, M., Pastor-Satorras, R., Díaz-Guilera, A., and Arenas, A., Models of social networks based on social distance attachment, *Phys. Rev. E* **70**, 056122 (2004).

[48] Bollobás, B., Riordan, O., Spencer, J., and Tusnády, G., The degree sequence of a scale-free random graph process, *Random Struct. Alg.* **18**, 279–290 (2001).

[49] Bonacich, P. F., Power and centrality: A family of measures, *Am. J. Sociol.* **92**, 1170–1182 (1987).

[50] Borgatti, S. P., Structural holes: Unpacking Burt's redundancy measures, *Connections* **20**(1), 35–38 (1997).

[51] Borgatti, S. P., Centrality and network flow, *Soc. Networks* **27**, 55–71 (2005).

[52] Borodin, A., Roberts, G. O., Rosenthal, J. S., and Tsaparas, P., Finding authorities and hubs from link structures on the World Wide Web, in V. Y. Shen, N. Saito, M. R. Lyu, and M. E. Zurko, eds., *Proceedings of the 10th International World Wide Web Conference*, pp. 415–429, Association of Computing Machinery, New York (2001).

[53] Boyd, D. M. and Ellison, N. B., Social network sites: Definition, history, and scholarship, *Journal of Computer-Mediated Communication* **13**, 210–230 (2008).

[54] Brandes, U., Delling, D., Gaertler, M., Görke, R., Hoefer, M., Nikoloski, Z., and Wagner, D., On finding graph clusterings with maximum modularity, in *Proceedings of the 33rd International Workshop on Graph-Theoretic Concepts in Computer Science*, no. 4769 in Lecture Notes in Computer Science, Springer, Berlin (2007).

[55] Brin, S. and Page, L., The anatomy of a large-scale hypertextual Web search engine, *Comput. Netw.* **30**, 107–117 (1998).

[56] Broder, A., Kumar, R., Maghoul, F., Raghavan, P., Rajagopalan, S., Stata, R., Tomkins, A., and Wiener, J., Graph structure in the web, *Comput. Netw.* **33**, 309–320 (2000).

[57] Broido, A. and Claffy, K. C., Internet topology: Connectivity of IP graphs, in S. Fahmy and K. Park, eds., *Scalability and Traffic Control in IP Networks*, no. 4526 in Proc. SPIE, pp. 172–187, International Society for Optical Engineering, Bellingham, WA (2001).

[58] Burlando, B., The fractal dimension of taxonomic systems, *J. Theor. Bio.* **146**, 99–114 (1990).

[59] Burt, R. S., Network items and the General Social Survey, *Soc. Networks* **6**, 293–339 (1984).

[60] Burt, R. S., *Structural Holes: The Social Structure of Competition*, Harvard University Press, Cambridge, MA (1992).

[61] Caldarelli, G., Pastor-Satorras, R., and Vespignani, A., Structure of cycles and local ordering in complex networks, *Eur. Phys. J. B* **38**, 183–186 (2004).

[62] Callaway, D. S., Newman, M. E. J., Strogatz, S. H., and Watts, D. J., Network robustness and fragility: Percolation on random graphs, *Phys. Rev. Lett.* **85**, 5468–5471 (2000).

[63] Cano, P., Celma, O., Koppenberger, M., and Buldú, J. M., Topology of music recommendation networks, *Chaos* **16**, 013107 (2006).

[64] Carvalho, R., Buzna, L., Bono, F., Gutierrez, E., Just, W., and Arrowsmith, D., Robustness of trans-European gas networks, *Phys. Rev. E* **80**, 016106 (2009).

[65] Catania, J. A., Coates, T. J., Kegels, S., and Fullilove, M. T., The population-based AMEN (AIDS in Multi-Ethnic Neighborhoods) study, *Am. J. Public Health* **82**, 284–287 (1992).

[66] Chen, Q., Chang, H., Govindan, R., Jamin, S., Shenker, S. J., and Willinger, W., The origin of power laws in Internet topologies revisited, in *Proceedings of the 21st Annual Joint Conference of the IEEE Computer and Communications Societies*, IEEE Computer Society, New York (2002).

[67] Chung, F. and Lu, L., The average distances in random graphs with given expected degrees, *Proc. Natl. Acad. Sci. USA* **99**, 15879–15882 (2002).

[68] Chung, F., Lu, L., and Vu, V., Spectra of random graphs with given expected degrees, *Proc. Natl. Acad. Sci. USA* **100**, 6313–6318 (2003).

[69] Clarkson, G. and DeKorte, D., The problem of patent thickets in convergent technologies, in W. S. Bainbridge and M. C. Roco, eds., *Progress in Convergence: Technologies for Human Wellbeing*, no. 1093 in Annals of the New York Academy of Science, pp. 180–200, New York Academy of Sciences, New York (2006).

[70] Clauset, A., Moore, C., and Newman, M. E. J., Hierarchical structure and the prediction of missing links in networks, *Nature* **453**, 98–101 (2008).

[71] Clauset, A., Newman, M. E. J., and Moore, C., Finding community structure in very large networks, *Phys. Rev. E* **70**, 066111 (2004).

729

[72] Clauset, A., Shalizi, C. R., and Newman, M. E. J., Power-law distributions in empirical data, *SIAM Review* **51**, 661–703 (2009).

[73] Cohen, J. E., *Ecologists' Co-operative Web Bank, Version 1.0: Machine-Readable Data Base of Food Webs*, Rockefeller University, New York (1989).

[74] Cohen, R., Erez, K., ben-Avraham, D., and Havlin, S., Resilience of the Internet to random breakdowns, *Phys. Rev. Lett.* **85**, 4626–4628 (2000).

[75] Cohen, R. and Havlin, S., Scale-free networks are ultrasmall, *Phys. Rev. Lett.* **90**, 058701 (2003).

[76] Cohen, R., Havlin, S., and ben-Avraham, D., Efficient immunization strategies for computer networks and populations, *Phys. Rev. Lett.* **91**, 247901 (2003).

[77] Cole, B. J., Dominance hierarchies in *Leptothorax* ants, *Science* **212**, 83–84 (1981).

[78] Coleman, J. S., Katz, E., and Menzel, H., The diffusion of an innovation among physicians, *Sociometry* **20**, 253–270 (1957).

[79] Colizza, V., Barrat, A., Barthélemy, M., and Vespignani, A., The role of the airline transportation network in the prediction and predictability of global epidemics, *Proc. Natl. Acad. Sci. USA* **103**, 2015–2020 (2006).

[80] Connor, R. C., Heithaus, M. R., and Barre, L. M., Superalliance of bottlenose dolphins, *Nature* **397**, 571–572 (1999).

[81] Cormen, T. H., Leiserson, C. E., Rivest, R. L., and Stein, C., *Introduction to Algorithms*, MIT Press, Cambridge, MA, 2nd ed. (2001).

[82] Cover, T. M. and Thomas, J. A., *Elements of Information Theory*, John Wiley, New York (1991).

[83] Cox, R. A. K., Felton, J. M., and Chung, K. C., The concentration of commercial success in popular music: an analysis of the distribution of gold records, *J. Cult. Econ.* **19**, 333–340 (1995).

[84] Crovella, M. E. and Bestavros, A., Self-similarity in World Wide Web traffic: Evidence and possible causes, in B. E. Gaither and D. A. Reed, eds., *Proceedings of the 1996 ACM SIGMETRICS Conference on Measurement and Modeling of Computer Systems*, pp. 148–159, Association of Computing Machinery, New York (1996).

[85] Danon, L., Duch, J., Diaz-Guilera, A., and Arenas, A., Comparing community structure identification, *J. Stat. Mech.* P09008 (2005).

[86] Davis, A., Gardner, B. B., and Gardner, M. R., *Deep South*, University of Chicago Press, Chicago (1941).

[87] Davis, G. F. and Greve, H. R., Corporate elite networks and governance changes in the 1980s, *Am. J. Sociol.* **103**, 1–37 (1997).

[88] Davis, G. F., Yoo, M., and Baker, W. E., The small world of the American corporate elite, 1982–2001, *Strateg. Organ.* **1**, 301–326 (2003).

[89] de Castro, R. and Grossman, J. W., Famous trails to Paul Erdős, *Math. Intell.* **21**, 51–63 (1999).

[90] De Vries, H., Finding a dominance order most consistent with a linear hierarchy: A new procedure and review, *Anim. Behav.* **55**, 827–843 (1998).

[91] De Vries, H., Stevens, J. M. G., and Vervaecke, H., Measuring and testing the steepness of dominance hierarchies, *Anim. Behav.* **71**, 585–592 (2006).

[92] Dobson, I., Carreras, B. A., Lynch, V. E., and Newman, D., Complex systems analysis of series of blackouts: Cascading failure, critical points, and self-organization, *Chaos* **17**, 026103 (2007).

[93] Dodds, P. S., Muhamad, R., and Watts, D. J., An experimental study of search in global social networks, *Science* **301**, 827–829 (2003).

[94] Dodds, P. S. and Rothman, D. H., Geometry of river networks, *Phys. Rev. E* **63**, 016115, 016116, & 016117 (2001).

[95] Dorogovtsev, S. N., Goltsev, A. V., and Mendes, J. F. F., Ising model on networks with an arbitrary distribution of connections, *Phys. Rev. E* **66**, 016104 (2002).

[96] Dorogovtsev, S. N. and Mendes, J. F. F., Scaling behaviour of developing and decaying networks, *Europhys. Lett.* **52**, 33–39 (2000).

[97] Dorogovtsev, S. N. and Mendes, J. F. F., Language as an evolving word web, *Proc. R. Soc. London B* **268**, 2603–2606 (2001).

[98] Dorogovtsev, S. N. and Mendes, J. F. F., Evolution of networks, *Adv. Phys.* **51**, 1079–1187 (2002).

[99] Dorogovtsev, S. N., Mendes, J. F. F., and Samukhin, A. N., Structure of growing networks with preferential linking, *Phys. Rev. Lett.* **85**, 4633–4636 (2000).

[100] Dorogovtsev, S. N., Mendes, J. F. F., and Samukhin, A. N., Giant strongly connected component of directed networks, *Phys. Rev. E* **64**, 025101 (2001).

[101] Drews, C., The concept and definition of dominance in animal behaviour, *Behaviour* **125**, 283–313 (1993).

[102] Dunne, J. A., Williams, R. J., and Martinez, N. D., Food-web structure and network theory: The role of connectance and size, *Proc. Natl. Acad. Sci. USA* **99**, 12917–12922 (2002).

[103] Ebel, H., Mielsch, L.-I., and Bornholdt, S., Scale-free topology of e-mail networks, *Phys. Rev. E* **66**, 035103 (2002).

[104] Eckmann, J.-P. and Moses, E., Curvature of co-links uncovers hidden thematic layers in the world wide web, *Proc. Natl. Acad. Sci. USA* **99**, 5825–5829 (2002).

[105] Erdős, P. and Rényi, A., On random graphs, *Publicationes Mathematicae* **6**, 290–297 (1959).

[106] Erdős, P. and Rényi, A., On the evolution of random graphs, *Publications of the Mathematical Institute of the Hungarian Academy of Sciences* **5**, 17–61 (1960).

[107] Erdős, P. and Rényi, A., On the strength of connectedness of a random graph, *Acta Mathematica Scientia Hungary* **12**, 261–267 (1961).

[108] Erickson, B., Some problems of inference from chain data, in K. F. Schuessler, ed., *Sociological Methodology 1979*, pp. 276–302, Jossey-Bass, San Francisco (1978).

[109] Estoup, J. B., *Gammes Stenographiques*, Institut Stenographique de France, Paris (1916).

[110] Eubank, S., Guclu, H., Kumar, V. S. A., Marathe, M. V., Srinivasan, A., Toroczkai, Z., and Wang, N., Modelling disease outbreaks in realistic urban social networks, *Nature* **429**, 180–184 (2004).

[111] Faloutsos, M., Faloutsos, P., and Faloutsos, C., On power-law relationships of the internet topology, *Comput. Commun. Rev.* **29**, 251–262 (1999).

[112] Fararo, T. J. and Sunshine, M., *A Study of a Biased Friendship Network*, Syracuse University Press, Syracuse (1964).

[113] Feld, S., Why your friends have more friends than you do, *Am. J. Sociol.* **96**, 1464–1477 (1991).

[114] Fernholz, D. and Ramachandran, V., The diameter of sparse random graphs, *Random Struct. Alg.* **31**, 482–516 (2007).

[115] Ferrer i Cancho, R., Janssen, C., and Solé, R. V., Topology of technology graphs: Small world patterns in electronic circuits, *Phys. Rev. E* **64**, 046119 (2001).

[116] Ferrer i Cancho, R. and Solé, R. V., The small world of human language, *Proc. R. Soc. London B* **268**, 2261–2265 (2001).

[117] Ferrer i Cancho, R. and Solé, R. V., Optimization in complex networks, in R. Pastor-Satorras, J. Rubi, and A. Díaz-Guilera, eds., *Statistical Mechanics of Complex Networks*, no. 625 in Lecture Notes in Physics, pp. 114–125, Springer, Berlin (2003).

[118] Fiedler, M., Algebraic connectivity of graphs, *Czech. Math. J.* **23**, 298–305 (1973).

[119] Fields, S. and Song, O., A novel genetic system to detect protein–protein interactions, *Nature* **340**, 245–246 (1989).

[120] Fisher, M. E. and Essam, J. W., Some cluster size and percolation problems, *J. Math. Phys.* **2**, 609–619 (1961).

[121] Flack, J. C., Girvan, M., de Waal, F. B. M., and Krakauer, D. C., Policing stabilizes construction of social niches in primates, *Nature* **439**, 426–429 (2006).

[122] Flake, G. W., Lawrence, S. R., Giles, C. L., and Coetzee, F. M., Self-organization and identification of Web communities, *IEEE Computer* **35**, 66–71 (2002).

[123] Flory, P. J., Molecular size distribution in three dimensional polymers. I: Gelation, *J. Am. Chem. Soc.* **63**, 3083–3090 (1941).

[124] Fortunato, S., Community detection in graphs, *Phys. Rep.* **486**, 75–174 (2010).

[125] Fowler, J. H. and Jeon, S., The authority of Supreme Court precedent, *Soc. Networks* **30**, 16–30 (2008).

[126] Fowler, J. H., Johnson, T. R., Spriggs II, J. F., Jeon, S., and Wahlbeck, P. J., Network analysis and the law: Measuring the legal importance of Supreme Court precedents, *Political Analysis* **15**, 324–346 (2007).

[127] Frank, O., Estimation of population totals by use of snowball samples, in P. W. Holland and S. Leinhardt, eds., *Perspectives on Social Network Research*, pp. 319–348, Academic Press, New York (1979).

REFERENCES

[128] Freeman, L. C., A set of measures of centrality based upon betweenness, *Sociometry* **40**, 35–41 (1977).

[129] Freeman, L. C., *The Development of Social Network Analysis*, Empirical Press, Vancouver (2004).

[130] Freeman, L. C., Borgatti, S. P., and White, D. R., Centrality in valued graphs: A measure of betweenness based on network flow, *Soc. Networks* **13**, 141–154 (1991).

[131] Freeman, L. C., Freeman, S. C., and Michaelson, A. G., On human social intelligence, *J. Soc. Biol. Struct.* **11**, 415–425 (1988).

[132] Freeman, L. C., Freeman, S. C., and Michaelson, A. G., How humans see social groups: A test of the Sailer–Gaulin models, *J. Quant. Anthropol.* **1**, 229–238 (1989).

[133] Fronczak, A., Hołyst, J. A., Jedynak, M., and Sienkiewicz, J., Higher order clustering coefficients in Barabasi–Albert networks, *Physica A* **316**, 688–694 (2002).

[134] Galaskiewicz, J., *Social Organization of an Urban Grants Economy*, Academic Press, New York (1985).

[135] Gastner, M. T., *Spatial Distributions: Density-Equalizing Map Projections, Facility Location, and Two-Dimensional Networks*, Ph.D. thesis, University of Michigan (2005).

[136] Gastner, M. T. and Newman, M. E. J., Diffusion-based method for producing density equalizing maps, *Proc. Natl. Acad. Sci. USA* **101**, 7499–7504 (2004).

[137] Gastner, M. T. and Newman, M. E. J., Optimal design of spatial distribution networks, *Phys. Rev. E* **74**, 016117 (2006).

[138] Girvan, M. and Newman, M. E. J., Community structure in social and biological networks, *Proc. Natl. Acad. Sci. USA* **99**, 7821–7826 (2002).

[139] Gleiser, P. and Danon, L., Community structure in jazz, *Adv. Complex Syst.* **6**, 565–573 (2003).

[140] Gleiss, P. M., Stadler, P. F., Wagner, A., and Fell, D. A., Relevant cycles in chemical reaction networks, *Adv. Complex Syst.* **4**, 207–226 (2001).

[141] Goldstein, M. L., Morris, S. A., and Yen, G. G., Problems with fitting to the power-law distribution, *Eur. Phys. J. B* **41**, 255–258 (2004).

[142] Grandy, Jr., W. T., *Foundations of Statistical Mechanics: Equilibrium Theory*, Reidel, Dordrecht (1987).

[143] Grant, T. R., Dominance and association among members of a captive and a free-ranging group of grey kangaroos (*Macropus giganthus*), *Anim. Behav.* **21**, 449–456 (1973).

[144] Grassberger, P., On the critical behavior of the general epidemic process and dynamical percolation, *Math. Biosci.* **63**, 157–172 (1982).

[145] Grossman, J. W., The evolution of the mathematical research collaboration graph, *Congressus Numerantium* **158**, 202–212 (2002).

[146] Grossman, J. W. and Ion, P. D. F., On a portion of the well-known collaboration graph, *Congressus Numerantium* **108**, 129–131 (1995).

[147] Grujić, J., Movies recommendation networks as bipartite graphs, in M. Bubak, G. D. Albada, J. Dongarra, and P. M. A. Sloot, eds., *Proceedings of the 8th International Conference on Computational Science*, no. 5102 in Lecture Notes in Computer Science, pp. 576–583, Springer, Berlin (2008).

[148] Guardiola, X., Guimerà, R., Arenas, A., Diaz-Guilera, A., Streib, D., and Amaral, L. A. N., Macro- and micro-structure of trust networks, Preprint cond-mat/0206240 (2002).

[149] Guare, J., *Six Degrees of Separation: A Play*, Vintage, New York (1990).

[150] Guimerà, R. and Amaral, L. A. N., Functional cartography of complex metabolic networks, *Nature* **433**, 895–900 (2005).

[151] Guimerà, R., Sales-Pardo, M., and Amaral, L. A. N., Modularity from fluctuations in random graphs and complex networks, *Phys. Rev. E* **70**, 025101 (2004).

[152] Gupta, S., Anderson, R. M., and May, R. M., Networks of sexual contacts: Implications for the pattern of spread of HIV, *AIDS* **3**, 807–817 (1989).

[153] Gutenberg, B. and Richter, R. F., Frequency of earthquakes in california, *B. Seismol. Soc. Am.* **34**, 185–188 (1944).

[154] Harary, F., On the notion of balance of a signed graph, *Mich. Math. J.* **2**, 143–146 (1953).

[155] Harary, F., *Graph Theory*, Perseus, Cambridge, MA (1995).

[156] Hethcote, H. W., The mathematics of infectious diseases, *SIAM Rev.* **42**, 599–653 (2000).

[157] Holland, P. W. and Leinhardt, S., An exponential family of probability distributions for directed graphs, *J. Amer. Stat. Assoc.* **76**, 33–50 (1981).

[158] Holme, P., Edling, C. R., and Liljeros, F., Structure and time-evolution of an Internet dating community, *Soc. Networks* **26**, 155–174 (2004).

[159] Huberman, B. A., *The Laws of the Web*, MIT Press, Cambridge, MA (2001).

[160] Huxham, M., Beaney, S., and Raffaelli, D., Do parasites reduce the chances of triangulation in a real food web?, *Oikos* **76**, 284–300 (1996).

[161] Jaffe, A. and Trajtenberg, M., *Patents, Citations and Innovations: A Window on the Knowledge Economy*, MIT Press, Cambridge, MA (2002).

[162] Jeh, G. and Widom, J., SimRank: A measure of structural-context similarity, in *Proceedings of the 8th ACM SIGKDD International Conference on Knowledge Discovery and Data Mining*, pp. 538–543, Association of Computing Machinery, New York (2002).

[163] Jenks, S. M. and Ginsburg, B. E., Socio-sexual dynamics in a captive wolf pack, in H. Frank, ed., *Man and Wolf*, pp. 375–399, Junk Publishers, Dordrecht (1987).

[164] Jeong, H., Mason, S., Barabási, A.-L., and Oltvai, Z. N., Lethality and centrality in protein networks, *Nature* **411**, 41–42 (2001).

[165] Jeong, H., Néda, Z., and Barabási, A.-L., Measuring preferential attachment in evolving networks, *Europhys. Lett.* **61**, 567–572 (2003).

[166] Jeong, H., Tombor, B., Albert, R., Oltvai, Z. N., and Barabási, A.-L., The large-scale organization of metabolic networks, *Nature* **407**, 651–654 (2000).

[167] Jones, J. H. and Handcock, M. S., Sexual contacts and epidemic thresholds, *Nature* **423**, 605–606 (2003).

[168] Kansky, K. J., *Structure of Transportation Networks: Relationships Between Network Geometry and Regional Characteristics*, University of Chicago, Chicago (1963).

[169] Katz, L., A new status index derived from sociometric analysis, *Psychometrika* **18**, 39–43 (1953).

[170] Kennedy, J. W., Quintas, L. V., and Syslo, M. M., The theorem on planar graphs, *Historia Math.* **12**, 356–368 (1985).

[171] Kernighan, B. W. and Lin, S., An efficient heuristic procedure for partitioning graphs, *Bell System Technical Journal* **49**, 291–307 (1970).

[172] Kernighan, B. W. and Ritchie, D. M., *The C Programming Language*, Prentice Hall, Upper Saddle River, NJ, 2nd ed. (1988).

[173] Killworth, P. D. and Bernard, H. R., Informant accuracy in social network data, *Hum. Organ.* **35**, 269–286 (1976).

[174] Killworth, P. D. and Bernard, H. R., The reverse small world experiment, *Soc. Networks* **1**, 159–192 (1978).

[175] Killworth, P. D., Johnsen, E. C., Bernard, H. R., Shelley, G. A., and McCarty, C., Estimating the size of personal networks, *Soc. Networks* **12**, 289–312 (1990).

[176] Kleinberg, J. M., Authoritative sources in a hyperlinked environment, *J. ACM* **46**, 604–632 (1999).

[177] Kleinberg, J. M., Navigation in a small world, *Nature* **406**, 845 (2000).

[178] Kleinberg, J. M., The small-world phenomenon: An algorithmic perspective, in *Proceedings of the 32nd Annual ACM Symposium on Theory of Computing*, pp. 163–170, Association of Computing Machinery, New York (2000).

[179] Kleinberg, J. M., Small world phenomena and the dynamics of information, in T. G. Dietterich, S. Becker, and Z. Ghahramani, eds., *Proceedings of the 2001 Neural Information Processing Systems Conference*, MIT Press, Cambridge, MA (2002).

[180] Kleinberg, J. M., Kumar, S. R., Raghavan, P., Rajagopalan, S., and Tomkins, A., The Web as a graph: Measurements, models and methods, in T. Asano, H. Imai, D. T. Lee, S.-I. Nakano, and T. Tokuyama, eds., *Proceedings of the 5th Annual International Conference on Combinatorics and Computing*, no. 1627 in Lecture Notes in Computer Science, pp. 1–18, Springer, Berlin (1999).

[181] Kleinfeld, J., The small world problem, *Society* **39**(2), 61 (2002).

[182] Klovdahl, A. S., Urban social networks: Some methodological problems and possibilities, in M. Kochen, ed., *The Small World*, Ablex Publishing, Norwood, NJ (1989).

[183] Klovdahl, A. S., Potterat, J. J., Woodhouse, D. E., Muth, J. B., Muth, S. Q., and Darrow, W. W., Social

networks and infectious disease: The Colorado Springs study, *Soc. Sci. Med.* **38**, 79–88 (1994).

[184] Knuth, D. E., *The Stanford GraphBase: A Platform for Combinatorial Computing*, Addison-Wesley, Reading, MA (1993).

[185] Kohli, R. and Sah, R., Market shares: Some power law results and observations, Working paper 04.01, Harris School of Public Policy, University of Chicago (2003).

[186] Korte, C. and Milgram, S., Acquaintance links between White and Negro populations: Application of the small world method, *J. Pers. Soc. Psychol.* **15**, 101–108 (1970).

[187] Krapivsky, P. L. and Redner, S., Organization of growing random networks, *Phys. Rev. E* **63**, 066123 (2001).

[188] Krapivsky, P. L. and Redner, S., A statistical physics perspective on Web growth, *Comput. Netw.* **39**, 261–276 (2002).

[189] Krapivsky, P. L., Redner, S., and Leyvraz, F., Connectivity of growing random networks, *Phys. Rev. Lett.* **85**, 4629–4632 (2000).

[190] Krapivsky, P. L., Rodgers, G. J., and Redner, S., Degree distributions of growing networks, *Phys. Rev. Lett.* **86**, 5401–5404 (2001).

[191] Krebs, V. E., Mapping networks of terrorist cells, *Connections* **24**, 43–52 (2002).

[192] Lakhina, A., Byers, J., Crovella, M., and Xie, P., Sampling biases in IP topology measurements, in *Proceedings of the 22nd Annual Joint Conference of the IEEE Computer and Communications Societies*, Institute of Electrical and Electronics Engineers, New York (2003).

[193] Landauer, T. K., Foltz, P. W., and Laham, D., An introduction to latent semantic analysis, *Discourse Processes* **25**, 259–284 (1998).

[194] Leicht, E. A., Clarkson, G., Shedden, K., and Newman, M. E. J., Large-scale structure of time evolving citation networks, *Eur. Phys. J. B* **59**, 75–83 (2007).

[195] Leicht, E. A., Holme, P., and Newman, M. E. J., Vertex similarity in networks, *Phys. Rev. E* **73**, 026120 (2006).

[196] Lewis, K., Kaufman, J., Gonzalez, M., Wimmer, A., and Christakis, N., Tastes, ties, and time: A new social network dataset using Facebook.com, *Soc. Networks* **30**, 330–342 (2008).

[197] Liljeros, F., Edling, C. R., and Amaral, L. A. N., Sexual networks: Implication for the transmission of sexually transmitted infection, *Microbes Infec.* **5**, 189–196 (2003).

[198] Liljeros, F., Edling, C. R., Amaral, L. A. N., Stanley, H. E., and Åberg, Y., The web of human sexual contacts, *Nature* **411**, 907–908 (2001).

[199] Lloyd, A. L. and May, R. M., How viruses spread among computers and people, *Science* **292**, 1316–1317 (2001).

[200] Lorenz, M. O., Methods of measuring the concentration of wealth, *Pub. Am. Stat. Assoc.* **9**, 209–219 (1905).

[201] Lotka, A. J., The frequency distribution of scientific production, *J. Wash. Acad. Sci.* **16**, 317–323 (1926).

[202] Lowry, O. H., Rosebrough, N. J., Farr, A. L., and Randall, R. J., Protein measurement with the Folin phenol reagent, *J. Biol. Chem.* **193**, 265–275 (1951).

[203] Lu, E. T. and Hamilton, R. J., Avalanches of the distribution of solar flares, *Astrophys. J.* **380**, 89–92 (1991).

[204] Lueg, C. and Fisher, D., eds., *From Usenet to CoWebs: Interacting with Social Information Spaces*, Springer, New York (2003).

[205] Lusseau, D., The emergent properties of a dolphin social network, *Proc. R. Soc. London B (suppl.)* **270**, S186–S188 (2003).

[206] MacKinnon, I. and Warren, R., Age and geographic inferences of the LiveJournal social network, in E. Airoldi, D. M. Blei, S. E. Fienberg, A. Goldenberg, E. P. Xing, and A. X. Zheng, eds., *Statistical Network Analysis: Models, Issues, and New Directions*, vol. 4503 of *Lecture Notes in Computer Science*, pp. 176–178, Springer-Verlag, Berlin (2007).

[207] Mariolis, P., Interlocking directorates and control of corporations: The theory of bank control, *Soc. Sci. Quart.* **56**, 425–439 (1975).

[208] Maritan, A., Rinaldo, A., Rigon, R., Giacometti, A., and Rodríguez-Iturbe, I., Scaling laws for river networks, *Phys. Rev. E* **53**, 1510–1515 (1996).

[209] Martinez, N. D., Artifacts or attributes? Effects of resolution on the Little Rock Lake food web, *Ecol. Monographs* **61**, 367–392 (1991).

[210] Martinez, N. D., Constant connectance in community food webs, *Am. Nat.* **139**, 1208–1218 (1992).

[211] Maslov, S., Sneppen, K., and Zaliznyak, A., Detection of topological patterns in complex networks: Correlation profile of the internet, *Physica A* **333**, 529–540 (2004).

[212] May, R. M. and Anderson, R. M., The transmission dynamics of human immunodeficiency virus (HIV), *Philos. Trans. R. Soc. London B* **321**, 565–607 (1988).

[213] McCarty, C., Killworth, P. D., Bernard, H. R., Johnsen, E. C., and Shelley, G. A., Comparing two methods for estimating network size, *Hum. Organ.* **60**, 28–39 (2001).

[214] McMahan, C. A. and Morris, M. D., Application of maximum likelihood paired comparison ranking to estimation of a linear dominance hierarchy in animal societies, *Anim. Behav.* **32**, 374–378 (1984).

[215] Medus, A., Acuña, G., and Dorso, C. O., Detection of community structures in networks via global optimization, *Physica A* **358**, 593–604 (2005).

[216] Menger, K., Zur allgemeinen Kurventheorie, *Fundamenta Mathematicae* **10**, 96–115 (1927).

[217] Meyer, C. D., *Matrix Analysis and Applied Linear Algebra*, Society for Industrial and Applied Mathematics, Philadelphia (2000).

[218] Mézard, M. and Montanari, A., *Information, Physics, and Computation*, Oxford University Press, Oxford (2009).

[219] Milgram, S., The small world problem, *Psychol. Today* **2**, 60–67 (1967).

[220] Milo, R., Kashtan, N., Itzkovitz, S., Newman, M. E. J., and Alon, U., Subgraphs in networks, *Phys. Rev. E* **70**, 058102 (2004).

[221] Milo, R., Shen-Orr, S., Itzkovitz, S., Kashtan, N., Chklovskii, D., and Alon, U., Network motifs: Simple building blocks of complex networks, *Science* **298**, 824–827 (2002).

[222] Mitzenmacher, M., A brief history of generative models for power law and lognormal distributions, *Internet Mathematics* **1**, 226–251 (2004).

[223] Mollison, D., Spatial contact models for ecological and epidemic spread, *J. Roy. Stat. Soc. B* **39**, 283–326 (1977).

[224] Molloy, M. and Reed, B., A critical point for random graphs with a given degree sequence, *Random Struct. Alg.* **6**, 161–179 (1995).

[225] Moody, J., Race, school integration, and friendship segregation in America, *Am. J. Sociol.* **107**, 679–716 (2001).

[226] Moore, C., Ghoshal, G., and Newman, M. E. J., Exact solutions for models of evolving networks with addition and deletion of nodes, *Phys. Rev. E* **74**, 036121 (2006).

[227] Moore, C. and Mertens, S., *The Nature of Computation*, Oxford University Press, Oxford (2010).

[228] Moreno, J. L., *Who Shall Survive?*, Beacon House, Beacon, NY (1934).

[229] Moreno, Y., Pastor-Satorras, R., and Vespignani, A., Epidemic outbreaks in complex heterogeneous networks, *Eur. Phys. J. B* **26**, 521–529 (2002).

[230] Neukum, G. and Ivanov, B. A., Crater size distributions and impact probabilities on Earth from lunar, terrestial-planet, and asteroid cratering data, in T. Gehrels, ed., *Hazards Due to Comets and Asteroids*, pp. 359–416, University of Arizona Press, Tucson, AZ (1994).

[231] Newman, E. I., A method of estimating the total length of root in a sample, *J. Appl. Ecol.* **3**, 139–145 (1966).

[232] Newman, M. E. J., Models of the small world, *J. Stat. Phys.* **101**, 819–841 (2000).

[233] Newman, M. E. J., Clustering and preferential attachment in growing networks, *Phys. Rev. E* **64**, 025102 (2001).

[234] Newman, M. E. J., Scientific collaboration networks: I. Network construction and fundamental results, *Phys. Rev. E* **64**, 016131 (2001).

[235] Newman, M. E. J., Scientific collaboration networks: II. Shortest paths, weighted networks, and centrality, *Phys. Rev. E* **64**, 016132 (2001).

[236] Newman, M. E. J., The structure of scientific collaboration networks, *Proc. Natl. Acad. Sci. USA* **98**, 404–409 (2001).

[237] Newman, M. E. J., Assortative mixing in networks, *Phys. Rev. Lett.* **89**, 208701 (2002).

[238] Newman, M. E. J., Ego-centered networks and the ripple effect, *Soc. Networks* **25**, 83–95 (2003).

[239] Newman, M. E. J., Mixing patterns in networks, *Phys. Rev. E* **67**, 026126 (2003).

[240] Newman, M. E. J., Properties of highly clustered networks, *Phys. Rev. E* **68**, 026121 (2003).

[241] Newman, M. E. J., The structure and function of complex networks, *SIAM Rev.* **45**, 167–256 (2003).

[242] Newman, M. E. J., Analysis of weighted networks, *Phys. Rev. E* **70**, 056131 (2004).

[243] Newman, M. E. J., A measure of betweenness centrality based on random walks, *Soc. Networks* **27**, 39–54 (2005).

[244] Newman, M. E. J., Power laws, Pareto distributions and Zipf's law, *Contemp. Phys.* **46**, 323–351 (2005).

[245] Newman, M. E. J., Finding community structure in networks using the eigenvectors of matrices, *Phys. Rev. E* **74**, 036104 (2006).

[246] Newman, M. E. J., Modularity and community structure in networks, *Proc. Natl. Acad. Sci. USA* **103**, 8577–8582 (2006).

[247] Newman, M. E. J., Random graphs with clustering, *Phys. Rev. Lett.* **103**, 058701 (2009).

[248] Newman, M. E. J., Forrest, S., and Balthrop, J., Email networks and the spread of computer viruses, *Phys. Rev. E* **66**, 035101 (2002).

[249] Newman, M. E. J. and Girvan, M., Mixing patterns and community structure in networks, in R. Pastor-Satorras, J. Rubi, and A. Díaz-Guilera, eds., *Statistical Mechanics of Complex Networks*, no. 625 in Lecture Notes in Physics, pp. 66–87, Springer, Berlin (2003).

[250] Newman, M. E. J. and Girvan, M., Finding and evaluating community structure in networks, *Phys. Rev. E* **69**, 026113 (2004).

[251] Newman, M. E. J., Moore, C., and Watts, D. J., Mean-field solution of the small-world network model, *Phys. Rev. Lett.* **84**, 3201–3204 (2000).

[252] Newman, M. E. J. and Park, J., Why social networks are different from other types of networks, *Phys. Rev. E* **68**, 036122 (2003).

[253] Newman, M. E. J., Strogatz, S. H., and Watts, D. J., Random graphs with arbitrary degree distributions and their applications, *Phys. Rev. E* **64**, 026118 (2001).

[254] Newman, M. E. J. and Watts, D. J., Scaling and percolation in the small-world network model, *Phys. Rev. E* **60**, 7332–7342 (1999).

[255] Newman, M. E. J. and Ziff, R. M., Fast Monte Carlo algorithm for site or bond percolation, *Phys. Rev. E* **64**, 016706 (2001).

[256] Ng, A. Y., Zheng, A. X., and Jordan, M. I., Stable algorithms for link analysis, in D. H. Kraft, W. B. Croft, D. J. Harper, and J. Zobel, eds., *Proceedings of the 24th Annual International ACM SIGIR Conference on Research and Development in Information Retrieval*, pp. 258–266, Association of Computing Machinery, New York (2001).

[257] Ogielski, A. T., Integer optimization and zero-temperature fixed point in Ising random-field systems, *Phys. Rev. Lett.* **57**, 1251–1254 (1986).

[258] Onnela, J.-P., Saramäki, J., Hyvönen, J., Szabó, G., Lazer, D., Kaski, K., Kertész, J., and Barabási, A.-L., Structure and tie strengths in mobile communication networks, *Proc. Natl. Acad. Sci. USA* **104**, 7332–7336 (2007).

[259] Padgett, J. F. and Ansell, C. K., Robust action and the rise of the Medici, 1400–1434, *Am. J. Sociol.* **98**, 1259–1319 (1993).

[260] Park, J. and Newman, M. E. J., Solution of the 2-star model of a network, *Phys. Rev. E* **70**, 066146 (2004).

[261] Park, J. and Newman, M. E. J., Solution for the properties of a clustered network, *Phys. Rev. E* **72**, 026136 (2005).

[262] Pastor-Satorras, R., Vázquez, A., and Vespignani, A., Dynamical and correlation properties of the Internet, *Phys. Rev. Lett.* **87**, 258701 (2001).

[263] Pastor-Satorras, R. and Vespignani, A., Epidemic dynamics and endemic states in complex networks, *Phys. Rev. E* **63**, 066117 (2001).

[264] Pastor-Satorras, R. and Vespignani, A., Epidemic spreading in scale-free networks, *Phys. Rev. Lett.* **86**, 3200–3203 (2001).

[265] Pastor-Satorras, R. and Vespignani, A., *Evolution and Structure of the Internet*, Cambridge University Press, Cambridge (2004).

[266] Pelletier, J. D., Self-organization and scaling relationships of evolving river networks, *Journal of Geophysical Research* **104**, 7359–7375 (1999).

[267] Perlman, R., An overview of PKI trust models, *IEEE Network* **13**, 38–43 (1999).

[268] Pitts, F. R., A graph theoretic approach to historical geography, *The Professional Geographer* **17**, 15–20 (1965).

[269] Plischke, M. and Bergersen, B., *Equilibrium Statistical Physics*, World Scientific, Singapore, 3rd ed. (2006).

[270] Pool, I. de S. and Kochen, M., Contacts and influence, *Soc. Networks* **1**, 1–48 (1978).

[271] Pothen, A., Simon, H., and Liou, K.-P., Partitioning sparse matrices with eigenvectors of graphs, *SIAM J. Matrix Anal. Appl.* **11**, 430–452 (1990).

[272] Potterat, J. J., Phillips-Plummer, L., Muth, S. Q., Rothenberg, R. B., Woodhouse, D. E., Maldonado-Long, T. S., Zimmerman, H. P., and Muth, J. B., Risk network structure in the early epidemic phase of HIV transmission in Colorado Springs, *Sex. Transm. Infect.* **78**, i159–i163 (2002).

[273] Press, W. H., Teukolsky, S. A., Vetterling, W. T., and Flannery, B. P., *Numerical Recipes in C*, Cambridge University Press, Cambridge (1992).

[274] Price, D. J. de S., Networks of scientific papers, *Science* **149**, 510–515 (1965).

[275] Price, D. J. de S., A general theory of bibliometric and other cumulative advantage processes, *J. Amer. Soc. Inform. Sci.* **27**, 292–306 (1976).

[276] Radicchi, F., Castellano, C., Cecconi, F., Loreto, V., and Parisi, D., Defining and identifying communities in networks, *Proc. Natl. Acad. Sci. USA* **101**, 2658–2663 (2004).

[277] Rapoport, A. and Horvath, W. J., A study of a large sociogram, *Behavioral Science* **6**, 279–291 (1961).

[278] Ravasz, E. and Barabási, A.-L., Hierarchical organization in complex networks, *Phys. Rev. E* **67**, 026112 (2003).

[279] Rea, L. M. and Parker, R. A., *Designing and Conducting Survey Research: A Comprehensive Guide*, Jossey-Bass, San Francisco, CA, 2nd ed. (1997).

[280] Redner, S., How popular is your paper? An empirical study of the citation distribution, *Eur. Phys. J. B* **4**, 131–134 (1998).

[281] Reichardt, J. and Bornholdt, S., Statistical mechanics of community detection, *Phys. Rev. E* **74**, 016110 (2006).

[282] Ripeanu, M., Foster, I., and Iamnitchi, A., Mapping the Gnutella network: Properties of large-scale peer-to-peer systems and implications for system design, *IEEE Internet Computing* **6**, 50–57 (2002).

[283] Roberts, D. C. and Turcotte, D. L., Fractality and self-organized criticality of wars, *Fractals* **6**, 351–357 (1998).

[284] Rodríguez-Iturbe, I. and Rinaldo, A., *Fractal River Basins: Chance and Self-Organization*, Cambridge University Press, Cambridge (1997).

[285] Rothenberg, R., Baldwin, J., Trotter, R., and Muth, S., The risk environment for HIV transmission: Results from the Atlanta and Flagstaff network studies, *J. Urban Health* **78**, 419–431 (2001).

[286] Sade, D. S., Sociometrics of Macaca mulatta: I. Linkages and cliques in grooming matrices, *Folia Primatologica* **18**, 196–223 (1972).

[287] Sailer, L. D. and Gaulin, S. J. C., Proximity, sociality and observation: The definition of social groups, *Am. Anthropol.* **86**, 91–98 (1984).

[288] Salganik, M. J., Dodds, P. S., and Watts, D. J., Experimental study of inequality and unpredictability in an artificial cultural market, *Science* **311**, 854–856 (2006).

[289] Salganik, M. J. and Heckathorn, D. D., Sampling and estimation in hidden populations using respondent-driven sampling, *Sociol. Methodol.* **34**, 193–239 (2004).

[290] Salton, G., *Automatic Text Processing: The Transformation, Analysis, and Retrieval of Information by Computer*, Addison-Wesley, Reading, MA (1989).

[291] Schaeffer, S. E., Graph clustering, *Comp. Sci. Rev.* **1**, 27–64 (2007).

[292] Schank, T. and Wagner, D., Approximating clustering coefficient and transitivity, *J. Graph Algorithms Appl.* **9**, 265–275 (2005).

[293] Scott, J., *Social Network Analysis: A Handbook*, Sage, London, 2nd ed. (2000).

[294] Sen, P., Dasgupta, S., Chatterjee, A., Sreeram, P. A., Mukherjee, G., and Manna, S. S., Small-world properties of the Indian railway network, *Phys. Rev. E* **67**, 036106 (2003).

[295] Shuzhuo, L., Yinghui, C., Haifeng, D., and Feldman, M. W., A genetic algorithm with local search strategy for improved detection of community structure, *Complexity* **15** (2010).

[296] Sibani, P. and Littlewood, P. B., Slow dynamics from noise adaptation, *Phys. Rev. Lett.* **71**, 1482–1485 (1993).

[297] Simkin, M. V. and Roychowdhury, V. P., Read before you cite, *Complex Systems* **14**, 269–274 (2003).

[298] Simkin, M. V. and Roychowdhury, V. P., Stochastic modeling of citation slips, *Scientometrics* **62**, 367–384 (2005).

[299] Simon, H. A., On a class of skew distribution functions, *Biometrika* **42**, 425–440 (1955).

[300] Smith, M., Invisible crowds in cyberspace: Measuring and mapping the social structure of USENET, in M. Smith and P. Kollock, eds., *Communities in Cyberspace*, Routledge Press, London (1999).

[301] Smith, R. D., Instant messaging as a scale-free network, Preprint cond-mat/0206378 (2002).

[302] Solé, R. V., Pastor-Satorras, R., Smith, E., and Kepler, T. B., A model of large-scale proteome evolution, *Adv. Complex Syst.* **5**, 43–54 (2002).

[303] Solomonoff, R. and Rapoport, A., Connectivity of random nets, *B. Math. Biophys.* **13**, 107–117 (1951).

[304] Stauffer, D. and Aharony, A., *Introduction to Percolation Theory*, Taylor and Francis, London, 2nd ed. (1992).

[305] Stoica, I., Morris, R., Karger, D., Kaashoek, M. F., and Balakrishnan, H., Chord: A scalable peer-to-peer lookup service for Internet applications, in *Proceedings of the 2001 ACM Conference on Applications, Technologies, Architectures, and Protocols for Computer Communications (SIGCOMM)*, pp. 149–160, Association of Computing Machinery, New York (2001).

[306] Strauss, D., On a general class of models for interaction, *SIAM Rev.* **28**, 513–527 (1986).

[307] Strogatz, S. H., *Nonlinear Dynamics and Chaos*, Addison-Wesley, Reading, MA (1994).

[308] Stutzbach, D. and Rejaie, R., Characterizing today's Gnutella topology, Technical Report CIS-TR-04-02, Department of Computer Science, University of Oregon (2004).

[309] Szabó, G., Alava, M., and Kertész, J., Structural transitions in scale-free networks, *Phys. Rev. E* **67**, 056102 (2002).

[310] Thompson, S. K. and Frank, O., Model-based estimation with link-tracing sampling designs, *Surv. Methodol.* **26**, 87–98 (2000).

[311] Travers, J. and Milgram, S., An experimental study of the small world problem, *Sociometry* **32**, 425–443 (1969).

[312] Turner, T. C., Smith, M. A., Fisher, D., and Welser, H. T., Picturing Usenet: Mapping computer-mediated collective action, *J. Comput.-Mediat. Commun.* **10**(4), 7 (2005).

[313] Tyler, J. R., Wilkinson, D. M., and Huberman, B. A., Email as spectroscopy: Automated discovery of community structure within organizations, in M. Huysman, E. Wenger, and V. Wulf, eds., *Proceedings of the First International Conference on Communities and Technologies*, Kluwer, Dordrecht (2003).

[314] Udry, J. R., Bearman, P. S., and Harris, K. M., National Longitudinal Study of Adolescent Health. This work uses data from Add Health, a program project designed by J. Richard Udry, Peter S. Bearman, and Kathleen Mullan Harris, and funded by a grant P01–HD31921 from the Eunice Kennedy Shriver National Institute of Child Health and Human Development, with cooperative funding from 23 other federal agencies and foundations. Special acknowledgment is due Ronald R. Rindfuss and Barbara Entwisle for assistance in the original design. Persons interested in obtaining data files from Add Health should contact Add Health, Carolina Population Center, 123 W. Franklin Street, Chapel Hill, NC 27516–2524 (addhealth@unc.edu). No direct support was received from grant P01–HD31921 for this analysis.

[315] Valverde, S., Cancho, R. F., and Solé, R. V., Scale-free networks from optimal design, *Europhys. Lett.* **60**, 512–517 (2002).

[316] van Hooff, J. A. R. A. M. and Wensing, J. A. B., Dominance and its behavioral measures in a captive wolf pack, in H. Frank, ed., *Man and Wolf*, pp. 219–252, Kluwer, Dordrecht (1987).

[317] Vázquez, A., Flammini, A., Maritan, A., and Vespignani, A., Modeling of protein interaction networks, *Complexus* **1**, 38–44 (2003).

[318] Vázquez, A., Pastor-Satorras, R., and Vespignani, A., Large-scale topological and dynamical properties of the Internet, *Phys. Rev. E* **65**, 066130 (2002).

[319] Wakita, K. and Tsurumi, T., Finding community structure in mega-scale social networks, in

P. Isaías, M. B. Nunes, and J. Barroso, eds., *Proceedings of the IADIS International Conference, WWW/Internet 2007*, pp. 153–162, IADIS Press, Lisbon (2007).

[320] Wasserman, S. and Faust, K., *Social Network Analysis*, Cambridge University Press, Cambridge (1994).

[321] Watts, D. J., *Small Worlds*, Princeton University Press, Princeton (1999).

[322] Watts, D. J., Dodds, P. S., and Newman, M. E. J., Identity and search in social networks, *Science* **296**, 1302–1305 (2002).

[323] Watts, D. J. and Strogatz, S. H., Collective dynamics of 'small-world' networks, *Nature* **393**, 440–442 (1998).

[324] West, D. B., *Introduction to Graph Theory*, Prentice Hall, Upper Saddle River, NJ (1996).

[325] West, G. B., Brown, J. H., and Enquist, B. J., A general model for the origin of allometric scaling laws in biology, *Science* **276**, 122–126 (1997).

[326] West, G. B., Brown, J. H., and Enquist, B. J., A general model for the structure and allometry of plant vascular systems, *Nature* **400**, 664–667 (1999).

[327] White, D. R. and Reitz, K. P., Measuring role distance: Structural, regular and relational equivalence, Technical report, University of California, Irvine (1985).

[328] White, J. G., Southgate, E., Thompson, J. N., and Brenner, S., The structure of the nervous system of the nematode *Caenorhabditis Elegans*, *Phil. Trans. R. Soc. London* **314**, 1–340 (1986).

[329] Wilf, H., *Generatingfunctionology*, Academic Press, London, 2nd ed. (1994).

[330] Willis, J. C. and Yule, G. U., Some statistics of evolution and geographical distribution in plants and animals, and their significance, *Nature* **109**, 177–179 (1922).

[331] Yeomans, J. M., *Statistical Mechanics of Phase Transitions*, Oxford University Press, Oxford (1992).

[332] Yook, S. H., Jeong, H., and Barabási, A.-L., Modeling the Internet's large-scale topology, *Proc. Natl. Acad. Sci. USA* **99**, 13382–13386 (2001).

[333] Yule, G. U., A mathematical theory of evolution based on the conclusions of Dr. J. C. Willis, *Philos. Trans. R. Soc. London B* **213**, 21–87 (1925).

[334] Zachary, W. W., An information flow model for conflict and fission in small groups, *J. Anthropol. Res.* **33**, 452–473 (1977).

[335] Zanette, D. H. and Manrubia, S. C., Vertical transmission of culture and the distribution of family names, *Physica A* **295**, 1–8 (2001).

[336] Zipf, G. K., *Human Behaviour and the Principle of Least Effort*, Addison-Wesley, Reading, MA (1949).

INDEX

Page numbers in **bold** denote definitions or principal references.